OF THE ELEMENTS
Radioactive Isotopes

Selected Radioactive Isotopes

Naturally occurring radioactive isotopes are designated by a mass number in blue (although some are also manufactured). Letter m indicates an isomer of another isotope of the same mass number. Half-lives follow in parentheses, where s, min, h, d, and y stand respectively for seconds, minutes, hours, days, and years. The table includes mainly the longer-lived radioactive isotopes; many others have been prepared. Isotopes known to be radioactive but with half-lives exceeding 10^{12} y have not been included. Symbols describing the principal mode (or modes) of decay are as follows (these processes are generally accompanied by gamma radiation):

- α alpha particle emission
- β− beta particle (electron) emission
- β+ positron emission
- EC orbital electron capture
- IT isomeric transition from upper to lower isomeric state
- SF spontaneous fission

Isotope	Half-life & decay
91Pa 231	(3.28×10⁴ y) α
92U 233	(1.59×10⁵ y) α
234	(2.44×10⁵ y) α
235	(7.04×10⁸ y) α
236	(2.34×10⁷ y) α
238	(4.47×10⁹ y) α
93Np 236	(1.1×10⁵ y) EC, β⁻
237	(2.14×10⁶ y) α
239	(2.346 d) β⁻
94Pu 238	(87.75 y) α
239	(2.41×10⁴ y) α
240	(6.54×10³ y) α
242	(3.8×10⁵ y) α
244	(8.3×10⁷ y) α
95Am 241	(432 y) α
243	(7.37×10³ y) α
96Cm 242	(163.2 d) α
244	(18.12 y) α
247	(1.55×10⁷ y) α
248	(3.5×10⁵ y) α, SF
97Bk 247	(1.4×10³ y) α
98Cf 249	(351 y) α
251	(900 y) α
99Es 252	(472 d) α
253	(20.47 d) α
254	(276 d) α
100Fm 255	(20.1 h) α
257	(100.5 d) α
101Md 258	(55 d) α
102No 259	(58 min) α
103Lr 260	(3.0 min) α
104 261	(65 s) α
105 262	(40 s) α
106 263	(0.9 s) α

Periodic Table Entries

VIII
2 4.00260 He — Helium
4.215 / 0.95 (at 26 atm) / 0.1787*
1s²

IIIB
5 10.81 ±4,3 B — Boron
4275 / 2300 / 2.34
1s²2s²p¹

IVB
6 12.011 ±4,2 C — Carbon
4470* / 4100* / 2.62
1s²2s²p²

VB
7 14.0067 ±3,5,4,2 N — Nitrogen
77.35 / 63.14 / 1.251*
1s²2s²p³

VIB
8 15.9994 −2 O — Oxygen
90.18 / 50.35 / 1.429*
1s²2s²p⁴

VIIB
9 18.998403 −1 F — Fluorine
84.95 / 53.48 / 1.696*
1s²2s²p⁵

10 20.179 Ne — Neon
27.096 / 24.553 / 0.901*
1s²2s²p⁶

13 26.98154 3 Al — Aluminum
2793 / 933.25 / 2.70
[Ne]3s²p¹

14 28.0855 4 Si — Silicon
3540 / 1685 / 2.33
[Ne]3s²p²

15 30.97376 ±3,5,4 P — Phosphorus
550 / 317.30 / 1.82
[Ne]3s²p³

16 32.06 ±2,4,6 S — Sulfur
717.75 / 388.36 / 2.07
[Ne]3s²p⁴

17 35.453 ±1,3,5,7 Cl — Chlorine
239.1 / 172.16 / 3.17*
[Ne]3s²p⁵

18 39.948 Ar — Argon
87.30 / 83.81 / 1.784*
[Ne]3s²p⁶

IB / IIB
28 58.70 2,3 Ni — Nickel
3187 / 1726 / 8.90
[Ar]3d⁸4s²

29 63.546 2,1 Cu — Copper
2836 / 1357.6 / 8.96
[Ar]3d¹⁰4s¹

30 65.38 2 Zn — Zinc
1180 / 692.73 / 7.14
[Ar]3d¹⁰4s²

31 69.72 3 Ga — Gallium
2478 / 302.90 / 5.91
[Ar]3d¹⁰4s²p¹

32 72.59 Ge — Germanium
3107 / 1210.4 / 5.32
[Ar]3d¹⁰4s²p²

33 74.9216 ±3,5 As — Arsenic
876 (subl.) / 1081 (28 atm) / 5.72
[Ar]3d¹⁰4s²p³

34 78.96 −2,4,6 Se — Selenium
958 / 494 / 4.80
[Ar]3d¹⁰4s²p⁴

35 79.904 ±1,5 Br — Bromine
332.25 / 265.90 / 3.12
[Ar]3d¹⁰4s²p⁵

36 83.80 Kr — Krypton
119.80 / 115.78 / 3.74*
[Ar]3d¹⁰4s²p⁶

46 106.4 2,4 Pd — Palladium
3237 / 1825 / 12.0
[Kr]4d¹⁰

47 107.868 1 Ag — Silver
2436 / 1234 / 10.5
[Kr]4d¹⁰5s¹

48 112.41 2 Cd — Cadmium
1040 / 594.18 / 8.65
[Kr]4d¹⁰5s²

49 114.82 3 In — Indium
2346 / 429.76 / 7.31
[Kr]4d¹⁰5s²p¹

50 118.69 4,2 Sn — Tin
2876 / 505.06 / 7.30
[Kr]4d¹⁰5s²p²

51 121.75 ±3,5 Sb — Antimony
1860 / 904 / 6.68
[Kr]4d¹⁰5s²p³

52 127.60 −2,4,6 Te — Tellurium
1261 / 722.65 / 6.24
[Kr]4d¹⁰5s²p⁴

53 126.9045 ±1,5,7 I — Iodine
458.4 / 386.7 / 4.92
[Kr]4d¹⁰5s²p⁵

54 131.30 Xe — Xenon
165.03 / 161.36 / 5.89*
[Kr]4d¹⁰5s²p⁶

78 195.09 2,4 Pt — Platinum
4100 / 2045 / 21.4
[Xe]4f¹⁴5d⁹6s¹

79 196.9665 3,1 Au — Gold
3130 / 1337.58 / 19.3
[Xe]4f¹⁴5d¹⁰6s¹

80 200.59 2,1 Hg — Mercury
630 / 234.28 / 13.53
[Xe]4f¹⁴5d¹⁰6s²

81 204.37 3,1 Tl — Thallium
1746 / 577. / 11.85
[Xe]4f¹⁴5d¹⁰6s²p¹

82 207.2 4,2 Pb — Lead
2023 / 600.6 / 11.4
[Xe]4f¹⁴5d¹⁰6s²p²

83 208.9804 3,5 Bi — Bismuth
1837 / 544.52 / 9.8
[Xe]4f¹⁴5d¹⁰6s²p³

84 (209) 4,2 Po — Polonium
1235 / 527 / 9.4
[Xe]4f¹⁴5d¹⁰6s²p⁴

85 (210) ±1,3,5,7 At — Astatine
610 / 575
[Xe]4f¹⁴5d¹⁰6s²p⁵

86 (222) Rn — Radon
211 / 202 / 9.91*
[Xe]4f¹⁴5d¹⁰6s²p⁶

Lanthanides

64 157.25 3 Gd — Gadolinium
3539 / 1585 / 7.89
[Xe]4f⁷5d¹6s²

65 158.9254 3,4 Tb — Terbium
3496 / 1630 / 8.27
[Xe]4f⁹6s²

66 162.50 3 Dy — Dysprosium
2835 / 1682 / 8.54
[Xe]4f¹⁰6s²

67 164.9304 3 Ho — Holmium
2968 / 1743 / 8.80
[Xe]4f¹¹6s²

68 167.26 3 Er — Erbium
3136 / 1795 / 9.05
[Xe]4f¹²6s²

69 168.9342 3,2 Tm — Thulium
2220 / 1818 / 9.33
[Xe]4f¹³6s²

70 173.04 3,2 Yb — Ytterbium
1467 / 1097 / 6.98
[Xe]4f¹⁴6s²

71 174.967 3 Lu — Lutetium
3668 / 1936 / 9.84
[Xe]4f¹⁴5d¹6s²

Actinides

96 (247) 3 Cm — Curium
—/ 1340 / 13.511
[Rn]5f⁷6d¹7s²

97 (247) 4,3 Bk — Berkelium
[Rn]5f⁹7s²

98 (251) 3 Cf — Californium
900
[Rn]5f¹⁰7s²

99 (252) 3 Es — Einsteinium
[Rn]5f¹¹7s²

100 (257) Fm — Fermium
[Rn]5f¹²7s²

101 (258) Md — Mendelevium
[Rn]5f¹³7s²

102 (259) No — Nobelium
[Rn]5f¹⁴7s²

103 (260) Lr — Lawrencium
[Rn]5f¹⁴6d¹7s²

NOTES:

Based upon carbon-12. ( ) indicates most stable or best known isotope.

Entries marked with asterisks refer to the gaseous state at 273 K and 1 atm and are given in units of g/l.

The A & B subgroup designations, applicable to elements in rows 4, 5, 6, and 7, are those recommended by the International Union of Pure and Applied Chemistry. It should be noted that some authors and organizations use the opposite convention in distinguishing these subgroups.

* Estimated Values

Courtesy Sargent-Welch Scientific Company

JOHN E. LACKEY
University of Southern Mississippi

JERRY L. MASSEY
MERLIN D. HEHN

SOLID STATE ELECTRONICS

HOLT, RINEHART AND WINSTON
New York • Chicago • San Francisco • Montreal
Toronto • Sydney • Tokyo • Mexico City • Rio de Janeiro • Madrid

Production Manager: Paul Nardi
Book Production Services: The Publisher's Network, Morrisville, PA
Cover Design: Maricarol Blanco Cloak

Copyright© 1986 by CBS College Publishing
All rights reserved
Address correspondence to:
383 Madison Avenue
New York, NY 10017

Library of Congress Cataloging in Publication Data

Lackey, John E.
  Solid state electronics.

  Includes index.
  1. Solid state electronics. I. Massey, Jerry L.
II. Hehn, Merlin D. III. Title.
TK7871.85.L23   1986   621.381   85-21970
ISBN 0-03-071882-1

Printed in the United States of America

Published simultaneously in Canada

678   039   987654321

CBS COLLEGE PUBLISHING
Holt, Rinehart and Winston
The Dryden Press
Saunders College Publishing

Dedication

*In memory of our colleague Jerry L. Massey.
It was our misfortune to lose an excellent coauthor
during the preparation of this book.
Jerry, you are greatly missed!*

Acknowledgments

Whenever a project of this scope is undertaken, many persons and corporations deserve recognition and appreciation. I would like to express my appreciation to all individuals and industries that assisted in any way, and I sincerely hope that no one feels left out.

My heartfelt thank you to my wife, Anne, for her support throughout the development of the manuscript. Her belief in the value of the project has encouraged me and served to keep me on track.

It has been my pleasure to again work with Jerry Massey and Merlin Hehn. Their assistance, throughout this project, was invaluable. Jerry's death during this time has created a huge vacuum. My sincere thanks to both coauthors for their dedication and support.

The support of the University of Southern Mississippi is acknowledged with pleasure. Without the support of this institution, this project would have been much more difficult.

Appreciation is expressed to Heath Company for the photographs used in the Introduction. The entire manuscript was prepared using a Heathkit H-100 microcomputer and a Diablo 630 printer. Word-processing software developed by Software Toolworks— PIE, TEXT, and SPELL—served the project well.

Sergeant-Welch Scientific provided the Periodic Table of Elements used inside the front cover. Texas Instruments Incorporated and International Electronic Research Corporation (IERC) supplied materials that are contained in the appendicies. RCA Corporation allowed us to use materials in the introduction and in an appendix. I wish to express my sincere thanks to these industries.

Within CBS Educational and Professional Publishing, Paul Becker, H. L. Kirk, and Deborah Moore have been especially helpful. To all of those persons who reviewed manuscript material and assisted in other ways, I express my appreciation.

Connie Kellner and Maricarol Blanco Cloak of The Publisher's Network have been pillars of support throughout the final production process. Through them, I would like to express my appreciation to copy editors and others who have assisted with the production process.

John E. Lackey

Contents

Contents

	Preface	xvii
	Introduction	xviii
	What is Solid State Electronics? xix	
Chapter 1	**Physics for Semiconductor Materials**	1

Objectives 1
Introduction 2
Atomic Structure 2
    Energy Bands 5
    Covalent Bonding 6
Self-Check 7
Germanium and Silicon Crystals 8
    Doping Germanium and Silicon Crystals 8
    Donor Impurity Bound in a Silicon Crystal 9
    Acceptor Impurity Bound in a Silicon Crystal 9
    Conduction Within Semiconductor Materials 10
    Charges in N- and P-Type Materials 11
    Current Flow in N-Type Material 12
    Current Flow in P-Type Material 12
Self-Check 13
Summary 14
Review Questions and/or Problems 14

Chapter 2	**The PN Junction Diode**	18

Objectives 19
Introduction 20
The PN Junction 20
Self-Check 24
Junction Barrier Functions 24
Biased PN Junctions 26
    Forward Bias 27
    Reverse Bias 29
Self-Check 31
PN Junction Characteristics 31
    Characteristic Curves 33
    Avalanche (Zener) Current 34
Self-Check 35
Bias Voltage Requirements 35
    Forward Bias Circuits 37
    Reverse Bias Circuits 38
    Modes of Operation 40
    Diode Load Lines 40
    Checking Diode Condition with an Ohmmeter 43
Self-Check 44
Summary 44
Review Questions and/or Problems 45

Chapter 3	**Bipolar Transistors**	50

Objectives 51
Introduction 52
The Bipolar Transistor 52
    Transistor Elements 53
Self-Check 55

Transistor Biasing Requirements   56
Self-Check   60
Transistor Currents   60
Self-Check   64
Leakage Current   64
Biasing Arrangements   66
Self-Check   68
Summary   68
Review Questions and/or Problems   69

Chapter 4   Transistor Configurations - Common-Emitter Amplifier   72
Objectives   73
Introduction   74
Transistor Circuit Configurations   74
The Common-Emitter Configuration   76
   Fixed Bias   76
   Self-Bias   78
   Voltage Divider Bias   78
   Dual-Source Bias   80
Self-Check   80
Transistor Current Relationships   81
Effect of Input on Output   82
Voltage Distribution in a Common-Emitter Amplifier   84
Approximation Method   87
Self-Check   88
Approximation Method with Fixed Bias   89
Approximation Method with Self-Bias   90
Transistor Characteristic Curves   91
DC Load Lines for the CE Amplifier   93
Determining Operation (Q) Point   96
Self-Check   98
Equations Used in DC Analysis of a CE Amplifier   99
Self-Check   106
Summary   106
Review Questions and/or Problems   107

Chapter 5   Transistor Configurations - Common-Base and Common-Collector Amplifier   112
Objectives   113
Introduction   114
The Common-Base Configuration   114
Current and Voltage Gain in Amplifiers   115
DC Analysis of a Common-Base Circuit   117
DC Load Lines for the Common-Base Circuit   120
Effect of Changes in Input Voltage on Output Voltage   122
Self-Check   124
Common-Collector (Emitter Follower) Configuration   125
   Common-Collector Amplifier Operation   126
   Phase Relationships in a Common-Collector Circuit   127
   DC Analysis of a Common-Collector Circuit   128
Self-Check   130

Summary 131
Review Questions and/or Problems 131

Chapter 6 Special Solid State Devices — 134
Objectives 135
Introduction 136
Zener Diodes 136
Tunnel Diode 139
The Varactor Diode 141
Light-Emitting Diode (LED) 143
Liquid Crystal Display (LCD) 145
The Photosensitive Devices 146
PIN Diodes 147
Hot Carrier (SCHOTTKY) Diodes 147
Self-Check 148
Four-Layer (Latching) Devices 148
Silicon-Controlled Rectifiers (SCR) 150
Silicon-Controlled Switch (SCS) 153
Silicon Bilateral Switch (SBS) 154
TRIACS and DIACS 155
Self-Check 156
Summary 156
Review Questions and/or Problems 157

Chapter 7 Diode Circuits and Power Supply Filters — 160
Objectives 161
Introduction 162
Rectifiers 163
Half-Wave Rectifiers 164
   Peak Inverse Voltage (PIV) 167
   Ripple Frequency 168
   Average DC Output Voltage 168
Self-Check 169
Full-Wave Rectifiers 169
   Conventional Full-Wave Rectifiers 169
   Ripple Frequency 171
   Peak Inverse Voltage (PIV) 171
   Average DC Output Voltage 172
Self-Check 173
   Full-Wave Bridge Rectifiers 173
   Peak Inverse Voltage 174
   Average DC Output Voltage 175
Self-Check 177
Filters 178
Filters Convert Pulsating DC to a Steady DC Output 178
   Simple Capacitive Filter 179
   Simple Inductive Filter 180
   L-Type Filters 182
   Pi-Type Filters 183
   Operational Characteristics 185
Self-Check 190
Troubleshooting Rectifier Circuits 190

Half-Wave Rectifier Troubleshooting 191
Conventional Full-Wave Rectifier Troubleshooting 192
Full-Wave Bridge Rectifier Troubleshooting 194
Self-Check 195
Limiters and Clampers 196
  Limiters 196
  Biased Limiter Circuits 197
  Biased Shunt Positive Limiters 197
  Biased Shunt Negative Limiters 198
  Double-Diode Limiters 199
  Double-Zener Diode Limiters 199
  Other Types of Limiters 200
Self-Check 200
  Clampers 200
  Positive Clampers 201
  Negative Clampers 202
  Other Clampers 202
Self-Check 203
Summary 203
Review Questions and/or Problems 205

Chapter 8　Transistor Amplifiers　　　　210
Objectives 211
Introduction 212
Effect of Changes in Resistance 212
Input Impedance 214
Self-Check 220
Frequency Response 220
  Upper Frequency Cut Off 222
AC Voltage Gain 223
Self-Check 226
AC Load Lines 226
Self-Check 231
Output Impedance 231
Signal Behavior 233
Self-Check 239
The Common-Collector Amplifier 240
  Input Impedance 240
The Common-Base Amplifier 245
Self-Check 248
Summary 249
Review Questions and/or Problems 250

Chapter 9　Amplification and Voltage Regulation　　　　254
Objectives 255
Introduction 256
Amplifier Review 256
Phase Splitters 257
Differential Amplifiers 258
Self-Check 262
Classes of Operation 263
  Class A Operation 264

Class B Operation   265
Class C Operation   265
Class AB Operation   265
Overdriven Amplifiers   266
Self-Check   266
Temperature Stabilization   267
Emitter Resistor Stabilization   267
Thermistor Stabilization   269
Diode Stabilization   271
Amplifier Comparisons   273
Amplifier Summary   275
Self-Check   276
Electronic Voltage Regulators   276
Zener Diode Regulators   277
Electronic Voltage Regulator (EVR) Circuits   278
Self-Check   281
Summary   281
Review Questions and/or Problems   283

Chapter 10 Amplifier Applications and Coupling 292
Objectives   293
Introduction   294
Amplifier Coupling   295
Direct Coupling   295
RC (Resistive-Capacitive) Coupling   299
Impedance (LC) Coupling   303
Transformer Coupling   305
Self-Check   306
Cascaded Amplifiers   307
Troubleshooting Amplifiers   309
Classification of Amplifiers   310
Classification By Frequency   310
Classification By Type of Operation   311
Classification By Use   311
Self-Check   311
Power Amplifiers   312
Single-Ended Power Amplifiers   314
Self-Check   315
Double-Ended (Push-Pull) Power Amplifier   316
Class A Push-Pull Power Amplifiers   317
Class B Push-Pull Power Amplifiers   318
Class AB Push-Pull Power Amplifiers   319
Even Harmonic Cancellation   320
Phase Splitter Input to Push-Pull Amplifiers   322
Complimentary-Symmetry Amplifiers   325
Compensated Common-Symmetry Circuits   327
Single-Source Complimentary-Symmetry Amplifiers   330
Troubleshooting a Push-Pull Amplifier   330
Self-Check   331
Summary   332
Review Questions and/or Problems   333

Chapter 11	**Narrow-Band and Wideband Amplifiers**	340

    Objectives 341
    Introduction 342
    Applications of Narrow and Wideband Amplifiers 342
    Narrow-Band Amplifiers 344
        Signal-To-Noise Ratio 349
        Troubleshooting Solid-State Narrow-Band Amplifiers 350
    Self-Check 353
    Wideband Amplifiers 354
        Square Wave Characteristics 354
        Wideband Amplifier Characteristics 357
        Low Frequency Compensation 360
        Combined Low and High Frequency Compensation 361
        Stagger-Tuned Wideband Frequency 362
        Troubleshooting Wideband Amplifiers 364
    Self-Check 366
    Summary 366
    Review Questions and/or Problems 367

Chapter 12	**Wave Generation**	370

    Objectives 371
    Introduction 372
    Classification of Wave Generators 372
    Requirements for Oscillation 373
    Self-Check 375
    Sine-Wave Generators 376
        RC Networks as Frequency-Determining Networks 376
        LC Tank Circuits as Frequency-Determining Networks 378
        Crystals as Frequency-Determining Devices 382
        Voltage-Controlled Oscillators 384
    Self-Check 384
    Nonsinusoidal Wave Generators 385
        Square and Rectangular Wave Generators 385
        Sawtooth-Wave Generators 388
        Trapezoidal-Wave Generators 390
    Self-Check 391
    Summary 391
    Review Questions and/or Problems 392

Chapter 13	**Sine Wave Oscillators**	396

    Objectives 397
    Introduction 398
    Basic Oscillator Operations 398
        The Amplifier 400
        The Frequency-Determining Device 400
        Regenerative Feedback 401
    Self-Check 402
    Armstrong Oscillators 403
        Troubleshooting the Armstrong Oscillator 409
    Self-Check 410
    The Series-Fed Hartley Oscillator 411
    The Buffer Amplifier 412

The Shunt-Fed Hartley Oscillator 413
Self-Check 416
The Colpitt's Oscillator 416
The Clapp Oscillator 419
The Butler Oscillator 421
Self-Check 422
The RC Oscillators 422
   The RC Phase Shift Oscillator 423
   The Wien Bridge Oscillator 425
Self-Check 430
Frequency Multipliers 430
Self-Check 433
Summary 433
Review Questions and/or Problems 435

Chapter 14 Nonsinusoidal Wave Generators 442
Objectives 443
Introduction 444
Pulsed Oscillators 444
Self-Check 450
Blocking Oscillators 451
Self-Check 458
Solid-State Multivibrators 459
   Multivibrator Waveforms 459
   Multivibrator Types 462
   The Astable Multivibrator 462
   The Monostable Multivibrator 464
   The Bistable Multivibrator 466
   The Schmitt Trigger 469
Self-Check 472
Sawtooth and Trapezoidal Wave Generators 473
   Sawtooth Wave Generation 474
   Trapezoidal Wave Generators 475
Self-Check 476
Summary 476
Review Questions and/or Problems 477

Chapter 15 The Junction Field-Effect Transistor (JFET) 490
Objectives 491
Introduction 492
JFET Construction 492
   JFET Operational Characteristics 493
   JFET - Bipolar Transistor Comparisons 495
   Operation with $V_{GS} = 0$ Volts 497
   Characteristic Curves 500
   Transconductance 505
Self-Check 505
Biasing JFET Transistors 506
   Fixing Bias JFET Circuits 506
   Self-Biased JFET Circuits 510
   Design of a JFET Circuit Using Graphic Analysis 516
Self-Check 519

JFET Amplifiers   520
   The Common-Source Amplifier   520
   Voltage Gain   521
   Impedance Values   522
   Load Line Analysis of a Common-Source Amplifier   523
   The Common-Drain Amplifier   525
   The Common-Gate Amplifier   526
Advantages and Disadvantages of JFET Circuits   528
The P-Channel JFET   528
The Use of BI-FET Circuitry   529
Self-Check   530
Summary   531
Review Questions and/or Problems   532

Chapter 16 MOSFETs—Metal Oxide Semiconductor Field Effect Transistors   536
Objectives   537
Introduction   538
Enhancement- and Depletion-Mode MOSFETs   538
Operation of an E-MOSFET   540
Operation of a DE-MOSFET   542
MOSFET Characteristic Curves   544
   E-MOSFET Characteristic Curves   544
   DE-MOSFET Characteristic Curves   546
Self-Check   547
MOSFET Amplifiers   548
   Common-Source Amplifiers   548
   Input and Output Impedance (Resistance)   549
   Bipolar-JFET-MOSFET Comparisons   550
V-MOS Devices   550
CMOS Construction   551
Self-Check   552
Summary   553
Review Questions and/or Problems   554

Chapter 17 Selected Solid State Devices Operational Characteristics   558
Objectives   559
Introduction   560
The Silicon Controlled Rectifier   560
   The SCR as a Switch   561
   DC Power Control Using an SCR   563
   AC Power Control Using an SCR   564
   SCR Summary   565
Self-Check   567
The Unijunction Transistor   568
   The UJT Sawtooth Generator   573
   The Programmable UJT (PUT)   575
Self-Check   577
The Tunnel Diode   577
   The Tunnel Diode Oscillator   579
   The Tunnel Diode Amplifier   579
Self-Check   580

Optoelectronic Devices 581
   The Photodiode 582
   The Phototransistor 583
   Optoelectronic Couplers 584
   The LASER Diode 584
Self-Check 585
Summary 585
Review Questions and/or Problems 586

Chapter 18   Integrated Circuits Fabrication and Applications   590
Objectives 591
Introduction 592
Summation of Events and Definitions 592
   Discrete Microcomponents 593
   Thick-Film or Ceramic Printed Circuits 593
   Thin-Film Fabrication 594
   Silicon-Integrated Circuits 594
   Hybrid Microcircuitry 594
Ceramic Printed Circuitry 595
Self-Check 599
Thin-Film Circuits 600
Self-Check 606
Integrated Circuits 606
   Reliability 607
   Cost 607
   Size and Weight 607
   Power 607
Fabrication of the Integrated Circuit (IC) 608
   Substrate Preparation 608
   Photoengraving 610
   Diffusion 611
   Oxidation 613
   Epitaxy 614
Self-Check 615
   Interconnection, Lead Attachment, and Encapsulation 616
   Quality Control 618
Multichip Circuitry 618
Self-Check 619
Summary 620
Review Questions and/or Problems 620

Chapter 19   Operational Amplifiers and Timers   624
Objectives 625
Introduction 626
Operational Amplifier Characteristics 626
   Input Circuits 627
   Frequency Response 631
   Slew Rate 632
   Large Signal Bandwidth Limitations 633
Self-Check 633

Sample Operational Amplifier Applications   635
　　The Inverting Amplifier   635
　　The Non-Inverting Amplifier   635
　　The Voltage Follower   636
　　The Summing Amplifier   636
　　The Subtractor   637
　　The Adder-Subtracter   637
　　The Differentiator   637
　　The Integrator   637
　　An Audio Amplifier   638
　　The Differential Amplifier   639
　　The Sine Wave Oscillator   639
Self-Check   640
The 555 Timer   641
Astable Multivibrator Operation   642
Monostable Multivibrator Operation   643
Self-Check   643
Summary   644
Review Questions and/or Problems   644

Glossary   648
Bibliography   667
Appendix A   670
Appendix B   676
Appendix C   680
Appendix D   686
Appendix E   688
Appendix F   694
Appendix G   704
Further Operating Considerations   708
Appendix H   710
Answers to Self-Check   717
Answers to Review Questions and/or Problems   721
Index   725

Preface

SOLID STATE ELECTRONICS was written for a wide audience, and is suitable for use by students ranging from high school to engineering school. A reading level has been maintained which should allow all potential students to understand the material.

This text was written with both teachers and students in mind. For the students, frequent Self-Checks have been included that the student can use as internal, and immediate, review. Answers to all Self-Check items are included at the back of the book. For the teacher and student, several items have been included at the end of each chapter that can be used in class discussion, homework, or as test questions. Answers for odd-numbered review items are included at the back of this book. Technical terms have been kept to a minimum, and those used have been fully explained.

Solid-state electronics can be a very complex subject. We believe, however, that the subject can be taught with a minimum of complexity to a majority of students enrolled in electronics programs throughout the world. We have, therefore, approached the subject with minimum technical and mathematical emphasis. With over 80 years of electronics experience between us, we have found that this approach can prepare excellent technicians.

This book incorporates most of the procedures found in our earlier book *Fundamentals of Electricity and Electronics*. Students should be able to move directly from that book to this one with little difficulty. We have tried to maintain the same reading level and presentation methods found in the earlier book.

In an effort to provide useful information, such as specification sheets, and characteristic curves, we have included several appendicies. Some of these are devoted entirely to specification sheets. One, however, discusses precautions that must be taken when working with MOS devices.

We have presented the material in a sequence which we feel best suits the subject. We have, however, gathered all information into self-supporting units to meet the needs of instructors who may prefer a different sequence.

Introduction

What is Solid State Electronics?

In the 1940s and 1950s few people could have imagined the effect that the 1948 development of the transistor would have on the field of electronics. As late as 1960 there was doubt, on the part of many, that solid-state devices would ever become widely used in electronic circuitry.

During this period the primary active device that was in use was the vacuum tube (Figure I-1). These devices required large spaces for their supporting circuitry, large amounts of power, and high voltages. This resulted in the need for large discrete components which made the circuits large and bulky.

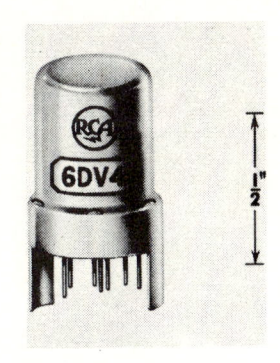

(Courtesy RCA Corporation)

Fig. I-1 Representative Vacuum Tubes

The one-inch marking beside each picture gives you an idea of the space required for each tube. When you consider the fact that each tube required heat in the range of several watts for operation, you can understand the problems associated with heat dissipation and circuit cooling. Using tubes, today's digital computer would require two or more rooms of space.

With the arrival of the space age in 1957, there was an immediate need for circuits that were smaller, weighed much less, and required much less power for their operation. Solid-state diodes and transistors were the devices immediately available. Figure I-2 shows the type of devices available at that time.

Fig. I-2 Representative Transistors

Circuitry designed for use with these components was still point-to-point discrete components. The lower voltage required for operation of solid-state devices allowed the finished circuits to be smaller and lighter in weight. Figure I-3 illustrates a circuit of this type. Notice the discrete components and the method used for connection. The circuit in Figure I-3 contains two transistor amplifiers. The board on which the circuit is mounted is approximately 2 × 3 inches. However, when mounted on edge, this assembly requires a relatively small space. When compared to the space required by two vacuum tube amplifiers, this is a marked decrease. Additionally, circuit power requirements are greatly reduced. This results in much less heat dissipation and reduces the need for circuit cooling.

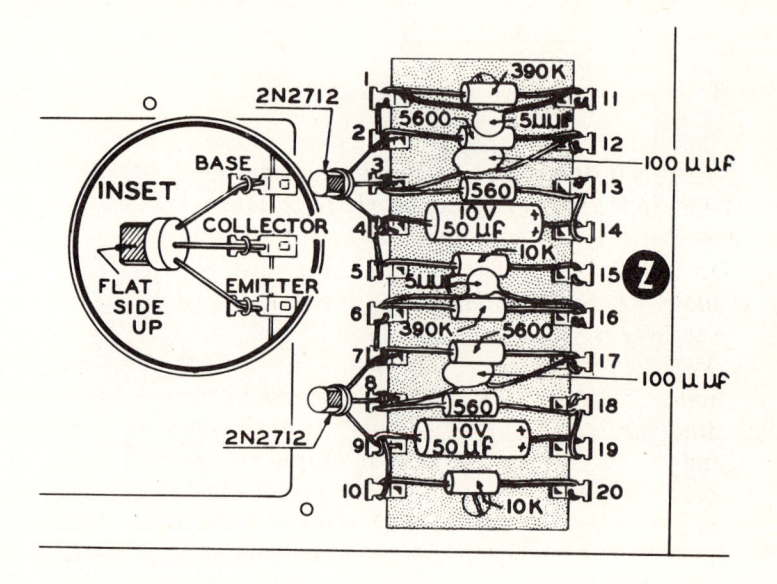

Fig. I-3 Dual-Transistor Amplifiers—Discrete Wired

There was, however, still a need for circuits that were smaller and still more power efficient. The next step toward miniaturization was to shorten the leads, reduce the size of the mounting board, and replace individual conductors with metal traces deposited on a phenolic board. The resulting boards were called printed circuit boards.

A completed PC board is shown in Figure I-4. The circuit shown includes 21 transistors and the associated discrete components required to fabricate a ten-stage shift register. This circuit was used in early solid-state digital computers. The PC board used for this circuit is approximately 3 × 6 inches. This provided more miniaturization than earlier fabrication techniques. A computer designed using these circuits could easily fit in one room and provide added capability over the vacuum tube type.

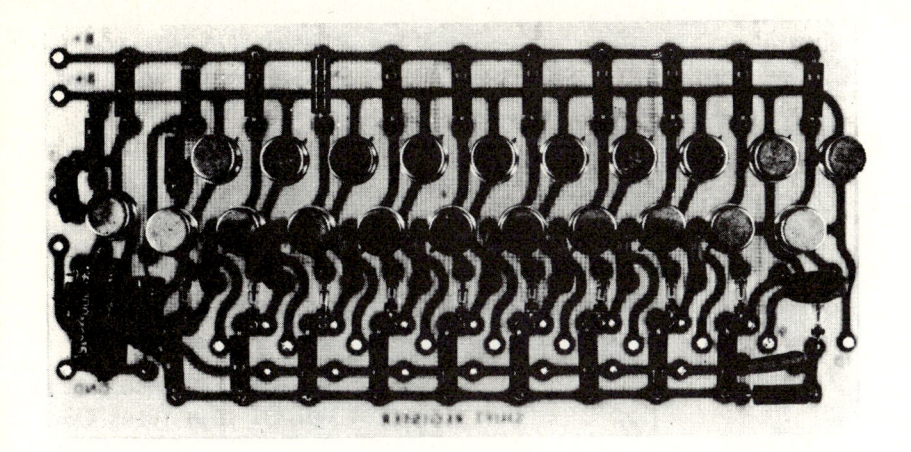

Fig. I-4 Circuit Board for Ten-Stage Shift Register

Improvements in circuit design have continued. Miniaturization has advanced to the point that the circuit shown in Figure I-4 can now be placed in an integrated circuit chip approximately 1/8 inch square, with much spare space. By the time this chip is mounted in a package that allows it to be interconnected into a larger circuit, it may occupy a space as large as 1/2 inch square.

With present day design, the same 1/8 inch integrated circuit chip, mentioned above, can be designed to contain large numbers of amplifier circuits. Using packaged ICs like those shown in Figure I-5, it is possible to design circuits that have different applications on a single PC board. The PC board shown in Figure I-5 probably contains the equivalent of several thousand ten-stage shift registers. Imagine the space required and the heat generated by a circuit this size that used vacuum tubes. This PC board is approximately 6 × 9 inches while a vacuum tube circuit of the same type would have to be housed in a large room. Consider the fact that this board can be, and often is, mounted on edge, taking up even less space. In fact, this board would probably occupy less space than a single vacuum tube amplifier.

Using these fabrication procedures, computers with much greater power than the earlier models can be designed. Figure I-6 shows a desktop computer of the type available today. Realize that the computer is contained in the unit below the display monitor. Compare this unit's size with the telephone beside it and notice how small the computer really is. This is even more remarkable when you realize that less than 50% of the unit is computer circuitry. The remainder consists of the keyboard and two internal disk drives that provide auxiliary storage. This computer can contain more memory and larger capacity than the large computers in use 10 to 15 years ago.

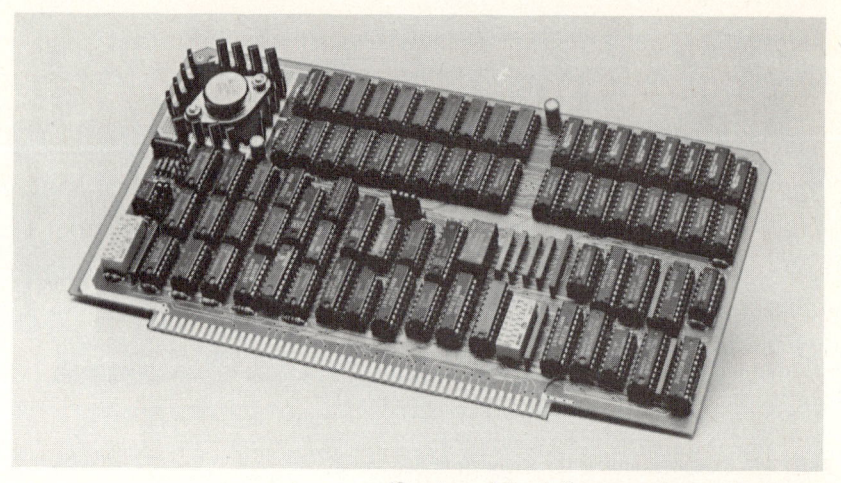

(Courtesy of Heath Company, Benton Harbor, MI)

Fig. I-5 High Density Printed Circuit Board

(Courtesy Heath Company, Benton Harbor, MI)

Fig. I-6 Desk-Top Personal Computer

Using the computer and printer shown in Figure I-7, the authors prepared the manuscript for this book. Using earlier circuit design techniques, both of these units would have been immense. In fact, size would have made them so expensive that only a few could afford them.

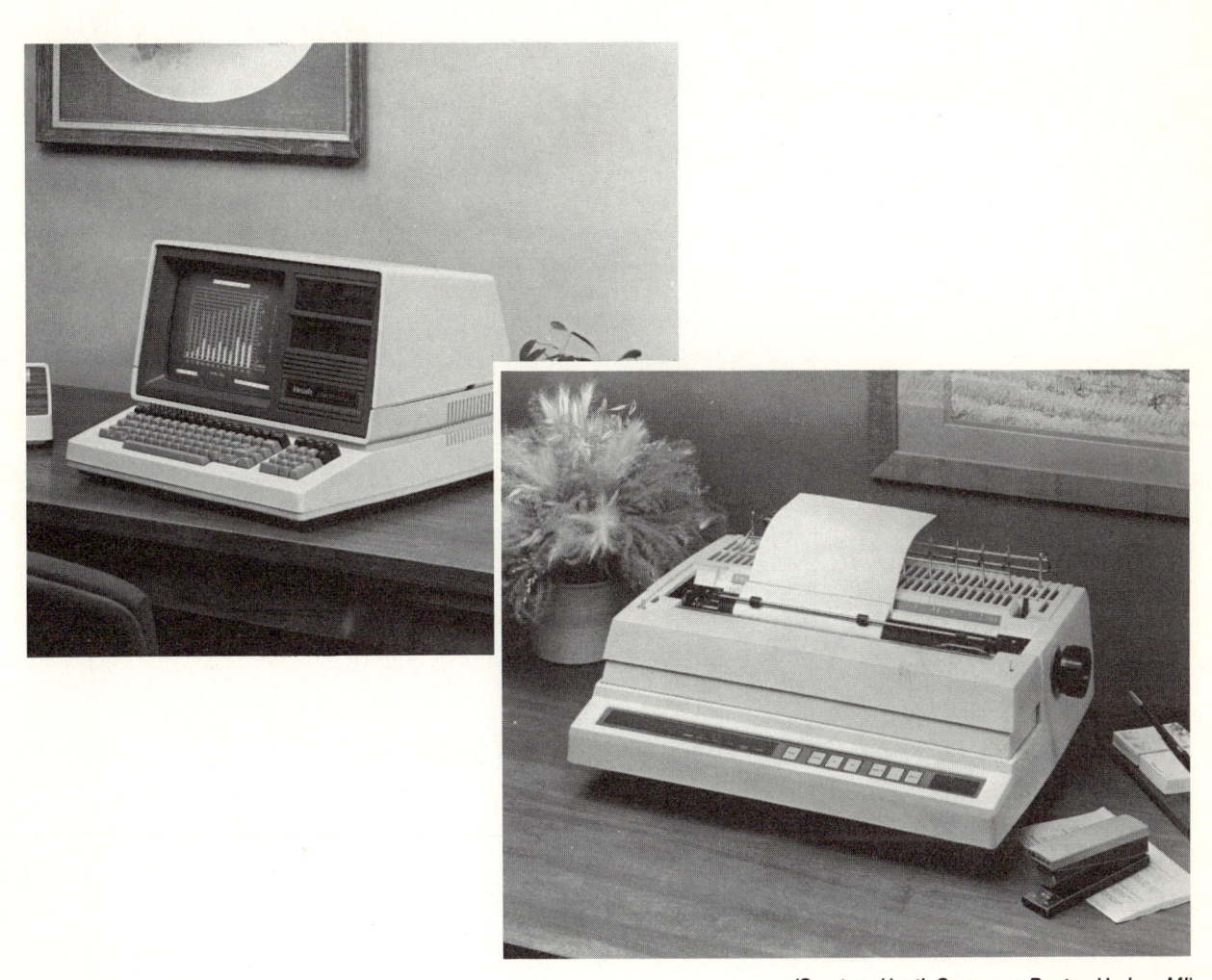

(Courtesy Heath Company, Benton Harbor, MI)

Fig. I-7   Desk-Top Computer and Letter-Quality Printer

Future technology will continue to miniaturize circuitry, even further. Those of you just entering the electronics field will be part of this advancement. A thorough understanding of the materials contained in this book will help to prepare you for those experiences.

1
Physics for Semiconductor Materials

Objectives

1. Define centrifugal and centripetal force and explain how each affects the electrons within an atom.
2. Define kinetic and potential energy and explain the difference between these energy forms.
3. Explain the relationship that exists between electron orbit and energy level.
4. Define orbit and shell and explain how each relates to the understanding of an atom's structure.
5. Define valence and explain the part that it plays in determining the chemical activity of an element.
6. Define energy level and explain how it is part of the structure of an atom.
7. Define conductors, insulators, and semiconductors and relate each to a group of elements that are classified by valence.
8. Define conduction band and free electrons and explain the relationship that each has with the other.
9. Explain the meaning and relationship that exists between valence band, forbidden band, and conduction band.
10. Explain the covalent bonding process.
11. Explain the part that electrons and holes play in semiconductor devices.
12. Identify two types of current that exist in a semiconductor device and explain how each reacts to a voltage applied to the semiconductor material.
13. Define the following terms and explain how they apply to semiconductor devices.

    a. Impurities
    b. Donor materials
    c. Acceptor materials
    d. N-type material
    e. P-type material
    f. Trivalent materials
    g. Pentavalent materials
    h. Majority carriers
    i. Minority carriers
    j. Hole flow

Introduction

The purpose of this chapter is to introduce all students to a set of physical properties that will assist them in learning the solid-state principles that follow. Knowledge of the physics of solid-state (semiconductor) materials is necessary in order to understand the operation of devices that range from the PN junction diode to the most advanced integrated circuit.

During the last 20 years, the advances that have occurred in the field of electronics have been remarkable. These advances can, to a great degree, be traced to the perfecting of semiconductor physics and its applications. As with other areas of science, it is necessary to start the study of these devices with a review of atomic structure.

Atomic Structure

To explain the structure of an atom it is probably best if we look at the simplest of all atoms — the hydrogen atom. The hydrogen atom contains one proton (positively charged particle) in its nucleus and one electron (negatively charged particle) that orbits its nucleus. As the electron revolves around the nucleus, it has two types of force affecting its position. Centrifugal force is a force that tends to cause the electron to be thrown into space. Centripetal force is an inward-moving (pulling) force that exists because of the attraction that exists between the electron and the proton. This force tends to pull the electron toward the nucleus. According to the energy level of the electron, it will assume an orbit at a position where the two forces, centrifugal and centripetal, are precisely balanced. The path that the electron follows around the nucleus is called the orbit or shell.

The electron in a hydrogen atom and the electrons in all other atoms contain two types of energy — kinetic and potential. Kinetic energy is the result of the electron's motion. Potential energy is determined by the electron's position with respect to the nucleus. The combined energy that results from these two types of energy determines the radius of the electron's orbit. If the electron's energy level is increased, the electron will move to a different orbit with a larger radius than before. If the electron's energy level decreases, the electron will move to a lower orbit with a smaller radius. Orbits or shells are designated by either numbers or letters as shown in Figure 1-1. The first shell (K) is the one closest to the nucleus, and its electrons are at the lowest energy level within the atom. Moving outward from the nucleus, the L, M, N, O, P, and Q shells each contain electrons with progressively higher energy levels.

These orbits or shells are called **permissible energy levels**. As this name implies, these are the levels at which electrons may orbit. Quantum physics theory states that:

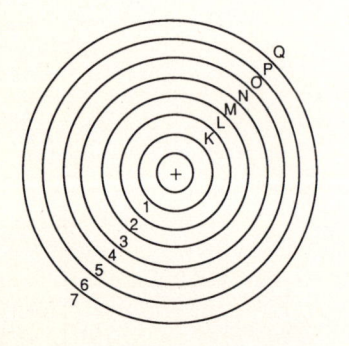

Fig. 1-1 Energy Shell Designations

<blockquote>An electron cannot exist in the space
between permissible energy levels.</blockquote>

Every electron that orbits the nucleus of an atom must exist at a permissible energy level. Application of heat to a hydrogen atom will cause the electron to absorb added energy. As the electron reaches a new energy level, it will jump to a higher permissible level. If the new energy is removed, the electron will return to the orbit where it existed before it absorbed the new energy. In nature, electrons always try to return to and occupy the lowest possible energy level. This means that the K shell is filled before the L and all other shells receive any electrons.

Figure 1-2 displays an energy level diagram that can be used to explain the position of shells in relation to the nucleus. The K (first) shell, the one nearest the nucleus, represents the lowest permissible energy level at which electrons can orbit. The L (second) shell is the second lowest energy level, and the energy level increases with each shell extending outward. Any electron that orbits the nucleus must be located in the narrow spaces represented by the energy shells.

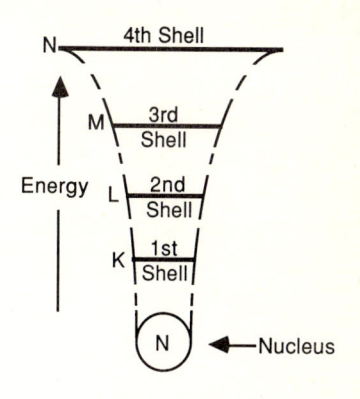

Fig. 1-2   Energy Shell Diagram

Electrons that orbit in the outermost shell of the atom are the ones which enter into chemical or electrical combinations with other atoms. These electrons are called **valence electrons** and the outermost shell is called the **valence shell**.

Each element is assigned an atomic number that represents its place within the table of elements. (A Table of Elements can be found in the front of this book). The atomic number represents the number of protons found in the nucleus of that element.

A hydrogen atom has one electron which orbits in the K shell; helium has two electrons and both orbit in the K shell, but the elements that have higher atomic numbers have the K, L, and other shells filled. The two elements that are widely used in semiconductor fabrication are silicon and germanium.

Silicon has an atomic number of 14 and its electrons orbit as follows:

K shell contains 2 electrons
L shell contains 8 electrons
M shell contains 4 electrons

Figure 1-3 is a pictorial diagram of the energy shells and the electrons contained in each shell of the silicon atom. Note that there are four valence electrons located in the outermost (M) orbit.

A germanium atom has an atomic number of 32. Electrons are located in shells K, L, M, and N. Electron and shell location is illustrated in Figure 1-4. Note that germanium also has four valence electrons, but in this case the valence electrons are located in the N shell.

In the case of both atoms (silicon and germanium) the valence is four. The valence electrons, however, are located at different energy levels. The number of electrons contained in the valence shell of an atom governs its chemical activity. Elements range in chemical activity

Si

Atomic Weight 28.08
Atomic Number 14

+14 2 8 4

The Silicon Atom
14 Electrons in orbit
14 Protons in Nucleus

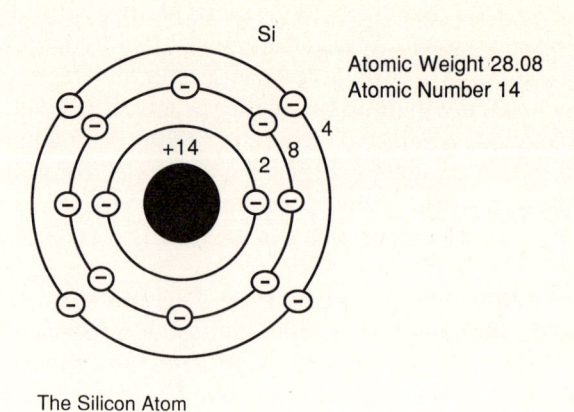

Fig. 1-3 Pictorial Diagram of a Silicon Atom

Ge

Atomic Weight 72.60
Atomic Number 32

+32 2 8 18 4

The Germanium Atom
32 Electrons in orbit
32 Protons in Nucleus

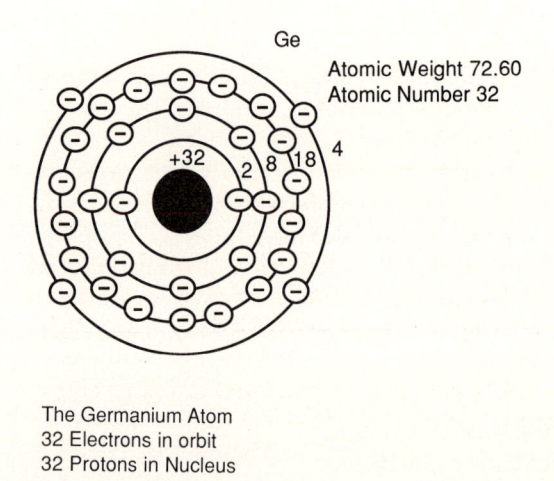

Fig. 1-4 Pictorial Diagram of a Germanium Atom

from highly active to stable. Element stability is defined as the condition where the valence orbit is filled and the atoms display very little chemical activity.

Atoms that have eight valence electrons are very stable and have little chemical activity. For this reason we call them insulators. Conductors, on the other hand, are made from elements whose atoms have a low valence. (Gold, silver, and copper each have a valence of 1.) The elements that exist in the valence ranges of 3, 4, and 5 are for the most part, neither good conductors nor good insulators. Carbon, silicon, and germanium have valences of four and all three are considered to be semiconductors. Permissible energy levels within an orbit exist such that an atom's outermost energy level will never contain more than eight valence electrons.

Energy Bands

Atoms that are brought near to each other will interact with each other. Atoms of the same element will have identical energy levels. Areas between energy levels are areas where no electrons may exist. These are called forbidden bands. Look at Figure 1-5 for a comparison of two isolated silicon atoms.

The two atoms are separated by a space large enough to prevent the shells from overlapping. Notice that the energy levels within the two atoms are identical. Figure 1-6 illustrates the effect of moving the two atoms closer together. In this case, the valence shells will actually overlap and the valence bands of the two atoms will act as if there was only one valence band. Note that only the outermost (valence) band has been affected. The inner shells (K and L) are unchanged. As more atoms are brought into the same area, they too will join valence bands with those already present. When enough silicon atoms are near each other, the result is a continuous permissible energy level. Conduction can be supported by this valence band.

The valence band and the next higher energy level (conduction band) join and form continuous paths through the semiconductor material. The valence and conduction bands are separated by a region which is called the forbidden band.

Electrons that are in the valence band are under the influence of the atom's nucleus; however, application of an external energy source can cause the energy level of an electron to be elevated to the point that it breaks away from the valence band and moves into the conduction band. The energy applied to the electron must be large enough to cause it to move all the way across the forbidden band. It is not possible for the electron to move part way and stop within the forbidden band. When the electron reaches the conduction band, it is considered to be free from all influence exerted by the nucleus and is now free to drift from place to place. The electron is now a **free electron**.

It is easier to create free electrons in some elements than it is in others. Gold, silver, and copper, when at room temperature, have many free electrons drifting from place to place. For this reason, these elements are used as conductors. On the other hand, rubber, glass, and plastic are substances that have very few free electrons. These are poor conductors and are called insulators because of their limited ability to conduct current. There is no well-defined line that separates conductors from insulators. Actually, all substances have some free electrons and will conduct current. Some materials make good conductors and

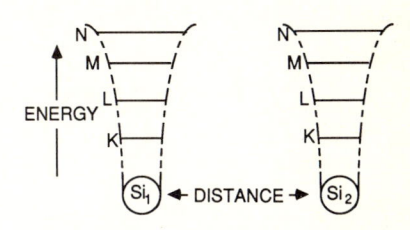

Fig. 1-5 Isolated Energy Diagrams of Silicon Atoms

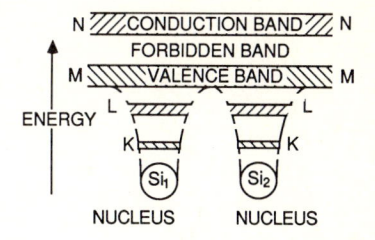

Fig. 1-6 Energy Diagram of Silicon Atoms Closely Spaced

others make good insulators; however, there is another group of elements that lie between these two extremes. These elements are called **semiconductors**. This group of elements is the one of greatest importance to us in this study of solid-state devices.

For a valence electron to be free from its shell and to enter the conduction band, energy must be gained from an external source. The amount of energy that is required to move the valence electron to free electron status is determined by the type of substance. Conductors require small amounts, insulators require large amounts, while semiconductors require a moderate amount of external energy to move their valence electrons into the conduction band.

Figure 1-7 illustrates the energy diagrams of each type. Compare the energy diagrams of the three. Notice that in the conductor, the valence and conduction bands overlap. In the semiconductor, the forbidden band is present but narrower than the one found in the insulator. In the conductor, all of the valence electrons exist in the conduction band and only a small amount of energy is required to make them move about. In the insulator, with its wide forbidden band, it takes much more energy to move an electron to the conduction band. Semiconductors fit somewhere in between these two extremes. Less energy is required to move the valence electron of a semiconductor into the conduction band than is required for the insulator, but more energy is required than for the conductor.

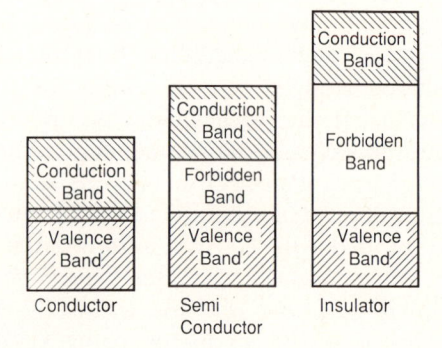

Fig. 1-7 Energy Diagrams Compared

Covalent Bonding

The **valence** of an element is a measure of its ability to combine with other elements to form molecules and chemical compounds. Hydrogen, for example, has a valence of 1 and will combine with other elements very easily. This is because its outer shell (K) requires two electrons to become stable. Oxygen has an atomic number of 8; 6

Physics for Semiconductor Materials

electrons in the L (valence) shell and 2 electrons in the K shell. It needs two additional electrons in its valence band to reach a stable state. One atom of oxygen combines with two atoms of hydrogen to form a molecule of water. Figure 1-8 illustrates this combination. Once combined in this manner, all three atoms appear to have complete outer shells. Under this condition, the outer shells are complete and a stable condition exists. When hydrogen and oxygen atoms combine to form water, the force that bonds them together is called **covalent bonding**.

A covalent bond is formed when valence electrons are shared by atoms.

When water is formed, the two hydrogen atoms share their valence electrons with the oxygen atom. The oxygen atom, in turn, shares one of its valence electrons with each of the hydrogen atoms. This allows each atom to satisfy its need for chemical stability. Now all three atoms appear to have a complete outer shell and the molecule they form is stable.

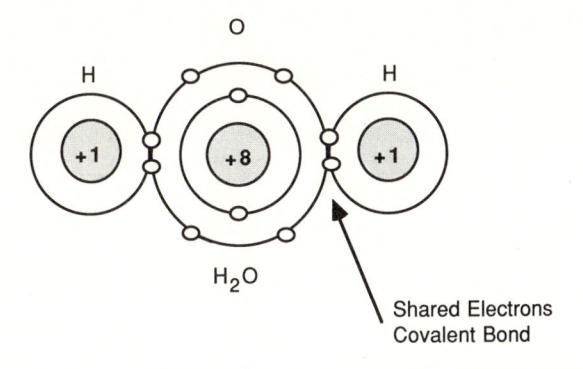

Fig. 1-8 Covalent Bonding to Form a Molecule

Self-Check

Indicate whether each statement is True or False.

1. _____ The materials from which conductors are made have valence electrons that exist at an energy level which makes them easily moved into the conduction band.
2. _____ Elements that have a high valence make good conductors.
3. _____ Covalent bonding refers to the sharing of valence electrons by atoms.

4. _____ Silicon and germanium both have a valence of four.
5. _____ Electrons that orbit in the M shell exist at a higher energy level than the electrons in the N shell.

Germanium and Silicon Crystals

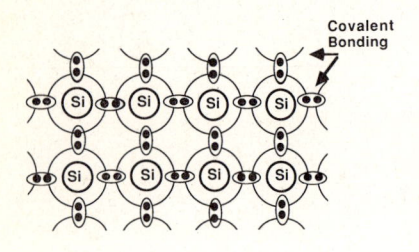

Fig. 1-9 Arrangement of Atoms in a Silicon Crystal

A CRYSTAL is a solid substance in which the atoms or molecules are arranged in definite, repeating patterns and the external surface has a symmetrical arrangement. This arrangement is called a **lattice structure**. Germanium and silicon are both crystalline substances. Remember, each silicon atom has four valence electrons and needs four more electrons in its valence band to become stable. To reach stability each silicon atom will align itself between its four neighbor atoms so that its spacing is equal in all directions. In this position each atom can share its valence electrons with the adjoining atoms. This type of arrangement is shown in Figure 1-9. Notice that each atom appears to have eight valence electrons. The covalent bonding that occurs between the five atoms results in a stable substance that is held together by the covalent bonding. When viewed under a microscope this substance looks like a lattice structure. Figure 1-9 illustrates this arrangement.

Doping Germanium and Silicon Crystals

In their pure (**intrinsic**) form, silicon and germanium crystals are of little use as semiconductor devices because they are actually insulators. When we add certain **impurities**, however, the crystal can be made to conduct current of sufficient amounts to be useful. To do this, the impurity that is added to pure silicon or germanium must be rigidly controlled as to quality and quantity. By adding impurities, the crystal is converted into a crystal that appears to have either an excess, or a deficiency of electrons.

Atoms that have an imbalance of electrons are called **ions**. **Positive ions** have a deficiency of electrons and **negative ions** have excess electrons. The type of impurity used will determine whether the excess is holes or electrons. After silicon or germanium has been mixed with the impurity, it is said to be **extrinsic** (not pure). The impurities used to mix with silicon and germanium are a type that will be bound into the crystal by covalent bonding in much the same way as atoms of the pure crystal were bound together. The process of mixing impurities with a pure semiconductor material is called **doping**.

Arsenic, phosphorus, antimony, and bismuth are elements that have five valence electrons. These are called **pentavalent** elements because of their five valence electrons. When a pentavalent material is mixed with pure silicon or germanium the results can be predicted. Figure 1-10 shows a crystal with one arsenic atom covalent bonded into a silicon crystal.

Donor Impurity Bound in a Silicon Crystal

Refer to Figure 1-10. When a pentavalent atom is doped into a pure silicon or germanium crystal, four of its electrons are used in the covalent bonding. This leaves one electron that is not bound into the stable crystal. The extra electron is allowed to move into the conduction band of the crystal and it becomes a **free electron**. Since a pentavalent impurity donates one electron to conduction, it is called a **donor** impurity. When donor impurities are added to crystals, the number of free electrons is increased. In other words:

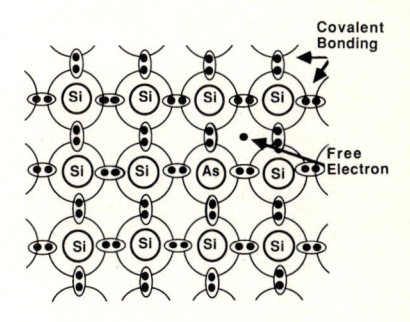

Fig. 1-10 Pentavalent Impurity Covalent Bonded with Silicon

A pure semiconductor crystal is doped with a donor impurity to increase the number of free electrons available to support current flow.

Because of the electron's negative charge, the material that results is said to be a **negative-carrier**, or **N-type material**. The more donor atoms that are added, the greater the number of free electrons that exist. In actuality, the doping rate is on the order of one impurity for each 10 million germanium or silicon atoms.

Semiconductor materials that are formed using donor impurities are N-type materials.

Acceptor Impurity Bound in a Silicon Crystal

Indium, boron, and gallium are **trivalent** (three valence electrons) impurities. When they are used as dopants the material that results is a **P-type semiconductor**. When a trivalent atom is bound into a crystal, as shown in Figure 1-11, one of the covalent bonds will be minus an electron needed to make the crystal stable. This leaves a vacancy at one position within the covalent bonding of each impurity atom. The vacancy is called a **hole**. For the material to be stable, each hole would have to *accept* an electron. For this reason, this type material is called an **acceptor** material.

As acceptor impurity atoms are added to silicon or germanium, the number of *positive carriers* (holes) is increased. To take this further, the hole acts the same as a positive electrical charge and therefore the acceptor material is called a **positive-carrier** or **P-type semiconductor**. When a free electron is created in the space near an acceptor material, the electron will be captured (accepted) by the P-type mate-

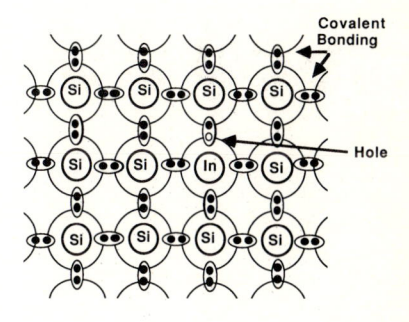

Fig. 1-11 Trivalent Impurity Covalent Bonded with Silicon

rial. The capture results from the tendency of all materials to exist in a stable state and they do this by forming complete covalent bonds. The acceptor atom accepts and "stores" an electron from the valence band of the crystal. The movement of an electron into an acceptor atom leaves a hole in the valence band. This "creates" a positive mobile carrier in the valence band as shown in Figure 1-11.

Conduction Within Semiconductor Materials

As mentioned earlier, a semiconductor's characteristics lie between those of a conductor and those of an insulator. A semiconductor's valence electrons can be converted to free electrons if a large enough energy source is applied. When valence electrons have absorbed energy and moved into the conduction band, the material takes on some of the characteristics of a conductor. Both silicon and germanium will release a few electrons to the conduction band at room temperature. Each electron that moves into the conduction band leaves behind it a "vacancy" that can accept one electron. The position vacated by the electron is called a "hole". A hole has the characteristics of a positive ion whose charge is opposite but equal to the electron's negative charge. When an electron breaks its bond and creates a hole, we say that an **electron-hole pair** has been generated.

If voltage is applied to a pure crystal of germanium or silicon, electrons will break their covalent bonds and either of two types of current flow is possible. Free electrons can move through the conduction band of the crystal in a path that supports movement from the negative to positive poles of the power source. Hole flow (movement of holes) can occur in a direction opposite to the flow of electrons. Hole movement is within the valence band and consists of the following: An electron moves from its covalent bond into and occupies a hole. The position from which the electron moved is now a hole. As another electron moves into and fills this hole, the hole's position has now moved to the location from which the electron moved. Therefore, holes seem to move in a direction that is opposite to the flow of electrons. This is to say:

Holes flow from the positive pole of a power source to the negative pole of the power source.

> **NOTE:** Hole flow is also called **conventional current flow.**

It is important to remember that hole flow exists only in the valence band of the semiconductor, and electron flow exists only in the conduction band. Current flow in the external circuit can be supported within a semiconductor by either hole flow or electron flow.

Figure 1-12 illustrates the movement of a hole through a crystal. Originally the hole was positioned at P_1. A valence electron, located at P_2 moves over to fill the hole. When the electron left P_2, a new hole was created there and the hole at P_1 disappeared. This would indicate that the hole has moved from P_1 to P_2. The action repeats over and over again as the hole moves from one atom to another within the crystal.

Charges in N- and P-Type Materials

When a donor impurity is combined with a silicon or germanium crystal, the donor atom forms covalent bonds with four atoms and donates one electron to the conduction band. However, when the donor atom loses an electron, it becomes a positive ion. The negative electrical charge that accompanies the electron is balanced by the positive charge of the donor positive ion. The net result is that the semiconductor crystal continues to have a neutral charge due to the fact that there are equal numbers of electrons and protons within the crystal. There will be, however, regions within the crystal where positive charges exist and other areas where negative charges exist. Remember, though, this type material is called an N-type because current is supported by negatively charged carriers.

P-type material has a similar structure. Suppose that silicon is doped with the acceptor impurity indium. Indium has three valence electrons. When it is used to dope silicon, a hole will exist in one of the covalent bonds. If an electron moves into and occupies this hole, the indium atom will have an excess electron in its valence band. In this way, the indium atom will become a negative ion. If the silicon atom releases an electron to the indium atom, the silicon atom will become a positive ion. Again, the net charge of the crystal continues to be neutral because the number of electrons and protons remains equal. However, positive charges exist in some portions of the crystal and negative charges exist in other portions.

The ionized (charged) atoms are equally distributed throughout the crystal. Remember, like charges repel each other. Therefore, positive ions repel positive ions and negative ions repel negative ions. This results in equal distribution. If large numbers of positive charges could exist in an area, that area would attract electrons from the areas that contained electrons and both would be neutralized in the process. The end result is that the charges are distributed equally throughout the material.

Both holes and electrons are current **carriers**. Holes are positive carriers and electrons are negative carriers. The carrier that is present in the largest amount is called the **majority** carrier and the other is called the **minority** carrier. In N-type semiconductor material, elec-

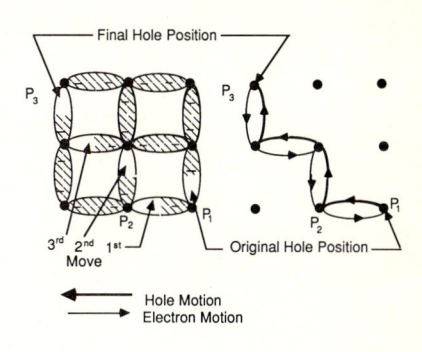

Fig. 1-12 Hole Flow in a Semiconductor

trons are the majority carriers and holes are the minority carriers. In P-type material, holes are majority carriers and electrons are minority carriers. Figure 1-13 shows the majority and minority carriers.

Current Flow in N-Type Material

Current flow in N-type material is illustrated in Figure 1-14. The conduction that exists in an N-type semiconductor is similar to the conduction that exists in a conductor. When voltage is applied to the material, conduction will exist through the crystal as shown. Remember, in this case conduction occurs as a result of carriers moving through the conduction band.

Some differences exist between N-type material and a copper conductor with respect to the effect of temperature. The semiconductor's temperature coefficient is such that an increase in its temperature causes its resistance to decrease. In a copper conductor, increasing the temperature increases the resistance of the conductor. A semiconductor material has a **negative temperature coefficient**. This means that as operating temperature increases, the resistance of the material will decrease.

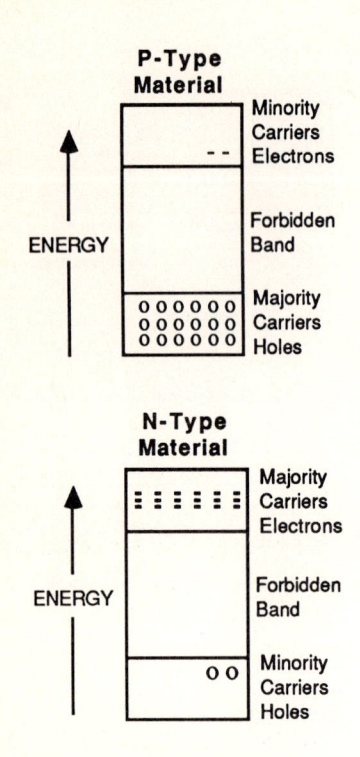

Fig. 1-13 Majority Versus Minority Carriers (energy level diagrams)

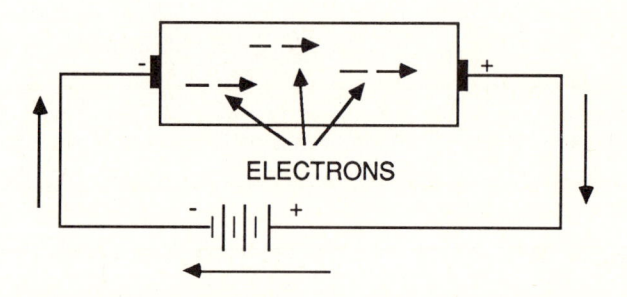

Fig. 1-14 Current Flow in N-Type Material

Current Flow in P-Type Material

Current flow in P-type material is illustrated by Figure 1-15. Remember, conduction occurs in P-type material because of positive carrier (hole) action within the valence band of the crystal. In order for the hole to move, an electron from a nearby atom must move into and occupy the hole. The position from which the electron moved is now a hole and, in effect, the hole has moved to a new location. In hole flow, the movement of electrons into a hole near the positive pole of the battery results in a continual movement of holes toward the negative terminal of the power source. The end result is that electron flow in the circuit outside the crystal is identical to that in the N-type crystal.

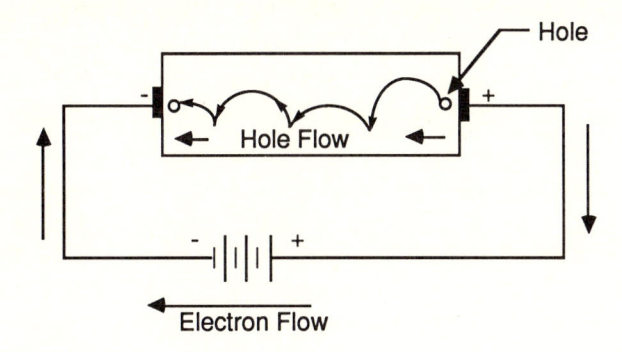

Fig. 1-15 Current Flow in P-Type Material

In both N-type and P-type materials, current flow in the external circuit consists of electrons moving out of the negative terminal of the battery and into the positive terminal of the battery. Both P- and N-type semiconductor materials have negative temperature coefficients. If the battery connections to the semiconductor crystal are reversed, current will flow in the opposite direction in the same amount that it flowed in the previous direction. This leads to the definition:

**A device that conducts current equally
well in both directions is called
a bidirectional device or a bilateral device.**

Self-Check

Indicate whether each statement is True or False.

6. ____Positive-current carriers are called holes.
7. ____N-type material results when a semiconductor is doped using a trivalent substance.
8. ____Hole flow is also referred to as conventional current flow.
9. ____One silicon atom forms covalent bonds with four neighbor atoms.
10. ____After a semiconductor material has been doped, it will have a neutral electrical charge.
11. ____Electron flow within a silicon crystal is conducted within the valence band.
12. ____P-type material contains an equal number of positive and negative ions.
13. ____A pentavalent atom that is covalently bonded into a silicon crystal has released an electron that becomes a free electron.

14. ____Silicon has a positive temperature coefficient.
15. ____Electron flow outside a semiconductor material is from the positive pole to the negative pole of the power source.

Summary

Semiconductor materials have a lattice structure. The atoms that are bound into the crystal are held in position by covalent bonding of their valence electrons. Natural materials such as silicon and germanium must be doped with an impurity before they are useable as semiconductor materials in electrical circuits. To create P-type material, the dopant atoms must be trivalent (three valence electrons). To create N-type material, the dopant must be pentavalent (five valence electrons). In the P-type material, a hole is created in the lattice structure for each impurity atom that is covalent bonded into the structure. These holes become the positive current carriers. In the N-type material, one free electron is released for each impurity atom that is covalent bonded into the crystal's structure. These electrons become the negative current carriers for the N-type semiconductor.

Review Questions and/or Problems

1. When an atom has more electrons than protons, it is called a ____.
   a. positive ion   c. positive atom
   b. negative ion   d. negative atom

2. A good conductor ____.
   a. opposes the movement of electrons
   b. has many free electrons
   c. must have very few electrons
   d. has many electrons

3. The path that an electron takes around the nucleus of an atom is called a/an ____.
   a. orbit   c. energy level
   b. shell   d. all of these

4. A good insulator _____.

   a. has many free electrons
   b. has few free electrons
   c. will stop all current flow
   d. all of the above

5. A semiconductor's resistance is ____ the resistance of an insulator.

   a. equal to
   b. less than
   c. greater than
   d. none of these

6. A P-type crystal will be created when silicon is doped with a ____ substance.

   a. pentavalent
   b. trivalent

7. The majority carriers in an N-type semiconductor are ____ and the minority carriers are _____.

   a. holes, electrons
   b. holes, holes
   c. electrons, holes
   d. electrons, electrons

8. In N-type semiconductor materials, current flow is supported by _____.

   a. positive ions
   b. electrons
   c. positive carriers
   d. holes

9. The energy diagram for an insulator shows a/an ____ forbidden region between the conduction and valence bands.

   a. overlapping
   b. wide
   c. narrow

10. Germanium and silicon have a valence of _____.

    a. 2
    b. 4
    c. 6
    d. 8

11. A hole has characteristics that are similar to those of a ____ ion.

    a. positive
    b. negative

12. A crystalline structure will exist when ____ atoms share covalent bonds.

    a. 3
    b. 4
    c. 5
    d. 6

13. In its natural state, silicon has many characteristics that are similar to a/an _____.

    a. conductor
    b. insulator

14. In a P-type semiconductor, current flow is supported by _____.

    a. electrons
    b. positive ions
    c. holes
    d. negative ions

15. When an atom has fewer electrons than protons, it is called a _____.

    a. positive ion
    b. negative ion
    c. positive atom
    d. negative atom

16. A semiconductor's resistance is ____ the resistance of a conductor.

    a. equal to
    b. less than
    c. greater than
    d. none of these

17. An N-type crystal will be created when silicon is doped with a ____ substance.

    a. pentavalent
    b. trivalent

18. The majority carriers in a P-type semiconductor are ____ and the minority carriers are _____.

    a. holes, electrons
    b. holes, holes
    c. electrons, holes
    d. electrons, electrons

19. The valence of an element indicates its chemical activity.

    a. True
    b. False

20. Conductors are made of elements that have ____ valence.

    a. high
    b. medium
    c. low

2
The PN Junction Diode

Objectives

1. Define the following terms:

   a. PN junction
   b. Diode
   c. Depletion region
   d. Recombination
   e. Junction
   f. Junction barrier
   g. Junction resistance
   h. Junction capacitance
   i. Cutoff
   j. Barrier region
   k. Bias
   l. Forward bias
   m. Reverse bias
   n. Peak inverse voltage (PIV)
   o. Characteristic curve
   p. Diode load line
   q. Saturation
   r. Active

2. Recognize and draw the schematic symbol for a PN junction diode.
3. Recognize and draw the schematic diagrams for PN junction diodes that are

   a. Reverse biased.
   b. Forward biased.

4. Predict the direction and amount of current that will flow when a diode is

   a. Reverse biased.
   b. Forward biased.

5. Describe the effect that forward (or reverse) bias has on

   a. Majority carriers.
   b. Minority carriers.

6. Name the three modes of diode operation and verbally describe the diode's condition in each mode.

Introduction

Now that you have been introduced to semiconductor physics, ways can be explored that show you how N- and P-type materials are used.

P- and N-type materials will be joined together to form a PN junction. The PN junction forms a device called a **diode** which has specific characteristics. Knowing these characteristics allows you to accurately predict the diode's operation. In this chapter we discuss the construction and use of diodes. Actually, the diode is a very simple solid state device, but by knowing how it reacts to current flow, we can relate that knowledge to other PN junction devices. The diode has two forces that combine to control its operation; an electrostatic field that exists across the junction, and voltage that is applied to the junction. A voltage that is used to control current flow is called **bias** voltage.

Before starting this study, let us review some facts regarding P- and N-type materials. **P-type material** is produced by doping silicon or germanium with a **trivalent** (3 valence electrons) element called an **acceptor** impurity. In the P-type material, **holes** are the **majority carriers** and the **electrons** are **minority carriers**. You should remember, though, that since the material has an equal number of electrons and protons, it has a neutral electrical charge.

N-type material is produced by doping germanium or silicon with a **pentavalent** (5 valence electrons) element called a **donor** impurity. In the N-type material, **majority carriers** are **electrons** and **minority carriers** are **holes**. This material, like the P-type, has an equal number of electrons and protons which assures that it retains its neutral electrical charge.

The PN Junction

It is possible to produce semiconductor materials that serve specific purposes. This is done by using a chemical process wherein donor impurities are doped into one part of a semiconductor crystal, and an acceptor material is doped into the other part. The result is a single crystal that has one P-type section and one N-type section. The region where the P- and N-type materials physically join is called a **PN junction**. For a PN junction to be electrically useful it must be one continuous crystalline structure. Metallic contacts are bonded to each end of the crystal. The result is a **PN junction diode** (Figure 2-1) that can be attached within the conductive path of a power source.

An isolated piece of N-type material is electrically neutral; that is, for every free electron in the conduction band there is a positively-ionized donor atom in the crystal. Even though there are many free electrons, each is balanced electrically by a positive ion. Therefore, the net charge of the crystal is neutral.

Figure 2-2 contains an electrical representation of isolated P- and N-type materials having balanced charges. Figure 2-2(a) uses energy levels to show the charged carriers. Figure 2-2(b) shows the even distribution of carriers and ions throughout the two crystals. The ions

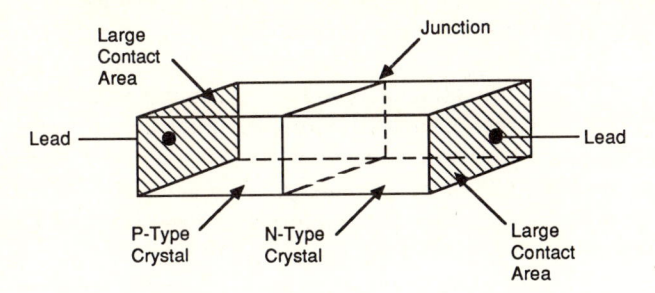

Fig. 2-1   PN Junction Pictorial Diagram

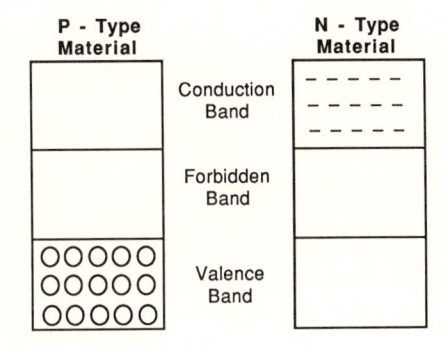

(a) Energy Level Diagram

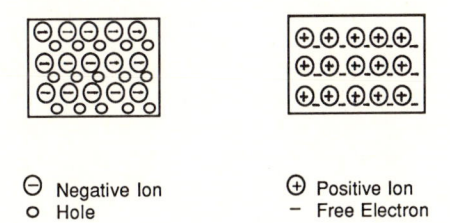

⊖ Negative Ion
o Hole

⊕ Positive Ion
− Free Electron

(b) Pictorial Diagram

Fig. 2-2   Majority Versus Minority Carriers (Energy Level Diagrams)

and carriers are placed side-by-side to indicate the balancing effect of positive and negative charges.

When P- and N-type semiconductor crystals are joined to form a junction, an interesting thing happens. Electrons in the N-type material begin to physically move across the junction into the P-type material. As the electrons move through the junction, they release their excess energy and are attracted into the valence band of the P-type material. Here they will be attracted into a hole. This process is called **recombination**. As recombination occurs one free electron and one hole are removed from the current-carrying process. When the electron leaves the N-type material, a positive ion is created. Remember, a positive ion is deficient one, or more, electrons. The N-type material is no longer neutral but has a positive electrical charge equal to the number of positive ions that are present. Much the same thing occurs in the P-type material. The elimination of a hole in the P-type material results in the P-type material's having one excess electron for each hole that recombines with an electron. This causes an **electrostatic** condition to occur in the region of the junction which causes the N-type section to have a positive static charge. This same condition exists in the P-type material, causing it to have a negative static charge. An **electrostatic field**, therefore, exists across the junction, as a result of recombination. The electrostatic field exerts its force from the positive charge (N-type material) to the negative charge (P-type material). This condition is depicted by the accumulation of holes and electrons that appear on either side of the junction, shown in Figure 2-3. The dark line in the middle of the figure represents the junction with the depletion area (set aside by the dashed lines).

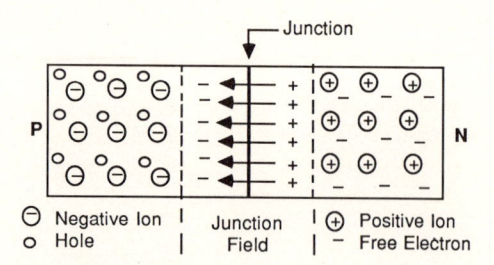

Fig. 2-3 PN Junction Field

The ions that are created by recombination are locked into the crystalline structure and cannot move; thus, they form a fixed electrical charge that exists on either side of the junction. In the N-type material, there is a layer of positive ions and in the P-type, there is a layer of negative ions. The result is an electrostatic field existing across the junction which results from oppositely-charged ions. Figure 2-3 illustrates the electrostatic field that exists. This is called the **junction field**. (The field shown in the diagram is greatly exaggerated for purposes of explanation and illustration.) As electrons move across the junction, the electrostatic field increases. Electrons continue to cross the junction until the electrostatic field is so strong that electrons can no longer overcome its charge. At this point, a balance exists and, for all practical purposes, the movement of electrons across the junction stops. Recombination occurs in a very small amount of time and is complete long before the crystal is used to form a useful operating device. Recombination affects only those carriers that are closest to the junction. Throughout the rest of the P- and N-type materials, carriers remain in their electron-hole pairs and the material remains electrically balanced.

Figure 2-3 also illustrates that there are no mobile carriers within the junction field. If a mobile carrier should appear within the field, it would be forced outward to the boundaries of the field. Electrons are attracted in a direction opposite the direction of the arrows, and holes are attracted in the direction indicated by the arrows. Because the junction field contains no mobile carriers, it is called a **depletion region, depletion layer**, or **barrier region**. It is possible that this area might also be referred to as the **space charge region, barrier**, or **junction barrier**.

A second way of illustrating the junction barrier is shown in Figure 2-4. An energy level diagram shows the free electrons in the conduction band of the N-type material and the holes in the valence band of the P-type material. Free electrons from the N-type material move across the junction into the conduction band of the P-type material. This electron joins with a hole and contributes to the building of the junction barrier. The energy diagram shows the barrier to have both

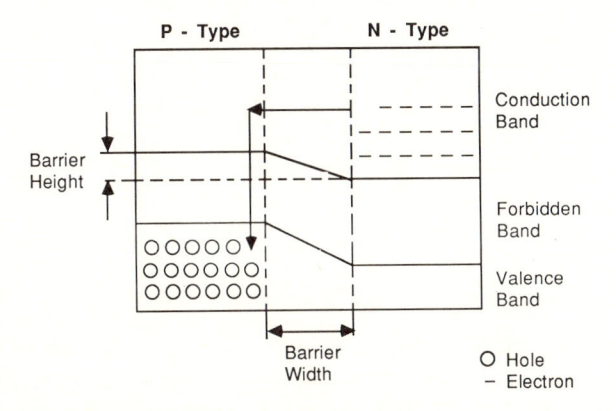

Fig. 2-4 Junction Barrier Formation Energy Level Diagram

height and width. The physical distance from one side of the barrier to the other side is called the barrier width. The width of the barrier, when no external potential is applied, depends on the number of mobile carriers that exist within the crystal. The barrier height is called the **potential hill**. The potential hill is the difference of potential that exists across the barrier. As recombination of holes and electrons continues, the barrier becomes wider and the potential hill becomes larger. When the barrier is large enough to prevent further recombination of electrons and holes, a balance will exist. In other words, the barrier prevents total recombination of electrons and holes across the PN junction. Those holes and electrons that remain in each section remain bound in electron-hole pairs.

Self-Check

Indicate whether each statement is True or False.

1. ____In P-type material, holes are majority carriers.
2. ____N-type material is produced by doping silicon or germanium with a trivalent substance.
3. ____PN junction diodes are formed by doping opposite ends of a crystal with the same impurity.
4. ____In N-type material, holes are the minority carriers.
5. ____Recombination occurs at the time that the junction is formed.
6. ____Across a PN junction, the two sections have opposite electrical charges.
7. ____It is possible to form a PN junction by physically connecting a piece of P-type material to a piece of N-type material.
8. ____A pentavalent substance is called a donor impurity.
9. ____An isolated piece of semiconductor material will have a neutral electrical charge.
10. ____The electrostatic charge that is created in the barrier will prevent the total recombination of electrons and holes.

Junction Barrier Formation

When an N-type crystal and a P-type crystal are joined to form a junction, the device that results is called a PN junction diode. Diodes of this type have some limitations. One limitation is that, as a result of increased heat, light, or electromagnetic radiation, electron-hole pairs are formed at random throughout the crystal.

When this happens, the electron-hole pairs that are generated are minority carriers. These minority carriers cause current to flow in the reverse direction during the time when we would prefer that the diode

The PN Junction Diode

did not conduct. Minority carriers are present in both sections; holes in the N-type material and electrons in the P-type. The direction of the junction field is such that minority carriers (electrons) are easily accelerated to the point that they cross the junction and recombine in the other crystal. Recombination causes the number of **holes** contained in P-type materials to decrease, and the number of **free electrons** available in the N-type material to decrease. Current flow that results from the movement of minority carriers is called **reverse current**. Reverse current results from hole pairs that are generated within the lattice crystal. Reverse current does *not* result from the flow of the carriers released by doping.

Figure 2-5 shows the electron-hole pairs that exist within an energy level diagram. When external energy is applied to the PN junction diode, some of the electrons that are bound into the lattice structure will absorb enough energy to break their covalent bonds.

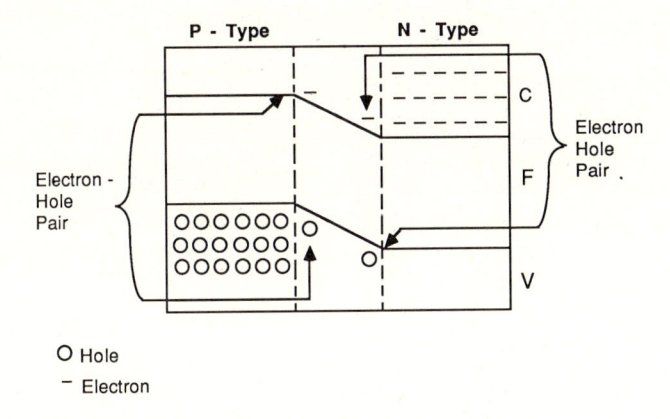

Fig. 2-5 Electron-Hole Pair Generation Energy Level Diagram

These electrons, within the N-type material, then move into the conduction band. In the **P-type** material, electrons are called **minority carriers**. In **N-type** material, the holes generated are the **minority carriers**. It is possible for the electron in the P-type conduction band to move into the N-type material and recombine with a hole. It is also possible for a hole to attract an electron from the N-type material into the P-type material, where they recombine. The current that results is called **minority current**. In later studies, minority current will be called **reverse** current.

After the recombination of minority carriers, the diode takes on different characteristics from those stated earlier. The electrons that move out of the N-type material leave holes behind. When electrons cross the junction and recombine with holes, excess electrons exist within the P-type material. This means that the N-type material has a positive charge and the P-type has a negative charge. The two opposite charges are concentrated within the area of recombination. Now the negatively charged P-type material will repel any electron that attempts to cross the junction.

The area around the junction, where recombination has occurred, will have free electrons in the P-type material and holes in the N-type material. The area where this occurs is called the **depletion region**. The depletion region is actually a barrier that will resist the movement of other electrons across the junction. This barrier can be represented by a battery as shown in Figure 2-6. Notice that the electrons (majority carriers) located in the N-type material feel a negative (repelling) force at the barrier and will not move across the junction.

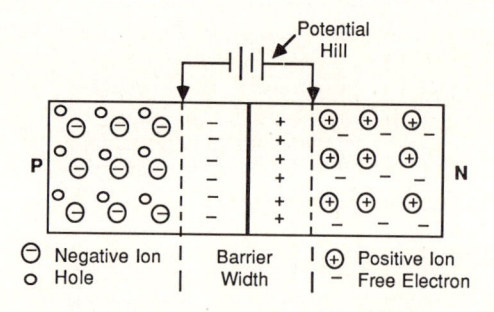

Fig. 2-6 Battery Representation of the Barrier Region

Biased PN Junctions

When a battery is connected across a PN junction, it causes the depletion region to react in a predictable way. Any voltage that is connected across a PN junction is referred to as **bias** voltage. When the battery is connected such that its energy opposes the barrier, current flow will be aided, and the junction is said to be **forward biased**. If the battery's potential aids the barrier, it makes current flow even more difficult, and the junction is said to be **reverse biased**. When the junction is forward biased it offers less opposition to the flow of electrons and either decreases the width of, or cancels the barrier entirely. When the junction is reverse biased it has more opposition to the flow of electrons than an unbiased junction. Reverse bias actually increases the barrier width. A reverse biased junction is a **high resistance** junction and a forward biased junction is a **low resistance** junction.

Forward Bias

To forward bias a junction, the battery must be connected as illustrated in Figure 2-7. Notice that the positive terminal of the battery is connected to the P-type material and the negative terminal of the battery is connected to the N-type material. A good way to remember this is to memorize the following:

"P" to "P" and "N" to "N" for FORWARD bias.

Note that forward bias opposes the electrostatic field that exists across the junction.

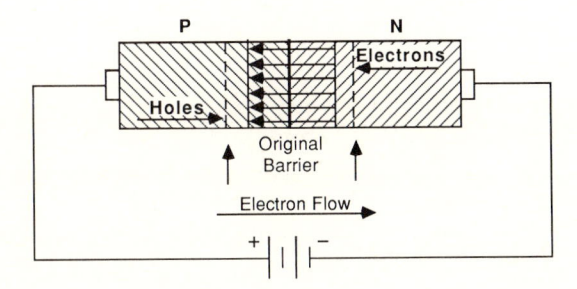

Fig. 2-7 Forward Biased PN Junction

With a positive connected to the P-type material, the positive carriers (holes) will be repelled away from the terminal and will move toward the junction. The negative, connected to the N-type material, repels free electrons and they move toward the junction. The combination of the two repelling forces act to narrow the barrier. As the barrier is narrowed, its height will be lowered. This creates a forward biased junction which will allow the flow of majority carriers.

Electron flow in a forward biased PN junction moves through the N-type material, into and through the P-type material and back to the battery. This conduction can be explained by the following description.

   An electron leaves the negative pole of the power source and moves to the terminal connected to the N-type material. The electron enters the N-type material where it is a majority carrier which is repelled to a point near the junction. As it approaches the accumulation of negative electrons near the junction, it meets a repelling force and begins to slow. As it slows, it releases part of its excess energy which is, in turn, absorbed by an electron nearer the junction. The added energy allows the electron to move across the barrier where it enters the P-type material. The entering electron is a minority carrier and will quickly fill a hole in the P-type material. As it stops moving, the excess energy that caused its movement is transferred to one of the valence electrons within the lattice structure. This electron is moved to the positive connection of the P-type material by hole flow. Once it arrives at the P-type material's positive connection, the high positive pressure (present as a result of the battery) attracts the electron out of the P-type material. Once it is in the conductor, the electron is free to return to the positive pole of the battery. Notice that the electron moves from negative to positive in the external circuit and from N to P within the junction diode.

It should be noted that minority carriers are moving away from the junction during the period of time when majority carriers are moving toward the junction.

You should remember that when a junction is forward biased, *majority carriers* are the source of current flow and the *minority carriers* are reverse biased. A forward biased junction will conduct quite heavily and current flow can be quite large. As forward bias is increased, the number of majority carriers that move through the barrier will increase, and total current will increase. There is a point, however, where increasing the forward bias no longer increases current flow. When this happens, the junction is said to be **saturated**. Operating a junction at saturation could possibly cause the junction to be destroyed by heat. This means that the device must be selected to meet the requirements of the operating circuit. Selection of diodes and other devices with the correct ratings for a desired application will be discussed in this and later chapters.

Figure 2-8 illustrates the effect of a forward bias on barrier width and height. With forward bias, both the height and width of the barrier will decrease. The decrease in width and height represents a lower resistance and allows more current to flow.

When forward bias exceeds the barrier potential of the junction, electrons cross the junction. When electrons cross the junction they recombine with a hole in the P-type section. For each recombination, an electron leaves the battery and enters the N-type material. At the same time, an electron is attracted away from the P-type material and returns to the battery. Notice that the majority carriers (electrons and holes) support the flow of current. For this reason, forward current is called **majority current**. With forward bias applied:

**Majority carriers move toward the junction.
Minority carriers move away from the junction.**

Zero Bias | Forward Bias

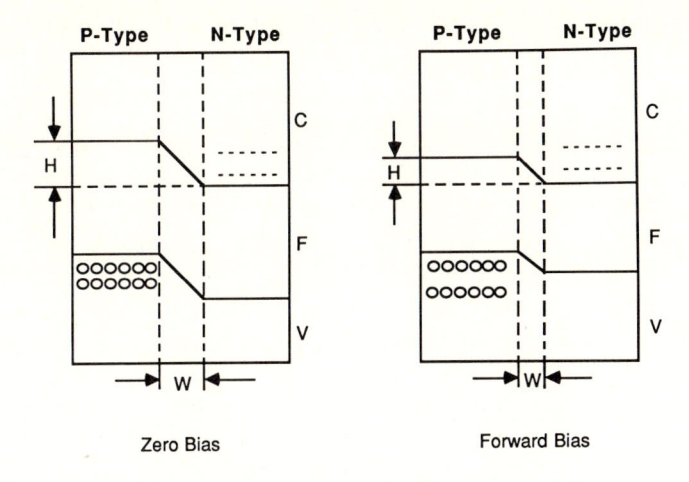

Fig. 2-8 Effect of Forward Bias on Barrier Height (H) and Width (W)

Reverse Bias

A junction that is connected:

"P" to "N" and "N" to "P" is reverse biased.

Reverse bias is connected in a way that aids the electrostatic field of the junction and increases the barrier width. When connected for reverse bias, the negative pole of the battery attracts the holes (majority carriers) away from the junction, and the positive pole attracts the electrons (majority carriers) away from the junction. The barrier width and height are both increased by reverse bias. This increases the resistance of the junction and means that majority current is cut off.

Figure 2-9 represents a reverse biased PN junction. Notice that the barrier has increased. Because the majority carriers are concentrated

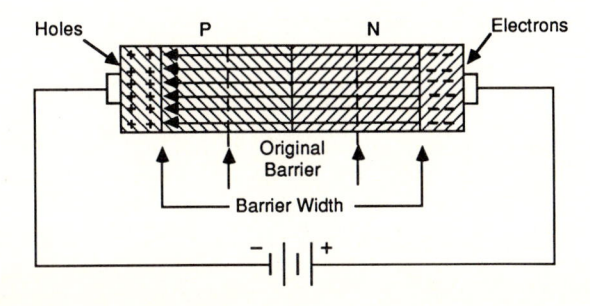

Fig. 2-9 Reverse Biased PN Junction

further away from the junction, majority current will not flow. The small amount of current that does flow is the result of minority carriers and is called **minority** or **reverse** current. With reverse bias applied:

Majority carriers move away from the junction.
Minority carriers move toward the junction.

Energy diagrams that explain the effect of reverse bias on a PN junction barrier are shown in Figure 2-10. You can see from the diagram that as reverse bias increases, the barrier increases. As the barrier increases, junction resistance increases.

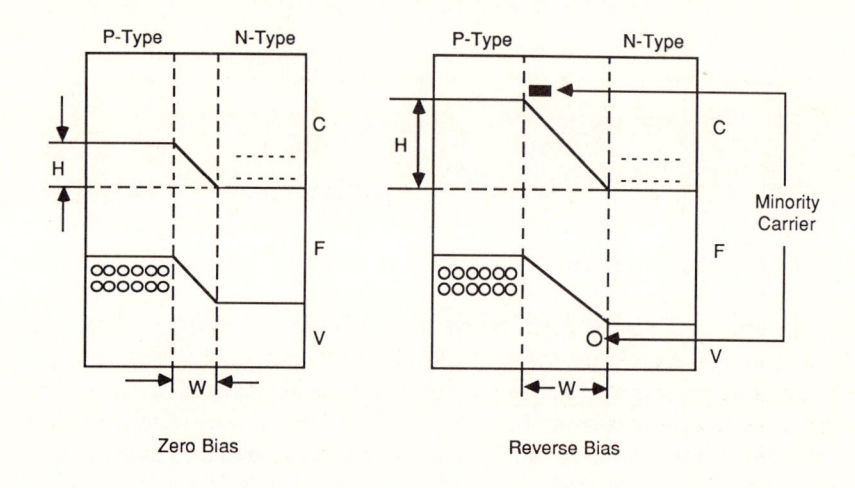

Fig. 2-10 Effect of Reverse Bias on Barrier Height (H) and Width (W)

In a reverse biased junction, majority carrier current is cut off. Minority carriers, however, are affected differently. In P-type material, the minority carriers (electrons) are repelled and move to the junction area as are the minority carriers (holes) in the N-type material. This sets up a condition where the minority carriers can, and will, cross the barrier in small amounts to provide a small reverse current. Minority carrier (reverse) current is of no great concern in most operating situations. However, in some cases, it is possible that a device which is highly sensitive to temperature must be operated at a point where reverse current becomes a problem. When this is the case, special precautions must be taken. Remember, as temperature increases the barrier width will narrow providing less resistance to current flow. Therefore, at high temperatures, reverse current can be quite high and possibly destroy the device.

Self-Check

Indicate whether each statement is True or False.

11. ____ A forward biased junction will have high resistance.
12. ____ A reverse biased junction has a positive voltage connected to its P-type material.
13. ____ As forward bias increases, the barrier width increases.
14. ____ In a forward biased PN junction, external current is supported by electrons.
15. ____ Minority current flows when reverse bias is present.

PN Junction Characteristics

Earlier, we mentioned the fact that a junction presents an opposition to current flow. This opposition consists of resistance for both DC and AC current. Capacitance is described as two conductors separated by a dielectric (insulator). Capacitive reactance (X_C) is the opposition that a capacitor offers to the flow of alternating current (AC). Two sections of semiconductor material that are separated by a junction meet the requirements for a capacitor. Therefore, in addition to resistance, a PN junction provides capacitive reactance (X_C) to AC current flow. In the circuits discussed here, X_C will not affect circuit operation. In more complex circuits, we must consider junction capacitance in our circuit designs.

The pictorial diagram and schematic symbol for a PN junction diode are shown in Figure 2-11. The straight line (bar) represents the

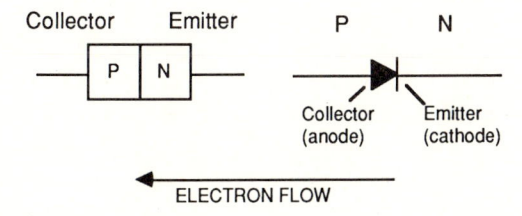

Fig. 2-11 PN Junction Diode Symbols

emitter (cathode) of the diode and the arrow represents the **collector** (anode). In a diode, electron current flows opposite to the direction indicated by the arrow.

**To forward bias a diode the anode must
be more positive than the cathode.**

**To reverse bias a diode the cathode
must be more positive than the anode.**

A numbering system that is used to identify diodes is one that starts with 1N. A common diode that finds its way into many circuits is the 1N4001. The first two places (1N) identify the device as diode. The remaining places identify the diode by type. Using this identification number, detailed characteristics of the diode can be checked for use in a circuit design.

A diode is rated according to the following characteristics:

Maximum Average Forward (Majority) Current
Peak Recurrent Forward Current
Maximum Surge Current
Peak Inverse (Breakdown) Voltage
Maximum Power Dissipation

Maximum average forward current is rated at a specific temperature (usually 25°C or 77°F) and refers to the maximum average current that the diode can conduct without damage. Typical amounts vary from a few milliamps to very high amperes.

Peak recurrent current is the maximum peak current that the diode can conduct when a varying current is present, as in AC operations. This is the maximum (peak) current that the diode can conduct on recurring applications of a peak forward bias.

Maximum surge current is the maximum current that can be permitted to flow in the forward direction for very short periods of time. This value will be larger than either of the two currents explained above. Care should be taken to assure that diode operational current is well below the surge current value.

Peak inverse voltage (PIV) is one of the most important characteristics of a diode. The PIV value indicates the maximum amount of reverse bias that can be applied to a diode without causing a break down of a diode junction. PIV may also be referred to as **breakdown** voltage or **peak reverse voltage (PRV)**.

Maximum Power Dissipation refers to the ability of the diode to dissipate power. Each diode is manufactured to meet specific applications. Care must be taken to assure that the diode used is capable of handling the power requirements of the circuit being designed. Silicon diodes are better able to withstand the high power dissipations that occur in many circuits and are used for these applications. Silicon diodes are used extensively in power supply circuitry. Germanium diodes do not have the high power capability of silicon but they do have

The PN Junction Diode

lower junction resistances. Germanium diodes are often used in signal handling applications.

Technicians must remember that all of these ratings vary at temperatures other than the ambient temperature stated as part of the diode specifications. If the diode is to be operated at a temperature higher than the stated ambient temperature, all of these ratings will be lower than those stated for the diode. It is good practice, therefore, to select a diode that has ratings well above the requirements of the circuit in which it will operate.

Characteristic Curves

Earlier, it was established that resistors, pieces of P-type crystal, pieces of N-type crystal and many other devices were **bidirectional** devices. The diode is a **unidirectional** device. Its design is such that it will allow practical amounts of current to flow in one direction only. Figure 2-12 shows the current-voltage characteristics of a 200 Ω resistor. Note

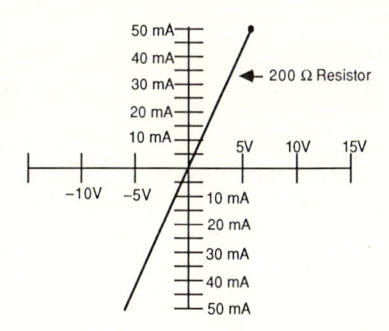

Fig. 2-12 Current Voltage Relationships for a 200 Ohm Resistor

that the voltage and current relationship is very linear and remains that way along the entire length of the line.

Figure 2-13 displays the voltage-current relationships of a diode. Some interesting differences are immediately visible when we compare Figures 2-12 and 2-13. Using Ohm's law:

$E = I \times R$ or

$I = \dfrac{E}{R}$

several operating points can be calculated. When these points are plotted for a resistor (Figure 2-12) the current voltage line is perfectly straight. The same slope is present regardless of the direction of current flow through the resistor.

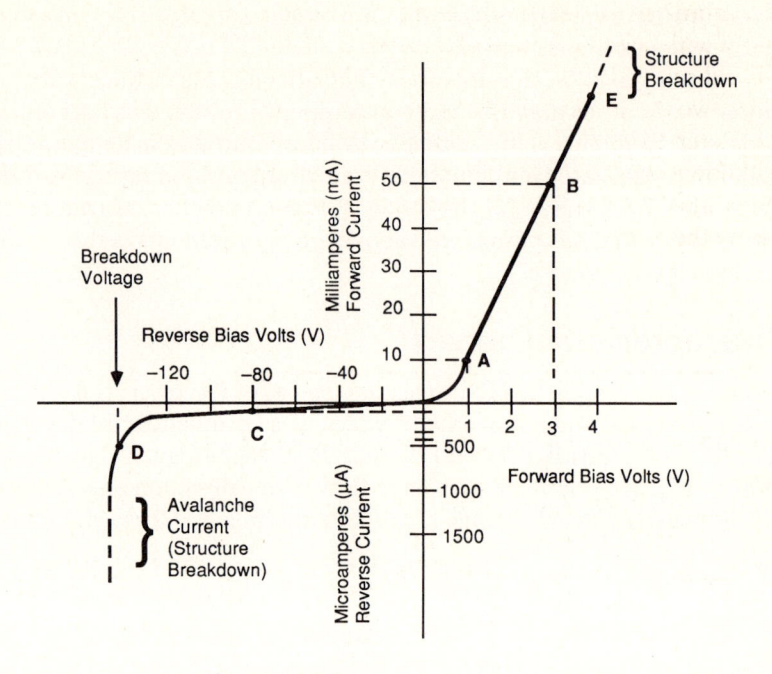

Fig. 2-13 Current Voltage Relationships for a Diode

For the diode, however, things are quite different. When the diode is forward biased, a small part of the voltage is used to overcome the barrier. Between points A and B (Figure 2-13), the diode conducts over a wide range with little change in junction resistance. At point B, forward bias voltage is about 3 V and current is approximately 50 mA. This means that the junction resistance is approximately 60 Ω. At other current levels the diode's resistance will be different.

When reverse bias is applied, however, the story is different. The junction presents a high resistance to current flow and will not begin to conduct significant amounts until high reverse bias is applied. At point C, the reverse bias voltage is about 80 V and the current is about 100 μ A. This means that the junction resistance is approximately 800 kΩ.

By comparing Figures 2-12 and 2-13, it is easy to see that the current-voltage plot for the resistor is much more linear than the diode's curve. For this reason we refer to a diode as being a *non-linear* device. In practical applications, we say that this non-linearity makes the diode a **unidirectional** device that allows current to flow in only one direction.

Avalanche (Zener) Current

If reverse bias is increased beyond a critical value, reverse current will begin to increase rapidly. The point where this happens is referred to as the **avalanche (breakdown)** point. At this point, current will increase greatly with a small change in forward bias. In the Zener diode, which will be discussed later, this is called the **Zener** current. In conventional diodes, we call it the **avalanche** current.

Refer to Figure 2-13. This characteristic curve shows that the breakdown (avalanche) begins at approximately 125 V. In conventional diodes like the one depicted here, avalanche current will probably destroy the diode.

Self-Check

Indicate whether each statement is True or False.

16. _____ A diode characteristic curve represents every condition at which the particular diode can operate.
17. _____ Peak inverse voltage (PIV) is the voltage at which the diode can operate efficiently.
18. _____ A diode is a unidirectional device.
19. _____ Maximum surge current is the maximum current at which a diode can operate on a long-term basis.
20. _____ Breakdown voltage and peak inverse voltage are terms that mean the same thing.

Bias Voltage Requirements

We have already established the fact that for current to flow through a junction, the electrostatic field (barrier) that exists across the junction must be cancelled. The first portion of forward bias voltage that is applied is used to overcome this electrostatic field. The amount of forward bias that is required to cancel the barrier may vary slightly from diode to diode. As a general rule, however, you can use 0.3 V for a germanium diode and 0.7 V for a silicon diode as the **average** voltage required to cancel the barrier. Refer to Figure 2-14(a) for a silicon diode's characteristic curve and to Figure 2-14(b) for a germanium diode's characteristic curve. As soon as the forward bias exceeds the barrier potential, the diode begins to conduct. You should remember that the voltage used to overcome the barrier remains a relatively constant 0.7 V as long as the diode conducts. If the applied voltage should decrease below barrier potential (0.7 V), the diode will cut off. You can expect that the voltage drop on a germanium junction will be 0.3 V, and on a silicon junction it will be 0.7 V during all periods that the junction is conducting.

36 Solid State Electronics

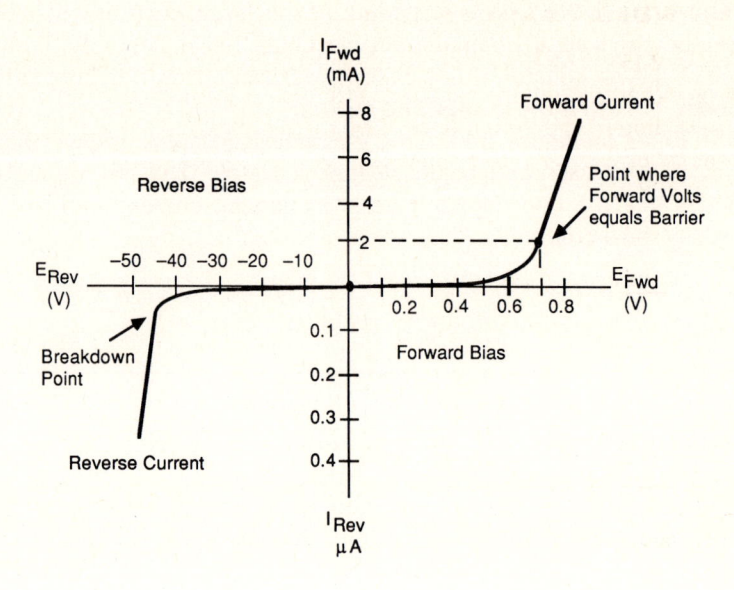

(a) Silicon Diode

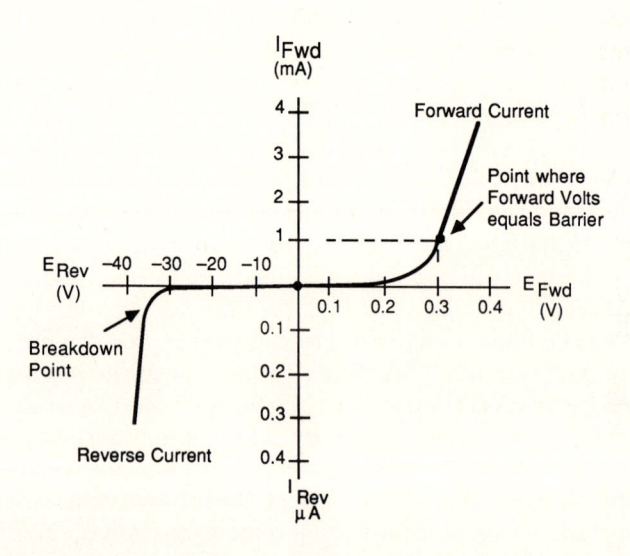

(b) Germanium Diode

Fig. 2-14 Diode Characteristic Curves

Forward Bias Circuits

Figure 2-15(a) depicts one type of circuit where a diode might be used. Notice that the battery has its negative pole connected to the **cathode** of the diode. This tells us the diode is forward biased and that if the voltage exceeds barrier potential (0.7 V) current will flow through resistor R_L.

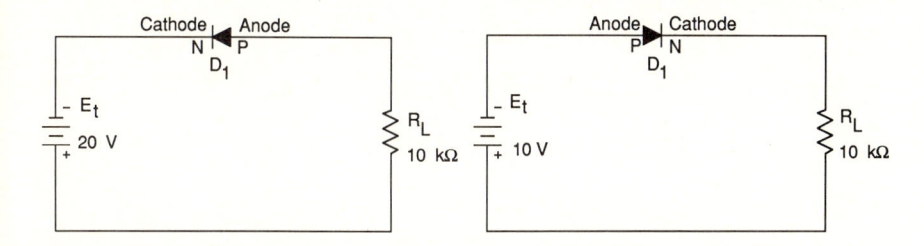

(a) Forward Biased Diode Circuit

(b) Reverse Biased Diode Circuit

Fig. 2-15

The circuit has two resistances (R_D and R_L) with R_D representing the diode's resistance. Applied voltage (20 V) will be divided across the two resistances with approximately 0.7 V on R_D and 19.3 V on R_L. Remember, the voltage required to cancel the barrier potential for a silicon diode is approximately 0.7 V. Therefore, V_{RL} can be found by using the formula:

$$V_{RL} = V_t - V_D = 20\text{ V} - 0.7\text{ V} = 19.3\text{ V}$$

If $R_L = 10\text{ k}\Omega$, then I_t can be calculated as follows:

$$I_t = \frac{V_{RL}}{R_L}$$

$$I_t = \frac{19.3\text{ V}}{10\text{ k}\Omega} = 1.93\text{ mA}$$

To calculate the diode's resistance, use the formula:

$$R_D = \frac{V_D}{I_t}$$

$$R_D = \frac{0.7\text{ V}}{1.93\text{ mA}} = 360\text{ }\Omega$$

Figure 2-15(b) contains the circuit diagram for a reverse biased PN junction diode. Reverse bias assures that the diode remains cut off, preventing forward current from flowing. The cutoff diode acts like an open and drops total voltage. Because zero current flows through R_L, V_{RL} will equal 0 V.

Reverse Bias Circuits

A reverse biased diode circuit is shown in Figure 2-16(a) and the diode's characteristic curve is shown in Figure 2-16(b). We know that excessive

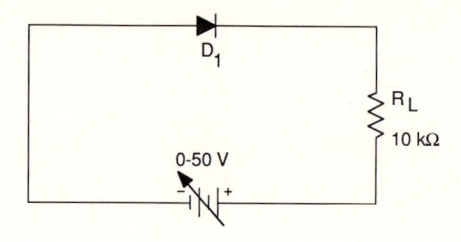

(a)

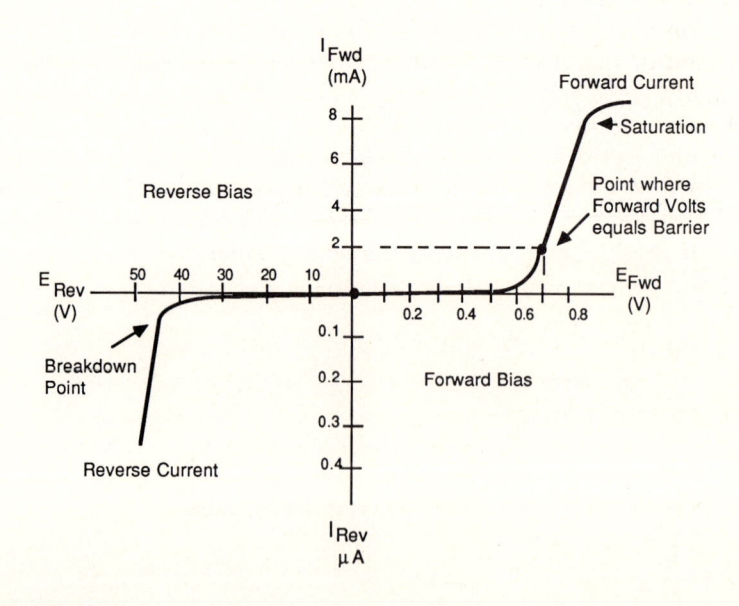

(b) Silicon Diode

Fig. 2-16

reverse voltage (PIV) will cause a diode to break down and probably be destroyed. At breakdown, the reverse current in a diode increases rapidly with small changes in bias voltage. This is called avalanche current. The resistance of the diode (R_D) is very large, past the breakdown point, and the power dissipation caused by the high reverse bias voltage and avalanche current will cause the diode to open. Breakdown voltage will be determined by the diode's design characteristics and construction.

Refer to Figure 2-16(a). To see how this could affect a circuit we will assume that D_1 has a REVERSE BREAKDOWN voltage of 45 V. The power source is a variable DC power supply delivering 0-50 V. As we begin to increase the reverse bias, minority current begins to increase slightly. Remember, though, there are very few minority carriers, but their number increases with increases in operating temperature. As reverse bias increases, reverse current increases, causing the operating temperatures to increase, which releases more minority carriers. This is not a problem until we approach breakdown. At 40 V of reverse bias, reverse current is approximately 0.05 μA. Past 40 V, however, reverse current begins to increase rapidly and by 45 V, it has increased to approximately 0.3 μA. Exceeding the PIV for a conventional diode would probably result in its being destroyed by excessive heat which could cause its junction to break down. As heat increases within a semiconductor junction, the junction's resistance decreases allowing junction current to increase. Increased junction current will cause junction temperature to increase. Increased heat will cause junction current to increase further. Increased current causes heat to increase. This sequence of events could, under the right circumstances, result in heat and current increases that would cause the junction to be destroyed by the high temperature. The added heat dissipation that results during this period is called "THERMAL RUNAWAY".

If the diode current is unknown, we can determine it by the following measurements and calculations:

1. Measure the voltage drop across R_L using a very accurate meter such as a digital multimeter or electronic VOM. Assume that your measurement is 5 V.
2. Calculate circuit current using the formula:

$$I_t = \frac{V_{RL}}{R_L}$$

$$I_t = \frac{5\,V}{10\,k\Omega} = 0.5\,mA$$

Diodes are designed and manufactured to operate within wide ranges. Most diodes, however, are not designed to operate under reverse-bias conditions past the avalanche point. Operating a conventional diode past the avalanche point will usually cause the diode to be destroyed. With specification sheets (see Appendix A for an example) it is possible to select the proper diode to meet your needs and to design the circuit that will do the job.

Modes of Operation

A diode can be operated in any of three modes; cutoff, active, or saturation. These terms are defined as follows:

1. **Cutoff** – Circuit is connected and power is applied but the diode does not conduct. Forward current is 0 mA.
2. **Active** – Circuit is connected, power is applied, and the diode is conducting in the forward direction. Forward current will be larger than zero but less than saturation current. Only in the active condition all of the laws (Ohms, Kirchhoff's, Power, etc.) apply.
3. **Saturation** – Circuit is connected, power is applied, and forward current flows. The diode is said to be saturated when forward current is so large that increasing forward bias has little effect on forward current.

The relationships that exist between each of the operational conditions and diode current are shown in Table 2-1.

Table 2-1 Mode of Operational Versus Current	
MODE OF OPERATION	**CURRENT LEVEL**
Saturation	Maximum
Active	Variable
Cutoff	Zero

Diode Load Lines

Load lines are not often used with diodes. They are, however, used extensively with other devices. A load line is:

Any line drawn on the characteristic curve for a device with the intent of establishing an operating or quiescent (Q) point for the device.

The characteristic curve for a diode is shown in Figure 2-16(b). The curved line on this figure shows currents ranging from saturation to avalanche. Notice that the right side of the graph is forward bias and the left side is reverse bias. Above the horizontal axis on the graph, forward current is plotted. Below the horizontal axis, reverse current is plotted. This curve provides all of the vital information concerning the diode. This same curve was used earlier to determine where the avalanche would begin and the point at which saturation would occur.

To plot a DC load line for the silicon diode, we need to use only the forward bias portion of the chart. Figure 2-17(a) depicts a diode circuit and Figure 2-17(b) shows a characteristic curve with the load line

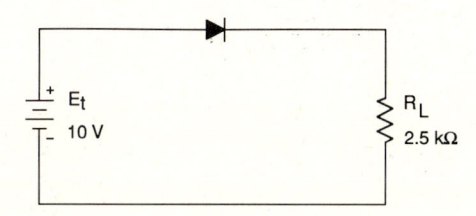

(a) Circuit Schematic

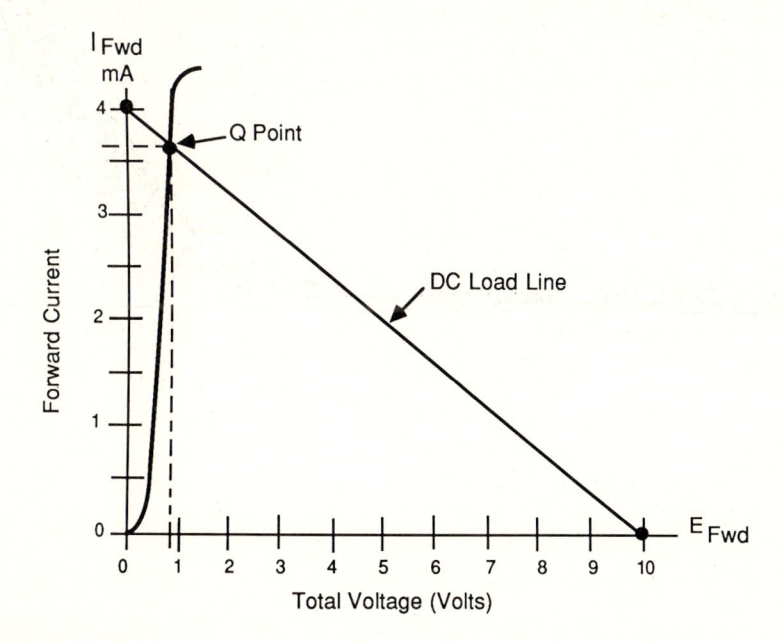

(b) Diode DC Load Line

Fig. 2-17

drawn. To plot the load line we must locate the point that represents the maximum voltage that the diode will be required to drop, and the point that represents the maximum current that the diode will be required to conduct. The maximum voltage drop for R_D will be V_t since the diode drops V_t when it is cut off.

From the schematic we find that V_t will be 10 V. Locate this point (10 V) along the horizontal axis labeled **total voltage** and place a dot at that position. Maximum current that could flow would exist only at such times that D_1 was a direct short (0 Ω). At that time R_t would equal R_L. Calculate maximum current as follows:

$$I_{Max} = \frac{V_t}{R_L}$$

$$I_{Max} = \frac{10 \text{ V}}{2.5 \text{ k}\Omega} = 4 \text{ mA}$$

Now that we found that I_{max} will equal 4 mA, locate the 4 mA position on the forward current (vertical) line. Once located, mark the 4 mA spot with a dot. Use a pencil and straightedge to connect the 4 mA and 10 V spots that you have marked. The line that results represents all of the possible operating (Q) points for this particular diode and this circuit. Because the only voltage involved is DC voltage, we call this a **DC load line**. AC load lines will be discussed when circuits that have AC as part of their operation are discussed. The point at which the load line crosses the characteristic curve is the Q point for this circuit. You should realize, though, that the Q point will change for each resistor (R_L) that is used.

Remember, a silicon diode drops approximately 0.7 V when conducting. Therefore, locate the 0.7 V position along the horizontal axis and project a dashed line up to the load line. From the point of intersection at the load line, move left (dashed line) to the vertical axis of the chart. This point (approximately 3.7 mA) represents the amount of current that flows through the diode at this Q point. The conditions that exist in the circuit are:

$I_t = 3.72$ mA
$V_D = 0.7$ V
$V_{RL} = 9.3$ V

Many types of diodes are available commercially. A few of these are shown in Figure 2-18.

The PN Junction Diode

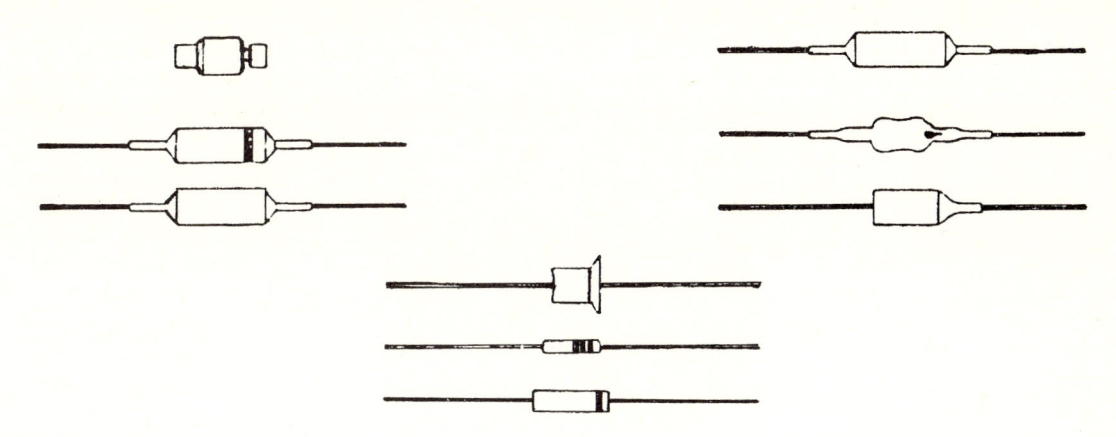

Fig. 2-18 Commercially Available Diodes

Checking Diode Condition with an Ohmmeter

It is possible to check most diodes for operational quality with an ohmmeter. Remember, the ohmmeter has its own internal power source. Using this source, it is possible to forward or reverse bias a diode in order to check its junction resistance. These checks can be done as follows:

1. To **forward bias** the diode, connect the positive lead of the ohmmeter to the anode of the diode and the negative lead to the cathode. See Figure 2-19(a). With the diode forward biased, the ohmmeter will indicate very low resistance.
2. To **reverse bias** the diode, connect the positive lead of the ohmmeter to the cathode of the diode and the negative lead to the anode. See Figure 2-19(b). With the diode reverse biased, the ohmmeter will indicate very high resistance.
3. A shorted diode indicates 0 ohms on the ohmmeter.
4. An open diode indicates **infinity** on the ohmmeter.

(a) Forward Bias   (b) Reverse Bias

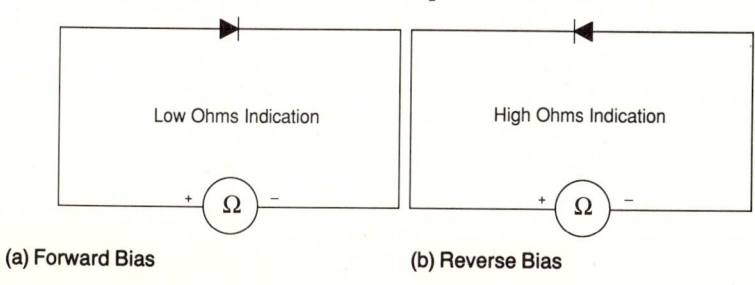

Fig. 2-19 Ohmmeter Checks

Two cautions are important at this point:

1. On some analog ohmmeters the common lead is not the negative lead. To determine which lead is positive, use a good diode and connect it to the ohmmeter in both polarities. The polarity that provides the least resistance indication is the forward biased connection. If it is other than the conventional common (negative) design, mark the meter to indicate this fact.
2. On newer digital and electronic multimeters, it is common to provide a constant current source instead of a common voltage source. In at least one case, a constant current of 1 mA is supplied. This current will destroy many diodes. When using this type meter, use the **low power ohms** position for checking diodes. In this position the meter supplies a much smaller current.

It is also possible that some digital ohmmeters may not be able to supply sufficient voltage to forward bias a PN junction. If this is the case, the ohmmeter cannot be used to check a diode's condition.

Self-Check

Indicate whether each statement is True or False.

21. ____The load line, constructed for a diode will pass through an infinite number of operating points represented by V_t and I_t.
22. ____A forward biased diode will be cutoff.
23. ____Reverse biased diodes have wide barrier regions.
24. ____Under normal conditions, minority current is *not* a problem.
25. ____Forward biased diodes have positive voltage connected to their cathodes.

Summary

Semiconductor materials have a lattice structure. The atoms that are bound into the crystals are held in position by covalent bonding of their valence electrons. To make the natural semiconductors (silicon, germanium, and others) usable in electronics circuitry, the semiconductor is "doped" with another element's atoms in specific quantities. To create P-type material, the dopant atoms must be trivalent (three valence electrons). To create N-type material, the dopant must be pentavalent (five valence electrons). In the P-type material, a hole is created in the lattice structure at each point where an impurity atom is covalent bonded into the structure. These holes become the positive current carriers. In the N-type material, a free electron is released each time an impurity atom is covalent bonded into the crystal's structure. These electrons become the negative current carriers for the N-type semiconductor.

When P-type and N-type crystals are joined, the area of connection is referred to as the junction. Recombination of electrons and holes across the junction causes a barrier that has both height and width to be formed in the junction area. One device that is formed in this way is called a diode.

A diode is a unidirectional device that allows current flow in one direction only. By forward biasing the diode's junction, the barrier can be cancelled and current will flow freely. By reverse biasing the junction, the barrier's height and width are increased and current flow is cutoff.

Diodes have three modes of operation within which they may operate: cutoff, saturation, and active. Diodes that perform work are operating in the active mode. Use of the characteristic curve for a diode allows us to predict the voltage at which the diode will enter or depart any of these modes. Adding a load line to the characteristic curve allows us to establish an operating point for the diode and any series resistor.

Review Questions and/or Problems

1. When P- and N-type materials are joined, the area where they join is called the _____.

    a. carrier
    b. recombination
    c. junction
    d. union

2. The impurity used as a dopant for creating N-type material is called a/an _____.

    a. donor material
    b. acceptor material
    c. negative ion
    d. positive ion

3. The two elements that form a PN junction are known as the _____.

    a. emitter and the collector
    b. anode and the cathode
    c. neither a nor b
    d. either a or b

4. In a forward biased diode, majority carriers move _____.

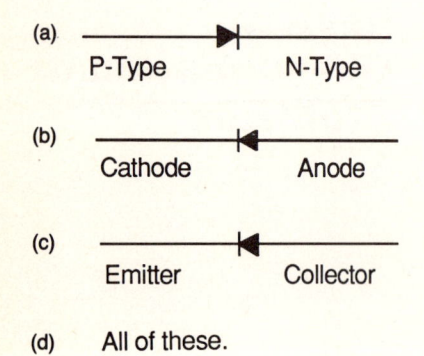

   a. toward the negative pole of the battery
   b. toward the positive pole of the battery
   c. toward the diode's junction
   d. away from the diode's junction

5. For a PN junction diode to be electrically useful _____.

   a. a continuous crystalline structure must exist
   b. P- and N-type materials must be placed side by side
   c. both materials must contain donor impurities
   d. none of the above

6. When P- and N-type materials are joined to form a junction, the area where recombination occurs is called _____.

   a. the barrier
   b. the depletion region
   c. the potential hill
   d. all of these

7. An electron crossing the junction within a diode _____.

   a. will enter the P-type material
   b. is attracted by a positive charge on the P-type material
   c. is repelled by a negative charge in the N-type material
   d. all of the above

8. Which of the symbols at the left is labeled correctly?

   (a) P-Type — N-Type
   (b) Cathode — Anode
   (c) Emitter — Collector
   (d) All of these.

9. Total recombination of holes and electrons across a PN junction is prevented by the _____.

   a. barrier
   b. space gap
   c. junction
   d. all of these

10. When no external energy is applied to a diode, the holes and electrons _____.

    a. move in circular paths
    b. are immobile until forward bias is applied
    c. drift through the crystal structure
    d. all of the above

The PN Junction Diode

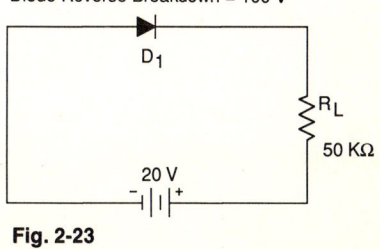

11. The diode in Figure 2-20 is _____.

    a. reverse biased
    b. forward biased

12. A forward biased PN junction has _____ junction resistance and _____ current flow.

    a. high, low
    b. low, high
    c. high, high
    d. low, low

Fig. 2-20

13. To forward bias the diode in Figure 2-21, you would connect the positive terminal of the battery at point _____ and the negative terminal at point _____.

    a. A, B
    b. B, A

14. A diode indicates 50 Ω on an ohmmeter. It is _____ biased.

    a. forward
    b. reverse

Fig. 2-21

15. The current that results from breakdown reverse bias is called _____ current.

    a. majority
    b. saturation
    c. avalanche
    d. cutoff

16. The arrow on a diode's schematic symbol points to _____.

    a. the collector
    b. the anode
    c. the cathode
    d. none of these

17. Refer to Figure 2-22. If D_1 is a silicon diode, how much voltage is dropped across R_L?

    a. 9.7 V
    b. 9.3 V
    c. 0.7 V
    d. 0.3 V

Fig. 2-22

18. Refer to Figure 2-23. How much current (I_t) is flowing in this circuit with the conditions shown?

    a. 8 mA
    b. 7.9 mA
    c. 4 mA
    d. 0 mA

Diode Reverse Breakdown = 100 V

Fig. 2-23

19. A diode is biased such that increasing forward bias will cause I_t to increase. The diode is operating in which mode?

   a. cutoff
   b. saturation
   c. active

20. The two parameters used to plot a DC load line are _____.

   a. voltage versus resistance
   b. voltage versus current
   c. voltage versus power
   d. current versus resistance

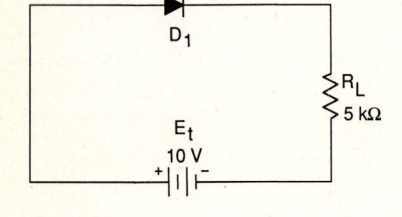

21. Refer to Figure 2-24. The germanium diode's voltage drop will equal _____.

   a. 9.3 V
   b. 9.7 V
   c. 0.7 V
   d. 0.3 V

Fig. 2-24

22. The term used to indicate maximum reverse bias that a diode can withstand is _____.

   a. breakdown voltage
   b. cutoff voltage
   c. peak inverse voltage
   d. both a and c

23. A diode is useful because it is a _____ device.

   a. unilateral
   b. bilateral

24. An ohmmeter indicates 500 kΩ when it is connected to a diode. The diode is _____.

   a. shorted
   b. open
   c. reverse biased
   d. forward biased

25. You are checking an open diode with your ohmmeter. Which of the following indications would you expect to receive?

   a. Zero ohms
   b. Infinity
   c. High Ohmic Indication
   d. Low Ohmic Indication

3
Bipolar Transistors

Objectives

1. Define the terms:

    a. Transistor
    b. NPN
    c. PNP
    d. Emitter
    e. Forward bias
    f. Gain
    g. Base
    h. Collector
    i. eb junction
    j. cb junction
    k. Reverse bias

2. Describe the construction of a bipolar transistor.
3. Relate the doping used for each element of a transistor to the part that element contributes to transistor operation.
4. Explain the current distribution that occurs in transistors.
5. Describe the biasing that is used with a transistor that is correctly biased for a specific application.
6. Given an assortment of transistor circuit schematics, select those schematics that show correctly biased transistors.

Introduction

The subject of discussion in Chapter 2 was the PN junction diode. In your study of diodes you learned the characteristics of the PN junction and the requirements for the current flow within a PN junction. The diode is constructed of a single piece of semiconductor crystal that has been doped with a different impurity at each end. The result is a device that allows current to flow in one direction only.

In this chapter, we begin the study of transistors. The transistor is a semiconductor that contains three sections of semiconductor material. These sections are placed such that two distinct junctions are formed within the device. Because they have two junctions, we refer to them as **bipolar transistors**. The term transistor was derived from the words "transfer" and "resistor". There are numerous types of bipolar transistors, but the theory that supports one transistor is suitable for understanding all types.

The transistor is another important electronic device. It makes use of the flow of current carriers within its semiconductor materials and the phenomena that accompany this current flow. The transistor that will be studied here has three sections and is used as a switching or an amplifying device.

The Bipolar Transistor

The **bipolar (junction) transistor (BJT)** is a **three-element** device. These three elements are called the **emitter, base,** and **collector**. The function of each element is:

1. The emitter (E) provides (emits) the majority carriers necessary to support current flow.
2. The base (B) is used to control the flow of majority carriers within all elements of the transistor.
3. The collector (C) supports the majority of the transistor's current flow. In most cases the work done by a transistor is accomplished by the current that flows through the collector.

During manufacture of the transistor, a metal lead is connected to each of the transistor's elements. In circuit assembly, these leads are used to connect the transistor into the circuit.

Bipolar transistors come in many shapes and packages. A few of these are shown in Figure 3-1. Notice that some of these have pins connected to each element, as indicated by the labels "E", "B", "C". Other packages use the metal case of the package as the collector lead. In this type, care must be taken to assure solid metallic contact between the case and the chassis on which the transistor is mounted. With some transistors, it is necessary to assure that their cases are fully insulated from the metal chassis. As you work with different types and packages, you will develop the skills necessary to recognize each type.

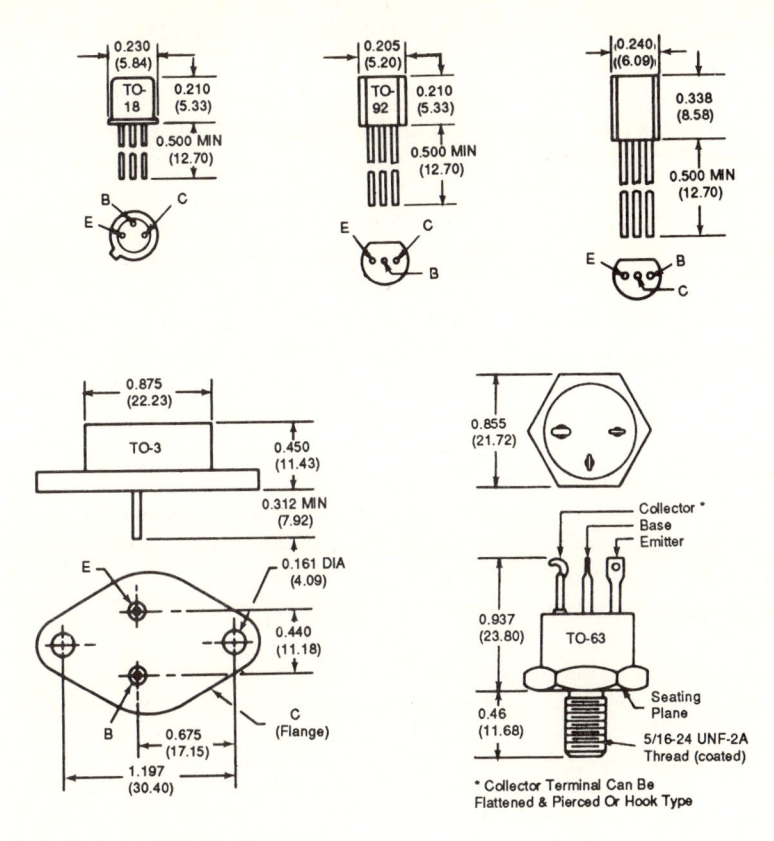

Dimensions in inches (millimeters).
All dimensions are max. unless otherwise indicated.

Fig. 3-1

Transistor Elements

As stated above, the bipolar transistor has three elements: (1) emitter, (2) base, and (3) collector. The base is a very thin strip of material which is very lightly doped and is located between the emitter and collector. Light doping assures that the number of majority carriers in the base will be limited. The purpose for the different levels of doping will become evident as we proceed. The emitter is a piece of semiconductor material that is more heavily doped than either of the other two elements. This element supports 100% of the current flow and it is said to "emit" the current carriers for that type transistor. The collector is usually larger than either of the other elements. In addition to its physical size, it has been doped with fewer impurities than the emitter. You should understand that the base is much smaller than either the emitter or collector and has very little doping when compared to them.

Two types of bipolar (junction) transistors are available. They are called **NPN** and **PNP** transistors. The NPN transistor is constructed using N-type material as the emitter and collector while the base is made of P-type material. The PNP transistor is constructed using P-type material as the emitter and collector while the base is made of N-type. Figure 3-2 is a pictorial diagram of the two types.

Figure 3-2 clearly shows that each transistor has two junctions which leads to its being called a **junction, bipolar,** and **bipolar-junction** transistor. The abbreviation "BJT" is sometimes used to refer to this device. The junction that separates the base and emitter is called the **emitter-base (eb)** junction, and the one separating the base and collector is called the **collector-base (cb)** junction. Remember, to be electrically useful, semiconductor devices must have a continuous crystalline structure. For this reason a single piece of material is used and the amount and type of dopant needed for each element is introduced at the right point and in the right amount. Therefore, even though we discuss each element as if it was a separate piece of material, you should understand that each transistor is actually one piece of crystalline material that has been doped to create the three elements.

Schematic symbols for each type transistor are shown in Figure 3-3. Note that each element is labeled. The emitter and collector join the base at approximately a 30° angle. The emitter is identified by an arrowhead. This arrowhead serves to tell us three things: (1) the location of the emitter; (2) the type of transistor that is being represented (the point of the arrowhead **always** points toward the N-type material. Therefore, if the arrowhead points outward, the transistor is an NPN type. If the arrowhead points toward the base, the transistor is a PNP type); (3) the direction that current will flow. Electron current will always move in a direction opposite the direction of the arrow. In the NPN transistor, electron current flows from the emitter into the base. In the PNP transistor, electron current flows from the base into the emitter.

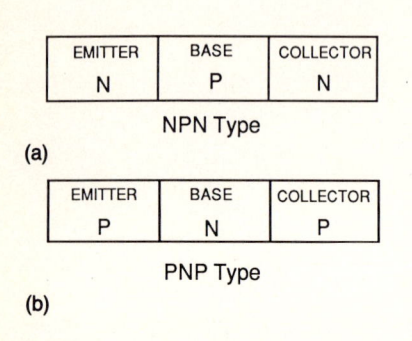

Fig. 3-2 Transistor Pictorial Diagrams

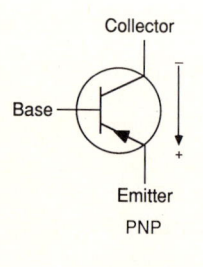

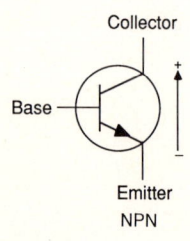

Fig. 3-3 Transistor Schematic Symbols

> **NOTE:** From this point on, any mention of current flow will refer to electron flow unless otherwise stated.

As with the junction diode, recombination across the junctions in a transistor cause a depletion region (potential hill, barrier) to exist at each junction. The electrostatic fields that result in an NPN transistor are shown in Figure 3-4.

In this figure, recombination has already occurred and barriers have been established. The base is a thin section of P-type material that has holes as its majority carriers. The emitter and collector are N-type material and have electrons as their majority carriers. Recombination

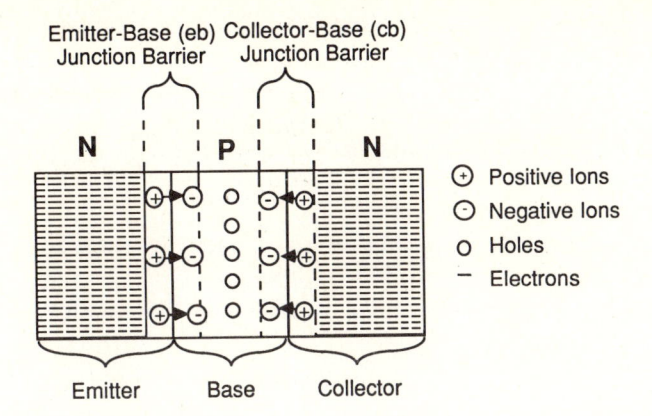

Fig. 3-4 Unbiased NPN Transistor

of some of the carriers has formed the junction barriers shown. The PNP transistor has identical characteristics except that its majority and minority carriers are opposite those of the NPN transistor.

In most cases where either of these transistors is used, the emitter-base junction is used as the input circuit. The collector-base junction is the output circuit. In some circuit designs, the output is taken from the emitter circuit. In other words, the signal we are working with will be coupled *into* the transistor on either the base or emitter. The signal will then be taken *out* of the transistor at either the collector or emitter. Inputs and outputs will be explained more fully in the next chapter.

Self-Check

Answer each item by inserting the word or words required to correctly complete each statement.

1. An NPN transistor has _____ element(s) made from N-type material and _____ element(s) made from P-type material.
2. A bipolar transistor contains three elements. These are?

    a. _____
    b. _____
    c. _____

3. The _____ element receives less dopant than the other two.
4. The arrowhead on a transistor symbol identifies the:

    a. _____
    b. _____
    c. _____

5. Of the emitter and collector, the _____ receives the heaviest amount of doping.
6. The process of _____ involves the movement of majority carriers within an unbiased junction.
7. In P-type material, majority carriers are the _____.
8. Current in a transistor moves in the _____ the arrowhead shown on its schematic symbol.
9. As a result of recombination, each junction in a transistor becomes a _____ to current flow.
10. The _____ element of a transistor is the smallest of the three elements.

Transistor Biasing Requirements

If you think back to the junction diode, you will remember that applying a voltage in the correct polarity will either forward bias or reverse bias the junction. If sufficient forward bias (approximately 0.3 V for germanium diodes or 0.7 V for silicon diodes) is connected across a junction, it cancels the junction barrier and causes the junction to conduct. Reversing the voltage polarity causes the barrier to widen and the diode to cutoff. The same thing is possible, and is used, in transistor operation. The voltage that is connected to a junction is called **bias voltage**. In order for a transistor to conduct properly, the following biasing must exist:

The EMITTER-BASE (eb) junction is FORWARD BIASED.
The COLLECTOR-BASE (cb) junction is REVERSE BIASED.

Forward bias at the emitter-base junction reduces or cancels that junction barrier. As forward bias to a junction is increased, the junction resistance decreases. When a transistor is conducting, the emitter-base junction offers a *very low resistance* to current flow. The reverse bias that is connected to the collector-base junction widens the junction barrier causing the collector-base junction to have a *very high resistance*. Figure 3-5 depicts both NPN and PNP transistors and the biasing arrangements they require for conduction. Notice that the batteries are labeled V_{CC} and V_{EE}. V_{CC} provides reverse bias for the collector-base junction and V_{EE} provides forward bias for the emitter-base junction. Notice that V_{EE} reduces the barrier width and decreases the junction's resistance (R_{eb}). V_{CC} increases the barrier width and increases the junction's resistance (R_{cb}). The results of biasing each transistor are identical, the only difference being the polarity for connecting the batteries.

Bipolar Transistors

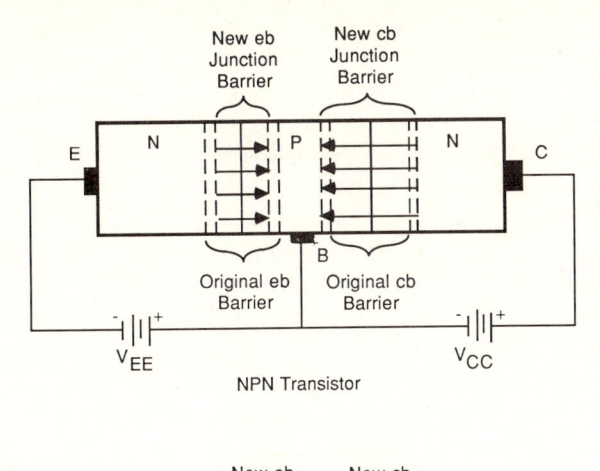

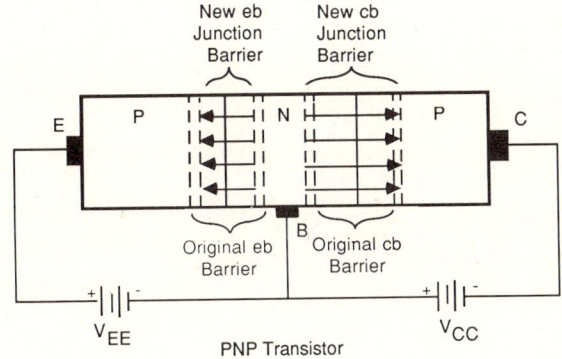

Fig. 3-5

Current paths that exist in each type transistor are shown in Figure 3-6. An NPN transistor is shown in Figure 3-6(a). The solid arrows indicate *electron* flow and the dashed arrows indicate *hole* flow. In the external circuit, using electron flow, the current will be *electrons*. (With conventional current flow, the external current would be *hole* flow and current, within the transistor, would flow *with* the emitter arrow). Emitter current (I_E) is shown leaving the emitter battery (V_{EE}) and flowing into the N-type emitter. In N-type material, the majority carriers are electrons. Therefore, the entering electrons add to the number of majority carriers. The extra electrons that enter the emitter have two forces acting on them: (1) the repelling force of the battery and (2) the attracting force of the holes in the base. Still having the energy they absorbed from these forces, the electrons move into the base where they are minority carriers and outnumber the holes. Remember, the base is P-type material, it is lightly doped and will have few holes (majority carriers). As many electrons as possible recombine with holes in the base and cause other electrons to leave the base lead and return to V_{EE}. The great majority of the electrons that enter the base do not have a hole with which they can recombine. The reverse bias applied across

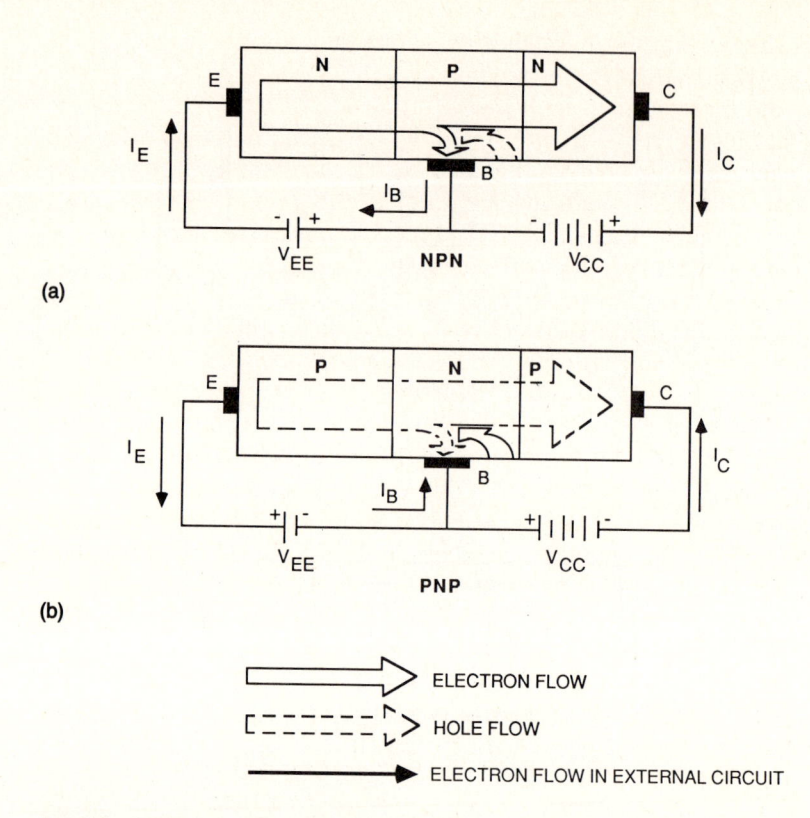

Fig. 3-6 Transistor Current Paths

the collector-base junction attracts the electrons to the end near V_{CC} and repels the minority carriers (holes) to the area of the junction. Remember, the collector is N-type material having electrons as its majority carriers. The extra electrons that are located in the base come under the influence of the concentration of holes (minority carriers) that exist near the collector-base junction. The attracting force that exists between the electrons and the holes will cause the electrons, which are already traveling at high speed, to pass through the reverse biased collector-base junction. As stated earlier, the base is very small and thin. This, plus the fact that electrons are exposed to a much greater collector surface than base-lead surface, plus the attracting force supplied by the holes, will assure that amounts ranging to more than 99% of all electrons entering the base pass through into the collector. Once in the collector, the electrons join the majority carriers (electrons) and come under the influence of the positive potential exerted by V_{CC}. This attracting force accelerates each electrons movement. It (the electron) easily moves from the collector into the external circuit and back to the battery V_{CC} as collector current (I_C).

The current paths in the PNP transistor are shown in Figure 3-6(b). Majority carriers in the emitter and collector are holes. In the base, the majority carriers are electrons. Within the PNP transistor, current flow is supported as follows: V_{EE} exerts a force on the emitter that causes a hole-pair to be generated in the P-type material. The electron from this pair is attracted out of the emitter and to V_{EE} as emitter current (I_E). The hole that remains is a majority carrier within the emitter and is repelled to the area near the junction. As it nears the holes that are near the junction, it begins to feel a repelling force which causes it to slow and to release its energy. This energy is absorbed by another hole, raising its energy level enough for it to cross the emitter-base junction. Once in the base, the hole becomes a minority carrier. A small number of the entering holes will combine with majority carriers (electrons) in the base. As an electron and a hole recombine in the base, an electron enters the base from the battery (V_{EE}) to replace the one used in recombination. Electrons that enter the base (amounts from less than 1% up) in this manner make up the base current (I_B). The great majority (amounts to more than 99%) of the holes entering the base from the emitter will be exposed to a strong electrostatic field located at the collector-base junction. This field and the concentration of electrons located in the collector will aid hole movement, allowing them to move easily through the collector-base junction and enter the collector. When a hole enters the collector, it joins the majority carriers already moving toward the connection with V_{CC}. For each hole that enters the collector, an electron is attracted away from the negative terminal of V_{CC} by the holes that are located in the collector. Electrons entering the collector are the ones that make up collector current (I_C).

In this transistor notice that current flows from the base to the emitter. In the NPN transistor, current flows from the emitter to the base. Although current flow in the PNP transistor is opposite that of the NPN transistor, notice that the majority carriers in both transistors move from the emitter to the collector.

Electrons and holes that recombine within the base are lost to the collector current. This means that the fewer recombinations that occur within the base, the more efficient the circuit's operation. This is the reason for the small, thin size and light doping of the base element. Since 100% of the current flows in the emitter, it must be able to support heavier carrier movement than the collector. For this reason it is doped more heavily than the collector or base. Normally, the collector element is larger than the emitter or base. This increases the probability that the minority carriers entering the base will be able to contribute to collector current. The larger size also gives the collector the ability to dissipate the larger amounts of heat that are generated by the carriers moving through the high resistance collector-base junction.

Self-Check

Answer each item by inserting the word or words required to correctly complete each statement.

11. In a bipolar transistor, the _____ junction is reverse biased and the _____ junction is forward biased.
12. To forward bias a junction, the positive pole of the battery must be connected to the _____ material and the negative pole to the _____ material.
13. Forward bias of approximately _____ is required to cancel the barrier potential of a silicon junction.
14. As the forward bias applied to the emitter-base junction of a transistor is increased, collector current will _____.
15. In an NPN transistor, the base is made of _____ material.

Transistor Currents

For you to be able to understand transistor operation, you must thoroughly understand transistor currents. At this point we will examine the different currents, their relative size, and how each is related to the other. We have already referred to three currents: emitter current (I_E), base current (I_B), and collector current (I_C).

Figure 3-7 shows an NPN transistor that has 0.7 V forward bias applied to the emitter-base junction by V_{EE} and 10 V reverse bias applied to the collector-base junction by V_{CC}.

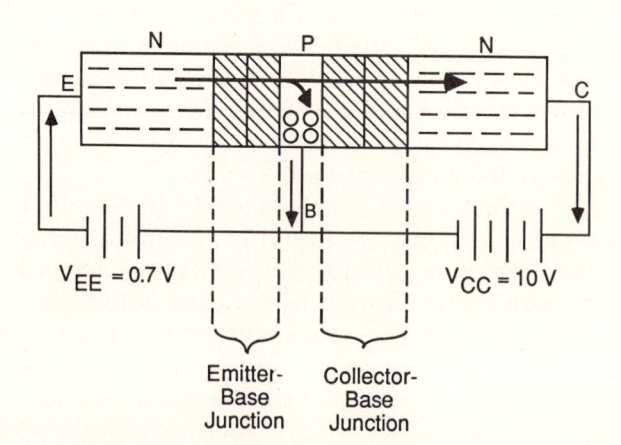

Fig. 3-7 Pictorial Diagram Biased NPN Transistor

The 0.7 V applied to the eb junction forward biases the emitter-base junction into conduction. Now, the eb junction has become a low resistance junction to the majority carriers in the emitter. At the same time, the 10 V applied to the collector-base junction has caused it to become a high resistance junction.

The N-type material in the emitter has a large concentration of majority carriers (electrons). The P-type base has only a few majority carriers (holes) because it is lightly doped, and the N-type collector has many majority carriers (electrons). The forward bias supplied by V_{EE} will cause some of the electrons in the emitter to enter the base. The emitter has "emitted" electrons which are now located in the base.

In the base element, all electrons are minority carriers. A few of the extra electrons do, however, manage to recombine with majority carriers (holes) in the base. For each electron that recombines, another electron's energy level raises to the point that it crosses into the external circuit and returns to V_{EE} as I_B. For an electron to flow as base current, it is necessary for an electron to recombine with a hole in the base. By lightly doping the base element, we limit the number of holes available for recombination. Also, the surface area available for entering the external circuit is relatively small when compared to the surface area that the collector-base junction presents to the electrons now located in the base.

Those electrons that cannot recombine in the base are under the influence of the electrostatic field that exists across the collector-base junction. The forward movement, already present in the electron, and the force of the junction field cause the electron to be moved into the collector quite easily. Once in the collector, the electron again becomes a majority carrier.

Approximately 92% to more than 99% of the electrons entering the base are forced to enter the collector depending on the transistor's construction. Once in the collector, the electrons are majority carriers and are under the influence of the positive pole of V_{CC}. This causes them to move through the collector, to leave the transistor, and to return to V_{CC} as I_C.

The amount of current that flows in each lead of the transistor can be discussed in terms of percentage. In either type of transistor, 100% of the current flows in the emitter. Base current and collector current combine to equal emitter current. This means that:

$$I_E = I_C + I_B$$

Depending on the transistor's construction, I_B ranges from less than 1% to 8% of I_E. However, with the transistors we use today, the base current is kept quite small. In these studies we will use 2% as the amount of I_E that flows as I_B. In the same transistor, 98% of I_E will flow as I_C. For our purposes, we will say that I_C equals 98% of I_E. For example:

If I_E equals 100 µA,
then I_B equals 2 µA,
and I_C equals 98 µA.

NOTE: These values are compromises. In actuality, both I_B and I_C can vary from one transistor to another. The values chosen here are used for convenience only and should not be considered as true for all bipolar transistors.

The fact that emitter-base voltage (V_{eb}) is relatively small (with a corresponding small base current) does not mean that these parameters are unimportant. In fact, the effect that V_{eb} has on I_B and the effect that I_B has on the other currents is considered to be the **controlling factor** for transistor operation. To illustrate the controlling effect that V_{eb} has on the transistor currents, refer to Figure 3-8.

Figure 3-8(a) shows an NPN transistor that has 0.6 V of forward bias and 10 V of reverse bias. Under these conditions, the current will flow from the emitter to the base (2%) and emitter to collector (98%). The width of the arrows represents the amounts of current that are flowing as I_E, I_B, and I_C. Figure 3-8(b) shows the effect that increasing forward bias to 0.7 V will have on the three currents. With the increase in forward bias, R_{eb} has decreased and all three currents have increased. Figure 3-8(c) shows the effect of decreasing forward bias to the point that it equals 0.5 V. In this case, R_{eb} has increased causing all three currents to decrease. Notice that relatively small changes in V_{eb} will cause changes in I_E, I_B, and I_C. By varying the bias voltage applied to the emitter-base junction, we can control base current. Any change in base current causes a corresponding change in both emitter current and collector current. From these effects we can make the following statements.

- Bias voltage is applied to a transistor to assure that the transistor is either conducting or cutoff.

- Once biased into conduction, transistors can be considered as current-controlled devices.

- Small changes in base current will cause large changes in emitter and collector currents.

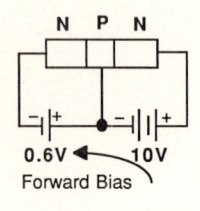

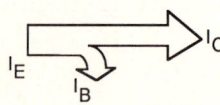

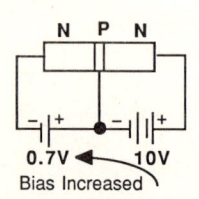

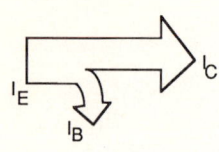

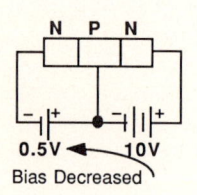

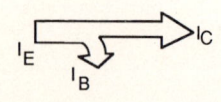

Fig. 3-8 Control Effect of Forward Bias (NPN)

The emitter-base voltage applied to a PNP transistor will have the same controlling effect as that discussed for the NPN. Figure 3-9(a) illustrates a PNP transistor that has 0.6 V forward bias applied to the emitter-base junction and 10 V reverse bias applied to the collector-base junction. The dashed arrows depict the currents that could be expected from this biasing arrangement. Increasing the emitter-base junction bias to 0.7 V will have the effect shown in Figure 3-9(b). Decreasing forward bias to 0.5 V will have the effect shown in Figure 3-9(c). If you compare Figures 3-8 and 3-9, you can see that the effect of small changes in forward bias are identical for the two transistors.

Although the base current in the transistor is controlled by the emitter-base voltage, the control is nonlinear. Figure 3-10 shows this relationship. Note that the relationship between the base-emitter voltage and the collector current changes much more drastically between 1.0 V and 1.1 V than it did between 0.9 V and 1.0 V. It is easy to see that the change is much greater with the higher forward bias. In one case, a 0.1 V change causes collector current to change by 2 mA. In the other case, the same change in voltage causes collector current to change 7 mA.

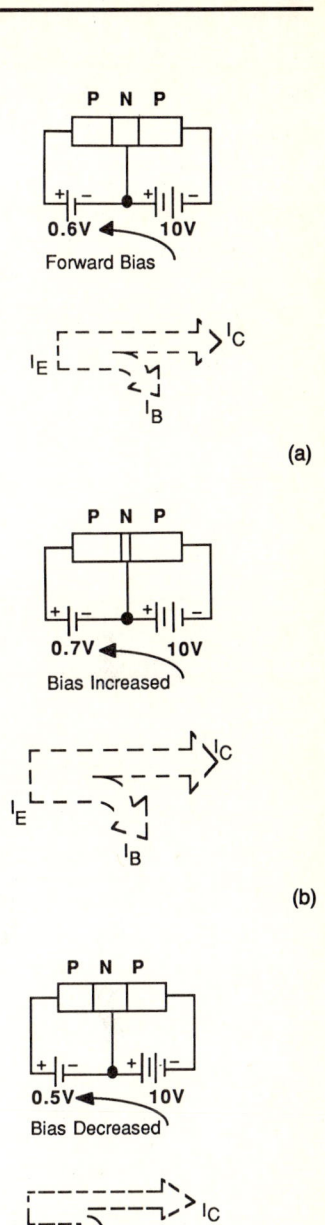

Fig. 3-9 Control Effect of Forward Bias (PNP)

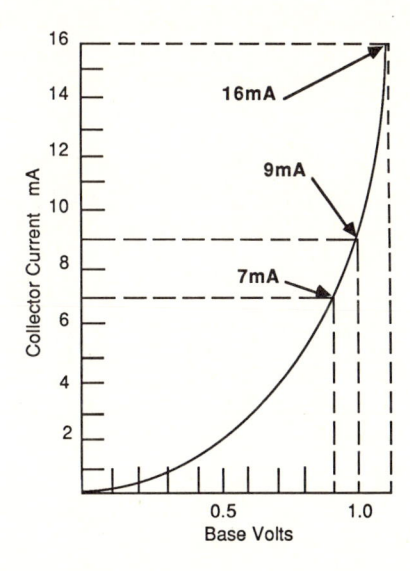

Fig. 3-10 Nonlinear Relationship (E_B to I_C)

In most operational circuits, any nonlinearity that occurs as a result of variations in V_{eb} is not desirable. It is more desirable to cause collector current changes by varying the base current.

Figure 3-11 shows the relationship of changes in I_C as compared to changes in I_B. Notice that I_B is plotted in microamperes and collector current is plotted in milliamperes. In this case the relationship between changes in I_B and I_C is practically linear. Therefore we can state that:

Changes in collector current are directly proportional to changes in base current.

Because of the linearity that exists between changes in I_B and I_C, transistors can be referred to as **current-controlled devices**.

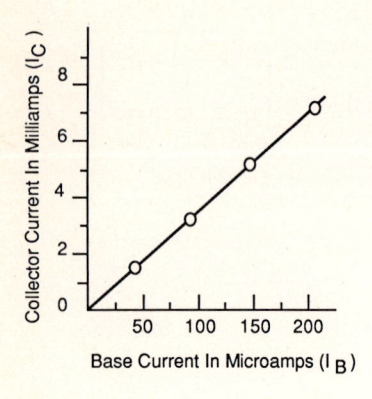

Fig. 3-11 Linear Relationship (I_B to I_C)

Self-Check

Answer each item by inserting the word or words required to correctly complete each statement.

16. Transistor current is distributed differently in each element. State these percentages below.

    a. _____
    b. _____
    c. _____

17. In an operating transistor, the emitter-base junction is a _____ resistance and the collector-base junction is a _____ resistance.
18. Majority current carriers within a transistor move toward the _____.
19. Transistors are _____-controlled devices.
20. Changes in collector current are _____ proportional to changes in base current.

Leakage Current

As you recall, the collector-base junction is reverse biased when a transistor is in operation. Opening the emitter lead would cause I_E to stop. Without the flow of I_E, there would be no current flow supported by the majority carriers. There would be, however, a small amount of current flowing in the collector-base junction as a result of the forward biased minority carriers. The current that flows from collector to base

Bipolar Transistors

under these conditions is called **leakage current** or I_{CBO}. (I_{CBO} stands for current from Collector to Base with emitter Open.) Even though the emitter lead is open, I_{CBO} will flow as long as the collector-base junction remains reverse biased.

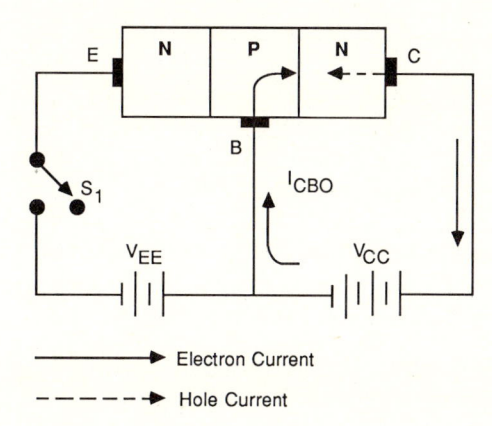

Fig. 3-12 I_{CBO} in a NPN Transistor

Figure 3-12 shows the circuit schematic for an NPN transistor that has its emitter open and its collector-base junction reverse biased. Notice that the direction of I_{CBO} is opposite to the direction of normal base current. Under normal operating conditions, the amount of I_{CBO} is of little importance. With the use of silicon transistors this phenomena will present little problem. When operating germanium transistors in a high temperature setting, I_{CBO} can be a problem. In fact, it is possible for the transistor to destroy itself because of **thermal runaway** as was discussed in diodes. Remember, as current increases, junction current increases causing junction temperature to increase. The increase in temperature causes more minority carriers to be released that can support I_{CBO} at higher levels. The increased current causes increased junction temperature and the cycle can continue until I_{CBO} becomes so heavy that the junction is destroyed. I_{CBO} is actually flowing when the emitter is connected. Since I_{CBO} flows through the base in the opposite direction to the desired current, it, in effect, cancels a portion of the normal base current. At normal operating ranges this portion is in the range of a very few microamperes. At higher temperatures, though, the increased I_{CBO} can actually decrease the effectiveness of base current's control over collector current.

As a general rule, we consider that I_{CBO} will double for each 8° to 10° (Celsius) increase in operating temperature. Since the amount of I_{CBO} is determined by heat, the junction temperature must be kept within specific ranges. Most transistors are specified for operation at

25° Celsius to prevent thermal runaway. Temperature changes above or below this level may affect circuit operation. The amount of I_{CBO} is dependent upon operating temperature and is relatively independent of V_{CC}.

Biasing Arrangements

Figure 3-13(a) shows the schematic diagram for a PNP transistor that is biased using two batteries. Notice the polarity of the batteries; V_{EE} forward biases the emitter-base junction and V_{CC} reverse biases the collector-base junction. Remember that for:

Forward bias – P-type material connects to the positive pole of the battery and N-type material connects to the negative pole.

Reverse bias – P-type material connects to the negative pole of the battery and N-type material connects to the positive pole.

In Figure 3-13(a), the forward bias applied to the emitter-base junction equals 0.3 V. An NPN transistor with the same biasing arrangement is shown in Figure 3-13(b). In each case, E_1 provides the forward bias for the emitter-base junction and E_2 reverse biases the

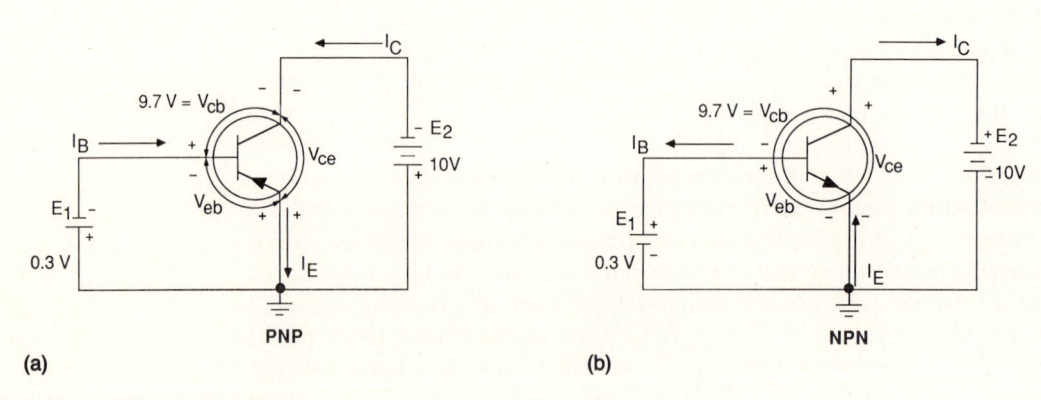

Fig. 3-13 Bias Polarities and Current Directions

collector-base junction. The main differences between Figures 3-13(a) and (b) is the fact that one is PNP and the other NPN. Also, the batteries are reversed in polarity, which causes current to flow in the opposite direction. These are conditions that could actually occur when using a germanium transistor except that batteries with 0.1, 0.2, and 0.3 V are nonexistent. Therefore, a circuit must be designed that will provide the required biases but do it using a conventional power supply.

Bipolar Transistors

A more convenient design is one that uses a resistive voltage divider which is connected to a single power source. By selecting the correct ratio between the divider's resistances, we can supply the needed voltages to both junctions. The use of a single power source is illustrated in Figure 3-14. In Figure 3-14(a) we see an NPN transistor that is connected to a single battery (V_{CC}) that supplies 10 V. By selecting resistors that have the proper value, this 10 V can be divided as shown. This circuit provides 0.3 V forward bias on the emitter-base junction and 9.7 V reverse bias on the collector-base junction. Remember, parallel branches have equal voltage drops. The arrows on Figure 3-14(a) show the direction that I_E, I_B, and I_C will flow.

Figure 3-14(b) is used to show a different way of connecting the circuit of Figure 3-14(a). Notice that the battery is included as a dashed line, but all currents and voltages are identical to the circuit discussed

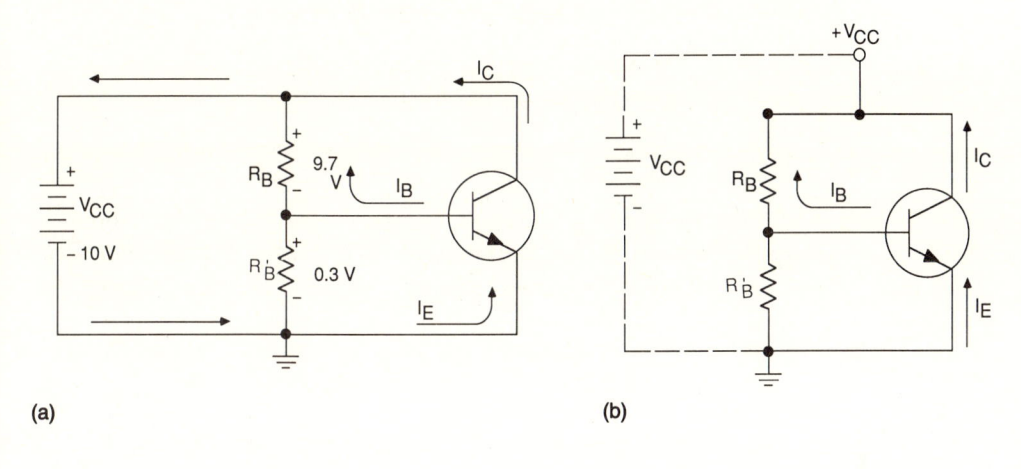

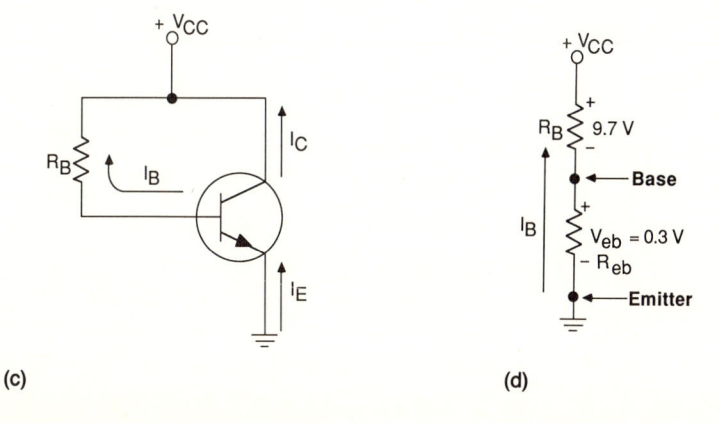

Fig. 3-14 Single Source Biasing Circuits

above. Since it is common to operate many circuits from a single power source, we do not show the battery in every operational circuit. Complex schematics have the value of V_{CC} noted in the legend, and conductors that connect the separate circuits are identified as V_{CC}. This type of marking is illustrated in Figure 3-14(c). When the collector connection is labeled $+V_{CC}$, we know that the positive pole of the battery is connected to this point, and current will flow from ground to $+V_{CC}$. If the point is labeled $-V_{CC}$, the negative pole is connected to the collector and current will flow from $-V_{CC}$ to ground.

In fact, many circuits will not use R_B' in the circuit. In these cases, the connection is like that shown in Figure 3-14(c). In this arrangement, R_{eb} is used to develop the forward bias for the emitter-base junction. The equivalent biasing circuit for this schematic is shown in Figure 3-14(d). Resistors R_B and R_{eb} form a voltage divider where R_B establishes the reverse bias for the collector-base junction, and R_{eb} develops the forward bias for the emitter-base junction. Because the emitter is at ground potential, any voltage dropped on R_{eb} will be positive at the base. A positive voltage at the base meets the requirements for a forward biased junction.

Self-Check

Answer each item by inserting the word or words required to correctly complete each statement.

21. Leakage current is supported by _____ carriers flowing through the _____ junction in the _____ direction.
22. The amount of I_{CBO} is controlled by _____.
23. For an NPN transistor, V_{CC} must have _____ polarity.
24. A single battery biasing arrangement has a _____ that develops the reverse bias applied to the collector-base junction.
25. When biasing an NPN transistor, the collector must be _____ with respect to the emitter.

Summary

The two types of bipolar Transistors are NPN and PNP. The schematic symbol for each type is shown in Figure 3-15. On these symbols, an arrowhead is used to represent three things:

1. the emitter of the transistor.
2. the type of transistor.
3. the direction of current flow.

As you can see from the symbols, each transistor has three elements. These are the emitter, base, and collector. The emitter and base elements form an emitter-base junction, and the collector forms a second junction with the base. The emitter-base junction acts as the input circuit for the transistor, and in most applications, the collector serves as the output connection. During the construction of transistors, care is taken to prepare each section of the transistor for the part it will play in the operational circuit. Each element of the transistor is doped differently from the others. The emitter and collector elements are both of the same type semiconductor material. The emitter, however, is doped more heavily than the collector but the collector usually has a larger physical size than the emitter. The base is formed in a very small and narrow region between the emitter and collector. Doping in the base is very light which limits the number of majority carriers that are present.

In an operating transistor the emitter-base junction is normally forward biased and the collector-base junction is reverse biased. Forward bias applied to the emitter-base junction causes a very heavy movement of majority carriers between the base and emitter. As a result, the base becomes flooded with minority carriers which will, in turn, cause collector majority carriers to support current flow within the external circuit. The end result is that the external current is always flowing from the negative to the positive pole of the battery. Inside the transistor, however, current flow is supported by the majority carriers located in the emitter and collector. As a result of the relations that exist within the transistor and the type bias applied to the junctions, the transistor is a current-controlled device whose operation is controlled by emitter current. This means that the output of the transistor can be controlled by the input.

In a bipolar transistor, emitter current equals 100% of the current. Emitter current divides between the base and collector at different ratios for different transistors. We, therefore, arbitrarily established percentages that would be used for explanations in this book. These are base current = 2% of I_E and collector current = 98% of I_E. You should realize, though, that these values were selected for convenience only. Within actual circuits, the division will vary from circuit to circuit.

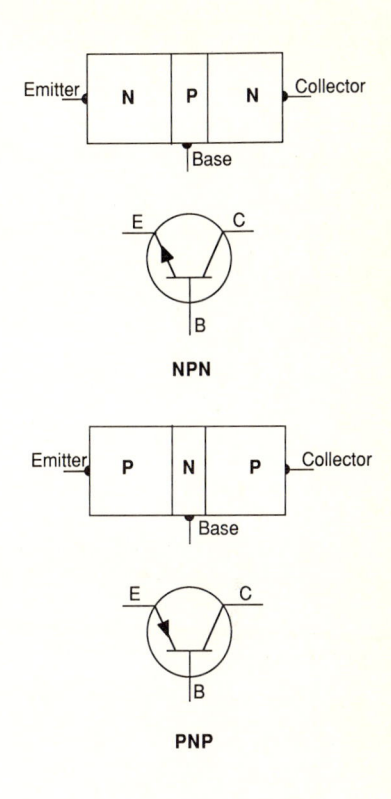

Fig. 3-15 Pictorial Diagrams and Schematic Symbols

Review Questions and/or Problems

1. Name the three elements of a junction transistor.

    a. _____
    b. _____
    c. _____

2. Name the two junctions found in a junction transistor.

    a. _____
    b. _____

3. The arrowhead on a transistor symbol identifies the _____ element and the _____ of transistor.

4. Draw the schematic symbols for NPN and PNP transistors and label each of the elements.

5. For normal operation of a transistor, the emitter-base junction is _____ biased and the collector-base junction is _____ biased.

6. Redraw Figure 3-16 and insert batteries that correctly bias each junction.

7. When a transistor is correctly biased, the emitter-base junction will have a _____ resistance and the collector-base junction will have a _____ resistance.

8. In the PNP transistor, the carrier that is used to support current flow is the _____.

9. Emitter current is equal to _____ percent of total current.

10. Base current is kept small by the fact that the base element is _____ doped and is very _____.

11. When the battery voltage (V_{CC}) is held constant, the voltage that controls the amount of I_C, I_E, and I_B is the _____.

    a. base-emitter voltage
    b. emitter-collector voltage
    c. base-collector voltage

12. Draw the schematic diagram of a properly biased NPN transistor circuit using a single power source. Show the correct power source polarity, current paths and direction. Label I_B, I_E, and I_C.

13. Increasing the forward bias applied to the emitter-base junction will cause I_B, I_E, and I_C to _____.

    a. increase
    b. decrease
    c. remain the same

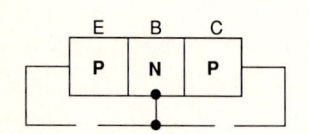

Fig. 3-16

14. In a transistor circuit, external current always flows from _____ to _____, and within the transistor, current flow is supported by the _____ carriers.

15. Redraw Figure 3-17 and insert batteries that correctly bias each junction.

16. In the NPN transistor, the carrier that is used to support current flow is the _____.

17. Draw the schematic diagram of a properly biased PNP transistor circuit using a single power source. Show the correct power source polarity, current paths and direction. Label I_B, I_E, and I_C.

Fig. 3-17

18. Collector current is equal to _____ percent of total current.

19. Which of the circuits in Figure 3-18 is properly biased for conduction?

(a) (b) (c)

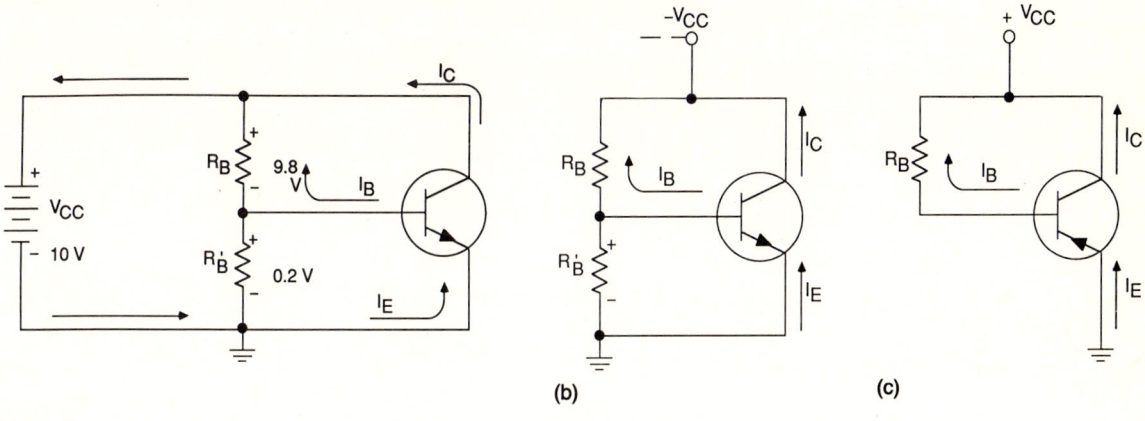

Fig. 3-18

20. Identify the circuit in Figure 3-19 that is properly biased for conduction.

(a) (b)

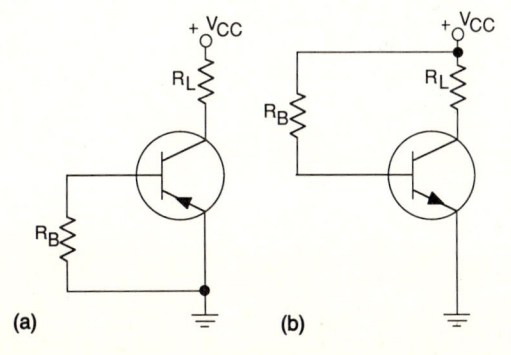

Fig. 3-19

4

Transistor Configurations - Common-Emitter Amplifier

Objectives

1. Define the terms:

    a. Common emitter
    b. Common base
    c. Common collector

2. Identify the schematic for a common-emitter configuration amplifier.
3. Describe the operation of a common-emitter amplifier.
4. Identify circuits that use:

    a. Self-bias
    b. Fixed bias
    c. Voltage divider bias

5. Explain the meaning of the terms *alpha* (α) and *beta* (β).
6. Compare changes of input current or voltage with the effect that it causes at the collector of a common-emitter amplifier.
7. Analyze circuits using the DC load line to establish

    a. maximum collector current (I_CMax).
    b. maximum collector voltage (V_CMax).

8. Use a completed DC load line to establish a circuit's operating (Q) point.
9. Calculate base current, collector current, and emitter current using circuit parameters and beta.
10. Apply approximation and equivalent resistance analyses to the common-emitter amplifier.

Introduction

You were introduced to the bipolar transistor in Chapter 3. At that point we were concerned with the characteristics of the two types (PNP and NPN) of bipolar transistors. You learned the characteristics of the three elements and two junctions that are part of the transistor's construction.

Biasing arrangements for both types of bipolar transistors were discussed. You learned that for a bipolar transistor to operate in the **active** region its junctions must be biased as follows:

The emitter-base junction must be forward biased.
The collector-base junction must be reverse biased.

To bias these junctions, the following rules apply:

- When a junction is connected such that its P-type material is at a more positive potential than the N-type material, it is forward biased.

- When a junction is connected such that its N-type material is at a more positive potential than the P-type material, it is reverse biased.

Circuits were shown and explained that could be used to bias these transistors. You learned that it was possible to bias a bipolar transistor using either one or two power supplies. Single source biasing was identified as the most commonly used biasing method. With single source biasing, a voltage divider is used to provide forward bias for the emitter-base junction. In some cases the emitter-base junction resistance will serve as part of the biasing voltage divider. The same voltage divider also supplies reverse bias for the collector-base (cb) junction.

In this chapter, we begin the study of operational transistor circuits. There are three basic amplifier configurations into which bipolar transistors are connected: these are **common-emitter (CE)**, **common-base (CB)**, and **common-collector (CC)**. You should understand that the *common* designation identifies the element that is common to both the input and output signal of an operational circuit.

Transistor Circuit Configurations

There are three circuit configurations into which bipolar transistor amplifier circuits are classified: these are the **common-emitter (CE)**, **common-base (CB)** and **common-collector (CC)** configurations. Figure 4-1(a) contains single and dual source circuits for the CE amplifier. Figure 4-1(b) contains the dual-source circuit diagram for the CB configuration. Figure 4-1(c) contains two circuits for the CC amplifier.

Transistor Configurations - Common-Emitter Amplifier

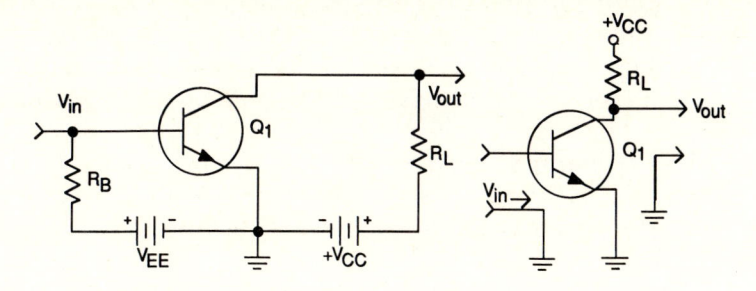

(a) Common-Emitter Configuration

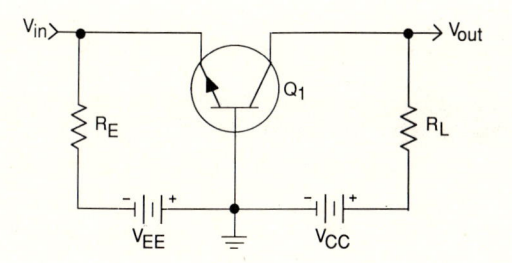

(b) Common-Base Configuration

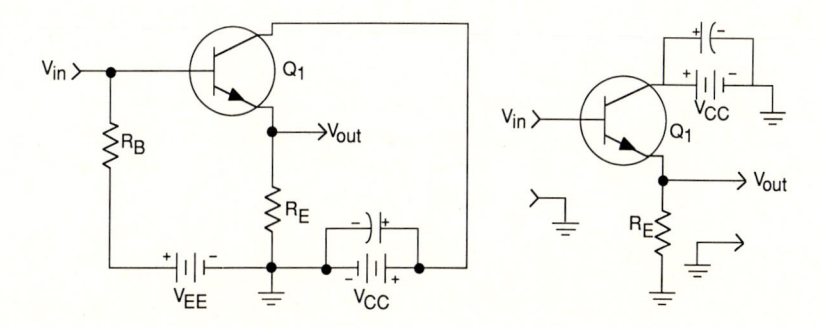

(c) Common-Collector Configuration

Fig. 4-1 Transistor Configuration

In the two common-emitter amplifier schematics, in Figure 4-1(a), notice that the *emitter is grounded*. Also notice that in the right schematic, the input and output circuits are referenced to ground. The fact that the emitter, the input signal, and the output signal all share the same ground tells us that they all have the ground *in common*. Thus, we use the name **common emitter**.

In Figure 4-1(b) the circuit has its base, input signal, and output signal all sharing a common ground. We call this circuit the **common-base** configuration.

In Figure 4-1(c) the collector, input signal, and output signal share a common ground. This circuit is called the **common-collector** configuration.

In this chapter we will concentrate on the study of the CE amplifier. A full explanation of this configuration will aid in understanding the CB and CC configurations which are discussed in Chapter 5.

The Common-Emitter Configuration

To better understand the different biasing arrangements that can be used to bias a transistor amplifier, we will discuss: (1) fixed bias, (2) self bias, (3) voltage divider bias, and (4) dual-source (two battery) bias.

Fixed Bias

Refer to Figure 4-2. This schematic depicts a common-emitter amplifier using **fixed bias**. Note that R_L is connected between V_{CC} and the collector. This resistor (R_L) may be called the collector resistor (R_C) in other textbooks. The authors believe, however, that the name that is becoming the standard is *load resistor* and will use the symbol R_L for all resistors that are used to develop the output signal. R_L is inserted at this

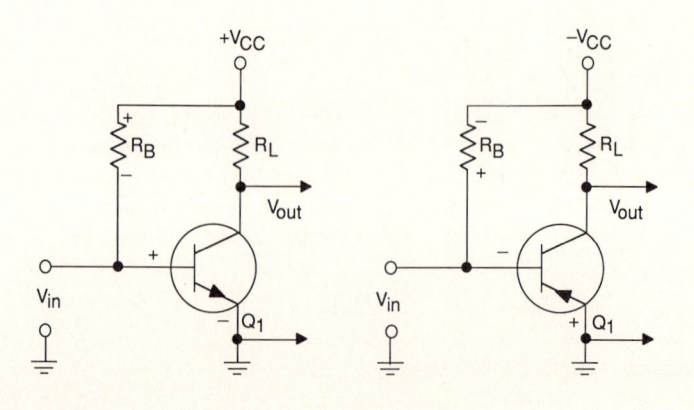

(a) NPN CE Amplifier (b) PNP CE Amplifier

Fig. 4-2 Common-Emitter Amplifier-Fixed Bias

location to limit the current flowing through the collector (I_C). This serves two purposes: collector current (I_C) is limited to a level that will not damage the transistor; and the voltage drop across R_L will cause the voltage drop at the collector (V_C) to vary. The size of R_L is critical to the desired circuit operation.

In the fixed bias circuit of Figure 4-2(a), R_B and the emitter-base resistor (R_{eb}) form a voltage divider to supply forward bias for the NPN transistor. Notice that V_{CC} is positive and is of the correct polarity to reverse bias the collector. With ground connected to the emitter, there are two paths for current flow. Current flows from ground, through the emitter and into the base. Once in the base, the majority carriers (electrons) will flow through either the base or the collector. R_B is connected between the base and V_{CC}. R_L is connected between the collector and V_{CC}. Remember from Chapter 3 that standards were established for use with this book. These standards are: approximately 2% of I_E will flow as I_B and approximately 98% will flow as I_C. When current flows through the emitter-base junction, a voltage is dropped across emitter-base resistance (R_{eb}). This voltage has the polarity that is shown on Figure 4-2(a). With negative voltage applied to the emitter and positive voltage to the collector, all polarities are correct for forward biasing an NPN transistor. The voltage drop across R_{eb} will be small because of the low junction resistance. In Chapter 3 you learned that forward bias for a silicon transistor is approximately 0.7 V and for a germanium transistor, approximately 0.3 V.

The forward bias that is applied to the emitter-base junction allows I_E to flow. The flow of I_E causes collector current (I_C) and base current (I_B) to flow. The voltage drop across the collector-base resistance (R_{cb}) plus R_L must equal the voltage drop across R_B. Notice that the two branches are in parallel. Because V_{CC}, R_L, and R_B are fixed values once the transistor begins to conduct, the bias applied across R_{eb} remains fixed for the entire time that the transistor continues to operate.

Figure 4-2(b) shows the schematic for a PNP CE amplifier. Notice that a $-V_{CC}$ is applied to this circuit. The $-V_{CC}$ has correct polarity for reverse biasing the collector-base junction. R_L, R_B, and V_{CC} are fixed values and are connected in the same way as the NPN transistor. The difference in operation is that current flows from the $-V_{CC}$ to ground. Current, again, has two paths: through R_B, into the base, through the emitter, and to ground; and through R_L, the collector, the base, the emitter, and to ground.

The current flowing through R_B and the base causes a voltage drop (negative to positive) across the emitter-base junction. This is a small voltage because of the small resistance of the emitter-base junction during conduction. It is, however, of the correct polarity and large enough to forward bias the emitter-base junction. The transistor will

Self-Bias

Circuits that are **self biased** are shown in Figure 4-3. Notice that both NPN and PNP types are shown. When R_B is connected in this way, we say that the transistor has self-bias applied. In some cases, you may hear this type bias referred to as **collector feedback bias**.

Operation of this circuit is very similar to that of the fixed bias circuits. In the NPN transistor circuit shown in Figure 4-3(a), current flows from ground to $+V_{CC}$. Leaving the base, two paths are present: through R_B, R_L, to V_{CC}; through R_{cb}, R_L, to V_{CC}. Current flowing from the base (I_B) and through R_B causes a voltage drop on the emitter-base junction (negative to positive) as shown on the schematic. This voltage is the correct polarity and has enough amplitude to forward bias the transistor. As long as all parameters remain constant, current flow in the circuit will continue at the same levels.

In the PNP circuit shown in Figure 4-3(b), operation is only slightly different. To reverse bias the collector-base junction, we must use $-V_{CC}$. Therefore current will flow from $-V_{CC}$ to ground. Leaving the bottom of R_L, current can take either of two paths: through R_B and through the emitter-base junction to ground; or through the collector-base junction and through the emitter-base junction to ground. The current that flows through R_B is large enough to forward bias the emitter-base junction by dropping a voltage across R_{eb} of the polarity shown on the drawing. The voltage across R_{eb} is enough to forward bias the emitter-base junction, which causes I_E, I_B, and I_C to remain constant as long as all other parameters remain constant.

Notice that in both of these circuits, R_B is in parallel with R_{cb}. This assures that the voltage drop on R_B and R_{cb} will be equal when the transistor is operating. Any change in I_C causes the voltage drop on R_L to change. As V_{RL} changes, the voltage drop across the transistor changes. Because R_B is in parallel with R_{cb}, the voltage drop on R_B will vary along with variations in V_{RL}. This causes forward bias to change which affects the amount of current flowing in the transistor. You can see that the transistor's operation controls its bias. The name **self-bias** is derived from this fact.

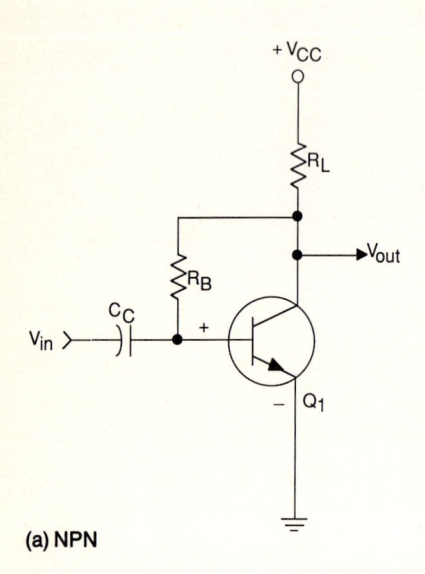

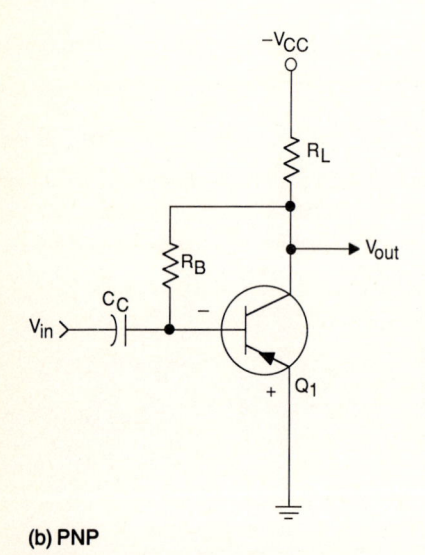

Fig. 4-3 Common-Emitter Amplifier -Self Bias

(a) NPN

(b) PNP

Voltage Divider Bias

Remember, a **voltage divider** is a resistive circuit that is used to divide a voltage into two or more voltages. Figure 4-4 contains both NPN and PNP common-emitter amplifier circuits where a voltage divider is used to forward bias the emitter-base junction. This method was discussed briefly in Chapter 3. It is covered in more detail here.

Transistor Configurations - Common-Emitter Amplifier

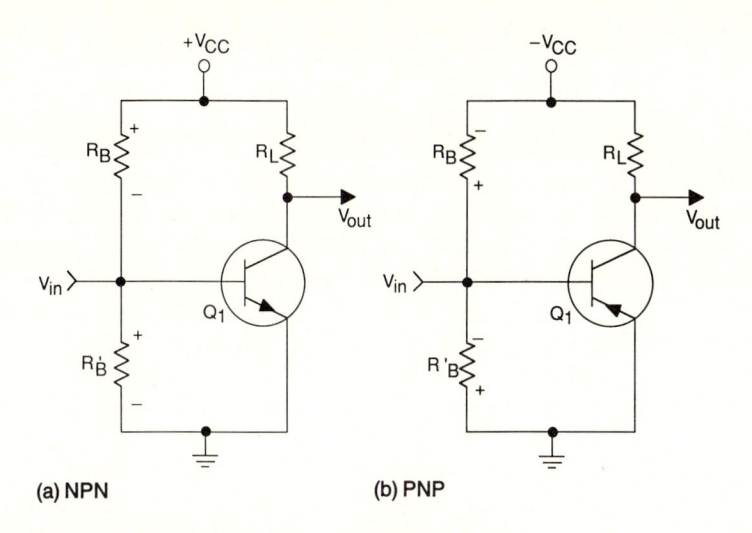

(a) NPN (b) PNP

Fig. 4-4 Voltage Divider Bias Common-Emitter Amplifier

Voltage divider biasing is probably the most commonly used of all biasing methods. Observe Figure 4-4(a). Notice that the voltage divider (R_B and R_B') forms a parallel branch with Q_1 and R_L between ground and V_{CC}. The voltage divider can and will operate independently of the transistor's branch. One branch for current flows from ground to $+V_{CC}$ through R_B' and R_B. Current flowing through this branch will drop voltages ($V_{RB'}$ and V_{RB}) whose sum equals $+V_{CC}$. The voltage drop on R_B' will have negative to positive polarity, as is shown on the schematic. This is the correct polarity for forward biasing the emitter-base junction of Q_1.

The PNP circuit shown in Figure 4-4(b) operates in a very similar way. Except for the fact that a $-V_{CC}$ is applied and current flows from $-V_{CC}$ to ground, the operation is the same. Again, the voltage drop on R_B' provides forward bias for the emitter-base junction.

Remember, forward bias applied to a bipolar transistor will range from approximately 0.3 V (germanium) to 0.7 V (silicon). These voltages are quite small and it is difficult to select resistor ratios that can drop these low voltages. To compensate for this fact, a resistor (R_E) is inserted between the emitter and ground. (See Figure 4-5). By careful selection, R_E values can be selected to allow it to drop a specific amount of the voltage applied to the base by $V_{RB'}$. Thus, V_{RE} equals $V_{RB'}$ less the voltage necessary to forward bias the emitter-base junction. Notice that R_{eb} and R_E are in parallel with R_B', and their voltage drops must equal $V_{RB'}$. In later circuits, we will discuss other advantages of using the emitter resistor (R_E).

A PNP transistor using R_E and voltage divider bias operates in the same way as the NPN type. PNP transistor current and voltage polarities are opposite to those discussed here.

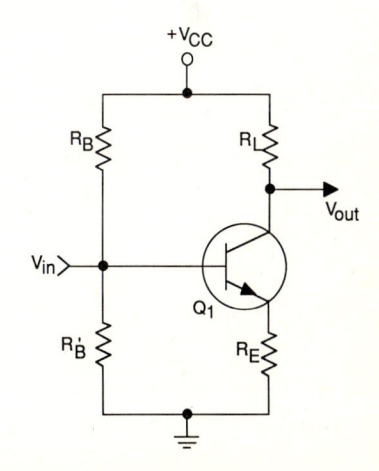

Fig. 4-5 Voltage Divider Bias with Emitter Resistor

Dual-Source Bias

A less often used method of transistor biasing is shown in Figure 4-6. Notice that this method uses two batteries. In fact, it is called **dual-source** bias. In some instances, you may also hear it called **dual-battery** or **dual-supply** bias. The batteries are labeled V_{EE} and V_{CC}.

In Figure 4-6, both the base and collector are positive with respect to the emitter. If the base is more positive than the emitter (with the NPN transistor) the emitter-base junction is forward biased and will conduct. In this circuit, R_E is selected to compensate for the relatively large voltage supplied by V_{EE}. R_E will drop the voltage supplied by V_{EE} that is not required for forward bias.

Power source (V_{EE}) and resistor R_E provide the forward bias that establishes the operating Q point. This allows emitter current (I_E), base current (I_B), and collector current (I_C) to begin. Once forward bias is applied, these currents are maintained by V_{CC}.

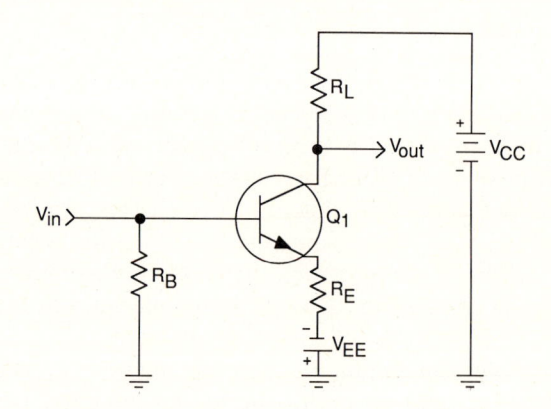

Fig. 4-6 Dual Source Bias Circuit

Self-Check

Indicate whether the statement is True or False.

1. ____The dual-source bias arrangement is the type most often used.
2. ____Using a fixed bias arrangement, forward bias is developed across R_B.
3. ____In a self-biased common-emitter amplifier, R_B and R_{cb} are in parallel.
4. ____When voltage divider bias is used, it is common to have an emitter resistor (R_E) in the circuit.

5. _____ A common-emitter amplifier has its input signal, output signal, and collector all referenced to a common ground.
6. _____ Current flows from V_{CC} to ground in an NPN transistor.
7. _____ Base current will flow through R'_B of the voltage divider when using voltage divider bias.
8. _____ I_B will flow in R_L of a self-biased common-emitter amplifier.
9. _____ Of the four types voltage divider bias is the most often used.
10. _____ R_B and R_{cb} have equal voltage drops in a fixed bias circuit.

Transistor Current Relationships

In Chapter 3 you were told that three currents can be identified within an operating transistor. These are: emitter current (I_E), base current (I_B), and collector current (I_C). You also learned that I_E = 100% of the current, with I_B and I_C dividing I_E such that:

$$I_E = I_B + I_C$$

Amounts that were selected for our use were 2% for I_B and 98% for I_C. The relationship that, in fact, exists between these three currents is governed by the amount of doping received by each element.

The relationship that exists between base and collector current is identified by the Greek letter beta (β). Beta can be calculated by the formula:

$$\beta = \frac{I_C}{I_B}$$

Using the amounts that were discussed above (2% and 98%), we will assume that a transistor is conducting 100 µA. This means that I_C = 98µA and I_B = 2 µA. By substituting these values in the formula shown above we can calculate the beta for this transistor as follows:

$$\beta = \frac{98\ \mu A}{2\ \mu A} = 49$$

In other words, I_C is 49 times as large as I_B and the transistor is said to have a beta of 49.

The relationship that exists between I_E and I_C is referred to by use of the Greek letter alpha (α). To calculate alpha use the following formula:

$$\alpha = \frac{I_C}{I_E}$$

When you substitute the values used above into this formula you can solve for alpha as follows:

$$\alpha = \frac{98\mu A}{100\,\mu A} = 0.98$$

Because I_E always equals 100% of current, and I_C must be less than 100%, alpha will always be less than one.

These equations and the ones listed below will take on more importance as you proceed through future lessons. Two formulas that are used to convert between alpha and beta are:

$$\alpha = \frac{\beta}{\beta + 1}$$

$$\beta = \frac{\alpha}{1 - \alpha}$$

Effect of Input on Output

For an electronic circuit to be useful it must be able to accept an input, to process that input, and to deliver a desired effect in its output. A CE amplifier is shown in Figure 4-7 that uses voltage divider bias with a switching arrangement connected to its input circuit. Notice that by changing the position of Sw_1, we can select normal forward bias, forward bias plus $+V$, and forward bias plus $-V$. Examine the circuit and you will see that the battery connected to the input is in series with the V_{in}, the emitter-base junction and R_E. The output is taken between the collector and ground, meaning that the voltmeter is connected in parallel with the series-connected collector-base junction, emitter-base junction and R_E.

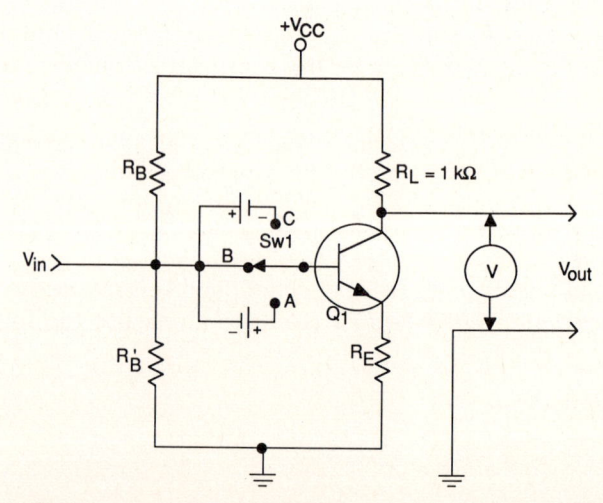

Fig. 4-7 Effect of Changes in Input on Output

Transistor Configurations - Common-Emitter Amplifier

Notice that Sw_1 is set to point B. This sets up the same condition that was discussed earlier: the transistor has normal forward bias and is conducting at a stable state. This is the **operating point** or **quiescent point** for the circuit. Normally the term "quiescent" means "at rest". In electronics we use quiescent point, abbreviated **Q point**, as the point at which a circuit operates when no input signal is present. In other words, this transistor is operating at its Q point with normal forward bias applied to its input. Assume that at the Q point I_C equals 5 mA. Therefore:

$$V_{RL} = I_C \times R_L = 5\,mA \times 1\,k\Omega = 5\,V$$

With $V_{RL} = 5\,V$, the voltmeter will indicate:

$$V_M = V_{CC} - V_{RL} = 10\,V - 5\,V = 5\,V$$

Now let's see what happens when the amount of forward bias changes, as would occur with an input signal. By switching Sw_1 to position A, a small positive voltage (+V) is added to the forward bias applied to the base of Q_1. This added voltage (+V) is divided across the emitter-base junction resistance and R_E in proportion to the size of the two resistances. The end effect is that the forward bias applied to the transistor has been increased. You know that an increase in forward bias will cause I_E, I_B, and I_C to increase. Assume that the increase in forward bias causes I_C to increase by 1 mA. Now I_C will equal 6 mA. What has happened to the voltages across R_L and at the collector? V_{RL} can be calculated as follows:

$$V_{RL} = I_C \times R_L = 6\,mA \times 1\,k\Omega = 6\,V$$

With $V_{RL} = 6\,V$, what will the voltmeter indicate?

$$V_M = V_{CC} - V_{RL} = 10\,V - 6\,V = 4\,V$$

Notice that the collector voltage (V_M) is less positive, which means that it is negative going. The voltmeter now indicates 4 V.

Now we will move Sw_1 to position C and determine what the effect will be. At point C a small −V is added to normal forward bias. This voltage (−V) will oppose the forward bias already present and will cause forward bias to decrease. If forward bias decreases, I_E, I_B, and I_C will also decrease. Assume, again, that the change in I_C equals 1 mA. Collector current (I_C) now equals 4 mA. The values for V_{RL} and V_M can be calculated as follows:

$$V_{RL} = I_C \times R_L = 4\,mA \times 1\,k\Omega = 4\,V$$

With $V_{RL} = 4$ V, what will the voltmeter indicate?

$$V_M = V_{CC} - V_{RL} = 10\text{ V} - 4\text{ V} = 6\text{ V}$$

You should be able to see that by applying a negative voltage to the input, the output voltage increases (goes positive) and the voltmeter indication changes to 6 V.

Notice that placing a positive input voltage on the base of Q_1, causes the voltmeter indication to decrease to 4 V. In other words, a positive going input causes a negative going output.

For a common-emitter amplifier, the input and output signals are 180° out-of-phase.

Note that with the NPN circuit discussed here, all voltages are positive with the output either going more positive or less positive. With a PNP transistor, the voltages will be negative. A $-V_{CC}$ will be applied and the output voltage used for the reference will be a negative voltage. Changes in input voltage (forward bias) will cause the output voltage to go more or less negative.

Voltage Distribution in a Common-Emitter Amplifier

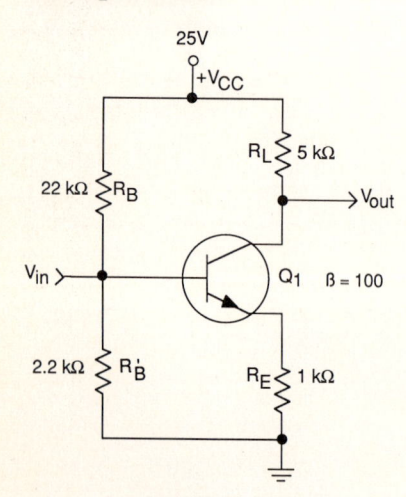

Fig. 4-8 Voltage Distribution in a CE Amplifier

A germanium NPN transistor common-emitter amplifier using voltage-divider bias is shown in Figure 4-8. To analyze this circuit some circuit parameters must be established. Notice that V_{CC} is $+25$ V. The beta of the transistor is assumed to be 100 and all component values are as shown on the schematic. Forward bias for the emitter-base junction is provided by the voltage divider formed by R_B and R'_B. Because the voltage divider is a separate branch connected to V_{CC}, it can be analyzed for its initial voltage and current values. Remember, the voltage divider is being analyzed as if the transistor branch did not exist. Total resistance in the voltage divider branch is calculated as follows:

$$R_{VD} = R_B + R'_B = 22\text{ k}\Omega + 2.2\text{ k}\Omega = 24.2\text{ k}\Omega$$

Then branch current will be:

$$I_{VD} = \frac{V_{CC}}{R_{VD}}$$

$$I_{VD} = \frac{25\text{ V}}{24.2\text{ k}\Omega} = 1.033\text{ mA}$$

You can now calculate the voltage drop on R'_B.

$$V'_{RB} = I_{VD} \times R'_B = 1.033\text{ mA} \times 2.2\text{ k}\Omega = 2.272\text{ V.}$$

Transistor Configurations - Common-Emitter Amplifier

Now the transistor will be reinserted into the circuit to see how the entire circuit operates. As soon as $+V_{CC}$ is applied, forward bias of 2.272 V is applied to the base of Q_1. However, when Q_1 begins to conduct, I_B must pass through R_B as it returns to V_{CC}. This means that I_{RB} is no longer 1.003 mA but is the sum of the currents $I'_{RB} + I_B$.

Assuming that Q_1 is a germanium NPN transistor, we will analyze the circuit to determine its characteristics. The forward bias for a germanium transistor is approximately 0.3 V. As soon as base current begins to flow, which is almost immediately, R_{eb} and R_e are in parallel with R'_B. R_B is also in parallel with R_L and R_{cb}. This means that the voltages developed by the voltage divider are now affected by the transistor's conduction. The result is that as transistor current changes, the distribution of voltage between R_B and R'_B will change.

To solve for the voltage drop on R_E we use:

$$V_{RE} = V'_{RB} - V_{Reb} = 2.272\,V - 0.3\,V = 1.972\,V$$

To solve for I_E we use:

$$I_E = \frac{V_{RE}}{R_E}$$

$$I_E = \frac{1.972}{1\,k\Omega} = 1.972\,mA$$

Notice on the schematic drawing (Figure 4-8) that beta (β) of the transistor is stated as 100. If you were using a stock transistor, you could determine the value for beta by using data (HFE) supplied on the transistor's specification chart. (See Appendix B for a sample specification sheet). Regardless of how beta is obtained, its value can be used to solve for alpha (α). Remember, alpha is the current gain ratio.

$$\alpha = \frac{\beta}{\beta + 1}$$

$$\alpha = \frac{100}{101} = 0.99$$

Using this value for alpha, I_C can be calculated using the following procedure:

$$I_C = \alpha \times I_E = 0.99 \times 1.972\,mA = 1.95\,mA$$

Then I_B can be calculated in either of the following ways:

$I_B = I_E - I_C = 1.972 \text{ mA} - 1.95 \text{ mA} = 22 \mu A$

$I_B = \dfrac{I_C}{\beta}$

$I_B = \dfrac{1.95 \text{ mA}}{100} = 19.5 \mu A$

> **NOTE:** The slight errors that exist between the two values for I_B result from the necessity to round off numbers during the mathematical calculations.

Once we know the values for I_E, I_B and I_C, we can complete the analysis of this circuit. These analyses are:

$V_{RL} = R_L \times I_C = 5 \text{ k}\Omega \times 1.95 \text{ mA} = 9.75 \text{ V}$

$V_C = V_{CC} - V_{RL} = 25 \text{ V} - 9.75 \text{ V} = 15.25 \text{ V}$

$V_{RE} = I_E \times R_E = 1.97 \text{ mA} \times 1 \text{ k}\Omega = 1.97 \text{ V}$

$V_{CE} = V_{CC} - (V_{RL} + V_{RE})$

$V_{CE} = 25 \text{ V} - (9.75 + 1.97) \text{ V}$

$V_{CE} = 25 \text{ V} - 11.72 \text{ V} = 13.28 \text{ V}$

Now we can calculate the resistance of each junction as follows:

$R_{cb} = \dfrac{V_{CE} - V_{EB}}{I_C}$

$R_{cb} = \dfrac{13.28 - 0.3 \text{ V}}{1.95 \text{ mA}}$

$R_{cb} = \dfrac{12.98}{1.95 \text{ mA}} = 6.656 \text{ k}\Omega$

$R_{eb} = \dfrac{V_{EB}}{I_E}$

$R_{eb} = \dfrac{0.3 \text{ V}}{1.97 \text{ mA}} = 152 \Omega$

As you can see, this system of analysis is time consuming and can result in mathematical errors. Rounding off of decimal quantities will cause errors even if no mistakes are made. An alternate method that recognizes the fact that precise accuracy is not necessary is the **approximation** method. This system has proven to be accurate enough to serve most design purposes because of the tolerance allowed in electrical circuit operation.

Use Figure 4-9 as a reference as we proceed through an approximation of the NPN amplifier's characteristics. The first thing that must be done is to calculate the current flow (I_{VD}) for the voltage divider that contains R_B and R'_B.

$$R_{VD} = R_B + R'_B = 15\ k\Omega + 5\ k\Omega = 20\ k\Omega$$

$$I_{VD} = \frac{V_{CC}}{R_{VD}}$$

$$I_{VD} = \frac{40\ V}{20\ k\Omega} = 2\ mA$$

Approximation Method

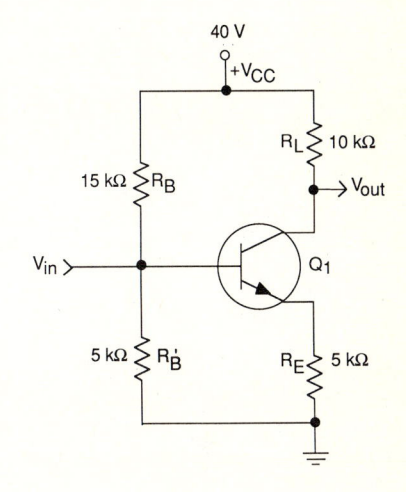

Fig. 4-9 Circuit Approximations - CE Voltage Divider Bias

Now that I_{VD} is known, the amount of voltage (V_B) that is applied to the base of Q_1 can be calculated. Base voltage (V_B) will be approximately equal to the voltage drop on R'_B. V'_{RB} is found as follows:

$$V'_{RB} = I_{VD} \times R'_B = 2\ mA \times 5\ k\Omega = 10\ V$$

Base voltage (V_B) can also be approximated without knowing I_{VD} as follows:

$$V_B = \frac{R'_B}{R_B + R'_B} \times V_{CC}$$

$$V_B = \frac{5\ k\Omega}{20\ k\Omega} \times 40\ V = 10\ V$$

This tells you that a $V_B \cong 10\ V$ is applied to the base of Q_1. Assuming that this is a germanium transistor, we can see that the 0.3 V required for forward bias is small compared to the 10 V V_B. Therefore, it is possible to ignore the 0.3 V when approximating I_E and assume that $V_{RE} \cong 10\ V$. Using this voltage and R_E, I_E can be approximated as follows:

$$I_E = \frac{V_{RE}}{R_E}$$

$$I_E = \frac{10\ V}{5\ k\Omega} = 2\ mA$$

Again, when I_B is compared to I_C and I_E, I_B can be ignored because of its small size. Therefore, it is possible to say that:

$$I_C \cong I_E$$

$$I_C \cong 2 \text{ mA}$$

Using $I_C \cong 2$ mA, we can solve for V_{RL}:

$$V_{RL} = I_C \times R_L = 2 \text{ mA} \times 10 \text{ k}\Omega = 20 \text{ V}$$

and for V_{CE}:

$$V_{CE} = V_{CC} - (V_{RE} + V_{RL}) = 40 \text{ V} - (10 + 20)\text{V} = 10 \text{ V}$$

Emitter-based resistance (R_{eb}) can be ignored because of its small size. Then collector-base resistance (R_{cb}) is calculated as follows:

$$R_{cb} = \frac{V_{CE}}{I_C}$$

$$R_{cb} = \frac{10 \text{ V}}{2 \text{ mA}} = 5 \text{ k}\Omega$$

This concludes the approximation analysis. Notice that it is much easier than the "precise" mathematical analysis and it results in data satisfactory for our purposes. For this reason, all future analyses will be done using the approximation method.

Self-Check

Answer each item by inserting the word or words required to make each statement true.

11. The Greek letter _____ is used to identify the ratio I_C divided by I_B.
12. In a common-emitter amplifier, as the input voltage goes in a positive direction, the output voltage goes in a _____ direction.
13. Adding a positive voltage to the input of a PNP common-emitter amplifier will cause V_{CE} to _____.
14. A common-emitter amplifier is operating with voltage divider supplied bias, the current flowing in R_B (I_{RB}) will equal _____.
15. The Greek letter _____ is used to identify the ratio I_C divided by I_E.
16. To correctly bias a PNP common-emitter amplifier, V_{CC} must have a _____ polarity.
17. When using the approximation method, we assume that I_C = _____ and _____ is so small that it can be ignored.
18. In Figure 4-8, R_E was inserted so that it would drop a voltage equal to _____ − _____.

19. In a common-emitter amplifier, the output signal is taken at the _____ and is referenced to _____.
20. In a common-emitter amplifier, the input and output voltages are _____ phase.

Approximation Method with Fixed Bias

The approximation method can also be applied to transistor circuits that use fixed bias and no R_E. Observe the fixed bias CE transistor amplifier in Figure 4-10. Because R_{eb} is very small, it can be ignored when we calculate I_B using the formula:

$$I_B = \frac{V_{CC}}{R_B}$$

$$I_B = \frac{10 \text{ V}}{200 \text{ k}\Omega} = 0.05 \text{ mA or } 50 \text{ }\mu\text{A}$$

From earlier discussions you know that it is possible to calculate I_C using:

$$I_C = \beta \times I_B = 50 \times 50 \text{ }\mu = 2500 \text{ }\mu\text{A or } 2.5 \text{ mA}$$

The voltage across R_L can be found by using:

$$V_{RL} = I_C \times R_L = 2.5 \text{ mA} \times 2 \text{ k}\Omega = 5 \text{ V}$$

Using these values, V_{CE} can be calculated:

$$V_{CE} = V_{CC} - V_{RL} = 10 \text{ V} - 5 \text{ V} = 5\text{V}$$

This gives us all the data needed to solve for R_{cb}:

$$R_{cb} = \frac{V_{CE}}{I_C}$$

$$R_{cb} = \frac{5 \text{ V}}{2.5 \text{ mA}} = 2 \text{ k}\Omega$$

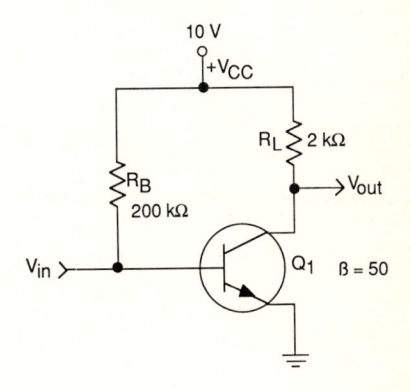

Fig. 4-10 Circuit Approximations - CE Fixed Bias

Notice that the approximation method was used to quickly solve for the major parameters of the circuit. As stated in the previous section, these data are accurate enough for most electronic design purposes.

Approximation Method with Self-Bias

The schematic diagram shown in Figure 4-11 is an NPN common-emitter amplifier with self-bias. Notice R_B is in parallel with R_{cb}. Because the two are in parallel, we already know certain things about them, such as:

1. $V_{RB} = V_{CB}$
2. Current division between the branches will be inversely proportional to the resistance contained in the branches.

It has already been established that:

$$I_C = \beta \times I_B$$

Remember, current and resistance are inversely proportional. Therefore, if $I_C = \beta \times I_B$, then the ratio between R_B and R_{cb} must have the opposite relationship, with R_B being larger than R_{cb}. The ratio of difference will be 50:1. The conclusion is that for **self-biased** circuits only:

$$R_B = \beta \times R_{cb}$$

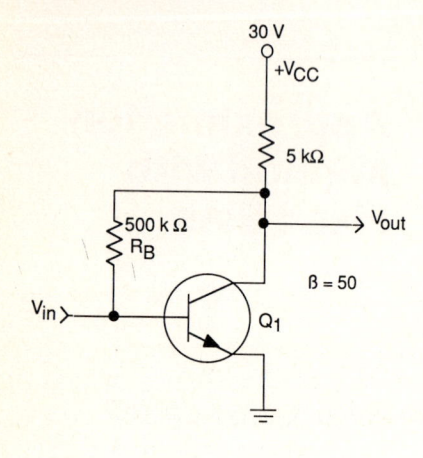

Fig. 4-11 Circuit Approximations - CE Self Bias

Now, to transpose this formula and calculate the size of R_{cb}:

$$R_{cb} = \frac{R_B}{\beta}$$

$$R_{cb} = \frac{500 \text{ k}\Omega}{50} = 10 \text{ k}\Omega$$

With R_{cb} known, the circuit approximation can be completed. Notice that R_B and R_{cb} are in parallel. The large size of R_B means the equivalent resistance of the network is approximately equal to R_{cb}. Therefore $R_{eq} \cong R_{cb}$. Total resistance of the circuit will be approximately equal to:

$$R_t = R_L + R_{cb} = 5 \Omega + 10 \text{ k}\Omega = 15 \Omega$$

$$I_C = \frac{V_{CC}}{R_t}$$

$$I_C = \frac{30 \text{ V}}{15 \text{ k}\Omega} = 2 \text{ mA}$$

$V_{RL} = I_C \times R_L = 2 \text{ mA} \times 5 \text{ K}\Omega = 10 \text{ V}$

$V_{CE} = V_{CC} - V_{RL} = 30 \text{ V} - 10 \text{ V} = 20 \text{ V}$

$I_B = \dfrac{I_C}{\beta}$

$I_B = \dfrac{2 \text{ mA}}{50} = 0.04 \text{ mA or } 40 \text{ }\mu\text{A}$

Again, you can see the advantages of using the approximation method. It yields quantities that are much easier to use for calculations than the decimal values derived in precise mathematical solutions. Further, analyses can be completed more quickly.

Transistor Characteristic Curves

Figure 4-12(a) contains a circuit that has a V_{CC} that is variable from 0-30 V, a bias battery, a current limiting rheostat, and a switch in its emitter circuit. Figure 4-12(b) contains a set of characteristic curves that depict all possible operating points for the transistor with V_{CC} ranging from 0-30 V.

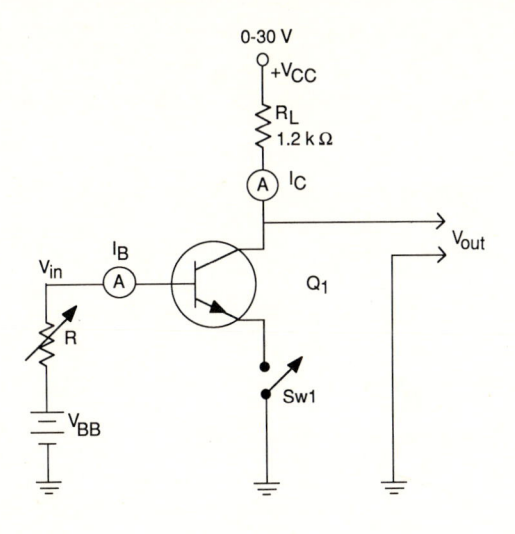

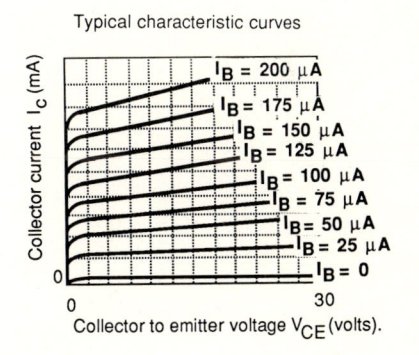

(a) Test Circuit

(b) Graph

Fig. 4-12 CE Amplifier Characteristic Curves

Using the circuit shown in Figure 4-12(a), the characteristic curve chart can be constructed as follows:

1. With Sw_1 open, V_{CC} is varied from 0-30 V. During this period the only current that flows is I_{CBO}, which flows from collector to base. The curve that is labeled $I_B = 0$ represents the I_{CBO} that is flowing. With forward bias applied and $I_B = 0$, there is a reverse current I_{CBO} flowing. This current (I_{CBO}) can be determined by comparing a point along the $I_B = 0$ line to I_C, represented by the left edge of the chart. Remember, I_{CBO} results from the movement of minority carriers and can, in most cases, be ignored.
2. With Sw_1 closed, V_{CC} can be adjusted to each voltage setting (2.5 to 30 V) in 2.5 V steps. With the voltage set to each of these steps, the rheostat R is adjusted for a base current of 25, 50, 75, 100, 125, 150, 175 and 200 μA. With voltage and base current set to stable values, the point that corresponds to I_C is marked along the vertical line. After all points have been plotted, the dots are connected to form the set of curves shown in Figure 4-12(b). These curves represent points at which the transistor can be operated.

As early as your study of DC, you learned that electronic devices could be destroyed by exposure to high heat (wattage). Transistors are no different. Each transistor is designed to operate under prescribed conditions which include maximum power dissipation capabilities. In Figure 4-13 you can see a dashed, curved line that represents the maximum power dissipation curve for a 12.5 watt transistor. It is possible to operate this transistor to the left of the power curve without it being damaged. Operating it at a point to the right of this curve will probably result in its being destroyed.

Two other things are shown on the graph in Figure 4-13. Note that an area along the bottom edge has been shaded, as has a vertical strip on the left side of the chart. The area at the bottom represents the conditions under which the majority current will be cutoff. The shaded area at the left represents the operating points at which the transistor will be saturated.

In most laboratories you will find at least one oscilloscope that is used as a curve tracer. By attaching it to a transistor and following the instructions provided, the screen will display the characteristic curves of the transistor. The maximum power curve will not be displayed, and when designing a circuit, you must calculate power to assure that the transistor is operated within its limits.

Transistor Configurations - Common-Emitter Amplifier

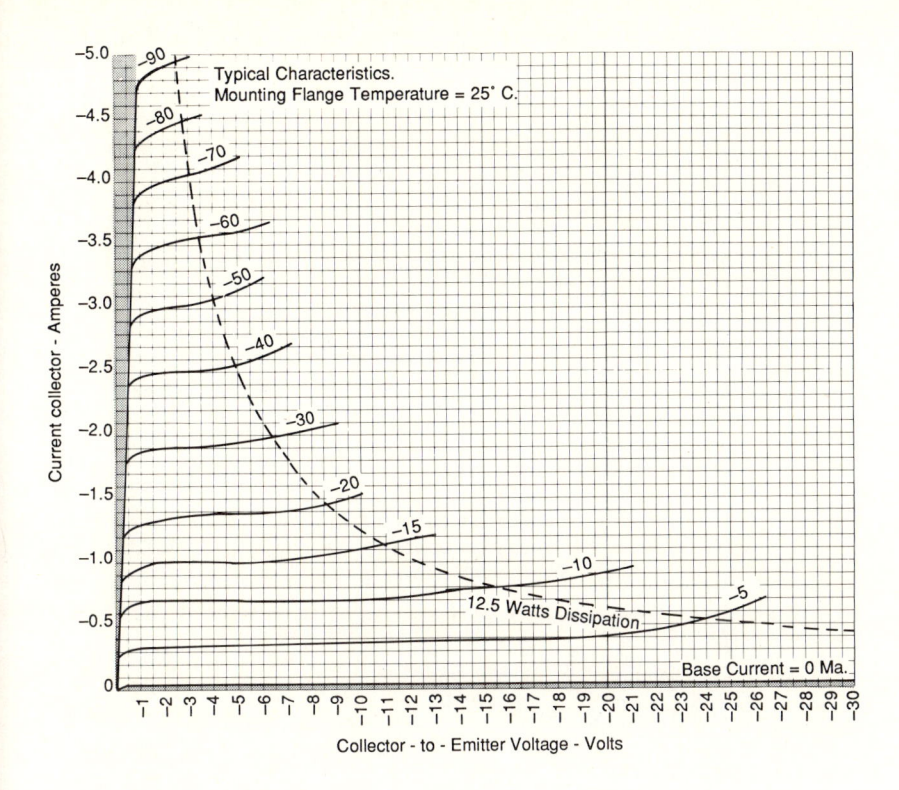

Fig. 4-13 Maximum Power Dissipation Curve

A transistor's DC load line is plotted much like those plotted for diodes (Chapter 2). The transistor DC load line involves plotting a voltage (V_{CE}) versus collector current (I_C). Using a circuit like the one shown in Figure 4-14 and a characteristic curve chart like the one shown in Figure 4-15, we proceed as follows:

1. Identify the collector supply voltage which, in this case, is +12 V V_{CC}. Locate +12 V along the bottom of the chart and place a dot at that location. This is the voltage that will be dropped between the collector and ground when the transistor is cutoff. See Figure 4-15(a).
2. Calculate the collector current that would flow if the transistor was shorted and V_{CC} was dropped across R_L. This is:

$$I_C \text{ Max} = \frac{V_{CC}}{R_L}$$

$$I_C \text{ Max} = \frac{12 \text{ V}}{1.2 \text{ k}\Omega} = 10 \text{ mA}$$

DC Load Lines for the CE Amplifier

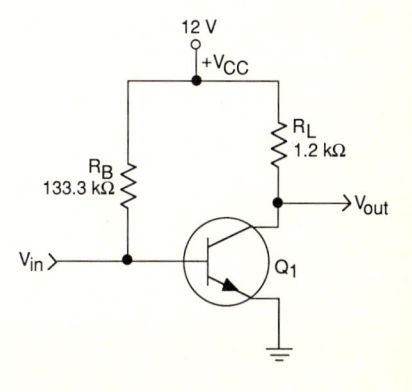

Fig. 4-14 Common-Emitter Amplifier

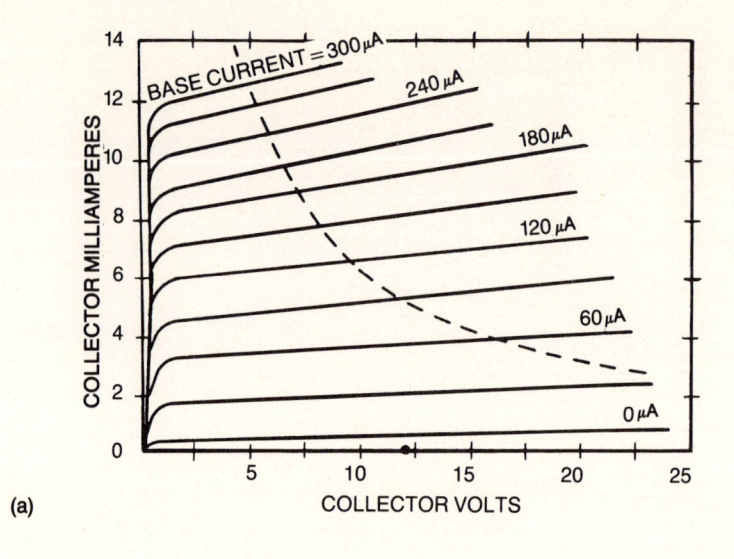

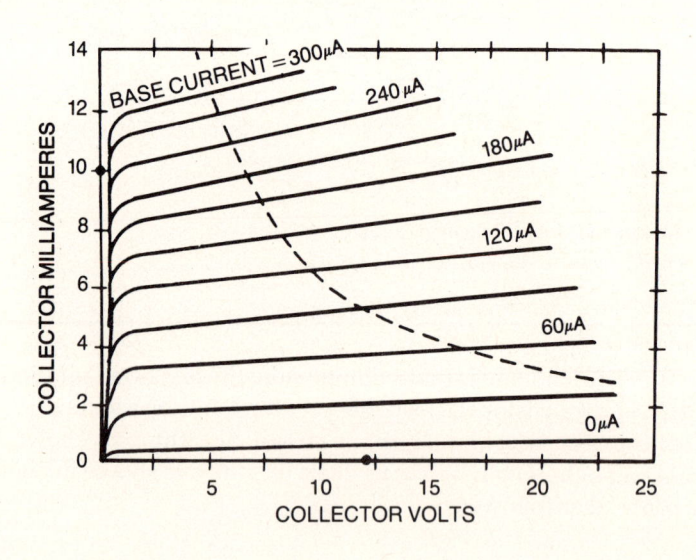

Fig. 4-15 Plotting a DC Load Line

3. Locate the 10 mA point along the vertical side of the chart and place a dot at this point. This is the saturation current that would flow through a shorted Q_1. See Figure 4-15(b).
4. Use a straight edge and connect the two dots (V_{CC} and I_C) that you placed on the chart. See Figure 4-15(c).
5. The diagonal line that you have drawn represents an infinite number of points at which Q_1 can be operated with a V_{CC} of +12 V and an R_L of 1200 Ω.

Transistor Configurations - Common-Emitter Amplifier

(c)

(d)

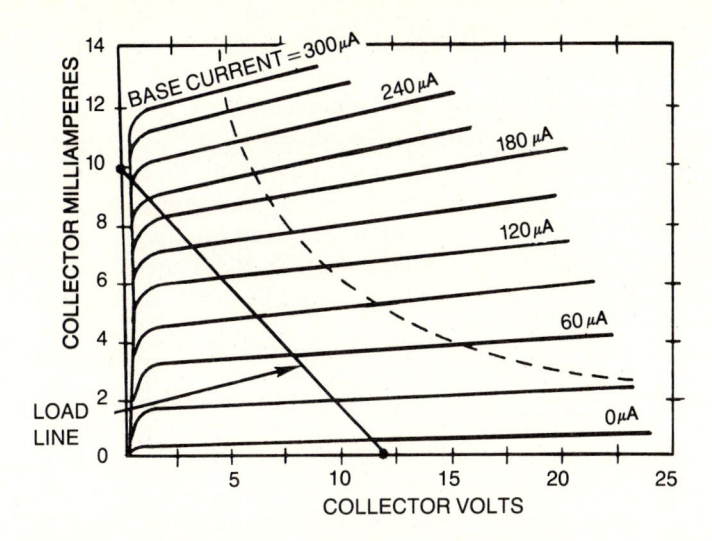

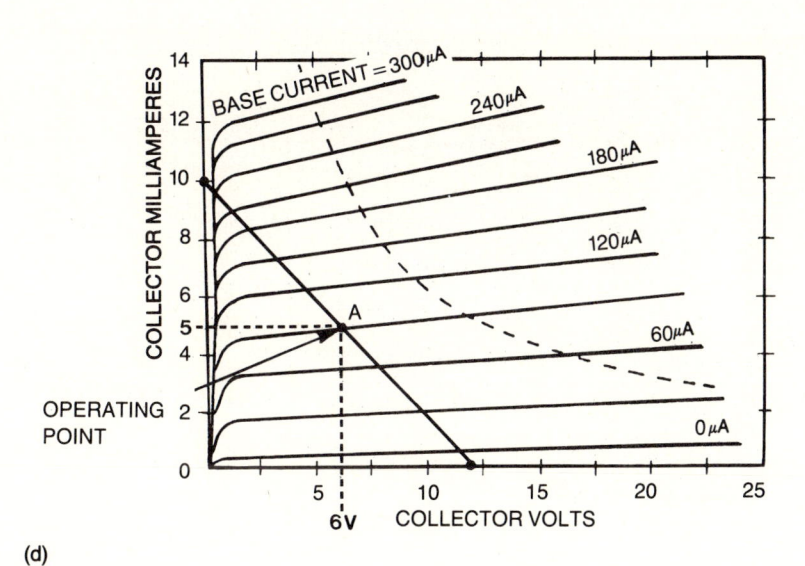

Fig. 4-15 (continued)

6. Notice on Figure 4-15(d) that a dot has been placed on the load line. This point is called the operating point (quiescent Q point). For this discussion the Q point was selected arbitrarily. Actual selection procedures will be discussed later. This Q point represents the point where forward bias applied to Q_1 causes the following conditions:

1. V_C to ground = 6 V
2. I_C = 5 mA

These points are located by constructing horizontal and vertical lines from the point where the base current (I_B) curve crosses the load line (Q point). The vertical line intersects the V_{CC} axis at 6 V. The horizontal line intersects the I_C axis at 5 mA.

Remember, this is only one operating point. By changing the size of R_B, we can change the operating point. By changing forward bias, the operating point can be varied along the load line between all points from cutoff to saturation. Should V_{CC} or R_L change, a completely different circuit would exist and it would be necessary to construct a new load line.

Determining Operating (Q) Point

Once a circuit is designed using specific voltages, resistors, and a transistor, its operation is limited to operating along the points that make up the circuit's DC load line. Once forward bias is applied, the circuit becomes an active circuit and will operate at some point along the load line. We will establish the Q point for a circuit using the PNP common-emitter amplifier circuit shown in Figure 4-16(a).

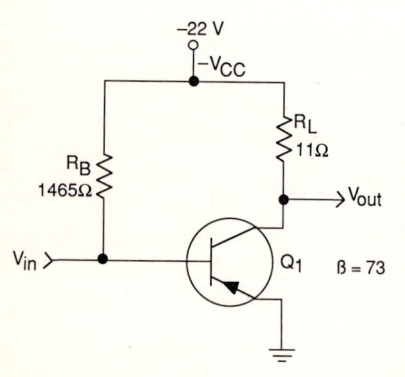

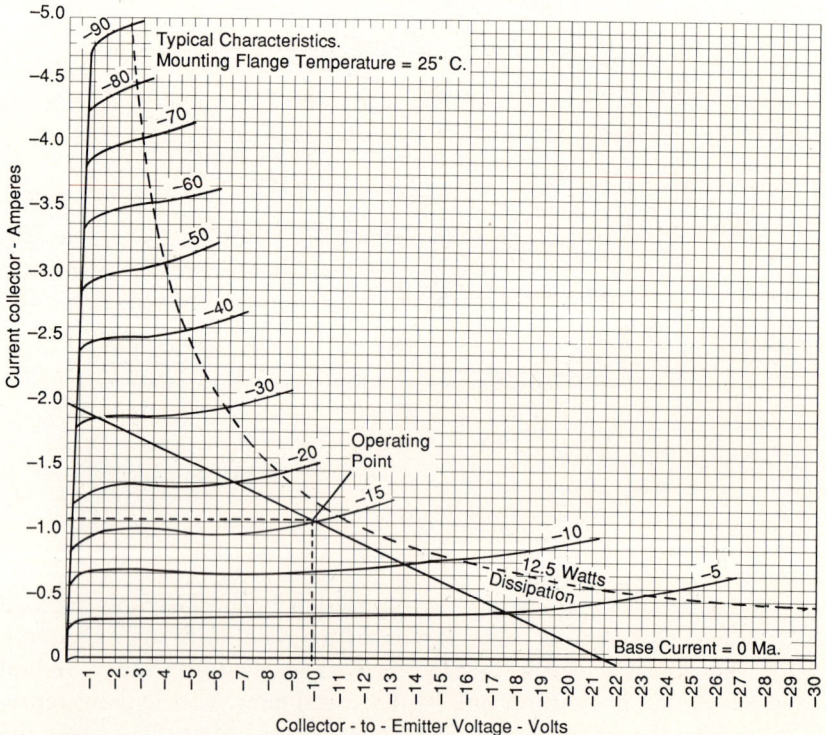

(a) circuit

(b) graph

Fig. 4-16 DC Load Line Analysis

Transistor Configurations - Common-Emitter Amplifier

The characteristic curves for this transistor are shown in Figure 4-16(b). First we must establish the extremes of our load line by identifying maximum I_C and maximum V_{CE}. With the transistor cutoff, $V_{CE} = V_{CC}$ or -22 V. With the transistor shorted, $I_C = V_{CC}$ divided by R_L or 2.0 A. Locate these points on the characteristic curve of Figure 4-16(b) and then use a straight edge to connect them with a line. Using the approximation method, you can calculate I_B as follows:

$$I_{RB} = \frac{V_{CC}}{R_B}$$

$$I_{RB} = \frac{-22 \text{ V}}{1465 \text{ }\Omega} = 15 \text{ mA}$$

Locate the point where the $I_B = 15$ mA curve intersects the load line. This is the Q point. Construct a dashed horizontal line as shown in Figure 4-16(b). Note that this line intersects the I_C axis where $I_C = 1.1$ A. Constructing the vertical dashed line (Figure 4-16(b)) yields a line that intersects the V_{CE} axis at the point where $V_{CE} = -9.9$ V. The following statements can be made about this circuit.

The point where the base current curve crosses the load line is the Q point.

At the Q point for this circuit: $I_C = 1.1$A
$V_{CE} = -9.9$ V

You can prove that $V_{CE} = -9.9$ V by calculation. If $I_C = 1.1$ A, then:

$$V_{RL} = I_C \times R_L = 1.1 \text{ A} \times 11 \text{ }\Omega = -12.1 \text{ V}$$

Therefore:

$$V_{CE} = V_{CC} - V_{RL} = -22 \text{ V} - (12.1) \text{ V} = -9.9 \text{ V}$$

A parameter not yet discussed is the ratio of voltage dropped across the transistor (V_{CE}) compared to V_{CC}. This parameter is called **gamma** and its symbol is the Greek letter gamma (γ). This ratio is referred to as *circuit gamma*. Gamma provides a quick and easy way of predicting the location of a circuit's Q point and requires few calculations. To calculate gamma, use the following formula:

$$\text{Gamma } (\gamma) = \frac{V_{CE}}{V_{CC}}$$

$$\text{Gamma } (\gamma) = \frac{9.9 \text{ V}}{22 \text{ V}} = 0.45 \text{ or } 45\%$$

> **NOTE:** Should an unbypassed emitter resistor (R_E) be present, its voltage would be treated as part of V_{RL}. This means that $V_{CE} = V_{CC} - V_{RL} - V_{RE}$.

Gamma is normally expressed as a decimal. In this case

Gamma $(\gamma) = 0.45$

Examine Figure 4-17. Note that the load line drawn on this chart is divided into 10 equal parts. Gamma is an expression of the percentage ratio of the operating point along the load line. Many amplifiers are designed to operate as near the center (0.5) of the load line as possible.

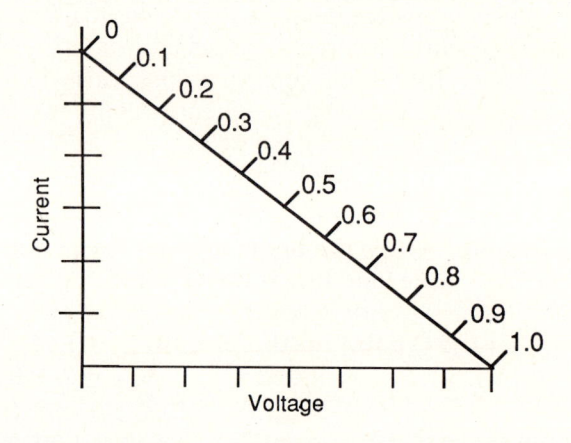

Fig. 4-17 Gamma Relationships

Except for this circuit, all discussions have centered on the NPN transistor. Remember, the only difference between NPN and PNP transistor circuits is the polarity of V_{CC} used and the direction of current flow. V_{CC} for the NPN transistor is positive polarity. For the PNP transistor, a $-V_{CC}$ is used. In the NPN transistor, current flows from ground to V_{CC}. In a PNP transistor, current flows from the $-V_{CC}$ to ground. All voltage and current values are identical except for polarity.

Self-Check

Answer each item by inserting the word or words required to make each statement a true statement.

21. The characteristic curve chart for a transistor can be used to design a circuit which uses that transistor at any of its operating _____.

Transistor Configurations - Common-Emitter Amplifier

22. A DC load line for a common-emitter amplifier connects all possible operating points between maximum _____ and maximum _____.
23. To change a DC load line for a given transistor, we must use a different _____ or a different _____.
24. On a characteristic curve chart, the area near the left edge of the chart represents _____.
25. A transistor has 20 V, V_{CC} applied and operates with an R_L that is 2.5 kΩ; what is maximum I_C?

When we analyze a transistor circuit for its DC parameters, we can convert the CE amplifier to its DC equivalent resistive circuit. When using the approximation method, the equivalent resistance circuit is especially easy to analyze. Figure 4-18 shows a fixed bias NPN CE amplifier and its equivalent resistive circuit. Notice that we have omitted R_{eb}. This was explained earlier and was justified by the fact that the conducting eb junction has such low resistance that its resistance has very little effect on the amount of current that flows. We also stated earlier that I_C and I_E are approximately equal. We, therefore, make the assumption that:

$$I_C \cong I_E$$

Equations Used in DC Analysis of a CE Amplifier

(a) CE Amplifier

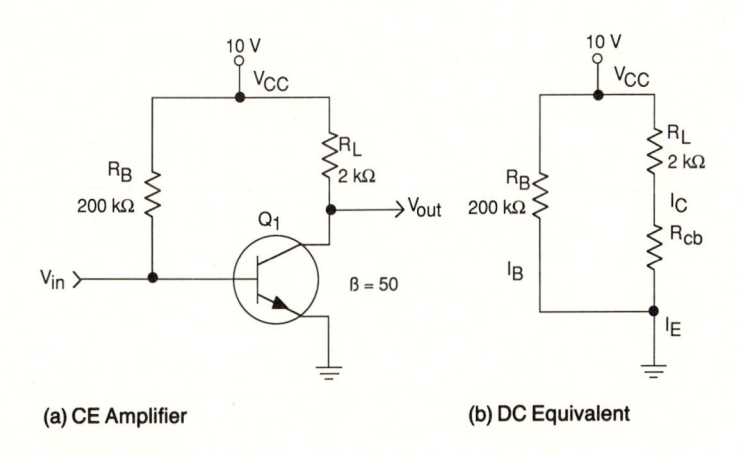

(b) DC Equivalent

Fig. 4-18 CE Amplifier Without R_E

When Figure 4-18(b) is examined you see that R_B is in parallel with the series connected R_L and R_{cb}. Therefore, I_B can be solved using:

$$I_B = \frac{V_{CC}}{R_B}$$

$$I_B = \frac{10 \text{ V}}{200 \text{ k}\Omega} = 50 \text{ }\mu\text{A}$$

On the schematic diagram, beta is shown to be 50. Beta can be used as follows:

$$I_C = \beta \times I_B = 50 \times 50 \text{ }\mu\text{A} = 2500 \text{ }\mu\text{A or 2.5 mA}$$

With beta and R_B we can transpose the following formula to calculate R_{cb}.

$$R_B = \beta \times (R_L + R_{cb})$$

Transposing this formula results in the following:

$$R_{cb} = \frac{R_B}{\beta} - R_L$$

$$R_{cb} = \frac{200 \text{ k}\Omega}{50} - 2 \text{ k}\Omega = 2 \text{ k}\Omega$$

After having solved for the value of R_{cb}, the resistance values for all resistors shown in Figure 4-18(b) are known. These values are shown on the drawing. Now it is possible to solve this circuit exactly like any other DC resistive circuit. These solutions are:

$$I_B = \frac{V_{CC}}{R_B}$$

$$I_B = \frac{10 \text{ V}}{200 \text{ k}\Omega} = 50 \text{ }\mu\text{A}$$

$$I_C = \frac{V_{CC}}{R_L + R_{cb}}$$

$$I_C = \frac{10 \text{ V}}{4 \text{ k}\Omega} = 2.5 \text{ mA}$$

$$I_E = I_C + I_B = 2.5 \text{ mA} + 0.05 \text{ mA} = 2.55 \text{ mA}$$

Figure 4-19 shows a fixed biased NPN CE amplifier and its equivalent circuits. Notice that the circuit contains an emitter resistor (R_E). **Note:** With R_E in the circuit the approximation method cannot be used.

Transistor Configurations - Common-Emitter Amplifier

When reduced to its equivalent resistance, we see that the circuit is a simple series-parallel circuit. In the circuit, though, R_B is in parallel with R_L and R_{cb}, as in the fixed bias circuit explained earlier. This means that you can use the same formula to calculate the value for R_{cb}:

$$R_{cb} = \frac{R_B}{\beta} - R_L$$

$$R_{cb} = \frac{200 \text{ k}\Omega}{100} - 1 \text{ k}\Omega$$

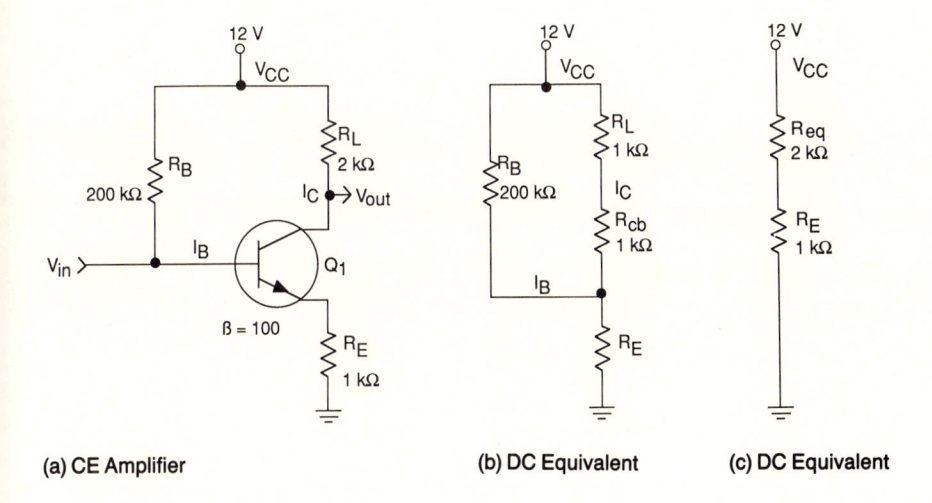

$$R_{cb} = 2 \text{ k}\Omega - 1 \text{ k}\Omega = 1 \text{ k}\Omega$$

(a) CE Amplifier

(b) DC Equivalent

(c) DC Equivalent

Fig. 4-19 CE Amplifier with Fixed Bias

This means that the R_L and R_{cb} branch has a total resistance of:

$$R_{br} = R_L + R_{cb} = 1 \text{ k}\Omega + 1 \text{ k}\Omega = 2 \text{ k}\Omega$$

The equivalent resistance of the parallel network can also be calculated as follows:

$$R_{eq} = \frac{R_B \times (R_L + R_{cb})}{R_B + (R_L + R_{cb})}$$

$$R_{eq} = \frac{200 \text{ k}\Omega \times (1 \text{ k}\Omega + 1 \text{ k}\Omega)}{200 \text{ k}\Omega + (1 \text{ k}\Omega + 1 \text{ k}\Omega)}$$

$$R_{eq} = \frac{400 \text{ M}\Omega}{202 \text{ k}\Omega} = 1.98 \text{ k}\Omega$$

Because 1.98 kΩ is so close to 2 kΩ, we say that $R_{eq} \cong 2 \text{ k}\Omega$.

Once the parallel network is reduced to its equivalent resistance, the circuit can be simplified to a two resistor equivalent as shown in Figure 4-19(c). By adding the two resistors we get:

$$R_t = R_{eq} + R_E = 2\,k\Omega + 1\,k\Omega = 3\,k\Omega$$

I_E can now be calculated using the formula:

$$I_E = \frac{V_{CC}}{R_t}$$

$$I_E = \frac{12\,V}{3\,k\Omega} = 4\,mA$$

V_{RE} can be found using the formula:

$$V_{RE} = I_E \times R_E = 4\,mA \times 1\,k\Omega = 4\,V$$

Remember, we assume that $I_C \cong I_E$. Therefore $I_C \cong 4\,mA$. Now V_{RL} can be calculated as follows:

$$V_{RL} = I_C \times R_L = 4\,mA \times 1\,k\Omega = 4\,V$$

then V_{CE} can be calculated using:

$$V_{CE} = I_C \times R_{cb} = 4\,mA \times 1\,k\Omega = 4\,V$$

to solve for V_{CC} use:

$$V_{CC} = V_{RL} + V_{Rcb} + V_{RE} = 4\,V + 4\,V + 4\,V = 12\,V$$

Now use these values to solve for I_B:

$$I_B = \frac{I_C}{\beta}$$

$$I_B = \frac{4\,mA}{100} = 0.04\,mA \text{ or } 40\,\mu A$$

This completes the resistive analysis of the fixed bias CE amplifier. Now we will analyze a self-biased NPN CE amplifier. Refer to Figure 4-20 for the schematic diagram and resistance equivalent for this type circuit. Notice that in this circuit, R_B is in parallel with R_{cb} *only*

Transistor Configurations - Common-Emitter Amplifier

(a) CE Amplifier

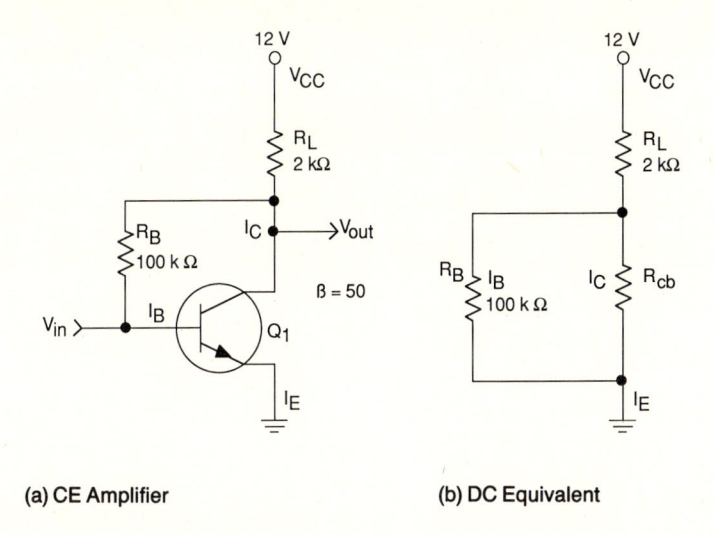

(b) DC Equivalent

Fig. 4-20 CE Amplifier with Self Bias

(Figure 4-20(b)). To analyze this circuit you must first solve for R_{cb}. This is as done as follows:

$$R_{cb} = \frac{R_B}{\beta}$$

$$R_{cb} = \frac{100 \text{ k}\Omega}{50} = 2 \text{ k}\Omega$$

This means that the 100 kΩ R_B is in parallel with a 2 kΩ R_{cb}. Therefore, $R_{eq} \cong 2 \text{ k}\Omega$. Now you can solve for R_t in this way:

$$R_t = R_L + R_{cb} = 2 \text{ k}\Omega + 2 \text{ k}\Omega = 4 \text{ k}\Omega$$

and then for I_E:

$$I_E = \frac{V_{CC}}{R_t}$$

$$I_E = \frac{12 \text{ V}}{4 \text{ k}\Omega} = 3 \text{ mA}$$

Remember, I_C is approximately equal to I_E, so use 3 mA as your value for I_C. To find the values for V_{RL}, V_{CE}, and I_B, use the following formulas:

$$V_{RL} = I_C \times R_L = 3\,\text{mA} \times 2\,\text{k}\Omega = 6\,\text{V}$$

$$V_{CE} = I_C \times R_{cb} = 3\,\text{mA} \times 2\,\text{k}\Omega = 6\,\text{V}$$

$$I_B = \frac{I_C}{\beta}$$

$$I_B = \frac{3\,\text{mA}}{50} = 0.06\,\text{mA or } 60\,\mu\text{A}$$

Now let's analyze a CE amplifier that uses an NPN transistor and voltage divider bias. The schematic and resistive equivalent circuits are shown in Figure 4-21. Again, R_{eb} is so small that it is ignored. Also, continue to assume that $I_C \cong I_E$. With these assumptions, the equivalent circuit can be solved using the same procedures as were used to solve resistive bridge circuits earlier in your studies. Figure 4-21(c) depicts this equivalent circuit.

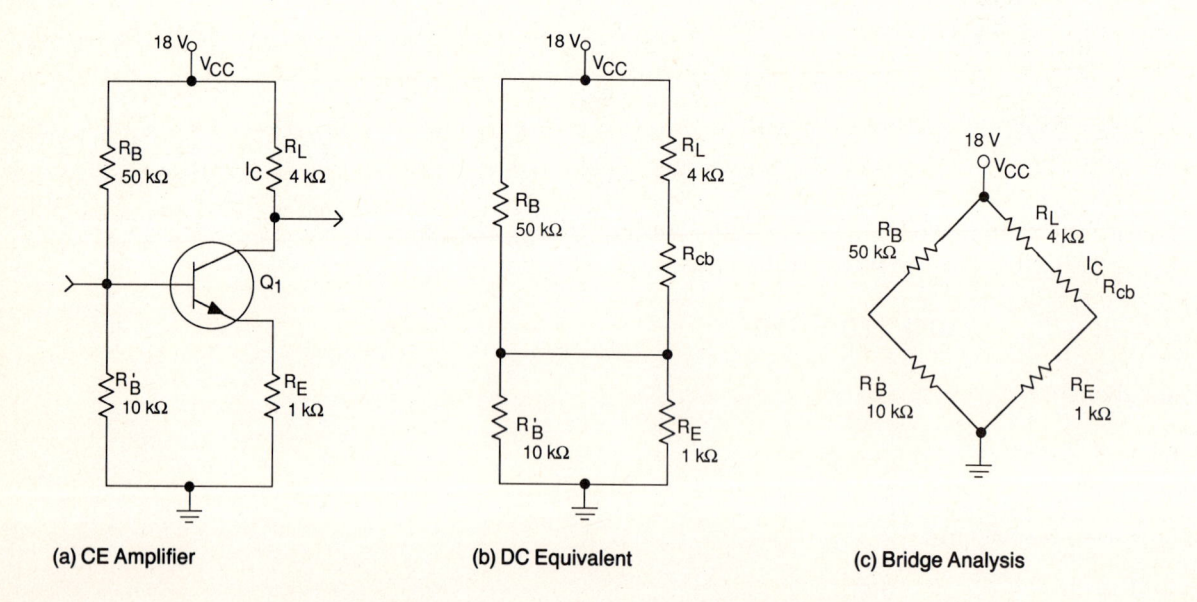

(a) CE Amplifier (b) DC Equivalent (c) Bridge Analysis

Fig. 4-21 CE Amplifier with Voltage Divider Bias

Transistor Configurations - Common-Emitter Amplifier

Treat this circuit as you would a balanced resistive bridge circuit. Then its relationship can be expressed by the ratio:

$$\frac{R_B}{R_B'} = \frac{R_L + R_{cb}}{R_E}$$

Cross multiplication of this ratio yields the following equation:

$$(R_L + R_{cb})(R_B') = (R_B)(R_E)$$

Dividing both sides of the equation by R_B' results in this formula:

$$R_L + R_{cb} = \frac{R_B \times R_E}{R_B'}$$

To isolate R_{cb} on one side of the equation, we must subtract R_L from both sides of the equation. This leaves the formula needed to solve for R_{cb}.

$$R_{cb} = \frac{R_B \times R_E}{R_B'} = -R_L$$

$$R_{cb} = \frac{50 \text{ k}\Omega \times 1 \text{ k}\Omega}{10 \text{ k}\Omega} - 4 \text{ k}\Omega$$

$$R_{cb} = 5 \text{ k}\Omega - 4 \text{ k}\Omega = 1 \text{ k}\Omega$$

The rest of the analysis can be done using Ohm's law. We will use it to solve for R_t, I_E, V_{RL}, V_{CE}, and V_{RE}:

$$R_t = R_L + R_{cb} + R_E$$

$$R_t = 4 \text{ k}\Omega + 1 \text{ k}\Omega + 1 \text{ k}\Omega = 6 \text{ k}\Omega$$

$$I_E = \frac{V_{CC}}{R_t}$$

$$I_E = \frac{18 \text{ V}}{6 \text{ k}\Omega} = 3 \text{ mA}$$

$$V_{RL} = I_C \times R_L = 3 \text{ mA} \times 4 \text{ k}\Omega = 12 \text{ V}$$

$$V_{CE} = I_C \times R_{cb} = 3 \text{ mA} \times 1 \text{ k}\Omega = 3 \text{ V}$$

$$V_{RE} = I_E \times R_E = 3 \text{ mA} \times 1 \text{ k}\Omega = 3 \text{ V}$$

This completes the explanation of DC analysis of CE amplifier circuits. Differing biasing arrangements have been discussed, but the techniques are similar for each type. In fact, many of the formulas are identical from one type to the next.

Self-Check

Answer each item by inserting the word or words required to make each statement a true statement.

26. The circuits shown in Figure 4-20 is a _____ biased circuit.
27. To analyze a voltage divider biased common-emitter amplifier we use the same procedure as we used for the resistive _____ circuit.
28. In a self-biased CE amplifier, to use the equivalent resistance method you must calculate the equivalent resistance of _____ in parallel with _____.
29. In a fixed biased CE amplifier, to use the equivalent resistance method you must calculate the equivalent resistance of _____ in parallel with _____.
30. Equivalent resistance and approximation methods of analysis (are) (are not) as accurate as mathematical analysis.

Summary

Bipolar junction transistors (BJTs) are used to design amplifiers that have three different configurations; common-base, common-collector, and common-emitter amplifiers. Both NPN and PNP transistors can be configured into each of these types.

Of the three types, the common-emitter amplifier is the one most often used. Its name comes from the fact that its emitter, input circuit, and output circuit share a common ground. Forward bias must be applied to a transistor's junction in order to place it into conduction. Four biasing arrangements are used to provide amplifier bias: self-bias, fixed bias, voltage divider bias, and dual-source bias. Of these arrangements, voltage divider bias is most often used. Each arrangement has different characteristics. Transistor characteristics vary from one type to another. To assist in the use of different transistors, transistor characteristic curve charts are available. These charts can be used to determine base current, collector current, collector voltage, cutoff, saturation, I_{CBO}, and maximum power dissipation. It is possible to use this chart to plot a DC load line for a specific transistor operating with a specific load resistor. Once the load line has been constructed, it can be used to select a point at which the transistor will operate. The operating point is also called the *quiescent point* or *Q point*.

Analysis of transistor amplifiers can be performed using either precise mathematical calculations or approximation methods. Analysis reveals that the common-emitter amplifier has low (medium) current gain and high voltage gain.

Transistor Configurations - Common-Emitter Amplifier

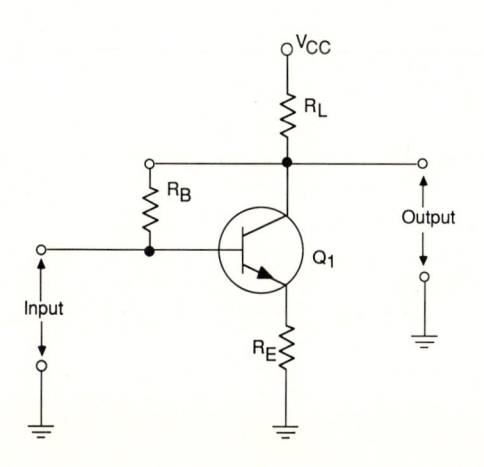

Review Questions and/or Problems

1. To forward bias a PNP transistor the collector must be more _____ than the emitter.

2. In the common-emitter amplifier, input signal, output signal, and the emitter share a common _____.

3. The circuit shown in Figure 4-22 uses a/an _____ transistor.

4. In Figure 4-22, the emitter is at _____ volts potential.

5. Forward bias for the circuit in Figure 4-22 is developed by _____.

6. Refer to Figure 4-23. In this circuit, V_{CC} must be a _____ voltage.

7. In this circuit (Figure 4-23) _____ bias is used.

8. Examine Figure 4-23. In this circuit, $V'_{RB} = V_{RE}$.

   a. True b. False

9. Refer to Figure 4-24. This circuit uses _____ bias.

10. In Figure 4-24, R_B is in parallel with R_{cb}.

    a. True b. False

11. Refer again to Figure 4-24. In this circuit, I_{RL} equals $I_{RB} + I_{cb}$.

    a. True b. False

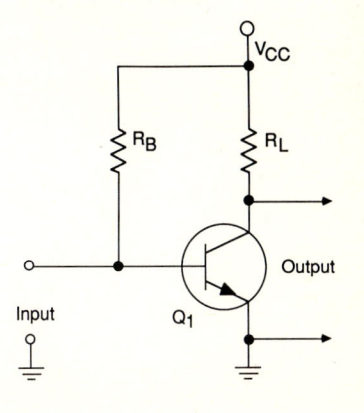

Fig. 4-22

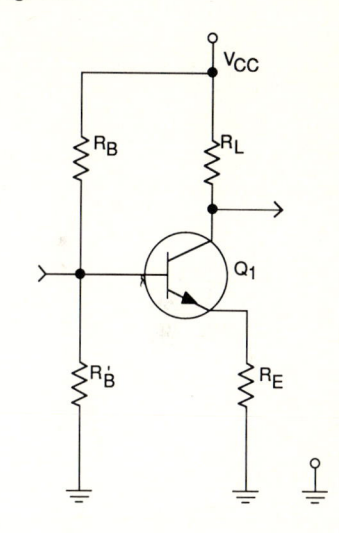

Fig. 4-23

Fig. 4-24

12. In Figure 4-24, $I_C = I_{RL}$.

    a. True   b. False

13. Refer to Figure 4-25. This circuit is an example of _____ bias.

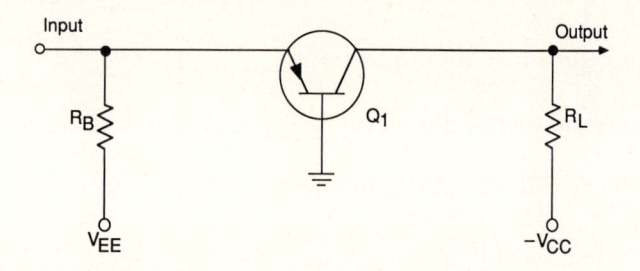

Fig. 4-25

14. The circuit in Figure 4-25 is a PNP common _____ amplifier.

15. In Figure 4-26, the dashed line that the arrow points to tells the transistor's _____.

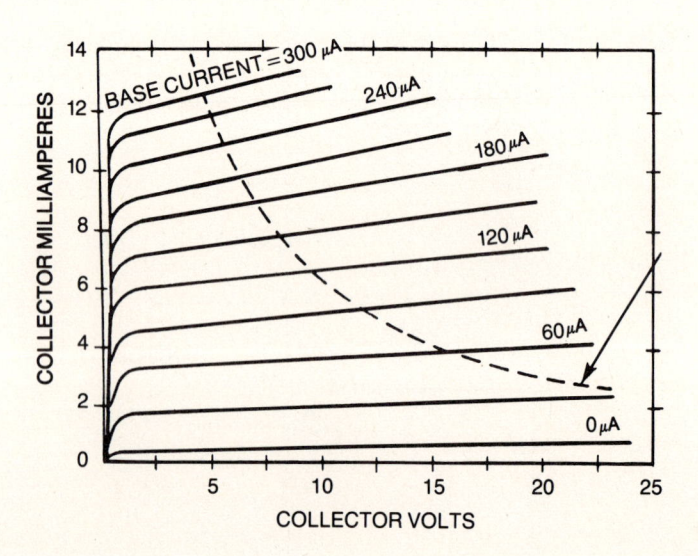

Fig. 4-26

Transistor Configurations - Common-Emitter Amplifier

16. The V_{CC} plot for the circuit shown in Figure 4-20(a) would be plotted at what point on Figure 4-27?

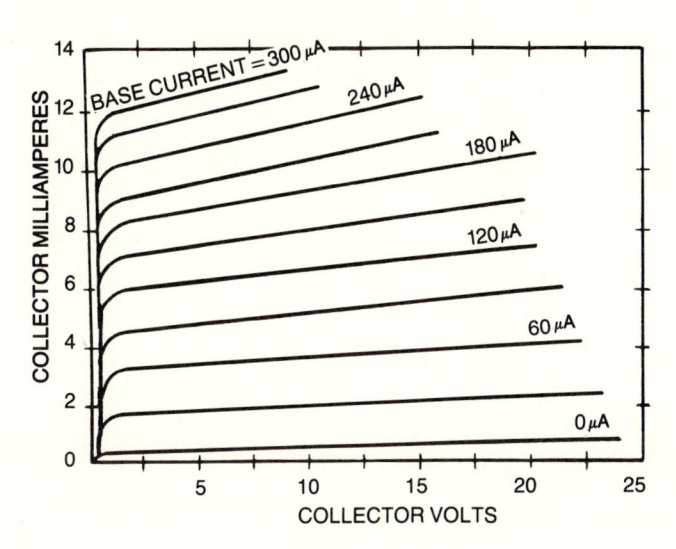

Fig. 4-27

17. Again using Figure 4-20(a) and 4-27, what would the transistor's maximum I_C equal, and where would you plot I_C on the chart?

18. Examine the load line drawn on the graph in Figure 4-28. What size R_L would be required for this load line to be accurate?

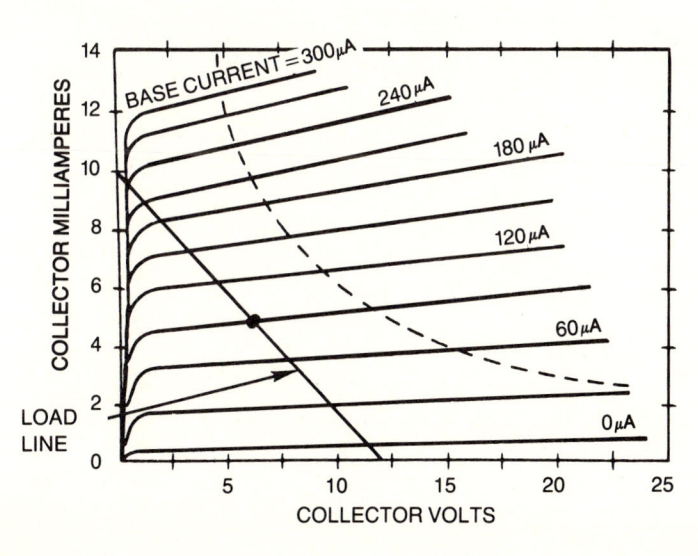

Fig. 4-28

Solid State Electronics

19. Assume that the transistor circuit for Figure 4-28 has been designed to operate with 90 µA I_B. What would the values for V_{CE} and I_C be at the operating point?

20. A circuit is designed such that its Q point is at point B on the load line shown in Figure 4-29. If reverse bias were applied to the circuit and it entered cutoff, what would be the value of V_{CE}?

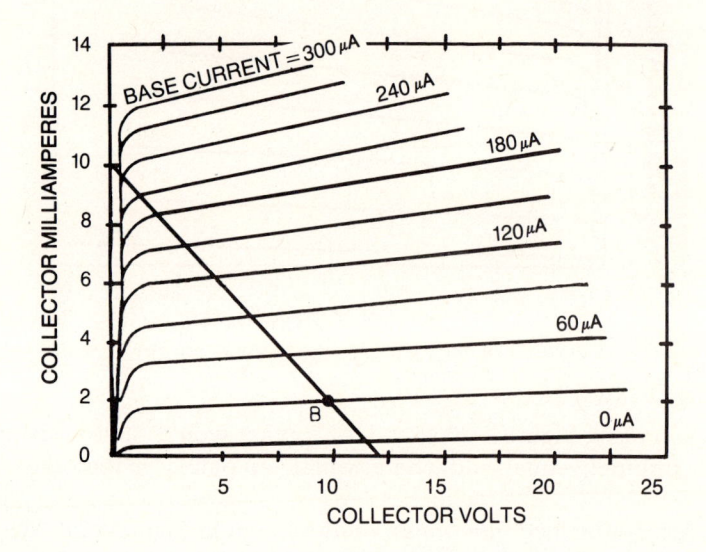

Fig. 4-29

5

Transistor Configurations - Common-Base and Common-Collector Amplifier

Objectives

1. Recognize the circuit schematics for common-base and common-collector amplifiers.
2. Discuss each of these amplifiers including the purpose of each resistor contained in the circuit.
3. Discuss the different biasing arrangements and how they affect circuit operation.
4. Describe the procedure used to construct a DC load line for a common-base amplifier.
5. Given the DC load line for a CB circuit, predict the values of V_{CE} and I_C for a given I_E.
6. Predict the effect that a change in input signal will have on the output voltage of the circuit.
7. Trace current through all paths of the NPN and PNP common-collector and common-base amplifier circuits.

Introduction

In Chapter 4 you were introduced to the three configurations into which bipolar transistor amplifiers are classified. These were the common-emitter, common-base, and common-collector configurations. The remainder of Chapter 4 was used to thoroughly introduce you to the common-emitter configuration (amplifier).

In this chapter the other two configurations are discussed. The common-base (CB) amplifier is used when good voltage gain is desired and input and output signals must be in phase. The common-collector (CC) amplifier is used where voltage gain is not important, but impedance matching is important. Note that in some cases the CC may be called an *emitter follower*. Each configuration will be analyzed using methods similar to those in Chapter 4.

The Common-Base Configuration

From the last chapter, you should remember that the **common-base** amplifier has its **input signal, output signal**, and **base** sharing **common ground**. This fact gives rise to the name **common-base amplifier**.

Figure 5-1 shows the schematic diagram for a common-base amplifier. Note that this schematic is for an amplifier that has dual-source bias which is supplied by $+V_{CC}$ and $-V_{EE}$. The input is applied to the *emitter* and the output is taken from the *collector*. Any signal that is applied to the input is developed across the emitter-base junction. With $-V_{EE}$ applied, the ground connected to the base is *less negative* (more positive) than the $-V_{EE}$. The base-emitter voltage that results meets all of the requirements for forward biasing the emitter-base junction. With these conditions, Q_1 will be conducting. With the base negative and the collector at $+V_{CC}$, the collector-base junction is reverse biased. This means that all conditions are right for a properly operating transistor circuit.

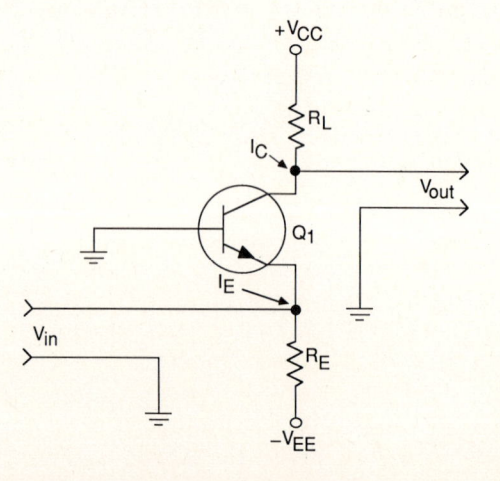

Fig. 5-1 Dual Source Common-Base Amplifier

Transistor Configurations - Common-Base and Collector Amplifier

Notice that Q_1 is an NPN transistor. This means that current within Q_1 flows from the emitter to the collector. When current flows in this direction both the base and collector are positive with respect to the emitter. This means that output voltage (V_C) is positive.

Using N-type material as the emitter requires that the eb junction be forward biased (positive at the base) for the transistor to conduct. Should the input signal go in a negative direction, the forward bias applied to Q_1 is increased, causing I_E, I_B, and I_C to increase. The increase in I_C causes the voltage drop on R_L to increase, which means that the voltage drop from the output (collector) to ground must decrease (become less positive or go negative). In other words, applying a negative going input causes a negative going output.

Should a positive voltage appear at the input, forward bias would decrease, causing all currents to decrease. With a decrease in I_C the voltage drop on R_L (V_{RL}) will decrease. The decrease in V_{RL} means that the voltage between the collector (V_{CB}) and ground must increase. In this case, a positive voltage applied to the input causes the output voltage to increase (go positive).

From these two examples the following conclusion can be drawn:

**In a common-base (CB) amplifier, the
input signal is in-phase with the output signal.**

This contrasts with the common-emitter amplifier where the two signals were 180° out of phase.

Current and Voltage Gain in Amplifiers

In the common-emitter amplifier, the input signal is applied to the base of the transistor and the output is taken from the collector. See Figure 5-2 for a comparison of the input and output circuits for a CE amplifier and a common-base amplifier. With the common-emitter amplifier, the input is applied between the base and ground. The input signal is developed by the emitter-base resistance (R_{eb} and R_E). Remember, R_{eb} is the resistance of a forward biased junction and is a low resistance. The output signal of a common-emitter amplifier is taken between the collector and ground (across R_{cb}, R_{eb}, and R_E). R_{cb} is the resistance contained in the reverse biased collector-base junction. In other words, the output is taken across the same components as the input except for the additional resistance of R_{cb}. With the larger resistance that is used to develop the output signal, we can expect that the voltage at the collector (output) will be higher than that at the base (input). If the output voltage is larger than the input voltage, there has been a voltage gain within the amplifier.

You have already learned that base current is much smaller than collector current. In fact, beta expresses the relationship that exists

Solid State Electronics

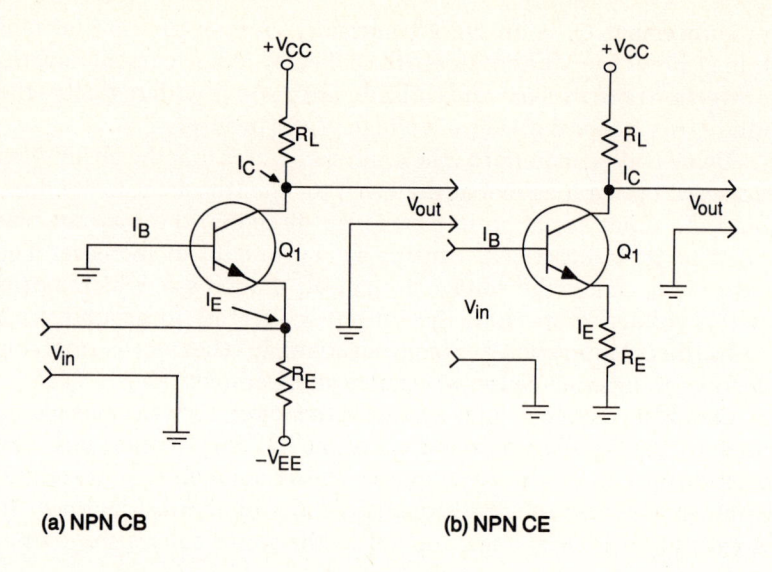

(a) NPN CB   **(b) NPN CE**

Fig. 5-2 Gain Comparisons in CB and CE Amplifier

between the two. If the output current (I_C) is larger than the input current (I_B), the transistor provides a current gain. The common-emitter amplifier supplies both voltage gain and current gain.

Now, to compare the common-base (CB) amplifier (Figure 5-2(a)) to the conditions that exist in a common-emitter (CE) amplifier (Figure 5-2(b)). In a CB amplifier the input signal is applied between the emitter and ground (R_{eb}). Because the emitter-base junction is forward biased, R_{eb} is a small resistance. The output is taken from the collector to ground, or across R_{cb}. The reverse biased condition that exists across the cb junction assures that R_{cb} is a large resistance. Because R_{cb} has a much higher resistance than R_{eb}, it will drop a much larger voltage; therefore, a CB amplifier can provide good voltage gain.

Remember, in all cases, gain (current or voltage) refers to the ratio that exists between the input signal and the output signal. In the common-emitter amplifier, the input was applied to the base and the output was taken from the collector. For this circuit voltage gain was calculated using the formula:

$$A_V = \frac{V_C}{V_{eb}}$$

and current gain using the formula:

$$A_I = \frac{I_C}{I_B}$$

Transistor Configurations - Common-Base and Collector Amplifier

For the common-base amplifier, the formula for voltage gain is:

$$A_V = \frac{V_C}{V_{eb}}$$

and current gain using the formula:

$$A_I = \frac{I_C}{I_E}$$

In the CB amplifier, you can see that the output voltage is produced by I_C. Because the input signal is developed across R_{eb}, I_E is the current that develops the input voltage. Remember, I_E is larger than I_C. Therefore, the common-base amplifier has a current gain of less than one. Actually, you learned earlier that the ratio of collector current to emitter current is the ratio for alpha (α). From this analysis we can state that a CB amplifier has:

A large voltage gain

A current gain of less than one

In other words, voltage will be amplified but current will be decreased when the input is compared to the output.

DC Analysis of a Common-Base Circuit

Common-base amplifier analysis can be performed by use of the equivalent resistance method similar to that used for the CE amplifier. Figure 5-3 contains the schematic diagram for a common-base amplifier and its equivalent resistive circuit. Ground is the common point for this circuit. Current flowing from $-V_{EE}$ to ground passes through two resistances; the forward biased eb junction (R_{eb}) and the 5 kΩ R_E. If we ignore the small resistance contained in R_{eb}, we can say that I_E flows through only 5 kΩ resistance (R_E). The amount of current flowing can be calculated as follows:

$$I_E = \frac{V_{EE}}{R_E}$$

$$I_E = \frac{10\ V}{5\ k\Omega} = 2\ mA$$

Using the approximation method:

$$I_C = I_E = 2\ mA$$

(a) NPN CB   **(b) Resistive Equivalent**

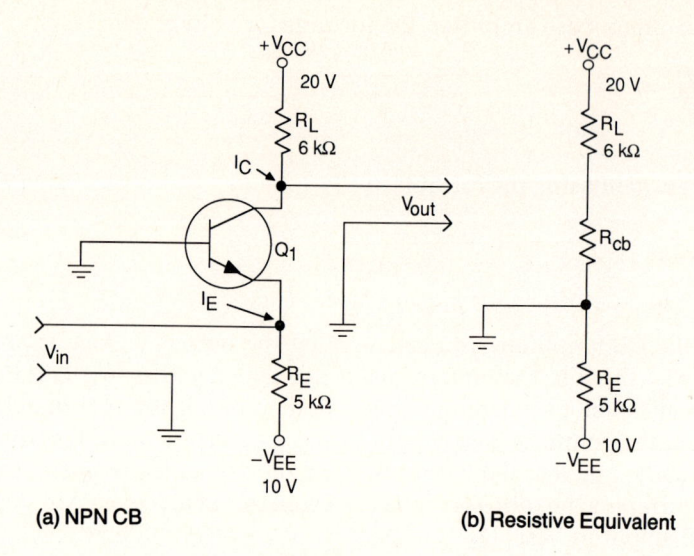

Fig. 5-3 Ground as the Common Point

Collector current flows from $-V_{EE}$, passes through R_E, R_{eb}, R_{cb}, R_L, and V_{CC}. To find V_{RL} use:

$$V_{RL} = I_C \times R_L = 2\,\text{mA} \times 6\,\text{k}\Omega = 12\,\text{V}$$

Using Kirchhoff's voltage law you can find V_{CB}:

$$V_{CB} = V_{CC} - V_{RL} = 20\,\text{V} - 12\,\text{V} = 8\,\text{V}$$

Now, Ohm's law can be used to solve for R_{cb}:

$$R_{cb} = \frac{V_C}{I_C}$$

$$R_{cb} = \frac{8\,\text{V}}{2\,\text{mA}} = 4\,\text{k}\Omega$$

As stated earlier, gamma (γ) is a convenient way of determining the position of a circuit's Q point along its DC load line. The circuit schematic of a common-base amplifier can be used to determine gamma (γ) and to locate the circuit's Q point. Examine Figure 5-4 as it will be used to develop these two points.

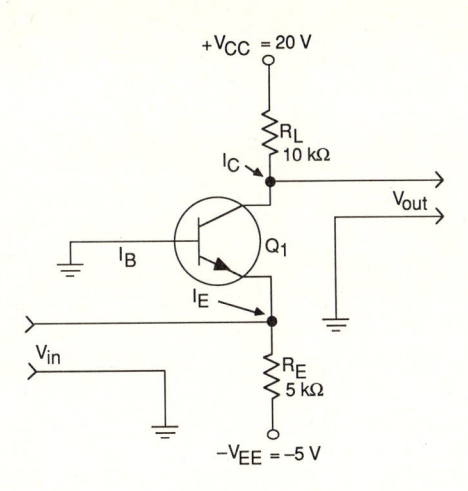

Transistor Configurations - Common-Base and Collector Amplifier

Fig. 5-4 Common-Base Amplifier

To locate the Q point, the emitter current (I_E) must be known. This is found by using the formula:

$$I_E = \frac{V_{EE}}{R_E}$$

$$I_E = \frac{5\,V}{5\,k\Omega} = 1\,mA$$

It has already been established that $I_C \cong I_E$. Therefore $I_C \cong 1\,mA$. Using this current you can calculate V_{RL}:

$$V_{RL} = I_C \times R_L = 1\,mA \times 10\,k\Omega = 10\,V$$

This means that V_{CB} equals:

$$V_{CB} = V_{CC} - V_{RL} = 20\,V - 10\,V = 10\,V$$

This provides the values:

$$I_C = 1\,mA$$
$$V_{CB} = 10\,V$$

which identify the conditions that exist at the Q point.

In a common-base amplifier gamma is the ratio of V_{CB} to V_{CC}.

$$\text{Gamma } (\gamma) = \frac{V_{CB}}{V_{CC}}$$

$$\text{Gamma } (\gamma) = \frac{10 \text{ V}}{20 \text{ V}} = 0.5$$

A gamma value of 0.5 tells you that the transistor is biased exactly in the center of its DC load line, midway between cutoff and saturation. A value less than 0.5 means that the transistor is biased closer to saturation than cutoff. A value larger than 0.5 means that the transistor is biased closer to cutoff than saturation.

DC Load Lines for the Common-Base Circuit

Figure 5-5 shows a family of characteristic curves for a typical common-base amplifier. While this set of characteristic curves may, at first glance, appear to be the same as the one discussed for the common-emitter amplifier, there are several important differences.

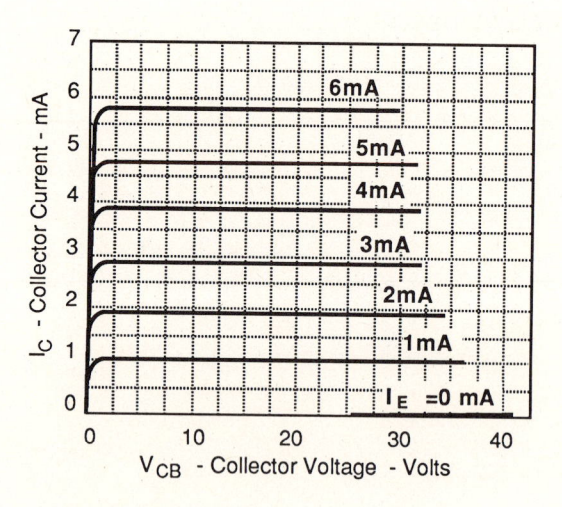

Fig. 5-5 Common-Base Output Characteristic Curves

1. The horizontal axis of the graph now represents collector-base voltage (V_{CB}). Remember, in this configuration the output is taken from the collector to ground (base).
2. The curves on this graph are labeled as emitter current (I_E). This is because the signal is applied to the emitter with respect to ground (base), and the current that develops the input signal is I_E.

Transistor Configurations - Common-Base and Collector Amplifier

3. The family of static curves shown here is developed in much the same way as the common-emitter characteristic curves discussed in Chapter 4. The transistor is connected to two variable power sources (V_{CC} and V_{EE}). V_{CC} is adjusted for a number of different voltages. With V_{CC} set to each voltage, I_E is then, adjusted for each of the currents represented on the chart. Dots are plotted for each combination of V_{CC} and I_E. After these plots have been made the curves are constructed and labeled.

The curves shown in Figure 5-5 show that V_{CB} has *no* effect on collector current (I_C). Changes in I_E do, however, have considerable effect on V_{CB}. From this fact we can state:

Emitter current (I_E) will have a controlling effect on collector current.

In the common-base amplifier the controlling effect that I_E has on I_C is called alpha (α). Two formulas for alpha are:

$$\alpha = \frac{I_C}{I_E}$$

$$\alpha = \frac{\beta}{\beta + 1}$$

Because collector current (I_C) is always less than emitter current (I_E), the value of alpha will always be *less than one (unity)*.

Figure 5-6 contains a common-base amplifier's characteristic curves. These curves can be used to calculate alpha for the transistor

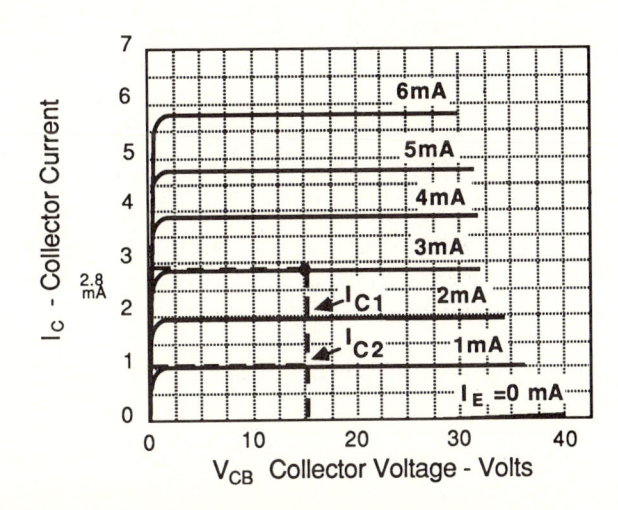

Fig. 5-6 Common-Base Output Characteristic Curves

they represent. Note that at the Q point on the graph, $I_C = 2.8$ mA and $I_E = 3.0$ mA. Using these figures, alpha can be calculated as follows:

$$\alpha = \frac{I_C}{I_E}$$

$$\alpha = \frac{2.8 \text{ mA}}{3.0 \text{ mA}} = 0.93$$

Part of emitter current (I_E) will flow through the base to ground. Because of this, collector current can never equal emitter current and, therefore, alpha can never be equal to one (1).

Alpha is generally referred to as being the current gain for a common-base amplifier since current gain is equal to the ratio of output current to input current. Again, we can see that in a common-base amplifier:

Alpha is always less than unity.

and

Current gain is always less than unity.

The common-base amplifier has its input applied to the emitter and the output taken from the collector. Because changes in I_C must be less than the changes in I_E, you can see that small changes in input (emitter) current result in smaller changes in output (collector) current. Thus, alpha (current gain) is small.

Effect of Changes in Input Voltage on Output Voltage

Use Figure 5-7 as a reference for the analysis that follows. Note that there is a switching network and batteries connected in series with the $-V_{EE}$. Follow along as we discuss the effect this has on the output voltage.

With the switch in position B, the amplifier operates in the active mode. Assume that with the switch set to point B, the circuit operates at point Q_{p1} on the characteristic curves chart shown in Figure 5-8. At this Q point, $V_C = 9$ V and I_C equals 1.9 mA.

Assume that when Sw_1 is moved to either of the other two positions, it causes I_E to change by 1 mA. Therefore, when the switch is set to position A, the two voltages are connected series aiding. This causes V_{EE} to increase. An increase in V_{EE} of Q_1 causes I_E to increase by 1 mA because of the increased forward bias. An increase in I_E of 1 mA means that the operating point is now at Q_{p3} on the graph. Now, $V_C = 3.5$ V and $I_C = 2.8$ mA. With the increased I_C, V_{RL} has increased, leaving less voltage to be dropped between the collector and ground (as V_{CB}).

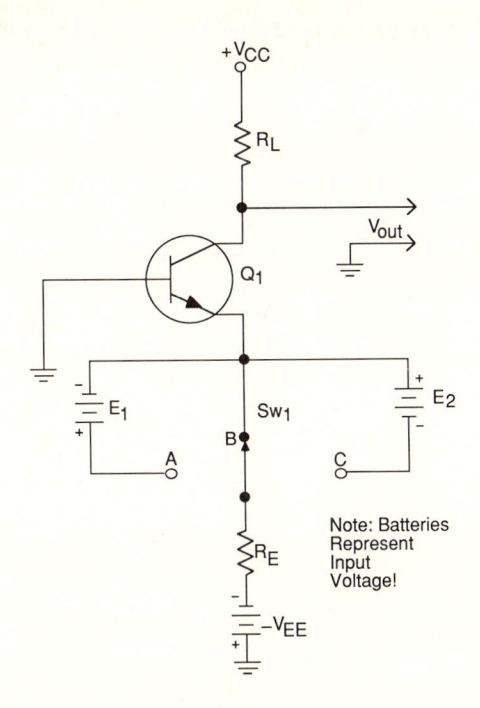

Transistor Configurations - Common-Base and Collector Amplifier

Fig. 5-7 Effect of Changes in Input Voltage

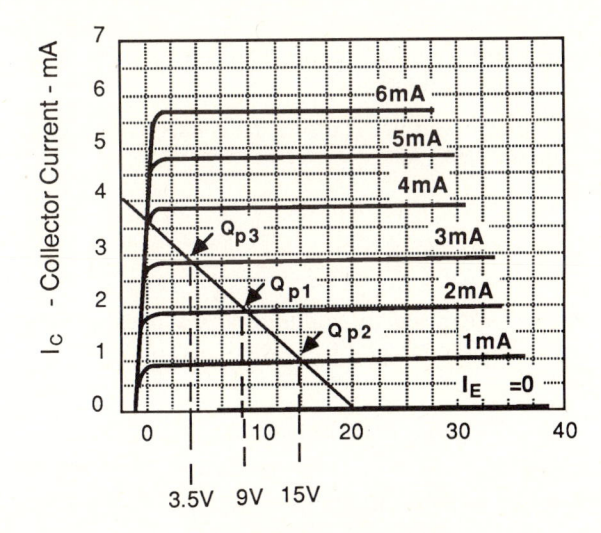

Fig. 5-8 Characteristic Curves Chart

Remember, a more negative potential was applied to the input which caused the output to go less positive. In other words, in a common-base amplifier:

Input voltage and output voltage are in phase.

When the switch is moved to position C, E_2 and V_{EE} are connected series opposing. This causes the voltage applied to the emitter to decrease. With the decrease in V_{EE}, I_E will also decrease. If I_E decreases I_C will decrease. Assuming that the decrease in I_E is 1 mA, the operating point will be at Q_{p2} on the characteristic curves chart. At this Q point, $V_C = 15$ V and $I_C = .95$ mA. With the decrease in I_C, V_{RL} will decrease, leaving a larger voltage to be dropped as V_{CB}. The statement made above is confirmed: causing the emitter (input) to be more positive causes the collector (output) to become more positive, and the two voltages are in phase.

From these analyses we can see that the changes in voltage and current distribution are:

$$V_C = V_{Qp3} - V_{Qp2} = 15\text{ V} - 3.5\text{ V} = 11.5\text{ V}$$

$$I_C = I_{Qp2} - I_{Qp3} = 2.8\text{ mA} - 0.95\text{ mA} = 1.85\text{ mA}$$

$$I_E = I_{Qp2} - I_{Qp3} = 3\text{ mA} - 1\text{ mA} = 2\text{ mA}$$

and that:

in a CB circuit input and output signals are in phase.

In a common-base amplifier, alpha (α) is used to identify current gain. The formula for alpha is:

$$\alpha = \frac{I_C}{I_E}$$

Note that the ratio used to calculate alpha uses I_C and I_E. The result represents the percentage of emitter current (I_E) that flows as collector current (I_C).

Self-Check

Answer each item by inserting the word or words required to make each statement true.

1. In the common-emitter amplifier, output signal is taken between _____ and _____.
2. In a common-base amplifier voltage gain is _____ and current gain is _____.

Transistor Configurations - Common-Base and Collector Amplifier

3. Output signal from a common-base amplifier is taken across the _____ resistance.
4. In a common-base amplifier, alpha (α) is _____ but in the common-emitter amplifier, beta (β) is _____.
5. In the common-base amplifier, output signal is taken from _____ to _____.
6. In the common-base configuration, current gain will equal _____.

   a. alpha   b. beta

7. Alpha (α) is the symbol used to represent the ratio _____.
8. In a common-base amplifier, current in the _____ will *always* be larger than the current through the _____, and these two currents will be used to calculate alpha.
9. The horizontal axis of a common-base amplifier characteristic curve chart represents _____.
10. The vertical axis of a common-base amplifier characteristic curve graph represents _____.
11. In a common-base amplifier, input and output voltages are _____ phase.
12. In the common-base amplifier, as I_E increases, I_C _____.
13. In the common-base configuration, a decrease in I_E will cause V_{Rcb} to _____.
14. A common-base amplifier is operating at its Q point. An increase in forward bias will cause its Q point to approach _____.

    a. cutoff   b. saturation

15. As the operating point of a transistor circuit approaches saturation, the amount of I_C is _____.

Common-Collector (Emitter Follower) Configuration

The third configuration that will be discussed is the **common-collector** configuration.

> Note: This type of circuit is often referred to as an emitter follower.

An NPN common-collector amplifier is shown in Figure 5-9. Note that the input is again applied to the base as it was in the common-emitter amplifier. The difference here is that the output is taken from the emitter to ground and that the circuit does not have a collector resistor. In fact, R_E acts as the load for a common-collector amplifier.

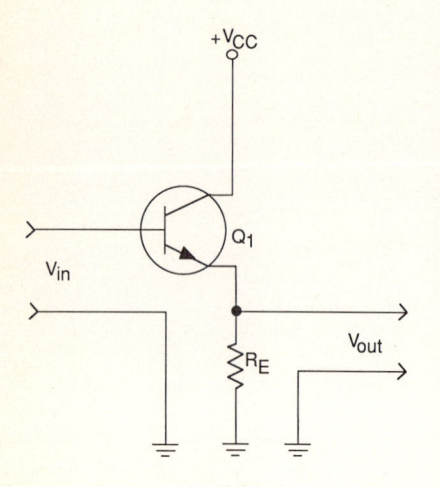

Fig. 5-9 Common-Collector Configuration

In this circuit, the collector is common to both the emitter and base. This means that it is common to both the input and the output signals.

Common-Collector Amplifier Operation

Use the circuit shown in Figure 5-10 as a reference for the following discussion. With Sw_1 in position B, the transistor is conducting at Q point, the eb junction is forward biased, and the cb junction is reverse biased. Note that the input and output voltages are both being developed across the eb junction and R_E. This results in an equivalent resistance like that shown in Figure 5-10(b). Notice that in this equivalent circuit, R_{eb} and R_E are in parallel with R'_B. From resistive circuits, you learned that the voltage drop on parallel branches must be equal. Therefore, if anything happens that affects the voltage on one branch, it has an equal effect on the other branch.

By switching Sw_1 to position A, a negative voltage is selected, which causes forward bias to decrease. The decrease in bias causes I_C, I_E, and V_{RE} to decrease. Remember, the voltage developed by R_E is the output voltage. Therefore, causing the input to go negative results in the output also going in a negative direction. With Sw_1 at position C, forward bias goes positive. The increased forward bias causes I_E, I_C, and V_{RE} to increase. With the increase of V_{RE}, the output goes more

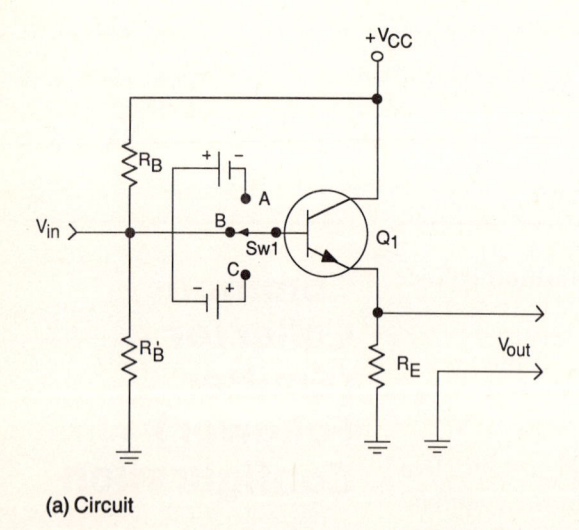

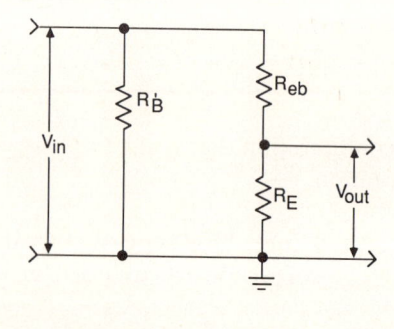

(a) Circuit

(b) Equivalent Resistive Circuit

Fig. 5-10 Common-Collector Configuration

Transistor Configurations - Common-Base and Collector Amplifier

positive. In this case, causing the input to become more positive causes the output to become more positive. We can conclude that:

$$V_{RE} = V_{input} - V_{Reb}$$

Granted, R_{eb} is a small resistance but the amount of voltage that it drops will determine how much of the input is passed to the output. In a common-collector circuit:

> **Output voltage is always less than input voltage.**

We can then draw the conclusion that:

> **Voltage gain within a common-collector amplifier is always less than unity.**

With Q_1 biased to operate at the mid-point of its operating range, it has the ability to develop either positive going or negative going changes in the output. Because V_{out} always follows the polarity of the input, and the output is taken from the emitter, we say that the emitter follows the input. Thus, the name **emitter follower**. Current gain of this circuit is the ratio:

$$\text{Gain} = \frac{I_{out}}{I_{in}}$$

Notice that the formula involves dividing the largest current (I_E) by the smallest current (I_B). This solution will result in a relatively large number for current gain. You should remember:

> **Of the three configurations, the common-collector configuration has the largest current gain.**

Phase Relationships in a Common-Collector Circuit

Now we will complete the analysis of the circuit shown in Figure 5-10. In its quiescent state, Q_1 is conducting. Current flows from ground through R_E, the eb junction, and to the base and collector. As current flows through R_E, a positive voltage is developed and is passed to the output. This positive voltage establishes a reference around which the output voltage will vary.

With the switch in position A, a negative going voltage is applied to the base of Q_1. This negative voltage is reverse bias and causes I_E to decrease. With the decrease in I_E, the voltage drop on R_E also decreases (goes in a negative direction). As a result, the output voltage also goes in a negative direction. Therefore, the output voltage is in phase with the input voltage.

With the switch set to position C, a positive going voltage is applied to the base of Q_1. This positive going voltage increases forward bias, which causes I_E to increase. With the increase in I_E, the voltage dropped on R_E will increase (go more positive). If the output goes more positive at the same time that the input goes more positive, the two voltages are in phase. Therefore, we can say that in a common-collector circuit the:

Input and Output Voltages are in phase.

If you think back and compare the three configurations, you will remember that input and output voltages have the following phase relationships:

- **Common-emitter** are 180° out of phase.
- **Common-base** are in phase.
- **Common-collector** are in phase.

DC Analysis of a Common-Collector Circuit

Figure 5-11 shows a common-collector circuit that has a voltage divider supplying its bias. The type of bias arrangement could be any of those we have covered, but since voltage divider bias is most often used we will discuss it here. Note that this circuit, like the last few that we have discussed, has its input applied to the base and its output taken from the emitter. Both input and output voltages are referenced to ground. Also, the collector is referenced to the same ground, leading to the name *common-collector*.

Remember, when analyzing the voltage divider biased circuit, we proceed as if we were analyzing a resistive bridge circuit. The ohmic value of R_{eb} is ignored and there is no collector resistor, so the ratio used is:

$$\frac{R_B}{R_B'} = \frac{R_{cb}}{R_E}$$

Transposing this formula allows you to solve for R_{cb}:

$$R_{cb} = \frac{R_B \times R_E}{R_B'}$$

$$R_{cb} = \frac{20 \text{ k}\Omega \times 5 \text{ k}\Omega}{5 \text{ k}\Omega} = 20 \text{ k}\Omega$$

Transistor Configurations - Common-Base and Collector Amplifier

(a) Circuit

(b) Resistive Equivalent

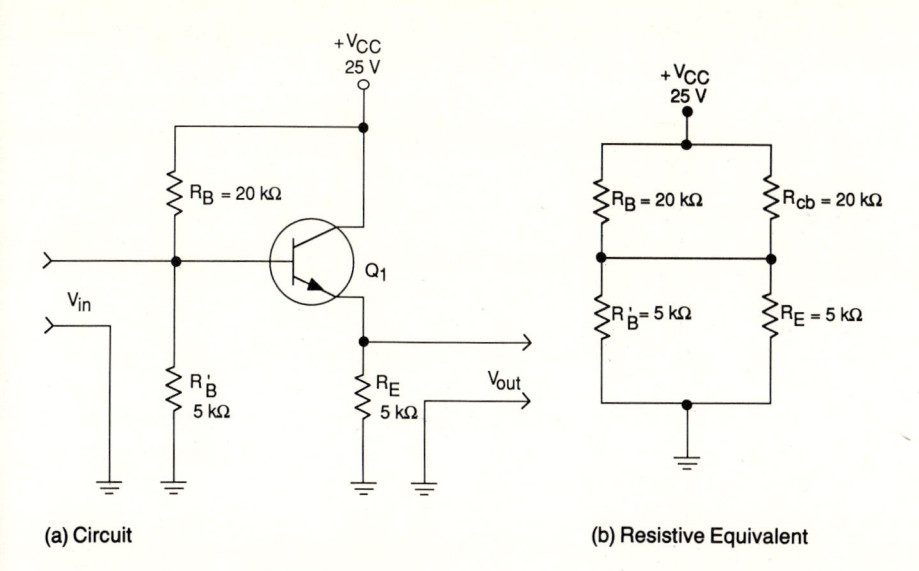

Fig. 5-11 Common-Collector Configuration

With R_{cb} known, you can redraw the equivalent circuit as shown in Figure 5-11(b). From this point the DC analysis is quite straightforward. First, analyze each branch with the understanding that in a balanced bridge, $V_{RB} = V_{cb}$ and $V'_{RB} = V_{RE}$. To analyze the voltage divider branch, we proceed as follows:

$$R_{VD} = R_B + R'_B = 20 \text{ k}\Omega + 5 \text{ k}\Omega = 25 \text{ k}\Omega$$

$$I_{VD} = \frac{V_{CC}}{R_{VD}}$$

$$I_{VD} = \frac{25 \text{ V}}{25 \text{ k}\Omega} = 1 \text{ mA}$$

$$V'_{RB} = I_{VD} \times R'_B = 1 \text{ mA} \times 5 \text{ k}\Omega = 5 \text{ V}$$

$$V_{RB} = I_{VD} \times R_B = 1 \text{ mA} \times 20 \text{ k}\Omega = 20 \text{ V}$$

You already know that $V_{RB} = V_{cb}$ and that $I_C \cong I_E$. Now you can solve for I_C, I_E, V_{cb}, and V_{RE} as follows:

$$V_{CB} = 20 \text{ V and } V_{RE} = 5 \text{ V}$$

$$I_C = \frac{V_{cb}}{R_{cb}}$$

$$I_C = \frac{20 \text{ V}}{20 \text{ k}\Omega} = 1 \text{ mA}$$

By neglecting the small amount of current that flows as I_B, we can assume that $I_E \cong I_C \cong 1\,\text{mA}$. I_E can be calculated as follows:

$$I_E = \frac{V_{RE}}{R_E}$$

$$I_E = \frac{5\,\text{V}}{5\,\text{k}\Omega} = 1\,\text{mA}$$

This method of analysis is used because it is quick and relatively accurate. In the actual circuit, however, calculated parameters will vary. For instance, I_C will actually equal $I_E - I_B$. Therefore, I_E and I_C cannot be equal. If I_C is less than 1 mA, then V_{CB} will be less than 20 V. These differences could go on and on. However, because of the allowable tolerances that exist in an electronic circuit, we can use an analysis of this type with the expectation that the results are close to actual circuit values.

Self-Check

Answer each item by inserting the word or words required to make each statement true.

16. In the common-collector circuit, the output is taken from the _____.
17. Voltage gain within a common-emitter circuit will always be _____.
18. Current gain in a common-collector amplifier will be equal to the ratio _____ over _____.
19. Of the three configurations, current gain is largest in the common-_____ amplifier.
20. In a common-collector circuit, the output voltage is _____ minus _____.
21. In order for the common-collector amplifier to handle both positive and negative input voltages it must be biased so that it operates at a point other than _____ or _____.
22. A common-collector circuit can be biased using the voltage divider arrangement ONLY. TRUE or FALSE.
23. In analyzing the common-collector amplifier we ignore the resistance of the _____ junction and assume that _____ equals _____.
24. Another name for the common-collector circuit is _____.
25. The common-collector has its input applied to the same point as the _____ circuit.

Transistor Configurations - Common-Base and Collector Amplifier

Common-base and common-emitter amplifiers have much in common. Both circuits can be biased using the same biasing methods.

A common-base amplifier gets its name from the fact that its *base*, input circuit, and output circuit all share a *common* ground. The input for a common-base amplifier is applied to the transistor's emitter and its output is taken from the collector. Analysis reveals that the common-base amplifier has low (less than 1) current gain and high voltage gain. Common-base amplifiers can be designed using characteristic curve charts similar to those used with the common-emitter circuit. Plotting of load lines and methods of analyses are similar. Both precise calculation and approximation methods can be used to analyze common-base amplifiers.

A common-collector amplifier has its *collector*, input circuit, and output circuit connected to a *common* ground. Quite often this circuit is called an *emitter follower*. The input for this amplifier is applied to the base and the output is taken from the emitter. Analysis of the circuit's operation reveals that current gain is quite high while voltage gain is low (less than 1). Characteristic curves are not normally used for the design of common-collector amplifiers but other design and analyses procedures are similar to those of the other two configurations. In fact, many of the assumptions made regarding this circuit are identical to those made about the other configurations.

Summary

Use the circuit shown in Figure 5-12 to answer review items 1-10.

1. The circuit shown in Figure 5-12 is a common-_____ configuration.

2. In the common-base configuration, the input is applied to the _____ and the output is taken from the _____.

3. By applying a positive going voltage to the input of a common-base circuit, we can expect that the change in output voltage will be _____.

4. If the forward bias applied to a common-base amplifier is decreased, what effect does this have on the collector current?

5. In the common-base circuit, output current flows in the _____ and input current flows in the _____.

6. Refer to Figure 5-12. Should Q_1 be changed to a PNP type, what other changes must be made in the circuit?

Review Questions and/or Problems

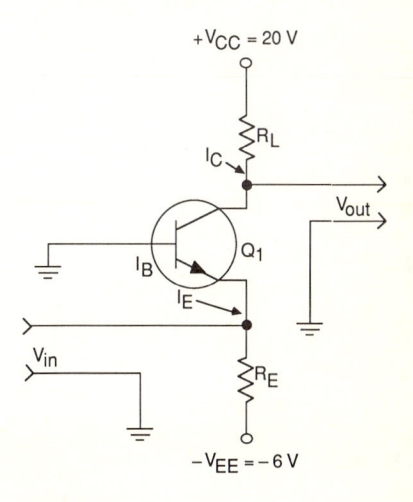

Fig. 5-12

7. Current gain, alpha (α), for a common-base circuit is calculated when _____ is divided by _____.

8. The symbol for current gain in a common-base circuit is (alpha – α). Using this symbol and the symbols for the two currents in question 7, write the current gain formula.

9. Given: $I_E = 12$ mA, $I_C = 11.4$ mA; calculate the value for alpha.

10. Voltage gain for a common-base circuit is calculated using the formula shown below.

    a. True   b. False

    Voltage Gain = $\dfrac{V_{out}}{V_{in}}$

NOTE: Use the circuit shown in Figure 5-13 to answer review items 11-20.

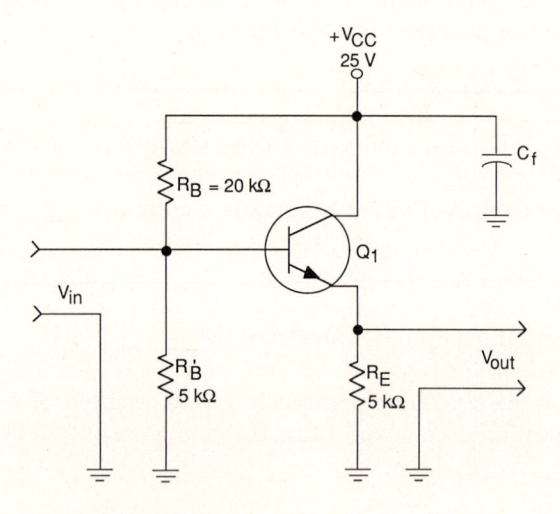

Fig. 5-13 Common-Collector Amplifier

11. The circuit shown in Figure 5-13 is a common-_____ configuration.

12. By applying a positive going voltage to the input of a common-collector circuit, we can expect that the change in output voltage will be _____.

Transistor Configurations - Common-Base and Collector Amplifier

13. In the common-collector configuration, the input is applied to the _____ and the output is taken from the _____.

14. In the common-collector circuit, output current flows in the _____ and input current flows in the _____.

15. Refer to Figure 5-13. Should Q_1 be changed to a PNP type, what other changes must be made in the circuit?

16. If the forward bias applied to a common-collector circuit is decreased, what effect does this have on the emitter?

17. Current gain for a common-collector circuit is calculated when _____ is divided by _____.

18. Given $I_E = 6$ mA, $I_B = 0.2$ mA; calculate the current gain for this common-collector circuit.

    GAIN = _____

19. In a common-collector amplifier, the input signal is applied across R_{eb} and R_E. Therefore, the voltage drop across R_E is _____ than the input voltage.

20. A common-collector circuit has a voltage gain of _____.

6
Special Solid State Devices

Objectives

1. Identify the schematic symbol for the following devices:

    a. Zener diode
    b. SCR
    c. Schottky diode
    d. Tunnel diode (Esaki diode)
    e. Light emitting diode (LED)
    f. Varactor diode
    g. Photo diode
    h. Four layer (Latching) diode
    i. Silicon controlled switch (SCS)
    j. Silicon bilateral switch (SBS)

2. Describe the operational characteristics of each device listed in objective 1.

Introduction

In this chapter we begin to study PN junctions that exhibit different characteristics from those discussed in previous chapters. Careful manufacturing processes allow the design of solid-state devices that can be used for very special purposes. During manufacturing, special attention is paid to the amount of doping, the properties of the depletion region(s), and the attachment of special electrodes which are used to control the conditions of one or more PN junctions.

An unlimited number of devices could be included in this category, but as you might guess, this is impractical. We will cover only those devices that appear most frequently in the design of circuits in use today. Since these devices are used in numerous ways, we must also limit the number of uses that are discussed for each device. Emphasis is placed on the operational aspects of the device. In some cases, device applications are explained more fully in later chapters.

Zener Diodes

One family of diodes is given the name **Zener**. To operate properly, the zener diode must:

1. be reverse biased.
2. operate in the **breakdown** (Zener) region
3. be insensitive to temperature change

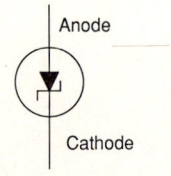

Fig. 6-1 Zener Diode Schematic Symbol

To perform these functions the Zener diode is designed of silicon material, giving it a high heat dissipation capability. Zener diodes operate in the **avalanche (Zener)** region of their characteristic curve. This means that the junction can withstand large variations of current with little change in reverse bias voltage. The schematic symbol for the Zener diode is shown in Figure 6-1.

A Zener diode characteristic curve is shown in Figure 6-2. With forward bias a Zener diode operates the same as a conventional diode. As with the conventional diode, reverse bias causes the junction barrier to widen. This means that the depletion region contains fewer majority carriers. For the minority carriers, this voltage is forward bias, which causes reverse current to flow. The flow of reverse current causes the junction's operating temperature to increase. The increased temperature causes more minority carriers to be produced. This means that additional carriers are available to support reverse current.

As reverse bias is increased, the flow of minority (reverse current) carriers continues to increase. At some point determined by the diode's characteristics, a voltage is reached that causes the reverse current to increase abruptly (avalanche). The point at which avalanche current begins to flow is called the **breakdown point**. In Figure 6-2, the breakdown point occurs with approximately 44 volts of reverse bias. You should realize that there are Zener diodes that have breakdown

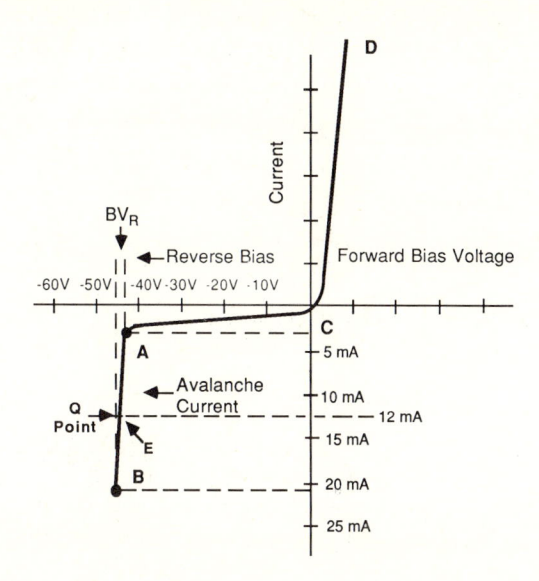

Fig. 6-2 Zener-Diode Characteristic Curve

voltages as low as a few tenths of a volt and as high as several hundred volts. Notice that between points A and B of the characteristic curve, a change of 2 V causes Zener current to change from approximately 3 mA to 21 mA. This means that the current can vary over a range of 18 mA while junction voltage varies only 2 volts. The region between points A and B is called the **Zener** (avalanche) region.

Assume that a reverse bias of −45 V is applied to the diode. This would place the operating point (point E) at approximately 12 mA. From point A, a change of 3 V in reverse bias would cause a change of 19 mA in current. From the operating point current could drop to 3 mA (a change of 9 mA) with a change in voltage of 1.5 V. The Zener diode operates between points A and B of this characteristic curve. It has the ability to undergo:

Large changes in current with small changes in bias voltage.

A voltage that is regulated to 45 V ± 1.5 V is a quite stable voltage.

Zener diodes can be destroyed by excessive current. As protection, a resistor is connected in series with the Zener diode. This resistor limits Zener current flow and drops the difference between applied voltage and the rated voltage of the diode. An example of this is shown

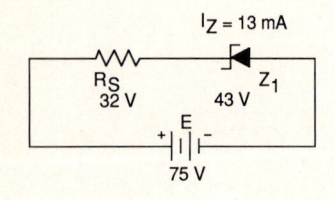

Fig. 6-3 Sample Zener Diode Bias Circuit

in Figure 6-3. Note that at the operating point where $E_Z = 43$ V and $I_Z = 13$ mA, the resistor must conduct 13 mA and drop 32 V. To determine the size of R_S use Ohm's Law as follows.

$$R_S = \frac{E_{RS}}{I_Z}$$

$$R_S = \frac{32 \text{ V}}{13 \text{ mA}} = 2460 \; \Omega$$

Remember, silicon diodes can withstand much higher currents and heat than germanium. For this reason Zener diodes are made of silicon. When designing a circuit that involves a Zener diode, you must select a Zener that is capable of dissipating the power it will be required to handle. Of course, the same goes for the series resistor and its power rating.

A major use of a Zener diode is that as a voltage regulator. A simple voltage regulator circuit is shown in Figure 6-4. Notice that the Zener is connected such that it is reverse biased. The battery voltage is 15 V and the Zener voltage (regulated voltage) is 6 V at 10 mA. R_L is a resistive load that draws 5 mA when connected to a 6 V source. This means that R_S must drop 9 V and will have 15 mA ($I_Z + I_{RL}$) of current flow. By using Ohm's law you can calculate the resistance for R_S. R_S must equal 600 Ω. A slight change in voltage across R_S, whether caused by the power source or a change in R_L, can be compensated for by a corresponding change in current through Z_1. The change in Zener current will act to maintain V_{RL} at 6 V even though total current has changed. Should the input voltage increase, bias applied to Z_1 will increase, causing I_Z to increase. The increase in I_Z causes I_t to increase, which causes V_{RS} to increase. This maintains the voltage drop on R_L at 6 V. If the current through R_L increases, the effect is the same. If the input voltage decreases, or I_{RL} decreases, the diode current will decrease, causing V_{RS} to decrease, maintaining V_{RL} at 6 V.

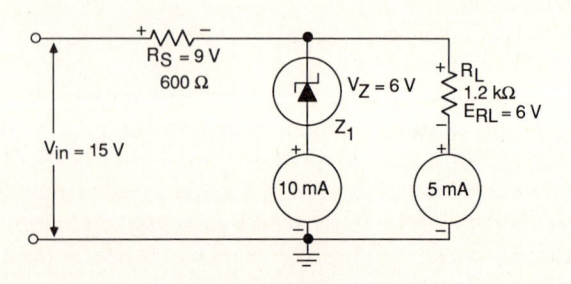

Fig. 6-4 Sample Zener Diode Regulator Circuit

Special Solid State Devices

Circuits that use the Zener diode are discussed in later chapters. Practical electronic voltage regulators, using a Zener diode, are discussed in Chapter 9.

Tunnel Diode

In earlier lessons you learned that current carriers are produced within semiconductor materials by a process called doping. The semiconductor that results is either P- or N-type material. N-type material has electrons as its (majority carriers) and P-type material has holes as its majority carriers. The number of majority carriers available within P- or N-type semiconductors is determined by the amount of impurity that is doped into the silicon or germanium.

By increasing the number of impurity atoms used in doping, we increase the availability of current (majority) carriers in the resultant material. A junction exists where P- and N-type materials are physically joined. The greater the number of majority carriers available in the P- and N-type materials, the less forward bias required to start current flow in the junction. A junction that has been formed using heavy doping has unique characteristics. One device that uses this characteristic is the **tunnel diode**.

The tunnel diode gets its name from the fact that electrons can suddenly disappear from one side of the junction and immediately reappear on the opposite side even though they (the electrons) do not have enough energy to overcome the barrier. When electrons behave in this manner it is referred to as the **tunnel effect**. The electron acts as if it can "tunnel" underneath the barrier.

A tunnel diode is a PN junction similar to the PN junction diode covered in Chapter 2. In the tunnel diode, however, the barrier width (depletion region) is designed to be very thin (less than a millionth of an inch) by the doping process. A comparison of the barrier widths is shown in Figure 6-5. The barrier of a tunnel diode is so thin that penetration (tunneling) of the junction is possible with a small forward bias.

In Figure 6-5(a) we see the pictorial diagram of a conventional diode. We know that its N-section has electrons as majority carriers and the P-section has holes. A tunnel diode pictorial diagram is shown in Figure 6-5(b). Note that this device has sections of P- and N-type material like other diodes. However, because of the heavy doping used in forming the materials, there is a large number of majority carriers in each section. The large number of majority carriers will cause the junction and the barrier region to be very thin. This is the condition that exists prior to the application of a forward bias to the tunnel diode junction, or at point 1 of Figure 6-6.

(a) Conventional Diode

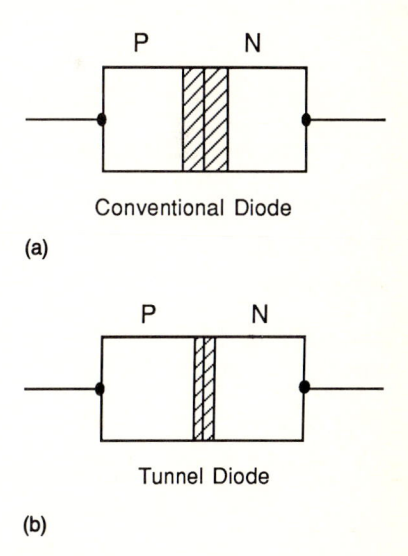

(b) Tunnel Diode

Fig. 6-5 Comparison of Barrier Widths

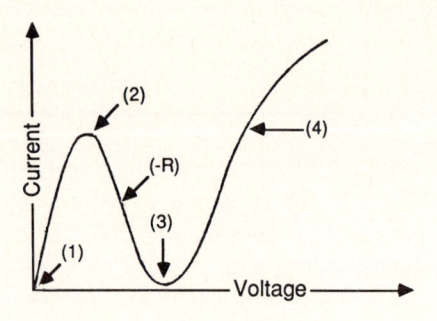

Fig. 6-6 Characteristic Curve - Tunnel Diode

As forward bias is increased, current continues to increase to a point (point 2) along the characteristic curve. At point 2 current slows, stabilizes, stops, and begins to decrease. Between points 2 and 3 the current decreases back to nearly zero. This region (points 2 to 3) is called the **negative resistance** region. Negative resistance is said to be present when:

An increase in voltage across a resistor causes current in the resistor to decrease.

When the current decrease reaches a point (point 3) much lower than when the decrease began, it again stabilizes and begins to increase. From point 3 to point 4, the tunnel diode operates very similar to a conventional junction diode. The schematic symbol for a tunnel diode is shown in Figure 6-7.

Figure 6-6 is used to summarize this discussion of tunnel diodes. Point 1 represents the zero bias point of the characteristic curve. As forward bias is increased to the **peak point** (2), current in the tunnel diode increases rapidly (because of tunneling) and then stabilizes. At point 2 an increase in forward bias causes current in the junction to decrease. This decrease will continue until current is near zero (point 3). The region where increases in bias voltage result in decreased junction current is called the **negative resistance region**. Point 3 is often called the **valley point**. As forward bias continues to increase, the junction current begins to increase. Past point 3 the diode junction acts much like a conventional diode.

To get some idea of the amounts of forward bias required to cause these conditions, refer to Figure 6-8. Notice that for the germanium tunnel diode, the entire process (point 1 to point 2) is caused by a change in forward bias of approximately 500 mV (0.5 V). The **negative resistance region** exists between forward bias levels of approximately 100 mV (0.1 V) to 500 mV (0.5 V), or for approximately 400 mV (0.4 V).

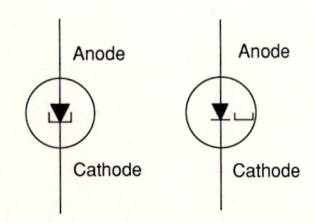

Fig. 6-7 Tunnel Diode Schematic Symbols

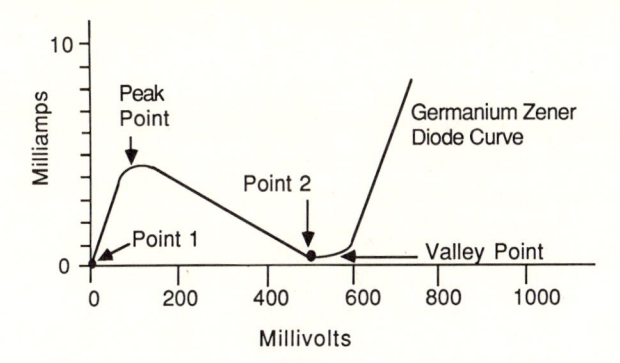

Fig. 6-8 Comparison of Bias Voltage and Negative Resistance

The Varactor Diode

The **varactor** is a voltage-controlled semiconductor capacitor. To understand how it operates we must review the PN junction. When a PN junction is reverse biased, electrons in the N-type material are attracted away from the junction. The holes in the P-type material are also attracted away from the junction. This means that no majority carriers are located near the junction (in the barrier region) as long as the junction is reverse biased. A reverse biased junction forms a capacitor. One plate exists in the N-type material and the other plate exists in the P-type material. The depletion region forms the separating dielectric. Figure 6-9(a) shows the most often used schematic symbols for varactors. Refer to Figure 6-9(b) for a comparison of the barrier widths that exist for the unbiased and reverse biased junctions. All current carriers are trapped in positions that leave the depletion region free of current carriers.

(a) Symbols

(b) Pictorial Diagrams

Fig. 6-9 Varactors

A capacitor is formed by:

two conductors separated by a dielectric.

Knowing this, you can see that the junction would now satisfy the definition of a capacitor. With the current carriers pulled away from the junction, the P-section and N-section serve as the capacitor plates and the barrier serves as the dielectric.

If the reverse bias applied to a PN junction is increased, the distance that separates the electrons of the N-section from the holes in the P-section is increased. As this distance is increased, the barrier increases, and the capacitance of the junction decreases. If reverse bias is decreased, the carriers move closer together and junction capacitance is increased. From this explanation you can see that by varying the reverse bias applied to a PN junction, the capacitance of the junction can be varied.

By using special doping, a PN junction can be designed for specific capacitance ranges. The device that results is called a **varactor**. The relationship that exists between the amount of reverse bias and capacitance is shown in Figure 6-10. Notice that V_R represents the reverse bias and capacitance is represented along the vertical axis. Notice that capacitance is largest when a small V_R is applied and it decreases rapidly over the first 30 V of V_R change. The upper limit of V_R is determined by the breakdown potential of the PN junction and the lower limit is restricted by 0 V.

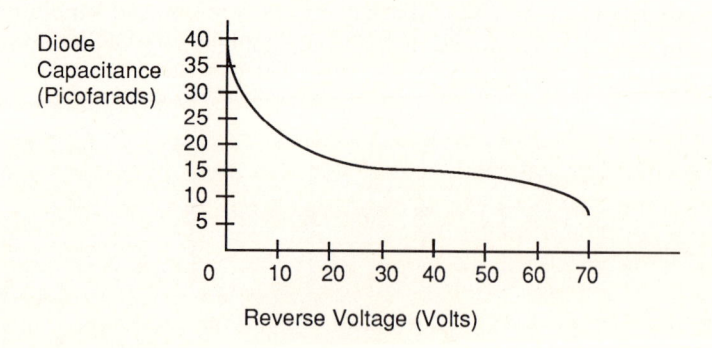

Fig. 6-10 Voltage/Capacitance Relationships - Varactor

The capacitance effect is present in all PN junctions. When a circuit is designed where capacitance values are critical the junction capacitances must be considered as part of total capacitance. A varactor can be used in many applications that require a variable capacitor.

Placing a varactor in parallel with a fixed capacitor allows variations in total capacitance that equal the sum of the parallel capacitors.

Connecting a fixed capacitor in series with a varactor provides variations that are smaller than the value of the smallest capacitor. Using these methods, varactors can be used to provide voltage-adjustable capacitances for many applications.

Light-Emitting Diode (LED)

A PN junction device that is used in many applications where a visual indication is needed is the **light-emitting diode (LED)**. Remember those little red lights that are used to display the time and date on many digital watches? Those lights are actually LEDs. They are placed in a pattern that allows all numbers 0-9 to be indicated by lighting assorted LEDs. Figure 6-11(a) contains a drawing that identifies the seven separate LEDs that are required to display all numerical digits (0-9). Discrete circuit components are shown in Figures 6-11(b) and (c). LED operation involves a completely different area of semiconductor physics from the devices studied earlier. In Chapter 1 you learned that electrons are the

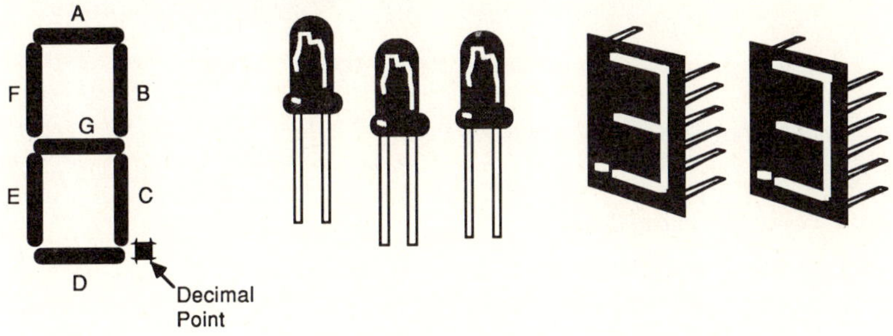

(a) Seven-Segment Display
(b) Discrete Components
(c) Two-Digit Readout

Fig. 6-11

majority carriers in N-type semiconductors, and holes are the majority carriers in P-type semiconductors. You also learned that electrons (in N-type material) exist in the conduction band and that the holes (P-type material) are bound into the valence band. From the energy diagram of these materials you learned that the conduction band represents a higher energy level than the valence band. Therefore, electrons in the conduction band have higher energy values than holes in a valence band. When electrons move across a PN junction and recombine with a hole, they release excess energy that is not needed for recombination. In the PN junction diode, this excess energy is dissipated as heat, and care has to be taken to assure that the diode used can dissipate the excess power. In some semiconductor materials this excess energy is released in the form of visual light. If the light-emitting junction is contained within a translucent material, the light that is emitted is visible to the human eye. The light that results from the conduction of these PN junctions is the reason for the name *light-emitting diode (LED)*.

The schematic symbol for an LED is shown in Figure 6-12(a). Notice the arrows that point away from the diode symbol. These arrows represent light that is being transmitted away from the junction. The actual light emitter is a very small crystal that is mounted on the cathode lead of the device. When forward bias (approximately 1.6 V) is applied, the LED will emit sufficient light to remain visible all of the time.

(a) Symbol (b) Cutaway View

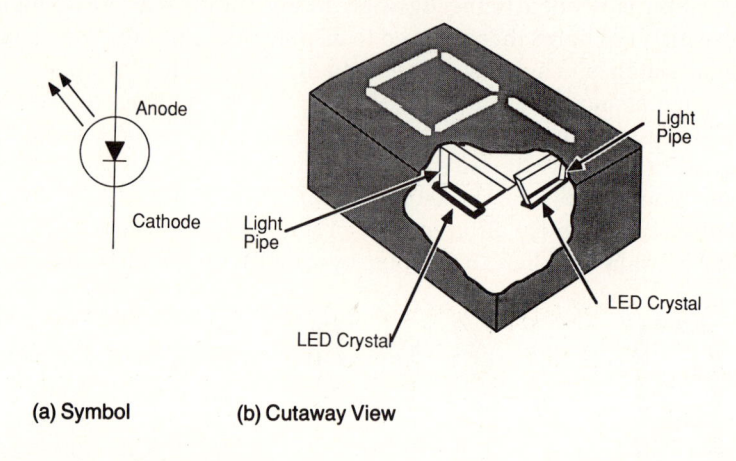

Fig. 6-12 Light-Emitting Diodes

Figure 6-12(b) shows a cutaway view of a seven-segment display that is made from seven LEDs. To spread the light that is emitted over enough space to make it useable as part of the seven-segment display, this light is conducted through *light pipes* as illustrated in Figure 6-12(b).

To this point, all of the devices studied have been made from either silicon or germanium that has been doped. There are, however, other elements that fall within the semiconductor classification that are used in the manufacture of semiconductor devices. Gallium is one of these. LEDs are made from doped gallium. A *gallium arsenide* (GaAs) will have a green glow, *gallium arsenide sulphide* (GaAsS) glows red, *gallium phosphide* (GaP) glows yellow and *gallium nitride* (GaN) glows blue.

The LED is a rugged device that can be operated in any position and under nearly all conditions. It emits a light that is visible under practically all conditions. Because of these facts it can serve as an excellent indicator in a readout arrangement or as a separate unit, as shown in Figure 6-11(b), for indicating power ON/OFF conditions. The major disadvantage that is associated with the LED is its high current drain. An average LED requires 20 mA for its operation. In a two-digit display, like the one shown in Figure 6-11(c), there are 14 LEDs. If all 14 of the LEDs are glowing, total current drain could exceed 280 mA (0.28 A). With this much current required to operate the display, plus that needed to operate the remainder of the circuit, battery life could be very short. In battery powered applications it is common practice to install a switch that is used to activate the display for short periods of time. Battery life can be increased by turning the display off when it is not needed.

In equipment where power consumption is not a problem, the LED makes an excellent display. It is capable of supplying continuous operation for long periods of time with few problems. Today we see LEDs used as displays for digital clocks, microwave ovens, stereo tuners, power on indicators, power level indicators on stereo amplifiers, and many other applications. As stated earlier, LEDs are available in the colors red, green, yellow, and blue.

The LED has forward bias requirements similar to those of the PN junction diode, but LEDs are not able to withstand reverse bias of even very small voltages. For this reason, it is necessary to assure that reverse bias is *never* applied to an LED.

Liquid Crystal Display (LCD)

The **liquid crystal display (LCD)** serves many of the same purposes as the LED. The LCD, however, has one big advantage over the LED; it requires only a small fraction of the current that is required by LED displays. This makes it especially adaptable to battery powered applications such as calculator displays and watch readouts. We will briefly discuss the theory of operation behind LCDs.

In ordinary liquids the molecules are randomly spaced and form no specific pattern. In a liquid crystal, however, all molecules are arranged in a definite pattern. When an electric current is applied to a liquid crystal, the molecules through which current flows become highly agitated.

Under zero current conditions, all molecules within the crystal are "at rest" and the crystal is transparent. When activated by current carriers, the molecules begin to vibrate and have the uneven surface required for reflection of light. This results in the activated areas appearing as a nontransparent area that is visible when under lighted conditions. The pattern of vibrated molecules can be such that the light reflected is similar to the LED displays, or any other graphical form desired.

Because the LCD operates on the principle of light reflection, it requires very little current for its operation. The only energy required to operate the display is the small current needed to agitate the molecules. The average current for an LCD is 20-25 μA. When compared to the 20 mA required for a single LED, the advantages of using the LCD in portable applications becomes obvious.

The LCD display can be designed for both large and small applications. Currently, LCD displays are used in portable computers, digital watches, pocket calculators, electronic typewriter displays and even small television screens.

Some of the newer portable personal computers have display terminals that are made from LCD materials. In one case, this display is capable of displaying 16 lines of 80-character print. Display quality approaches that available from CRT displays. The thinness and light weight of these displays makes them ideal for portable computers.

The Photosensitive Devices

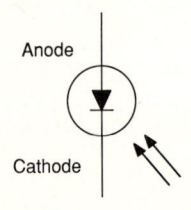

Fig. 6-13 Schematic Symbol - Photo Diode

We have already discussed a diode that emits light from its junction; the LED. A **photo diode** is a diode that has a light-sensitive junction. When this junction is exposed to light it will conduct reverse current in predictable amounts. Examine Figure 6-13. Notice that the arrows point inward (toward the diode) and indicate that this is a light-sensitive device.

To better understand the operation of the photodiode, we will review some of the lessons learned earlier. When a diode is reverse biased there is a small amount of reverse current in the junction that results from the minority carriers being forward biased. With a fixed reverse bias, reverse current varies in direct proportion to the operating temperature of the PN junction. By experiment, it was discovered that a diode's reverse current performed in the same manner when light was allowed to strike the junction. As the amount of light was increased, the amount of reverse current increased. If the light was decreased, the current decreased. In fact, the amount of reverse current that flows is directly proportional to the amount of light that strikes the junction. Additionally, once the junction is reverse biased, the amount of reverse bias has little effect on the amount of reverse current that flows.

It should be noted that all PN junctions display light-sensitive characteristics. To avoid light interference, diodes, transistors, and other devices are sealed in light-proof packages.

It was found that by using special construction methods, junctions could be made that were especially sensitive to light. Today, many uses are made of this phenomena. Transistors, diodes, and SCRs are some of the devices that receive wide use. (Schematic symbol for a photo diode is contained in Figure 6-13.) One method used to make photosensitive devices involves the diffusing of P-type silicon into an N-type silicon chip until the N-type material becomes so thin that it can be penetrated by light. When light is allowed to strike this device, the energy level of the valence electrons in the P-type material is raised to the conduction level of the N-type material. These electrons are then able to become part of the reverse current flowing across the junction. Variations in the intensity of the light striking the device results in the variation of reverse current.

Special Solid State Devices

Examine Figure 6-14. The top of this device has a lens built into it. This lens is the area through which visible light is allowed to strike the photosensitive material. When light increases, reverse current increases. When the light decreases, reverse current decreases. The stronger the light that is admitted, the heavier the reverse current that will flow. Use of a lens allows close control of the amount of light that is allowed to strike the junction.

The amount of current produced by a photosensitive junction is very small. Remember, it is reverse current. Photosensitive devices do, however, have the ability to respond to changes in light, which makes them valuable when a circuit must be switched on and off at a very fast rate. Another application allows the use of light-emitting devices and light-sensitive devices to be used as optional isolators.

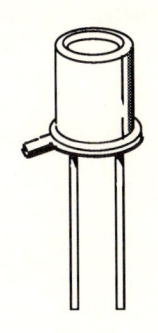

Fig. 6-14 Pictorial - Photo Diode

PIN Diodes

Another group of diodes are constructed using still a different method. In these diodes a strip of **intrinsic** semiconductor is inserted between the P- and N-type sections. (An intrinsic semiconductor is one that has not been doped.) This diode is called a **PIN Diode**. The conventional PIN diode is nothing more than a high value resistance that can be controlled by a DC current. Figure 6-15(a) contains the pictorial diagram of a PIN diode and Figure 6-15(b) contains the schematic symbol for the PIN diode.

When a lens is inserted over the junction of a PIN diode and it is exposed to light, the junction operates as a photodiode. The PIN diode is desireable as a photodiode because it responds to slow changes in light intensity better than the conventional photosensitive diode.

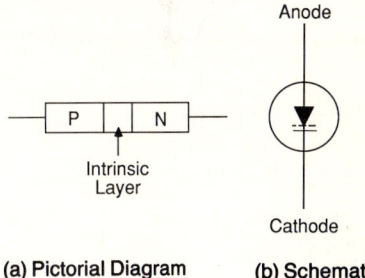

(a) Pictorial Diagram (b) Schematic Symbol

Fig. 6-15 PIN Diode

Hot Carrier (SCHOTTKY) Diodes

This diode carries two names. It was originally named the **SCHOTTKY** diode after the German scientist who discovered it. More recently it has become known as the **hot carrier diode** because its operation is very similar to a vacuum tube diode.

The pictorial diagram and schematic symbol for this type diode appear in Figure 6-16. The diode is constructed quite differently from the others we have discussed. In actuality, this is not a PN junction diode. It is constructed using a silicon N-type section, but the P-section is fabricated out of gold, silver or aluminum. This process forms a semiconductor-to-metal junction. The N-type material serves as the cathode for the device. Barrier potential is quite low, usually about 0.3 V. In addition there are very few minority carriers present. The Schottky (hot carrier) diode is able to respond to very fast variations in anode current.

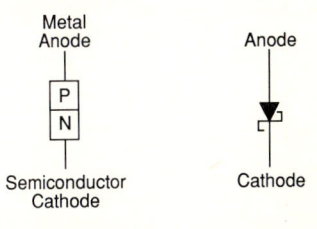

(a) Pictorial Diagram (b) Schematic Symbol

Fig. 6-16 Hot Carrier Diode

Self-Check

Answer each item by inserting the word or words required to make each statement true.

1. A diode that has its P- and N-sections separated by a layer of intrinsic semiconductor is a/an _____ diode.
2. A diode that is used as a voltage regulator is the _____ diode.
3. A device whose junction is sensitive to light is called a/an _____ diode.
4. An intrinsic semiconductor is one that has _____.
5. A device that has electrons moving back and forth across its junction, even when unbiased, is called a _____.
6. The varactor operates on the principle that a _____ bias causes a corresponding variation in _____.
7. The LCD is used for display purposes in _____ equipment because of its _____.
8. A Zener diode operates with _____ bias and within its _____ region.
9. When arrows appear as part of a schematic symbol they represent the effect of _____.
10. _____ diodes use an anode that is made of metal.

Four-Layer (Latching) Devices

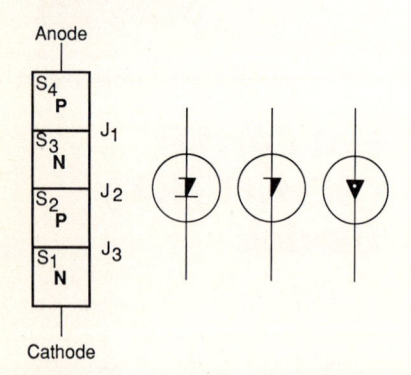

Fig. 6-17 Four-Layer Diode
(a) Pictorial Diagram
(b) Schematic Symbol

Just as P- and N-type semiconductor materials are formed into a two layer device (PN junction diode), other semiconductor devices are built that have three or more alternate layers of P- and N-type materials. The bipolar transistor is a three-layer (element) device having two junctions, but some other transistors use additional layers and have more than two junctions. A four-layer device is created by making a **PNPN** sandwich that has three junctions. Many applications have been found for four-layer devices as they have a completely different set of characteristics from diodes and bipolar transistors.

A device that is used in many circuits to control current is a semiconductor device that is constructed using four sections of doped material. These sections (layers) are formed as two P- and two N-sections. A pictorial diagram of **a four-layer (PNPN)** device is shown in Figure 6-17(a), and the most used schematic symbols are shown in Figure 6-17(b). Internally, the device has connections that allow it to work like two separate transistors.

These connections are illustrated in Figure 6-18(a) and the pictorial diagram for the connections is shown in Figure 6-18(b). By examining Figure 6-18, we can get a better idea of how the device operates. In Figure 6-18(a), you can see that one section of the N-type material and one section of P-type material are divided into two sections for operational purposes. One part of the split N-type acts as the base of a PNP transistor (Q_1), and the other part acts as the collector of an NPN

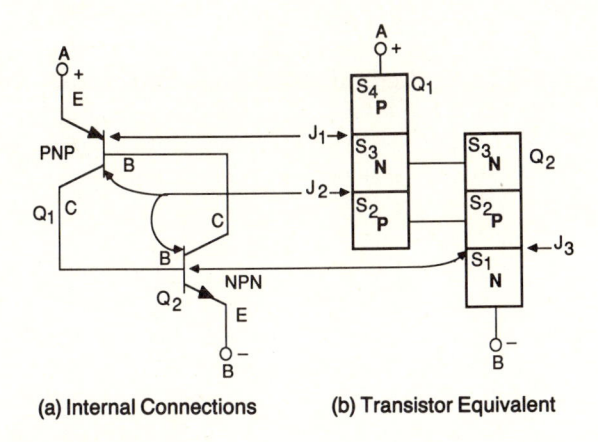

Fig. 6-18 Four-Layer Diode

transistor (Q_2). The split P-section acts as the collector for Q_1 and the base of Q_2.

When the positive terminal of a power source is connected to point A of the device and the negative terminal is connected to point B, the device is forward biased. Conduction, however, does not occur in the same way as other devices we have studied. Note that these conditions exist on each junction. The positive potential applied to point A will attract the minority carriers (electrons) in the P-section and will repel the majority carriers (holes) to the junction area. This causes the barrier at J_1 to be narrowed, as happens with forward bias. The negative potential at point B attracts the minority carriers (holes) and repels the majority carriers (electrons). This narrows the barrier at J_3 and forward biases J_3. It is possible to forward bias these two junctions and still not have heavy current. For the device to operate like a diode, the difference in potential between points A and B must be large enough to cause **breakover** in J_2. When the number of electrons entering section S_2 is sufficient to saturate the section, current carriers will be forced across J_2 in the reverse direction. This is called the **breakover point**. Once breakover has occurred, the device will continue to conduct current from point B to point A until the voltage applied to point A is removed.

Another way of explaining the initial conduction of the four-layer device is as follows: with forward bias applied, electrons will enter the emitter of Q_2 through point B. Current will flow through the base and collector of Q_2 and into the base of Q_1. Remember, in a transistor, current in the collector is much larger than base current. The larger current entering the base of Q_1 quickly saturates the base section. The amount of current that flows in the collector of Q_1 is even higher than the base current. Collector current of Q_1 flows from the base of Q_2, causing the base of Q_1 to saturate. After a short time, both transistors are saturated. At that point, the junctions act as if they were shorted and the device is said to be **latched on**. The term *latching* comes from the fact and once *turned on*, it remains *latched on* until the voltage applied to points A and B is removed. The term *diode* describes a device that will conduct in one direction only. Thus, the name we often hear used is **latching diode**.

Silicon-Controlled Rectifiers (SCR)

The **silicon-controlled rectifier (SCR)** is a four-layer (PNPN) semiconductor device having three electrodes: a **cathode**, an **anode**, and a **gate**. A pictorial diagram of this device is shown in Figure 6-19(a) and its schematic symbol in Figure 6-19(b). Notice that this device is identical to the four-layer diode discussed in the last section with the exception of the gate electrode.

Remember, the term **rectifier** means to:

allow current to flow in one direction only.

An SCR differs from the junction diode in that it does not conduct from cathode to anode until a specific minimum voltage is applied. The level of the required minimum voltage can be varied by application of a bias voltage to the gate. Using the gate as the controlling element, the SCR becomes a valuable device for controlling current within high power circuits.

Figure 6-20 shows an SCR that is connected to a circuit that can be used to control forward bias characteristics of the SCR. The supply voltage (E_t) is connected such that a positive potential is applied to the anode and the cathode is referenced to the negative pole of E_t. Connecting the SCR in this manner provides forward bias to the anode with respect to the cathode. In the P-type anode, the minority carriers (electrons) are attracted away from the junction and the majority carriers (holes) are repelled toward the junction. In the cathode, the minority carriers (holes) are attracted away from the junction but the majority carriers are repelled toward the junction. The forces that exist at J_1 and J_3 cause the majority carriers in S_3 and S_2 to be attracted toward the junctions J_1 and J_3. These conditions are those that exist across forward biased junctions, placing J_1 and J_3 in the forward biased state. At the junction between S_2 and S_3 (J_2), the minority carriers are near the junction while the majority carriers have moved further away. On J_2 this sets up a reversed biased condition. The reverse bias placed on J_2 prevents current from flowing from the cathode to the anode. Notice that as the wiper (moveable contact) on R_1 is moved upward, the

(a) Pictorial Diagram
(b) Schematic Symbols

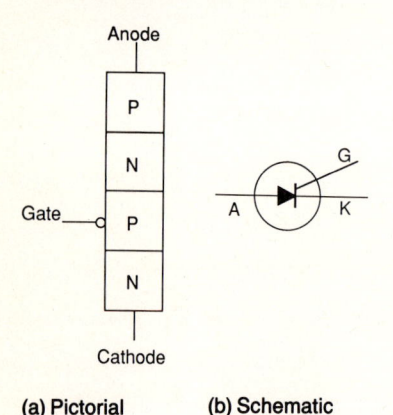

Fig. 6-19 Silicon-Controlled Rectifiers (SCRs)

Special Solid State Devices

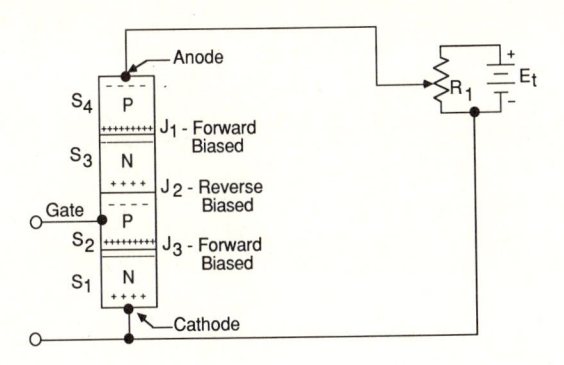

Fig. 6-20 Pictorial Diagram - SCR Biasing

positive voltage at the anode is increased. This makes the forward bias at J_1 even stronger and places the junction in a condition where it is even more ready to handle current flow.

In order for the SCR to enter the ON (active) mode, one of two things must happen: the anode/cathode voltage must increase to a certain point, or a small positive voltage must be applied to the gate. If the anode/cathode voltage is increased above a specific level, the junction barriers will break down and the device will start to conduct. The voltage level where this happens is called the **forward breakover potential**. Forward breakover is based on the following: electrons from within S_1 are attracted across the forward biased junction (J_3) by the positive voltage applied to the anode. If S_1 is more heavily doped than S_2 and the voltage applied to the anode is large enough, some of the electrons entering S_2 will not find holes with which to recombine. This sets up a state where electrons outnumber the holes. This state is referred to as **saturation**. The electrons that do not recombine act like minority carriers within S_2 and are forced to an area near J_2. As they approach the electrons already located in this area, they slow and return their energy to the mass. This released energy is absorbed by an electron near the junction, raising its energy level enough so that it can cross J_2 into S_3. Remember, J_2 is forward biased to minority carriers and the electrons approaching this junction see a forward biased junction. Once in the S_4, the electron is a minority carrier and will be attracted out through the anode and return to the battery as anode current. In S_2 electrons now greatly out-number holes, and sections S_1, S_2, and S_3 act as a single piece of N-type material. The anode (S_4) is made of P-type material and, therefore, acts as the anode for a device that operates much like the PN junction diode. In this case (J_1) is the only controlling junction within the SCR. The SCR remains in the ON state with high conduction until current flow drops to the point that the number of electrons entering S_2 is less than the number of holes available for recombination. When this happens all electrons will recombine with holes and no electrons will be available to support breakover current. The minimum current required to support breakover is called the **holding current**. As soon as the current across J_1 decreases below the holding current level, J_2 reverts to a reverse biased junction. Another way that conduction from cathode to anode can be started is by use of the **gate**.

If a small positive voltage is applied to the gate, it will cause conduction to start immediately. Notice that in Figure 6-21 the gate is connected to S_2. When a positive voltage is applied to S_2, electrons will be attracted from S_1 into S_2. The positive voltage applied to the gate aids the anode voltage, and because it is much nearer to J_3, it has much more of an effect (control) than anode voltage. Now, some of the electrons entering S_2 will combine with holes, will be conducted out of the gate as I_G and return to E_2 as I_G. Those electrons that do not recombine become available for and support anode current. In this condition the SCR has been gated ON. As the voltage applied to the gate is made more positive, the amount of anode to cathode voltage (selected by R_1) that will be required to support anode current will decrease. Figure 6-22 shows two curves; one with zero gate current ($I_{G1} = 0$) and one with positive gate voltage and gate current (I_{G2}) flowing. You can see that the breakover voltage point is lower with gate current than without gate current. The *Volts Forward* axis of the graph represents the anode to cathode voltage. As the gate current increases, the anode to cathode voltage steadily decreases. If the bias applied to the gate is increased to the point where S_2 becomes saturated, forward current will flow even though the anode to cathode voltage is very low.

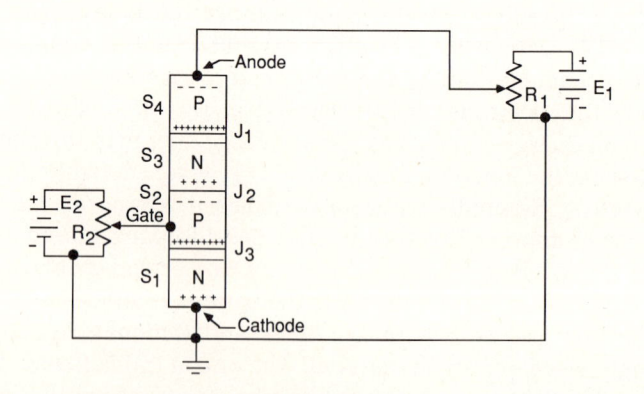

Fig. 6-21 Pictorial Diagram - SCR Biasing

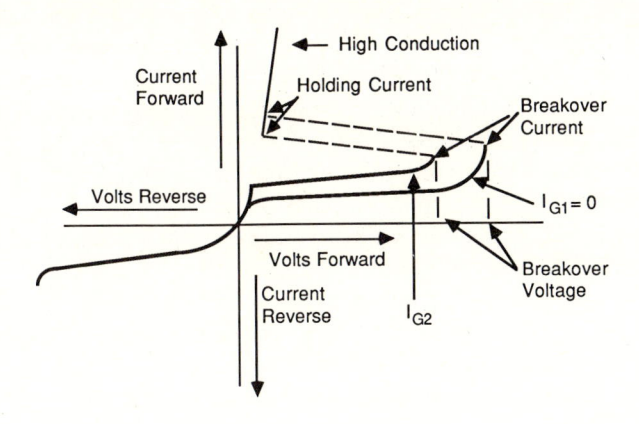

Fig. 6-22 Gate Current Versus Breakover Voltage - SCR

Once the gate voltage has turned the SCR ON, the current that flows through the SCR is not controlled by gate voltage. The SCR continues to conduct until forward current drops below the *holding current* level. Any action that causes the anode/cathode circuit to open, or that reverses the polarity of the voltage applied across the anode/cathode circuit, will cause forward current to stop.

On the other hand, if the SCR is allowed to conduct excessive current, it will be destroyed. For this reason, care must be taken in the selection of SCRs that fit the requirements of the circuit being designed. Because it is able to control current flow in one direction only, the SCR is classified as a unilateral device.

Silicon-Controlled Switch (SCS)

A **silicon-controlled switch (SCS)** is actually a low current SCR that has two gate terminals. See Figure 6-23 for a pictorial diagram and the schematic symbol for an SCS. Note that the SCS has both an **anode gate** and a **cathode gate**. Having two inputs allows the SCS to be triggered by either a positive or negative bias.

(a) Pictorial Diagram   (b) Schematic Symbol

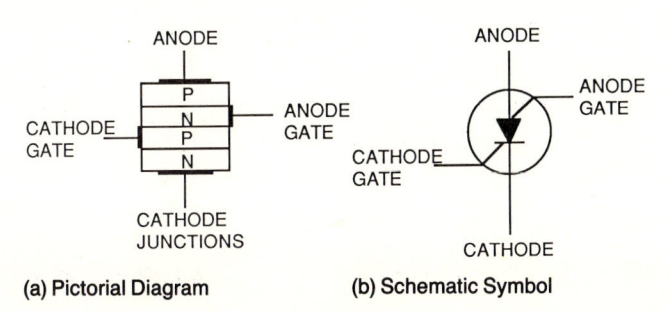

Fig. 6-23 Pictorial Diagram and Schematic Symbol

Internal connections of an SCS are shown in Figure 6-24. Notice the way that the NPN and PNP transistors are interconnected. A positive input to the base of Q_2 will forward bias Q_2 which will, by its conduction, turn Q_1 ON. Similarly, a negative input to the base of Q_1 will forward bias Q_1 and turn Q_2 ON. Placing pulses of opposite polarity to the gates (positive to cathode gate or negative to anode gate) will turn both transistors ON. We must emphasize that *only* one pulse is required to turn both transistors ON. This pulse may be applied to either gate as long as it has the correct polarity for the gate being triggered. Once ON, the SCS remains *latched on* until anode current drops below the *holding current level*. During the conduction of an SCS, current will flow from the cathode to the anode, as it does in other unilateral devices. Compared to currents that flow in the SCR, the SCS's current is very small.

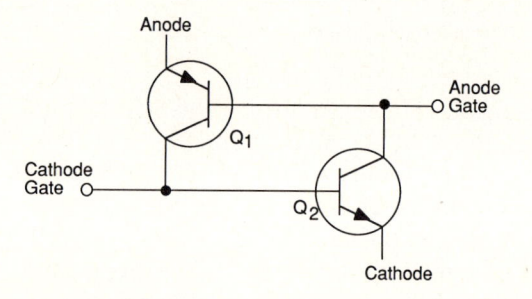

Fig. 6-24 Silicon Control Switch

Silicon Bilateral Switch (SBS)

A **silicon bilateral switch (SBS)** schematic symbol and pictorial diagram are shown in Figure 6-25. The operation of an SBS is similar to that of an SCS. The one major difference is that the SBS has two trigger points and can be used to control current in either of two directions, while the SCS can only control in one direction; thus the name **bilateral**. The SBS is very useful in the design of alternating current (AC) circuits.

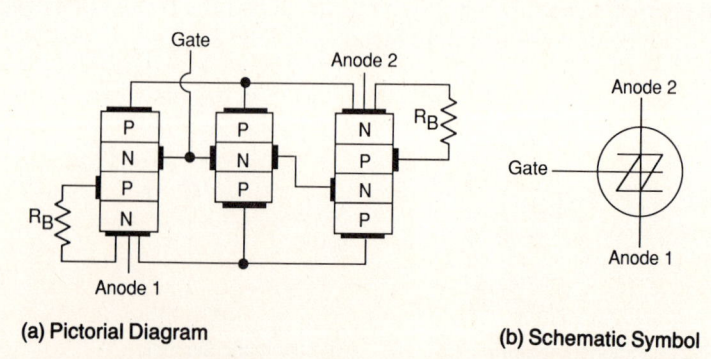

(a) Pictorial Diagram

(b) Schematic Symbol

Fig. 6-25 Silicon Bilateral Switch

TRIACS and DIACS

A **TRIAC** is simply two **SCRs** that have opposite but parallel connections. The TRIAC can be used to control current in either of two directions and is therefore a *bilateral* device. Remember, the SCR is a *unilateral device* that can only control current flowing from its **cathode to anode**. The TRIAC is the same as the SCR except that a second, and oppositely connected, SCR is also present as shown in Figure 6-26(a). In Figure 6-26(b) you see the schematic symbol, for a TRIAC and the internal connection is shown in Figure 6-26(c). Notice that the gate is connected such that when a voltage of either polarity is applied to it, one of the SCRs is triggered ON. To properly operate the TRIAC, both positive and negative triggers are applied to the gate. This allows each SCR to operate on command and current to flow in the desired direction. This is especially useful in AC circuits where current flows in each direction for a period of time.

(a) Pictorial Diagram   (b) Schematic Symbol   (c) Internal Connections

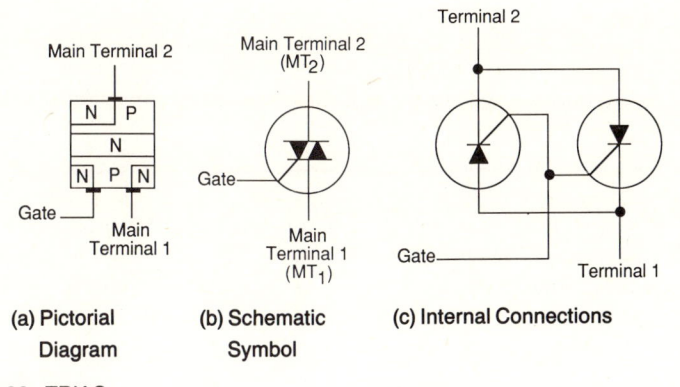

Fig. 6-26 TRIAC

Once triggered ON, the TRIAC is a *latched on* device whose conduction will stop only when the Anode voltage (E_{Mt2}) is either removed or has its polarity reversed. When its forward current drops below the *holding current* level, it can no longer support current.

The **DIAC** is simply a TRIAC that does not have a gate electrode. The DIAC schematic symbol is shown in Figure 6-27. Note that no gate is shown. For the DIAC to operate, it must be triggered by *breakover potential*. Once triggered ON, the DIAC will be *latched on* and will remain in conduction until the anode voltage is either removed or has its polarity reversed. With a reverse in polarity, the opposite SCR will be triggered ON and current will flow in the opposite direction. The result is that a DIAC is a *bilateral* device that is useful in AC circuits as it can be used to control current flow in either direction.

(a) Pictorial Diagram   (b) Schematic Symbol

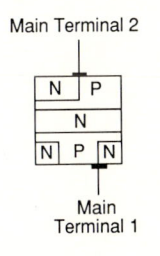

Fig. 6-27 DIAC

Self-Check

Answer each item by inserting the word or words required to make each statement true.

11. A four-layer diode is said to be _____ once current starts to flow.
12. In a four-layer device, current flows from the _____ to the _____.
13. In a four-layer device, the point where heavy current begins to flow is called the _____.
14. In order to turn OFF a four-layer device, we must remove _____ voltage.
15. An SCR is a _____ device.
16. Voltage applied to the gate of an SCR determines the amount of voltage that must be applied to the _____.
17. A silicon bilateral switch can control current flowing in _____ direction(s).
18. A four-layer diode is triggered by applying a voltage across its _____.
19. The SBS is actually two oppositely connected _____ that share the same _____.
20. A diode allows current to flow in _____.

Summary

Many semiconductor devices do not fit within the PN junction diode and bipolar transistor classifications; however, they are used widely and must be understood.

One such device is the Zener diode. This device is capable of supporting wide variations in current with small variations of reverse bias. When operating in this way, the Zener diode is said to be operating in its *breakdown* or *avalanche* region. Zener diodes are very useful as voltage regulating devices.

Tunnel diodes are heavily doped giving them the ability to support electron movement through the junction without the application of bias voltage. When properly biased the Tunnel diode will operate in a region where it has *negative resistance* characteristics.

Light-emitting diodes (LEDs) and liquid-crystal displays (LCDs) are devices that are used primarily for display purposes. The LED is a rugged, durable device that requires approximately 20 mA for its operation. When conducting, the LED's junction emits visible light. The LCD, however, uses a small current (approximately 20 µA) to agitate the surface of its crystal structure. Agitation causes the surface to be uneven, which allows it to reflect light from the surface areas that are uneven. Both devices make excellent displays; however, because of its current needs, the LED is usually found in applications where power consumption is not a problem.

Four-layer devices can be used to support unidirectional (one direction) or bidirectional (two-way) current flow. Several four-layer devices are used within electronic circuits. The most common is probably the silicon-controlled rectifier (SCR). The SCR will allow current to flow in one direction only. It is placed into conduction by the application of sufficient anode-to-cathode voltage to cause its junction to *breakdown*. Once in conduction the SCR can be turned off by either removal of the anode voltage or reversing its polarity. The SCR is called a rectifier because of its unilateral operation which is similar to that of the PN junction diode.

All four-layer devices can be turned ON by application of sufficient voltage to cause breakdown. Some devices, however, have gate electrodes that are used for triggering the device ON. The availability of a gate input allows the device to operate at low anode voltages.

Review Questions and/or Problems

1. A Zener diode is placed into its avalance conduction by applying _____ bias.

2. A tunnel diode receives (heavier) (lighter) doping than a conventional diode.

3. A varactor acts as a _____ when _____ bias is applied.

4. An SCR is made of _____ layers of semiconductor material.

5. The amount of anode voltage that must cause a four-layer diode to begin conduction is called the _____.

6. A PIN diode is made of three layers of material. These are:

   a. _____
   b. _____
   c. _____

7. Describe the effect that anode-to-cathode voltage has on an SCR after the device has been placed in conduction.

8. The minimum current that is required to maintain the conduction of a four-layer diode is the _____ current.

9. An SCR has anode-to-cathode voltage applied but has not yet had the gate voltage applied. What is the condition of each junction prior to breakover?

   a. Junction 1 is _____ biased.
   b. Junction 2 is _____ biased.
   c. Junction 3 is _____ biased.

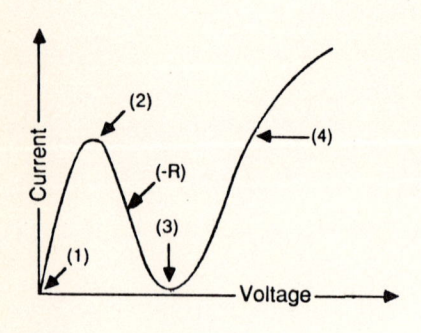

Fig. 6-28 Tunnel Diode Characteristic Curve

10. To increase the capacitance of a varactor's junction you must _____ bias.

11. Refer to Figure 6-28. The negative resistance region occurs between points _____ and _____.

12. Refer to Figure 6-29. Figure 6-29(a) is the schematic symbol for a/an _____.

13. Refer to Figure 6-29. Figure 6-29(b) is the schematic symbol for a/an _____.

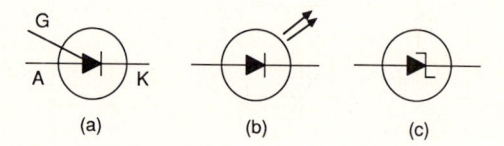

Fig. 6-29

14. Refer to Figure 6-29. Figure 6-29(c) is the schematic symbol for a/an _____.

15. An SCR conducts current in one direction only. Therefore it is classified as a/an _____ device.

16. The varactor represents its highest capacitance when its _____ bias is _____.

17. A seven-segment display is able to spread the light emitted by a/an _____ over relatively large areas by use of _____.

18. A/an _____ display operates on the principle of _____ reflection.

19. A hot carrier diode has a piece of _____ instead of an anode.

20. Applying a small positive voltage to the gate will cause an SCR to operate with _____ collector-to-anode voltage.

21. A TRIAC is capable of controlling current in _____ direction(s).

22. A device that has a lens built into its case is a _____.

23. A rectifier is a device that allows _____.

24. The DIAC is turned on by the application of a voltage to its _____.

25. Refer to Figure 6-30. The voltage regulation area of a Zener diode lies between points _____ and _____.

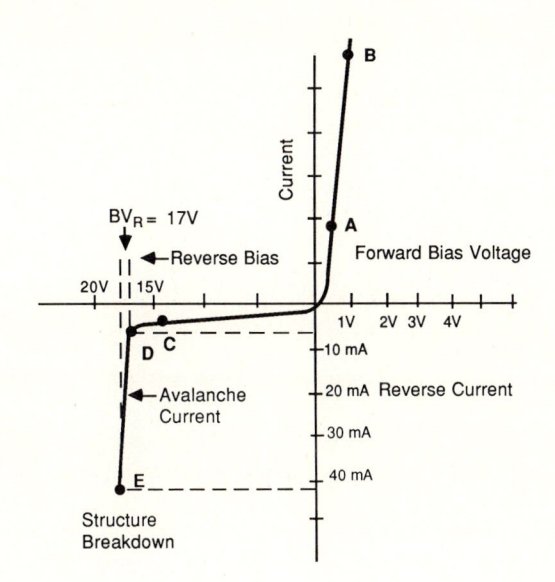

Fig. 6-30

7
Diode Circuits and Power Supply Filters

Objectives

1. Define the following terms.

    a. Input voltage
    b. Output voltage
    c. Ripple frequency
    d. Half-wave rectifier
    e. Peak output voltage
    f. Pi filter
    g. L filter
    h. Ripple voltage
    i. Diode limiters
    j. Full-wave rectifier
    k. Bridge rectifier
    l. Rectification
    m. Filter
    n. Average DC output
    o. Steady DC
    p. Pulsating DC
    q. Capacitive filter
    r. Diode clampers

2. Recognize and/or draw the schematic diagrams for:

    a. Half-wave rectifiers
    b. Full-wave rectifiers
    c. Bridge rectifiers
    d. Pi filters
    e. L filters
    f. Series limiters
    g. Shunt limiters
    h. Positive clamper
    i. Negative clamper

3. Explain, in writing, the operation of each circuit listed in objective 2.

4. When provided sufficient data, calculate the values for:

    a. Effective voltage
    b. Peak voltage
    c. Peak inverse voltage
    d. Average DC output voltage
    e. Peak output voltage
    f. Ripple frequency

5. Explain the difference between capacitive- and inductive-input filters.

6. Explain the purposes for and method of connection for capacitors and inductors in power supply filters.

7. Identify the input and output waveforms for:

    a. Series limiters
    b. Shunt limiters
    c. Positive clampers
    d. Negative clampers

Introduction

You are probably familiar with electronic equipment that can operate from either internal batteries or by being connected to a 120 V power source. You may have even wondered how it was possible to operate the same device from different types of power. It is easy to understand that the transistor radio can receive DC power from a battery, but how does it get DC power from an AC power source?

By definition, a **rectifier** is a device that changes *alternating current (AC)* into *pulsating direct current (PDC)*. Remember these three definitions:

1. Direct current is a current that flows in only one direction.
2. Alternating current is current that is continually changing in amplitude and periodically reverses direction.
3. The action of a rectifier diode that converts AC to pulsating DC is called rectification.

A PN Junction Diode allows current flow in one direction only. By controlling the *direction* of current flow a diode performs rectification. An ideal rectifier is one that offers an infinite impedance to current flow in one direction and zero impedance to current flow in the opposite direction. By using forward and reverse biasing, the PN junction diode can be made to approach these ideals. For this reason, the junction diode is the device most often used as the rectifier for converting AC to pulsating DC (PDC).

Once the incoming AC has been converted to PDC, it can be filtered using coils and capacitors. Filter coils and capacitors are usually selected to operate at the AC power frequencies of 50 Hz, 60 Hz, or 400 Hz. There are, however, some applications for which coil and capacitor sizes are selected for the specific application. High quality audio systems, for example, have filters designed for the minimum audio frequency to be amplified. Special power supplies that are used in certain scan applications have filter components selected to operate at the scanning frequency. A correctly designed output filter can, however, accept a PDC input and convert it into a relatively steady DC voltage.

The first part of this chapter is devoted to the types of circuits used to produce DC voltages. The PN junction diode is used as the rectifier that converts AC to PDC. Filter circuits are discussed that can be used to convert PDC into a steady DC voltage.

In later parts of the chapter, **diode limiters** and **diode clampers** are discussed. Many circuits are sensitive to *over voltage* conditions. Allowing a pulse exceeding a specific amplitude to be applied to the device's input can cause it to be destroyed. Limiters can be used to assure that protection is provided for devices of this type.

In other cases, a signal may need to have its DC reference level altered prior to application to a subsequent circuit. Clampers allow us to change the reference to any DC voltage level, positive or negative.

Diode Circuits and Power Supply Filters

The diode was introduced in chapters 1 and 2 of this book. At that time you learned that a diode consists of one section of P-type material and one section of N-type material. These sections are created by doping, in such a way that they form a continuous crystalline structure. The area where P- and N-type materials join is called the *junction*. When *forward bias* (Figure 7-1) is applied across the junction current flow will be heavy. When *reverse bias* (Figure 7-2) is applied across the junction very little current will flow.

Remember:

- **Forward bias** decreases the depletion region and decreases junction resistance. Current flow through the junction will *increase*.

- **Reverse bias** increases the depletion region and increases junction resistance. Current flow through the junction will *decrease*.

A diode that is connected in series with an AC power source and a load resistor (R_1), as shown in Figure 7-3, will be *forward biased* half of the time and will be *reverse biased* the other half. The type of bias

Rectifiers

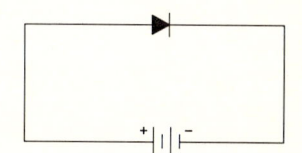

Fig. 7-1   Forward Biased Diode

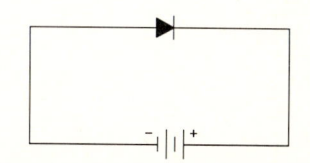

Fig. 7-2   Reverse Biased Diode

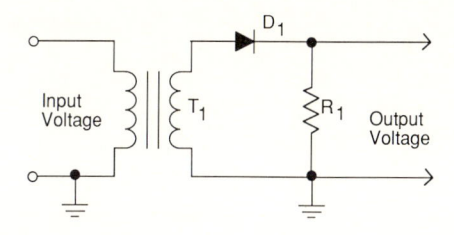

Fig. 7-3   Half-Wave Rectifier

applied changes with each alternation of the input AC signal. With each positive alternation of the AC voltage across the secondary of T_1, the diode is forward biased. If the junction is forward biased (positive voltage applied to the anode), current will flow through D_1 and the resistor (R_1). Because the diode only conducts during the positive alternation, current flow in the resistor will always be in the same direction. To flow from negative to positive, this current must leave ground, flow up through R_1, and through D_1, moving from right to left. Because current is flowing away from ground, we say that the voltage dropped on R_1 has positive polarity. Current will flow through R_1 in the same direction each time the diode conducts. Therefore, I_{R1} is *direct current*. In other words, the voltage applied to the diode-resistor combination is AC, but the output voltage taken across R_1 is DC. From this we can see that *rectification* has been accomplished. Because current flows in R_1 for only one half of the input cycle, we call this circuit a **half-wave rectifier**.

> **NOTE:** The fixed resistor (R_1) included in power supply circuits is used to provide protection for the diode circuit. It is also used to discharge reactive components when power is removed from the circuit. When filters are added to the output, this resistor is called a *bleeder* resistor. The circuit to be operated from the power source is connected in parallel with R_1.

Refer to Figure 7-3. The transformer provides an AC input to the circuit. It can be either a step-up or step-down transformer. The secondary of the transformer is connected to a diode which provides rectifier action for the AC input. In this circuit, R_1 serves two purposes: it limits the amount of current that can flow in the diode's junction to a safe level; it develops an output voltage that is a pulsating DC.

Half-Wave Rectifiers

Consider the half-wave rectifier shown in Figure 7-4. Assume that the polarity on the secondary is as shown in the figure, with positive at the top and negative at the bottom. This places a positive potential on the anode (left side) of the diode. Remember, the anode is made of P-type

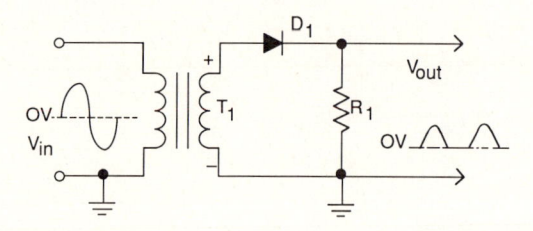

Fig. 7-4 Half-Wave Rectifier

material. The diode is now forward biased by the positive potential that is connected to the P-type material. During the positive alternation the diode will conduct, causing a voltage drop on R_1 that is directly proportional to the amount of current flow. As the voltage applied to the diode changes, so will the diode current (I_{D1}). The voltage drop on R_1 will change at the same rate as the pulses of I_{D1}. The output waveform across R_1 is shown in Figure 7-4.

Refer to Figure 7-5. Notice that the bottom of the transformer secondary is now positive and the top is negative. The negative potential is applied to the diode's anode and causes the diode to be reverse biased. When the diode is reverse biased, there is no current flow through the diode or R_1. Without current flow in the resistor, no output voltage is developed. The result is that the negative alternation is not passed to the output because it is blocked by the cut off diode.

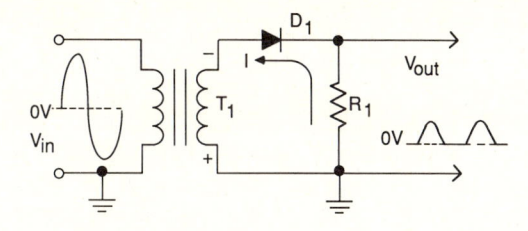

Fig. 7-5 Positive Half-Wave Rectifier

Remember, the AC input consists of an endless number of sine waves. Each positive alternation forward biases the diode and each negative alternation reverse biases the diode. The resulting waveshape that is developed on the resistor is a series of positive DC voltage pulses. Any device that is connected in parallel with R_1 will have the same pulsating DC dropped across it as is dropped across R_1.

Reversing the position of the diode's anode and cathode results in a negative pulsating DC output as is shown in Figure 7-6. Notice that the

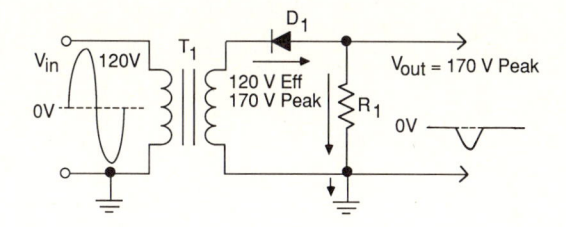

Fig. 7-6 Negative Half-Wave Rectifier

negative potential applied to the diode's cathode causes the junction to be forward biased, and a negative pulse is developed on R_1. When the positive pulse is applied, the diode is reverse biased and, therefore, cut off. No pulse is developed on R_1. The output now becomes a negative pulsating, half-wave DC output.

When an operating device is connected in parallel with R_1, a pulsating DC voltage is applied to that device. This causes pulsating DC to flow in the device. Remember, though, this current and voltage is pulsating and will have the same amplitude characteristics as the input voltage. If the device can operate on a pulsating DC voltage, then this could be an adequate power supply for its operation. In any event, we can say that the circuit shown in Figure 7-6 is a *DC power source* and that R_1 is the *load* to which power is supplied. Any current-carrying device that is connected in parallel with R_1 becomes part of the power supply's load.

A power source is loaded by the fact that it supplies current to a load. Maximum load (current) for a power supply (source) cannot exceed the maximum current available from the power supply. It is common, however, to refer to load as being:

Any device that draws current.

A load device that draws a small amount of current is said to be a *light load* and one that draws a large amount of current is a *heavy load*. When we speak of *load* we are speaking about the device that draws current from the power source. The power source is overloaded when the current demand placed on it exceeds its capability.

With this discussion, you can see how a diode can be used to control current flow and how an output voltage can be developed across R_1. This knowledge makes it possible for us to develop other facts regarding the output voltage.

As a way of review, look at the sine wave shown in Figure 7-7.

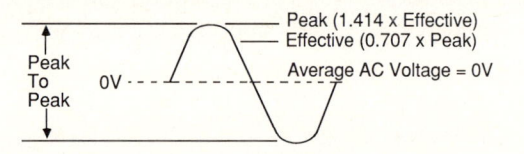

Fig. 7-7 Sine Wave Characteristics

Peak, peak-to-peak, and effective voltages of the sine wave are indicated. If the effective voltage (as at the wall receptacle of a convenience outlet) is known, the other values can be determined through use of certain formulas. These formulas are:

PEAK VOLTAGE	$E_{pk} = E_{eff} \times 1.414$
EFFECTIVE VOLTAGE	$E_{eff} = E_{pk} \times 0.707$
PEAK-TO-PEAK VOLTAGE	$E_{pk\text{-}pk} = E_{eff} \times 2.828$
PEAK-TO-PEAK VOLTAGE	$E_{pk\text{-}pk} = E_{pk} \times 2$
EFFECTIVE VOLTAGE	$E_{eff} = E_{pk\text{-}pk} \times 0.3535$
PEAK VOLTAGE	$E_{pk} = \dfrac{E_{pk\text{-}pk}}{2}$

With an understanding of the sine wave's voltage values you can learn how to analyze a half-wave rectifier's voltage characteristics. Refer to Figure 7-8 for this analysis.

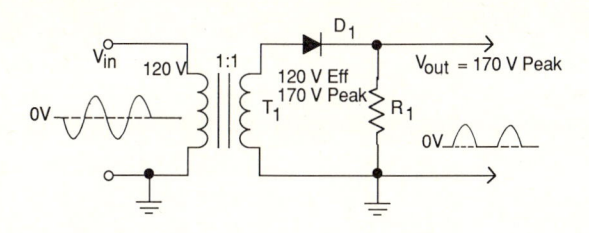

Fig. 7-8 Half-Wave Rectifier

From chapter 2 we know that the voltage drop on the diode junction is very small. In fact, this voltage drop is so small that we can ignore it in our practical analysis of the half-wave rectifier. The transformer shown in the schematic has a 1:1 windings ratio which tells us that the primary and secondary voltages are equal. This ratio tells us that peak and effective voltages are the same for primary and secondary windings. In the circuit schematic, the indicated effective voltage is 120 V and the peak voltage is 170 V. Two of the formulas listed above can be used to prove that $E_{pk} = 1.414 \times 120 \text{ V} = 170 \text{ V}$ and $E_{eff} = 0.707 \times 170 \text{ V} = 120 \text{ V}$.

Peak Inverse Voltage (PIV)

With a negative voltage at the top of the transformer secondary, the diode is reverse biased. This reverse bias keeps the diode cut off during the period that the input voltage is passing through the negative alternation. During this period, the junction acts as an open and must drop total voltage, which reaches 170 V at E_{peak}.

The maximum voltage that can be applied across a diode's junction is called PEAK INVERSE voltage. A good rule to remember is:

In a half-way rectifier, diode peak inverse voltage rating must be larger than the peak voltage of the transformer secondary.

You have already learned that the diode can withstand a limited amount of reverse bias voltage. In the half-wave rectifier the peak inverse voltage rating (PIV) must exceed the peak voltage of the secondary. Knowing the peak secondary voltage you can check the diode specifications sheet (Appendix A) to assure that the diode you place in the rectifier has a **PIV rating** that is larger than peak secondary voltage. One thing that you should keep in mind is that in step-up transformers, peak secondary voltage will be larger than peak primary voltage: therefore, the transformer's peak secondary voltage must be known prior to selecting a diode (rectifier).

Ripple Frequency

Also of concern is the **ripple frequency** of the rectifier output. That is, how many pulses of DC will flow through R_1 for each cycle of the input? As can be seen in Figure 7-8, there is one output pulse for each cycle of input. This means that:

> **Ripple frequency in a half-wave rectifier is equal to the frequency of the AC input.**

With an input frequency of 60 Hz, the output is 60 pulses per second (60 PPS).

Average DC Output Voltage

In a half-wave rectifier, the average DC voltage available in the output is relatively small due to the fact that current flows for only one-half of the time. To calculate the average DC output voltage, use either of the formulas:

$DC_{ave} = 0.318 \times E_{pk\ out}$

$DC_{ave} = 0.45 \times E_{eff}$

For the half-wave rectifier shown in Figure 7-8, we can calculate average DC output voltage as follows:

$DC_{ave} = 0.318 \times 170 = 54\ V$

$DC_{ave} = 0.45 \times 120 = 54\ V$

The advantage of using a half-wave rectifier is that it can be used to develop a relatively high peak output voltage. It is an economical circuit in that it requires only one diode. The half-wave rectifier is used where low voltage and low current are needed. An example is the battery charger for calculator power supplies and NiCad batteries.

When a sine wave is applied to a half-wave rectifier, only half of the input wave is reproduced in the output. This means that current flows in the load (R_1) for only one-half the time. During the other one-half cycle of the input, zero current flows in the load. The result is that when current is averaged across the entire time, average DC output is quite low.

Most of the circuits in use today have power requirements that are more constant. For this reason the half-wave rectifier is seldom used. The principles learned here do, however, apply to the other rectifier circuits that follow.

Self-Check

Answer each item by inserting the word or words required to correctly complete each statement.

1. A _____ allows current to flow in one direction only.
2. A _____ biased diode has a high junction resistance.
3. Current flow will be _____ in a forward biased diode.
4. For a half-wave rectifier, the average DC output voltage can be calculated by using the formula _____.
5. Output ripple frequency for a half-wave rectifier is _____ the frequency of the AC input voltage.

Full-Wave Rectifiers

There are two types of full-wave rectifiers that will be discussed. These are the **conventional full-wave** and **full-wave bridge rectifiers**.

Conventional Full-Wave Rectifiers

A conventional full-wave rectifier can be identified by the fact that it has a center-tapped transformer secondary as part of its construction. This rectifier also has two diodes, although this alone is not a distinguishing feature.

The conventional full-wave rectifier uses two diodes which provide paths for current flow, one for each alternation of the input AC. Examine the schematic in Figure 7-9. If you look closely, you can see that two separate half-wave rectifiers have been included in one circuit. Note that T_1 has a 1:2 step-up ratio. This means that $E_{pk\ sec} = 340$ V.

To understand how this circuit works refer to Figure 7-9(a). The polarity shown on the transformer secondary applies forward bias to D_1 and reverse bias to D_2. This causes D_1 to conduct and D_2 to remain cut off. When D_1 conducts, current will flow away from ground and through R_1. When current flows *away* from ground, a positive pulse is developed across R_1. Note that as current completes its path, it flows through *only* one-half of the secondary. In other words, only half of the peak secondary voltage is being used in the development of the output pulse. This means that in a conventional full-wave rectifier:

**Peak output voltage equals one-half
of the peak secondary voltage.**

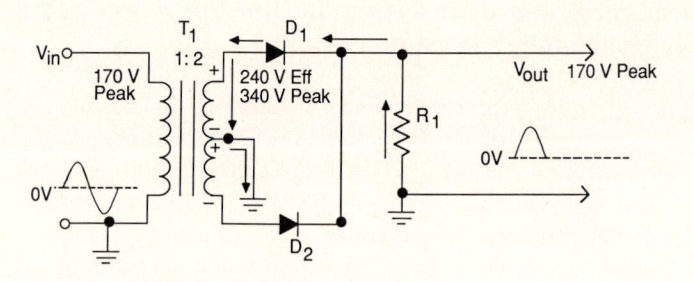

(a) Positive Full-Wave Rectifier/Positive Alternation Input

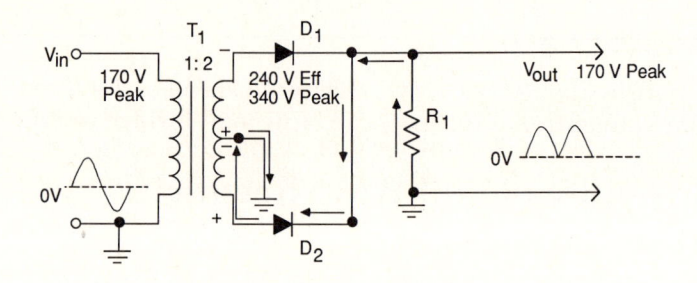

(b) Positive Full-Wave Rectifier/Negative Alternation Input

Fig. 7-9

In the rectifier shown above, peak output voltage is a positive 170 V.

Observe Figure 7-9(b). When the secondary voltage decreases to zero, both diodes are cut off for an instant. However, as secondary voltage reverses phase, the voltage applied to the diode will cause D_1 to enter cutoff and D_2 to conduct. Current is flowing away from ground through D_2. This will, again, develop a positive pulse in the output. Peak output voltage will again be +170 V, or one-half of the peak secondary voltage.

Note that when one diode is conducting, the other is cutoff. This means that current is directed so that it will always flow through R_1, in the same direction. If all current flow in R_1 is in the same direction, it is *direct current* and the output is a series of one polarity pulses (in this case positive).

When both diodes have their polarity reversed, the output voltage is a continuous series of negative pulses. A conventional negative full-wave rectifier is shown in Figure 7-10. In this rectifier, the peak output voltage equals −85 V. Note that the transformer has a 1:1 ratio with 170 V E_{peak} applied.

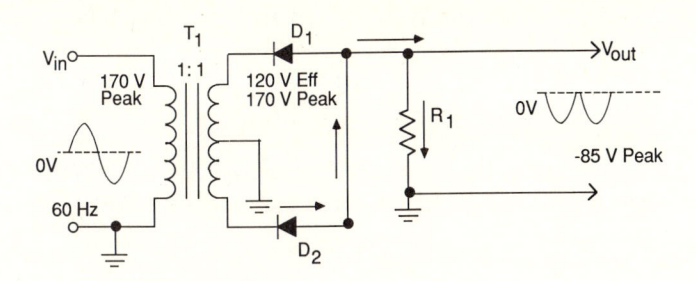

Fig. 7-10 Negative Full-Wave Rectifier

Ripple Frequency

In the half-wave rectifier, we found that the frequency of the ripple-voltage equaled the frequency of the AC input voltage. In the conventional full-wave rectifier, though, we see that each alternation of the AC input develops an output DC pulse: therefore, in the conventional full-wave rectifier:

Ripple Frequency = 2 × input AC frequency.

The ripple frequency of a conventional full-wave rectifier shown in Figure 7-10 can be calculated as follows:

$f_{ripple} = 2 \times f_{input}$

$f_{ripple} = 2 \times 60 \text{ Hz}$

$f_{ripple} = 120 \text{ PPS}$

Peak Inverse Voltage (PIV)

What about the peak inverse voltage rating of the diodes in Figure 7-10? When D_1 is forward biased, D_2 is reverse biased. If we assume that D_1 does not drop any voltage, we can calculate the PIV of D_2. Because D_1 is conducting, a negative voltage (-85 V) is developed at the top of R_1 and the right side of D_2. The left side of D_2 is connected to the bottom of T_1, which is $+85$ V with respect to ground. The difference in potential that exists across D_2 is:

$+85$ V $- (-85$ V$)$ or 170 V

Shown graphically, this difference in potential appears as depicted in Figure 7-11. The vector that illustrates the difference in potential is 170 V in magnitude.

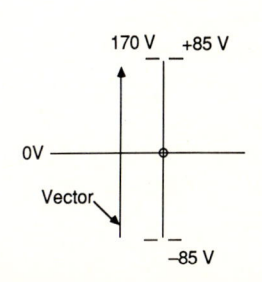

Fig. 7-11

For equal peak output voltages, the full-wave rectifier must have diodes that have a PIV rating twice as large as the half-wave rectifier. However, one rule can be used to remember their PIV requirements. This rule is:

Diodes used in full-wave rectifiers must have a PIV rating equal to or greater than peak secondary voltage.

Average DC Output Voltage

Remember, in a conventional full-wave rectifier, current flows through only one-half of the transformer's secondary. Current does, however, flow during each alternation of the input, creating one output pulse for each alternation of the input sine wave. This creates two output pulses for each sine wave of the input. This means that current flows in R_1 twice as long as in the half-wave rectifier. The current flowing through R_1 is direct current. To calculate the average DC output of a conventional full-wave rectifier, use one of the formulas:

$DC_{ave} = 0.636 \times E_{pk\ out}$

$DC_{ave} = 0.318 \times E_{pk\ sec}$

$DC_{ave} = 0.45 \times E_{eff\ sec}$

For the conventional full-wave rectifier shown in Figure 7-10, you can calculate average DC output voltage using any of the following formulas:

$DC_{ave} = 0.636 \times 85\ V = 54\ V$

$DC_{ave} = 0.318 \times 170\ V = 54\ V$

$DC_{ave} = 0.45 \times 120\ V = 54\ V$

Notice that this full-wave rectifier (Figure 7-10) supplies an average DC equal to the half-wave rectifier contained in Figure 7-8, even though its peak output voltage is only half that of the half-wave rectifier. An advantage of using a full-wave rectifier is that current flows approximately 100% of the time: therefore, greater amounts of power ($P = I^2R$) can be supplied by a full-wave rectifier. Conventional full-wave rectifiers are used in circuits where power requirements are moderate.

Use of a conventional rectifier results in:

$f_{ripple} = 2 \times f_{input}$

The higher ripple frequency allows the capacitors used in the filter to be selected for a higher frequency. Further, the diodes used in this type power supply can have lower current ratings. The result is that the circuit can be manufactured at a lower cost.

Self-Check

Answer each item by performing the action indicated.

6. One identifying feature of a conventional full-wave rectifier is its _____ transformer.
7. Another identifying feature of the conventional full-wave rectifier is that it contains _____ diodes.
8. A full-wave rectifier has a 400 Hz input frequency. What is the output ripple frequency?
9. The peak voltage across a transformer secondary is 300 V. If this transformer is connected to a conventional full-wave rectifier, what will the rectifier's average DC output voltage equal?
10. The transformer that is used with a conventional full-wave rectifier has a 1:4 step-up ratio. If the peak voltage on the primary is 100 V, what is the peak output voltage?

Full-Wave Bridge Rectifiers

The bridge rectifier is a modification of the full-wave rectifier. The bridge does not require a transformer that has a center-tapped secondary for its operation. It does, however, require two extra diodes, giving it a total of four.

For our discussion of the bridge rectifier, we will use Figure 7-12. When the transformer polarity is as shown (positive at the top of the secondary), D_3 and D_4 are forward biased and D_1 and D_2 are reverse biased. With D_3 and D_4 forward biased, current flows away from the

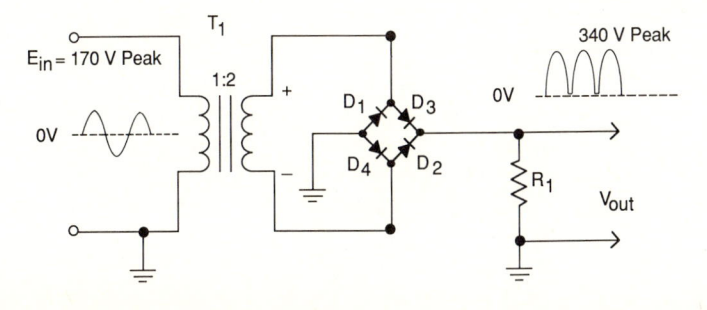

Fig. 7-12 Positive Full-Wave Bridge Rectifier

negative side of the transformer, through D_4 to ground, up through R_1, through D_3, and to the transformer secondary to complete the loop. Because the current flow through R_1 is flowing *away* from ground, the output pulse developed across R_1 is a *positive* pulse.

When the polarity across the secondary changes to that shown in Figure 7-13, D_3 and D_4 are reverse biased while D_1 and D_2 are forward

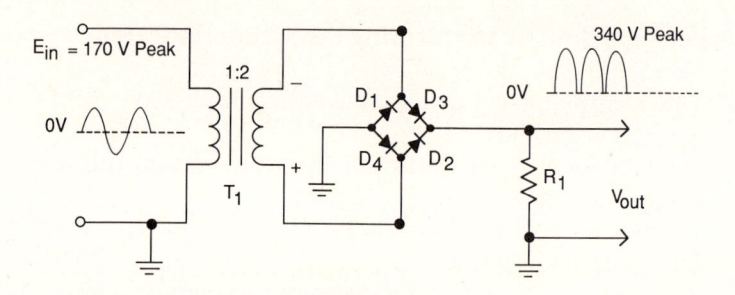

Fig. 7-13 Positive Full-Wave Bridge Rectifier

biased. Current flows away from the top of the transformer, through D_1, to ground, up through R_1, through D_2, and back to the bottom of the secondary to complete the loop. Again, current flow in R_1 is away from ground, and the output pulse developed by R_1 is positive.

The bridge rectifier is classified as a full-wave type because current flows in the load resistor on each alternation of the input AC. This means that the output ripple will be two times the input frequency, or 120 pulses per second when the input frequency is 60 Hz.

Peak Inverse Voltage

To evaluate the PIV felt by a cutoff diode, refer to Figure 7-14. When D_1 and D_2 are conducting, they have a very low junction resistance (shorted as shown in Figure 7-14), and drop very small amounts of

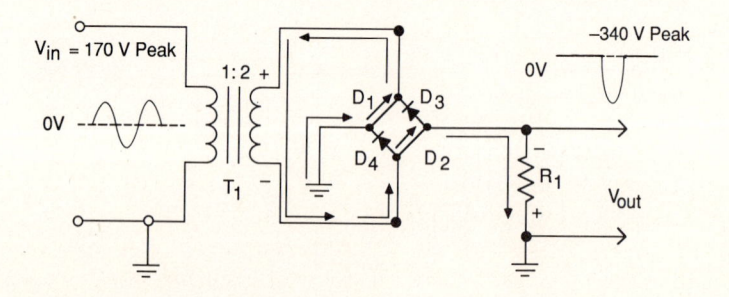

Fig. 7-14 Negative Full-Wave Bridge Rectifier/Positive Alternation Input

voltage. In effect, D_3 and D_4 are reverse biased and connected directly across the tranformer's secondary. This means that the PIV will equal E_{peak} of the secondary, which is 340 V for this circuit. Notice that the transformer is shorted to the ends of both diodes applying 340 E_{pk} reverse bias to each diode.

Average DC Output Voltage

Figure 7-15 illustrates a negative bridge rectifier. Note that the only difference between this and the positive bridge rectifier (Figure 7-13) is that all four diodes have had their polarity reversed. Another voltage

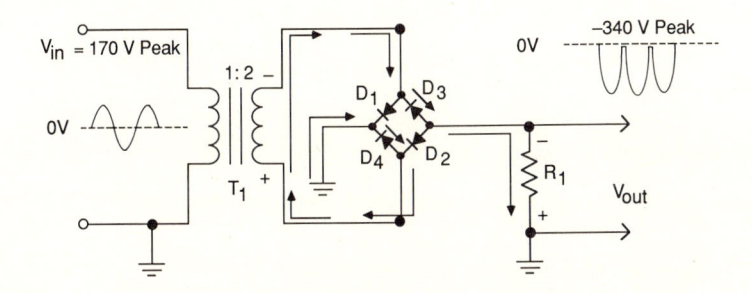

Fig. 7-15 Negative Full-Wave Bridge Rectifier/Negative Alternation Input

value to be considered is the **average DC output voltage**. Refer to Figure 7-15. For each alternation of the input voltage, a pulse is developed across R_1. Average DC output voltage can be calculated using either of the formulas shown below.

$DC_{ave} = 0.636 \times E_{pk\ out}$

Or:

$DC_{ave} = 0.90 \times E_{eff\ out}$

To compare the three rectifiers examine Figure 7-16. Notice that, in each case, the peak output voltage is equal to the peak voltage applied to any single diode.

The conventional full-wave rectifier requires a transformer having a center-tapped secondary. Using the same input voltage and identical transformers, each diode of the conventional full-wave rectifier has half as much peak output voltage applied as the diode in a half-wave rectifier. To maintain the trend, the peak output from the bridge rectifier is equal to that of the half-wave and double that of the conventional full-wave rectifier. Remember, the conventional full-wave

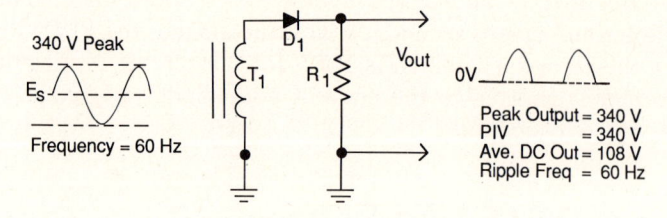

(a) Positive Half-Wave Rectifier

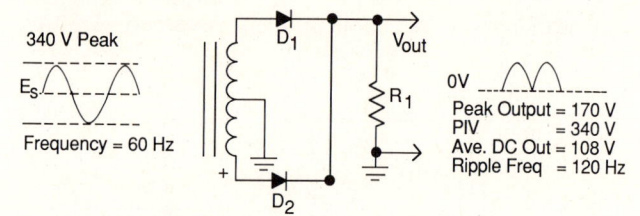

340 V Peak
E_s
Frequency = 60 Hz

Peak Output = 340 V
PIV = 340 V
Ave. DC Out = 108 V
Ripple Freq = 60 Hz

(b) Positive Full-Wave Rectifier

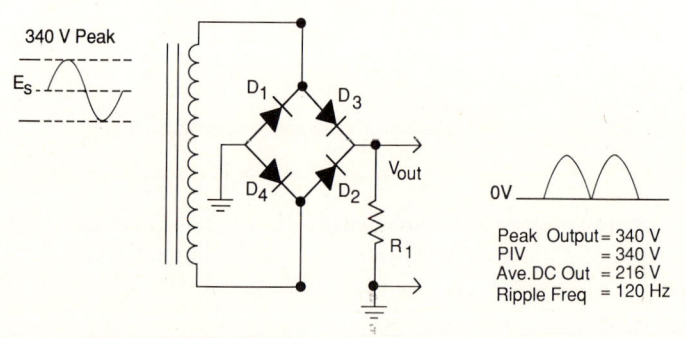

340 V Peak
E_s
Frequency = 60 Hz

Peak Output = 170 V
PIV = 340 V
Ave. DC Out = 108 V
Ripple Freq = 120 Hz

(c) Positive Full-Wave Bridge Rectifier

340 V Peak
E_s

Peak Output = 340 V
PIV = 340 V
Ave. DC Out = 216 V
Ripple Freq = 120 Hz

Fig. 7-16 Comparison of Three Types of Rectifier Circuits

rectifier's peak output voltage equals one-half the peak secondary voltage. This means that the peak voltage applied to the bridge rectifier's diodes is double that applied to the conventional full-wave rectifier's diodes.

The half-wave rectifier's average DC output voltage is equal to the average DC output of a full-wave rectifier when both transformers have equal secondary voltages. Because the bridge does not use the center-tapped secondary, its average DC output is double that of the full-wave and half-wave rectifiers when all three have equal AC voltages applied. Ripple frequency for the conventional and bridge full-wave rectifiers is double that of the half-wave because they develop pulses on each alternation of the input sine wave. In each case, the peak inverse voltage is equal to the peak voltage available across the entire secondary.

Diode Circuits and Power Supply Filters

Of the three circuits, it is more practical to use the bridge rectifier for today's electronics. The cost of a prefabricated bridge element like those shown in Figure 7-17 is quite small. In addition, the voltage and current capabilities of the bridge are much better than the other two. For these reasons you should consider using bridge rectifiers for any circuits you may design.

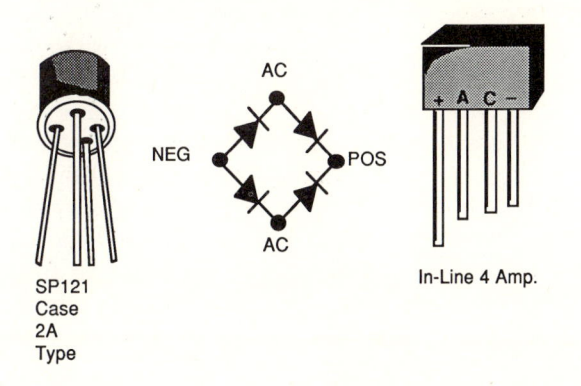

Full Wave 4 PIN DIP Bridge Rectifiers

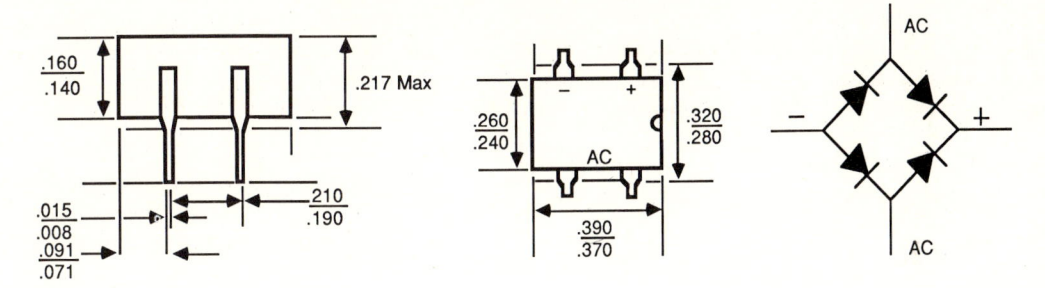

Fig. 7-17 Bridge Rectifiers

Answer each question by inserting the word or words required to correctly complete each statement.

Self-Check

11. A bridge rectifier requires _____ diodes.
12. Using the same transformers, full-wave bridge rectifiers will supply average DC output voltage that is _____ that of conventional full-wave rectifiers.
13. Ripple frequency in a bridge rectifier is _____ the AC input frequency.

14. Using the same transformer the PIV of the diodes used in a bridge rectifier will be _____ the PIV of the diodes used in a conventional full-wave rectifier.
15. Bridge rectifiers provide _____ voltage at _____ current levels.

Filters

To this point, we have discussed circuits that are used to convert AC to pulsating DC. In actuality, pulsating DC is of little use to those of us in electronics. The circuits that you work with require a very steady DC voltage that has minimum ripple. As you get into more complicated circuitry, the amount of ripple must be reduced to as low a level as possible. For most computer circuitry, even a small ripple voltage can cause errors to occur in the data processed. Ideally, the output of a DC power supply should be as stable as the output of a battery. It is impossible to obtain a completely ripple-free output from a power supply. It is possible, however, to approach that point through the use of **filter** circuits.

Filters Convert Pulsating DC to a Steady DC Output

A filter circuit reduces ripple voltage to a level that can be used by the circuits using power from that particular supply. Use of power supply filters increases the average DC output and decreases the AC (ripple) component present in the output.

The unfiltered output of a half-wave rectifier is shown in Figure 7-18(b). Note that the amplitude of the output pulses vary between 0 V and peak voltage. For this type rectifier, the average DC output equals 0.318 × peak output voltage and is represented by the solid line shown on the drawing. Average DC output is low when compared to peak

(a)

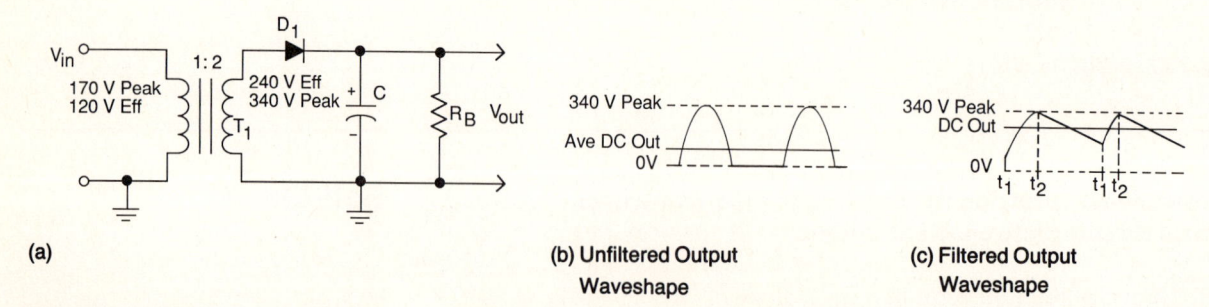

Fig. 7-18 Positive Half-Wave Rectifiers

(b) Unfiltered Output Waveshape

(c) Filtered Output Waveshape

output because current flows through R_B of the half-wave rectifier during one-half the time. When this is averaged over the total time, the average is low. The amount of voltage that the output pulse changes

Diode Circuits and Power Supply Filters

during each pulse is called the ripple voltage. In a half-wave rectifier, the ripple voltage equals the peak output voltage because voltage continually varies from 0 V to peak voltage. Ripple voltage is often referred to as the AC component of the output.

The pulsating DC output of any rectifier contains two components; an AC component and a DC component. The rectifier is used to convert AC into PDC. For the PDC to be converted to pure DC we must reduce the AC component (ripple voltage) to the minimum possible. This is done by use of a *filter* circuit.

Simple Capacitive Filter

The half-wave rectifier shown in Figure 7-18(a) contains a simple filter that consists of an electrolytic capacitor. The output of the rectifier is in the form of pulses. The capacitor filter stores energy during the time that the rectifier conducts. The stored energy is then returned to the circuit during the periods when the diode is cutoff.

Figure 7-18(b) shows the output of the half-wave rectifier without the capacitor in the circuit. With the capacitor (C) in the circuit, the following happens: When D_1 conducts, the capacitor is in series with current flow. The capacitor quickly charges to the peak output voltage. Resistor (R_B) is in parallel with C. Therefore, C must drop the same voltage as V_{RB}. Charging the capacitor results in the storage of energy in its electrostatic fields. When the current flow from the rectifier begins to decrease, the voltage drop on R_B tends to drop. The capacitor senses the change in voltage and immediately acts to oppose the change. The energy in the capacitor will be used to support current flow moving from the bottom of C to the bottom of R_B, upward through R_B, and back to the top of C. Instead of V_{RB} dropping at the same rate as the current flow in D_1, it now decreases more slowly if C and R_B form a relatively long time constant (TC). Remember:

$$1\,TC = R \times C$$

If the capacitance of C is large enough, current flows in R_B 100% of the time. With current flow existing for greater periods of time, the average DC output voltage increases. This causes a decrease in ripple voltage (AC component) and an increase in the DC component. Figure 7-18(c) illustrates the effect of connecting C into the circuit. Note that the duration of current flow in R_B and the average DC output have increased, and ripple voltage no longer drops to 0 V. The diode, however, only conducts during the short period between t_1 and t_2.

Figure 7-19(a) depicts the output wave of an unfiltered full-wave rectifier that has 170 V peak output. Note that it, too, contains a ripple voltage that varies from 0 V to peak voltage; however, current flow in R_B is present for double the amount of time as was true with the half-wave rectifier. The average DC output voltage for full-wave rectifiers is

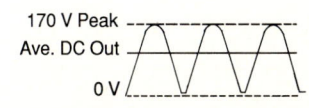

(a) Full-Wave Output (no filtering)

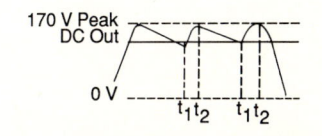

(b) Full-Wave Output (after filtering)

Fig. 7-19 Full-Wave Rectifier Comparisons

0.636 × peak output voltage and is depicted by the solid line in Figure 7-19. Connecting a capacitive filter in the output of a full-wave rectifier results in a waveshape like that shown in Figure 7-19(b). Note that the average DC output here is much greater for the same peak voltage and the same size capacitor than was true with the half-wave rectifier. This is because the capacitor has much less time to discharge and its voltage will not drop as low as in a half-wave rectifier.

After C has been charged, it acts as a "power supply" and the diode quickly becomes reverse biased once the input voltage decreases. The diode remains cut off until the voltage supplied by the transformer secondary exceeds the voltage that remains on the capacitor. Realizing this, we can see that in Figure 7-18(c) and 7-19(b), the diode only conducts for a portion of the positive going input pulse. Conduction occurs only during the periods identified between points t_1 and t_2. The diode starts to conduct at t_1 and cuts off at t_2. At all other times, E_C exceeds the input voltage and the diode is not conducting.

The output from a bridge rectifier is identical as it is also a full-wave rectifier. If the bridge rectifier has the same peak output voltage as the conventional full-wave rectifier, both output voltages will be identical. If the capacitive filter used for each rectifier is the same value, the waveshapes are identical to those shown in Figure 7-19.

In all three rectifiers, the average DC output is increased by the return of the energy to the circuit by C discharging. The result is that the output average is greater for the filtered circuit than it is for the unfiltered circuit. Because the discharge time of C depends upon the discharge time constant, we can increase the average DC output by increasing the size of either C or R_B. Increasing the size of these components does, however, have some limitations. For this reason, the simple capacitive filter is not sufficient for most power sources because it is not able to supply large currents to small load devices for long periods of time.

Simple Inductive Filter

An inductor (coil) can be used effectively in the filter circuit of a rectifier. The coil is a reactive device as is the capacitor. They both store energy in their fields; the coil in an electromagnetic field. During the part of the input alternation where load current would be decreasing, they both supply current to the load, thereby increasing the average DC output.

Figure 7-20(a) shows a half-wave rectifier with a coil in its output circuit. Notice that the coil is placed in series with the diode and the load resistor. Remember, the capacitor was placed in parallel with the load resistor. In Figure 7-20(b), the dashed lines represent the unfiltered output and the solid lines represent the inductively filtered output.

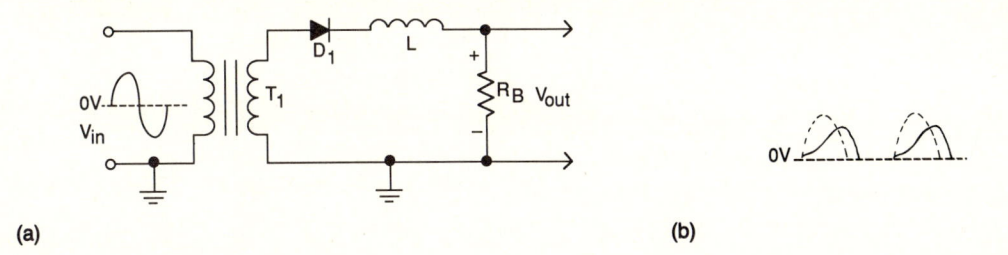

Fig. 7-20 Inductive Filter and Output Waveforms

The inductive filter uses the inductive reactance of the coil. As current increases in the coil, a magnetic field builds outward around the coil. This electromagnetic field stores electrical energy that is referred to as the **charge**. As current increases, the coil's charge is increasing. The coil's charge does not reach maximum, though, because of the lagging effect that exists in a coil. (Voltage leads current by 90° in a coil). Once the input voltage begins to decrease and the current through the load device tries to decrease, the coil acts to oppose the change in current flow. It does this through the collapse of its magnetic field which returns the energy that is stored in its magnetic field to the circuit.

To this point, you have studied simple capacitive and inductive filters. You have seen that each of these can affect the amount of current that flows in the load (R_B and devices in parallel with it). Each filter produced an increased output current through the bleeder resistor. Both store energy during times when the diode(s) conduct and then return stored energy during the period when the diodes are cutoff. Each type reactance has some value. Best filtering occurs when both capacitors and coils are used in the same circuit.

Remember:

In power supplies, filter coils are always connected in series with the load. Electrolytic capacitors are used in power supply output filters and are always connected in parallel with the load resistor.

The use of filter circuits provides a much more suitable voltage and current output from rectifier circuits than can be obtained otherwise. Their use is a form of regulation where regulation means to supply more constant current, voltage, or power to a circuit. In later chapters, we will add an electronic voltage regulator to these circuits to make them much more stable than with filters alone.

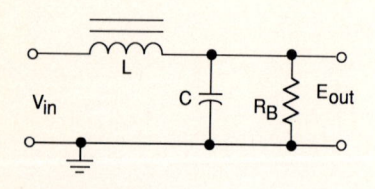

(a) Inductive Input

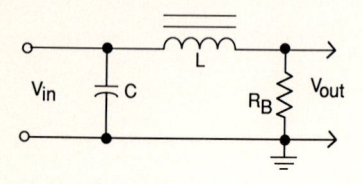

(b) Capacitive Input

Fig. 7-21 L-Type Filters

L-Type Filters

The use of capacitance and inductance as filters results in two circuits that are referred to as **L-type filters**. In one type, the coil is connected directly to the diode's output and is, therefore, called an **inductive input filter**. The other connects the capacitor directly to the diode's output and is called a **capacitive input filter**. In each case, the type input refers to the device connected to the diode's output.

An L-type inductive input filter is illustrated in Figure 7-21(a) and a capacitive input L-type filter is shown in Figure 7-21(b). Both receive their names from the fact that, when R_B is not present, they resemble the letter "L" lying on its side.

In Figure 7-22, an L-type inductive input filter is connected in the output of a full-wave rectifier. When AC is applied to the transformer primary, D_1 will conduct during the time that the voltage at the top of the transformer secondary is positive (+). Arrows show the two paths of current through C and R_B that combine to flow through L and D_1. As current begins to flow, C will charge quickly, and a magnetic field builds around L.

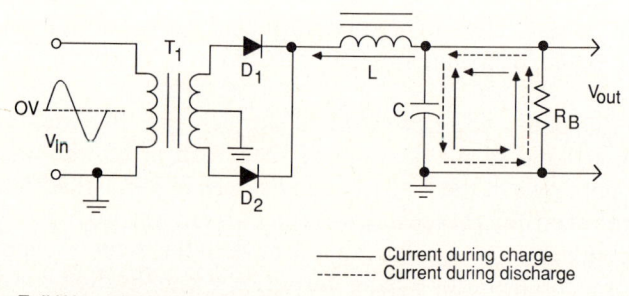

——— Current during charge
- - - - - Current during discharge

Fig. 7-22 Full-Wave Rectifier with L-Type Inductive Filter

D_1 places very little resistance in the charge path of C and L. However, the impedance of L is now in the charge path of C. The voltage dropped by L causes the voltage charged on C to be lower than with the simple capacitive filter. Because R_B is in parallel with C, its voltage will also be lower than before.

As the positive alternation on the secondary reaches peak and begins to decrease, D_1's current begins to decrease. The capacitor senses the change in voltage, begins to release its energy and maintains conduction in the bleeder resistor, as shown by the arrow. This tends to maintain voltage at its earlier level. The coil is, at the same time, acting to oppose the change in current that it senses by returning energy to the circuit from its collapsing field. The end result is that the conduction of the diode is extended for a longer period of time. This action provides a relatively constant output voltage and current to the load device. During the negative alternation of the input, the same sequence occurs except D_2 is now conducting. In this circuit, L maintains current flow in the diode long enough to support its discharge.

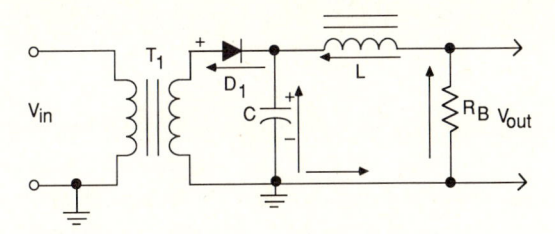

Fig. 7-23 Half-Wave Rectifier with L-Type Capacitive Input Filter

Figure 7-23 shows a **capacitive input L-type filter** connected in the output circuit of a half-wave rectifier. When AC is applied to this circuit, the positive voltage that is applied to the diode causes D_1 to conduct. As the diode conducts, the capacitor quickly charges to peak output voltage and the coil builds its magnetic field. Arrows depict the current paths. Notice that R_B and L are in series during the charge cycle. This means that the voltage drop on the load resistor is less than peak voltage by the amount dropped across L. During this alternation, both C and L are charged and store energy.

When voltage at the top of the transformer secondary begins to decrease, both C and L act to oppose the changes of current and voltage that are present across the load device. By returning their energy to the circuit, the voltage and current at the load remains relatively stable. In this circuit, L has a discharge path through the capacitor and resistor.

Pi-Type Filters

Pi-type filters receive the name from the fact that their circuit diagrams resemble the Greek letter pi. Two types of pi filters are used. These are the **RC** and **LC** types illustrated in Figure 7-24. As you might guess from these names, one type uses a coil and the other uses a resistor in conjunction with capacitors to perform the filtering action.

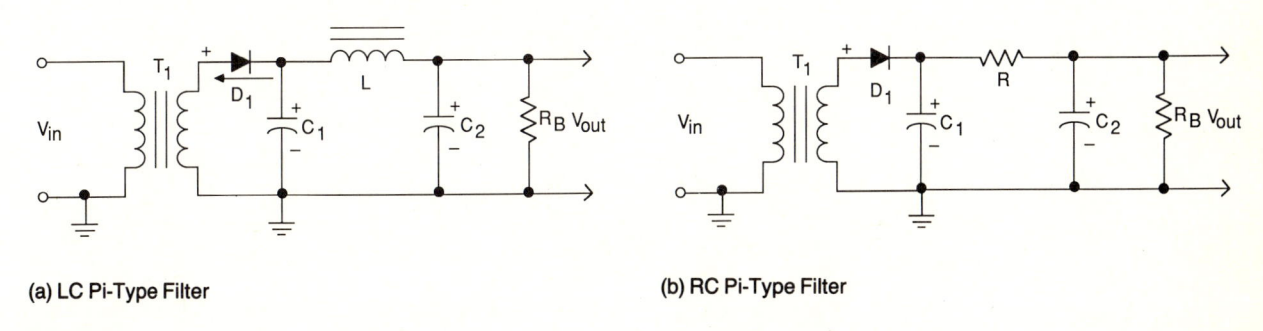

(a) LC Pi-Type Filter

(b) RC Pi-Type Filter

Fig. 7-24

The first pi-type filter that will be discussed is the **LC pi-type filter**. A half-wave rectifier with an LC pi-type filter connected to its output is shown in Figure 7-25. Components that make up the filter are C_1, C_2, and L.

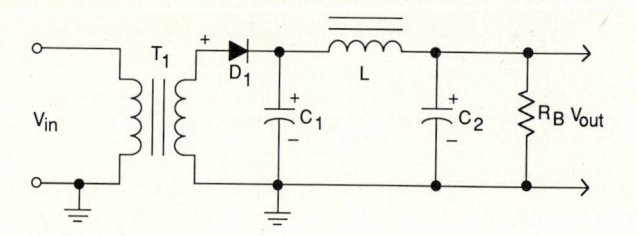

Fig. 7-25 LC Pi-Type Filter

When a sine wave is present at the secondary of the transformer, the positive alternation will cause the diode to conduct. At this time, C_1 and C_2 begin to charge. C_1 immediately charges to peak voltage, but C_2 is in series with L and charges to a slightly lower voltage.

During the negative alternation, the diode is reverse biased and cut off. This allows the coil and both capacitors to transfer their stored energy back to the circuit. The discharge path for all three is through the bleeder resistor and load device. This maintains the voltage drop across the load device at approximately the same level as peak voltage applied to the diode during the conduction alternation. With the LC pi-type filter, it is possible to obtain high output voltages with good control over voltage and current variations.

A less effective pi-type filter is the **RC type** shown in Figure 7-26. Replacing the coil with a resistor provides a filter that operates in a similar way to the one just discussed. The resistor, however, is not a reactive device and will not store energy; therefore, it cannot be used to help stabilize voltage and current during the time that the diode is cut off.

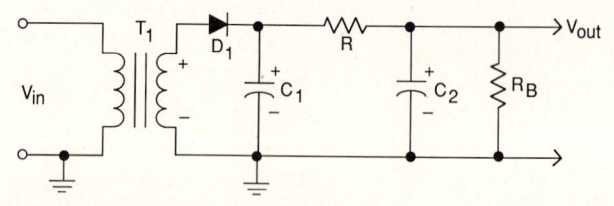

Fig. 7-26 RC Pi-Type Filter

In the LC type, the coil opposes any change in current. This tends to stabilize both current flow through, and the voltage drop across, the load device. With the RC type, the stabilization provided by the coil is not present. Without the coil's opposition to changes in current, voltage and current variations will be more drastic.

Operational Characteristics

To this point, we have discussed several different filtering techniques and how they affect the output of half- and full-wave rectifiers. Since all rectifiers have AC as their input voltage, the polarity induced into the transformer secondary is continually reversing, and the amplitude of the induced voltage is continually changing. The changes of voltage alternately forward and reverse bias the diodes used in the rectifier circuit. As you learned in earlier chapters, diodes have limited current-carrying capabilities. If too much forward current is allowed to flow, the diode can be destroyed.

When power is initially applied to a circuit, current flow will, quite often, **surge** through the circuit at levels much higher than normal. Certain methods are used within rectifier circuits to protect the diodes from damage due to surge currents. Devices that are used to protect a circuit from current surges are called **surge suppressors**. Figure 7-27 contains a few of these protective features.

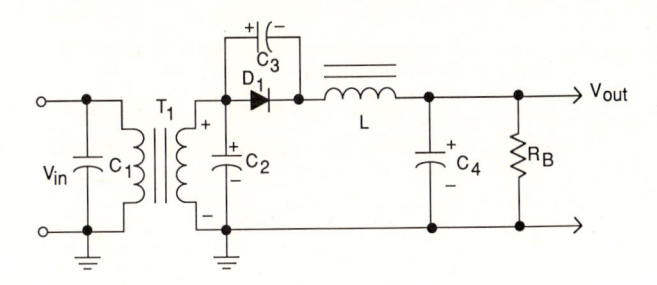

Fig. 7-27 LC Pi-Type Filter with Surge Suppressors

It is possible for transient currents to damage a diode during operation. To prevent this from happening, we must provide a path that has low opposition to the high frequency current flow caused by transient spikes. Remember, a capacitor can be selected whose X_C allows it to act as a short to all frequencies above a specific frequency. If we want to filter out all frequencies above 10 kHz, we can select a capacitor that has very low X_C to 10 kHz. By connecting the capacitor in parallel with the part of the circuit we want to protect, we can provide a low opposition path for currents at 10 kHz and above.

High frequency variations that occur in the transformer primary can be shorted by use of C_1. Capacitor C_3 is placed in parallel with the diode to provide a shunt path for high frequency currents that could otherwise destroy the diode. Connecting a capacitor across the secondary, as illustrated by C_2, will accomplish the same result by shunting high frequency, transient currents across the secondary prior to their reaching the diode.

When the diode is reverse biased, voltages that are too high can destroy the diode. C_3 is placed in parallel with the diode and will absorb any high voltage spikes that might occur across the diode. Remember, this capacitor was selected to have low opposition to high frequency AC transient voltages and currents. Because of the PIV rating of diodes, it is important that reverse bias be kept below their rated PIV. C_3 protects the diode against high peak inverse voltages that might cause it to be destroyed. A further advantage of C_3 is that it protects the diode from the high current surges that can occur when C_4 begins to charge. By conducting a portion of this current surge, C_3 can help prevent the diode from having its current rating exceeded.

Examine the half-wave rectifier, pi-filter circuit in Figure 7-28.

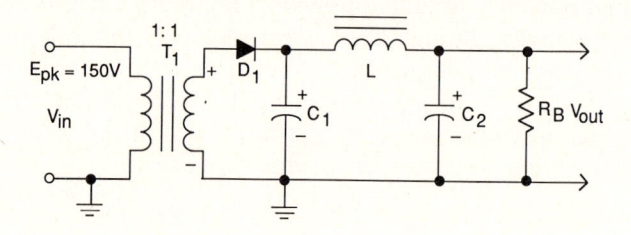

Fig. 7-28 PIV for a Filtered Half-Wave Rectifier

This circuit will be used to discuss the effect that connecting the pi filter to a rectifier circuit has on the peak inverse voltage applied to the diode. If the input sine wave has a peak voltage of 150 V, what is the PIV voltage applied to the diode? For this discussion, assume that the diode used was selected because of its 200 V PIV rating.

Since the transformer has a 1:1 ratio, the peak voltage across the secondary will be 150 V. When the positive alternation applies forward bias to the diode, C_1 charges to E_{peak} or +150 V. When the secondary changes polarity, a negative voltage is applied to the diode's anode. At this same time, the capacitor is placing +150 V on the diode's cathode. The result is that the diode has the −150 V peak applied by the transformer, plus +150, the charge of the capacitor applied as peak inverse voltage, for a total PIV of 300 V. In this case, PIV greatly exceeds the 200 V PIV rating used in selecting the diode. For a diode to operate in this circuit without damage, it should have a PIV of more than 300 V. A minimum PIV of 400 V would probably be a good selection.

A similar situation occurs with the full-wave rectifier when it is connected to an LC filter, as shown in Figure 7-29. Using the 150 V peak voltage and a 1:1 transformer, we have a maximum of 150 V peak secondary voltage. During conduction, each diode uses the voltage across one-half the secondary, or 75 V. The transformer voltage is limited to one-half $E_{pk\ sec}$, or 75 V. When the diode conducts for the first

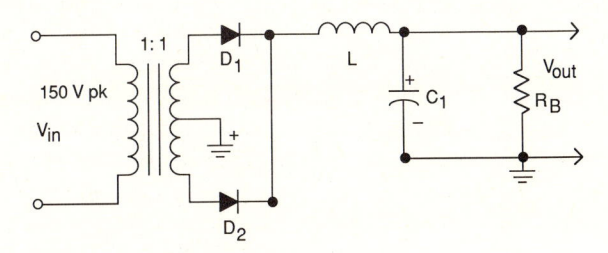

Fig. 7-29 PIV for a Positive Full-Wave Rectifier

time, however, it causes the capacitors in the filter to charge to peak output voltage, which is 75 V. As in the half-wave rectifier, a situation exists where the total PIV of the diode is the sum of peak output and capacitor voltages. Since each of these will be, at maximum, 75 V, the maximum reverse voltage that could be applied to the diode is 150 V. If the 200 V PIV rating of the diode selected above was used here, it should operate with no problem.

From our earlier discussion of the bridge rectifier, we can see that the PIV rating of these diodes must be greater than the peak voltage of the transformer secondary. Because the two diodes that are conducting act as shorts, the two diodes that are cut off are connected directly in parallel with the transformer secondary voltage; therefore, each diode must withstand the peak secondary voltage as its peak inverse voltage.

In power supply circuits, especially those with capacitive filters, it is common to see a fixed resistor connected in the position labeled as R_B in all these circuits. R_B serves two purposes:

1. it maintains a permanent load on the power supply during the periods when the operating device connected across the output is turned off. This protects the diode(s) from damage.
2. it provides a discharge path for the filter capacitors when the voltage is removed from the transformer.

In the last instance, R_B is referred to as a *bleeder* resistor because it "bleeds off" the charge on the capacitors and prevents what could be a dangerous situation from existing. Remember, a capacitor will retain its charge until it is provided with a path through which it can discharge. Because the bleeder resistor is not producing any useful work for the user, the current flow through it is kept quite low. In power supply design, it is common to restrict bleeder current to less than 10% of the total current load placed on the power supply. Allowing $I_{bleeder}$ to exceed 10% of total current decreases the power supply's efficiency. The loss in efficiency increases the cost of operating the circuit.

In some cases (Figure 7-30), the bleeder resistor may be two or more resistors connected in series. Note that each type rectifier is included. Each circuit has been modified to supply both positive and negative output voltages. Remember:

**Current flowing toward ground
develops negative voltage**

and:

**Current flowing away from ground
develops positive voltage.**

When the rectifier outputs are connected in this manner, the bleeder circuit forms a voltage divider that can be used to supply two or more output voltages. In each circuit, current flows from the bottom of the resistor network to the top. This means that the voltage dropped by each of the bottom resistors (R_{B2}) is *negative* and the voltage across each of the top resistor (R_{B1}) is *positive*. It is possible, by placing ground at a point between two resistors, to obtain both positive and negative voltages from the same power supply. When two or more resistors are used to divide the voltage care must be taken to assure that the ratio between the resistor values is correct to provide the voltages desired.

When using inductive **choke** filters, the amount of current flow is critical. A coil is considered to be a choke when it:

**Impedes the flow of pulsating direct
current by means of its self inductance.**

If current is not maintained above a critical level, the coil will fail to act as an inductive filter and will perform more like a capacitive filter. Current through the coil must be large enough to build a magnetic field that can maintain current for the period when the input voltage is decreasing or not present.

The size of the choke (coil) can be calculated using the following formula. This formula is used with either type of full-wave rectifier.

$$L = \frac{R_B}{1130}$$

> **NOTE:** It is not practical to use a choke in the half-wave circuit due to the fact that the diode current exists for such short periods of time.

Diode Circuits and Power Supply Filters

(a)

(b)

(c)

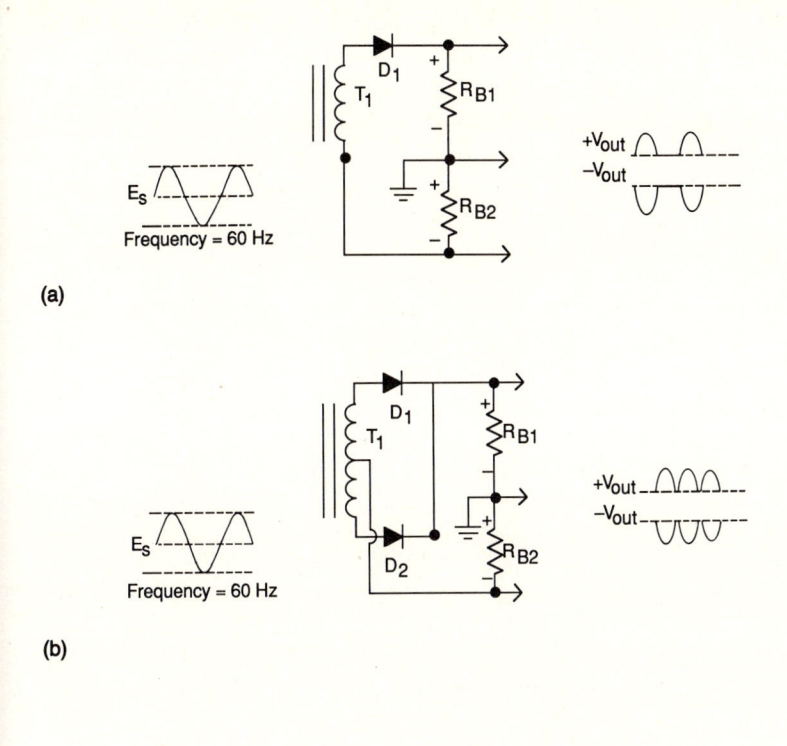

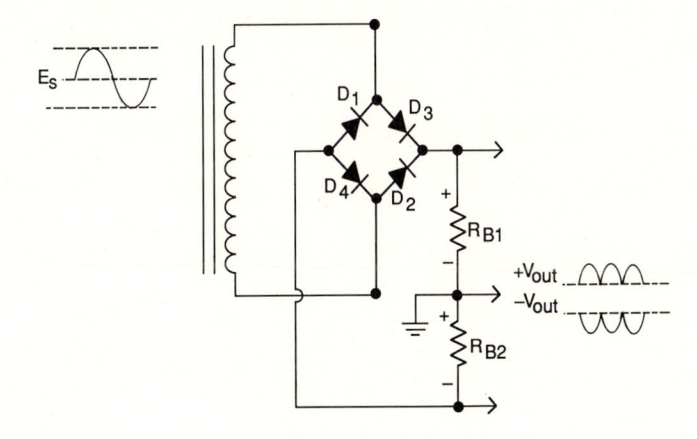

Fig. 7-30 Dual-Output Power Supplies

Once the size of L has been calculated, the value of the critical current required for it to operate as intended can be calculated using the formula shown below:

$$I \,(mA) = \frac{VOLTAGE}{L \,(Henries)}$$

If the circuit for which the power supply is being designed calls for 20 V output voltage and the choke is 20 mH, the minimum current must be 1 mA. To assure a flow of 1 mA at all times, the bleeder resistor must be 20 kΩ.

Self-Check

Answer each item by performing the action indicated.

16. What is the advantage of using a capacitor within a filter circuit?
17. What purpose is served by connecting a filter in the output of a rectifier circuit?
18. Name and describe the two types of pi filters. What is missing from the second type that is present in the first?
19. What is the advantage of using an inductor within a filter circuit?
20. What effect does the use of a filter in the output circuit of a rectifier have on peak output voltage?

Troubleshooting Rectifier Circuits

The process of troubleshooting is defined as follows:

Troubleshooting includes all of the actions taken to identify, locate, and repair an electrical circuit so it can be placed back in operation.

The ability to locate and replace components that malfunction in any operational circuit is an important skill for all technicians. Maintenance of rectifier and filter circuits is extremely important to most electrical equipment. Without the correct voltage and current, a circuit cannot operate correctly.

Two conditions cause most of the equipment malfunctions that you will encounter. These are *opens* and *shorts*. Some characteristics of these conditions are in order:

- An open component will have infinite resistance.

- An open component will stop all current flow through that part of the circuit.

- Anytime that a component opens in any circuit, total current within the circuit will decrease.
- A shorted component will have zero resistance.
- A shorted component will conduct all of the current within its part of the circuit.
- Anytime that a component shorts in any circuit, total current within the circuit will increase.

Half-Wave Rectifier Troubleshooting

Now we can take a look at how open and shorted components will affect the half-wave circuit. For this discussion refer to Figure 7-31.

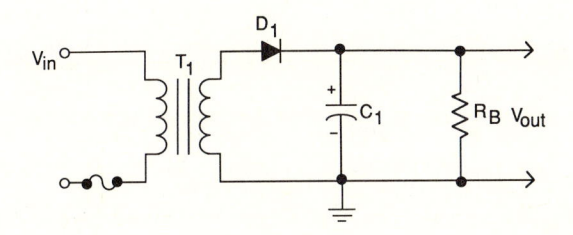

Fig. 7-31 Troubleshooting a Half-Wave Rectifier with Capacitive Input Filter

When the diode in this circuit shorts, rectification is not present and the AC signal is passed to the output. The output signal is a sine wave identical to the one on the secondary of the transformer. With a shorted diode, current flow is higher than normal. The higher current may cause the fuse to open. With an open fuse, all current flow stops and the output is zero. An open diode prevents all current from flowing in the output filter and, therefore, the output is zero.

A shorted bleeder resistor (R_B) places a zero resistance path across the transformer secondary when the diode is conducting. The result is that all of the current flows through the short and there is zero current in the operating device connected to the output. The excessive current that flows through the diode causes the diode's current to exceed the current rating and the fuse will be burned open. This results in zero current flow in the secondary circuit. Should the capacitor open, the output signal is a pulsating DC with the same characteristics as an unfiltered half-wave rectifier output. The loss of filtering causes ripple voltage to increase and average DC output to decrease. Table 7-1 summarizes the results of opening and shorting the diode or capacitors in a half-wave rectifier like the one shown in Figure 7-31.

Table 7-1
Half-Wave Rectifier Troubles and Symptoms

Trouble	Output Voltage	Ripple Amplitude	Ripple Frequency
diode open	zero current zero voltage	zero	zero
diode shorted	excess current zero output blown fuse	zero	zero
capacitor open	lower than normal no filtering	more than normal	no change
capacitor shorted	excess current zero output blown fuse diode open	zero	zero

Conventional Full-Wave Rectifier Troubleshooting

Now we will apply the same type of analysis to the conventional full-wave rectifier and filter circuit shown in Figure 7-32. Should either diode in this rectifier open, the output signal reverts to a half-wave

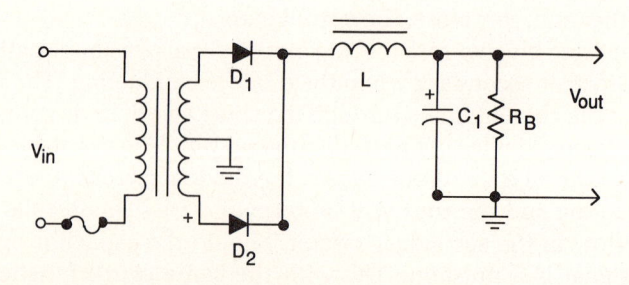

Fig. 7-32 Troubleshooting Full-Wave Rectifier with L-Type Filter

output. This results in a change in ripple amplitude and causes average DC output voltage to decrease. Figure 7-33 shows the output that can be expected from an unfiltered conventional full-wave rectifier when either diode is open.

Should either of the diodes short, the excessive current that results will cause the fuse to open, stopping all current. An open capacitor causes the filtering action to stop. The average DC output voltage then decreases while ripple voltage amplitude increases. Should the capacitor short, excessive current causes either/both the fuse or diode to open, stopping all current flow. If the coil opens, all output current stops. Should the coil short, everything will appear to be normal. The only change will be the loss of current filtering. Except in special situations, it is possible that the circuit would continue to operate with a shorted coil without any recognizable change in operation. Table 7-2 summarizes the results of opening or shorting of the diodes or capacitors in a full-wave rectifier like the one shown in Figure 7-32.

Fig. 7-33 Full-Wave Output with an Open Diode

Table 7-2
Conventional Full-Wave Rectifier Troubles and Symptoms

Trouble	Output Voltage	Ripple Amplitude	Ripple Frequency
diode open	half-wave out lower than normal	greater than normal	decreases by one-half
diode shorts	excess current zero output blown fuse	zero output	zero output
capacitor open	lower than normal decrease in filtering	greater than normal	no change
capacitor shorted	excess current zero output blown fuse diode open	zero	zero

Full-Wave Bridge Rectifier Troubleshooting

For an analysis of the full-wave bridge rectifier and pi filter, refer to Figure 7-34. If any one diode shorts in this circuit, the circuit will draw excessive current and the fuse will open. Should any one diode open, the circuit reverts to half-wave operation with ripple frequency decreasing by one-half, ripple amplitude increasing, and a large decrease in average DC output. If either of the capacitors should short, the diode conducts heavily and the fuse will blow. Should either capacitor open, filtering decreases with a corresponding decrease in DC output and an increase in ripple amplitude. Table 7-3 summarizes the results of opening or shorting diodes or capacitors in a bridge rectifier like the one shown in Figure 7-34.

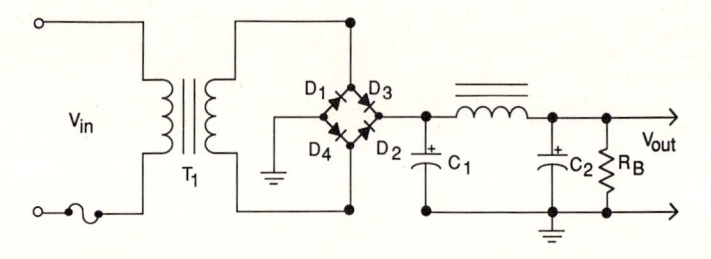

Fig. 7-34 Troubleshooting Bridge Rectifier with LC Type Filter

---

A Special Note Regarding Power Supply Troubleshooting

It is rare that power supply trouble involves only one trouble. Usually the diode or fuse, or both, will open if any component shorts. The rule is:

Always check for multiple troubles.

For example, if R_B or a filter capacitor shorts, excessive current can flow through the diode(s) and the tranformer. This will probably result in the diodes and/or fuse being destroyed. This could then result in several defective components from one initial failure.

Diode Circuits and Power Supply Filters

Table 7-3
Bridge Rectifier Troubles and Symptoms

Trouble	Output Voltage	Ripple Amplitude	Ripple Frequency
diode open	half-wave output lower than normal	greater than normal	decreases by one-half
diode shorts	excess current zero output blown fuse	zero output	zero output
capacitor open	lower than normal decrease in filtering	greater than normal	no change
capacitor shorted	excess current zero output blown fuse diode open	zero	zero

Self-Check

21. In the half-wave rectifier, an open diode will have what effect on the output voltage and frequency?
22. A shorted diode in a full-wave rectifier will have what effect on the average DC output voltage and frequency?
23. A shorted filter capacitor will have what effect on a rectifier circuit?
24. An open filter capacitor will have what effect on the output voltage and frequency of a full-wave bridge rectifier?
25. In the inductive input L-type filter, an open coil will have what effect on output voltage and frequency in a conventional full-wave rectifier circuit.

Limiters and Clampers

Often it is necessary to change the reference level of a signal as it passes from one circuit to another. In this section, we will explain how limiters (clippers) and clampers can be used to do this.

Limiters

A **limiter** is defined as:

> **Any circuit that prevents (limits) some characteristic of a waveform from exceeding a specified value.**

We will discuss circuits that can limit either or both amplitudes of an AC signal.

Limiter circuits are used for two primary purposes: (1) waveshaping and (2) circuit protection. Limiters that are used for waveshaping limit (clip) or modify the waveshape of a signal. When used as protective circuits, limiters prevent a voltage from exceeding a value that could damage the circuit that follows. A limiter which removes a portion of the negative half-cycle of a signal is called a *negative* limiter. A limiter that removes part of the positive half-cycle is called a *positive* limiter. Block diagrams and circuits for these two limiters are shown in Figure 7-35. Notice that all drawings have sine wave inputs but provide half-wave outputs. Each type can be modified to clip (limit) any portion of the alternation that is being limited.

(a) Negative Limiter

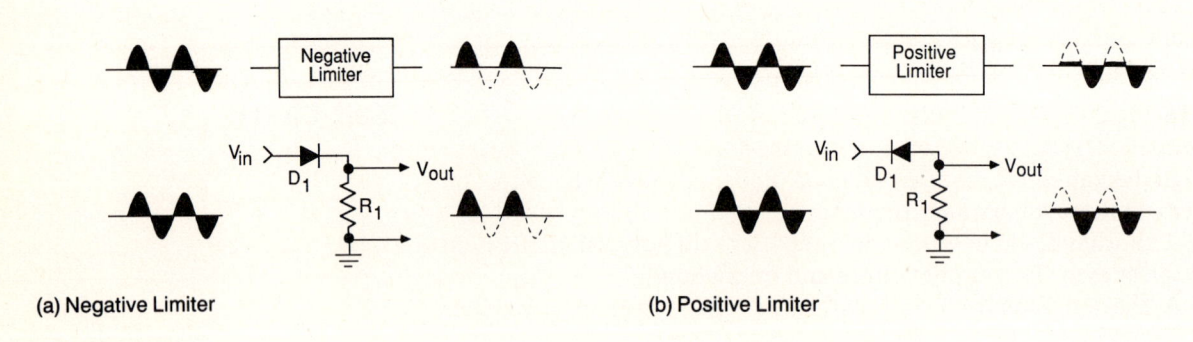

(b) Positive Limiter

Fig. 7-35 Limiters

The circuits shown in Figure 7-35 are called **series limiters** because the diode is in series with the output load. In Figure 7-35(a), the limiter is called a **series negative limiter** and in 7-35(b), the circuit is a **series positive limiter**. In either circuit, the alternation that is passed will forward bias the diode causing it to conduct. This drops voltage across the resistor that can be coupled to a subsequent circuit. The arrival of the opposite polarity reverse biases the diode, cutting off output current. This prevents voltage from being dropped on the output resistance and therefore, there is no voltage to be coupled to the next circuit.

A second type of limiter places the diode in parallel (shunt) with the input of the next circuit. These are called **shunt limiters**. Figure 7-36 contains schematics for **shunt positive** and **shunt negative limiters**.

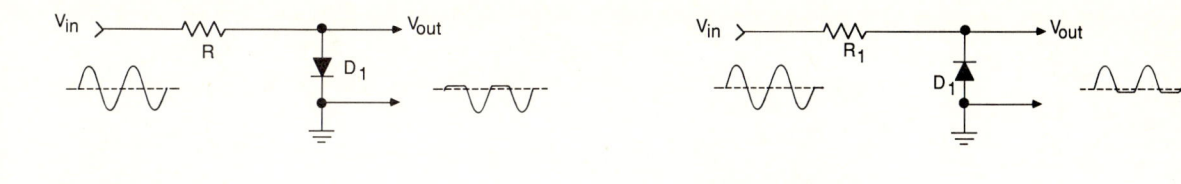

(a) Shunt Positive

(b) Shunt Negative

Fig. 7-36 Limiters

Notice that in both circuits, the limited alternation is not completely removed. This is because of the reverse current that flows in the junction of a reverse biased diode.

Biased Limiter Circuits

In order for limiter circuits to have a broader range of limiting, fixed bias is placed in the circuit. This bias determines the point where the diode begins to conduct and the duration of conduction. In the shunt limiters discussed above, limiting removed approximately 50% of the input sine wave. With bias, limiting can be referenced to any percent of the input signal ranging from 1% to 99%.

Biased Shunt Positive Limiters

Schematic diagrams with input and output waveshapes are contained in Figure 7-37. Notice that one circuit has the positive terminal of the battery connected to the diode's cathode, and the other has the negative

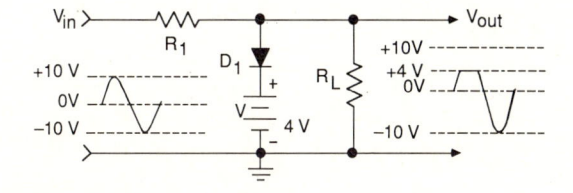

(a) Shunt Positive with Positive Bias

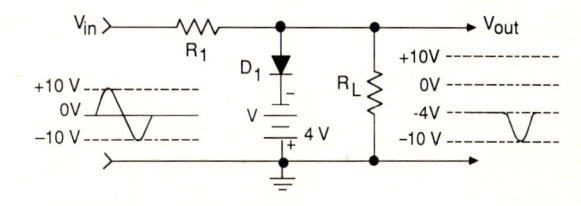

(b) Shunt Positive with Negative Bias

Fig. 7-37 Limiters with Bias

terminal connected. The amount of bias applied in this manner determines the conduction and cutoff point for the diode. In Figure 7-37(a), note that only a small portion of the positive alternation is limited. This is because the input signal amplitude must exceed +4 V before the diode conducts and begins to short out the signal. With negative bias applied, Figure 7-37(b), the input signal must decrease to −4 V before the diode is cutoff. Once in cutoff, the diode branch is open, forcing current to flow through R_L. Notice that less than 50% of the input signal is applied to R_L because of the limiting that occurs by the conducting diode.

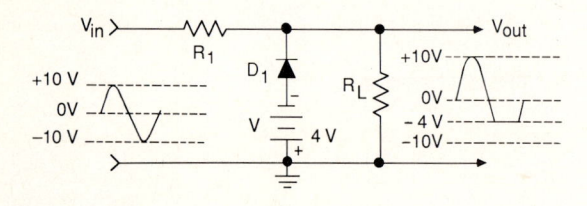

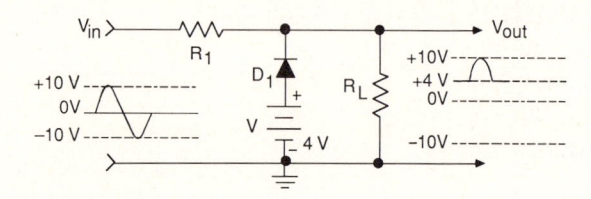

(a) Shunt Negative with Negative Bias

(b) Shunt Negative with Positive Bias

Fig. 7-38 Limiters with Bias

Biased Shunt Negative Limiters

Figure 7-38 contains shunt negative limiters that have positive and negative bias applied. These circuits operate similarly to the circuits discussed above. The difference is that negative bias results in 50% to 100% of the input signal being passed to the output. Positive bias results in the output being limited to less than 50% of the input signal.

Remember, shunt limiters perform their limiting when the diode conducts. The conducting diode provides a shorted path which bypasses the output. When the diode is cutoff, current flows in R_L and is not limited.

With the shunt positive limiter:

Positive bias can be used to limit the output signal between 50% and 100%.

and:

Negative bias can be used to limit the output signal between 0% and 50%.

With the shunt negative limiter:

> **Positive bias can be used to limit the output signal between 0% and 50%.**

and:

> **Negative bias can be used to limit the output signal between 50% and 100%.**

Double-Diode Limiters

A circuit that can be used to limit the peaks of both half-cycles of an AC input signal is the **double-diode biased limiter**. A circuit diagram for this circuit is shown in Figure 7-39. Note that this circuit contains a

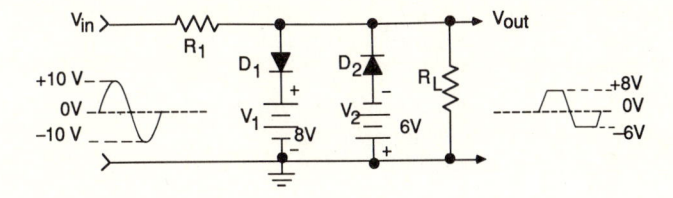

Fig. 7-39 Double Diode Limiter

shunt positive limiter with positive bias and a shunt negative limiter with negative bias. During the time that no signal is present, neither diode is forward biased. When either alternation of the input signal exceeds 8 V, however, one of the diodes will conduct, shorting the input signal. During the positive alternation, D_1 will conduct when input voltage exceeds +8 V. With the negative alternation of the input, D_2 will conduct when voltage exceeds −6 V.

Double-Zener Diode Limiters

A limiter circuit that uses two Zener diodes for its limiting is shown in Figure 7-40. Remember that a Zener diode operates like a conventional

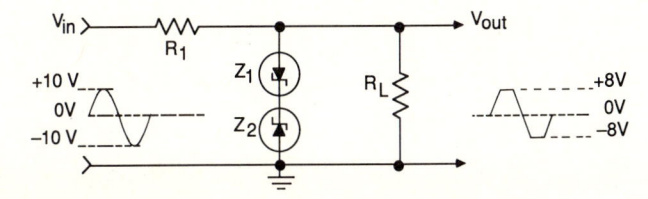

Fig. 7-40 Double Zener Diode

diode when forward biased. It will, however, regulate voltage to its designed value when reverse biased and operating in the avalanche (regulating) region. As each diode is reverse biased and reaches its regulating region, it limits the output to +8 V when Z_1 conducts and to −8 V when Z_2 conducts.

Other Types of Limiters

Bipolar transistors, JFETs, and MOSFETs can be designed to operate as limiters. To do this, bias must be applied to the device that will provide the type limiting desired. Limiting occurs when the device is cut off or saturated. Limiting of both peaks occurs in an overdriven amplifier.

Self-Check

Answer each item by performing the action indicated.

26. To identify a shunt limiter, we would look for the diode to be in _____ with the load.
27. The diode of a series positive limiter is _____ biased by the positive half-cycle of the input signal.

   a. reverse b. forward

28. The diode of a shunt positive limiter is _____ biased by the negative half-cycle of the input signal.

   a. reverse b. forward

29. A double-diode limiter limits _____ peak(s) of the input sine wave.

   a. either   c. both
   b. one

30. It is possible to use which of the following for limiting purposes?

   a. Zener diodes   c. bipolar transistors
   b. JFETs   d. all of these

Clampers

Before discussing **clampers** we must review RC circuits. Remember, series RC circuits are used extensively for coupling two electronic circuits. In the amplifier, we use a capacitor to block the DC present in the preceeding circuit from over-biasing the next stage. This was a form of clamping. With the amplifier input referenced to 0 V, the output is referenced to the value of collector voltage (V_C) that is used to establish the Q point. By using different time constants for the RC coupling circuit, we can determine how closely the shape of the output signal will be to the input. We must select a capacitor for the circuit that is capable of passing all the frequencies we wish to couple.

There are certain applications in electronics that require a signal be "clamped" to a specific DC level for use with a circuit. Clamping is:

> **The act of establishing a signal (either peak) to a preselected DC voltage level.**

Clamping does not change the shape or amplitude of the input signal. It only assures that the signal is referenced to the desired DC voltage level.

Positive Clampers

Figure 7-41 contains the schematic diagram for a positive diode clamper. The resistor (R_1) provides a discharge path for the capacitor (C_1).

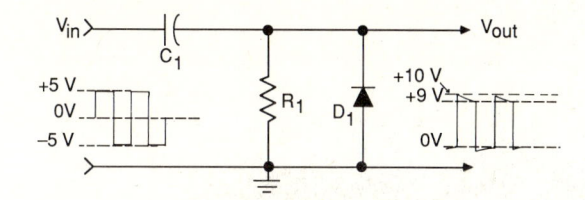

Fig. 7-41 Positive Clamper with Waveforms

The resistance of R_1 is large, making the time required for C_1 to discharge very long. The diode, however, provides a quick charge path for C_1. Once charged, the capacitor (C_1) acts as a power source.

Notice that the input applied to the circuit in Figure 7-40 is a square wave with a 0 V reference. This means that each alternation varies 5 V above or below the reference. The positive alternation swings to +5V and the negative −5V.

When the circuit is first turned on, C_1 quickly charges to −5 V with the first negative alternation. From this point, the capacitor voltage is in series with the +5 V of the positive alternation. When the positive alternation is present, the capacitor is discharging through R_1. The discharge that occurs during the time that the positive alternation is present is approximately 1 V, as shown on the positive peak of the output signal. During this period, the diode is reversed biased and does not conduct. This allows the entire 10 V to be coupled to the output, with the exception of the 1 V that is discharged through R_1. When the signal again goes negative, it is opposed by the 4 V of charge present on C_1. The diode is now forward biased and C_1 quickly charges to 5 V again, canceling out the -5 volts present at the input. This effectively clamps the output to a reference of 0 V with a peak amplitude of $+10$ V.

Variation of the time constant of R_1 and C_1 causes the amount of discharge of capacitor C_1 to vary during the period when the positive alternation of the input is present. In some cases, a quicker discharge may be desirable. Allowing C_1 to discharge faster does, however, cause the output signal to have greater distortion.

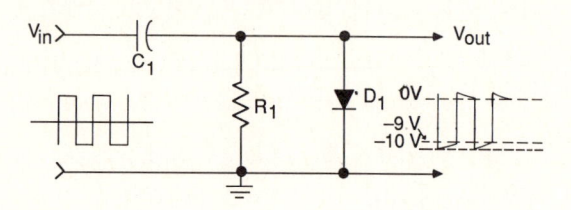

Fig. 7-42 Negative Clamper with Waveforms

Negative Clampers

In Figure 7-42, you see the circuit diagram for a **negative clamper**. Compare this with the positive clamper just discussed. Notice that the diode has its polarity reversed with respect to ground. Again, resistor R_1 supplies capacitor C_1 with a discharge path that requires a long discharge time. The diode, again, provides a fast charge path for C_1. After C_1 charges, it acts as a power source, which helps to determine the maximum and minimum voltages present at the output. In the case of the negative clamper, the output signal is clamped to 0 V but the signal at the output swings from 0 V to -10 V.

Other Clampers

As with the limiters discussed earlier, it is possible to bias the diode used in a clamper circuit and to "clamp" the output signal to some voltage other than 0 V. Figure 7-43 contains the schematic diagram for positive and negative clampers that have $+10$ V bias applied. Note that in each case the output signal is "clamped" to $+10$ V. For the positive clamper, the swing is from $+10$ V to $+60$ V, and for the negative clamper, the swing is from $+10$ V to -40 V. In each case, the input and output signals have the same amplitude and are "clamped" to the same DC level. The difference is that one provides a positive output signal and the other provides a negative output signal.

It is possible to reverse the bias applied to these two circuits and provide negative bias. When this is done, both signals are "clamped" to

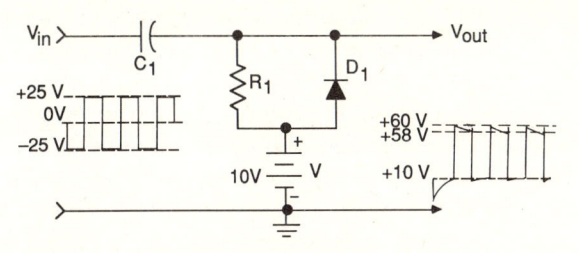

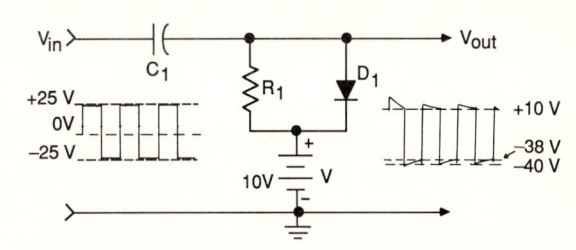

(a) Positive Clamper with 10 V Positive Bias

(b) Negative Clamper with +10 V Positive Bias

Fig. 7-43

−10 V. The positive clamper's output swings from −10 V to +40 V while the negative clamper provides a swing from −10 V to −60 V. From this you should be able to see that by selecting a specific bias, it is possible to clamp a signal to any DC reference.

Self-Check

31. Clampers are used to change the DC _____ voltage of a signal.
32. An unbiased positive clamper establishes its output signal at a _____ volt reference.
33. A negative clamper has +10 V bias applied. Its output signal will be "clamped" to _____ volts.
34. In a diode clamper, the capacitor discharges through the _____ and charges through the _____.
35. To have an output signal that has minimum distortion, the RC time constant of R and C must be such that the capacitor discharge takes a _____ time.

Summary

Rectification is the process of *converting an AC sine wave into pulsating direct current* (PDC). PN junction diodes are used as *rectifiers*. These diodes are unidirectional devices which allow current to flow in one direction only. If a diode is connected in series with an AC sine wave input, the diode's output current will be direct current that pulsates in the same polarity as the input alternation. This DC pulse can then be filtered to a state that approaches a straight line DC.

Three types of rectifier circuits are available. These are the half-wave rectifier, the conventional full-wave rectifier, and the full-wave bridge rectifier circuits.

A half-wave rectifier has one diode. The output of this diode is connected to some type of load device and the current that flows in the load device is pulsating DC. The polarity of the diode's connection determines whether the output pulse is negative or positive. The diode used in a half-wave rectifier will conduct during one alternation of each sine wave of the input. One output pulse is produced for each input sine wave, meaning that the ripple frequency equals the input frequency. Peak output voltage of a half-wave rectifier will equal the peak voltage applied to the diode's input. Half-wave rectifiers are used where high voltage is needed but current demands are low.

A conventional full-wave rectifier requires a center-tapped transformer secondary and two diodes for its operation. It produces one output pulse for each alternation of the input sine wave. This means that the ripple frequency of a full-wave rectifier is double the frequency of the input sine wave. Because of the center-tapped transformer secondary that is connected to the conventional full-wave rectifier, the voltage applied to each diode equals one-half of the peak voltage across the transformer secondary. Therefore, peak output voltage of the rectifier is one half of the peak voltage of the transformer secondary. Conventional full-wave rectifiers are used where voltage and current requirements are both moderate. Output pulse polarity is determined by the polarity of the diodes that are connected to the transformer.

The full-wave bridge rectifier does not require a center-tapped transformer but it can be identified by its four (4) diodes. Its peak output voltage is equal to peak secondary voltage. Ripple frequency is double the input frequency. This rectifier is used where both current and voltage requirements are high. Polarity of diode connection again determines the output polarity.

Filter circuits that contain capacitors and/or coils are used to reduce or remove the ripple component in the output of a rectifier. The result is a DC voltage that approaches the same level at all times. In fact, the filtered output of these rectifiers is suitable for most applications encountered in the electronic circuits of today. Filter circuits that were studied in this chapter were the L and Pi types. Of the two, the Pi filter provides the most stable output.

When voltage is first applied to a circuit, it causes a surge of current to flow through the circuit that could possibly damage the circuit's components. To reduce the possibility of such damage, surge suppressor modifications are added to the rectifier circuit.

Diodes are also used when voltages must have their amplitude limited or where the DC level of a pulsating signal must be changed. Limiters operate much like a rectifier except that they can be used to block or pass the peaks of all input pulses whose peaks exceed a predetermined value. Clampers have a capacitor connected into the circuit that charges to a DC voltage which is then used to provide a new DC reference for all arriving pulses. Both limiters and clampers can be operated with fixed bias applied and as either series or parallel connected circuits.

Review Questions and/or Problems

Answer each item by performing the action indicated.

1. A rectifier is a device that is capable of converting _____ to _____.

2. The characteristic that allows a diode to be used as a rectifier is that a diode will:

    a. not allow current to flow through it into a resistor.
    b. allow current to flow only when it is reverse biased.
    c. allow only hole current to flow.
    d. allow current to flow in one direction only.

3. Increasing forward bias across a diode:

    a. decreases both junction resistance and current flow.
    b. increases both junction resistance and current flow.
    c. increases junction resistance and decreases current flow.
    d. decreases junction resistance and increases current flow.

4. Refer to Figure 7-44. Which waveshape represents the output of this circuit?

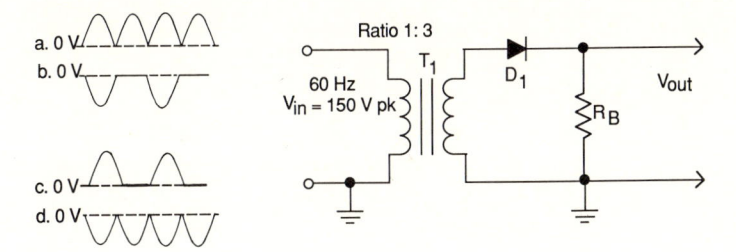

Fig. 7-44

5. Ripple frequency for the circuit shown in Figure 7-44 will be _____ PPS.

6. To reverse the output polarity of the rectifier circuit shown in Figure 7-44 you would _____.

7. Refer to Figure 7-44. The unfiltered average DC output voltage of this circuit equals _____ volts.

8. Refer to Figure 7-44. Peak output voltage of this circuit equals _____ volts.

9. Refer to Figure 7-44. Peak secondary voltage of this circuit equals _____ volts.

(a)

(b)

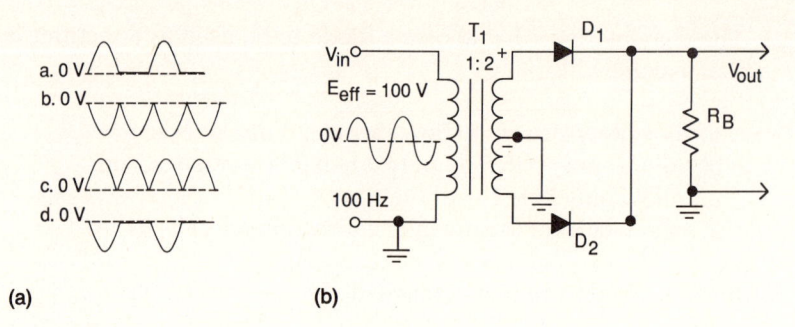

Fig. 7-45

10. Refer to Figure 7-45. Which waveshape represents the output of this circuit?

11. Ripple frequency for the circuit shown in Figure 7-45 will be _____ PPS.

12. To reverse the output polarity of the rectifier circuit shown in Figure 7-45, you would _____.

13. Refer to Figure 7-45. Average DC output voltage of this circuit equals _____ volts.

14. Refer to Figure 7-45. Peak output voltage of this circuit equals _____ volts.

15. Refer to Figure 7-45. Peak secondary voltage of this circuit equals _____ volts.

16. A diode opens in the circuit shown in Figure 7-45. Describe the output of the circuit.

17. Refer to Figure 7-46. Which waveshape represents the output of this circuit?

18. Ripple frequency for the circuit shown in Figure 7-46 is _____ PPS.

19. To reverse the output polarity of the rectifier circuit shown in Figure 7-46, you would _____.

Diode Circuits and Power Supply Filters

(a)

a. 0 V
b. 0 V
c. 0 V
d. 0 V

(b)

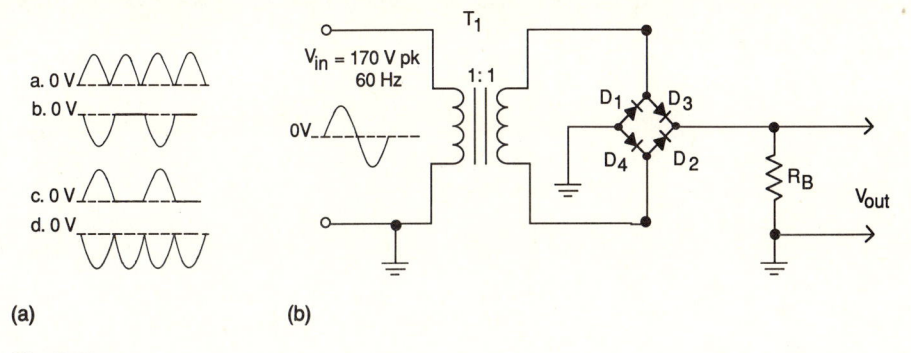

V_{in} = 170 V pk, 60 Hz, 1:1 transformer T_1, bridge rectifier with D_1, D_2, D_3, D_4, load R_B, V_{out}

Fig. 7-46

20. Refer to Figure 7-46. Average DC output voltage of this circuit equals _____ volts.

21. Refer to Figure 7-46. Peak output voltage of this circuit equals _____ volts.

22. Refer to Figure 7-46. Peak secondary voltage of this circuit equals _____ volts.

23. A diode opens in the circuit shown in Figure 7-46. Describe the output of the circuit.

24. A full-wave rectifier is supplying half-wave output voltage. What is the trouble?

    a. One diode is open
    b. Loose ground connection
    c. Both a and b
    d. Neither a nor b

25. A full-wave rectifier has zero output. This would be caused by which of the following?

    a. T_1 primary is open.
    b. D_1 is open.
    c. A load resistor is open.
    d. None of these.

26. A pi filter is a _____ input filter.

27. A capacitive filter will oppose any change in output _____.

28. Of the two LC filters discussed, the _____ filter provides the best filtering action.

29. What is the likely result if a diode shorts in one of the rectifier circuits?

30. Draw the output waveshape for the circuit shown in Figure 7-46. State the ripple frequency, then mark and label the following; zero volts, peak voltage, and average DC output voltage.

31. A negative limiter removes all or a part of the _____ half-cycle of the input. A positive limiter removes all or part of the _____ half-cycle.

32. The key to identification of a series limiter is the fact that the diode is _____ with the output signal.

33. In a series positive limiter, the diode is _____ biased by the positive half-cycle of the input.

34. In a shunt negative limiter, the diode is _____ biased by the positive half-cycle of the input.

35. The double-diode limiter includes a series of positive and a shunt negative limiter.

   a. True   b. False

36. All limiters are designed using diodes.

   a. True   b. False

37. A positive clamper without bias clamps the _____ peak of the output to 0 volts, and a negative clamper clamps the _____ peak to 0 volts.

   a. positive, positive   c. negative, negative
   b. positive, negative   d. negative, positive

38. A positive clamper with positive bias clamps the _____ peak of the output to a _____ voltage.

   a. positive, positive   c. negative, negative
   b. positive, negative   d. negative, positive

39. A _____ clamper with _____ bias clamps the negative peak of the output to a negative voltage.

    a. positive, positive
    b. positive, negative
    c. negative, negative
    d. negative, positive

40. A _____ clamper with _____ bias clamps the positive peak of the output to a positive voltage.

    a. positive, positive
    b. positive, negative
    c. negative, negative
    d. negative, positive

8 Transistor Amplifiers

Objectives

1. Define the terms *input impedance* and *output impedance*, and explain the effect that each has on the operation of transistor amplifiers.
2. Explain the actions that must be taken to assure that a transistor amplifier has a **frequency response** suitable for the signal that is to be amplified.
3. Define the term *AC load line* and explain the part that it plays in the design of a transistor amplifier.
4. Compare the following parameters for *common-emitter, common-base* and *common-collector* transistor amplifiers.

   a. Input impedance
   b. Output impedance
   c. Voltage gain
   d. Frequency response
   e. Power gain
   f. Current gain
   g. Phase shift
   h. Low frequency cut off
   i. High frequency cut off

Introduction

In Chapters 4 and 5 we discussed the DC characteristics of transistor amplifiers. At that time, we discussed the effect that application of DC voltages have on a transistor's operation.

Three transistor amplifier configurations–*common-base, common-collector* and *common-emitter*–were discussed. These circuits were analyzed from the DC aspects of their operation. All configurations were analyzed using the *approximation* method and methods were developed that allowed us to predict the transistor's operation.

At that point, we intentionally avoided discussing the effect of AC signals on transistors. Actually, very few transistor amplifier circuits are used that do not amplify AC waves. In electronics, we call the AC wave that we work with the *signal* (i.e., input signal, output signal, etc.).

In this chapter it is assumed that each student has completed the study of DC circuits and is ready to examine AC phenomena. This chapter will, therefore, concentrate on the effects of AC signals on transistor amplifiers which are biased to a specific Q point.

Effect of Changes in Resistance

By way of an introduction to amplification, let us discuss the effect that changes in resistance can have on voltage and current. Observe Figure 8-1. Notice that R_1 is a rheostat and R_2 is a fixed resistor. They form a voltage divider that has 25 V applied. With R_1 adjusted as shown in Figure 8-1, R_t of the circuit equals 25 kΩ and I_t equals 1 mA. The voltage between point D and ground is 10 V. When the arm of R_1 is moved upward, to point B, total resistance has decreased to the value of R_2, and $I_t = 1.67$ mA. In this case, the voltage at point D with respect to ground has decreased to zero. You should be able to see that by changing the resistance of R_1, the voltage at point D can be varied through a range of 0–10 V. If R_1 is adjusted so that the voltage at point D is 5 V (Q point), it is possible to vary the voltage 5 V positive to 5 V negative, when using the Q point as the reference. The variations that occur are directly proportional to the resistance value of R_1. As the resistance of R_1 increases, the voltage at point D increases. When the resistance of R_1 decreases, the voltage at point D decreases. In Figure 8-2, R_1 has been replaced by a transistor.

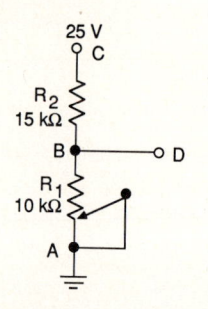

Fig. 8-1 Equivalent Resistive Circuit

Remember, in a transistor a small change in base current (I_B) causes a large change in collector current (I_C). This large change in I_C results in a large change of the internal resistance of the transistor. In other words, I_B is used to control the resistance of the transistor much like you would manually control the rheostat in Figure 8-1. A transistor operates much like a variable resistor. If I_B is small, the resistance of the transistor is large and the voltage at point B is large. If I_B is very large, the voltage at point B will be very small. This again shows that a change in I_B controls the voltage at point B. Should a base current be selected that sets the voltage at point B to mid-range between 0 V and V_{CC} (25 V),

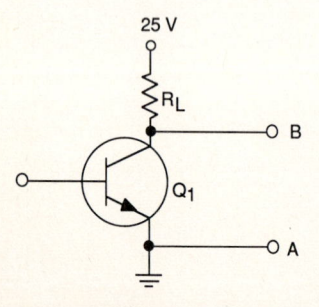

Fig. 8-2 Simplified Amplifier Circuit

then variations in I_B will cause the voltage at point B to vary above and below its stable (quiescent) voltage.

Another way to look at the effect that base current (or voltage) has on the circuit can be shown as follows. In this circuit, (Figure 8-3) three different voltages can be applied as forward bias. With the switch at A,

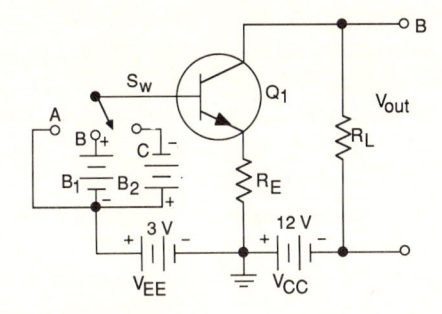

Fig. 8-3 Simplified DC Bias Circuit

the forward bias applied to the eb junction is $(3\text{ V} - V_{RE})$, as applied by V_{EE}. When the switch is moved to position B, V_{EE} and B_1 are connected series aiding. This increases the forward bias applied to the base of Q_1 causing I_B, I_E, and I_C to increase. When I_C increases, R_L will drop more voltage leaving less to be dropped between point B and ground; therefore, the output voltage has decreased. If the switch is moved to C, V_{EE} and B_2 are connected series opposing. This causes forward bias to decrease which, in turn, causes I_B, I_E, I_C, and V_{RE} to decrease. When I_C decreases, V_{RL} decreases, which leaves more voltage to be dropped between point B and ground. The result is that the output voltage has increased. Note that as the bias voltage is increased, the output voltage is decreased, giving a 180° phase shift between the base of the transistor and its collector output. Other types of amplifiers will not have 180° phase shift, but their operation is similar.

In Figure 8-4, placing the switch at A causes the transistor to conduct at a desired DC level. Current flow will be steady DC as long as

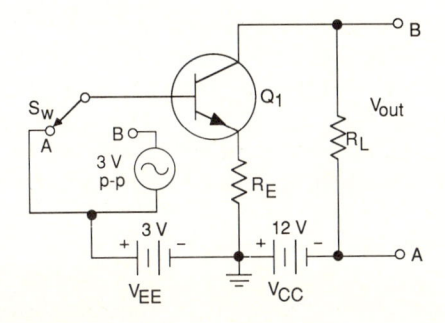

Fig. 8-4 Simplified AC/DC Circuit

the forward bias does not change. When Sw_1 is moved to position B, you not only select the DC voltage that sets the DC conduction level, but you also select an AC signal that varies forward bias above and below the stable (quiescent) bias level. During the signal's positive alternation, V_{EE} and the generator are series aiding, and forward bias is larger than the fixed DC bias. During this time, I_B, I_E, and I_C are conducting more heavily than with DC bias only. This causes the voltage at point B to decrease. When the negative alternation of the generator is present, forward bias decreases causing I_B, I_E, and I_C to decrease. This causes the voltage at point B to increase. Since the AC input voltage (or current) is constantly changing, so is the voltage at point B. In fact, if the forward bias is selected correctly and the output of the generator is controlled very closely, the output voltage will be an exact reproduction of the input voltage except that it will change over a much greater range of voltages and will be 180° out of phase with the input. In this way, we can apply a small AC signal to the base of the transistor and amplify it to a much larger AC of the same frequency in the output.

Figure 8-5 represents the same circuit as Figure 8-4 except that the transistor quiescent point is stabilized by a single battery biasing network. The generator supplying the input voltage or current most likely would be another amplifier. This is the way in which amplification is accomplished. Amplifier circuits are designed in one of three configurations: common-emitter, common-base, and common-collector. Capacitor (C_C) is placed in the input to isolate the different DC voltages that are present at different points within the circuit.

The remainder of this chapter is used to expand on the AC characteristics of transistor amplifiers. Remember, when AC is used, inductive and/or capacitive reactances enter the picture. With both AC and DC, we now have to consider frequency response and impedance.

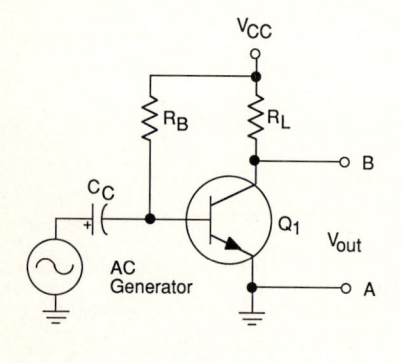

Fig. 8-5 Amplifier Circuit

Input Impedance

During your study of alternating current (AC) fundamentals, you learned the characteristics of *impedance*. You found that when AC is applied to a circuit, other factors (X_C and X_L) must be considered. Total opposition to current flow now results from R, X_C, and X_L. Remember:

Inductors oppose all changes in current

and:

Capacitors oppose all changes in voltage.

You also know that:

As frequency increases, inductive reactance (X_L) increases.

and that:

As frequency increases capacitive reactance (X_C) decreases.

Capacitance is present within a circuit any time two current-carrying conductors are separated by an insulator. Because air is an insulator, capacitance is distributed throughout an operating circuit. Inductance refers to the magnetic characteristics of a circuit. For inductance to occur, it is necessary to have a magnetic field, a conductor, and relative motion between the two. Any circuit that has variations in current flow will have inductance. All DC circuits that have variations in current flow contain both capacitance and inductance.

Impedance is defined as the total opposition that a circuit offers to the flow of alternating current. Impedance is a combination of *resistance, capacitive reactance*, and/or *inductive reactance*. From this explanation you can see that:

**Impedance is present in all circuits
that have variations in current flow.**

In transistor amplifier circuits, impedance analysis can be divided into two categories. A signal arriving at the input of the circuit meets an opposition that is present within the circuit. This opposition is referred to as **input impedance** (Z_{in}). The signal is processed by the amplifier and then presented to the output circuit. The object is to couple the signal to another circuit or device that will process it; an amplifier or speaker, for example. The characteristics of this output circuit are such that they present a completely different impedance to the next stage. This impedance is called **output impedance** (Z_{out}). When you buy a stereo system, the specifications tell you to connect it to speakers of specific ohmic value (often 8 Ω). The final amplifier has been designed to have an *output impedance* of approximately 8 Ω. The speaker that you buy has been designed to have an *input impedance* of 8 Ω.

Remember:

**For maximum power to transfer, the input impedance of the
load must equal the output impedance of the source.**

Theoretically, to obtain maximum efficiency from our circuits we must match (as nearly as possible) the output of one circuit (or device) to the input impedance of the next circuit (or device). However, the cost of operation of solid-state amplifiers is quite low. Because of this, efficiency of operation is sacrificed to obtain better performance in other areas. Amplifier impedance is important, however, and is discussed in this section.

First, we will discuss amplifier input impedance. Later we will discuss amplifier output impedance.

Figure 8-6 contains a common-emitter (CE) amplifier that has voltage divider bias. A capacitor (C_1) serves as the coupling capacitor.

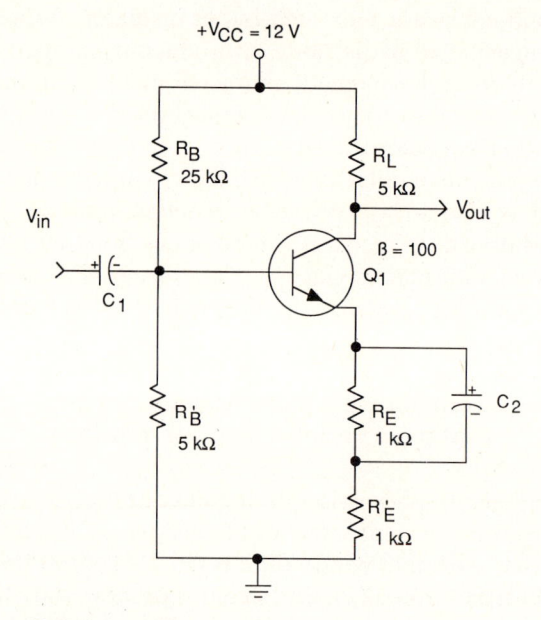

Fig. 8-6 Common-Emitter Amplifier

The transistor is an NPN transistor with a load resistor (R_L), two emitter resistors (R_E and R_E'), and an **emitter-bypass** capacitor (C_2).

The two capacitors that appear in this circuit have not yet been discussed. Coupling capacitor (C_1) is placed in the position shown to block any DC voltage that might be present at V_{in} from being applied to the base of Q_1 as bias. The presence of this voltage could saturate the transistor, causing it to stop acting as an amplifier. Using this arrangement, the voltage divider provides the bias necessary to establish the circuit's Q point. C_1 will charge to the DC difference in potential that exists between its right and left sides. Assume that the quiescent voltage on the left side is 4 V, and the bias applied at the right side is 1.5 V. The capacitor charges to 4 V − 1.5 V, or 2.5 V. The difference in potential is 2.5 V and the capacitor will charge to 2.5 V. After charging, C_1 acts like:

An open to DC, but will pass AC.

C_2 will charge to the voltage drop of R_E during the first instant of operation. After that, all DC current will flow through the resistor as the capacitor acts like an open to DC. When a change in current flow appears at the eb junction, the capacitor will attempt to charge or discharge to a new voltage level. If the capacitor is selected to have a low X_C to the frequency of operation, it will act like a short to that frequency. The capacitor allows the changing current to conduct and prevents the voltage drop on R_E from changing. R_E' is not bypassed with a capacitor and it will conduct all of the current change; therefore, $V_{R_E'}$ will vary with changes in current while V_{RE} remains relatively stable.

Capacitor C_1 blocks all DC voltage present at V_{in} and provides a path for AC. This is done by selecting a capacitor (C_1) whose X_C will be low to all desired input frequencies. Other characteristics of C_1 will be discussed in later lessons.

As explained above, any AC signal arriving at the left of C_1 sees an input impedance (Z_{in}) that opposes its movement. Assume that capacitor C_1 was selected because it has low reactance to all frequencies that are to be amplified. The discussion that follows will assume that $X_{C1} = 0\ \Omega$. The effect of having C_1 in the input circuit will be discussed later. Arriving at the right side of C_1, the AC signal has three paths through which it can flow:

1. through R_B to V_{CC}
2. through R_B' to ground
3. through the emitter-base junction to ground.

Remember, the emitter-base junction is forward biased, which causes it to reach a stable condition level as soon as the Q point is established. Once conduction starts, the eb junction continues to conduct until forward bias decreases below the cutoff level.

The first of the AC paths to be examined is the one through the emitter element. The AC resistance of this element is referred to by the symbol (r_e). Note: small letters (r_e) represent AC resistance.

The mathematical calculations involved in determining the r_e of a transistor are quite difficult and complicated. In our studies to this point, we have used approximations for our calculations. For approximation purposes, these mathematical analyses have been used to develop standards for transistors that allow us to approximate r_e. This is possible because transistors have much the same physical characteristics. You should, however, understand that this is a rough approximation that is suitable for circuits that operate at room temperature. In actual circuits, r_e varies with changes in current and temperature. To be absolutely correct, a separate r_e should be calculated for each circuit using AC load lines and input signals. Characteristic curves are not always available, though, and the calculations are not possible.

We will use 25 mV as our standard for the voltage drop across the AC resistance of the emitter.

With this voltage, the AC resistance of a transistor's emitter can be determined by use of the formula:

$$r_e = \frac{0.025 \text{ V}}{I_E}$$

In the r_e formula I_E represents the emitter current that flows from ground to emitter. This path (Figure 8-6) contains a capacitor bypassed R_E, with C_2 having low reactance to the frequency being amplified. We can ignore R_E when considering the AC path. R'_E, however, is not bypassed which means that R'_E is in the path of all current, AC and DC. To calculate total resistance that would be in the path of the signal current, we must first find the r_e of the transistor. The formula listed above, can be used as follows:

> **NOTE:** All resistances that exist in the emitter circuit of a CE amplifier that are not bypassed by a capacitor appear at the base as beta + 1 larger than their actual value. The reason for this is that emitter current is equal to beta plus one times base current $I_E = (\beta + 1)(I_B)$.

To complete the solution for r_e, we need values for V_E and I_E. Remember: $V_B \cong V_E$; therefore:

$$V_B = \frac{R'_B}{R_B + R'_B} \times V_{CC}$$

$$V_B = \frac{5 \text{ k}\Omega}{30 \text{ k}\Omega} \times 12 \text{ V} = 2 \text{ V}$$

Therefore, $V_E = 2$ V.
To find I_E:

$$I_E = \frac{V_E}{R_E}$$

$$I_E = \frac{2 \text{ V}}{2 \text{ k}\Omega} = 1 \text{ mA}$$

Now we know $I_E = 1$ mA, we can solve r_e as follows:

$$r_e = \frac{25 \text{ mV}}{I_E}$$

$$r_e = \frac{0.025 \text{ V}}{1 \text{ mA}} = 25 \text{ }\Omega$$

You can see that the application of this formula results in r_e of approximately 25 Ω. Now that we have r_e, we can solve for input resistance (R_{ib}).

$R_{ib} = (\beta + 1)(r_e + R_E')$

$R_{ib} = (101)(1025) = 103{,}525\ \Omega$ or $103.5\ k\Omega$

The AC input signal divides and flows through two other paths. One path is through R_B and the other is through R_B'. If the current divides to flow through two paths, the paths must be treated as parallel branches.

To calculate the equivalent resistance (R_{eq}) of this circuit, we use the same formula as always. This formula and its solution is as follows:

$R_{eq} = \dfrac{R_B \times R_B'}{R_B + R_B'}$

$R_{eq} = \dfrac{5\ k\Omega \times 25\ k\Omega}{5\ k\Omega + 25\ k\Omega}$

$R_{eq} = \dfrac{125\ M\Omega}{30\ k\Omega} = 4.16\ k\Omega$

Now we can calculate Z_{in} using R_{eq}, R_{ib} and this formula:

$Z_{in} = \dfrac{R_{eq} \times R_{ib}}{R_{eq} + R_{ib}}$

$Z_{in} = \dfrac{4.16\ k\Omega \times 103.5\ k\Omega}{4.16\ k\Omega + 103.5\ k\Omega}$

$Z_{in} = \dfrac{430.56\ M}{107.56\ k}$

$Z_{in} = 3999\ \Omega \cong 4\ k\Omega$

This is the **input impedance** (Z_{in}) when part of R_E (1 kΩ) is not bypassed with a capacitor.

Should this same circuit be operating with both emitter resistors being bypassed, the input impedance can be calculated as follows: (Note that r_e is the only resistance in the path of I_E.)

$R_{ib} = (\beta + 1)(r_e)$

$R_{ib} = (101)(25) = 2525\ \Omega$

$R_{eq} \cong 4.16\ k\Omega$ was solved above; therefore:

$$Z_{in} = \frac{R_{eq} \times R_{ib}}{R_{eq} + R_{ib}}$$

$$Z_{in} = \frac{4.16 \text{ k}\Omega \times 2.525 \text{ k}\Omega}{4.16 \text{ k}\Omega + 2.525 \text{ k}\Omega}$$

$$Z_{in} = \frac{10.504 \text{ M}\Omega}{6.685 \text{ k}\Omega}$$

$$Z_{in} = 1571 \, \Omega$$

Self-Check

Answer each question by performing the action indicated.

1. Small changes in I_B will cause a large change in ____ and ____.
2. The internal resistance of a conducting transistor is ____ proportional to I_C.
3. When the forward bias applied to a transistor is varied, the transistor acts much like a variable ____.
4. When I_C, in a common-emitter amplifier increases, the output (V_C)

   a. increases.   b. decreases.

5. Refer to Figure 8-6. What purpose does the inclusion of C_1 and C_2 serve in the operation of this circuit?
6. To calculate the approximate AC resistance of an emitter, you would divide 25 mV by ____.
7. Any resistance that appears in the emitter circuit of an amplifier, and that is not capacitor bypassed, will appear at the base as ____ larger than their actual value.
8. To calculate the input impedance (Z_{in}) of a common-emitter amplifier, you must find the equivalent resistance of ____, ____ and ____ in parallel.

Frequency Response

As previously explained, when capacitance and inductance are present in a circuit, that circuit has different characteristics from the DC circuits discussed earlier. Inductance and capacitance react to each frequency differently. We know that for a given capacitor:

The higher the frequency, the smaller X_C becomes.

Transistor Amplifiers

Examine Figure 8-7. Note that this circuit operates identically to the circuit in Figure 8-6. Two fixed capacitors have been placed in this circuit. When these capacitors were selected, it was necessary to take into consideration how they would react to the input signal. Remember, capacitors act like an open to any DC component. The main concern is how they will affect the AC signal. At low frequencies, a capacitor's X_C is high but at some higher frequency, X_C becomes so low that it can be considered a short. Capacitors are selected that have little or no effect on the frequency of operation (f_o). This assures that DC operation is maintained and that the AC signal meets minimum opposition.

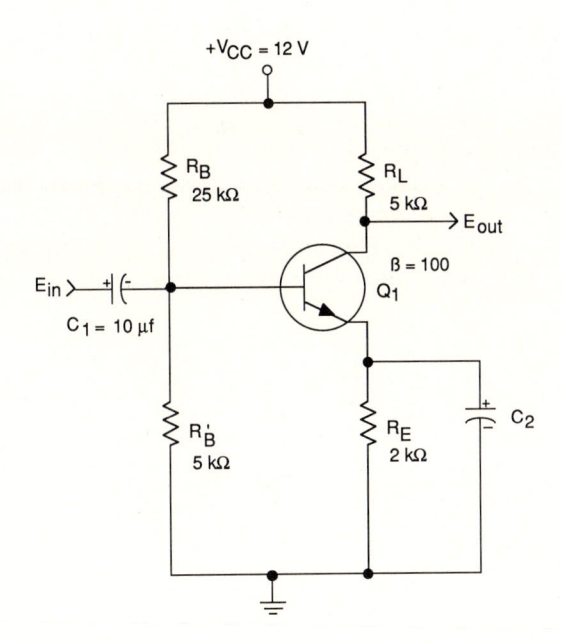

Fig. 8-7 Common-Emitter Amplifier

When calculating a circuit's low frequency response or low frequency cutoff (f_{co}), we use the following standard:

Low frequency cutoff occurs when $X_{C1} = Z_{in}$.

In the last section, we found that the Z_{in} for the completely bypassed R_E was approximately 1571 Ω. Using that value and the formula that follows, we can calculate f_{co} of this circuit.

$$f_{co} = \frac{1}{2 \times \pi \times z_{in} \times C_1}$$

Then:

$$f_{co} = \frac{1}{2 \times 3.14 \times 1571 - 0.00001}$$

$$f_{co} = \frac{1}{9.871 \times 10^{-2}}$$

$$f_{co} = 10.1 \text{ Hz}$$

Using a 10 μf capacitor as C_1, this circuit can be expected to amplify signals that have a frequency above 10.1 Hz.

Changing C_1 to a .01 μf capacitor will change the low frequency cut off to 10.1 Hz. An audio system that would not reproduce sounds below 10.1 Hz would be practically useless. You can see the importance of selecting the correct capacitor for use as the input capacitor.

Upper Frequency Cut Off

At some high frequency, the circuits distributed (parallel) capacitance reaches a frequency where its X_C becomes low. When the combined impedance at the output causes the signal amplitude to decrease to 0.707 × peak output, the signal is no longer usable. This point is called the upper-half power point.

When you select a transistor for use and examine the manufacturer's specifications, you have available to you a specification that can be used to determine upper frequency cut off. This parameter is identified as f_α for a common-base amplifier or f_β for a common-emitter amplifier. Another parameter that is supplied as part of the specifications is AC beta (β_{AC}). To convert between these two parameters, use the formulas that follow:

$$f_\alpha = \beta_{AC} \times f_\beta$$

or:

$$f_\beta = \frac{f_\alpha}{\beta_{AC}}$$

Transistor Amplifiers

Veatch (1977) and Tocci (1982) go even further in their explanations of upper frequency cut-off (Uf_{co}). They state that the upper frequency is also influenced by the voltage gain (A_V) of the amplifier. You should also remember that A_V is influenced by whether R_E is capacitor bypassed, or not. The procedure for calculating upper frequency limit using this approach is:

Assume $f_\alpha = 1\,\text{MHz}$, $\beta_{AC} = 50$, and $A_V = 40$.

$$f_\beta = \frac{f_\alpha}{\beta_{AC}} = 20\,\text{kHz}$$

Then:

$$Uf_{co} = f_\beta + \frac{f_\alpha}{A_V}$$

$$Uf_{co} = 20\,\text{kHz} + \frac{1\,\text{MHz}}{40} = 45\,\text{kHz}$$

Using this approach and the conditions stated above, the actual upper frequency limit is 45 kHz for this common-emitter amplifier.

AC Voltage Gain

We can approximate how much the input signal will be amplified by use of the formula:

$$A_V = \frac{R_L}{r_e + R_E}$$

> **NOTE:** R_E represents all unbypassed emitter resistance.

> **NOTE:** Amplification is also called *gain* and voltage gain is indicated by the symbol A_V. This approximation uses only those emitter resistors that are not bypassed with a capacitor. Also, R_L refers to the collector (load) resistor and r_e (for a germanium transistor) equals 0.025 (25 mV) divided by I_E.

The circuit in Figure 8-8 does not have an emitter resistor. In this circuit AC voltage gain equals:

$$A_V = \frac{R_L}{r_e}$$

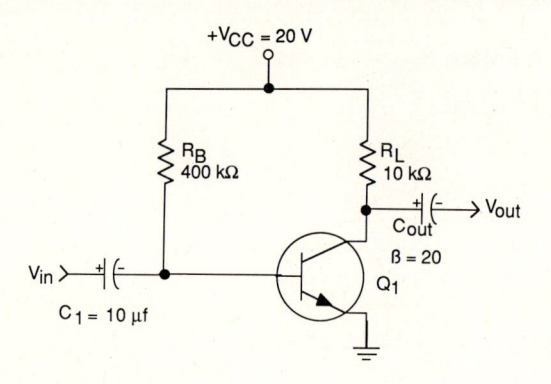

Fig. 8-8 Common-Emitter Amplifier

To solve for A_V, we must first calculate I_E. This can be calculated as follows:

$$I_B = \frac{V_{CC}}{R_B} = \frac{20\,V}{400\,k\Omega} = 0.05\,mA$$

Remember, $I_C = \beta \times I_B$

$$I_C = \beta \times I_B = 20 \times 0.05\,mA = 1\,mA$$

You have already learned that $I_E \cong I_C$. From this we can assume that $I_E = 1\,mA$ and then use this value to solve for r_e:

$$r_e = \frac{25\,mV}{I_E} = \frac{25\,mV}{1\,mA} = 25\,\Omega$$

Then:

$$A_V = \frac{R_L}{r_e}$$

$$A_V = \frac{10\,k\Omega}{25\,\Omega} = 400$$

Note that the circuit shown in Figure 8-8 has an AC voltage gain of 400 to 1. This means that for each 1 mV applied to the input, the output voltage changes by 400 mV.

Figure 8-9 contains the schematic diagrams for two common-emitter amplifier circuits. Each of these will be compared to the circuit in Figure 8-8. Figure 8-9(a) is biased by a voltage divider arrangement.

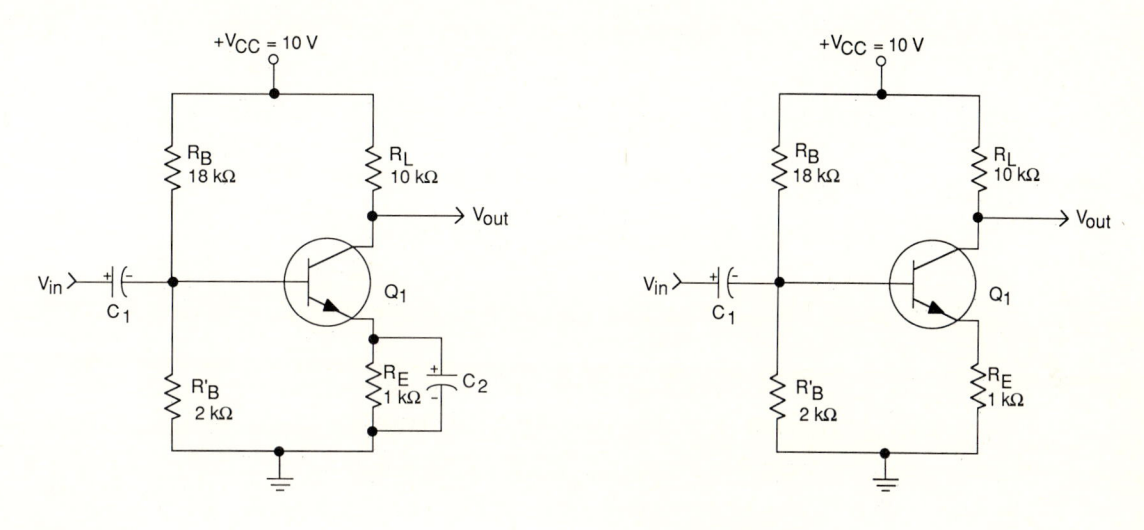

(a) R_E Bypassed

(b) R_E not Bypassed

Fig. 8-9 Comparison of Voltage Gains

In this circuit, all emitter resistors are bypassed by capacitors, leaving the AC resistance (r_e) as the only resistance in the input circuit. A DC analysis reveals that this circuit has an emitter current of 1 mA. Using this current and the procedures shown below, we can solve for the r_e of this circuit.

$$r_e = \frac{25\text{ mV}}{I_E} = \frac{25\text{ mV}}{1\text{ mA}} = 25\,\Omega$$

then:

$$A_V = \frac{R_L}{r_e}$$

$$A_V = \frac{10\text{ k}\Omega}{25\,\Omega} = 400$$

In Figure 8-9(b), the emitter resistor is not bypassed; therefore, the calculation for voltage gain is:

$$r_e = \frac{25\text{ mV}}{I_E} = \frac{25\text{ mV}}{1\text{ mA}} = 25\,\Omega$$

and:

$$A_V = \frac{R_L}{r_e + R_E} = \frac{10\text{ k}\Omega}{1025\,\Omega} = 9.75$$

When the calculations are compared for the two circuits, you can see that a bypassed R_E has a large effect on the voltage gain of the circuit. In this example, voltage gain decreased from 400 (R_E bypassed) to 9.75 for the unbypassed R_E.

Self-Check

Answer each item by inserting the word or words required to make each statement a true statement.

9. At high frequencies, the coupling capacitor will act as a/an _____ to the incoming signal.
10. To calculate the low frequency cut off of an amplifier, you must use the circuit's input _____ and the _____ of the coupling capacitor.
11. In audio systems, the upper frequency cut off must be at least _____ kHz.
12. AC voltage gain can be increased by placing a _____ capacitor around any emitter _____.
13. A common-emitter amplifier is operating with a 2 V peak-to-peak input signal, and the amplifier has a gain of 50. The output peak-to-peak amplitude equals _____.
14. AC voltage gain of an amplifier can be increased by increasing the size of _____.

AC Load Lines

In Chapter 4, we discussed DC load lines and the part they play in understanding transistors and transistor amplifiers. Remember, the V_{CC} point was located along the bottom axis of a transistor characteristic curve chart and the maximum I_C point was located along the left vertical axis of the chart. These two points were connected by a line that represented an infinite number of operation points for the circuit operating with this transistor, V_{CC} and I_C. We, then, located the point along the line where collector voltage (V_C) and collector current (I_C) intersected. This point represented the quiescent (operating or Q) point for that circuit. You also learned that the circuit would operate indefinitely with the same I_C and V_C. To change these parameters it was necessary to change either the size of R_L or the amount of V_{CC}.

A DC load line serves one purpose only: to identify the Q point that exists for a specific amount of I_C and V_C. However, an amplifier would be of little use if it operated with the same amount of current all of the time. As stated earlier in this chapter, electronic signals (audio and others) have alternating current characteristics. When the AC signal is applied to the input of an amplifier that is operating at its DC Q point, forward bias present at the eb junction is varied. Variations in V_{eb} cause base current (I_B), collector current (I_C), collector load resistor voltage (V_{RL}) and collector voltage (V_C) to change.

When an AC signal is applied to the transistor amplifier's input, several changes occur. To better understand these changes an **AC load line** is constructed. Constructing an AC load line is easier if the DC load line is constructed first. The DC load line can then be used in the construction of the AC load line.

Figure 8-10 shows a CE amplifier that is operating with fixed bias. For this explanation, a resistor has been added that represents the input resistance of the next stage or device. This is the impedance seen by the signal arriving at the output of the amplifier. This representation is for explanation purposes only and will not appear in schematics.

Your first job is to establish the Q point for this circuit. To do that, you must construct a DC load line. To assist you with understanding

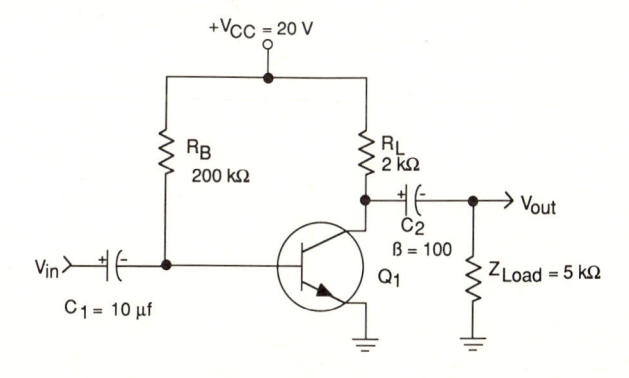

Fig. 8-10 Common-Emitter Amplifier-Circuit for a Load Line

transistor parameters, see Appendix B. This appendix contains specification sheets for NPN and PNP transistors suitable for use with this circuit. The DC load line is constructed as follows: (Use Figure 8-11 as a reference).

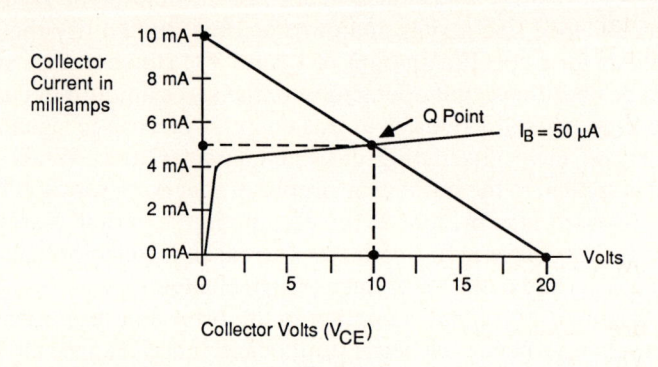

Fig. 8-11   Common-Emitter Amplifier DC Load Line

1. Identify V_{CC}. For the circuit in Figure 8-10, V_{CC} equals 20 V. Locate the 20 V position along the horizontal axis of Figure 8-11.
2. Now calculate I_C MAX:

$$I_{CMax} = \frac{V_{CC}}{R_L}$$

For our example:

$$I_C \text{ Max} = \frac{20 \text{ V}}{2 \text{ k}\Omega} = 10 \text{ mA}$$

Now locate the point along the vertical axis that corresponds to 10 mA I_C.

3. Connect the two points that you have identified. The DC load line is now complete.

To identify the circuit's Q point, you must determine the amount of I_B that is actually flowing. Earlier in this chapter, you learned that base current I_B can be calculated as follows:

$$I_B = \frac{V_{CC}}{R_B}$$

$$I_B = \frac{20 \text{ V}}{400 \text{ k}\Omega} = .05 \text{ mA}$$

Now that you know I_B, its value can be used to determine I_C and V_C. Refer to the Q point on Figure 8-11. The Q point marks the point where the I_B bias line crosses the load line. Note that two dashed lines extend away from the Q point. The horizontal line intersects the vertical (I_C) axis, indicating that $I_C = 5$ mA. The vertical line intersects the V_C (horizontal) axis at the $V_C = 10$ V point. Using this value for V_C, you can calculate V_{RL}:

$$V_{RL} = V_{CC} - V_C = 20 \text{ V} - 10 \text{ V} = 10 \text{ V}$$

You now know that the DC operating conditions are such that $I_C = 5$ mA and $V_{CE} = 10$ V. Notice that dashed lines for these values are shown on Figure 8-11, as well as the Q point.

When an AC input is applied to the circuit, there is a slightly different load line that represents the AC operation of this circuit (Figure 8-10). While this circuit is operating with the DC applied, C_2 charges to V_{CE} (10 V) with the polarity shown on the drawing. If an AC input signal drives the transistor into saturation, 10 mA will flow as a result of the circuit's DC characteristics (V_{CC} and R_L). When the transistor saturates, it acts as if it is a short. With Q_1 saturated, its low resistance presents little opposition to the capacitor's (C_2) discharge. The discharge path for C_2 is: from its negative side, down through Z_{Load} up through Q_1 and to the positive plate of C_2. As the capacitor discharges, the current that flows becomes part of I_C. The amount of this current can be calculated by dividing the amount of charge on C_2 (10 V) by the Z_{Load}. With the transistor shorted, Z_{Load} is the only opposition present in the discharge path of C_2. Therefore:

$$I_{ZLoad} = \frac{V_C}{Z_{Load}}$$

$$I_{ZLoad} = \frac{10 \text{ V}}{5 \text{ k}\Omega} = 2 \text{ mA}$$

Remember, maximum DC current equals 10 mA. Now we have found that it is possible for the collector current to increase as much as 2 mA due to the discharge of C_2. This means that maximum I_C, with AC applied, can be:

$$I_C = I_{DC} + I_{Load} = 10 \text{ mA} + 2 \text{ mA} = 12 \text{ mA}$$

This current (12 mA) is the maximum current used for plotting the AC load line. Use Figure 8-12 as a reference for the following explanation.

First, we will calculate the maximum voltage that could be dropped across Q_1, with an AC signal applied. Remember, Q_1 drops its largest voltage when cut off. With Q_1 *cut off*, $V_C = V_{CC}$ and $I_C = 0$.

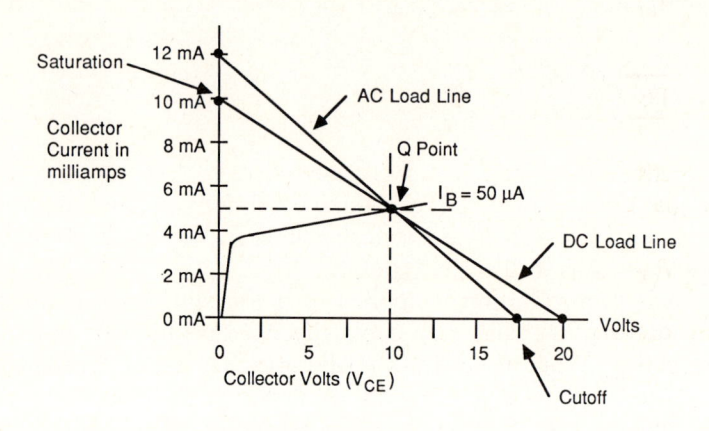

Fig. 8-12 Common-Emitter Amplifier Load Lines

Therefore, any current that flows must pass through Z_{Load}, C_2, R_L and to V_{CC}. Two voltages are present in this path, V_{CC} and V_{C2}. Note that V_{CC} is positive and that V_{C2} is charged to 10 V with its left side being positive (nearest to V_{CC}). With $+V_{CC}$ and V_{C2} applied to the same capacitor, we have two opposing voltages. The opposition of these two voltages means that we actually have 10 V to support current. This calculation is shown below:

$$V_{CE} = V_{CC} - V_{C2} = 20\,V - 10\,V = 10\,V$$

This means that the path through C_2, actually has 10 V applied. Voltage (V_{C2}) must be divided across R_L and Z_{Load}. To calculate the amount of voltage that each will drop, we must first calculate R_t.

$$R_t = R_L + Z_{Load} = 2\,k\Omega + 5\,k\Omega = 7\,k\Omega$$

We can now calculate I_{RL} as follows:

$$I_{RL} = \frac{V_C}{R_t}$$

$$I_{RL} = \frac{10\,V}{7\,k\Omega} = 1.43\,mA$$

Remember:

None of this current flows through the transistor.

Using I_{RL} we can calculate V_{RL}.

$V_{RL}^* = I_{RL} \times R_L = 1.43 \text{ mA} \times 2 \text{ k}\Omega = 2.86 \text{ V}$

> **NOTE:** * means AC value.

Subtracting this voltage from V_{CC}, we find that the maximum voltage drop that can occur on the collector is:

$V_C = V_{CC} - V_{RL}^* = 20 \text{ V} - 2.86 \text{ V} = 17.14 \text{ V}$

After locating this point on the horizontal axis of Figure 8-12, we can connect the two points (I_C^* and V_{CE}^*) and complete the AC load line. Both the DC and AC load lines are shown on Figure 8-12. Note that both load lines have the same Q point.

Self-Check

Answer each item by performing the action indicated.

15. To plot the DC load line for a specific amplifier circuit you must first calculate ____ and ____.
16. An AC and a DC load lines will share a common ____.
17. The Q point of an AC load line is the point where ____ and ____ intersect.
18. When compared to the DC load line, the I_C Max for an AC load line will be slightly ____ than that of the DC load line.

   a. higher   b. lower

Output Impedance

In an earlier section of this chapter, we discussed the fact that a circuit presents an impedance to any input signal. We referred to this as **input impedance**. We also reviewed the importance of impedance matching and restated the law:

> For maximum power to be transferred, load impedance must equal the internal impedance of the source.

In many electrical applications, the source and the load are both amplifier circuits. For best efficiency, therefore, it is necessary to match the impedance of the two amplifiers to assure that maximum power is transferred from one to the other. Unless maximum power is transferred, the unit's efficiency is less than desired. This decrease in efficiency increases operating costs. In many circuits, though, operating costs are so low that operating efficiency is ignored in favor of other operating conditions.

An amplifier not only has an input impedance (Z_i) but it also has an output impedance (Z_o). In this section we will discuss this parameter and how it can be calculated. **Output impedance** is defined as:

**Impedance present at the output of a circuit,
looking backwards into the circuit,
when the output path is open, and all
input signals and voltages are present.**

Remember, impedance is the total opposition that a circuit presents to an AC signal. Use Figure 8-13 as a reference for the following explanation. Capacitors that are connected in series with the input and output signals are selected to provide low X_C to the signals they must pass. We will therefore, assume that C_2 has zero capacitive reactance to the signal. This means that C_2:

**Acts as an open to DC and, therefore, blocks the DC
of one circuit from reaching the other circuit.**

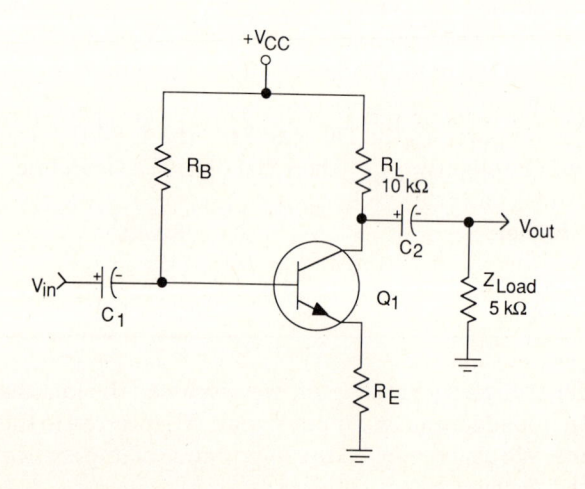

Fig. 8-13 Common-Emitter Amplifier-Output Impedance

Transistor Amplifiers

When an AC signal is applied to the input of the amplifier, everything performs exactly as we have been discussing. Upon arriving at the collector this signal appears as a change in collector current. If changing current is present, we have AC characteristics present. At the collector the signal (I_C) has two paths through which it can flow: path 1 is through R_L to V_{CC}; path 2 is through C_2, through Z_{Load} and to ground. NOTE: the voltage on Z_{Load} changes in proportion to the voltage at the collector of Q_1. Z_{Load} represents the input impedance of the next stage; therefore, any change on Z_{Load} is a bias change that is applied to the next circuit.

AC signal action within the two paths described above is identical to that of parallel circuits. Realizing this, we can use the *product over sum* parallel resistance formula to calculate the load impedance (Z_{Load}) that changing collector current meets upon reaching the collector. **AC load impedance** (Z_{Load}) can be calculated as follows:

$$Z_{AC} = \frac{R_L \times Z_{Load}}{R_L + Z_{Load}}$$

$$Z_{AC} = \frac{10 \text{ k}\Omega \times 5 \text{ k}\Omega}{10 \text{ k}\Omega + 5 \text{ k}\Omega}$$

$$Z_{AC} = \frac{50 \text{ M}\Omega}{15 \text{ k}\Omega}$$

$$Z_{AC} = 3.33 \text{ k}\Omega$$

Again, let us emphasize that this is opposition to the AC signal, not opposition to DC. All DC is blocked by the presence of C_2; therefore, the output opposition to DC is infinite. You should understand that Z_{out} of the unloaded transistor equals R_L.

Signal Behavior

When referring to **signal behavior**, we are actually discussing the affect that the application of an AC signal has on the collector of a transistor amplifier. Quite often, we refer to amplifiers of the type discussed here as being **small signal amplifiers**. By this, we mean that the effect of the input signal causes a voltage variation, at the collector that is a very small portion of V_{CC}. In **large signal amplifiers**, it is possible for the variation at the collector to be approximately the amplitude of V_{CC}.

Each amplifier that is designed is intended to be used for a specific job. Many of these amplifiers are used in circuits where the output signal must have identical shape to the input signal. If an identical reproduction of the shape of the input signal is reproduced in the output circuit, the amplifier is said to have *high fidelity*. In the design of a high

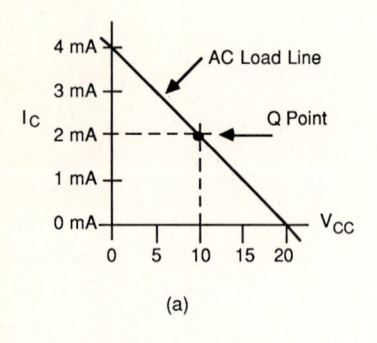

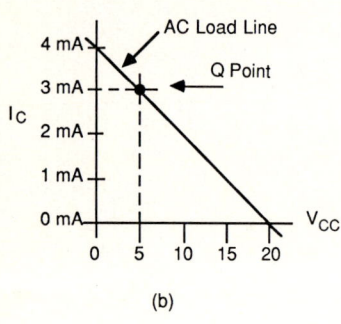

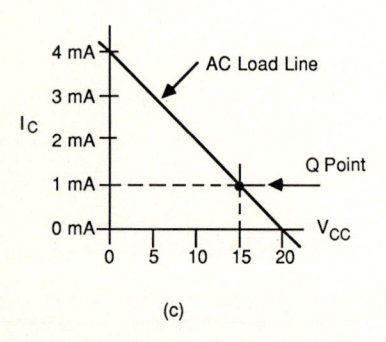

Fig. 8-14 Load Lines that Illustrate Saturation and Cutoff

fidelity amplifier, the transistor must be biased such that it operates at a point on its AC load line where the input signal will not drive it into either saturation or cutoff. Refer to Figure 8-14(a) for an example of this.

By biasing the amplifier at the $I_C = 2$ mA and $V_C = 10$ V operating (Q) point, it is possible for V_C to drop to 0 V or increase to 20 V. In other words, the output voltage could have a 20 V swing (from 0 V to 20 V). With the transistor biased to this Q point, the output would be symmetrical. The amplifier could be classified as a high fidelity amplifier.

If the transistor is biased such that its operating point is at the point shown in Figure 8-14(b), its output will be considerably different. At this Q point, $V_C = 5$ V and $I_C = 3$ mA. In other words, the transistor is operating near its saturation point. If a large signal is applied to the transistor, the collector voltage (V_C) can swing between 0 V and 20 V. The output signal will not however, be symmetrical. With an AC signal input, the positive alternation causes the transistor to saturate. After only 5 V of swing in the negative direction, the transistor becomes saturated. A saturated transistor cannot have a current change when forward bias is increased. This means that, during the positive alternation of the input signal, all of the signal past the point where the transistor becomes saturated, will not be developed in the output. In this case, the change in V_C could go to 5 V below the Q point ($V_C = 0$ V) and 15 V above ($V_C = +20$ V). The output signal from a circuit biased in this manner is not symmetrical. In the output signal, the negative alternation will not be fully reproduced.

A similar situation occurs with the biasing illustrated by Figure 8-14(c). With the Q point set to $I_C = 1$ mA and $V_C = 15$ V, the transistor is biased close to its cutoff point. In this case, the output voltage (V_C) can swing between 20 V and 0 V. The difference is that the negative alternation of the input causes the transistor to enter cutoff. With this being the case, part of the negative alternation of the input signal can not be developed in the output. With this circuit, the output signal is nonsymmetrical and the positive alternation of the output is not fully reproduced.

You should understand, though, that a transistor biased to any one of these points can be used as a symmetrical (high fidelity) amplifier. If the signal applied to the input is small enough to *not* saturate or cutoff the transistor, the output will be a faithful reproduction of the input signal.

You should realize that even a saturated transistor has some voltage drop between its collector and emitter (V_{CE}). With the collector saturated, V_{CE} is approximately equal to the voltage drop across the emitter-base junction. The presence of V_{CE} prevents the transistor's voltage drop from decreasing to 0 V at saturation. Care must be taken to assure that this fact is considered when designing amplifiers that are intended to work near saturation.

Earlier, voltage gain of the CE amplifier was discussed. We found that for the circuit used, the *unbypassed* gain equaled 9.75:1 (approximately 10:1) and the *bypassed* gain was 400:1. Let's discuss the effect that these gains have on the output swing of a CE amplifier. To calculate the amount of swing that is necessary at the input to obtain a desired swing for each amplifier at its output, we will proceed as follows:

$$\text{Input Swing} = \frac{\text{Desired Output Swing}}{A_V}$$

$$\text{Input Swing} = \frac{10 \text{ V}}{400} = 25 \text{ mV}$$

The amplifier that has a gain of 400 needs a peak-to-peak input signal of 25 mV to produce 10 V peak-to-peak in the output. Now, what about the amplifier whose gain is 10:1?

$$\text{Input Swing} = \frac{\text{Desired Output Swing}}{A_V}$$

$$\text{Input Swing} = \frac{10 \text{ V}}{10} = 1 \text{ V}$$

> **NOTE:** This formula can be used with any gain to determine input signal amplitude.

From these calculations you can see that a bypassed emitter resistor (R_E) results in the output signal having much greater amplitude.

Now you know how to calculate the maximum input swing that is needed to provide a given output when the AC voltage gain is known. We will use the swings (changes) that occur in the transistor to analyze the circuits. Also, because the NPN transistor was used as the basis for earlier explanations, we will change gears. For this analysis we will use a PNP transistor.

PNP and NPN transistors that have the same specifications and identical operating parameters are called **complementary transistors**.

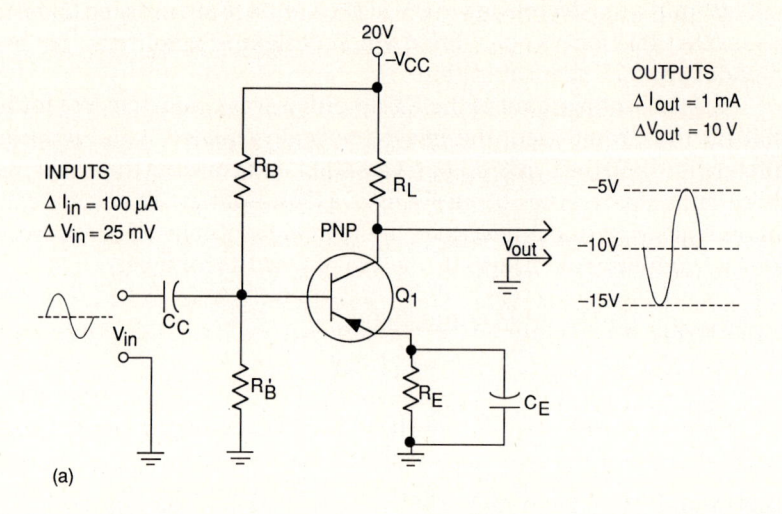

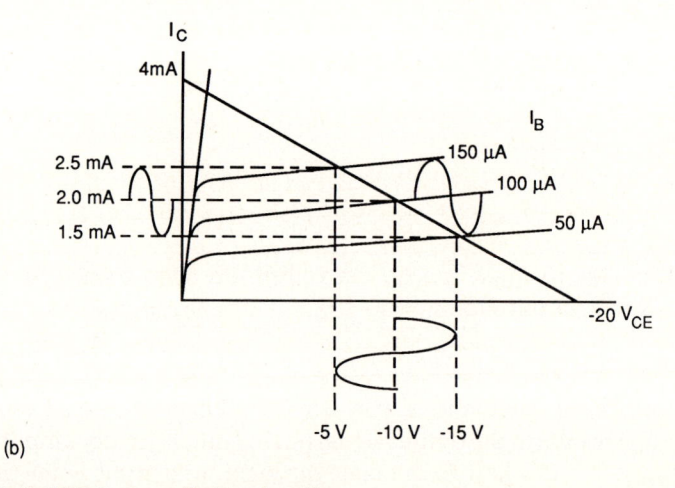

Fig. 8-15 PNP Common-Emitter Amplifier

Figure 8-15(a) contains a PNP common-emitter amplifier. This circuit is used for the explanation that follows. Note that V_{CC} for the circuit is -20 V. To select a Q point near the middle of the circuit's operating range we will assume:

$V_{CE} = -10$ V

$I_B = 100 \, \mu A$

$I_C = 2$ mA

Input Swing = 25 mV

AC Gain = 400

Transistor Amplifiers

With these conditions existing, (Figure 8-15(b)) a change in base current of 100 μA will cause the collector voltage to change 10 V and I_C to change 1 mA.

When the positive alternation of the input is applied to the base of Q_1, forward bias decreases (positive to N-type is reverse bias). With the decrease in forward bias, I_B and I_C decrease. I_C is now 1.5 mA and I_B = 50 μA. The decrease in I_C causes V_{RL} to decrease (go positive). This means that the negative voltage at the collector increases and the output goes negative. Note: For the output of a PNP transistor to go negative, the collector voltage must go more negative. With a positive input signal, the output goes negative, giving the 180° phase shift expected of a common-emitter amplifier.

When the negative alternation of the input arrives at Q_1 base, the following changes will occur. Application of a negative potential causes forward bias to increase. As forward bias increases, I_C and I_B also increase. As I_C increases, V_{RL} will increase. An increase in V_{RL} means that R_L is dropping a larger portion of the negative voltage. The voltage that remains at the collector (V_C) is decreasing (going less negative or positive). With V_C going positive, the output must also go positive. Again, we can see that the input and output signals are 180° out-of-phase. The voltage changes can be measured using either a VOM, a DMM or an oscilloscope. In addition, the waveshapes can be viewed and measured using an oscilloscope.

In electronics it is common to represent a change in any parameter by use of the Greek letter **delta**. The symbol for delta is Δ and it means a *small change in value*.

Using the changes (swings) that occur in the circuit, we can calculate several important parameters for this circuit. Remember, the two inputs have swings of 25 mV and 100 μA. To calculate the input power (P_{in}) of the circuit, use:

$P_{in} = \Delta I_B \times \Delta V_{in}$

$P_{in} = 100 \, \mu A \times 25 \, mV$

$P_{in} = 2.5 \, \mu W$

Using the same procedure, we can solve for the circuit's output power as follows:

$P_{out} = \Delta I_C \times \Delta V_{out}$

$P_{out} = 1 \, mA \times 10 \, V$

$P_{out} = 10 \, mW$

The amount of voltage gain, current gain, and power gain supplied by an amplifier can be calculated using one of the following ratios:

Voltage Gain $(A_V) = \dfrac{\Delta V_{out}}{\Delta V_{in}}$

Power Gain $(A_P) = \dfrac{P_{out}}{P_{in}}$

Current Gain $(A_I) = \dfrac{\Delta I_C}{\Delta I_B}$

Using these formulas and the results of the calculations completed earlier we can calculate the power gain (A_P) for the circuit.

Power Gain $(A_P) = \dfrac{P_{out}}{P_{in}}$

Power Gain $(A_P) = \dfrac{10\ mW}{2.5\ \mu W}$

Power Gain $(A_P) = 4000$

This tells us that output power is 4000:1 times greater than input power. To prove this as the correct ratio, do this:

$P_{out} = P_{in} \times A_P$

$P_{out} = 2.5\ \mu W \times 4000 = 10\ mW$

To calculate current gain, proceed as follows:

Current Gain $(A_I) = \dfrac{\Delta I_C}{\Delta I_B}$

Current Gain $(A_I) = \dfrac{1\ mA}{100\ \mu A}$

Current Gain $(A_I) = 10$

Collector current changes by a value that is 10 times greater than the change in base current. The circuit has a *current gain* of 10.

To calculate the voltage gain using the change (delta) values, do the following:

Voltage Gain $(A_V) = \dfrac{\Delta V_{out}}{\Delta V_{in}}$

Voltage Gain $(A_V) = \dfrac{10\ V}{25\ mV}$

Voltage Gain $(A_V) = 400$

Collector voltage changes by a value that is 400 times greater than the change in base voltage. The circuit has a *voltage gain* of 400.

An alternate method for calculating power gain is:

$A_P = A_V \times A_I$

$A_P = 400 \times 10$

$A_P = 4000$

Observe Figure 8-15(b). Changes (swings) in input current or voltages are plotted on a characteristic curve chart. Using these values, the circuit can be analyzed for its different gains. When you can do these calculations and understand their meaning, you understand common-emitter amplifiers.

Self-Check

Answer each item by performing the action indicated.

19. To assure that maximum power is transferred from an amplifier to the next stage or device, it is necessary to match the _____ between them.
20. Output impedance is calculated using _____ and R_L.
21. To amplify large signals, the collector voltage of a transistor must be able to vary an amount that approximately equals _____.
22. If the output signal is an exact reproduction of the input signal, the amplifier is said to have _____.
23. During operation, one alternation of the input signal causes the transistor to saturate. In this case the output signal would not be _____.
24. When calculating the maximum input swing, we divide the desired output swing by _____.
25. The Greek letter delta (symbol Δ) is used to represent the _____ of a parameter.
26. When a positive going signal is applied to the base of a PNP transistor, I_C will _____.

   a. increase   b. decrease

27. Power gain (A_P) can be calculated by multiplying _____ and _____.
28. A PNP transistor uses $-V_{CC}$. This means that its collector current must _____ for the output to go positive.

   a. increase   b. decrease

The Common-Collector Amplifier

The circuit shown in Figure 8-16 is a common-collector amplifier. It has a bypass capacitor connected to its collector, and it does *not* have a collector resistor. Placing the capacitor at the collector provides an AC path for the signal to be returned to ground. If this capacitor, was not there, the AC signal would be superimposed on the DC power supply voltage. This would cause variations in the V_{CC} that is applied to all other circuits, making it impossible to accurately predict circuit operation.

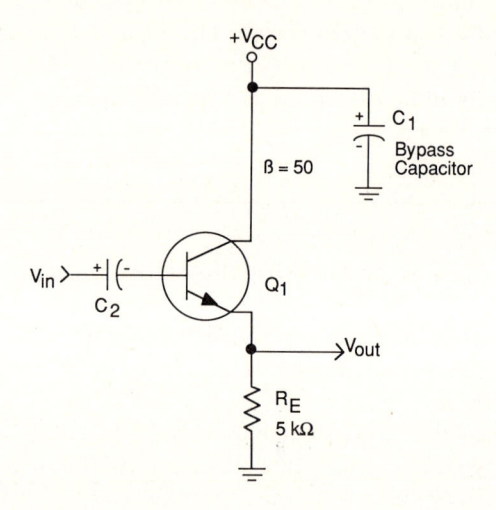

Fig. 8-16 Common-Emitter Amplifier

Because the output is taken from the emitter to ground, we cannot bypass R_E. To do so would prevent R_E from developing the output signal, and the output signal would be shorted to ground.

Input Impedance

Input impedance (Z_{in}) for this circuit is calculated in the same way as the common-emitter amplifier discussed above. Remember: the capacitor (C_2) blocks DC and passes AC; emitter current is ($\beta + 1$) times the base current; also, as we look into the base, the total resistance (r_{ib}) is ($\beta + 1$) times the emitter resistance. The same formula that was used previously is used to find R_{ib}.

$$R_{ib} = (\beta + 1) \times (r_e + R_E)$$

To find r_e we assume that a transistor has an $I_E = 1\,mA$.

$$r_e = \frac{25\,mV}{I_E}$$

$$r_e = \frac{25\,mV}{1\,mA}$$

$$r_e = 25\,\Omega$$

Then:

$$R_{ib} = (\beta + 1) \times (r_e + R_E)$$

$$R_{ib} = (50 + 1) \times (5025)$$

$$R_{ib} = 256{,}275\,\Omega \text{ or approximately } 256\,k\Omega$$

You can see that when compared to the R_{ib} of the common-emitter amplifier, this is a very high resistance. In fact, the common-collector amplifier is recognized for the fact that it has a *high input* impedance and a *low output* impedance. For this circuit, we can say that Z_{in} is approximately 256 kΩ.

Earlier, you learned that the voltage gain of the common-collector amplifier is less than unity and that the current gain of the circuit is $\beta + 1$. Note that this amplifier will faithfully reproduce the input signal. Because the majority of the input signal is dropped across R_E and because R_E develops the bias for the circuit, current will flow continuously. This circuit is used as the output amplifier in circuits where good fidelity and impedance matching are important. With its low output impedance, it is possible to insert a resistor (R_E) that can be connected directly in parallel with a 4 Ω, 8 Ω, or 16 Ω speaker.

Now we will analyze the common-collector amplifier using the changes in voltage, current, and power that are present at the input and output. There is not, normally, a characteristic curve chart for a common-collector amplifier. This will not, however, affect our ability to use the changes (deltas) for analysis.

Refer to Figure 8-17(a) for a PNP common-collector amplifier. Again, we use this amplifier to expose you to both NPN and PNP circuits. Refer to Figure 8-17(b) for the discussion that follows: Observe that ΔV_{in} equals 10 V and that ΔI_{in} equals 50 μA.

The circuit is operating at a Q point that has:

$I_B = 100\ \mu A$

$V_{RE} = -10\ V$

$I_E = 1\ mA$

and when the input is applied, the changes in output will be:

$\Delta I_E = 1\ mA$

$\Delta V_{out} = 9\ V$

$\Delta I_B = 50\ \mu A$

(a)

(b)

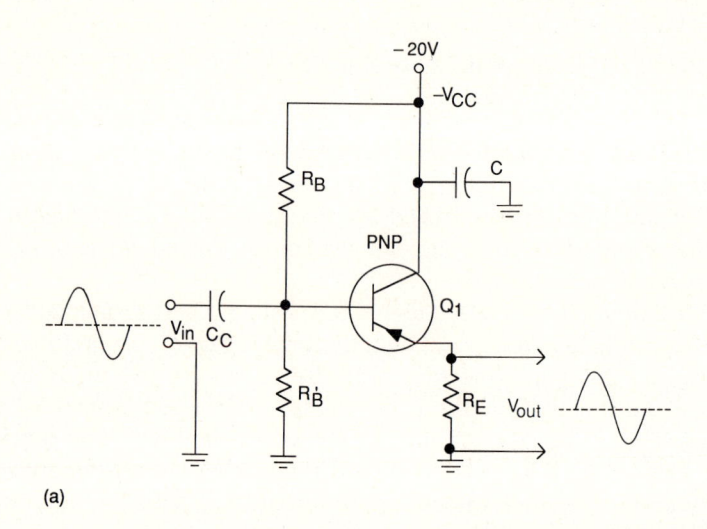

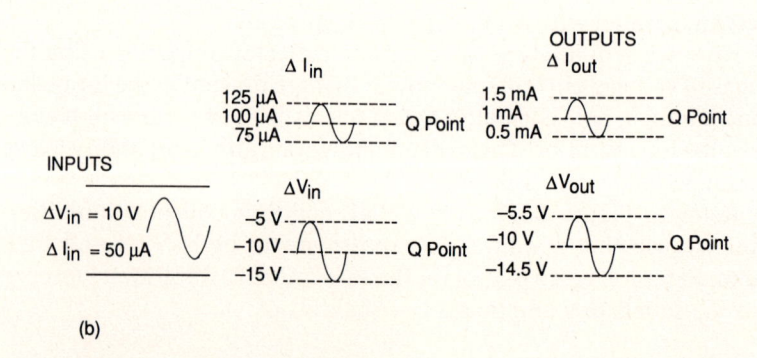

Fig. 8-17 PNP Common-Collector Amplifier

Transistor Amplifiers

Follow along on the circuit schematic as we discuss the circuit's operation and the three gains (A_P, A_V, and A_I).

When the positive alternation of the input is applied to the base of Q_1, the forward bias on Q_1 is reduced. As forward bias decreases, I_E decreases to 0.5 mA, and I_B decreases to 75 µA. The decrease in I_E causes V_{RE} to decrease to -5.5 V (go less negative). A voltage that becomes less negative is a positive going voltage. With a positive at the input and a positive output, we can see that the two signals are in phase. Remember, the common-collector amplifier has input and output voltages that are in phase, so this confirms that the circuit operates as expected.

During the time that the negative alternation is applied, forward bias is increasing. An increasing forward bias causes I_E to increase to 1.5 mA and I_B to increase to 125 µA. With the increase in I_E, the voltage drop on R_E increases (goes negative to -14.5 V). When compared to the -10 V reference, we can see that the output voltage is more negative than before. From this we can see that the output taken across R_E is going negative and is in phase with the input signal.

Now we will use the changes in current, voltage, and power to analyze this circuit. To calculate input power, we use the following procedures:

Input Power $(P_{in}) = \Delta I_B \times \Delta V_{in}$

Input Power $(P_{in}) = 50\ \mu A \times 10\ V$

Input Power $(P_{in}) = 500\ \mu W$

To calculate output power:

Output Power $(P_{out}) = \Delta I_E \times \Delta V_{RE}$

Output Power $(P_{out}) = 1\ mA \times 9\ V$

Output Power $(P_{out}) = 9\ mW$

To calculate power gain:

Power Gain $(A_P) = \dfrac{P_{out}}{P_{in}}$

Power Gain $(A_P) = \dfrac{9\ mW}{500\ \mu W}$

Power Gain $(A_P) = 18$

In this circuit, power gain equals 18, meaning that output power is 18 times larger than input power.

Current gain is again calculated using ΔI_{in} and ΔI_{out}, where $\Delta I_{in} = \Delta I_B$ and $\Delta I_{out} = \Delta I_E$. To calculate current gain, we must do the following:

Current Gain $(A_I) = \dfrac{\Delta I_E}{\Delta I_B}$

Current Gain $(A_I) = \dfrac{1\,mA}{50\,\mu A}$

Current Gain $(A_I) = 20$

In the common-collector configuration, the largest current (I_E) is compared to the smallest current (I_B). This results in the common-collector amplifier having the highest current gain of the three configurations. For this circuit we find that the current gain is at a 20:1 ratio.

To find the circuit's voltage gain we proceed as follows:

Voltage Gain $(A_V) = \dfrac{\Delta V_{out}}{\Delta V_{in}}$

Voltage Gain $(A_V) = \dfrac{9\,V}{10\,V}$

Voltage Gain $(A_V) = 0.9$

Notice that we again compared the change in input voltage to the change in output voltage. The result is a voltage gain (A_V) of 0.9. One of the characteristics of the common-collector amplifier is that it has a voltage gain of *less than unity*.

The alternate version for calculation of power gain is:

$A_P = A_V \times A_I$

$A_P = 0.9 \times 20$

$A_P = 18$

As we stated at the end of the common-emitter amplifier analysis, the test of how well you understand the common-collector amplifier depends upon how well you know and can apply the principles discussed here. To be fully competent as a technician, you must be able to supply these principles to circuit design and troubleshooting. The use of test equipment in confirming these parameters is an important part of the technician's job.

The Common-Base Amplifier

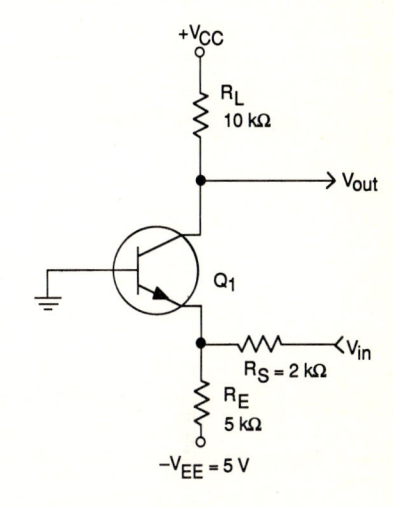

Figure 8-18 contains the schematic diagram for a **common-base** amplifier. Notice that it has a resistor in its input circuit and that the input is applied to the emitter. This resistor is placed here to increase the input resistance of the amplifier. It is called a **signal** resistor.

Notice that if the signal resistor was not present, the signal would be applied across R_{eb} only. In this circuit, r_e is much too small to effectively develop the input signal. By inserting R_S, we increase the input resistance of the circuit but also reduce the amplitude of the input signal. The addition of R_S allows the transistor to better develop the input signal. With R_S in the circuit R_S must be added to r_e.

To calculate voltage gain for this circuit, we use the formula:

$$A_V = \frac{R_L}{R_S + r_e}$$

From Figure 8-18 we can approximate $I_E = 1$ mA. To solve for r_e we must use:

$$r_e = \frac{25 \text{ mV}}{1 \text{ mA}} = 25 \ \Omega$$

Because r_e is so small when compared to R_S we can ignore it when calculating A_V.

$$A_V = \frac{R_L}{R_S}$$

$$A_V = \frac{10 \text{ k}\Omega}{2 \text{ k}\Omega}$$

$$A_V = 5$$

Therefore, for each 1 V change at the input, there is a 5 V change at the output.

Now we will analyze a PNP common-base amplifier using the changes that occur in the input and output. The formulas and calculations are quite similar to those studied for the other two configurations.

Refer to Figure 8-19(a) for a PNP common-base amplifier schematic diagram. Observe the circuit diagram and you will see that the input is applied to the emitter. This means that I_E is the *control* current for this amplifier. In the other two configurations, I_B was the *control* current.

When the positive alternation of the input signal is applied to the emitter, forward bias on Q_1 is increased. (Notice that we are applying a positive voltage to the P-type emitter). As forward bias increases, it

Fig. 8-18 Common-Base Amplifier

causes I_C to increase (I_C = 2.35 mA) and I_E to increase (2.5 mA) as shown in Figure 8-19(b). The increase in I_C causes V_{RL} to increase (drop more negative voltage). Because V_{RL} increases, V_C must decrease (go positive). V_{out} now equals -7.5 V.

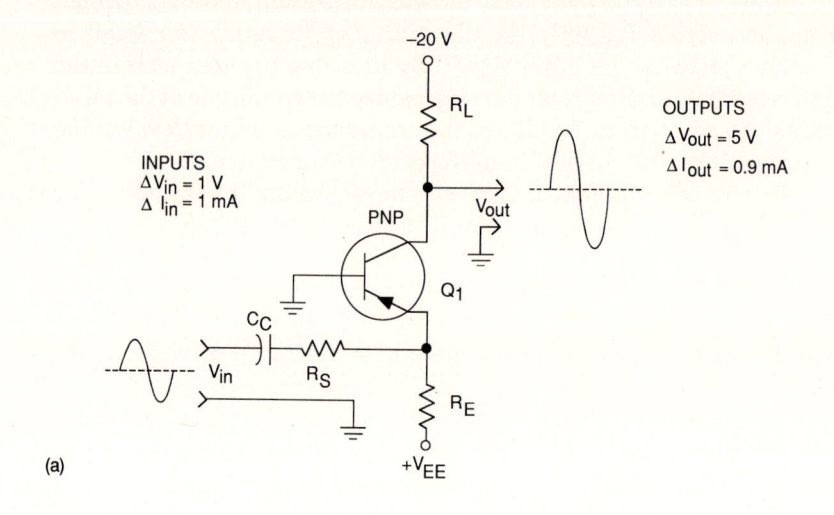

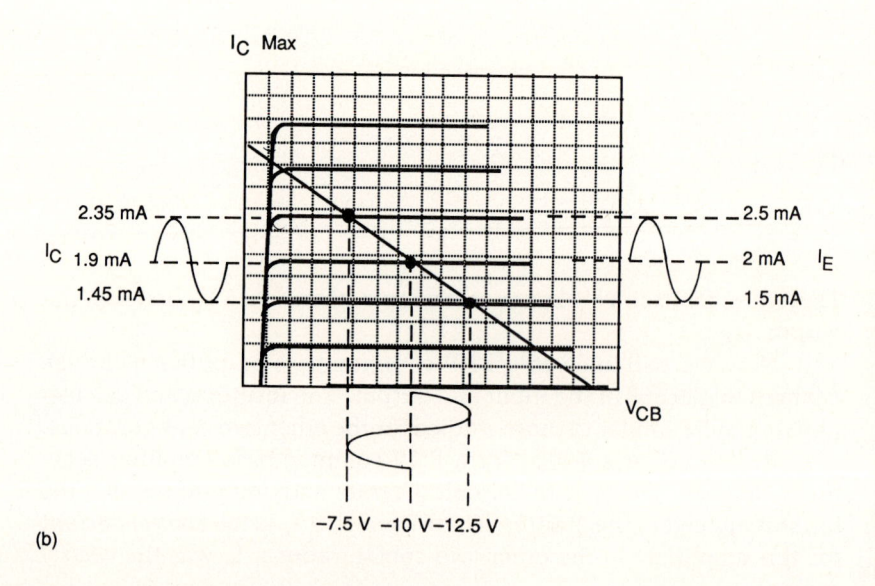

Fig. 8-19 PNP Common-Base Amplifier

With the arrival of the negative alternation, forward bias is decreased. The decrease in forward bias causes I_E to decrease (1.5 mA) and I_C to decrease (1.45 mA). When I_C decreases, it causes V_{RL} to decrease (go less negative). With V_{RL} dropping less voltage, the collector voltage (V_C) is more negative; therefore, the output voltage is also more negative (-12.5 V).

You should have observed that when the input went negative, the output went negative. When the input went positive, the output went positive. The fact that input and output signals are in phase is one characteristic of the common-base amplifier.

Now we will analyze the circuit using the changes to calculate A_P, A_I, and A_V. Power gain is calculated using the following procedures:

Input Power $(P_{in}) = \Delta I_E \times \Delta V_{in}$

Input Power $(P_{in}) = 1\,mA \times 1\,V$

Input Power $(P_{in}) = 1\,mW$

To calculate output power:

Output Power $(P_{out}) = \Delta I_C \times \Delta V_{CB}$

Output Power $(P_{out}) = 0.9\,mA \times 5\,V$

Output Power $(P_{out}) = 4.5\,mW$

Then, to calculate power gain:

Power Gain $(A_P) = \dfrac{P_{out}}{P_{in}}$

Power Gain $(A_P) = \dfrac{4.5\,mW}{1\,mW}$

Power Gain $(A_P) = 4.5$

current gain equals:

Current Gain $(A_I) = \dfrac{\Delta I_C}{\Delta I_E}$

Current Gain $(A_I) = \dfrac{0.9\,mA}{1\,mA}$

Current Gain $(A_I) = 0.9$

Notice that for the common-base amplifier, current gain is less than *unity*.

Voltage Gain $(A_V) = \dfrac{\Delta V_{out}}{\Delta V_{in}}$

Voltage Gain $(A_V) = \dfrac{5\,V}{1\,V}$

Voltage Gain $(A_V) = 5$

This yields a voltage gain ratio of 1:5; therefore:

$A_P = A_V \times A_I$

$A_P = 5 \times 0.9$

$A_P = 4.5$

You should realize that in operational common-base amplifiers, voltage gain and power gain, will be much higher than in this example.
    We repeat, your ability to use the data discussed in these last few sections will determine how successful you will be in electronics. It is impossible to know too much about the three amplifier configurations and how these parameters affect their design and operation.

Self-Check

Answer each item by performing the action indicated.

29. In the common-collector amplifier, a capacitor is placed in the collector circuit that places _____ at the collector.
30. The common-collector amplifier has an input impedance that is _____ than the input impedance of the common-emitter or common-base amplifiers.
31. One advantage gained from using the common-collector amplifier is that it has a _____.
32. In the common-collector amplifier, the _____ gain is low but the _____ gain is high.
33. The common-base amplifier is noted for its low _____ impedance and its high _____ impedance.
34. A signal resistor is added to the input circuit of a common-base amplifier to _____.
35. The change in output voltage for a PNP transistor common-collector amplifier is _____ than the change in input voltage.

36. The output current (I_E) of a common-collector amplifier will go _____ when the input signal goes positive.

    a. positive   b. negative

37. A common-collector amplifier has a current gain that is _____ than either the common-emitter or common-base amplifiers.
38. When a positive signal is applied to the emitter of a PNP common-base amplifier, the output will swing in a _____ direction.

    a. positive   b. negative

39. In the common-base amplifier, _____ current controls the circuit's operation.
40. The output signal is taken from the _____ in a common-base amplifier.
41. The collector current of a common-base amplifier is _____ than base current (I_B).

    a. larger   b. smaller

Summary

For maximum efficiency, amplifier circuits should have exact impedance matching between stages. Operating costs for solid-state devices is so low, though, that in most applications, efficiency is sacrificed for better overall operations.

The parameters Z_{in} and Z_{out} are important when considering the matching of impedance between stages. Ways to calculate and/or approximate these two parameters are available and were demonstrated in this chapter. The AC resistance (r_e) of a transistor's emitter element is an important factor in determining voltage gain. A standard exists that allows us to use 25 mV as voltage drop that exists across r_e in any transistor. Using this voltage and emitter current, we can calculate the value of r_e using the formula:

$$r_e = \frac{25 \text{ mV}}{I_E}$$

AC voltage gain is another valuable parameter. Knowing its value allows us to predict the exact amount that a signal will be amplified by a given amplifier. Parameters needed for this calculation are change (swing) in input voltage amplitude and change (swing) in amplitude of the output signal. The formula for AC voltage gain is:

$$A_V = \frac{\Delta V_{out}}{\Delta V_{in}}$$

As part of the calculation of input impedance and voltage gain, r_e is again an important parameter.

In order for you to better understand the AC parameters of an amplifier, we constructed and used an AC load line. Once constructed, the AC load line can be compared to the DC load line. The two load lines share a common Q point, but all other points along the two lines are different. Using the AC load line, the AC parameters of the circuit can be determined.

Important considerations of amplifier operation are: voltage gain, current gain, power gain, phase shift, input signal amplitude and phase, and output amplitude and phase. The ability to determine each of these characteristics is important. This means that you must devote the time necessary to learn them if you are to become proficient with bipolar transistor amplifiers.

Review Questions and/or Problems

1. What is the purpose of the capacitor in the circuit shown in Figure 8-20?

    a. It develops the output signal.
    b. It bypasses AC to ground and isolates the DC power from the AC variations.
    c. It assures that collector to emitter voltage will vary with the input voltage.
    d. It blocks DC and couples input signals above a specific frequency.

2. Refer to Figure 8-20. When emitter current (I_E) increases, the internal resistance of Q_1 is _____.

    a. increasing   b. decreasing

3. Check Figure 8-20. When the forward bias that is applied to Q_1 increases, the voltage drop on R_L _____.

    a. increases   b. decreases

Transistor Amplifiers

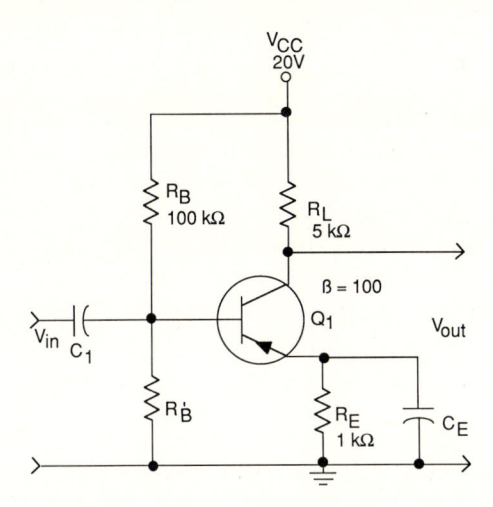

Fig. 8-20

4. Observe Figure 8-20. For this circuit, $V_{CC} = 20$ V and $R_L = 4$ kΩ. How much collector current flows if Q_1 is saturated?

5. Check Figure 8-20. Assume that $R_B = 100$ kΩ, $R_B' = 12$ kΩ, and $V_{CC} = 20$ V. When conducting at the Q point, I_B equals _____ A.

6. A common-emitter amplifier has an unbypassed emitter resistor (R_E). This causes voltage gain to be _____ than if R_E were bypassed.

   a. more   b. less

7. A germanium transistor is being used as Q_1, and emitter current (I_E) equals 1.25 mA. Calculate the AC resistance (r_e) for this circuit.

8. Check Figure 8-20. This circuit has an $R_{ib} = 2.5$ kΩ, $R_B = 15$ kΩ and R_B' of 5 kΩ. The circuit's input impedance equals _____.

9. Refer to Figure 8-20. In this circuit, $C_1 = 5$ μf. Using this capacitance and the Z_{in} calculated in question 8, solve for the circuit's low-frequency cutoff.

10. Observe Figure 8-20. An AC signal is applied to the input of Q_1 that has a peak value of 50 mV. This causes a change of 5 V at the output. The voltage gain for this circuit is _____.

11. Refer to Figure 8-20. Current gain for this circuit is 10:1. Using the voltage gain (A_V) from question 10, calculate power gain (A_P) for the circuit.

12. Refer to Figure 8-21. This circuit is a/an _____ amplifier.

13. Check Figure 8-21. For this circuit to operate as an amplifier, the input signal must be applied to _____.

   a. J_1 b. J_2

14. Observe Figure 8-21. Power requirements for this circuit are:

   a. $+V_{CC}, +V_{EE}$ c. $-V_{CC}, -V_{EE}$
   b. $+V_{CC}, -V_{EE}$ d. $-V_{CC}, +V_{EE}$

15. Check Figure 8-21. The current at Q_1 input varies 1 mA and input voltage varies 25 mV. This causes V_{out} to change 10 V and I_{out} to change 0.9 mA. The circuit's power gain (A_P) equals _____.

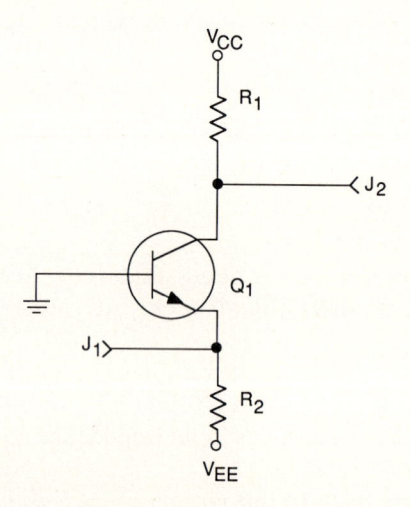

Fig. 8-21

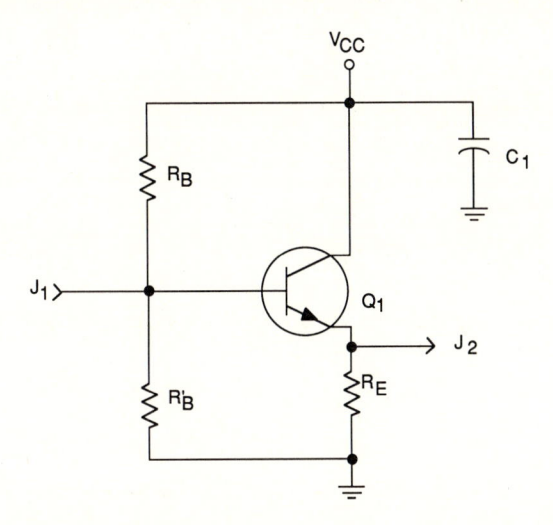

Fig. 8-22

16. Use the values provided in question 15 to calculate the circuit's current gain (A_I).

17. Refer to Figure 8-22. What is the purpose of C_1 in this circuit?

18. Check Figure 8-22. V_{CC} for this circuit must have a _____ polarity.

19. Observe Figure 8-22. In the circuit shown here, the output is taken from _____.

    a. J_1 b. J_2

20. Refer to Figure 8-22. This circuit has a voltage gain that is _____ than the circuit shown in Figure 8-21.

    a. larger b. smaller

9

Amplification and Voltage Regulation

Objectives

1. Explain why an amplifier is called a single-ended amplifier.
2. Discuss the operation of a phase splitter and describe some of its uses.
3. Define the term *differential amplifier* and explain what purpose this type amplifier serves.
4. Identify circuits that have stabilization provisions and explain how different stabilization circuits operate.
5. Describe what is meant by Class A, Class AB, Class B, and Class C operation and identify the type output wave that can be expected from each type operation.
6. Explain how a transistor amplifier is designed to operate in any of the four classes listed in objective 5.
7. Define the terms:

    a. Overdriven
    b. Saturation
    c. Bias stabilization
    d. Gain
    e. Distortion

Introduction

Specific characteristics of amplifiers have already been discussed. These discussions were intended to supply the building blocks upon which a thorough understanding of amplifiers can be developed. In this chapter we begin to explore more fully the operation of amplifiers that use the characteristics of selected Q point, capacitor coupling, and specific purpose.

You should understand that amplifiers are designed to operate in many different applications, a few of which will be discussed in this chapter. Prior to our discussion of specific types, we will take time to review some of the material presented earlier.

Amplifier Review

Refer to Figure 9-1 for this discussion. Notice that this circuit is an NPN common-emitter amplifier. It was chosen because it has current flowing from ground to $+V_{CC}$. Remember that PNP circuits operate identically to NPN circuits except that current flows from $-V_{CC}$ to ground. When we examine the circuit we see that it contains two capacitors (C_1 and C_2). We will assume that these capacitors were selected to operate at the desired frequency range, and therefore will act as a short to the frequencies being amplified. Remember, this does not change the fact that the capacitor acts as an open to DC.

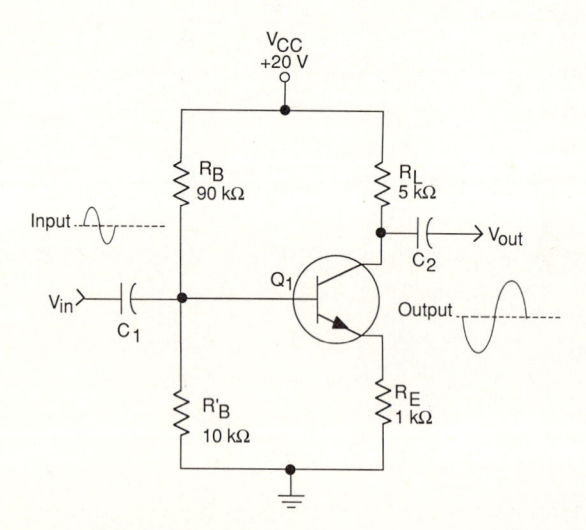

Fig. 9-1 NPN Common-Emitter Amplifier

To calculate the voltage gain (A_V) of this circuit we must use the formula:

$$A_V = \frac{R_L}{r_e + R_E}$$

In order to solve this formula, we must find r_e:

$$r_e = \frac{25 \text{ mV}}{I_E}$$

$$r_e = \frac{25 \text{ mV}}{2 \text{ mA}} = 12.5 \, \Omega$$

Note that r_e is very small when compared to the size of R_E. Because it is so small, we can ignore it in our calculation of A_V. Now we have:

$$A_V = \frac{5 \text{ k}\Omega}{1 \text{ k}\Omega} = 5$$

Notice that R_E is not bypassed with a capacitor. In the last chapter you learned that circuits which operate with R_E unbypassed have a small gain compared to those whose R_E is bypassed. This circuit has a voltage gain of 5:1, as calculated above. This tells you that for each input of 1 volt (peak-to-peak), you can expect an output of 5 V peak-to-peak. Notice that the circuit has a single output. In other circuits we may have more than one output. To identify this amplifier as one having a single output, it is often called a **single-ended amplifier**.

Phase Splitters

The circuit shown in Figure 9-2 is called a **phase splitter**. Note that it has two outputs; one taken from the collector and one taken from the emitter. In operation, this amplifier acts like two; one common-emitter and one common-collector. You know that a common-emitter amplifier has input and output signals that are 180° out of phase, and in the common-collector amplifier, input and output signals are in phase.

Observe the two output signals on Figure 9-2 and you will see that they are 180° out of phase. This is the reason that the circuit is called a **phase splitter**. The circuit receives one input signal and provides two output signals that are 180° out of phase. In other words, the circuit splits the signal into two phases. This circuit amplifies the signal at Output$_1$ exactly as it would in a conventional common-emitter amplifier having the same components. Output$_2$ has the same characteristics as the input signal but will be slightly smaller in amplitude than the input signal amplitude. The same voltage gain characteristic that we

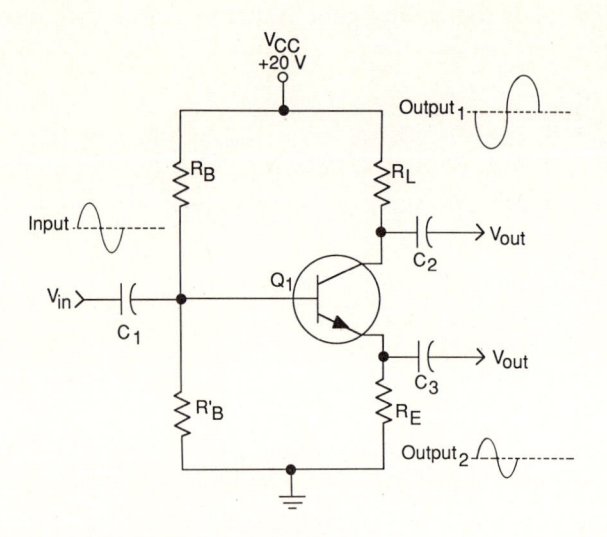

Fig. 9-2 NPN Common-Emitter Phase Splitter

discovered in the common-collector amplifier is present in the base-emitter circuit of this amplifier; therefore, Output$_2$ will be smaller in amplitude than the input signal. You can see from this that Output$_1$ will have more amplitude than Output$_2$. It is possible, however, to obtain output signals that have equal amplitude by using resistors that have equal values for R_L and R_E.

Phase splitter circuits are used to provide input signals to circuits that perform other functions. By designing circuits for specific use, the two outputs can be equal in amplitude and 180° out of phase. This is a necessity when we use the phase splitter as the input driver for a push-pull amplifier acting as the final amplifier of an audio system. Another use is in the color demodulation stage of a color television.

Differential Amplifiers

Another circuit that is used quite often is the **differential amplifier**. Figure 9-3 contains the schematic drawing for this circuit. Notice that the circuit contains two transistors that share one emitter resistor. Sharing the same R_E means that the current that flows through R_E must divide, and part flows through each transistor. If care is used in the selection of the resistors and transistors, the current flow in the branches will be equal. The common-emitter amplifier is used in differential amplifier circuits and either NPN and PNP transistors can be used.

Amplification and Voltage Regulation

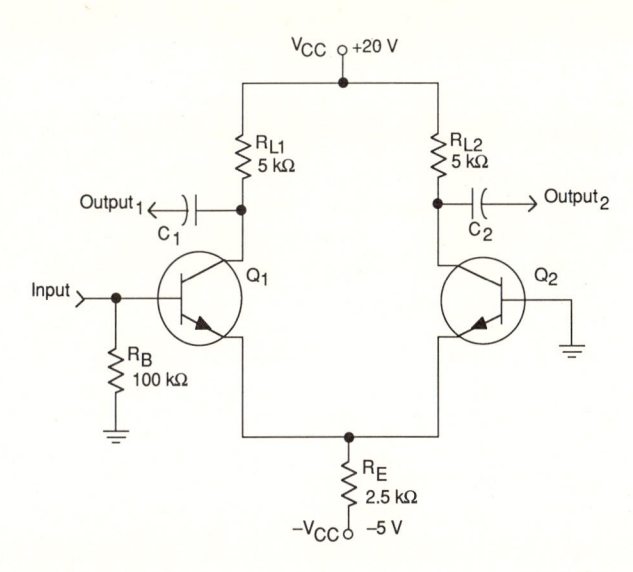

Fig. 9-3 Differential Amplifier

For the circuit shown, we can calculate R_E current flow using the formula:

$$I_{RE} = \frac{V_{EE}}{R_E}$$

$$I_{RE} = \frac{5\,V}{2.5\,k\Omega}$$

$$I_{RE} = 2\,mA$$

As stated above, the current that flows through R_E divides to provide I_E for each of the transistors. If current divides equally, each transistor has an emitter current (I_E) of 1 mA. We already know that I_C is approximately equal to I_E. Therefore we consider I_C to be 1 mA in each transistor.

With 1 mA flowing in each collector and each R_L being 5 kΩ, we can calculate V_{RL} as follows:

$$V_{RL} = I_C \times R_L = 1\,mA \times 5\,k\Omega = 5\,V$$

This tells us that the voltage drop on each load resistor will be:

$$V_{RL1} = 5\,V$$
$$V_{RL2} = 5\,V$$

Then, to find the voltage drop from the collector of each transistor to ground, we proceed as shown below:

$$V_C = V_{CC} - V_{RL} = 20\text{ V} - 5\text{ V} = 15\text{ V}$$

From this calculation, we can see that:

$V_{C1} = 15\text{ V}$

$V_{C2} = 15\text{ V}$

The output from this stage (input to the next stage) will be the difference in potential that exists between the two collector voltages. One possible combination of collector voltages is equal output voltages as illustrated above. In that case, the *difference* in potential is 0 V and the combined output is 0 V. Remember, voltage refers to difference in potential.

> **NOTE:** Each transistor has a quiescent output voltage of +15 V and when their amplitudes are compared, the result is:
>
> $+15\text{ V} - (+15\text{ V}) = 0\text{ V}$

This circuit is designed to assure that both Q points are a fixed voltage (+15 V) and their difference in potential is 0 V. Because components are often hard to match, it is common for a differential amplifier to have a balancing device. Resistor R_C (see Figure 9-4) has been inserted for this purpose. Adjustment of R_C will compensate for slight differences in the branch circuits by varying the amount of V_{CC} that is applied to each circuit.

The name *differential amplifier* is derived from the fact that the circuit has one or two inputs, connected so as to respond to the difference between two voltages or currents, but it effectively suppresses like voltages or currents. With the Q point set for 0 V output, each transistor is conducting 1 mA and has 15 V at its collector. Follow along as we discuss what happens when an input voltage is applied to the base of Q_1.

When a positive voltage is applied to the input of Q_1, it causes forward bias, I_E, and I_C of Q_1 to increase. With the increase in I_C, V_{RL1} increases. As V_{RL1} increases, less voltage is dropped between the collector and ground, causing Output$_1$ to go less positive (in a negative direction). Assume that the change equals 1 V: this causes Output$_1$ voltage to decrease to +14 V.

Amplification and Voltage Regulation

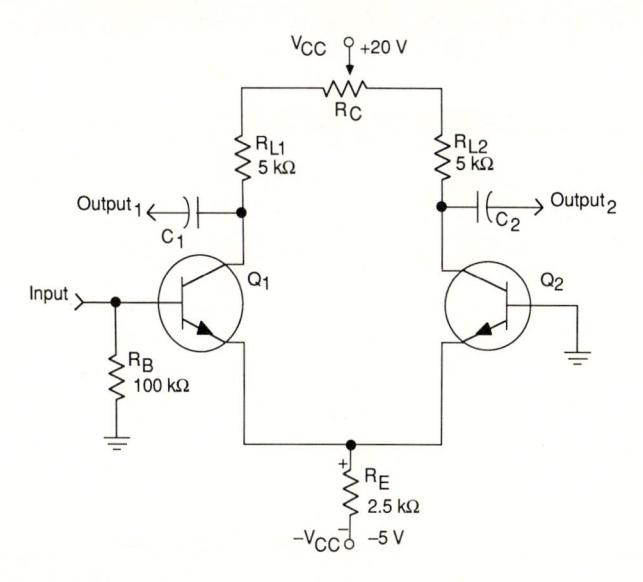

Fig. 9-4 Differential Amplifier

Now we will take a look at what happens to the branch that contains Q_2. When the current in Q_1 increases, the current in R_E has to increase. The increase in I_{RE} causes V_{RE} to increase (become more negative). The increased negative voltage has the same effect as a positive voltage at the top of R_E, which causes the forward bias on Q_2 to decrease. Decreased forward bias results in a decrease of I_E and I_C for Q_2. Decreasing I_C causes V_{RL2} to decrease and V_{C2} to increase. In other words, the voltage at the collector of Q_2 goes positive (opposite to that of Q_1).

Assume that this causes Output$_1$ voltage to decrease to +14 V. At the same time, the output voltage for Q_2 increases to +16 V. Comparison of the two output voltages reveals that:

$$+16 \text{ V} - (+14 \text{ V}) = +2 \text{ V}$$

In this case, the difference in potential is +2 V.

From this analysis you can see that collector voltages of Q_1 and Q_2 change in opposite directions. In the analysis above, V_{C1} becomes less positive (15 V to 14 V) and V_{C2} goes more positive (15 V to 16 V). The difference between the two outputs equals the sum of the two voltage changes. Notice that, in all cases, the base of Q_2 is grounded. The only way that Q_2 bias can be changed is by changes in the voltage drop on R_E (V_{RE}).

When the voltage applied to the base of Q_1 goes negative, the opposite sequence occurs. Forward bias decreases and causes I_E, I_C, I_{RE}, and V_{RE} to decrease. The decrease in I_C causes V_{RL1} to decrease and V_{C1} to increase (go more positive). The decreased (less negative) voltage dropped across R_E causes Q_2 forward bias to increase. With the increase in forward bias, V_{RL2} increases and V_{C2} decreases (goes less positive). V_{C1} is going more positive, V_{C2} is going less positive, and assuming that the total change is 2 V, we have:

$$+14 \text{ V} - (+16 \text{ V}) = -2 \text{ V}$$

You can see that the difference in potential between the two outputs is opposite to the difference discussed above. The difference in potential that results from comparing the two outputs is in phase with the input voltage. Application of a positive input causes a positive difference, and application of a negative input causes a negative difference.

Remember:

Input and output signals are in phase in CB and CC amplifiers.

and:

Input and output signals are 180° out of phase in a CE amplifier.

An understanding of this circuit will help you to learn a similar circuit that will be discussed in the chapter on operational amplifiers.

Self-Check

Answer each item by inserting the word or words required to correctly complete each statement.

1. The single-ended CE amplifier has one input and provides _____ output that has a _____ phase shift as compared to the input signal.
2. A circuit that can accept one input signal and provide two equal but opposite-phase outputs is called a/an _____.
3. A circuit that accepts a single input and then compares it to a second input is called a/an _____ amplifier.
4. In a differential amplifier, if the voltage at the collector of Q_1 goes positive, the voltage at the collector of Q_2 must go _____.
5. The secret to differential amplifier operation is that the base of one amplifier is _____ and the other base has an input signal applied.

Amplification and Voltage Regulation

Before going further, we will take a look at some definitions.

Fidelity – the faithful reproduction of an input signal in the output of an amplifier.

Distortion – any undesired change in the waveform of the original signal, resulting in an unfaithful reproduction of the input signal.

Amplitude Distortion – any change in waveshape that causes its output amplitude to not be proportional to the input amplitude.

Class of Operation – a classification of amplifiers that is based on the conduction time and output signal of different types of amplifiers as they compare to the input signal. These are:

> **CLASS A** – an amplifier that conducts 100% of the time during which an input signal is present. The transistor is never allowed to enter cutoff or saturation. Any AC sine wave applied to the input results in variation of I_C during 360° of the input. This assures that the output will contain the entire input signal. A Class A amplifier is often biased near the center of its load line.
>
> **CLASS B** – an amplifier that is biased at cutoff so that with no input, there is no conduction. The circuit will, however, conduct during exactly 50% of the time that an AC input signal is applied.
>
> **CLASS AB** – an amplifier that is biased such that it will conduct more than 50%, but less than 100%, of the time that an input signal is present. The Class AB amplifier is biased to operate between the Class A and Class B amplifiers. Without an AC input signal, Class AB amplifiers conduct 100% of the time.
>
> **CLASS C** – an amplifier that is biased below cutoff and that will conduct *less* than 50% of the time during which an AC input signal is applied.

Cutoff – an amplifier that is in the state of not conducting. Cutoff may be the result of fixed bias or may occur during the period when one alternation of the input signal causes the device to have reverse bias sufficient to stop conducting. During cutoff, an amplifier has $V_C = V_{CC}$ and $I_C = 0$.

Saturation – an amplifier that is biased such that it has maximum collector current. When the transistor is saturated, V_C is minimum and I_C is maximum. Further increases in forward bias will not cause I_C *to increase*.

Overdriven – a transistor that is biased such that one alternation of the input drives it into saturation and/or the other alternation drives it into cutoff.

Classes of Operation

The four classes of amplifier operation are: **Class A, Class B, Class AB,** and **Class C**. Class of operation is determined by the amount of bias that is applied to the eb junction, the amplitude of the input signal and the percent of time that the transistor conducts as compared to the input signal. These will, in turn, determine the percentage of time that the amplifier conducts.

Refer to Figure 9-5 for illustrations of each class of operation.

> **NOTE:** The waveshapes used for this figure are for the NPN common-emitter amplifier only. Other configurations and PNP circuits will have different waveshapes.

The solid line, in the output shows the time that the amplifier is conducting. The dashed line shows the part of the input signal that is lost due to saturation or cutoff of the transistor.

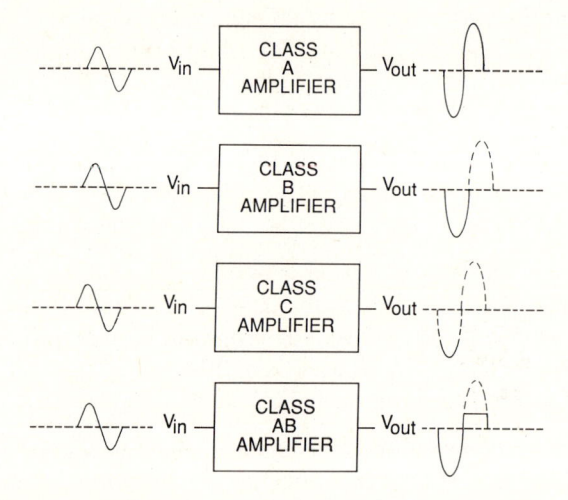

Fig. 9-5 Classes of Amplifier

Class A Operation

All of the amplifiers we have discussed to this point have been *Class A* amplifiers. This type of amplifier has changing current flow for 100% of the time during the complete 360° of an input signal. Note that the output signal is an exact reproduction of the input and has *high fidelity*.

To bias an amplifier for Class A operation and to provide linear output, we select a Q point where the input signal cannot drive the transistor into either saturation or cutoff. Biased to the mid-point of its load line, the amplifier can amplify equal positive and negative swings of the input without cutting off or saturating. A Class A audio amplifier provides the highest fidelity possible.

Class B Operation

Class B amplifiers (Figure 9-5) are biased such that I_C will flow for only 50% of the time that an AC sine wave is present as an input signal. Note that the output signal, in this example, produces only a negative alternation. The positive alternation is lost during the time that the transistor enters cutoff. *Fidelity* of this output is poor.

Class B amplifiers are designed by selecting an operating point that sets forward bias to the exact point of cutoff. With this condition existing, the alternation that forward biases the transistor drives the amplifier into conduction, and the entire alternation is amplified before the transistor is again cut off. The opposite alternation is not amplified because the transistor remains cut off during the time it is being applied to the input. Common-base and common-collector NPN amplifiers that are operating Class B will produce only positive output pulses.

For PNP transistors, operation is the same. The difference is that output amplitudes are opposite those discussed for NPN types.

Class C Operation

Class C amplifiers are biased to conduct for something less than 50% of the time. Collector current flows for less than one-half of the input AC signal. Note that only the most negative peak of the output is developed. The transistor is cut off for the rest of the time.

To design a Class C amplifier, you must select a reverse bias that holds the transistor in the cutoff condition when no input signal is present. When an input signal that increases forward bias is applied, the transistor's reverse bias decreases until it is overcome by the input signal. At that time the transistor has forward bias applied and will conduct. It continues to conduct until the input signal decreases (drops) below the reverse bias voltage level. When the input voltage drops low enough, reverse bias overcomes the forward bias and cuts the transistor off. For the transistor to conduct again, a new signal that has sufficient amplitude to forward bias the junction must be applied to the input.

Class AB Operation

Class AB amplifiers are biased to conduct for something more than 50% of the time but less than 100%. Collector current flows for more than one-half of the input AC signal. Note that only the most positive peak of the output is lost. The transistor conducts during the rest of the time. This class operates in the region between Class A and Class B.

To design this amplifier, we must select an operating point that biases the transistor so that the output signal's reference voltage is more than 50% of V_{CC}. With this condition existing, one alternation of the input signal is amplified. The other alternation drives the transistor into cutoff at some point before its peak voltage is reached. The transistor remains cut off until the input signal decreases to the point that it can no longer cancel the fixed forward bias. At that time, the transistor again begins to conduct.

Overdriven Amplifiers

One cause of amplitude distortion is the input of a signal with larger amplitude than the amplifier is capable of handling. Figure 9-6 is an illustration of this condition. With the NPN common-emitter amplifier that is illustrated here, you can see that one alternation causes the transistor to become saturated and the other alternation causes it to enter cutoff. In the common-emitter amplifier, the output signal has been shifted 180° with respect to the input. For common-base and common-collector amplifiers, the signal would not be shifted.

Fig. 9-6 An Overdriven Amplifier

PNP amplifiers that are overdriven produce signals that are opposite in polarity to those discussed for the NPN types. In addition, the polarity of V_{CC} and the direction of current flow are opposite.

Any class of amplifier can be driven into saturation by large input signals. Saturation is not, however, a normal operating condition. An overdriven amplifier is not operating Class A. Any amplifier that is driven into cutoff is operating in one of the other three classes (B, C, or AB) of operation. The class of operation is based on the percentage of time of conduction, not whether part of that time is spent in saturation.

Self-Check

Answer each item by inserting the word or words required to correctly complete each statement.

6. A Class ____ amplifier provides a high fidelity output.
7. A Class ____ amplifier conducts 50% of the time.
8. A/an ____ amplifier distorts both the positive and negative alternations of the output signal.
9. An amplifier is cut off 25% of the time: it is operating Class ____.
10. Of the four classes of amplifiers, Class ____ most severely distorts the output signal.

Temperature Stabilization

Many transistors are very sensitive to changes in operating temperature. Remember, heat is a by-product of voltage and current that is called **power**. Heat generated within the transistor affects the transistor's operation. Heat within the room where equipment operates can affect the operation of solid state components. Methods used with transistor circuitry to prevent the effects that result from temperature changes are called **temperature stabilization**.

Solid state devices operate with a **negative temperature coefficient**. This means that as temperature increases, junction resistance decreases. This decrease in resistance allows junction current to increase, causing an increase in temperature. This causes the entire cycle to begin again. This phenomena is called **thermal runaway**. Unless these effects are controlled, the transistor may become very hot and may even destroy itself.

In order to stabilize the circuit's operation, we must modify the circuit so that the base current changes are small when compared to changes in R_{eb} caused by temperature changes. What follows is a discussion of some of the circuit modifications used to stabilize base current in common-emitter amplifier circuits.

Emitter Resistor Stabilization

An NPN common-emitter amplifier whose emitter is grounded is shown in Figure 9-7. When the operating temperature of Q_1 increases, the resistance of the collector-base junction (R_{cb}) decreases. As R_{cb} decreases, I_E and I_C increase. The increased current causes the junction's operating temperature to increase. As the junction's temperature increases, R_{cb} decreases, I_C increases and more heat is generated.

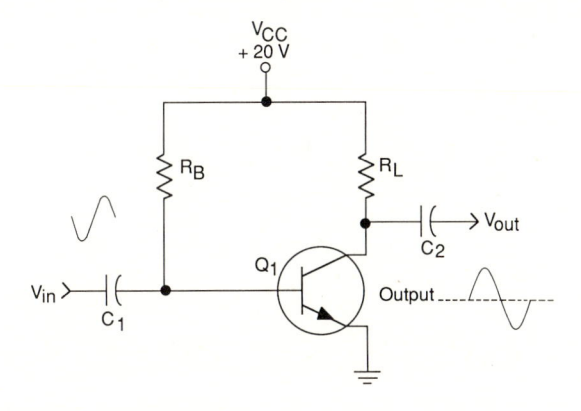

Fig. 9-7 Unstabilized CE Amplifier

If this action is allowed to continue, it is possible that this type operation could result in the transistor's self destruction. At best, the transistor would continue the heat, resistance, current, and temperature cycle until it (the transistor) became saturated.

Notice that in this example, I_C changes as a result of temperature, not because of a change in forward bias. The increase in I_C causes V_{RL} to increase and V_C to decrease. Should this continue, a point is reached where Q_1 is saturated (V_C = minimum and I_C = maximum). Note that the increase in temperature causes a change in gain. As temperature increases, gain decreases. This cannot be allowed to happen, so temperature stabilization must be provided.

Figure 9-8 shows the same CE amplifier with an emitter resistor (R_E) inserted. This circuit has a different set of characteristics. With current flowing through R_E, a positive voltage is present at the top of R_E. As operating temperature increases in Q_1, emitter-base resistance (R_{eb}) decreases, causing I_E to increase. The increased current flow causes V_{RE} to increase.

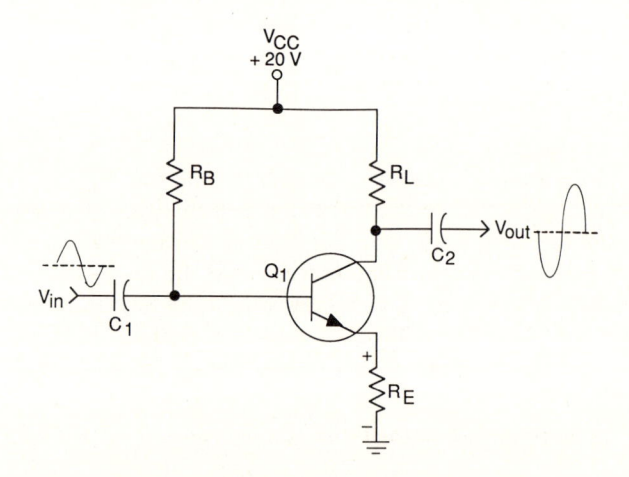

Fig. 9-8 Emitter Resistor Stabilization

When this happens, the larger positive voltage present at Q_1's emitter actually reduces forward bias. Remember, application of a positive voltage to N-type material is a form of reverse bias. In effect, the increase in temperature has actually reduced forward bias, which causes I_E to decrease. The reduced I_E causes less voltage to be dropped on R_E which results in I_C staying very nearly stable.

Emitter-resistor stabilization takes a relatively long time to react. It is, however, a quite satisfactory method for use with many circuits. The one disadvantage is that the emitter resistor (R_E) and R_{eb} of the junction resistance form a voltage divider. As a signal is applied to the input, the increased current flow that occurs in R_E prevents the full effect of the input signal from being applied as forward bias. This effect is called **degeneration**.

Degeneration can be eliminated by the use of a capacitor connected in parallel with R_E. Observe Figure 9-9, where capacitor C_E has been added to form this type circuit. Remember, a capacitor blocks DC but passes AC at and above specified frequencies. If C_E is selected such that its X_C is very low to the lowest frequency that will be amplified, it (C_E) acts like a short to all frequencies above that point. Once C_E has been selected and installed, we have, in effect, placed AC ground at the emitter of Q_1. Now we have a DC path through R_E, but all AC variations are shunted around R_E to ground. In this way, it is possible to maintain a stable DC voltage at the emitter of Q_1 while allowing the input signal to be applied as bias.

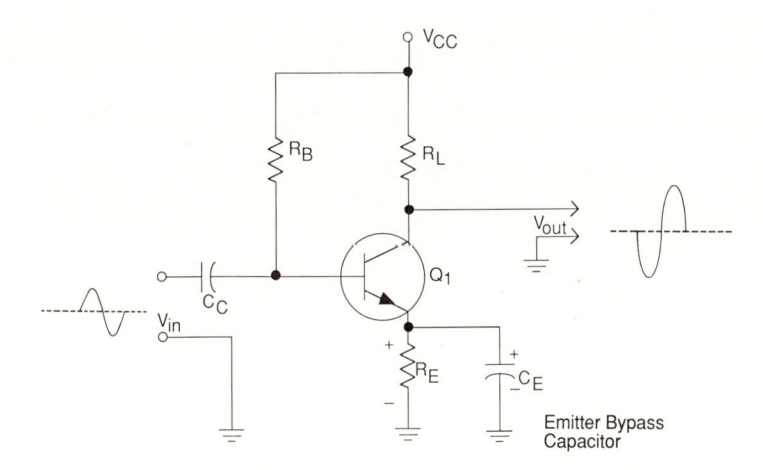

Fig. 9-9 Temperature Stabilization Without Degeneration

Thermistor Stabilization

A **thermistor** is a resistor that has been designed to be temperature sensitive. As current flows through the thermistor, its resistance changes. Actually, the word *thermistor* is short for thermal resistor, indicating that its resistance changes with any change in operating temperature. (Note the schematic symbol for this device in Figure 9-10). As used in this circuit, the thermistor has a **negative temperature coefficient**. In other words, as operating temperature increases, the thermistor's resistance decreases. Figure 9-10 contains a circuit with a thermistor connected between the base and ground.

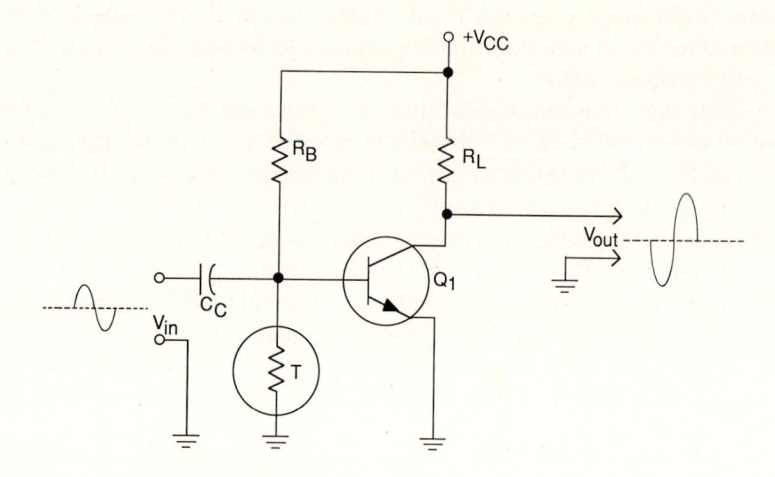

Fig. 9-10 Thermistor Stabilization

> **NOTE:** Although the thermistor is electrically located in parallel with R_{eb}, it is physically located such that its temperature reacts in the same manner as that of Q_1.

In this circuit, when temperature increases, I_C begins to increase and the thermistor's resistance decreases. Note that the thermistor is connected in parallel with R_{eb}, meaning that as its resistance decreases, the equivalent resistance of the parallel network decreases. This causes a decrease in forward bias which serves to keep I_C relatively stable.

In Figure 9-11 you see a comparison of circuits that use resistor, thermistor, and unstabilized emitter circuits. Curve X shows the

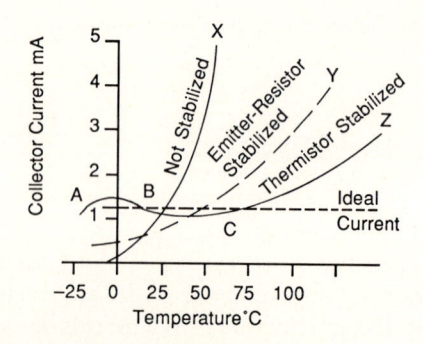

Fig. 9-11 Comparison of Stabilization Effects

unstabilized circuit, curve Y the emitter-resistor stabilization, and curve Z the thermistor effect. Notice the improvement in stability that is accomplished with thermistor stabilization (curve Z). The thermistor (curve Z) maintains I_C more closely to *ideal* current which is represented by a dashed line. Thermistor stabilized circuits maintain a relatively close control of I_C over a larger operating range. The actual current, however, is equal to the ideal current at points A, B, and C only. This is due to the difference in the variation of resistance that occurs across the thermistor and the base-emitter resistance. Because of the different materials used to make the two devices (transistors and thermistors), their resistances do not react in the same way.

Diode Stabilization

Figure 9-12 contains a circuit in which a diode is used for temperature stabilization. Diodes and transistors are made from the same materials. Both depend upon biased junctions for their operation. Because of this, current flow in this circuit is much nearer to ideal than those discussed above. In all other ways, the operating of this circuit is like the thermistor operation. This is **forward-biased diode stabilization**.

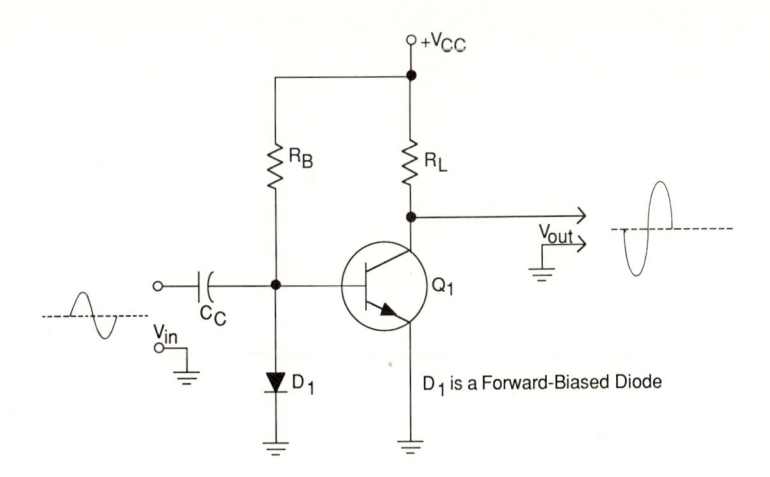

Fig. 9-12 Forward-Biased Diode Stabilization

Another type of diode stabilization uses a **reverse biased diode** as is shown in Figure 9-13. This diode works because it has minority current during the time it is reverse biased. Reverse current at temperatures less than 50° Celsius (120° Farenheit) is very small. If operating temperature increases above this level, however, the resistance of R_{cb} in the transistor decreases allowing I_{CBO} to increase. The same

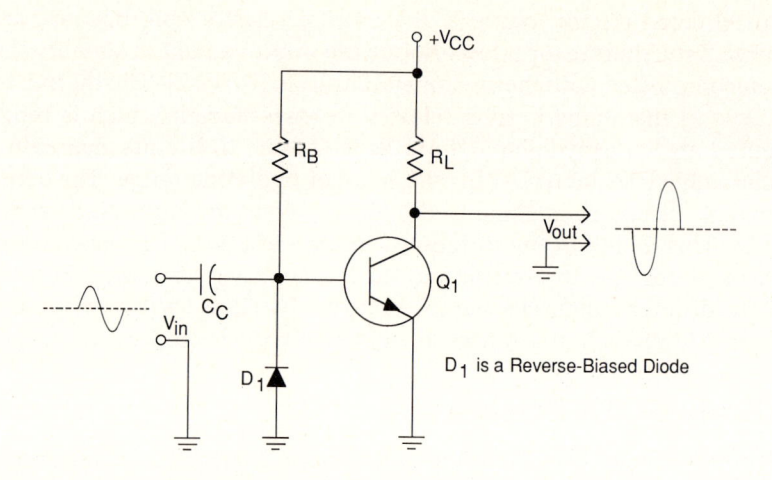

Fig. 9-13 Reverse Biased Diode Stabilization

change in operating temperature affects the junction of D_1 and causes its resistance to decrease. The decrease in diode junction resistance is enough to reduce the forward bias applied to the transistor. As a result, collector current decreases which stabilizes I_C. A reverse biased diode compensates for variations in minority current (I_{CBO}) in the transistor that result from increased temperatures.

It is not uncommon to find both forward and reverse biased diodes in the same amplifier circuit as shown in Figure 9-14. This arrangement is called **double-diode stabilization**. The forward biased diode compensates for, or stabilizes, the transistor's forward current, and the reverse biased diode stabilizes the transistor's reverse current.

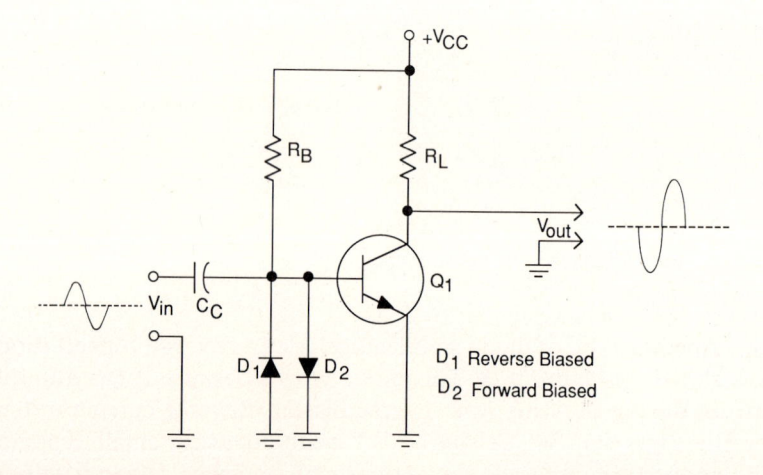

Fig. 9-14 Double Diode Stabilization

Amplification and Voltage Regulation

Figure 9-15 shows a comparison of the three types of diode stabilization. Curve A represents the unstabilized circuit. Curve B illustrates the effect of single (forward) biased stabilization. Curve C illustrates the effect of double-diode compensation. Note that double-diode stabilization provides very close to ideal current at temperatures approaching 110° Celcius (230° F).

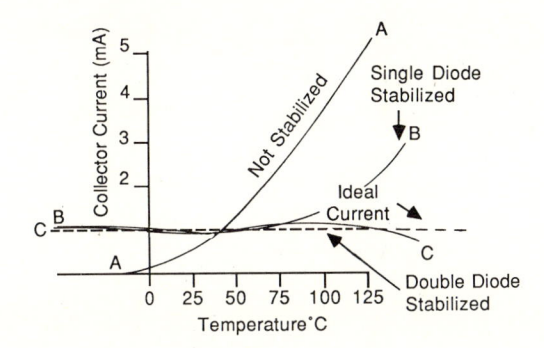

Fig. 9-15 Diode Stabilization Curves

Amplifier Comparisons

In Chapter 4 you were introduced to transistor amplifiers. Since that time we have covered several variations of those circuits. At this point, it seems only right that we take time to pull all the important parts of transistor amplifier theory together.

We have discussed three types of gain for transistor amplifiers. These are **current gain, voltage gain,** and **power gain**. We have stressed over and over again the fact that power is the product of current and voltage as expressed by the formula:

$$P = I \times E$$

therefore, it is reasonable to expect that:

Power Gain = Current Gain × Voltage Gain

The symbol "A" is used to identify **amplification factor**, a term that means the same as gain. Using this symbol, we can state the three gains as follows:

Current Gain = A_I

Voltage Gain = A_V

Power Gain = A_P

To properly design transistor amplifiers, it is necessary to have an understanding of **input** and **output resistance** for the different amplifier configurations. To transfer maximum power from one stage to the next, it is necessary to match the output of one stage to the input of the next stage as nearly as possible.

Table 9-1 contains information pertaining to R_{in}, R_{out}, A_V, A_I, and A_P for each of the three transistor amplifier configurations.

Table 9-1

Type of Circuit	R_{in}	R_{out}	A_V	A_I	A_P
Common Base CB	Low 50 - 100 Ω	High 300 k - 500 kΩ	Medium	Less Than One	Medium
Common Emitter CE	Medium 500 - 3.0 kΩ	Medium 30 k - 50 kΩ	High	Medium	High
Common Collector CC	High 20 k - 500 kΩ	Low 25 Ω - 1 kΩ	Less Than One	High	Medium

NOTE: R_{in} represents average impedance as found on specification sheets and R_{out} equals the reciprocal of output conductance as noted on specification sheets.

Another thing that must be considered regarding amplifiers is the relationship that exists between their input and output. Remember that the common-emitter amplifier has a 180° phase shift, and the others have input and output signals that are in phase. To assure that the output signal is in correct phase, you must decide which type to use. Remember, if you need two stages of amplification, you can use two cascade-connected common-emitter amplifiers and get a 360° phase shift between input signal and output. Table 9-2 shows a comparison of the three configurations with respect to phase relationship between input signal and output signal.

Table 9-2		
Type of Amplifier	Input Waveform	Output Waveform
Common Base CB	(sine wave, in phase)	(sine wave, in phase)
Common Emitter CE	(sine wave)	(inverted sine wave)
Common Collector CC	(sine wave, in phase)	(sine wave, in phase)

By now you should realize that selecting the amplifier that you need to do the job is a matter of considering the amount of gain (current, voltage, or power) needed, the correct configuration to provide the right input and output impedances, and the phase required in the output. As stated earlier, compromises in impedance matching can be, and are made because of the low cost of operating solid-state devices.

One other consideration is critical; *temperature* and *temperature stabilization*. Back in Chapter 1, we discussed the fact that semiconductor materials have a negative temperature coefficient. This means that as temperature increases, the resistance of a PN junction decreases. In some cases, the effect of high operating temperatures can result in the transistor destroying itself. For this reason, the location and temperature where the device is likely to be used must be considered. Types of temperature stabilization are discussed in this chapter that, when used, aid in the protection of transistor amplifier circuits.

Amplifier Summary

Throughout this discussion we have used NPN transistor circuits. Stabilization of PNP transistor circuits requires the same procedures and provides the same results. Care should be taken regarding polarities and connections.

Circuits that are designed to use PNP transistors have $-V_{CC}$. In addition, current flow in the transistor is from $-V_{CC}$ to ground. This means that current enters the collector and exits the emitter.

The circuits discussed in this chapter are NPN common-emitter amplifier circuits which have a 180° phase shift between input and output signals. If NPN common-base or NPN common-collector amplifiers are used, input and output signals are in phase.

Self-Check

Answer each item by inserting the word or words required to correctly complete each statement.

11. Solid-state devices have a _____ temperature coefficient.
12. Of the circuits studied, the _____ stabilized circuit provides a more stable current over greater variations in current.
13. A _____ is used to bypass R_E in a common-emitter amplifier to place _____ at the emitter.
14. A coupling or bypass capacitor must be selected that presents a low opposition to the _____ frequency contained in the signal being amplified.
15. The emitter bypass capacitor is necessary to overcome the _____ feedback that is induced when emitter resistor stabilization is used.
16. A Class A amplifier conducts _____ percent of the time.
17. Of the three types of transistor amplifiers, the _____ configuration provides the lowest voltage gain.
18. The _____ configuration provides a 180° phase shift between its input and its output.
19. The _____ amplifier is often used as an output stage to match impedance.
20. Of the three types of transistor amplifiers, the _____ configuration provides the highest voltage gain.
21. The _____ has the highest input resistance of the three configurations.
22. The _____ has the lowest output resistance of the three configurations.
23. The _____ has the highest power gain (A_P) of the three configurations.
24. The common-base amplifier has a current gain of _____.
25. Of the three types of transistor amplifiers, the _____ configuration provides the highest current gain.

Electronic Voltage Regulators

In earlier chapters, we discussed rectifiers and filters. At that point, we briefly mentioned voltage regulation. When discussing filter circuits, we developed the facts that capacitors serve to regulate voltage and coils serve to regulate current. We did, though, find that it was impossible to remove all ripple voltage, and therefore current variations, by use

Amplification and Voltage Regulation

of coils and capacitors. In this section, we will discuss an additional type of regulation that has wide application in the electronics of today; the **electronic voltage regulator (EVR)**.

You should understand that the output of a power supply is affected in one of two ways; either variation of the input *or* when current is drawn from its output circuit. Many electronic circuits can operate with moderate variations in voltage and current. Some circuits, however, have very critical requirements. Computer circuitry can malfunction with only slight, and very short, variations. It is, therefore, very important to have extremely stable power sources for those circuits that require them. The EVR circuit allows this.

Zener Diode Regulators

Zener diodes were discussed in an earlier chapter. **Zener** is the name given to a family of diodes that are designed to operate with **reverse breakdown** current. These diodes operate in the avalanche region of their characteristic curves and are not damaged by the reverse current.

The characteristic curve for a Zener diode is very similar to that of a PN junction diode. When forward biased, the two diodes operate exactly the same (note Figure 9-16, points A to B). The difference occurs when the Zener diode is reverse biased (points D to E). Notice that between these two points, current varies over a broad range with very

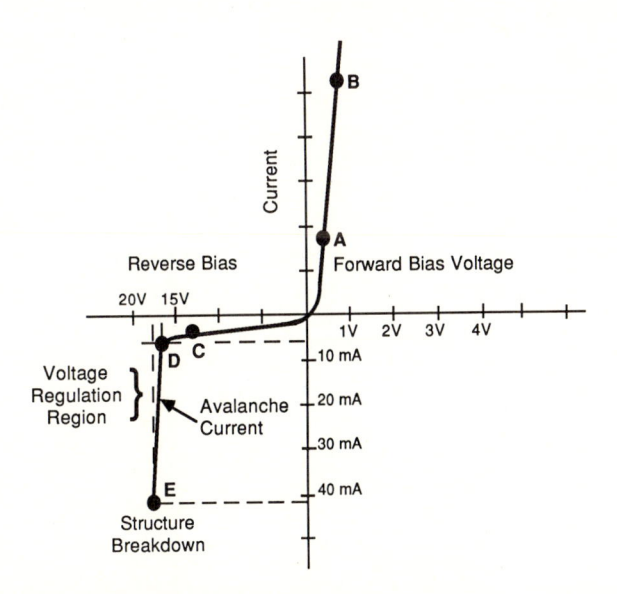

Fig. 9-16 Zener Diode Characteristic Curve

little change in voltage. In the example shown in Figure 9-16, the reverse bias voltage changes from 17 V to 17.5 V while current varies from 7 mA to 40 mA. This region is referred to as the **voltage regulating region** because of the small change in bias voltage that will occur from one current limit to the other. This constant voltage is a regulated voltage. The diode will, however, be destroyed in the event that too much current is allowed to flow. At that current level, the diode will undergo structural breakdown and will be of no further use. To prevent structural breakdown, we connect a resistor in series with the Zener diode. This resistor will limit Zener diode current to safe levels. A resistor (R_S) and a Zener diode having $V_Z = 17$ V are connected in series with a 30 V power source are shown in Figure 9-17. R_S drops the extra 13 V and limits current to safe levels for the Zener. We can use the Zener diode to regulate voltage between 17 V and 17.5 V.

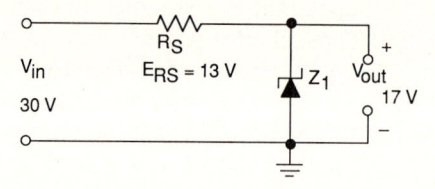

Fig. 9-17 Zener Diode Current

Electronic Voltage Regulator (EVR) Circuits

A more advanced circuit is the **electronic voltage regulator (EVR)**. It is designed to keep output voltage almost constant regardless of changes in either input voltage or output load. The EVR can be compared to a series resistive circuit where one of the resistors is adjustable. See Figure 9-18 for an illustration. The load device presents an impedance to the power source. The ideal power supply must be able to provide varying amounts of current at a constant voltage over long periods of time. With variations of input voltage, however, this is not possible. By placing a device (transistor) that is capable of having its impedance changed in the position of R_1, we can cause its opposition to change and maintain a relatively constant current in the load.

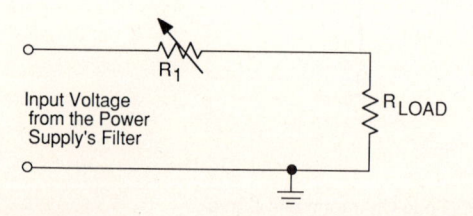

Fig. 9-18 Equivalent EVR Circuit

Amplification and Voltage Regulation

Before going further, we will discuss how this circuit could be used to regulate voltage. Should the voltage from the power supply increase without R_1 being present, the current and voltage at the load would increase. With an adjustable R_1, this change in load voltage and current could be reduced by increasing R_1 enough to drop the amount of voltage increase. If R_1 drops all of the increase, this will leave V_{LOAD} constant. If R_1 should decrease with any decrease in input voltage, the load voltage will, again, be maintained constant.

Figure 9-19 contains the circuit diagram for an *electronic voltage regulator*. Q_1 and Q_2 are both Class A amplifiers. These two amplifiers are used to detect and regulate any changes in input voltage that may occur. Note that Q_1 is in series with the load, and all of the power supply's output current must flow through this transistor. This places it in the same position as R_1 in Figure 9-18.

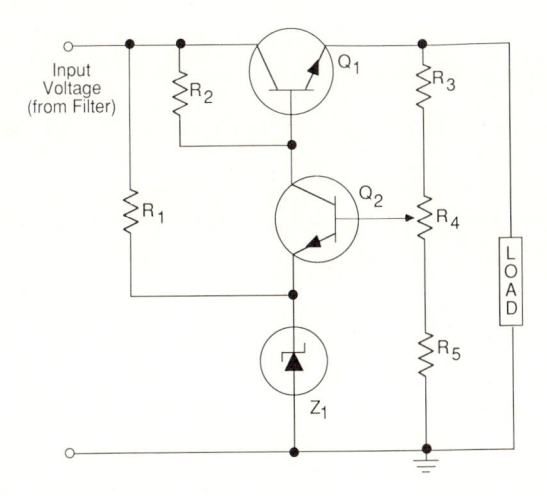

Fig. 9-19 Electronic Voltage Regulator Circuit

You already know that it is possible to control the resistance of a transistor by controlling the forward bias applied to its eb junction. Base current for Q_1 is controlled by the collector voltage of Q_2. R_2 is the collector resistor for Q_2 and also the forward bias resistor for Q_1. R_1 is the current limiting resistor for the Zener diode. R_3, R_4, and R_5 form a voltage divider, with R_4 being a variable resistor that can be used to adjust the output (load) voltage. Q_2 is forward biased by the voltage selected at the wiper of R_4 and the voltage established by Z_1 at its emitter. The Zener voltage is referred to as the **reference voltage**. Using this design, the voltage present at the wiper of R_4 is continuously sampled. If the output voltage should try to increase, the voltage at the wiper (R_4) would also increase. The increased voltage at the base of Q_2 changes the forward bias of the transistor.

To see how the circuit operates, follow along with this discussion. We will discuss two conditions that may occur in any operating circuit; a change in input voltage, or a change in the current in the load.

Any increase in input voltage will be felt across the voltage divider network of R_3, R_4, and R_5. The voltage drop at the wiper arm of R_4 will also increase. The increased voltage at R_4's wiper is applied to the base of Q_2 as an increase in forward bias. The increased bias causes Q_2 to conduct more heavily. This means that I_B, I_C, and I_E also increase. The voltage drop across R_2 increases, making the voltage at the base of Q_1 less positive. The less positive voltage at the base of Q_1 causes forward bias to decrease, currents I_B, I_E, and I_C to decrease, and the internal resistance of Q_1 to increase. With its increased resistance, Q_1 drops more voltage (collector to emitter). The increased voltage drop is almost equal to the change in input voltage.

If the input voltage *decreases*, each of these actions is the opposite of those explained above. The end result is the canceling of the decrease in input voltage. Using this mode of operation, the EVR acts (instantaneously) to cancel any variations that may occur in the input voltage.

If load current changes (another amplifier is turned on, for example) a similar chain of events is set up. An increased load current (decrease in load impedance) means that the voltage dropped by the load devices decreases. The change in load voltage is felt across the parallel network of R_3, R_4, and R_5. The decrease in voltage across the load causes the voltage at the wiper of R_4 to decrease which will, in turn, cause Q_2's forward bias to decrease. Q_2 then conducts less and causes the voltage drop on R_2 to decrease. This means that the base of Q_1 becomes more positive, causing its forward bias to increase. The increased bias causes I_C to increase and the transistor's internal resistance to decrease. Q_1 now drops less voltage, allowing the load voltage to increase back to where it was before the change in load current. Acting in this way, the EVR acts to effectively cancel out any changes in input voltage and load current that fall within its operating range.

Q_1 is referred to as the **series regulator**. The voltage on R_4 is referred to as the **error voltage**, and Q_2 is called the **error amplifier**. The Zener diode operates exactly like the Zeners discussed earlier in this section, with R_1 acting as the *limiter resistor*. The Zener diode is used to establish the *reference voltage* to which Q_2 is biased. Output voltage of a circuit of this type is very closely regulated.

The design of this circuit has been greatly simplified by the availability of integrated circuits that provide regulation for wide ranges of voltages. Sample regulators are shown in Figure 9-20.

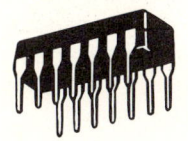

Fig. 9-20 Commercially Available EVR Packages

As mentioned earlier, the output voltage of these EVRs can be adjusted to a specific voltage value using R_4. Moving the wiper arm of R_4 up causes the forward bias on Q_2 to increase. This causes Q_2 to conduct more heavily, causing R_2 to drop more voltage. The forward bias on Q_1 has now decreased, causing its internal resistance to increase and its voltage drop to increase. The increased voltage drop on Q_1 causes the voltage present at the output to decrease. Adjusting the wiper arm of R_4 downward causes the output voltage to increase. In this manner, the output voltage can be adjusted for best operating value after the load device has been connected. The EVR then regulates the voltage around the selected voltage valve.

Self-Check

Answer each item by inserting the word or words required to correctly complete each statement.

26. The simplest voltage regulator discussed here consists of a _____ diode and a limiting _____.
27. The Zener diode regulator provides a relatively stable _____ over a wide range of _____.
28. In the electronic voltage regulator (Figure 9-19), Q_1 acts as the _____ regulator.
29. In an EVR, the error voltage is detected by _____.
30. An increase in forward bias on Q_1 within an EVR causes the voltage drop on the load device to _____.

Summary

Amplification of AC signals is an important function in electronics. Many solid-state devices can be used for amplification. Bipolar junction transistors are one type.

Amplifiers are classified according to frequency response and the percentage of time the transistor conducts. Frequency ranges are DC, audio frequency, video frequencies, and radio frequencies. Percentage of time of conduction classifications are:

CLASS A – conducts 100% of the time during which an input signal is applied to its input. It will never enter cutoff or saturation.

CLASS AB – is biased such that one alternation of the input causes it to cutoff. When a signal that provides is applied, the transistor supports current flow more than 50%, but less than 100%, of the time.

CLASS B – is biased at cutoff. With the arrival of a signal, the forward biasing alternation causes the transistor to conduct, and it will continue to conduct for the entire alternation. The opposite alternation is reverse bias and the amplifier cannot conduct.

CLASS C – is biased well below cutoff. Application of a forward biasing voltage must first overcome the reverse bias that is applied. Once the transistor begins to conduct, it will continue to conduct until forward bias decreases to the point that it is canceled by the fixed reverse bias.

Because of the way it is constructed and the material from which it is constructed, the bipolar transistor is very sensitive to heat. In fact, the transistor can be destroyed by heat. Several circuit modifications are used to protect a transistor from the possibility of heat damage. Methods of circuit modification that are discussed in this chapter are: emitter resistor, shunting the base junction with a thermistor, one diode, or two diodes. Each modification uses characteristics of the component added to provide temperature stabilization. When R_E is added, it also severely limits the circuit's voltage gain. By placing a capacitor in shunt with R_E, it is possible to have both temperature stabilization and high gain.

Input resistance, output resistance, voltage gain, current gain, and power gain for each of the three amplifier configurations are important. Calculations and/or methods that can be used to determine these parameters are discussed in this chapter.

The common-emitter amplifier shifts the phase of an AC input signal by 180° before applying it to the circuit's output. Common-base and common-collector amplifiers do not provide a phase shift, meaning that input and output signals are in phase. Common-collector amplifiers have low voltage gain and can have very low output impedance. For this reason they are often used to match impedance with a speaker or other device. Common-collector amplifiers are also called emitter followers.

Amplification and Voltage Regulation

The voltage applied to many electronic circuits must be free of amplitude variations. To provide this type DC voltage from an AC operated power supply, we often connect an electronic voltage regulator (EVR) in the power supply's output. An EVR can sense any change in input or load voltages. Once a change is sensed, the EVR acts to immediately cancel the change, keeping current and voltage stable at the load. This is done by use of transistors, a Zener diode, and a resistive voltage divider.

Review Questions and/or Problems

NOTE: Figure 9-21 will be used to answer the next five questions.

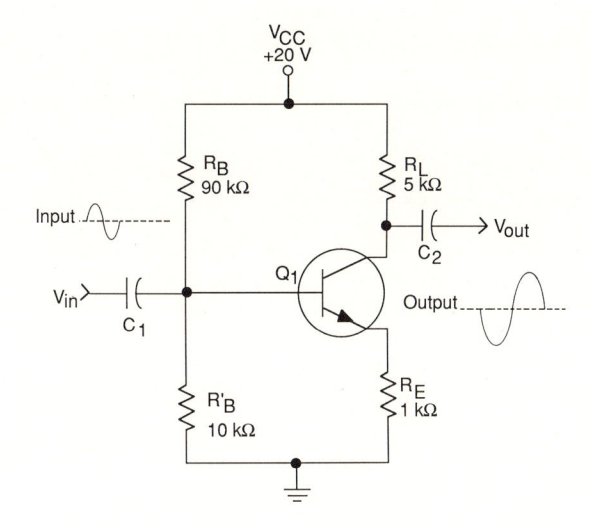

Fig. 9-21

1. Refer to Figure 9-21. In a common-emitter amplifier operating Class A _____.

   a. I_C flows 100% of the time
   b. the base is common to both input and output signals
   c. the output is taken across R_E
   d. the input signal is applied to the emitter

2. In Figure 9-21, what effect would increasing the size of R_L have on circuit voltage gain?

   a. increase
   b. decrease
   c. remain the same

3. Examine Figure 9-21. Bias is established by R_B and _____.

4. Check Figure 9-21. Capacitor C_1 is placed in this circuit for what purpose?

5. In Figure 9-21, what effect would increasing the size of R_E have on circuit voltage gain?

    a. increase
    b. decrease
    c. remain the same

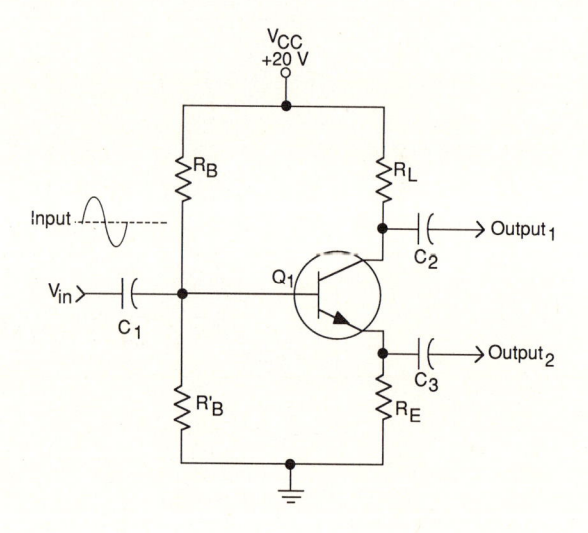

Fig. 9-22

NOTE: Figure 9-22 will be used to answer the next five questions.

6. This circuit is called a/an _____.

7. How does the phase of $Output_2$ compare to the phase of the input signal?

8. What effect does increasing the size of R_B have on the Q point of Q_1?

9. How does the phase of Output₁ compare to the phase of the input signal?

10. The amplitude of Output₂ will be _____ than the input signal's amplitude.

    a. larger   b. smaller

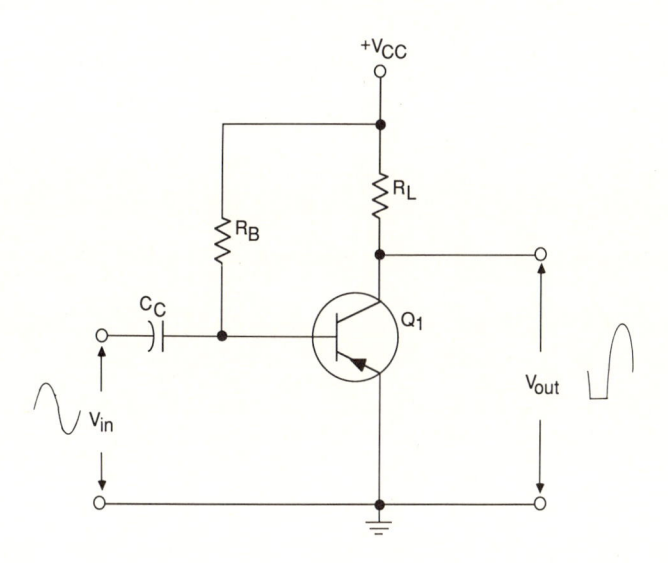

Fig. 9-23

NOTE: Figure 9-23 will be used to answer the next two questions.

11. Check Figure 9-23. This circuit is a common-_____ configuration.

12. Examine Figure 9-23. This circuit is operating Class _____.

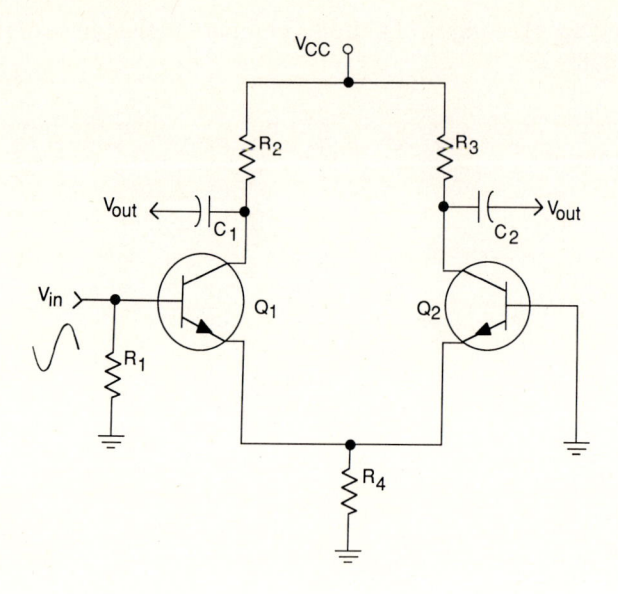

Fig. 9-24

NOTE: Figure 9-24 will be used to answer the next five questions.

13. This circuit is a/an _____.

14. Resistor R_4 serves as a common _____.

15. The output from this circuit is taken from _____.

16. If necessary, R_5 could be installed in this circuit for balancing purposes. Where would it be inserted and what purpose would it serve?

17. Explain why the base of Q_2 is grounded.

NOTE: Figure 9-25 will be used to answer the next three questions.

18. What is the purpose of C_1 in this circuit?

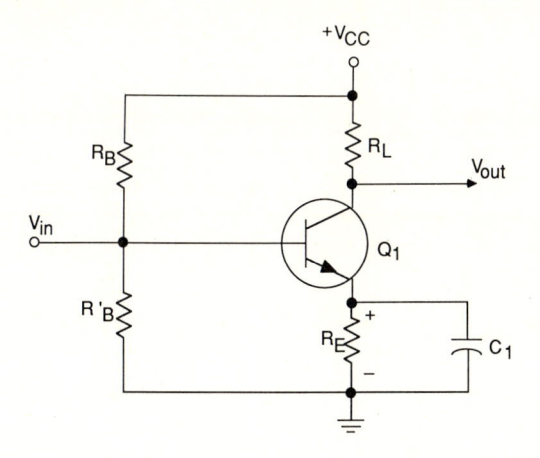

Fig. 9-25

19. What is the phase relationship between the input signal and the output signal?

20. How could the base of Q_1 be isolated from any DC voltage that might be present at the circuit's input?

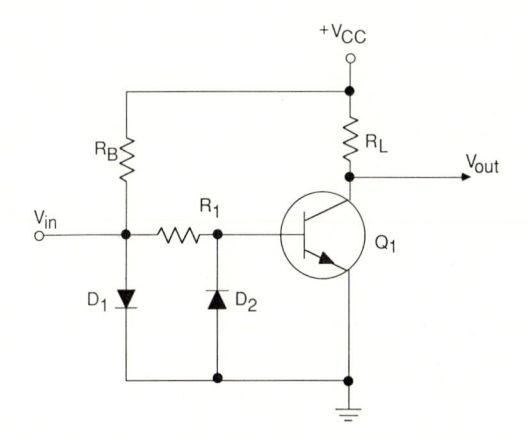

Fig. 9-26

> **NOTE:** Figure 9-26 will be used to answer the next two questions.

21. What type stabilization is shown here?

22. In this circuit, which diode is forward biased.

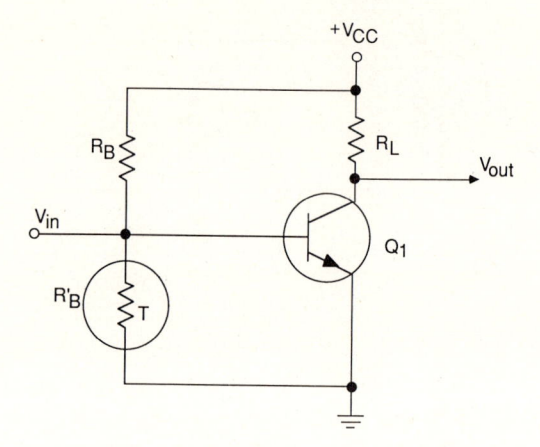

Fig. 9-27

> **NOTE:** Figure 9-27 will be used to answer the next three questions.

23. This circuit has _____ stabilization.

24. A thermistor has a _____ temperature coefficient.

25. Compare Figures 9-26 and 9-27. Which circuit has the best temperature stabilization and why?

> **NOTE:** Figure 9-28 will be used to answer the next five questions.

Amplification and Voltage Regulation

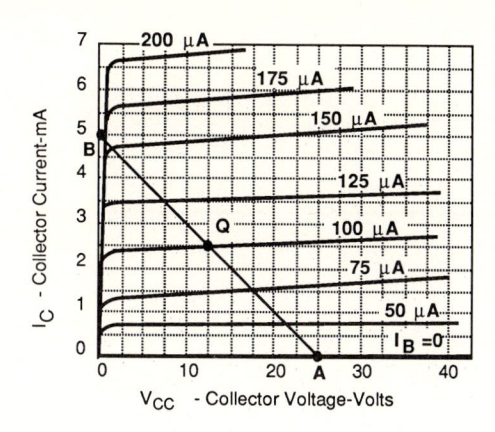

Fig. 9-28

26. The load line shown here is for a common-emitter amplifier. How much voltage is present at the collector of the amplifier when the transistor is cut off?

27. An AC input of 50 μA peak-to-peak is applied to the input of the transistor represented by this load line. What is the peak-to-peak swing of collector voltage?

28. When the transistor is cut off, collector current equals _____.

29. The transistor is saturated. What is collector current?

30. For the load line shown here, what size R_L is represented?

> **NOTE:** Figure 9-29 will be used to answer the next three questions.

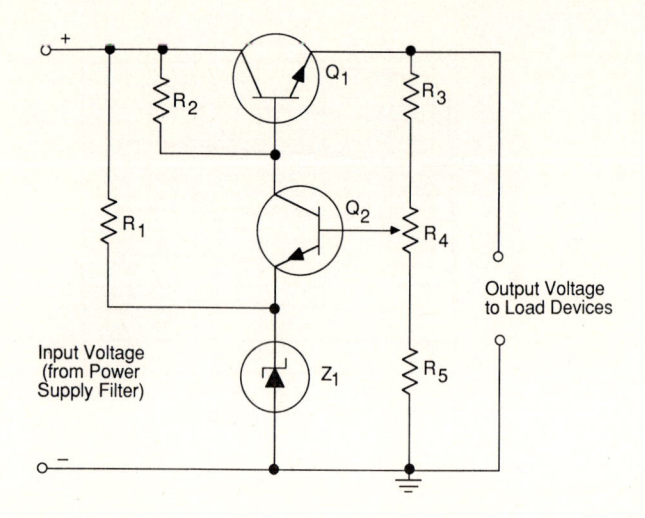

Fig. 9-29

31. Which transistor conducts total load current?

    a. Q_1 b. Q_2

32. Which transistor amplifies the error signal?

    a. Q_1 b. Q_2

33. In the EVR circuit (Figure 9-29), where is the reference voltage established?

    a. Across Q_1 c. At the arm of R_4
    b. Across Q_2 d. At the top of Z_1

34. The power supply output voltage increases slightly. What effect does this have on the EVR's operation?

35. The EVR is set to provide 10 V output. To adjust the output for 12 V, you would move the wiper of R_4 _____.

    a. up   b. down

10
Amplifier Applications and Coupling

Objectives

1. Define each of the following terms:

   a. Voltage gain
   b. Current gain
   c. Power gain
   d. Direct coupling
   e. RC coupling
   f. Power amplifier
   g. Output stage
   h. Impedance matching
   i. Impedance coupling
   j. Transformer coupling

2. Explain the part that each of the above plays in the design and operation of an electronic system.
3. Relate each type coupling to the applications that it is best suited to serve.

Introduction

It is rare that an electrical system will be able to operate with a single amplifier. In fact, it is more likely that several amplifiers will be used. These amplifiers must be connected so that the output from one amplifier serves as the input to the next. In this type system, a small signal is applied to the first (input) amplifier. Once amplified, the signal is then applied to the input of the second amplifier. This continues until the output signal from the final amplifier has sufficient amplitude to meet the needs for which it was designed. In audio, the final amplifier's output signal is used to drive a speaker. In a video system, the signal might be used to apply the picture to the screen of a television or a computer terminal. For broadcasting, the signal might be applied to an antenna for transmission over the air waves.

The purpose of each amplifier within an operating system is to receive the signal, to amplify it, and to provide it as an output suitable for acceptance and use by the next stage. Earlier, you learned that for maximum power to be passed from one circuit to another, the two impedances must be equal. This means that there must be ways of coupling two amplifiers together that provide for good power transfer. The methods that are used to *couple* amplifiers together will be discussed in the first part of this chapter.

Regardless of whether a system contains a few or many amplifiers, each amplifier has a specific job to do. Each stage must present an input impedance to the signal that allows the signal to be processed with best results. Each amplifier must process the signal by increasing its amplitude or power, and then present it at the stage's output. The last stage of the system is designed to develop the signal's power and/or amplitude to the level needed to drive the next circuit or device.

In general, the last stage of a series of amplifiers is called the **power amplifier**. Other names may be **output stage** or **final amplifier**. The power amplifier differs from the preceeding stages in that it is usually designed to produce maximum power gain instead of maximum voltage or current gain. Remember, power is the product of voltage and current. The power amplifier operates such that its output supplies the maximum possible power gain, or a combination of voltage gain and current gain that results in maximum power. This can be stated as:

$$A_{power} = A_{voltage\ gain} \times A_{current\ gain}$$

In the earlier chapters, we discussed the three transistor amplifier configurations and their classes of operation. Brief mention was made of voltage gain and biasing effects on those amplifiers. In this chapter, you will be given information pertaining to the different ways that amplifiers can be used, and how a circuit can be designed to achieve a desired effect.

Amplifier Coupling

As stated in the introduction, it is necessary that care be taken when stringing amplifiers together to assure that maximum efficiency of operation is being achieved. Four basic coupling methods are available; direct, resistance-capacitor (RC), impedance (LC), and transformer coupling. Each method has unique characteristics that make it more suitable for some applications than other methods. These are the items we will concentrate on in our discussion of *coupling methods*.

Direct Coupling

Direct coupling is one method used to connect amplifiers. Notice that Figure 10-1 contains two NPN common-emitter transistor amplifiers.

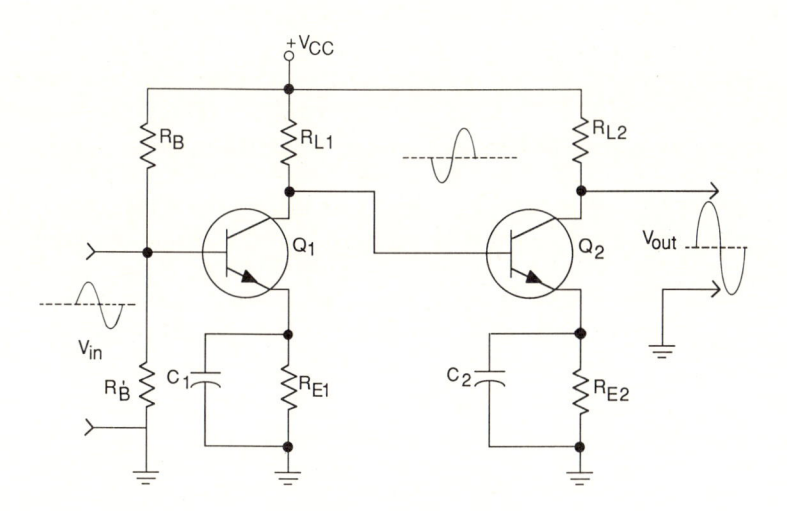

Fig. 10-1 Direct Coupled NPN Common-Emitter Amplifier

Notice that the collector of Q_1 is connected to the base of Q_2 by nothing more than a conductor. When two transistors are connected by a conductor they are said to be joined using **direct coupling**. In this arrangement, R_{L1} serves two purposes; as collector load resistor for Q_1, and as base resistor for Q_2.

When the positive alternation of the input signal arrives at the base of Q_1, it causes forward bias to increase. This increase in forward bias causes I_E, I_C, and V_{RL1} to increase. When V_{RL1} increases, the voltage dropped between Q_1 collector and ground goes negative (becomes less positive). When the positive alternation reaches peak and begins to decrease, I_E, I_C, and I_{RL1} begin to decrease. This causes the voltage drop at V_C of Q_1 to go more positive. As the input signal continues to decrease in amplitude, it will reach zero and then go into a negative alternation. From the positive peak of the positive alternation to the peak of the negative alternation, the input voltage is going negative, but the output voltage is going positive. At negative peak, the input signal begins to go positive and the voltage at the output of Q_1 beings to go negative. Note that the signal at the output (V_C) of Q_1 has been amplified and has undergone a phase shift of 180°.

Notice that both CE amplifiers are using bypassed emitter resistors thereby operating with higher gain. The signal developed at the collector of Q_1 is applied directly to the base of Q_2. When the output of Q_1 goes negative, it appears at the base of Q_2 as a reverse bias. This bias causes I_E and I_C to decrease. The decreasing I_C causes R_{L2} to drop less voltage. With V_{RL2} decreasing, the output voltage (V_C of Q_2) is increasing (going positive). This results in a positive going output at the collector of Q_2. When the output of Q_1 begins its positive alternation, the signal at the collector of Q_2 is going negative. Across the CE amplifier containing Q_2, the signal is again amplified and its phase is shifted by 180°. The end result is that the input signal at the base of Q_1 has been amplified by two stages of amplification and has had its phase shifted by 360°. We now have a signal that is considerably larger in amplitude but still has the same phase characteristics as the original input.

The power supply requirements for a circuit like that shown in Figure 10-1 are large when compared to other types of coupling. Each succeeding stage requires a higher DC voltage for its quiescent bias voltage. For the cascading of only a few stages, V_{CC} must be larger than 12 V.

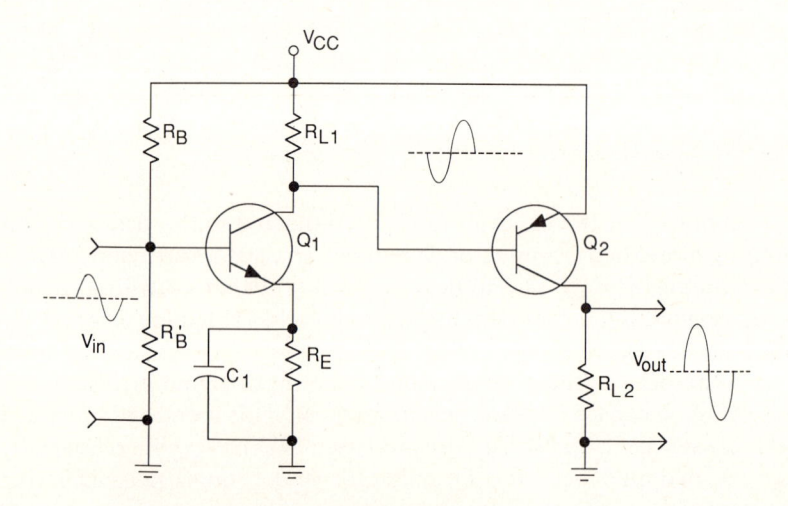

Fig. 10-2 Direct Coupled NPN and PNP CE Amplifiers

Figure 10-2 presents another type of direct coupling. In this circuit, Q_1 is an NPN and Q_2 is a PNP transistor. This arrangement allows us to use both type transistors with a single power source. Remember, we have used $+V_{CC}$ for NPN transistors and $-V_{CC}$ for PNP transistors. By connecting the transistors as shown here, they can both operate from a $+V_{CC}$ power source. This arrangement also eliminates the need for C_2 and R_{E2} shown in Figure 10-1. We will use the circuit shown in Figure 10-2 for this discussion.

Each transistor has its input signal applied to its base and provides an output at its collector. Q_1 is forward biased by resistor R_B and R_B'. R_{L1} serves as the collector load resistor for Q_1, and it also establishes the bias that is present at the base of Q_2. R_{L2} serves as the collector load resistor for Q_2. The signal is coupled out of each transistor from the collector, with reference to ground. With the circuit designed as shown and with no input signal, both amplifiers conduct at their Q point. In this state, each collector has a positive DC voltage present. Notice, though, that Q_2 has its emitter connected to the $+V_{CC}$. This means that any positive going voltage present at the collector of Q_1 is negative with respect to the emitter of Q_2. The positive voltage applied to the base of Q_2 is actually a forward bias voltage. In other words, the DC voltage present at the collector of Q_1 serves as the bias voltage for Q_2 base and forward biases Q_2, establishing its Q point.

When the positive alternation of the input signal arrives at the base of Q_1, it causes the transistor's forward bias to increase. The increased forward bias causes I_E, I_C, and V_{RL1} of Q_1 to increase. When V_{RL1} goes positive, the voltage from Q_1 collector to ground becomes less positive (goes negative). The output of Q_1 is coupled directly to the base of Q_2. This negative going voltage causes the forward bias on Q_2 to increase. An increase in forward bias causes I_E, I_C, and V_{RL2} for Q_2 to increase. When V_{RL2} increases, the voltage at the collector of Q_2 must go positive, providing a positive going output.

When the negative alternation of the input signal arrives at the base of Q_1, it causes the transistor's forward bias to decrease. The decrease in forward bias causes I_E, I_C, and V_{RL1} of Q_1 to decrease (go less positive). When V_{RL1} decreases, the voltage from Q_1 collector to ground increases (goes positive). This is a 180° phase shift of the input signal. The output of Q_1 is coupled directly to the base of Q_2. The positive going signal causes the forward bias on Q_2 to decrease. A decrease in forward bias causes I_E, I_C, and V_{RL2} to decrease. When V_{RL2} decreases, the voltage at the collector of Q_2 goes less positive. This provides a negative going output signal which gives the second 180° phase shift of the input signal.

Again, we have considerable amplification of the original input signal and a 360° phase shift. In each direct coupled amplifier, we achieve the same results. One advantage of the type shown in Figure 10-2 is that it can be used with smaller voltages (V_{CC}) than the one in

Figure 10-1. Another advantage is that the circuit can operate without components C_2 and R_{E2} that are necessary for the circuit shown in Figure 10-1.

Direct coupled amplifiers can be constructed with a minimum of parts and labor. This results in a low-cost product that serves many useful purposes. When circuits are coupled end-to-end, they are said to be *cascaded*. The number of stages that can be cascaded using *direct coupling* is limited because any undesired change that occurs in Q_1 is amplified in each succeeding stage. One example is the effect of changes in operating temperature. An effect felt by Q_1 is amplified by Q_2 and any other stages that might be connected. The result is a large change in the I_C of later stages resulting from small changes of temperature in early stages.

Amplifiers must be designed to operate at specific frequency bands. For example, an amplifier that is used for an *audio amplifier* must be able to amplify all frequencies between 20 Hz and 20 kHz equally well. The direct coupled amplifier has two distinct advantages; it can be used as a DC amplifier, and it can be used to amplify low frequency AC signals. Figure 10-3 contains the *frequency response curve* for a direct coupled amplifier. For a given input signal amplitude, the gain of the circuit remains constant from DC, through the audio range and into the lower radio frequencies. Radio frequencies range from 20 kHz to 300 GHz. Notice that the circuit has a *flat response* throughout the AF range (vertical dashed lines). Remember, capacitive reactance decreases as frequency increases. In this circuit, loss of amplification at higher frequencies is due to the *distributed (stray) capacitance* of the circuit. This is represented in Figure 10-3 by the dashed line that represents the frequency response.

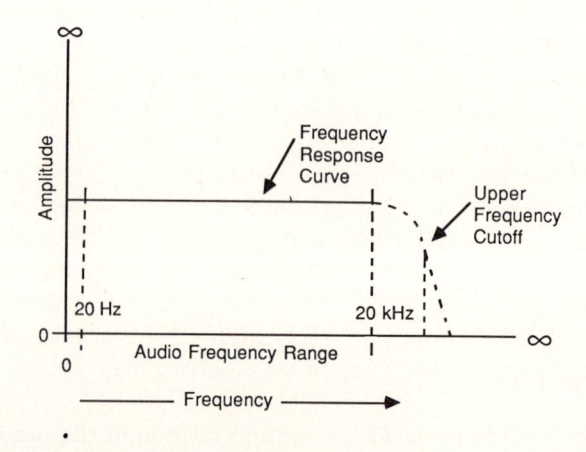

Fig. 10-3 Frequency Response Curve (Direct Coupling)

Amplifier Applications and Coupling

Figure 10-4 shows the *interelement capacitance* that is present in a transistor. Although the capacitances C_{eb}, C_{ce}, and C_{cb} are shown as external capacitors, they are actually the capacitance that exists in the junction areas of the transistor. The collector and base form the plates for C_{cb}, and the collector-base junction depletion region serves as the dielectric. As the frequency of the input signal increases, the X_C of the interelement capacitors decreases. At some frequency, X_C of the interelement capacitances is so low that the capacitances act as shunt (parallel) paths for the AC signal. These low reactances short both the input and output signals to ground, resulting in a loss of amplitude in the output. Once the output amplitude drops 50% below its peak power, the output signal is of little use.

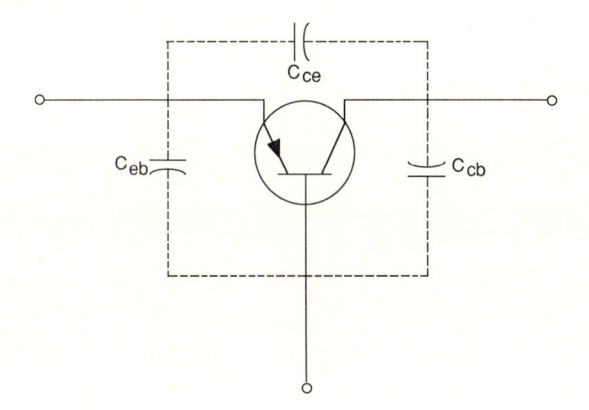

Fig. 10-4 Interelement Capacitance

We discussed distributed capacitance in an earlier chapter, but let us review this subject again. **Distributed (stray) capacitance** is the capacitance that exists between circuit components and wiring, plus the interelement capacitance present at the transistor junctions. The distributed capacitance between two conductors could cause output signal amplitude to decrease, as it is in parallel with the output of all transistors. Remember, total impedance will decrease as the reactance of any branch decreases. In audio amplifiers, distributed capacitance is not a problem. The frequencies being amplified are quite low (20 kHz maximum) and at this frequency X_C is still quite high. The effects of distributed capacitance become greater at higher frequencies.

RC (Resistive-Capacitive) Coupling

In this type coupling, we use a combination of capacitors and resistors for the coupling of stages and for the input signal. A circuit that uses **RC coupling** appears in Figure 10-5. In this circuit, RC coupling connects two NPN common-emitter amplifiers. Two load resistors are present

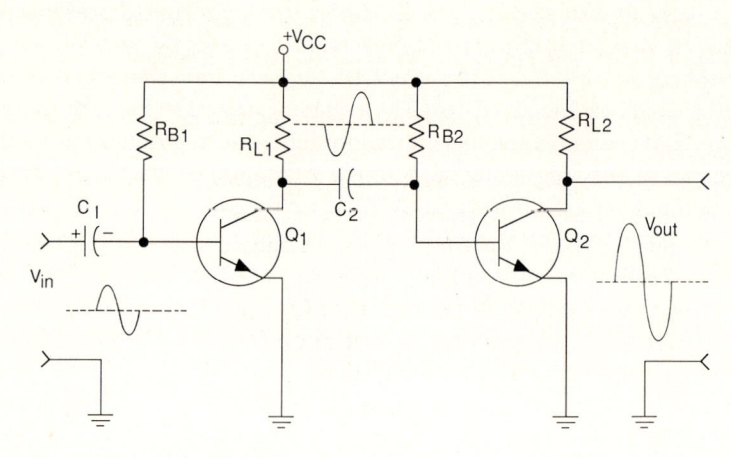

Fig. 10-5 RC Coupled Amplifiers

and are labeled R_{L1} and R_{L2}. Capacitors C_1 and C_2 are **coupling capacitors**. R_{B1} provides the bias for Q_1, and R_{B2} provides bias for Q_2.

If this circuit is operated with DC voltage and no input signal, the following conditions exist. R_{B1} and R_{eb} establish a forward bias on Q_1. The conduction of Q_1 causes R_{L1} to drop a voltage, which causes the collector voltage (V_{C1}) of Q_1 to be less than V_{CC}. This voltage (V_{C1}) is applied to the left plate of C_2. At the same time this has been happening, Q_2 has been forward biased by the conduction of junction resistance (R_{eb}) and R_{B2}. The small (0.7 V silicon, 0.3 V germanium) forward bias voltage is applied to the right side of C_2. This means that C_2 has a relatively high voltage (assume 10 V) on its left plate and a low voltage (0.7 V) on its right plate. Capacitor C_2 will charge to a voltage that equals the difference between these two voltages. In this case, the charge on C_2 is:

$$10\text{ V} - 0.7\text{ V} = 9.3\text{ V}$$

Forward bias applied to Q_2 causes it to conduct and causes R_{L2} to drop some voltage. When V_{RL2} increases, the voltage at the collector of Q_2 decreases to a less positive voltage. Assume that the DC voltage felt at Q_2 output is +10 V.

With no AC signal input, we have two forward biased amplifiers that have approximately +0.7 V applied to their base and a V_C of approximately +10 V. Capacitor C_2 is charged to +9.3 V and is effectively blocking the high DC (+10 V) present at Q_1 collector from being applied as bias to the base of Q_2. The left side of C_2 has +10 V applied and the right side has 0.7 V applied. To maintain this difference, C_2 must charge when Q_1 collector voltage increases or discharge when Q_1 collector voltage decreases. In either case, the charge or discharge has a path through R_{B2} to V_{CC}. The effect of the capacitor's charge and discharge causes the forward bias of Q_2 to increase or decrease in proportion to the change at the collector of Q_1. In effect, C_2 is:

Blocking the DC component but passing the AC component.

Now we will discuss what happens when an AC signal is applied to C_1. When the input signal goes positive, C_1 charges through R_{eb} of Q_1. The increased current flowing in the junction serves to increase Q_1 forward bias, which causes I_E, I_C, and V_{RL1} of Q_1 to increase. As V_{RL1} increases, the voltage at the collector of Q_1 decreases (goes negative). The positive voltage applied to the left side of C_2 is the same voltage as Q_1 collector voltage, and therefore, it must also decrease. C_2 reacts to this voltage change and tries to discharge the amount of voltage that is represented by the change of Q_1 collector voltage (V_{C1}). It must discharge through R_{B2} to V_{CC}. This applies a negative voltage to the base of Q_2, which decreases forward bias and I_C of Q_2. As Q_2 conduction (I_C) decreases, V_{RL2} also decreases (goes less positive).

With the negative going portion of the input signal, the process is identical except that the changes in current and voltage cause C_2 to charge to a new value. As C_2 charges, the forward bias on Q_2 increases. Increased forward bias at Q_2 causes V_{RL2} to go more positive and the output (V_C) of Q_2 to decrease (go less positive).

From this explanation, you should be able to see that any change appearing at the collector of Q_1 is coupled to the base of Q_2 by the charge and discharge of C_2. This allows us to use capacitors in our coupling circuits at points where we want to prevent the DC voltage of one circuit from affecting the bias of the next circuit. The use of the capacitor assures us that the AC (change) component of the input signal will be passed.

At some high frequency, the distributed capacitance affects output power and it drops below the half-power level. This point is called the upper frequency cutoff. This point identifies the highest frequency that the amplifier can effectively amplify.

The manufacturer's specification sheet provides us with one of two specifications:

1. The **alpha cutoff frequency** identifies upper frequency cutoff for a *common-base amplifier*. This parameter is identified with the symbol f_α.
2. The **beta cutoff frequency** identifies the upper frequency cutoff of a *common-emitter amplifier*. This parameter is identified by the symbol f_β.

Regardless of which specification is listed, one can be converted to the other by use of one of the following formulas:

$$f_\alpha = \beta \times f_\beta$$
$$f_\beta = \frac{f_\alpha}{\beta}$$

For example, assume that the specifications given are: $\beta = 50$ and $f_\alpha = 700$ kHz. To solve for f_β, do the following:

$$f_\beta = \frac{f_\alpha}{\beta}$$

$$f_\beta = \frac{700 \text{ kHz}}{50}$$

$$f_\beta = 14 \text{ kHz}$$

The low frequency response of a circuit is controlled by the size of the coupling capacitors and the input impedance of the stage. By changing the value of either one, you can change the lower frequency cutoff.

As a general rule, we state that low frequency cutoff occurs when the coupling capacitor's X_C equals the input impedance of the next stage. To determine the frequency where low frequency cutoff will occur, use the formula:

$$f_{Low} = \frac{1}{2\pi RC}$$

This formula is used with R and C representing the resistance and capacitance found in the coupling circuit.

The load connected to the circuit's output will have some effect on frequency response, but the above formulas provide a good "rule-of-thumb" for our use.

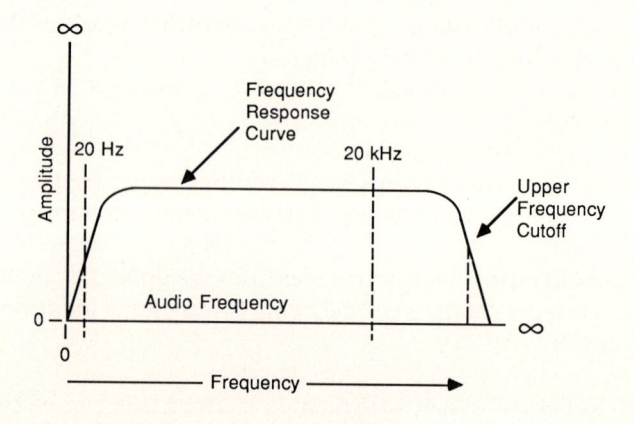

Fig. 10-6 Frequency Response Curve (RC Coupling)

Figure 10-6 contains a frequency response curve for an RC coupled audio amplifier. Assume that the high reactance of the coupling capacitor causes a loss of amplitude at the low frequency end of the curve, with output being usable above 15 Hz. The upper frequency cutoff occurs as stated above. To determine the size of C_1 and C_2, we proceed as follows:

Assume that $Z_{input} = 25 \, \Omega$.

We know that Z_{in} must $= X_C$ at 15 Hz.

Therefore, $X_C = 25 \, \Omega$

and lower frequency cutoff $= 15$ Hz.

Substitute these values in the transposed X_C formula shown below:

$$C = \frac{1}{2\pi f X_C}$$

$$C = \frac{1}{2(3.1415927)(15)(25)}$$

$$C = \frac{1}{2356.2}$$

$$C = 425 \, \mu f$$

You can see that this would be a very large capacitor. Except for very sophisticated audio systems, the selection of coupling capacitors is at best a compromise. In systems that operate with higher frequencies, this is not a problem. We must realize, however, that using a transistor having a higher input impedance overcomes some of this. One solution is to insert an unbypassed resistor in the emitter of all transistors that operate at audio frequencies. You should remember that all unbypassed emitter resistors appear as $\beta+1$ larger to the input. Their installation increases the circuit's input impedance. The addition of unbypassed emitter resistors also decreases the voltage gain for each amplifier that receives this added resistor.

Impedance (LC) Coupling

Impedance coupling is similar, in some ways, to RC coupling. In this type, we use the increasing reactance (X_L) of a coil to overcome some of the frequency loss that occurs at the high end of the frequency response curve. Figure 10-7 illustrates this type circuit. Note that the two load resistors have been replaced by L_1 and L_2. Impedance coupling is used in circuits that operate above the audio range. This is because:

As frequency increases, X_L increases.

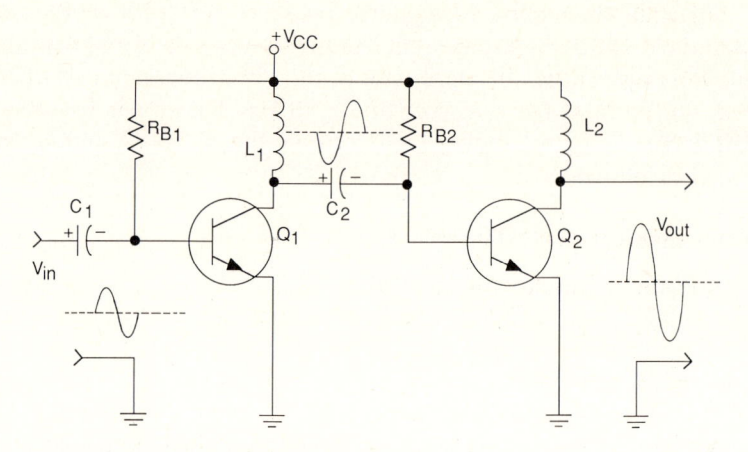

Fig. 10-7 Impedance Coupled Amplifiers

As the X_L of the collector-load coil increases, the voltage gain of the amplifier increases. At higher frequencies, X_L is large enough to significantly increase the voltage gain of the amplifier. The advantage gained from use of LC coupling is increased gain, which compensates for some of the signal lost as a result of distributed capacitance.

The main disadvantage of impedance coupling is that it is limited to high frequency applications. This is because of the large value inductors that are required to obtain high X_L at low frequencies. At low frequencies, X_L is so low that the amplifier's voltage gain is very low. Figure 10-8 contains the frequency response curve for an impedance coupled circuit. Notice the low voltage gain that occurs at low frequencies.

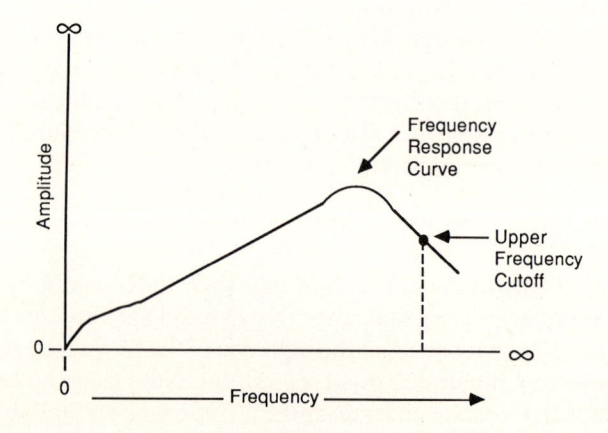

Fig. 10-8 Frequency Response Curve (Impedance Coupling)

Amplifier Applications and Coupling

At high frequencies, the output amplitude drops quickly due to capacitance added to the distributive capacitance by the space and insulation separating the coil's windings. The peak in the response curve results from the coil and the circuit's distributed capacitance resonating.

Transformer Coupling

The use of transformers for coupling of stages has definite advantages and disadvantages. Figure 10-9 contains the schematic diagram for two NPN common-emitter amplifiers that are connected using **tranformer coupling**. Notice that the primary winding of each transformer serves as the collector load device for a transistor. The secondary winding of T_1 couples the signal to the base of Q_2. R_{B1} and R_{eb} develop the forward bias for Q_1, while R_{B2} and R_B' provide forward bias for Q_2. The low DC resistance of T_1 secondary is so small that we can ignore its effect on Q_2 bias. C_1 places AC ground at the bottom of T_1 secondary. Placing an AC ground at this point assures that the secondary will couple the AC signal, which is then applied to the base of Q_2 as the input signal.

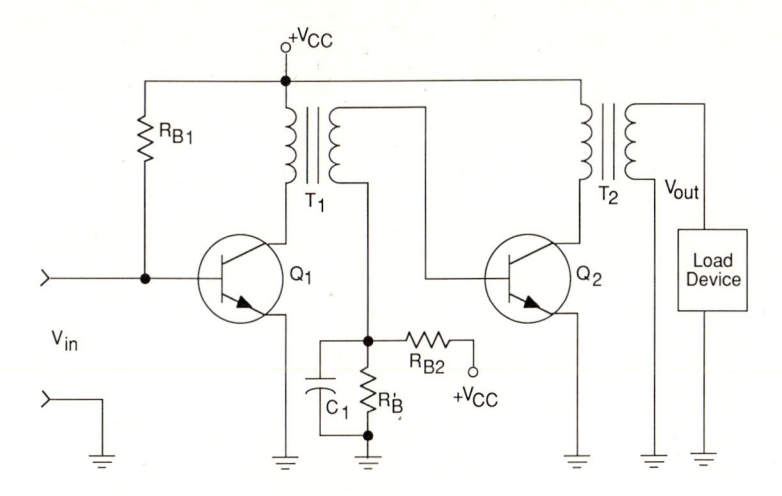

Fig. 10-9 Transformer Coupled Amplifiers

Because there is no collector load resistor to dissipate power, transformer coupled amplifiers are very power efficient. For this reason, transformer coupled amplifier circuits are used in many pieces of portable equipment where battery power is used.

Transformers provide a method for matching the impedance of one amplifier's output to another amplifier's input. This is a way of assuring that maximum power is transferred.

Transformers have an advantage: they can be designed into circuits that provide good selectivity of desired frequencies. It is possible by use of tuned transformers, to couple very narrow bands of high frequency signals. These applications are discussed in chapter 11.

The main disadvantage encountered in the use of transformer coupling is cost. When compared to other types of coupling, transformer coupling is quite costly. Also, the use of transformers adds bulk and weight to the circuit, and in many cases, this is a definite disadvantage.

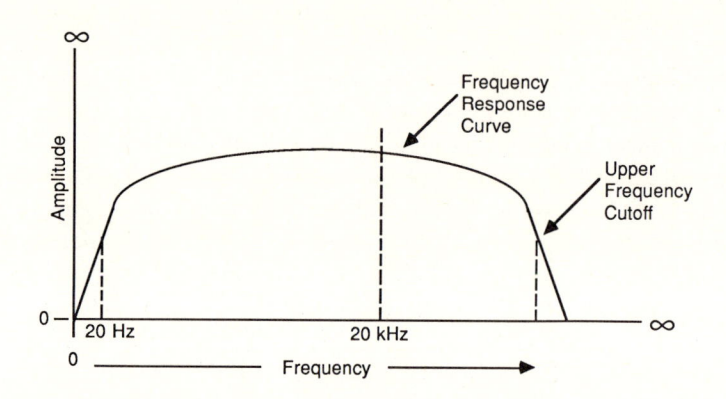

Fig. 10-10 Frequency Response Curve (Transformer Coupling)

A representive frequency response curve for transformer coupled amplifiers is shown in Figure 10-10. The low reactance of the transformer windings at low frequencies causes low frequency response to decrease. At high frequencies, the response is reduced by the distributed capacitance of the circuit. The capacitance between the turns of the transformer windings are included in the distributed capacitance.

Self-Check

Answer each item by inserting the word or words required to make each statement a true statement.

1. Amplifiers that use _____ coupling are used to amplify direct current.
2. For maximum power to be transferred, it is necessary to match the _____ impedance of the one stage to the _____ impedance of the next stage.

3. Of the four types of coupling, which provides the best low frequency amplification?

   a. RC
   b. Direct
   c. Impedance
   d. Transformer

4. Transformer coupling can be used to select and couple very narrow bands of frequencies.

   a. True
   b. False

5. When using RC coupling, the size of the input _____ and the _____ determines the low frequency response of the circuit.

6. Of the following, which type of coupling is used to extend the upper frequency response of a circuit?

   a. RC
   b. Direct
   c. Impedance
   d. Transformer

7. A disadvantage of direct coupling is the fact that the bias applied to the base of Q_2 is _____ than would normally be expected.

   a. higher
   b. lower

8. When using direct and RC coupling, the loss of gain that occurs at high frequencies results from the _____.
9. High frequency compensation is accomplished by the insertion of _____ into the circuitry.
10. Of the following, which type of coupling provides the best selection of a desired band of frequencies?

    a. RC
    b. Direct
    c. Impedance
    d. Transformer

Cascaded Amplifiers

When amplifiers are connected so that the output of one amplifier is applied as input to the next amplifier, the amplifiers are said to be **cascaded**. Whether direct, RC, impedance, or transformer coupling is used, the amplifiers are said to be connected in cascade (connected in line).

The gain of cascaded amplifiers can be determined by multiplying the gains of each stage. That is, multiply the gain of stage 1 by the gain of stage 2, by the gain of stage 3, and so on. Figure 10-11 illustrates this.

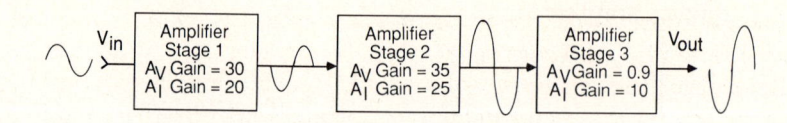

Fig. 10-11 Common-Emitter Amplifiers Three-Stage Block Diagram

Voltage gain (A_V) of stage 1 = 30 and current gain (A_I) = 20. Remember:

$$A_{power}\ (A_P) = A_V \times A_I$$

therefore:

$$A_P = 30 \times 20 = 600$$

For stage 2, the gains are: $A_V = 35$ and $A_I = 25$. In this circuit, power gain (A_P) equals:

$$A_P = A_V \times A_I$$

therefore:

$$A_P = 35 \times 25 = 875$$

For stage 3, the gains are: $A_V = 0.9$ and $A_I = 10$. In this circuit, power gain (A_P) equals:

$$A_P = A_V \times A_I$$

therefore:

$$A_P = 0.9 \times 10 = 9$$

Now that we know the voltage, current, and power gains of each stage, we can determine the total gain of the three stages of multiplication.

$$A_V = A_{V1} \times A_{V2} \times A_{V3} = 30 \times 35 \times 0.9 = 945$$

$$A_I = A_{I1} \times A_{I2} \times A_{I3} = 20 \times 25 \times 10 = 5000$$

$$A_P = A_{P1} \times A_{P2} \times A_{P3} = 600 \times 875 \times 9 = 4{,}725{,}000$$

This tells us that with 1 mV input, we can expect 0.945 V output. Or, with 1 μA input, we can expect 5 mA output. Or with 2 μW input power, we can expect 9.45 watts at the output of stage 3.

Troubleshooting Amplifiers

The malfunction of a component in an amplifier circuit produces a specific symptom. These symptoms "tell" the technician what is wrong. Before the technician can repair a circuit, he/she must have a thorough knowledge of the circuit and how it operates under normal conditions. Refer to Figure 10-12 for the discussion that follows.

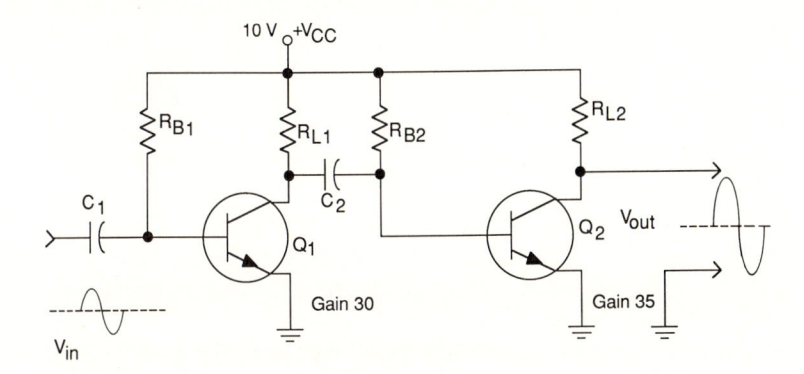

Fig. 10-12 Common-Emitter Amplifiers Two-Stage Circuit

During normal operations, the following DC voltages are approximations of what could be expected at the locations noted:

From the base of each transistor to ground $\cong +0.7$ V

From the collector of each transistor to ground $\cong +5$ V

To measure these voltages you must have a good quality multimeter and/or an oscilloscope, preferrably both. AC signals within the circuit can be observed and measured with the oscilloscope. An observation of the input to Q_1 should reveal a small AC waveshape (assume 1 mV). At the collector of Q_1, the scope should show a larger AC waveshape of approximately 30 mV that is 180° out of phase with the input. The AC signal at the base of Q_2 should have nearly the same amplitude as the collector of Q_1. We will assume they are equal at 30 mV. After being amplified by Q_2, the output voltage should have an approximate amplitude of 1 V when observed on the scope. This signal should be 180° out of phase with the input to Q_2. After having gone through a phase shift of 360°, the signal is now back in phase with the input to Q_1. When these signals are viewed and the DC voltages mentioned above are measured, we know that the circuit is working properly.

A common symptom of malfunction is the absence of an output. If you are listening to your stereo and the sound stops, it is very easy to recognize the symptom of no output, meaning no sound. Other circuits speak just as plainly although they *may not* be as easily understood.

Once the symptom appears, your job is to locate the problem and repair the defect.

Should the emitter-base junction of either transistor short, there would be no output signal. All signals would be shorted to ground through the shorted eb junction. If either transistor is open, the signal is blocked and cannot be coupled to the output. If R_{L1} opened, collector current for Q_1 would be blocked and no output could be developed for application to Q_2. If either C_1 or C_2 should open, the path for the signal would be open and no output could be present. If C_2 should short, the collector voltage from Q_1 would be applied to the base of Q_2. This high forward bias would drive Q_2 into saturation, and Q_2 could not amplify the incoming signal. If, however, Q_2 is not driven into saturation, the signal present at output will be very small and badly distorted.

No signal at the base of Q_2 could be another problem. This tells us that the problem must be between the input and the right side of C_2. The number of components that can be suspect is limited considerably. The problem must lie with the transistor (Q_1) being open or shorted, R_L open, or either C_1 or C_2 open.

Another symptom might be that when you measured the voltage between Q_1 collector and ground, the meter indicates V_{CC}. In this case, the trouble must be either an open Q_1 or a shorted R_{L1}.

To accurately analyze and troubleshoot transistor circuits, you must consider all possible current paths. Within each current path, you must understand the effect of each component and how the applied voltage (V_{CC}) is divided. It is possible for the collector-base junction of a transistor to open while the emitter-base junction still conducts through the biasing resistor R_B. Remember, anytime that current flows through a resistive device, there must be a voltage drop. This is not to say, however, that any time we measure voltage that current must be flowing. You should recall that an open drops the total voltage applied to that portion of the circuit.

Classification of Amplifiers

Amplifiers are classified (grouped) in several ways. The commonly used classifications are **frequency range, class of operation,** and **use.**

Classification By Frequency

Quite often, we refer to amplifiers by the frequency or range of frequencies that they are designed to amplify. A **DC amplifier** is one that is able to amplify *direct current* (0 Hertz). DC amplifiers are used in many devices, the oscilloscope is one. Another amplifier is designed to amplify all frequencies from approximately 20 Hz to 20 kHz. Because these are the frequencies that fall within the *audible* range, we call this an **audio amplifier.** Other amplifiers are designed to operate at frequencies well above 20 kHz and are referred to as **RF (radio frequency) amplifiers**. Still others are used to amplify signals that carry the picture in a television signal. These are called **video amplifiers**. Radio and video amplifiers are further divided into **narrow-band** and **wideband** classifications. These amplifiers are discussed in chapter 11.

Classification By Type of Operation

There are four classes of operation. These are *Class A, Class AB, Class B* and *Class C*. The class of operation is determined by the percentage of time that the amplifier conducts. These classes were discussed in Chapter 9.

Classification By Use

Amplifiers are quite often grouped into categories that pertain to how they are used in a circuit. Remember – current, voltage and power gains of a circuit are dependent on the design of that particular amplifier. Things that affect these gains are configuration, the size of the load resistor, and transistor type. An amplifier that is designed for high current gain is called a **current amplifier**. One that provides a high voltage amplification is called a **voltage amplifier**. A third type provides large power amplification and is called a **power amplifier**. The latter type is the subject that is discussed in the remainder of this chapter.

Self-Check

Answer each item by performing the action indicated.

11. Cascaded amplifiers have the _____ of one amplifier connected to the _____ of the next amplifier.
12. In cascaded amplifiers, total voltage gain can be calculated by _____.
13. A transistor has opened. A check of its collector voltage indicates _____.
14. Refer to Figure 10-12. C_2 is open. What symptom would you recognize as indicating this problem?
15. Refer to Figure 10-12. Q_2 is shorted. This will cause the output voltage to be _____.
16. Audio frequency amplifiers must be able to amplify all frequencies from _____ to _____ equally well.
17. A voltage amplifier is designed so that it will provide an output signal that has _____ voltage.
18. One way in which amplifiers are classified is by type of operation. This classification system is concerned with the _____ of time that the transistor conducts.

19. Refer to Figure 10-12. When operating normally, the sum of the voltage drops in each branch of this circuit equals V_{CC}.

   a. True   b. False

20. Radio Frequency (RF) amplifiers operate at frequencies above _____.

Power Amplifiers

One limitation of bipolar transistors is the amount of power that they can dissipate. The **maximum power dissipation** (P_{max}) rating of a transistor refers to its maximum permissible power. To exceed this power rating is to run the risk of destroying the transistor.

Maximum collector dissipation is the maximum power that can be safely dissipated by the collector of a transistor. This operating limit was discussed earlier when you were introduced to the transistor characteristic curve chart. Figure 10-13 is a sample chart. In our previous discussion, we identified the power dissipation rating as being a curved line (P_{max} curve) that connected the right ends of all base current lines on the chart. On our sample, this line is labeled *Transistor Total Dissipation (Watts)* = 5. We will refer to this as the P_{max} rating. Understand that P_{max} is the product of the DC quantities I_C and V_{CE} such that:

$$P_{max} = I_C \times V_{CE}$$

If the transistor manufacturer lists a P_{max} of 2 watts, this means that the product of the collector voltage and collector current for any base current *must not* exceed 2 watts. When operating with an AC input, it is possible for total power dissipation to exceed the 2 watts rating for short periods of time. The average power dissipation, though, must not exceed 2 watts.

In Figure 10-13, we are dealing with a silicon transistor. Notice that the lines terminate on the P_{max} curve, which represents 5 watts. Assume that we wanted to operate this transistor at 500 mA: we could determine the amount of V_{CE} that would be required as follows:

$$V_{CE} = \frac{5 \text{ Watts}}{500 \text{ mA}} = 10 \text{ V}$$

Now, if we locate the 500 mA line on the chart, then move right until we reach the P_{max} curve, and then move down to the bottom of the chart, we see that we intersect the horizontal line at 10 V. Another point is 250 mA and 20 V, another is 125 mA and 40 V, and another is 100 mA

Amplifier Applications and Coupling

and 50 V. Note that the product of each set is 5 watts. This confirms the fact that this curve represents the P_{max} characteristics of the transistor. You should recall that the load line for a transistor must be constructed to the *left* of the P_{max} curve or the transistor will be destroyed.

Typical Collector Characteristics

Common-emitter circuit, base input.
Mounting-base temperature = 25°C.

Transistor total Dissipation (Watts) = 5

Base milliamperes = 2

Collector - To - Emitter Volts

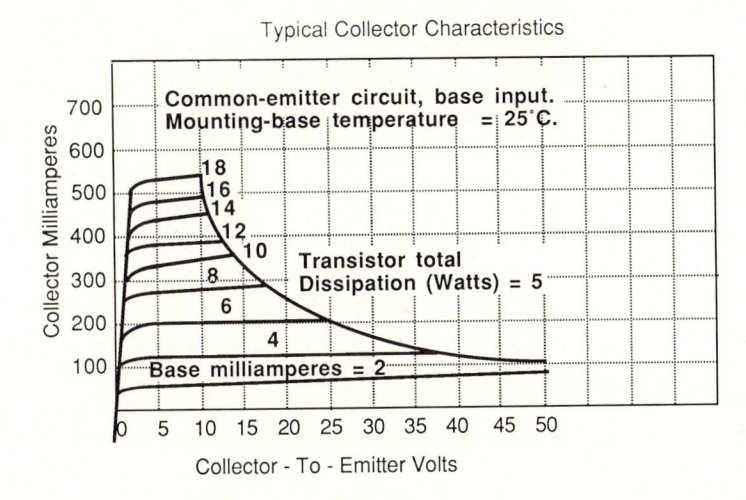

Fig. 10-13 Typical Transistor Curve Chart

A power amplifier is designed to operate at high power levels. When operating in this manner, the heat that is generated inside the transistor can be a problem. Remember, the gain and stability of a transistor depends on its operating temperature. To assist the transistor in the dissipation of the high heat that it dissipates, we often install a device called a **heat sink**.

Figure 10-14 contains pictures of typical power transistors and heat sinks. Typical power transistors are shown in Figure 10-14(a) and two drawings that depict a heat sink are shown in Figure 10-14(b). The

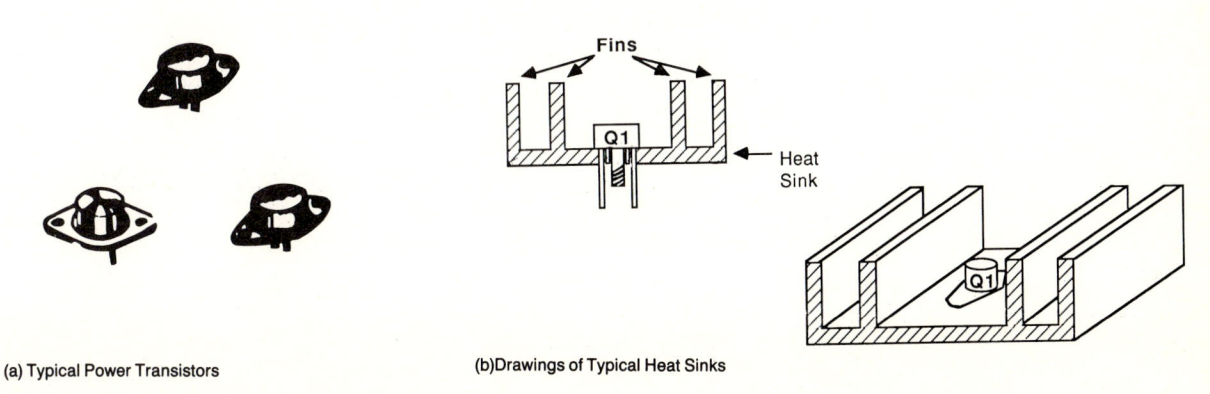

(a) Typical Power Transistors   (b) Drawings of Typical Heat Sinks

Fig. 10-14 Power Transistors and Heat Sinks

heat sink is made of metal and has fins that aid in spreading the heat over large areas so that it can radiate into the air more easily. In some cases, heat is such a problem that fans are installed that circulate air through the heat sink fins, removing the heat even faster. Appendix D contains data regarding heat dissipation (*heat sink*) devices.

In many cases, it is difficult to tell from a schematic diagram whether the circuit is designed for high power operation or not. Quite often, the low and medium power amplifier schematics are identical to the high power ones. The major difference is the installation of heat sinks, different mounting methods, and in some cases, the addition of a blower that aids in heat removal.

Another difference aside from heat is the fact that the power amplifier is the last stage of a system. In this position, it is designed to provide sufficient power to the device that receives the output power. To produce high power, the amplifier has a smaller load impedance than the preceeding stage. To have a maximum transfer of power from source to load, the output impedance and load impedance must be equal. When these conditions exist, we say that we have *matched impedance*. Remember, power is a function of current and resistance that can be represented by the formula:

$$P = I^2 \times R$$

To provide for high current, the value of R is kept quite small.

Single-Ended Power Amplifiers

A **single-ended amplifier** is one that normally uses only one transistor. Figure 10-15 contains the schematic diagram for a single-ended, NPN common-emitter power amplifier. All components in this circuit perform exactly as we described in chapter 9. Bias is developed by R_B and R'_B. Resistor R_B controls the amount of base current (I_B) which, in turn, establishes the Q point for the transistor. The emitter capacitor (C_E) is an AC bypass capacitor that places AC ground at the emitter of Q_1. The coupling capacitor (C_1) blocks DC and passes the AC input signal. R_E and R'_B provide bias stabilization for the transistor. R_L and V_C provide the output voltage variations that result from variations in I_B, I_E, and I_C.

The circuit is designed to use a transistor whose characteristics allow the production of high power at the output. R_L will probably have a smaller ohmic value than in previous single-ended amplifiers. It is possible that this transistor will be mounted on a heat sink. If not on a heat sink, it will almost definitely be secured firmly to the metal chassis that houses the circuit.

Amplifier Applications and Coupling

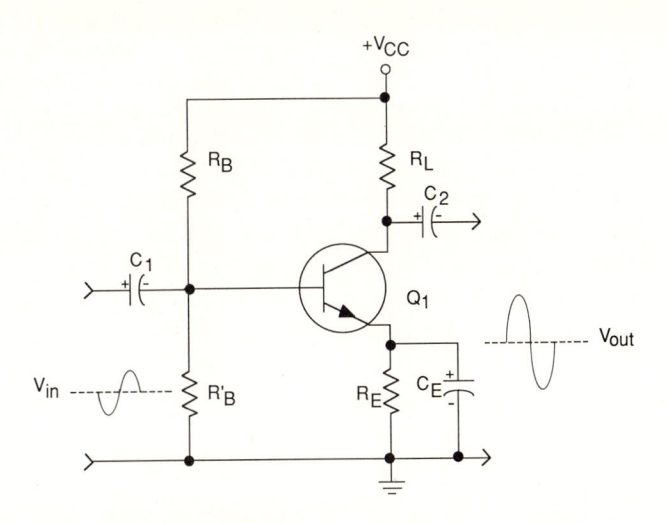

Fig. 10-15 Single-Ended Power Amplifier

Using this basic amplifier, it is possible to design multistage circuits that increase voltage and current through a succession of stages until the last stage (power amplifier) is reached. The input provided to the power amplifier is then converted into a signal that has sufficient power to perform the job needed. In an audio system, this would be the power needed to drive the speaker. Many systems require more power than can be provided by a single-ended amplifier. More power amplifiers are discussed in the section that follows.

Self-Check

Answer each item by performing the action indicated.

21. When using transistor characteristic curves to design a circuit, the load line must be constructed to the _____ of the P_{max} curve.

    a. right b. left

22. Power amplifiers can dissipate large amounts of power. To assist in the removal of this heat from the operating area of the transistor, we will install _____.
23. Power amplifiers are usually the _____ stages of an operational system.
24. A single-ended power amplifier has _____ transistor(s).
25. To get high-power output from a single-ended power amplifier, R_L will have a _____ ohmic value than before.

    a. smaller b. larger

Double-Ended (Push-Pull) Power Amplifier

A circuit that is called a **push-pull amplifier** is used in many applications where high power is required. The push-pull amplifier is said to be a **double-ended** amplifier because it has two amplifiers that operate 180° out of phase. The output currents from the two amplifiers add algebraically. This tends to increase the desired characteristics of the signal while canceling some of the less desireable characteristics. A push-pull amplifier consists of two transistors and the supporting circuitry shown in Figure 10-16. Note that this circuit has two transformers (T_1 and T_2) that have center-tapped windings. T_1 is called an **interstage** transformer, and T_2 is called an **output** transformer.

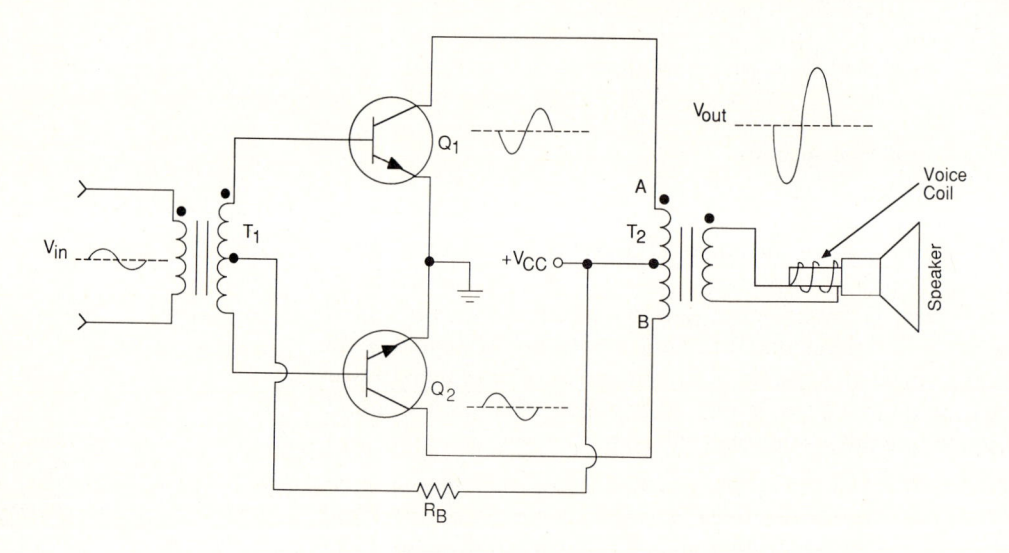

Fig. 10-16 Class A Push-Pull Power Amplifier

In this push-pull amplifier, a transformer is used to supply the input to Q_1 and Q_2. The center-tapped secondary of T_1 supplies two equal amplitude, 180° out of phase input signals to Q_1 and Q_2. This is a simple way of performing the job. Transformer (T_2) acts as the inductive load for each amplifier and couples the output to a speaker or other device.

The use of transformers, however, is expensive, plus it adds weight and bulk to the finished product. In some cases, these limitations will rule out the use of transformers for input and output coupling.

Resistor R_B provides forward bias for both Q_1 and Q_2. Application of this forward bias establishes the Q point for each transistor. One half of the transformer (T_2) primary provides the output load impedance for Q_1, and the other half provides the output load impedance for Q_2.

Transformer (T_2) is designed such that it provides impedance matching for the speaker, assuring that sufficient power is transferred from the source (amplifier) to the load (speaker). This is especially important when we realize that the output impedance of transistor circuits is usually very high, and the speaker's impedance is very low.

With no AC signal applied, no coupling occurs across T_1, and the circuit operates with DC to establish its Q point. With balanced transistors, both Q_1 and Q_2 will conduct equally. Current flows from ground, through the eb junction of each transistor, through one half of T_1 secondary, and through R_B to establish I_B for each transistor. This base current establishes the Q point. As a result, Q_1 conducts collector current from the emitter, to the collector, to point A on T_1, to the center tap, and to V_{CC}. Conduction for Q_2 is identical except that I_C enters T_2 at point B. Remember, induction only occurs when current in the primary is changing; therefore, no signal is induced into the secondary of T_2 when the circuit is conducting at its Q point.

Class A Push-Pull Power Amplifiers

Assume that the circuit operates Class A (conducts 100% of the time). Because the secondary of T_1 is center-tapped, two equal amplitude but 180° out of phase signals are present on T_1 secondary. One of these signals is applied to the base of Q_1 and the other is applied to the base of Q_2.

When the signal at the primary of T_1 goes positive, the following sequence of events will occur. When a dot is located at the top of each winding of a transformer, this tells us that there is *zero* phase shift between the primary and secondary (see Figure 10-16). This means that when the top of T_1 primary is positive, so is the top of the secondary, and the positive is applied to the base of Q_1. At the same time, the bottom of each winding and the base Q_2 will go negative. Since both transistors are NPN, forward bias on Q_1 increases, and Q_2 forward bias decreases. Increased bias on Q_1 causes I_C to increase, and point A will become less positive (go negative). The decreased forward bias at Q_2 causes I_C of Q_2 to decrease and point B to go more positive.

With a negative going voltage at point A and a positive going voltage at point B, we have two 180° out of phase currents flowing in the primary of T_2. These currents are added algebraically and when induced into the secondary, act as one large magnetic field. The result is that with T_2 having in-phase windings, the negative pulse developed at T_2 primary is coupled to the secondary as a negative pulse.

With the negative alternation of the input signal, the reverse happens. Q_1 conduction decreases, causing point A to go positive while Q_2 conduction increases and point B goes negative. Again, the two currents add algebraically, giving the effect of having one large positive pulse applied to T_2 primary. This large signal is coupled to the secondary and to the speaker.

The power output from this Class A amplifier can be more than double that available from a single-ended power amplifier operating Class A. An added advantage is that this circuit cancels even harmonics (provided the transformer circuits are balanced). Variations (noise) present in the V_{CC} applied to the circuit will affect both transistors. Because the primary of T_2 is center-tapped, equal variations cause equal voltage changes at points A and B which, in push-pull operation, will be canceled.

The Class A push-pull power amplifier is used when minimum distortion is desirable, but a Class A push-pull power amplifier operates with low efficiency because both transistors consume power even when no signal is applied to the input.

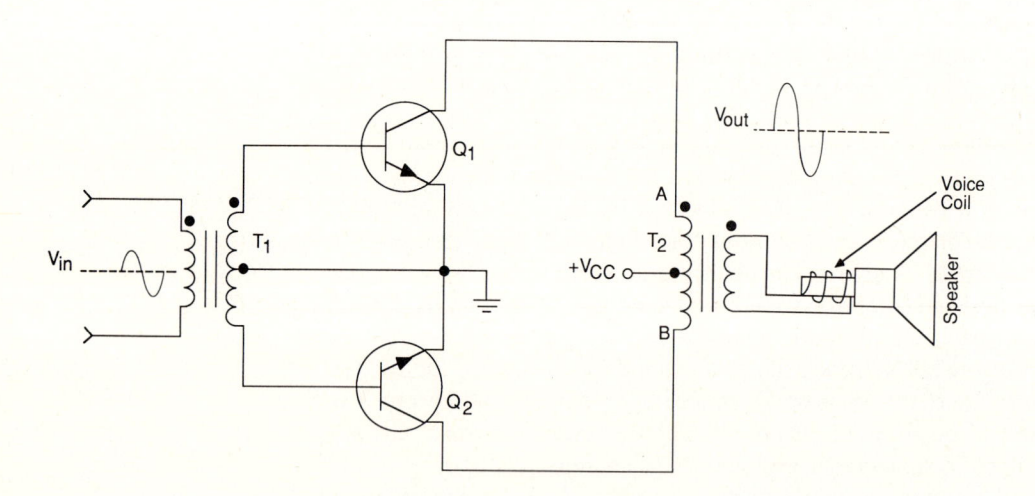

Class B Push-Pull Power Amplifiers

Another push-pull amplifier circuit that is used as a power amplifier is one in which each transistor conducts 50% of the time. This circuit is called a **Class B push-pull power amplifier**. Figure 10-17 contains a simplified schematic for an amplifier of this type.

Fig. 10-17 Class B Push-Pull Power Amplifier.

A Class B push-pull amplifier can be identified by the fact that there is no forward-bias network for the transistors. With no input signal, both transistors are cut off. Each transistor will conduct *only* when a positive signal is applied to its base. This occurs during half of each input cycle, causing each transistor to conduct half of the time and to be cut off the other half.

A positive at the top of T_1 causes Q_1 to conduct, and the negative at the bottom of T_1 causes Q_2 to remain cut off. During the negative alternation, Q_2 is biased on and Q_1 is biased to cutoff. At T_2 current flows through one half of the primary on each alternation of the input. These signals are coupled to the secondary and the speaker by T_2. To the speaker, the current that flows has the same effect as if the output had come from the conduction of a single transistor.

The Class B push-pull power amplifier operates with much greater efficiency than the Class A type. Notice that there is never a time when both transistors are conducting. This means that the power consumed to develop the output power is much less than that used in the Class A circuit, resulting in greater efficiency.

A disadvantage of this type circuit is that it has more distortion than the Class A push-pull power amplifier. Much of this distortion results from the fact that at the point where the alternations change from positive to negative and negative to positive, both transistors cut off momentarily. This is called **crossover distortion**.

Class AB Push-Pull Power Amplifiers

By *slightly* modifying the circuit shown in Figure 10-17, the amount of crossover distortion can be reduced significantly. To do this, a small bias is applied to the base of each transistor, which prevents the transistor's cutting off prior to crossover. The conduction of each transistor increases to slightly more than 50% of the time and results in **Class AB** operation. This circuit is called a **Class AB push-pull power amplifier**. A schematic diagram for this circuit is shown in Figure 10-18.

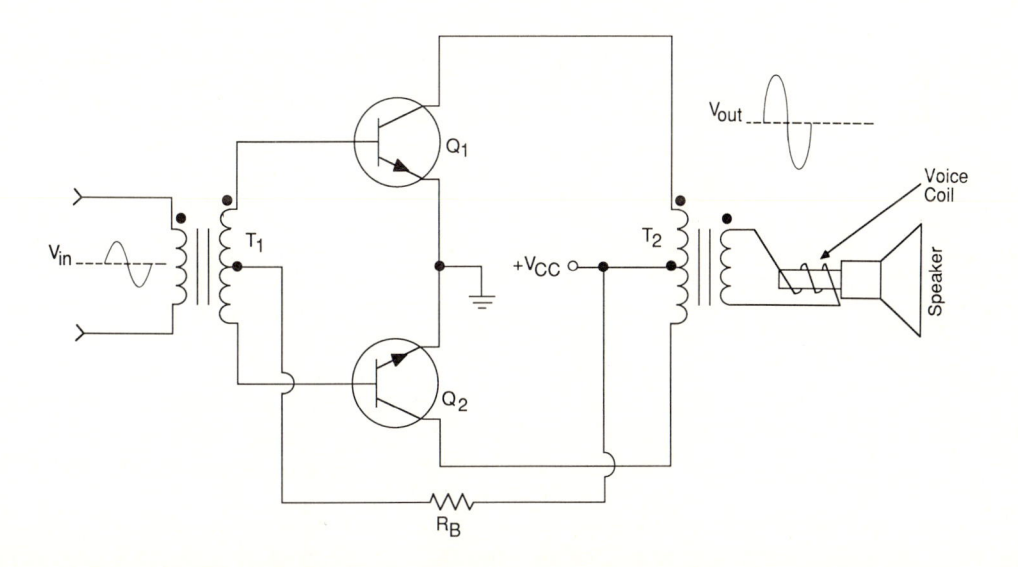

Fig. 10-18 Class AB Push-Pull Power Amplifier

The class AB push-pull amplifier operates very much like the Class B type. The exception is that the insertion of R_B applies a small forward bias to Q_1 and Q_2. This causes each transistor to conduct during the crossover from one alternation to the other. Once the opposite transistor begins increasing in conduction, the reverse bias applied to the opposite transistor biases it OFF. In this way, we cancel the distortion caused by crossover and sacrifice little in the way of power efficiency.

Even Harmonic Cancellation

When algebraically summing two signals across a device, harmonic distortion can present a problem. To better understand this, we will discuss fundamental and harmonic waves. Remember:

Any sine wave that is used as a reference is referred to as the fundamental wave at the sine wave's frequency.

and that:

Each fundamental wave has an infinite number of harmonics (multiples).

Assume a fundamental frequency of 1 kHz. All even multiples (2 kHz, 4 kHz, 6 kHz, etc.) are even harmonics. All odd multiples (3 kHz, 5 kHz, 7 kHz, etc.) are odd harmonics. When using push-pull power amplifiers, harmonics are allowed to add algebraically, and the even harmonics create distortion. Observe Figure 10-19 for an example. Note that when the fundamental wave (a) and the second harmonic (b) are allowed to add, severe distortion results in the output wave (c).

A Class A power amplifier operates in the linear portion of its characteristic curve and generates few harmonics. When an amplifier operates on the nonlinear portion (near cutoff and saturation) of its characteristic curve, harmonics are generated. Both Class B and AB amplifiers operate in this way. Note that the positive alternation is larger than the negative alternation. When the fundamental (a) and the second harmonic (b) are added, the resultant negative alternation (c) is highly distorted. The waveshape in Figure 10-19(c) illustrates the output from a single-ended power amplifier operating Class AB.

Balanced push-pull amplifiers provide cancellation of the even harmonics. An illustration of this is shown in Figure 10-20. Note the relationship that exists between the fundamental (F) and second harmonic (2F) waves. The fundamental wave at the collector of Q_1 is used as the reference. Note that the fundamental wave at the collector of Q_2 is 180° out of phase. At Q_1 the second harmonic is in phase with F. At Q_2

Amplifier Applications and Coupling

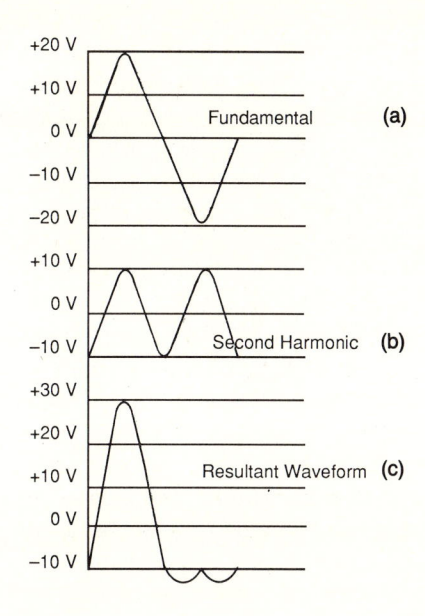

Fig. 10-19 Harmonics in Single-Ended Amplifiers

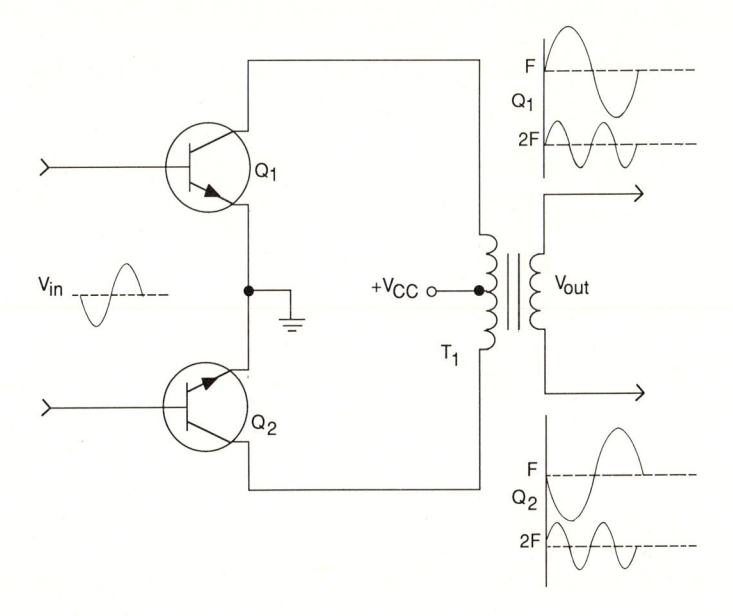

Fig. 10-20 Even Harmonic Cancellation

the second harmonic and fundamental waves are also in phase because the two transistor circuits are identical. However, the harmonic frequencies (2F) appear at opposite ends of the output transformer (T_1). With the top of T_1 positive (first alternation of 2F at Q_1) and the bottom of T_1 positive (first alternation of 2F) at the same time, there is zero difference in potential (between two 2F waves) across the primary of T_1. In this case, the 2F waves cancel and no output is created. During the negative alternation of the 2F waves, both go negative and continue to cancel. The end result is that the even harmonics cancel. Odd harmonics, however, will not cancel.

Since it is practically impossible to select two transistors that are absolutely identical, provisions are made within the circuit to assure balance. This is done by inserting a balance resistor in the emitter circuits of the transistors. In Figure 10-21, R_1 serves this purpose. By placing ground on the resistor's wiper, we can adjust R_1 to assure that both sides of the circuit are conducting at precisely the same level. If Q_1 has slightly more resistance than Q_2, R_1 can be adjusted to compensate for the difference.

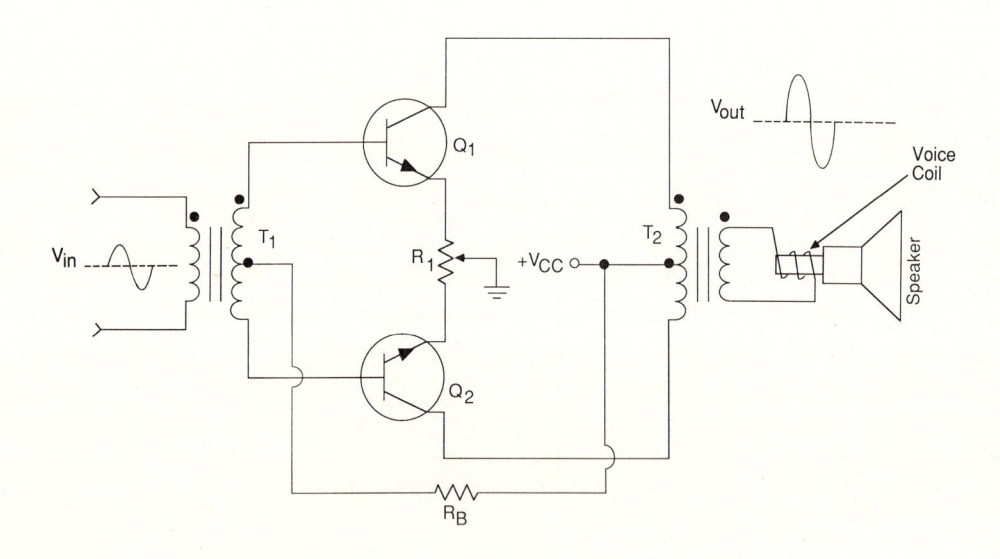

Fig. 10-21 Balanced Push-Pull Power Amplifier

Phase Splitter Input to Push-Pull Amplifiers

In many cases, a **phase splitter** is used to provide the input to the two transistors of a push-pull power amplifier. When used in this way, the phase splitter circuit is referred to as the **driver** stage because it is used to provide the "driving" or input signal. Phase splitters were discussed in the last chapter. At that time, you were told that the circuit could be

Amplifier Applications and Coupling

used to produce two outputs of equal amplitude that were 180° out of phase. This type of circuit is shown in Figure 10-22.

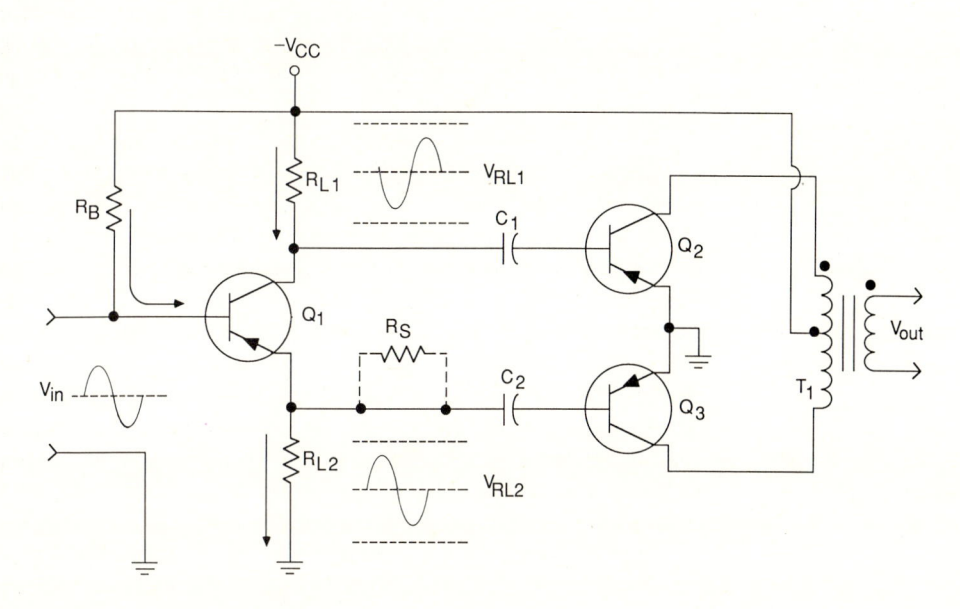

Fig. 10-22 Phase-Splitter Driven Push-Pull Amplifier

In some applications, you may see this circuit referred to as a **split-load phase inverter** or a **paraphase amplifier**. A phase splitter develops the two signals, equal in amplitude and 180° out of phase, that are applied to the bases of Q_2 and Q_3 through coupling capacitors C_1 and C_2. Resistor R_B and R_{eb} establish base current and the Q point for Q_1. Resistors R_{L1} and R_{L2} operate as a split load resistor that divides the gain of the stage equally between the emitter circuit and the collector circuit.

When the input signal at the base of Q_1 goes negative (note the PNP transistor), I_E and I_C increase. This increase in current is approximately the same in both circuits causing V_{RL1} to go negative and voltage (V_C) to go positive. At the same time, the emitter voltage is going more negative.

With the positive input, the reverse happens. V_C goes negative and V_E goes positive. Assuming that R_{L1} and R_{L2} are equal in ohmic value, the amplitude of the two signals will be equal. Because they are taken from opposite ends of the transistor, their phases are different by 180°.

There is a problem, the collector output impedance is much larger than that of the emitter. This can be overcome by inserting a series resistor in the coupling circuits of Q_3. This resistor is shown as a dashed line on Figure 10-22. (Note: This resistor would actually replace the short indicated by the solid line.) The value of the series resistor is selected such that its insertion adjusts the output emitter impedance of Q_1 to the input impedance of Q_3. Of course, connection of the series resistor creates another problem, it reduces the signal amplitude that is applied to Q_3 and results in a lower output. To compensate for this, the size of R_{L2} is made larger than R_{L1}.

Use of the unbypassed emitter resistor (R_{L2}) is necessary to develop the second output. Because of the unbypassed emitter resistor, the voltage gain of Q_1 is reduced to less than *unity*. The phase splitter will pass a much wider frequency band than was possible with the transformer. Construction costs are greatly reduced and the weight of the end product is much less when the phase splitter is used as the driver.

In a Class B push-pull amplifier, the use of coupling capacitors C_1 and C_2 creates a problem. The capacitors must have a discharge path. Because of the transistor action, the discharge path for the capacitors is through a cut off transistor. Figure 10-23 contains a modified circuit diagram that can be used to provide discharge paths for the capacitors. Note diodes D_1 and D_2. These diodes are called **discharge diodes** and are connected across the base-emitter circuit of each transistor. Now, the discharge path for each capacitor is through a forward biased diode junction.

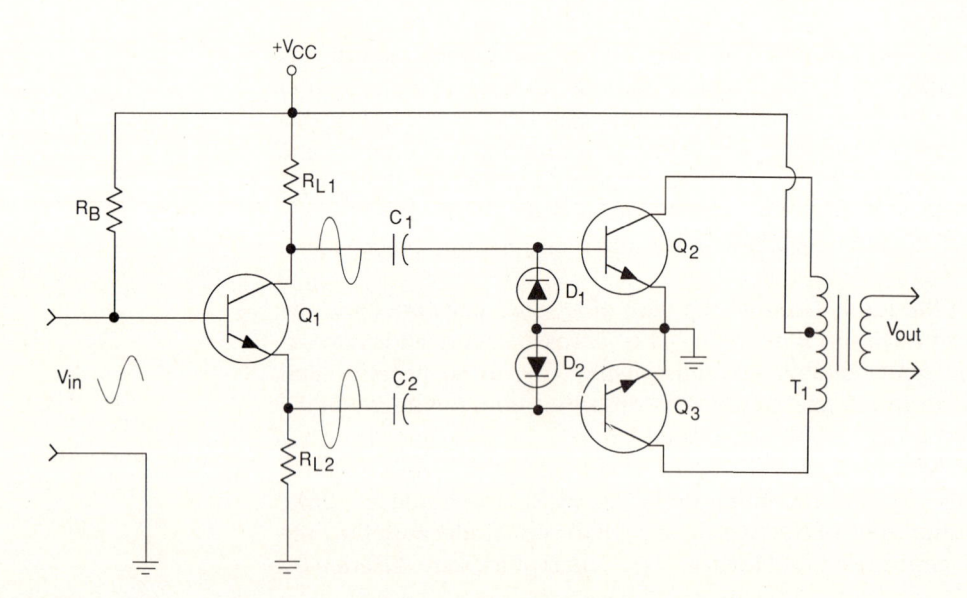

Fig. 10-23 Discharge Diodes

Complimentary-Symmetry Amplifiers

It is possible to purchase **matched pair transistors**. Manufacturers test and select transistors whose characteristics are close enough to operate identically, and they are matched in pairs. Matched pairs will consist of one PNP and one NPN transistor. Realize that in these two types, emitter-base current flows in opposite directions within each transistor.

When these transistors are connected into a single stage, the DC electron path is through both transistors. Figure 10-24 contains the circuit diagram for another type of power amplifier. Placing PNP and NPN transistors in series like this is referred to as a **complimentary-symmetry** circuit. This circuit acts like a push-pull amplifier that does not require either a phase splitter or transformer for its input.

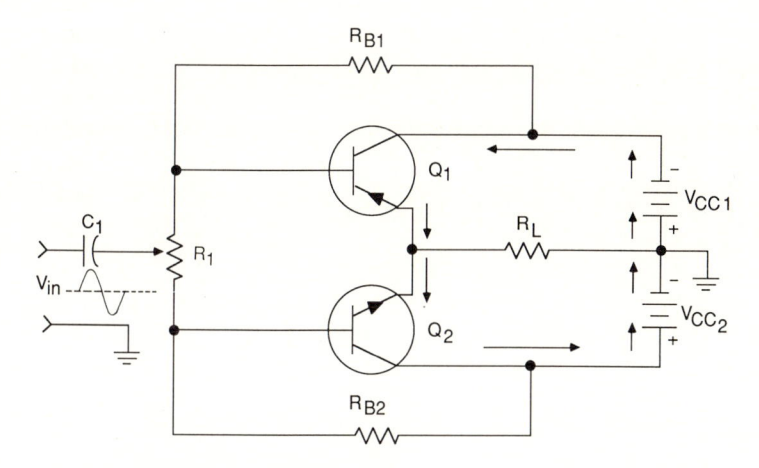

Fig. 10-24 Complimentary-Symmetry Circuit

Observe Figure 10-24. When a negative going signal arrives at the input (C_1), it increases forward bias on Q_1, causing it to conduct harder. A positive input signal forward biases Q_2 and causes it to conduct harder. When one transistor conducts, the other transistor is cut off because the same signal that forward biases one transistor reverse biases the other.

In the output circuit, this results in an action that can be understood if we analyze Figure 10-25. This is a simplified version of the output circuit of Figure 10-24. The internal emitter-collector resistances are included as variable resistors R_1 and R_2. With Class B operation, no current flows in either transistor and the variable resistors have infinite value. At this time, there is zero current through R_L. With a positive input signal, Q_2 conducts, causing the resistance of R_2 to decrease toward zero (point 3). At this time, Q_1 is cut off and R_1 remains infinite.

Current now flows from V_{CC2}, through R_L, through Q_2 (variable resistor R_2), and back to V_{CC2}. The amount of current that flows depends on the amplitude of the input signal. This current is represented by the dashed line. The voltage drop on R_L is in the polarity shown.

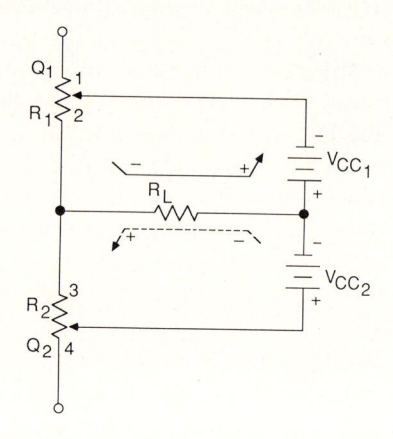

Fig. 10-25 Simplified Complimentary-Symmetry Output Circuit

With a negative input signal, Q_1 conducts and Q_2 is cut off. This means that the ohmic value of R_1 approaches zero (point 2) and R_2 is infinite. Current will now flow through V_{CC1}, through Q_1 (R_1), through R_L, and back to V_{CC1}. This current is indicated by the solid line. The voltage drop on R_L is in the polarity indicated. Notice that, in this case, the current path through R_L is opposite that for the conduction of Q_2.

To operate this circuit Class A, a voltage divider network must be connected $-V_{CC2}$ to $+V_{CC1}$. The result is that both transistors receive forward bias and conduct 100% of the time (Class A operation). In this case, neither of the transistors will ever cut off and therefore, the variable resistors (R_1 and R_2) shown in Figure 10-25 will never approach infinite values. In the quiescent state, current flows out of $-V_{CC2}$, through the series aiding $+V_{CC1}$, through both R_1 and R_2, and back to $-V_{CC2}$. Notice that no current flows in R_L. Actually, the output circuit can now be considered a balanced bridge. When a negative input arrives, the resistance of R_1 decreases and R_2 increases. This unbalances the bridge and causes current to flow in R_L in the direction of the solid arrow. When a positive signal arrives, R_1 increases in value and R_2 decreases. Again, the bridge is unbalanced. This time current flows through R_L in the direction indicated by the dashed arrow.

All current through R_L results from the current that flows as a result of a changing input signal. It is possible that R_L would actually be the voice coil of a speaker. In some cases, Class AB operation may be used because Class B operation is affected by crossover distortion.

Compensated Common-Symmetry Circuits

The circuit described above operates Class B. This causes crossover distortion to be present each time the transistors switch which leads to serious deficiencies. These are:

1. With both transistors unbiased when crossover occurs, a portion of each alternation must be used to raise forward bias to the level where the transistor begins to conduct. The use of the input signal for forward bias purposes severely limits the amplitude of the output signal. The greater the amount of input signal that is used for biasing, the greater crossover distortion becomes. Another result is that the amplifier cannot respond to signals that are smaller than the amplitude required for forward biasing the transistor. Assuming that both transistors are made of silicon, they require approximately 0.7 V amplitude for biasing. This means that 1.4 V of each input signal is lost in biasing the transistors into conduction.
2. The period during which both transistors are cut off is a period when no current flows in the load. This means that output is *not* present during 100% of the time that input signal is present. The greater the input signal amplitude that is required for biasing, the greater the amount of crossover distortion.

In your studies of push-pull amplifiers, you learned that it is possible to modify the circuit and reduce the amount of crossover distortion. A similar modification can be made to the *complimentary-symmetry amplifier*.

Figure 10-26 contains the schematic diagram for a modified two-source circuit that can be used to reduce crossover distortion to a low level. Note that the biasing voltage divider contains four resistors. Voltage drop for each resistor is noted on the circuit diagram. V_{R2} and V_{R3} each drop 0.6 V in their quiescent condition. At this time, both transistors are biased near the 0.7 V required for conduction. When an AC signal arrives at the input, the first 0.1 V is required to forward bias one transistor and reverse bias the other. A positive input turns Q_1 on and Q_2 off. The conduction of Q_1 causes current to flow in the speaker, developing the positive alternation of the output. A negative input causes the opposite to happen. Now, current flows in the opposite direction through the speaker, developing the negative alternation.

This circuit processes most of the input signal and minimizes crossover distortion. In this circuit, only 0.2 V is required for switching as opposed to 1.4 V in the uncompensated circuit. This circuit is able to respond to much smaller signals than the uncompensated circuit.

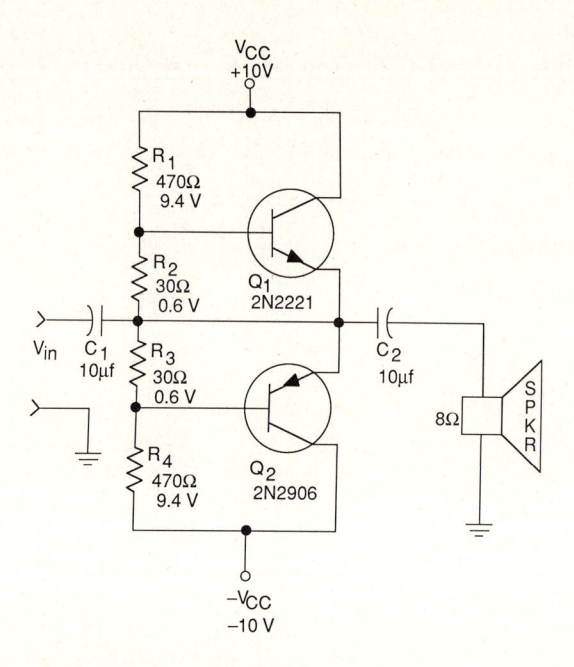

Fig. 10-26 Crossover Distortion Compensation

We have indicated on the schematics for the complimentary-symmetry circuits that two DC sources are required for their operation. The second power source is required to power the second transistor. It is quite common in solid state circuitry to use power supplies that provide both positive and negative DC voltages. Remember, we established in chapter 7 that:

> **When current flowing toward ground passes through a resistor, it drops a negative voltage.**

and:

> **When current flowing away from ground passes through a resistor, it drops a positive voltage.**

If we connect a voltage divider in the output of a power supply, we can provide two polarity DC for use in these circuits.

Observe Figure 10-27. Note that it has two load resistors that are center-tapped and the center-tap becomes circuit ground. Current flow in R_{L1} is away from ground and is positive. Current flow in R_{L2} is

Amplifier Applications and Coupling

toward ground and is negative. Using this arrangement we have a positive voltage (with respect to ground) at point A and a negative voltage (with respect to ground) at point B. If R_{L1} and R_{L2} have equal ohmic values, the power source is suitable for powering complimentary-symmetry circuits.

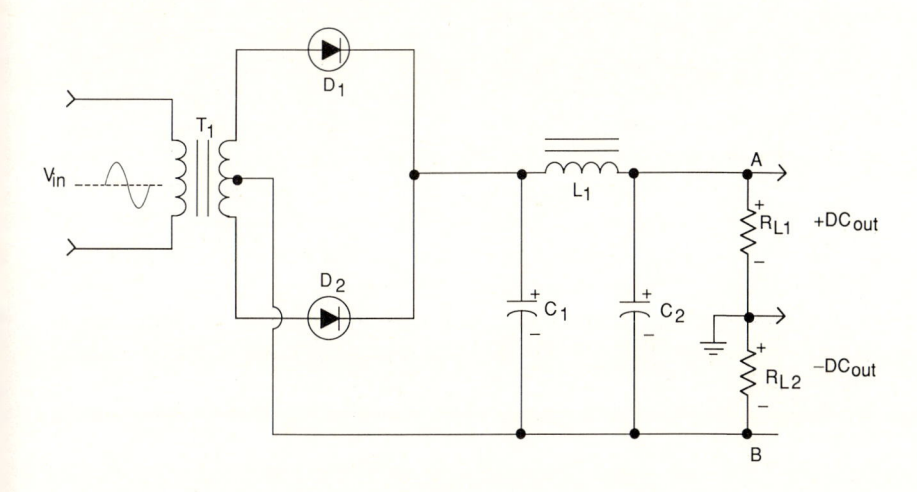

Fig. 10-27 Complimentary Symmetry Circuit Power Supply

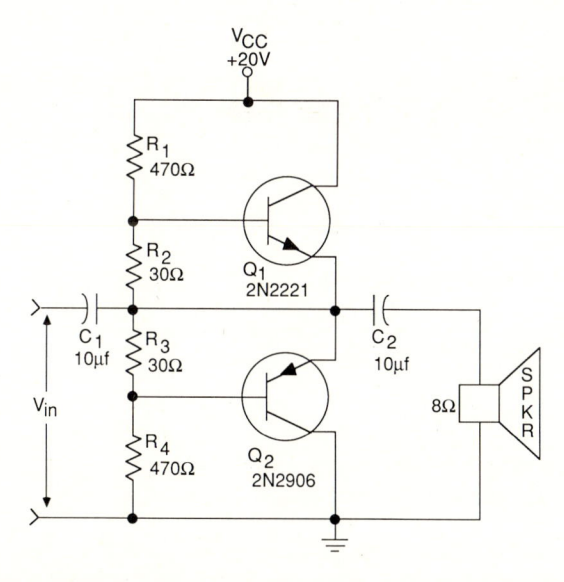

Fig. 10-28 Single-Source Circuit

Single-Source Complimentary-Symmetry Amplifiers

Another complimentary-symmetry circuit that is available for use is shown in Figure 10-28. Notice that this circuit is powered by a single power source. Biasing for this circuit is identical to that explained for the circuit in Figure 10-26. The insertion of R_2 and R_3 into the biasing network reduces crossover distortion.

NPN and PNP matched pair transistors are used as amplifiers Q_1 and Q_2. In their quiescent state, both transistors are biased slightly below full conduction. The application of a positive signal turns Q_1 on and Q_2 off. The conduction of Q_1 produces the positive alternation of the output in the speaker. When the negative alternation arrives, Q_2 is turned on and Q_1 is turned off. The conduction of Q_2 develops the negative alternation of the output.

This provides a circuit that can reproduce sine wave input signals of low amplitude. Crossover distortion has also been reduced.

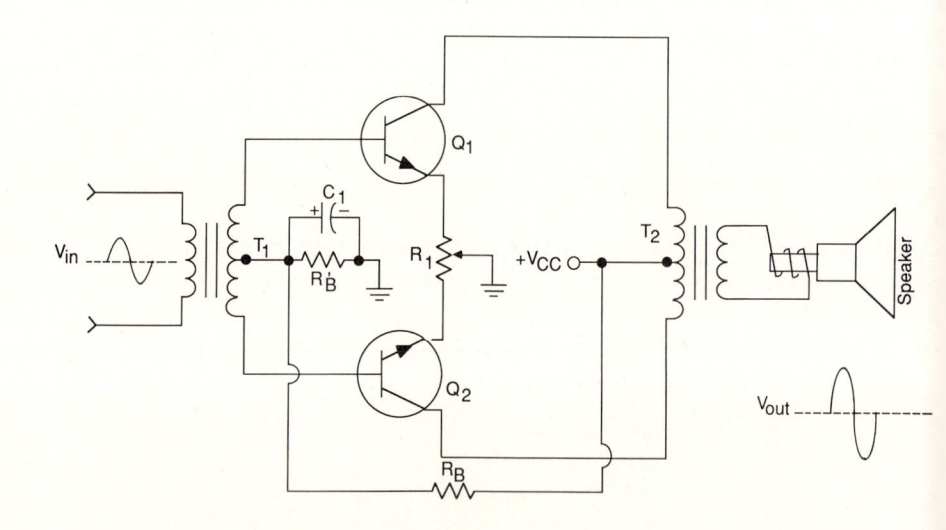

Fig. 10-29 Balanced Push-Pull Power Amplifier

Troubleshooting a Push-Pull Amplifier

Refer to Figure 10-29. This circuit is biased to operate as a Class A, balanced, push-pull power amplifier. R_B and R_B' form a biasing voltage divider that forward biases both transistors on. C_1 provides an AC path to ground, bypassing R_B' and thereby preventing degeneration.

If the primary of T_1 opens, there is no signal present at the secondary of T_1 or applied to either transistor. All other conditions are

Amplifier Applications and Coupling

normal. Both transistors conduct and all DC voltages will be normal. Use of an oscilloscope would reveal the absence of the AC signal. If the primary of T_1 is shorted, it is possible that the same indications might be present. It is more likely, however, that the increased current caused by the short will trip a circuit breaker or otherwise interrupt all DC operation. If half of T_1 secondary opens, or if either transistor opens, half of the output signal will be missing, making the signal applied to the speaker very weak. Measurement of DC voltages would reveal that one stage has abnormal voltages while the other is normal. Observation on the scope reveals that only half of the signal is present across the output. If R'_B opens, all indications (DC and AC) will be abnormal and there will be no output. If R'_B shorts, the circuit operates Class B and all indications will be abnormal. If C_1 opens, degeneration will occur across R'_B and will reduce output amplitude. If R_B opens, operation will be Class B. Shorting R_B will either destroy the transistors or blow the power supply's fuse. Opening R_1 will cause only one transistor to operate, according to the position of the wiper arm. Opening one half of T_2 primary causes the output signal to be weak. If the secondary of T_2 opens, the amplifiers will operate normally, and the only symptom is loss of sound. The trouble will probably have to be located using an oscilloscope. It is also possible that an open T_2 secondary will cause other damage in the circuit because of overload.

Self-Check

Answer each item by inserting the word or words required to make each statement a true statement.

26. Double-ended power amplifiers use _____ transistors.
27. Two methods are used to apply the inputs to a push-pull power amplifier, these are:

    a. _____
    b. _____

28. In a Class A push-pull amplifier, each transistor conducts _____% of the time.
29. Dots that appear on transformer symbols tell us the amount of _____ that the transformer provides.
30. When compared to single-ended power amplifiers, the push-pull amplifier can supply an output that is _____ that of the single ended type.
31. A Class B push-pull amplifier can be identified by the absence of a _____ network.
32. A Class B push-pull amplifier is much more _____ than the Class A circuit.

33. One disadvantage of the Class B push-pull amplifier is the presence of _____ distortion.
34. To overcome this distortion and maintain efficiency, the circuit can be modified to operate Class _____.
35. Complimentary-symmetry amplifiers differ from push-pull amplifiers in that they have _____ transistors.
36. A fundamental wave is actually any sine wave that is used as a _____ for its frequency.
37. Why are discharge diodes inserted into the circuitry of some power amplifiers?
38. Only Class _____ push-pull power amplifiers are used as audio amplifiers.
39. As used in power amplifiers, the phase splitter provides _____ outputs that have the _____ amplitude and a _____ phase difference.
40. A push-pull amplifier whose transistors each conduct 50% of the time is called a _____ power amplifier.

Summary

The earlier sections of this chapter were used to discuss amplifier coupling. Direct, RC, impedance, and transformer coupling were discussed. You learned that direct coupling can be used to amplify direct current. RC coupling has loss of gain at both low and high frequencies. Coupling capacitors are the problem at low frequencies and distributed capacitance causes loss of gain at high frequencies. Impedance coupling is used to extend the gain at high frequencies and to delay the effect of distributed capacitance. Transformer coupling has the ability to select specific bands of frequencies.

A push-pull amplifier consists of two transistors connected in parallel. Two input signals are required that have equal amplitude and that are 180° out of phase. In the Class B push-pull amplifier, crossover distortion is a problem. Class AB amplifiers overcome this distortion because both transistors are never cut off at the same time. The input to push-pull amplifiers can be provided by either transformers or phase splitters.

Complimentary-symmetry amplifiers use a matched pair of NPN and PNP transistors that operate together. Only one input signal is required to provide Class A output. Crossover distortion is present in these circuits. Complimentary-symmetry amplifiers can be designed that operate from either one or two power sources. Modification of circuit biasing networks can greatly reduce the effect of crossover distortion.

Review Questions and/or Problems

1. Amplifiers are used to:

   a. increase voltage gain.
   b. increase current gain.
   c. increase power gain.
   d. any of the above.

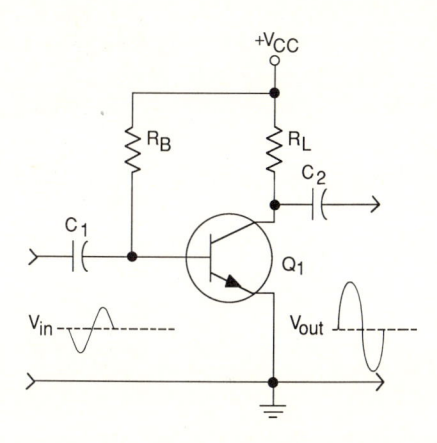

Fig. 10-30

2. Refer to Figure 10-30. This schematic represents a/an:

   a. double-ended amplifier.
   b. single-ended amplifier.
   c. push-pull amplifier.
   d. complimentary-symmetry amplifier.

3. Refer to Figure 10-30. Low frequency cutoff for this circuit occurs when _____ of C_1 equals _____ of the transistor.

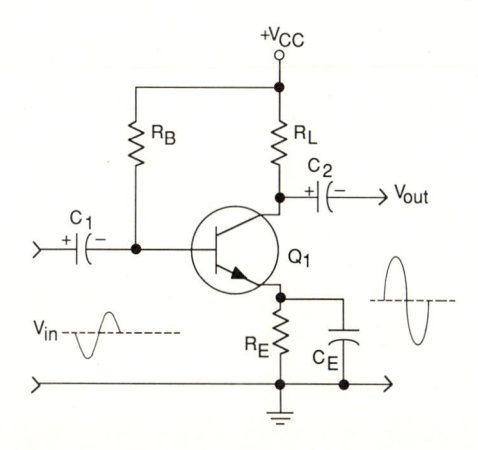

Fig. 10-31

4. Refer to Figure 10-31. The addition of R_E and C_E to the circuit:

   a. places AC ground at the emitter of Q_1.
   b. adds emitter-resistor temperature stabilization.
   c. provides separate paths for DC current and the AC signal.
   d. all of the above.

5. What modifications must be made in the circuit of Figure 10-31 to convert it to a DC amplifier?

   a. Remove C_E.
   b. Remove C_1.
   c. Remove C_2.
   d. all of these.

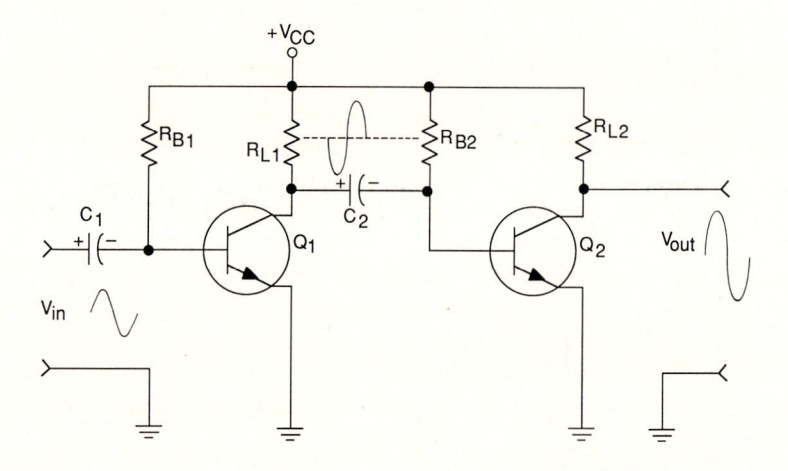

Fig. 10-32

6. Refer to Figure 10-32. This circuit is called a/an _____ coupled amplifier.

7. Check Figure 10-32. Q_1 has $A_V = 15$ and Q_2 has $A_V = 20$. Overall voltage gain of the circuit equals _____.

   a. 5:1
   b. 15:1
   c. 20:1
   d. 300:1

8. Refer to Figure 10-32. This circuit has a $Z_{in} = 50\ \Omega$ and we want to use it to amplify all frequencies above 100 Hz. What is the minimum value for C_1?

   a. 3.18 µf
   b. 31.8 µf
   c. 3.18 pf
   d. 31.8 pf

Amplifier Applications and Coupling

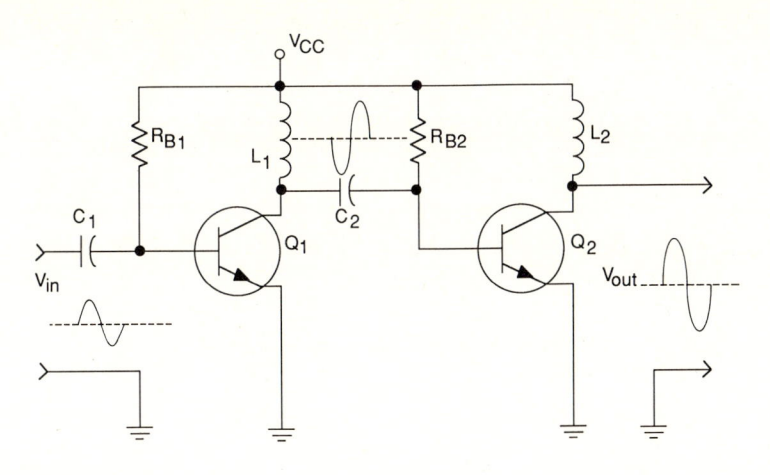

Fig. 10-33

9. Refer to Figure 10-33. This circuit uses _____ coupling.

   a. direct
   b. RC
   c. impedance
   d. transformer

10. L_1 and L_2 were inserted in Figure 10-33 to compensate for losses at _____ frequencies.

    a. low
    b. high

11. At low frequencies, Figure 10-33 provides _____ gain.

    a. high
    b. low

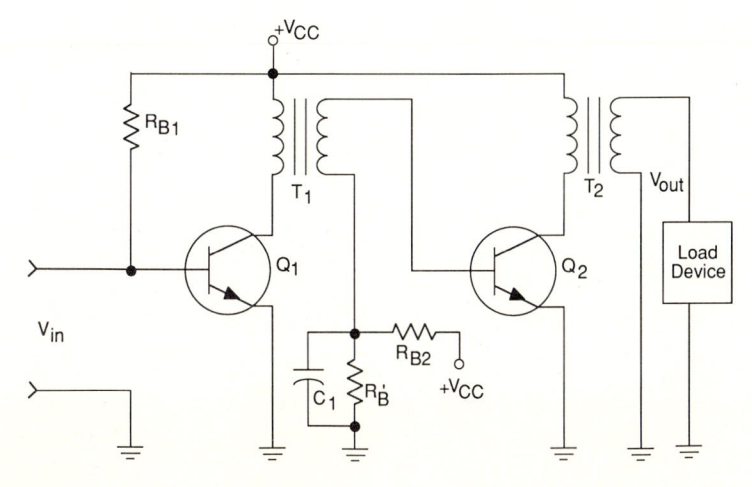

Fig. 10-34

12. Figure 10-34 shows the schematic diagram for a/an _____ coupled amplifier.

    a. direct
    b. RC
    c. impedance
    d. transformer

13. In Figure 10-34, the total phase shift is _____?

    a. 90°
    b. 180°
    c. 270°
    d. 360°

14. A circuit like the one shown in Figure 10-34 has which of the following advantage(s)?

    a. excellent power efficiency
    b. able to select a narrow band of frequencies
    c. can be used to amplify wide bands of frequencies
    d. both a and b

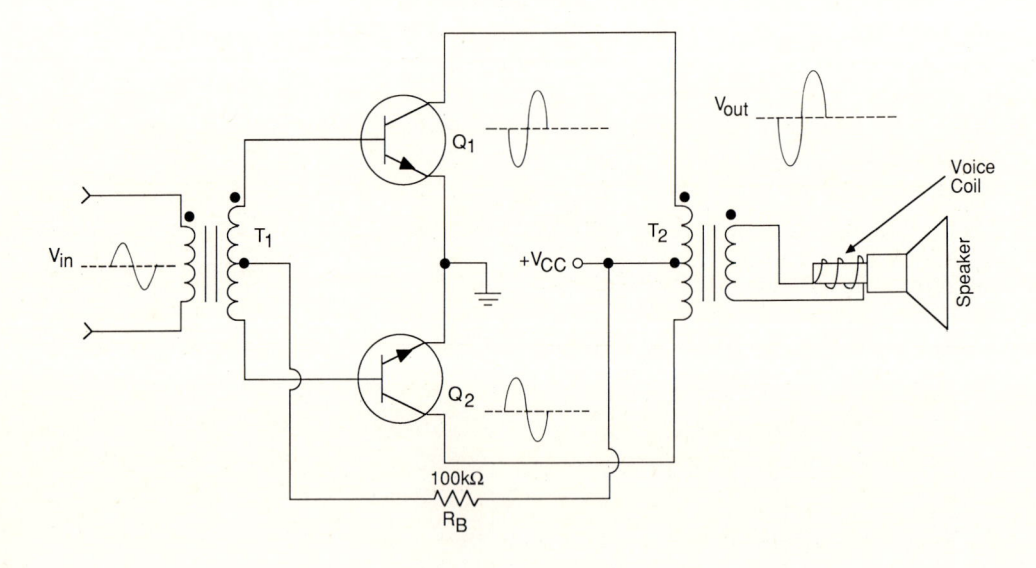

Fig. 10-35

15. The circuit shown in Figure 10-35 operates Class _____.

    a. A
    b. B
    c. C
    d. AB

16. The dots that appear on the transformer symbols of Figure 10-35 mean that there is a 180° phase shift across each transformer.

    a. True
    b. False

Amplifier Applications and Coupling

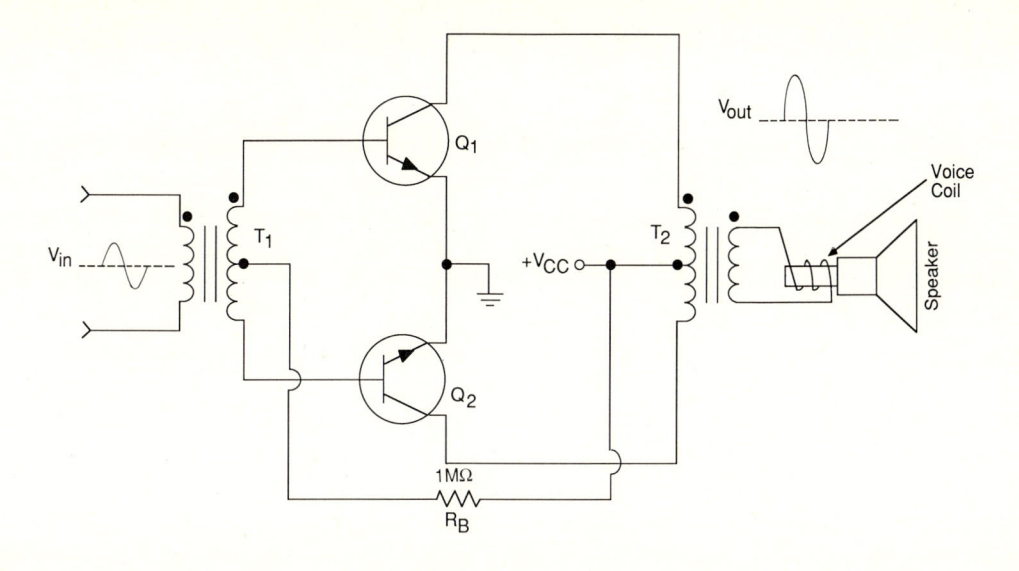

Fig. 10-36

17. Refer to Figure 10-36. This circuit operates Class AB. How is it different from Figure 10-35?

    a. The type of transistors is different.
    b. It uses different transformers.
    c. The amount of bias applied to Q_1 and Q_2 is different.
    d. There is no difference.

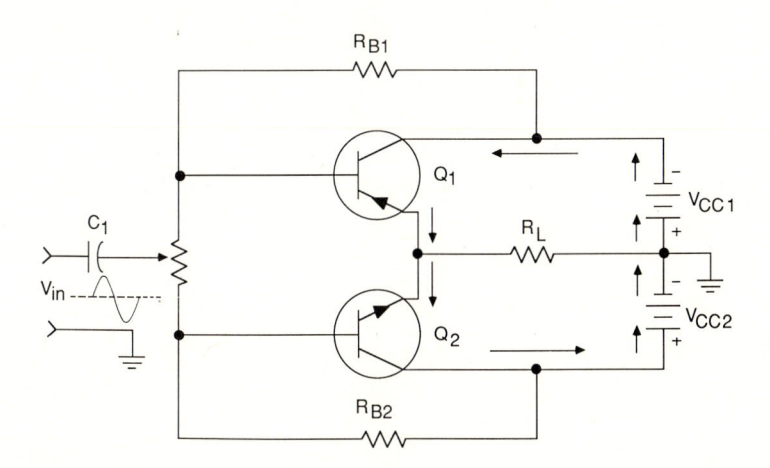

Fig. 10-37

18. Refer to Figure 10-37. This circuit is a/an _____ amplifier.

   a. Class A push-pull
   b. Class B push-pull
   c. Class AB push-pull
   d. complimentary-symmetry

19. What feature is present in Figure 10-37 that allows us to identify its type?

   a. no transformers
   b. two power sources
   c. one NPN and one PNP transistor
   d. both b and c

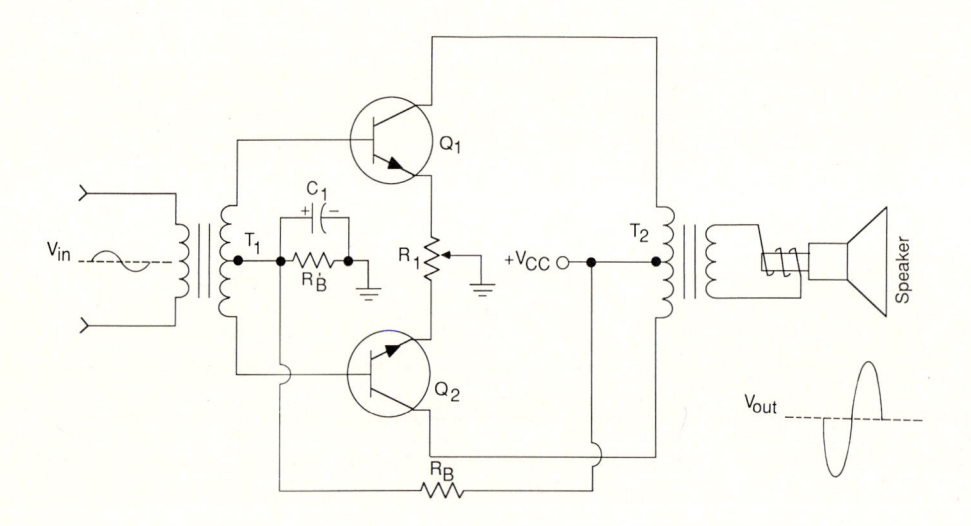

Fig. 10-38

20. Refer to Figure 10-38. What is the purpose for R_1 in this circuit? (Explain what R_1 is, why it is used, and how it affects circuit operation).

11
Narrow-Band and Wideband Amplifiers

Objectives

1. Define the terms:

   a. Narrow-band amplifier
   b. Wideband amplifier
   c. Sensitivity
   d. Selectivity
   e. Nonsinusoidal wave
   f. Tuned circuit
   g. Cascade tuned circuit
   h. Double tuned transformers
   i. Square wave
   j. Fundamental wave
   k. Series compensation
   l. Shunt compensation
   m. Series-shunt compensation
   n. Low frequency compensation
   o. High frequency compensation
   p. Response curves
   q. Tapped windings
   r. Unilateralized tuning
   s. Harmonics

2. Describe the methods used to design circuits with good selectivity and sensitivity.
3. Explain the frequency composition of a square wave.
4. Explain the difference between narrow-band and wideband amplifiers.

Introduction

In past chapters, we discussed the DC and AC characteristics of amplifiers. Class A, Class AB, Class B, and Class C amplifiers are classified according to the percentage of time that they conduct as compared to the time an input is present. The amplifiers were also studied to determine how faithfully an input sine wave was reproduced in the output. You learned that an audio amplifier operates Class A (in the linear portion of the characteristic curve) and provides faithful signal reproduction. Power amplifiers were discussed that could be used to provide good fidelity and high power for use in audio systems. Each amplifier's operation was compared to the audio frequency range (20 Hz to 20kHz).

In this chapter, amplifiers are discussed that operate at other frequency ranges. Each circuit is examined to determine how frequency affects its design and its operation. Two types of amplifiers are discussed, **narrow band** and **wideband**. These amplifiers are defined as follows:

Narrow-band amplifier – an amplifier designed for best operation over a narrow band of frequencies. A narrow-band amplifier provides maximum gain to a center frequency and then limits output to a few frequencies on either side of center.

Wideband amplifier – an amplifier that is capable of passing a wide range of frequencies with equal gain. The wideband amplifier has a bandpass that is large enough to pass all of the input frequencies (lowest to highest) that are contained in an expected input.

Using these definitions, the audio amplifiers discussed in Chapter 10 are classified as wideband amplifiers. They are designed to pass a wide range of frequencies while amplifying each frequency an equal amount.

Each type circuit will be discussed more fully in this chapter.

Applications of Narrow and Wideband Amplifiers

The first circuit that we will discuss is the narrow-band amplifier. These circuits are the heart of all communications systems. They are used in both the transmitters and receivers of AM, FM, shortwave, two-way, and CB radio systems. To give you a better understanding of why these amplifiers are needed, we will take time to verbally describe a typical AM radio station and the radio that uses its signal.

The station is assigned an operating frequency by the Federal Communications Commission. Let us assume that the station we are discussing is assigned the operating frequency of 1000 kHz (1 MHz.) Along with this authorization, the station is assigned a power level at which it can transmit and a bandwidth of 10 kHz within which it must operate. This means that the signal being transmitted from the station's antenna must be contained within the narrow band of 10 kHz. When we compare this to the assigned frequency (carrier frequency) of the station, we realize that the transmitted signal must be 1000 kHz ± 5 kHz. In other words, this signal must be contained between the two extremes of 995 kHz and 1005 kHz. An amplifier that is capable of this response is classified as a narrow-band amplifier. Once the station has developed the signal to the power level authorized, the signal is fed to the antenna. At the antenna, this signal causes radio-frequency electromagnetic waves to be radiated into the air.

Unless intentionally restricted, this wave radiates into a 360° area. Because of its movement through the air and the large area it covers, the signal intensity at any given point is relatively low.

Once the wave is transmitted into the air, two of the requirements for induction (a magnetic field and motion) are present. When this wave strikes a conductor, current is induced into the conductor. Any antenna will have voltage and current induced into it and this induction is a true reproduction of the signal that was used to radiate the RF wave. In most AM radios, the antenna is a ferrite rod/coil arrangement called either a *loopstick* or *ferrite-rod antenna*. A ferrite-rod antenna consists of a coil wound around a ferrite rod. The signal that strikes this antenna can be selected by another narrow-band amplifier and processed into an audible signal within an AM radio.

All radio waves that strike the antenna will cause current to be induced into the metal parts of the antenna. For AM signals, these waves are at the AC (carrier) frequency that the station is assigned by the FCC, ±5 kHz. Stop for a moment and think of all the AM, FM, TV, CB, and two-way radio systems that are located close enough to hear their signals if they were detected and amplified. The number is staggering. A radio receiver must do two things: first, it must select the station that we want to receive from all those present at the antenna; second, it must be sensitive enough to take a very small signal and amplify it for application to the speaker. In the case of an AM station, the voltage that is induced into the receiving antenna is probably no more than 10 μV. As you can see, this is a very small voltage. The fact that the radio can take this signal and use it is referred to as the receiver's *sensitivity*. **Sensitivity** is defined as:

> **The minimum input signal required in a radio receiver or similar device, to produce a specified output signal having a specified signal-to-noise ratio**.

In other words, sensitivity is the ability to receive small signals.

Once a receiver has been designed that is capable of receiving a signal of 10 μV or less, the next problem is the selection of the desired frequency. **Selectivity** is defined as:

The ability to choose one narrow band of frequencies while rejecting all other frequencies.

Circuits that have the sensitivity and selectivity needed to choose a specific radio signal are called **narrow-band amplifiers**. These circuits are discussed in the first part of this chapter.

Other circuits have completely different applications. Many of them must amplify very wide frequency bands. These are called **wideband amplifiers**. Many of the signals that must be amplified are not sine wave signals. Signals that are other than a pure sine wave are called **nonsinusoidal signals**. A nonsinusoidal wave is one that does not vary at the sine wave rate. Examples are the sawtooth wave that is used in the oscilloscope sweep circuits, square and rectangular waves that are used for timing circuits, and the peaked waves that are used as trigger pulses. Each of the nonsinusoidal waves consist of a *fundamental* wave and a large number of harmonics. For an amplifier to provide an output that is an exact reproduction of a nonsinusoidal wave, it must be capable of amplifying the fundamental wave and each harmonic equally well. This type device is called a wideband amplifier and is discussed in the last section of the chapter.

Narrow-Band Amplifiers

Before beginning our analysis of narrow-band amplifiers, we will discuss the schematic diagram and operation of a typical RF amplifier.

Figure 11-1 contains a schematic diagram for a PNP transistor RF amplifier. Note that the PNP transistor is connected in the common-emitter configuration.

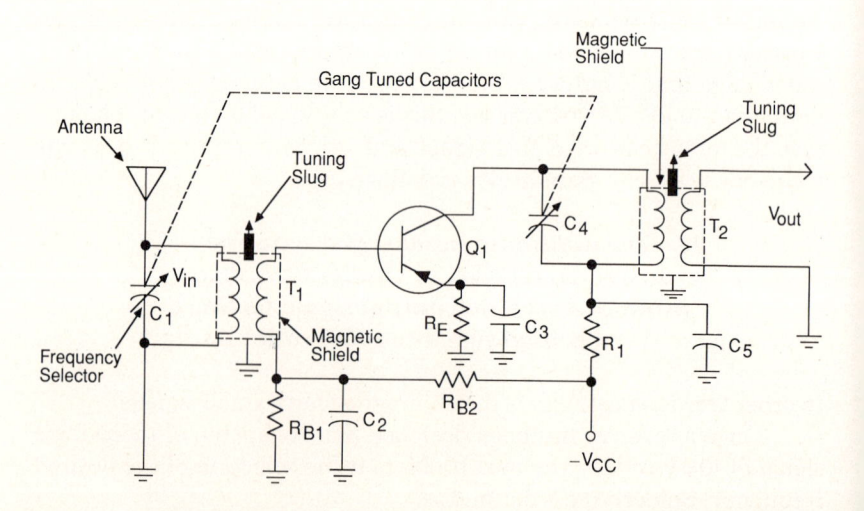

Fig. 11-1 Transistor RF Amplifier

Narrow-Band and Wideband Amplifiers

Note that the Q point for Q_1 is established by the voltage divider R_{B1} and R_{B2}. Resistors R_{B1} and R_E provide stabilization for the amplifier, and C_3 reduces degeneration by placing AC ground at the emitter. C_2 and C_5 act as *decoupling* capacitors that isolate the RF signal from the power supply. Remember, we discussed RF bypass capacitors earlier. The process of RF bypass is also referred to as "decoupling." R_1 is part of Q_1 collector load. Both parallel resonant tank circuits are tuned to resonate at the carrier frequency of the AM station being received. Note that C_1 and C_4 are *ganged* together. This indicates that when one capacitor is tuned, both capacitors are tuned, and both tanks continually operate at the same resonant frequency. The fact that the two capacitors are mechanically connected is illustrated by the dashed line that connects them. By tuning both the input and output of the amplifier, better selectivity is provided than if the input alone is tuned.

The capacitor connected across T_1 primary operates with the primary winding to form a parallel (resonant) tank circuit. Figure 11-2 shows the schematic of this tank circuit.

Notice that the tank sits between the antenna and ground. Of all the radio stations whose radio waves cut the antenna, only the single station whose frequency is identical to the frequency selected by the tuned resonant tank circuit will induce a usable voltage into T_1 secondary. Tuning C_1 changes the tank's resonant frequency. This allows a desired signal (station) to be selected.

In Figure 11-1, the signal (current) that flows in T_1 primary causes voltage and current to be induced into the secondary of T_1. This voltage and current is applied to the base of Q_1 where it causes forward bias to vary. The variation in forward bias that is applied by T_1 secondary causes I_C of Q_1 to vary at the same rate as the input signal. At the collector of Q_1, the signal has been amplified and is 180° out of phase with the input signal. Transistor (Q_1) collector current (I_C) flows in the parallel resonant tank circuit formed by C_4 and the primary of T_2. Voltage is coupled to the secondary of T_2 and is available for application to subsequent circuits.

The signal arriving at the output (T_2 secondary) is an exact reproduction of the signal transmitted (1000 ± 5kHz) by the AM station that we assumed in the introduction. This signal can be coupled to the next stages where it is amplified and converted back into the audio signals that we hear.

One amplifier stage does not usually provide enough amplification to do the job. Usually, two or more RF amplifiers are connected in cascade to give the higher gain that is desired. Using two or more stages also increases selectivity, which is an advantage.

The block diagram contained in Figure 11-3 illustrates a two-stage RF amplifier with *ganged tuning*. To simplify the explanation, we will assume that each amplifier has a gain of one (1). The two cascaded amplifiers require three tuned circuits (T_1, T_2, and T_3) to allow for their

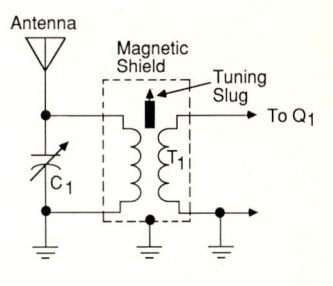

Fig. 11-2 Tuned Tank Circuit

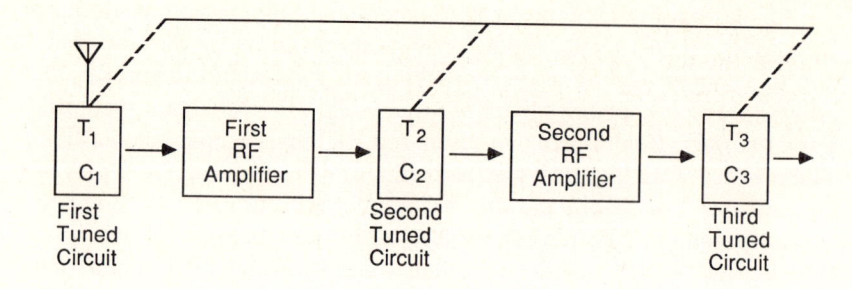

Fig. 11-3

inputs and outputs. Capacitors (C_1, C_2, and C_3) are ganged together. When two capacitors or resistors are mounted on a single shaft that is used to adjust their values at the same time, they are said to be *gang tuned*. The use of ganged tuning allows all three capacitors to be adjusted at the same time. This assures that all tank circuits operate at the same frequency.

Figure 11-4 depicts the frequency response for this circuit as used in AM radio circuitry. Note that the curve has good amplitude and narrow bandwidth (BW). BW is $F_r \pm 5kHz$. The more tuned circuits through which the signal is coupled, the narrower the bandwidth becomes. For AM radio, three tuned stages is enough.

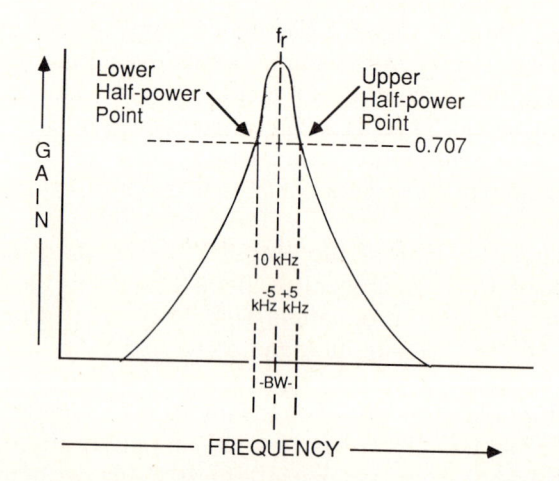

Fig. 11-4 Frequency Response Curve

Narrow-Band and Wideband Amplifiers

Suppose that three signals, all having equal amplitude (10V), arrive at the antenna at the same instant. With the RF amplifier tuned to the center frequency (F_o), the other two frequencies (F_L and F_U) appear above and below the resonant frequency. This is illustrated in Figure 11-5.

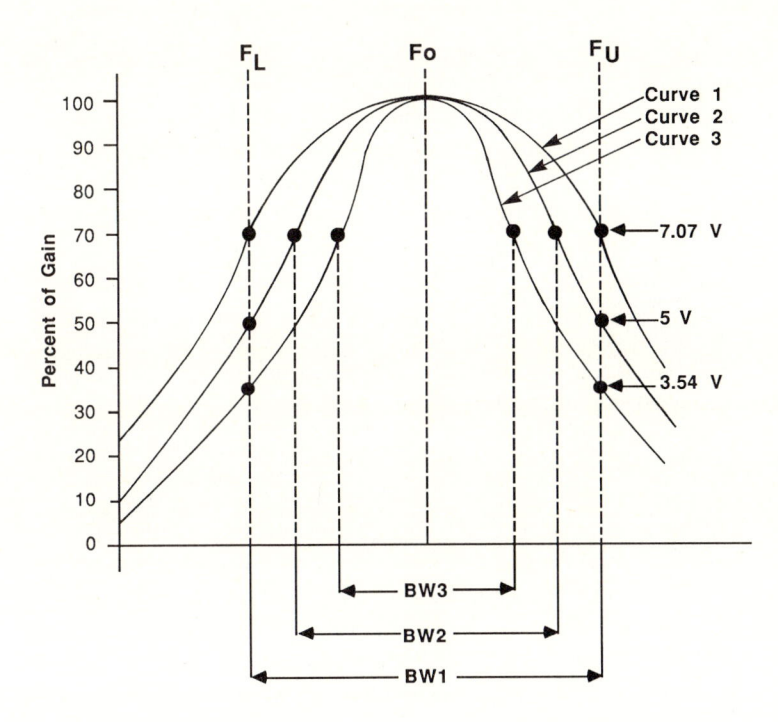

Fig. 11-5

This figure shows the response curves that result from the cascading of the two RF amplifiers. Curve 1 represents the response curve of the first tuned circuit. Curve 2 represents the response of the first RF amplifier which includes the first and second tuned circuits. Curve 3 represents the response curve for both RF amplifiers connected in cascade, including the three tuned circuits. The maximum voltage developed across the tuned circuits at the resonant frequency is considered to equal 100% for reference purposes. The symbols F_L (lowest frequency), F_U (highest frequency), and F_o (middle frequency) identify each input frequency. Remember, frequencies F_L and F_U are above and below F_o and fall at the half-power points marked on F_o. This means that they (F_L and F_U) coincide with the half-power points of the first tuned circuit, which is represented by curve 1.

If the maximum voltage that is developed across the first tuned circuit is 10 volts, the voltage developed by F_L and F_U, at their half-power points, equals:

$0.707 \times 10\,V = 7.07\,V$

The bandwidth of the first tuned circuit is noted by BW_1 on Figure 11-5. All three frequencies (F_o, F_L, and F_U) are amplified by the first RF amplifier and are applied to the primary of T_2. Remember, we assumed a gain for each stage of 1, therefore there will be no change in amplitude between input and output. This means that the center frequency (F_o) has an amplitude of 10 V, or 100% of the input amplitude. Even though the other frequencies are amplified by a gain of 1, they are now 7.07 V in amplitude and arrive at tuned circuit 2 as 7.07 V signals. This is because they are above and below the tuned frequency of tuned circuit 1, and only 70.7% of their peak entered the circuit.

Now, input amplitudes for the three signals are; $F_o = 10\,V$, F_L and $F_U = 7.07\,V$. At tuned circuit 2, the half-power point for F_o is again 7.07 V. The half-power points for F_L and F_U equal:

$0.707 \times 7.07\,V = 5\,V$

This situation is represented by curve 2 of Figure 11-5. Notice that the bandwidth of tuned circuit 2 is narrower. The narrower bandwidth for F_o is shown by a comparison of BW_2 to BW_1. The half-power points for F_L and F_U coincide with the vertical lines used to represent the half-power point of BW_1 and are identified as 5 V.

With another gain of 1 in the second RF amplifier, F_o appears at tuned circuit 3 with an amplitude of 10 V. F_L and F_U, however, arrive at the tuned circuit with an amplitude of 5 V. At tuned circuit 3, the half-power point for $F_o = 7.07\,V$ but those for F_L and F_U now equal:

$0.707 \times 5\,V = 3.54\,V$

This point is noted in Figure 11-5 as 3.54 V. Any signal that falls below the half-power point of the tuned frequency is said to be rejected by the tuned stages. You can see that both F_U and F_L are rejected at the second tuned circuit.

Observe that the bandwidth of F_o is even narrower at the third tuned circuit. This bandwidth is represented by BW_3. Notice that the resonant frequency (F_o) has maintained its peak voltage (10 V) in all tuned circuits. F_L and F_U have, however, decreased in amplitude for each tuned circuit. The result is that at T_3 we have a resonant frequency signal whose amplitude is 10 V and two other frequencies whose amplitude has been attenuated by the circuit's action. Notice that for each stage of amplification, the amplitude of all frequencies other than the resonant frequency is decreased. In the case discussed here, the two frequencies (F_L and F_U) have decreased in amplitude from 10 V at tuned circuit 1 to 3.54 V at tuned circuit 3. You can see that the bandwidth narrows as the signal passes through each stage that contains a tuned circuit. Adding more stages results in a narrower bandwidth, steeper slopes on the response curve, and an increase in the selectivity of the circuit.

The circuit that is represented by the block diagram of Figure 11-3 is shown in Figure 11-6. Remember, a parallel resonant tank circuit has maximum impedance to resonant frequency and has a lower impendance to any other frequency. For this reason, the impedance of the primary tank circuits of T_1, T_2, and T_3 will be less to all frequencies other than resonant frequency. The two-stage, three-tuned circuit amplifier shown here results in a condition that passes the resonant frequency and blocks all other frequencies. The effect of the three-tuned circuits is identical to that explained above.

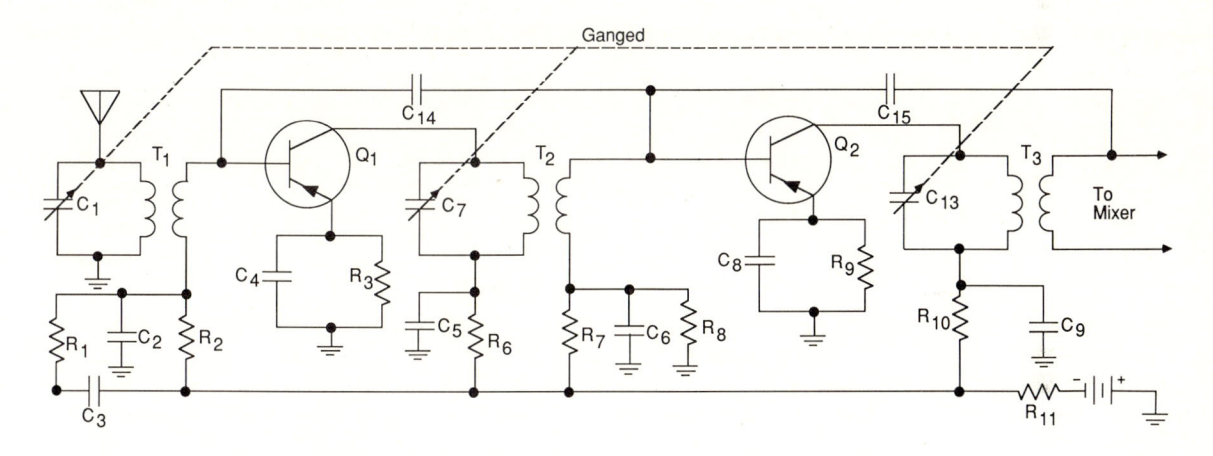

Fig. 11-6 Two Stage Tuned RF Amplifier

Signal-To-Noise Ratio

At the same time radio waves are transmitted, atmospheric and other electrical systems are generating noises at the same frequencies as the station's carrier frequency. These noises are called **static**. An example of this is the static that you hear on your radio during thunderstorms. Any signal arriving at the antenna that has the same frequency as the tuned frequency is amplified and reproduced by the speaker. This means that the station's signal must be large when compared to the noise signal. Like transmitted signals, any noise whose frequency falls outside the RF amplifier's bandwidth is ignored and is not coupled to the RF amplifier. The effect that noise voltages have can be decreased by operating with a narrow (decreased) bandwidth. You should realize, however, that decreasing the bandwidth by too much can cause the loss of part of the desired frequency. The ratio of the magnitude of the signal to noise present is the **signal-to-noise ratio**.

Troubleshooting Solid-State Narrow-Band Amplifiers

The procedures used to troubleshoot these circuits are no different from those discussed for other circuits. First you must recognize the symptoms. Next you must know the purpose of each component in the circuit. An understanding of how each component reacts to both direct current and the AC signal is very important. Refer to Figure 11-7 for an explanation of troubleshooting techniques. You have already studied the troubles that result from problems in bias circuits like R_{B1} and R_{B2}, from the loss of V_{CC}, and from problems with other DC components. Now we will discuss the AC signal problems associated with the tuned circuits used to select and couple the AC signal. Notice that the circuit shown in Figure 11-7 contains two tuned circuits, one in the input and the other in the output. As we discuss these problems, keep in mind the relationship that exists between bandwidth, selectivity, gain, and signal-to-noise ratio.

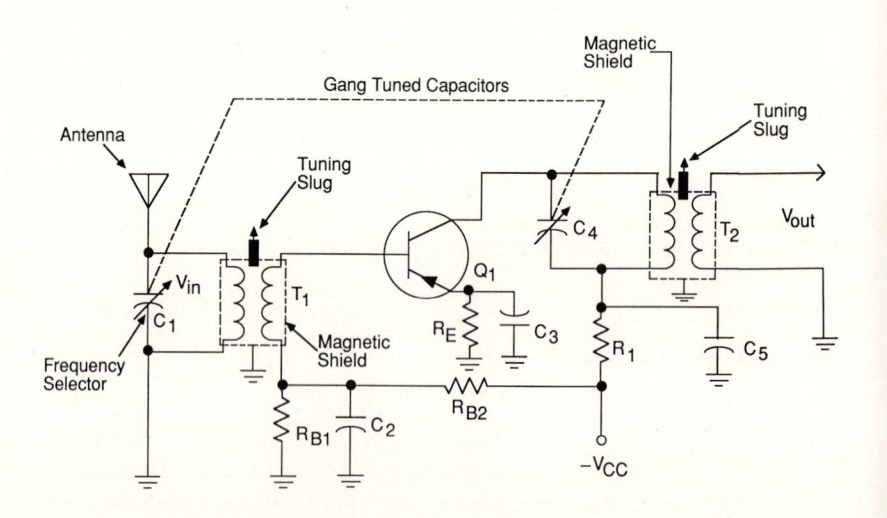

Fig. 11-7 Transistor RF Amplifier

Capacitor C_1 open. Symptoms are:
    Output signal amplitude is lower than normal.
    Bandwidth is increased.
    Decrease in signal-to-noise ratio (more noise).
    Bias and collector DC voltages are normal.

Reasons for symptoms:
    Total Q of the circuit decreases causing BW to increase.
    Increased BW causes poorer selectivity.
    Increased BW causes poorer signal-to-noise ratio.

Primary of T_1 open. Symptoms are:
    No output signal.
    Bias and collector voltages are normal.

Reasons for symptoms:
    No coupling of signal at T_1.
    Open T_1 primary does not open a DC path, therefore all DC voltage measurements are normal.

Capacitor C_4 open. Symptoms are:
    Output signal amplitude is lower than normal.
    Bandwidth is wider than before.
    Signal-to-noise ratio has decreased (more noise).

Reasons for symptoms:
    Opening C_4 causes circuit Q to decrease.

Primary of T_2 open. Symptoms are:
    No signal present at the output.
    Voltage measured at Q_1 collector equals 0 V.
    Q_1 does not amplify the input signal, as viewed on an oscilloscope.

Reasons for symptoms:
    T_2 primary being open stops all collector current for Q_1. Without collector current, Q_1 is of no value to the circuit and no output signal is developed on the primary of T_2.

Short in either T_1 primary or C_1. Symptoms are:
    No output signal present at Q_1 collector or T_2 secondary.
    DC bias and collector voltages for Q_1 are normal.

Reasons for symptoms:

All input signals are shorted to ground.

This does not affect the DC conduction of Q_1, therefore, all DC voltage measurements are normal.

Short in either T_2 primary or C_4. Symptoms are:

No output signal present at the secondary of T_2.

Collector voltage of Q_1 is slightly lower than normal.

Reasons for symptoms:

Short across T_2 primary causes zero output signal to be developed across T_2 primary. Without signal being developed on the primary, it is impossible to couple a signal.

Shorting of T_2 primary resistance causes a slight decrease in Q_1 load resistance. This causes Q_1 collector current to increase and V_{RL} to increase. An increase in V_{RL} leaves less voltage to be dropped between Q_1 collector and ground.

T_1 secondary shorted. Symptoms are:

No output signal.

Forward bias applied to Q_1 increases slightly.

Q_1 collector voltage decreases slightly.

Reasons for symptoms:

With T_1 secondary shorted, it is impossible for a signal to be coupled to Q_1 base.

Shorting T_1 primary removes a small DC resistance (coils winding resistance) from the base circuit, which causes I_B and I_C to increase.

With the increase in I_C, V_{RL} increases.

When V_{RL} increases, less voltage is dropped between Q_1 collector and ground.

T_2 secondary shorted. Symptoms are:

No output signal.

Transistor bias and collector voltages are normal.

Reasons for symptoms:

Signal developed on T_2 primary cannot be coupled to a shorted secondary; therefore, zero output signal.

Primary of T_2 completes all DC circuitry for Q_1; therefore, all DC voltages and currents are normal.

Q₁ open. Symptoms are:

No output.

Collector voltage of $Q_1 = V_{CC}$.

Reasons for symptoms:

Open transistor will not conduct and cannot amplify the input signal.

Open components drop total voltage and with the transistor open, the problem is between the collector and ground; hence, V_C measures V_{CC}.

Q₁ shorted. Symptoms are:

With Q_1 shorted, I_C is maximum, V_C is minimum, and Q_1 is saturated.

No output signal.

Reasons for symptoms:

With Q_1 shorted, the ability to vary I_C is gone. This means that V_{RL} cannot vary and thereby cause collector voltage on Q_1 to vary. Since the variations in Q_1 collector voltage form the output, there is no output. A short drops minimum voltage: therefore, the voltage dropped between Q_1 collector and ground is minimum.

Without changes in I_C, it is impossible to develop an output signal.

In this section we have discussed some of the theoretical problems that could occur in a transistor narrow-band amplifier. Much emphasis was placed on the effect that would result from problems in the tank circuits and transformer secondaries. An open or short in R_L, R_B and other components common to other amplifiers will result in the same symptoms as those discussed earlier. If you need to review these problems, refer back to previous troubleshooting explanations.

Answer each item by inserting the word or words required to correctly complete each statement.

Self-Check

1. The antenna used in the modern AM radio is often called a _____ antenna or a _____ antenna.
2. _____ refers to a receivers ability to choose one frequency or band of frequencies from all those present at an antenna.
3. The bandwidth of the third tuned circuit in Figure 11-3 is _____ than the bandwidth of the first tuned circuit.

4. An RF amplifier that is used to tune an AM station has a maximum bandwidth of _____ kHz.
5. To increase the selectivity of a tuned RF amplifier, we can add more _____.
6. Adding more stages and tuned circuits to an RF amplifier causes the circuit's bandwidth to _____.
7. RF amplifiers are joined together by _____ coupling.
8. Radio frequency (RF) amplifiers are coupled using _____ coupling so that the input and output circuits can be _____ to resonance.
9. A comparison of the amplitude of desired signals to noise present at the input to a tuned RF amplifier is referred to as the _____.
10. When two or more resonant circuits are tuned by the same mechanical shaft, they are said to be _____ tuned.
11. Describe the effect on the output signal that would result from the secondary of either T_1 or T_2 opening. Check Figure 11-7.
12. Examine Figure 11-7. Resistor (R_{B1}) opens in this circuit. This will cause Q_1 to:

   a. saturate
   b. cutoff
   c. operate normally

13. What effect would shorting the secondary of T_1 have on the output signal at the secondary of T_2? Check Figure 11-7.
14. A check of I_C shows that Q_1 is saturated. What would cause this? Check Figure 11-7.
15. Examine Figure 11-7. What is the purpose of C_3?

Wideband Amplifiers

The ability of an amplifier to amplify a nonsinusoidal waveshape is an important part of amplifier design. A nonsinusoidal wave does not vary at the same rate as does a sine wave. This wave is composed of a fundamental frequency and a large number of harmonics. An amplifier that is designed to amplify a nonsinusoidal wave must amplify not only the fundamental wave, but it must also amplify all harmonics equally well. A circuit that can do this is called a **wideband amplifier**.

Square Wave Characteristics

To understand the need and purpose of a wideband amplifier, we must understand the make-up of a nonsinusoidal wave. Because a square wave is a representative of this group of waves, we will take time to discuss its composition. By definition, a **square wave** is a:

Periodic wave which alternately assumes two fixed values for equal lengths of time, the transition time being negligible in comparison with the duration of each fixed value.

Each cycle of a square wave is composed of one positive going alternation and one negative alternation. A perfect square wave has vertical sides (leading and trailing edges), and both top and bottom are perfectly flat horizontal lines.

The sides of each alternation are actually fast voltage changes that occur in nanoseconds of time. These voltage changes represent the high frequency components of the square wave. The flat tops and bottoms are caused by the low frequency components.

The square wave is composed of a fundamental frequency and an infinite number of odd harmonics. Odd harmonics have a specific phase and amplitude relationship to the fundamental frequency. The algebraic summation of the fundamental and odd harmonics serves to create a square wave that fits the description above. A square wave is shown in Figure 11-8.

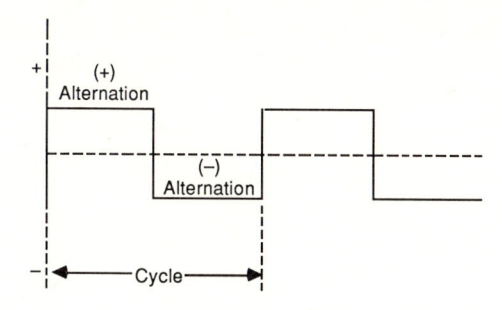

Fig. 11-8 Square Wave

Note that one cycle of the square wave consists of two alternations. One cycle of the square wave's fundamental frequency will occur during the same time as one cycle of a sine wave having the same frequency.

The effect of only a few harmonics being added (algebraically) to the fundamental is illustrated in Figure 11-9. The fundamental frequency is illustrated by the sine wave in Figure 11-9(b). The duration (time between the start of one positive going alternation and the start of

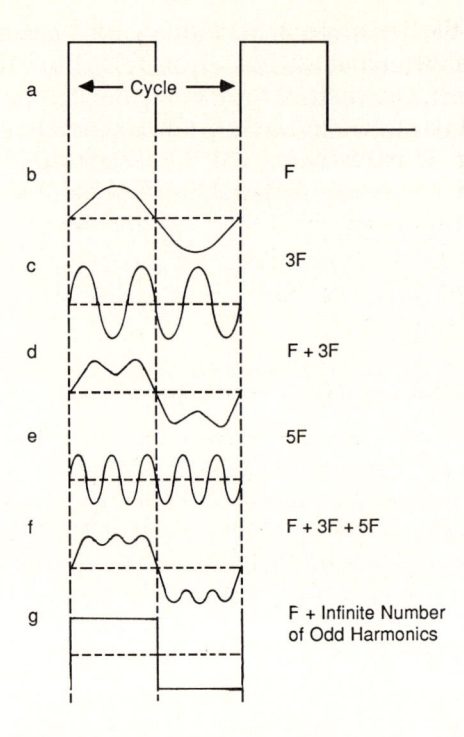

Fig. 11-9

the next positive going alternation) represents the duration of the fundamental frequency. To convert this time to frequency use the formula:

$$f = \frac{1}{t}$$

f = frequency in Hertz
t = time in seconds

The third harmonic of the fundamental frequency is shown in Figure 11-9(c). Note that it has a frequency three times that of the fundamental. When these two waves (fundamental and 3rd harmonic) are summed the result appears like Figure 11-9(d). The 5th harmonic is shown in Figure 11-9(e). Figure 11-9(f) represents the results of summing the fundamental, 3rd harmonic, and 5th harmonic. Note that the resultant wave is beginning to take on the characteristics of a square wave. Figures 11-9(a) and 11-9(g) are square waves that contain a fundamental frequency and an infinite number of odd harmonics.

In addition to phase, for the square wave to have proper size, the amplitude of each harmonic must meet certain requirements. The 3rd harmonic's amplitude is one-third the fundamental frequency amplitude, the 5th harmonic is one-fifth, the 7th harmonic one-seventh and so on. Also, the fundamental and all odd harmonics must have the same phase when the fundamental goes through zero. Notice that with each odd harmonic, the change-over point for each cycle finds the fundamental and the harmonic having similar voltage changes. When the vertical edge of the fundamental is going positive, so are all odd harmonics. When the vertical edge of the fundamental is going negative, so are all odd harmonics. By observing Figures 11-9(b), 11-9(c), and 11-9(e), you can see that all three waves are in-phase at the zero points. As long as these amplitude and phase relationships exist, the result will be a square wave.

Wideband Amplifier Characteristics

Before any amplifier can be used to amplify wideband signals, we must overcome two frequency response limitations. Refer to Figure 11-10 for an example of these limitations. In this circuit, high frequency response is limited by C_o and C_i. C_o represents the distributed capacitance and the output capacitance of Q_1. C_i represents the input capacitance to Q_2. Remember, the capacitive reactance of C_o and C_i will decrease as operating frequency increases. This decrease in reactance results in a decrease of gain at high frequencies, resulting in a circuit gain that falls below an acceptable level. Low frequency response is limited by the coupling capacitor (C_C). At low frequencies, the capacitive reactance of C_C approaches infinity. The high reactance that is connected in series with the transistor input acts to limit the amount of signal that is applied to the base of Q_2. With a smaller signal applied to its base, Q_2 output signal decreases below a usable level. The discussion that follows presents some of the methods used to overcome these frequency limitations in order to increase the bandwidth of the amplifier.

The frequency range of a video amplifier (a typical wideband amplifier) must extend from a few hertz to several megahertz. To be more specific, we will say that we need an amplifier that is capable of amplifying all frequencies from 10 Hz to 4 MHz.

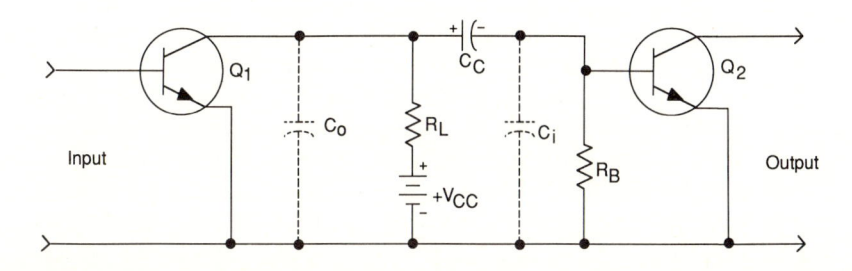

Fig. 11-10

One means of obtaining a wide bandwidth is to reduce the size of R_L for Q_1. (See the simplified schematic in Figure 11-11(a).) Figure 11-11(b) shows the frequency response that could be expected from this circuit with either a large or small R_L.

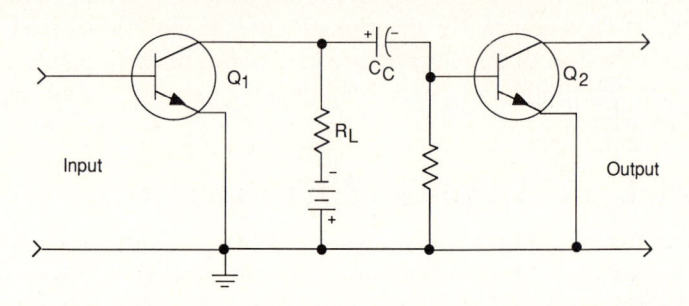

(a)

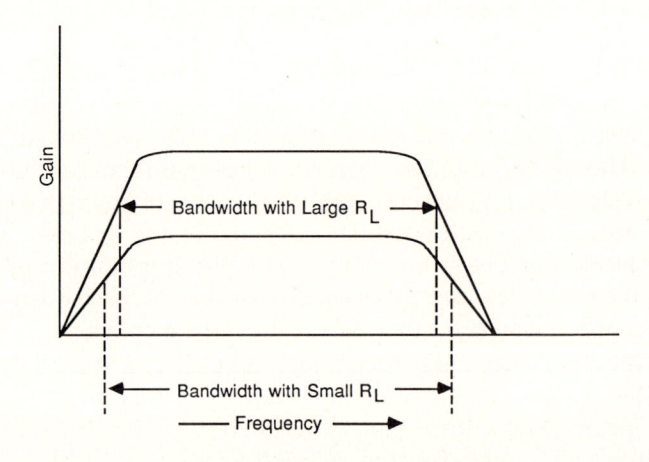

(b)

Fig. 11-11

When R_L is large, the gain of Q_1 is much higher than with a smaller R_L. The half-power points (0.707 × Peak Output) are closer together for the larger R_L. When the size of R_L is decreased, the half-power points are further apart, meaning that the bandwidth has increased. To increase bandwidth in this manner, we must sacrifice gain, but for some circuits, decreasing the size of R_L increases bandwidth enough to meet circuit needs.

Two other methods are used to extend high frequency response (to 4 MHz in our example). Both methods fall under the broad classification of **high frequency compensation**. The two methods are called

shunt and **series** compensation. These names serve to identify the position at which a compensation component is located within the circuit. Figure 11-12(a) contains the schematic diagram for a simplified circuit using shunt compensation.

For shunt compensation, coil (L_1) is connected in shunt (parallel) with the output capacitance (C_o) of Q_1 and the input capacitance (C_i) of Q_2. Notice that L_1 is connected in series with R_L. In this position, collector current (I_C) must flow through L_1. At low frequencies, L_1 will act as a short; but as operating frequency increases, the inductive reactance (X_L) of the coil will increase. Remember, at high frequencies the reactance of C_o and C_i is so low that they begin to short the signal to ground. By placing L_1 in parallel with these capacitances, a parallel resonant circuit is formed whose resonant frequency is near the frequency where C_o and C_i have low X_C. At its resonant frequency, a parallel resonant tank circuit has maximum impedance. By adding L_1 in the shunt position, a high impedance replaces the low reactance that would otherwise be present because of C_o and C_i. The tank now presents a high impedance to the output of Q_1. This causes a larger portion of the signal to be applied to Q_2. Because L_1 is in series with a relatively large fixed resistance (R_L), the Q of the tank is low, causing it to have a wide bandwidth. If care is taken in the selection of all components, a circuit of this type can be designed that will easily pass the 4 MHz upper frequency we selected for our discussion. Because L_1 is connected in shunt with the signal path, this is called **shunt compensation**. It is possible that you will also hear this coil (L_1) called **shunt peaking coil**.

Figure 11-12(b) contains a schematic for a series compensated circuit. In this circuit, coil L_2 is connected in series with Q_2 input capacitance C_i. Coil L_2 can be placed on either side of C_C. The circuit then operates as follows: L_2 and C_i form a series resonant circuit that resonates at the high end of the uncompensated circuit's frequency response curve. C_C can be ignored because at high frequencies its X_C is so low that it acts like a short. Within a series resonant circuit, circuit impedance is minimum and circuit current is maximum. This places a high voltage across C_i. The higher voltage is applied to the base of Q_2, which maintains the gain at a high level. This compensates for the loss of gain that is present in the uncompensated circuit. When components are carefully selected, it is possible to operate this circuit at frequencies well above 4 MHz. Because L_2 is connected in series with the signal, this is called **series compensation**. You may also hear it referred to as a **series peaking coil**.

In some applications, both series and shunt compensation are used in a single circuit. Figure 11-12(c) contains the schematic diagram for a circuit of this type. Note that both L_1 (shunt compensation) and L_2 (series compensation) are present. The explanations stated above apply to this circuit.

(a)

(b)

(c)

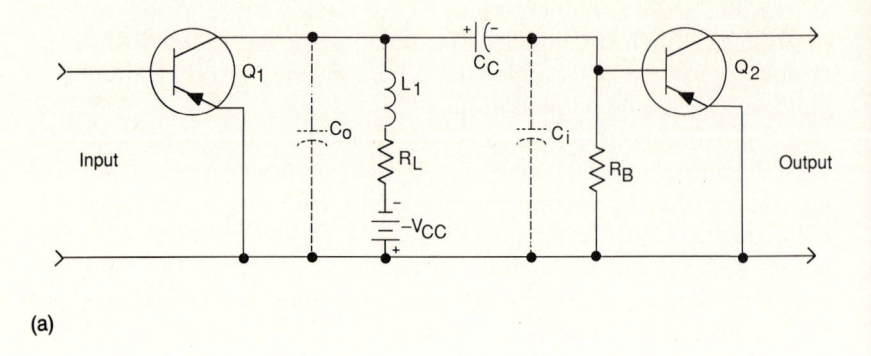

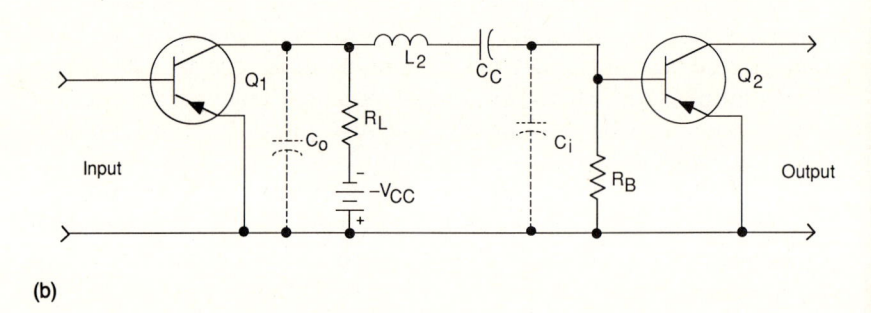

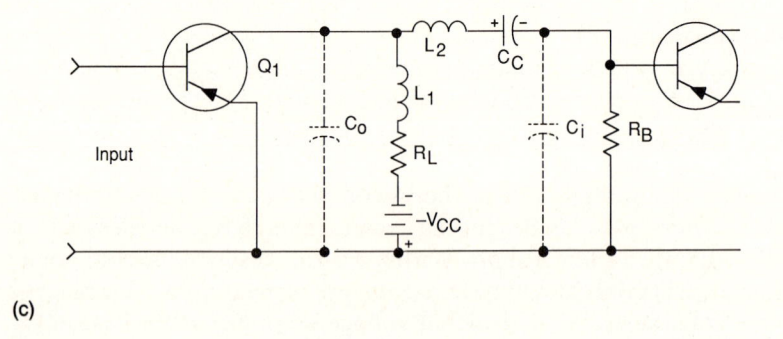

Fig. 11-12

Low Frequency Compensation

At lower frequencies, the X_C of C_o and C_i is so high that these capacitances have *no* effect on circuit operation. Low frequency response is affected by the reactance of C_C and the input resistance (R_{in}–made up of R_B and R_{eb}) of the transistor. For low frequency response, the reactance of C_C must be low as compared to R_{in} of the transistor. At the point where X_C becomes larger than R_{in}, the output signal is too low for practical use.

To reduce the amount of distortion and loss of gain that can result at low frequencies, a filter circuit is placed in series with R_L. Figure 11-13 is a simplified schematic that contains a filter of this type. This filter consists of R_F and C_F.

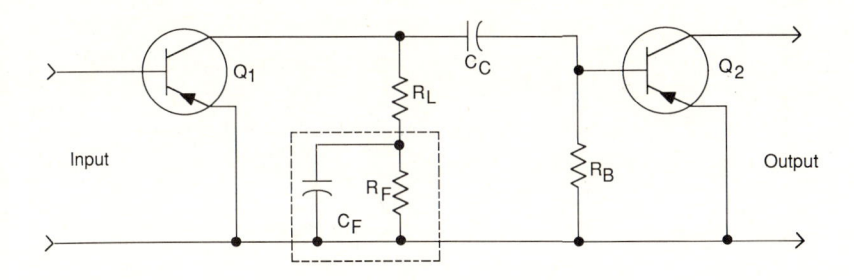

Fig. 11-13

Remember, at low frequencies X_{CF} will be high, and as operating frequency increases, X_{CF} decreases. This increases the collector load impedance at low frequencies, which increases Q_1 gain. This places a larger voltage on the left side of C_C. The increased voltage is coupled to Q_2 where it is amplified and is available at the output. Low frequency response is increased to the point where very low frequencies can be passed. At higher frequencies, capacitive reactance of C_F decreases to the point where the collector load is equal to R_L only; thus, the low frequency compensation filter has no effect on high frequency response. **Low frequency** compensation is accomplished by the use of a filter circuit inserted in series with R_L.

Combined Low and High Frequency Compensation

To get the full wide-band response required for video amplifiers, it is common to use multiple types of frequency compensation. Figure 11-14 contains the schematic diagram for an amplifier that is capable of wide band response (10 Hz to 4 MHz). Note that both high and low frequency compensation is used in this circuit. You should realize that both types of compensation work separately; therefore, each can supply the compensation for which it is designed. The compensation is then removed from the circuit by its low reactance at the opposite end of the frequency response curve. At low frequencies, the X_L of L_1 is very low, and the coil acts as a conductor and has no effect on the collector load. The X_L of L_2 is very small at low frequencies and, therefore, has no effect on the coupling circuit. For low frequencies, C_2 and C_4 have such large X_C that their being in parallel with R_3 and R_7 will increase *low frequency* gain. You can see that the components inserted for high frequency compensation have no effect at low frequencies. At high frequencies, however, the X_C of C_2 and C_4 is so small that the capacitors act like shorts, removing R_3 and R_7 from the collector loads of Q_1 and Q_2. You should be able to see how C_2 and C_4 increase low frequency gain; L_1, L_2, L_3, and L_4 increase high frequency gain.

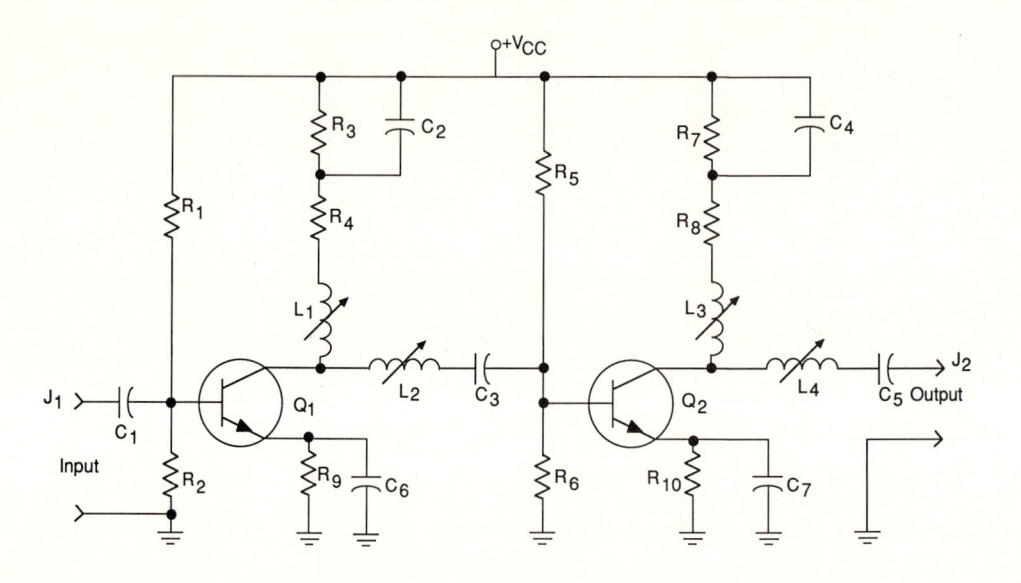

Fig. 11-14

Stagger-Tuned Wideband Amplifiers

Another method used to extend the bandwidth of an amplifier circuit is **stagger-tuned transformer coupling**. To accomplish wideband operation, the tuned circuits are alternately tuned slightly above or below the center frequency that is to be coupled.

Figure 11-15 contains a simplified schematic diagram (power sources and biasing networks have been omitted) for a two-stage, stagger-tuned RF amplifier.

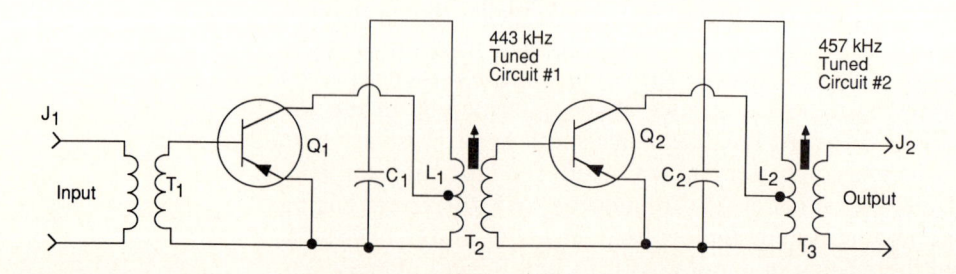

Fig. 11-15

Notice that even though the center frequency desired in the output is 450 kHz, neither transformer is tuned to this frequency. T_2 (L_1 and C_1) is tuned to 443 kHz, and T_3 (L_2 and C_2) is tuned to 457 kHz. Figure 11-16(a) shows the response curves for each amplifier.

Circuit 1 is tuned to 7 kHz below the desired output center frequency (450 kHz) and has a tuned resonant frequency of 443 kHz. Circuit 2 is tuned 7 kHz above the center frequency and has a resonant frequency of 457 kHz. Each amplifier has a bandwidth of 14 kHz. This makes them share 450 kHz as a common half-power point. For circuit 1, 450 kHz occurs at the upper-half power point, and for circuit 2, the lower half-power point occurs at 450 kHz.

(a)

(b)

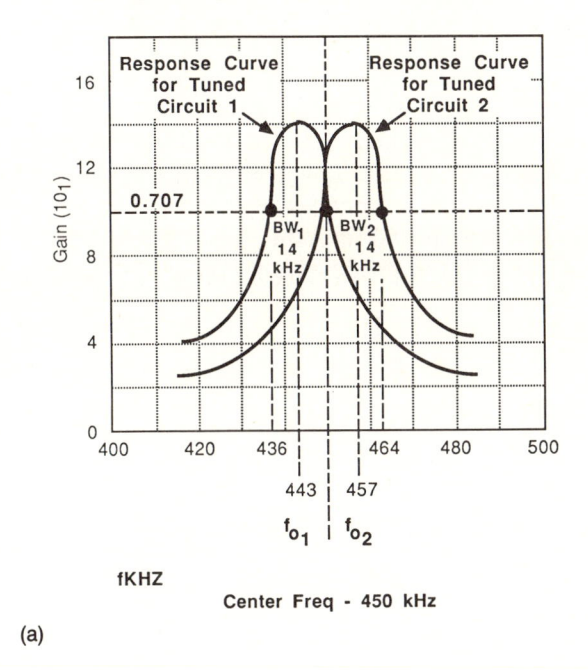

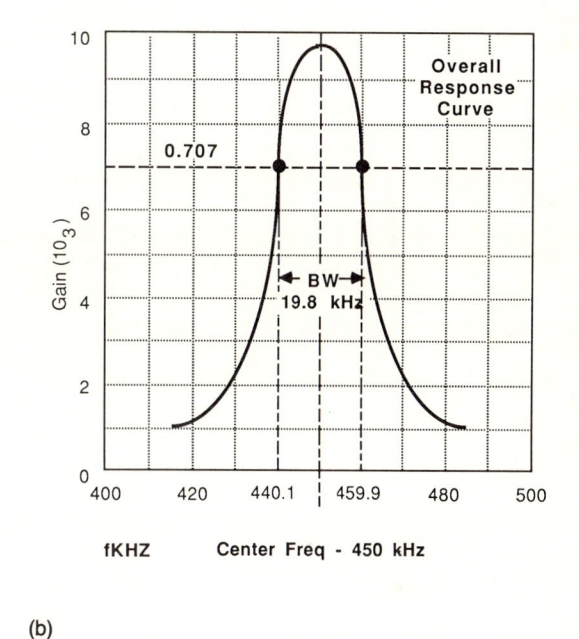

Fig. 11-16

It might appear that the overall bandwidth of the stagger-tuned pair would be 28 kHz (14 kHz + 14 kHz), but that is not the case. One method of determining the overall response curve of two or more stages is to apply a number of frequencies (for example, 415 kHz to 485 kHz) to the input of Q_1 and then plot the output amplitude at T_2 secondary for each frequency. Signal generators are presently available that can be set to sweep a band of frequencies for this purpose. When all frequencies have been applied and their outputs plotted, you can connect the plotted points to determine the actual response curve of gain versus frequency. Refer to Figure 11-16(b) and note that when all the amplitude points of Figure 11-16(a) are plotted, the resultant response curve has a bandpass from 440.1 kHz to 459.9 kHz or a bandwidth of 19.8 kHz with a center frequency of 450 kHz. The end result is that stagger tuning provides a wider bandwidth than if each circuit was tuned to the same frequency. Remember, when this circuit had all stages tuned to the same frequency, the bandwidth became narrower.

Troubleshooting Wideband Amplifiers

To locate trouble within a circuit, you must first understand how that circuit operates. Figure 11-17 contains the waveshapes that could be expected, if viewed on an oscilloscope, when checking the circuit of Figure 11-14 for normal operation and/or changes in compensation. Figure 11-17(a) shows a typical input square wave, and Figure 11-17(b) shows the output signal that is present when the circuit is operating normally. Note that the schematic (Figure 11-14) has its input labeled as J_1 and output as J_2. These test points are intentionally installed during manufacture to make it easy to check the circuit's operation. Notice that two common-emitter amplifiers are used. This means that the input and output signals will be in phase. Since the square wave in Figure 11-17(b) is an amplified and undistorted reproduction of Figure 11-17(a), you can tell that the circuit is operating normally when these waveshapes are viewed at J_1 and J_2.

(a) Input

(b) Normal Output

(c) Low Frequency Loss — Increased Gain

(d) High Frequency Loss — Over Compensation

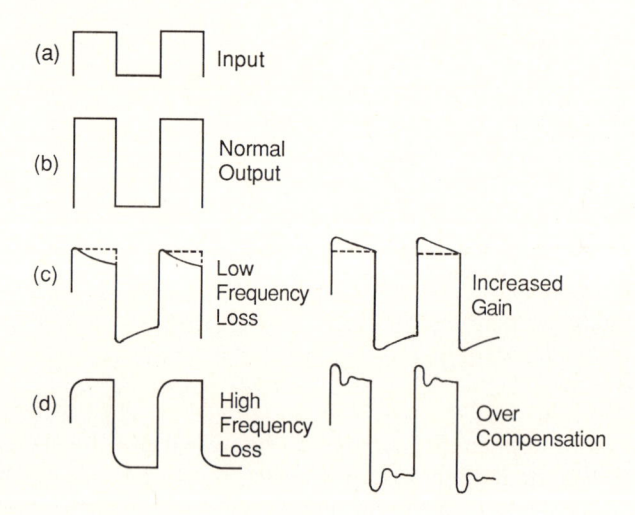

Fig. 11-17

If you observed the correct input signal at J_1 (Figure 11-14) but failed to see an output signal at J_2, you would know that some trouble had stopped the signal from being coupled to the output. As discussed earlier, many things could cause this to happen. Troubles like transistors that are open or shorted, open coupling capacitors, and open components—such as L_1, R_3, and L_2—can cause loss of the output signal.

Observe Figure 11-17(c) and note that it is labeled "LOW FREQUENCY LOSS". The signal observed is no longer a square wave. Both horizontal (peaks) lines are no longer perfectly straight but are "tilted." Remember, the horizontal lines in a square wave represent the low frequency component; therefore, any distortion present here is a result of poor low frequency compensation. The "tilted" appearance tells us that an RC time constant has decreased, and the output waveshape is approaching differentiation. Some possible causes are: (Refer to Figure 11-14.)

R_3, R_7, C_2, or C_4 shorted

or:

C_2 or C_4 open.

Observe the waveshape labeled "INCREASED GAIN" and you can see the difference between C_2 and C_4 opening or shorting. If either of these capacitors shorts, the tilted wave is the result; but if either capacitor opens, the result is increased gain, as shown in the waveshape labeled "INCREASED GAIN". Opening either of these capacitors (C_2 or C_4) not only affects the low frequency compensation, but also changes the gain of the stage. An open C_2 or C_4 means that the resistor in parallel with that capacitor becomes part of the collector load at all frequencies. The larger the collector load, the higher the circuit gain. The opening of either of these capacitors results in two symptoms: reduced low frequency compensation and increased gain.

Figure 11-17(d) contains a waveshape that shows the effect of high frequency distortion. The waveshape is labeled "HIGH FREQUENCY LOSS." Notice that it is approximately the correct amplitude, but its leading edges are "rounded off." Again, you should remember that the fast changing vertical edges represent the high frequency component of a square wave. Any "rounding" of these edges results from a loss of high frequency compensation. When this waveshape is viewed at J_2, it means that some of the high frequency harmonics are being lost. A short in L_1, L_2, L_3, or L_4 could be the cause of this problem.

In the practical circuits that technicians repair, these coils are often adjustable. The technician is responsible for aligning the circuit to assure correct operation. If these coils are not correctly adjusted, either loss of compensation or over compensation can result. Figure 11-17(d) contains a waveshape that could be viewed should these coils be adjusted to the point of *over compensation*.

Self-Check

Answer each item by performing the action indicated.

16. Loss of upper frequency amplification is caused in an RF amplifier by _____ of a circuit.
17. Typical video amplifiers have a bandwidth that extends from approximately 10 Hz to _____.
18. Placing a coil in series with R_L in the collector load of a transistor causes gain at high frequencies to _____.
19. Placing a resistive-capacitive filter in series with R_L causes low frequency gain to _____.
20. Stagger-tuned, two-stage RF amplifiers develop twice the bandwidth of a single tuned amplifier.

    a. True   b. False

21. An open coupling capacitor (C_3) (Figure 11-14) causes a loss of output signal.

    a. True   b. False

22. Placing a resistive-capacitive filter in series with R_L causes high frequency gain to _____.
23. A square wave consists of a fundamental wave and _____.
24. A coil is placed in parallel with the output signal. It is referred to as a _____ compensation network.
25. Placing a coil in series with R_L in the collector load of a transistor causes gain at low frequencies to _____.

Summary

Narrow-band amplifiers are used to amplify a specific group of frequencies. This band of frequencies is normally selected by use of transformer coupling. The coupling circuit is tuned to resonate at the center frequency of the band to be amplified. The upper and lower half-power points establish the highest and lowest usable frequencies that will be coupled. Q of the tuned circuit determines the bandwidth. Higher Q results in a circuit having narrower bandwidth and smaller bandpass. Transformer coupling can also be used to aid with impedance matching between stages.

Wideband amplifiers are used to amplify nonsinusoidal waveforms. A nonsinusoidal wave consists of a fundamental sine wave plus harmonics. The type of harmonics, odd or even, and the number of harmonics will determine the shape of a nonsinusoidal wave. Wideband amplifiers must be able to amplify the fundamental wave plus all of the harmonic frequencies that make up the waveform.

Narrow-Band and Wideband Amplifiers 367

Frequency compensation circuits are used to extend the frequency range of a circuit. At low frequencies, gain is affected by coupling capacitance. To amplify low frequencies, the input capacitance to a circuit must have low X_C to the lowest frequency to be amplified. High frequency amplification is affected by distributed (stray), capacitance which is in parallel with the output. At high frequencies, the small value of stray capacitance will have a low X_C. The low X_C acts as a short between the signal path and ground, shorting the signal. To compensate for high frequency loss, a coil is placed in the circuit. It is possible for the coil to be connected in either series or parallel with the signal. This gives rise to the names *shunt compensation* and *series compensation*. Using a combination of capacitors and coils it is possible to extend both low and high frequency gain of an amplifier.

Review Questions and/or Problems

1. An AM radio operates with RF amplifiers whose bandwidth is tuned for _____ kHz.

2. Wideband amplifiers are used to amplify _____.

   a. sinusoidal waveshapes
   b. nonsinusoidal waveshapes
   c. sine waves
   d. all of these

3. Selectivity is defined as the ability to _____.

   a. receive and amplify small signals
   b. transmit small signals
   c. transmit a narrow band of frequencies
   d. choose a frequency or small band of frequencies

4. A square wave is composed of a fundamental wave and an infinite number of _____.

   a. harmonics
   b. even harmonics
   c. odd harmonics

5. To isolate tuned circuits that operate in the RF band from magnetic interference, we enclose them within a/an _____.

6. When two components are mounted on the same shaft and are both tuned at the same time we say that they are _____ tuned. This is shown in the schematic diagram by a _____.

7. Adding tuned stages to an RF amplifier will increase _____.

   a. selectivity
   b. bandwidth
   c. sensitivity
   d. all of these

8. The ability of a circuit to receive and amplify small signals is referred to as _____.

   a. selectivity
   b. bandwidth
   c. sensitivity
   d. all of these

9. Refer to Figure 11-10. In this simplified circuit, C_i and C_o are the result of _____ capacitance.

10. Using the same circuit, C_C will cause loss of gain at _____ frequencies.

11. Refer to Figure 11-10. Increasing the size of R_L will cause bandwidth to _____.

    a. increase
    b. decrease
    c. remain the same

12. Refer to Figure 11-10. Increasing the size of R_L will cause circuit gain to _____.

    a. increase
    b. decrease
    c. remain the same

13. Refer to Figure 11-12. In these circuits, L_1 and L_2 are placed in the circuit to _____ high frequency response.

    a. increase
    b. decrease

14. Refer to Figure 11-13. In this circuit, R_F and C_F are placed in the circuit to _____ low frequency response.

    a. increase
    b. decrease

15. Refer to Figure 11-14. In this circuit C_2 and C_4 will have _____ X_C at high frequencies assuring that R_3 and R_7 are _____ from the circuit.

16. Refer to Figure 11-14. In this circuit L_1, L_2, L_3, and L_4 will have _____ X_L at high frequencies assuring that high frequency gain is _____.

17. In stagger-tuned circuits, tuned circuit 1 is tuned _____ resonance and tuned circuit 2 is tuned _____ resonance.

    a. above, below
    b. above, above
    c. below, below
    d. below, above

18. Refer to Figure 11-17. Distortion of the top of the trailing edge of a waveshape indicates _____.

    a. low frequency loss
    b. over comparison
    c. high frequency loss
    d. all of these

19. Refer to Figure 11-17. Rounding off on the top of the output waveshape indicates _____.

    a. low frequency loss
    b. over compensation
    c. high frequency loss
    d. all of these

20. Audio amplifiers that are used in a radio are classified as _____ amplifiers.

    a. narrow band
    b. wideband

12
Wave Generation

Objectives

1. List the two broad classifications into which wave generators can be grouped.
2. Name the three classes into which sinusoidal oscillators are grouped.
3. Discuss the nonsinusoidal oscillators and identify their waveshapes.
4. List the three requirements for oscillation and discuss the part that each plays in the generation of an output wave.
5. Explain the need for amplitude and frequency stability.
6. Define the following terms:

    a. RC oscillator
    b. Flywheel effect
    c. Piezoelectric effect
    d. Feedback
    e. Feedback network
    f. Sawtooth wave
    g. Jump voltage
    h. Fall or Decay
    i. Damping
    j. LC tank circuit
    k. Frequency-determining device
    l. Square wave
    m. Rectangular wave
    n. Trapezoidal wave
    o. Slope
    p. Time base generator
    q. Sinusoidal
    r. Nonsinusoidal

Introduction

The need to generate electronic waves of specific frequency and shape plays an important part in the field of electronics. Wave generators of various types are used to generate everything from a sine wave to square and peaked waves that range in frequency from a few Hertz to several thousand gigahertz (1×10^9). Many different circuits are used as wave generators.

One of these circuits is called an **oscillator**. An oscillator circuit is simply an amplifier that supplies its own input. It is, therefore, only natural that the study of oscillators should follow amplifiers. As we proceed through this chapter, we will classify oscillators according to *waveshape* and will establish the *requirements for oscillation*. We will present the most common types of wave generators and the output waveshape that can be expected from each type.

Classification of Wave Generators

It is possible to classify wave generators into two broad categories: **sinusoidal** and **nonsinusoidal**. In these classifications, the generators are grouped according to the type of output wave they generate.

A wave generator that produces a sine wave in its output is called a **sinusoidal** oscillator. Ideally this output has constant amplitude and constant frequency. In actuality, something less than this must be accepted, as perfection is not obtainable. The degree to which oscillator designs approach perfection depends on several factors: the class of operation, the characteristics of the amplifier used, frequency stability, and amplitude stability.

Sine-wave generators produce sine wave signals at frequencies that range from a low audio frequency to extremely high frequencies of 30–50 gigahertz and higher. To do this, the circuits use various methods to establish the oscillator's operating frequency. Simple resistor-capacitor circuits can be used to create low frequency audio signals. Radio frequency sine wave oscillators use LC resonant tank circuits, and extremely high frequency oscillators operate using tuned resonant cavities. Still other oscillators use crystal-controlled circuitry to create and regulate the circuit's output frequency.

Nonsinusoidal oscillators generate complex waveforms: square, rectangular, sawtooth waves, and/or trigger pulses. Because their output is usually characterized by a sudden change from one position to another, these oscillators are called **relaxation** oscillators. The operating frequency of these amplifiers is usually controlled by the charge and/or discharge of a capacitor connected in series with a resistor. Some types do, though, contain coils which affect output frequency. In other words, like the sinusoidal oscillators discussed above, nonsinusoidal oscillators can use both RC and LC frequency determination networks. Individual circuits are called multivibrators, pulsed and blocking oscillators, and sawtooth generators.

Wave Generation

Many of these circuits are designed to generate an output pulse only when told to do so (triggered). Triggered oscillators can be either sinusoidal or nonsinusoidal. To achieve this, the transistor is biased such that at the Q point it is either cut off or it is conducting at a specific level. Application of a pulse or gate to the oscillator's input causes the transistor's bias to change in the direction and amount necessary to generate the output signal.

Requirements for Oscillation

For oscillators to operate correctly, certain conditions must exist. What characteristics do we look for in a good oscillator? How can these characteristics be improved? Why would we choose one type oscillator over another? These are questions that must be answered before a detailed analysis of the circuitry can begin. For an oscillator to operate satisfactorily over long periods of time, the following three requirements must be met:

1. amplification
2. feedback in proper phase and amplitude
3. a frequency-determining device or network

By definition an oscillator is:

A device that converts DC power into AC power at some predetermined frequency.

An oscillator can be thought of as an amplifier that is capable of providing its own input signal. Oscillators are used, primarily, to generate a desired waveform that has *constant amplitude* at a *constant frequency*. The challenge is to design circuits that can do this and maintain signal stability over long periods of time.

An oscillator circuit, like an amplifier, requires a power source (V_{CC}) for its operation. Remember, the power present at the output of an amplifier is much higher than that required at the input. It is possible, therefore, to feed a small portion of the output power back to the base of the transistor for use as the input. The amount of power that is fed back must be strong enough to continually drive the amplifier, causing it to operate as a signal generator. Not only must the feedback be strong enough to drive the amplifier, it must be in phase with the signal already present. Feedback that is in phase with the signal that is already present is called **regenerative feedback**.

Refer to Figure 12-1 where an amplifier and its feedback path are shown. Earlier we discussed *degenerative feedback*, which acts to decrease the gain of an amplifier. Degenerative feedback consists of a signal that is 180° out-of-phase with the normal signal. *Regenerative*

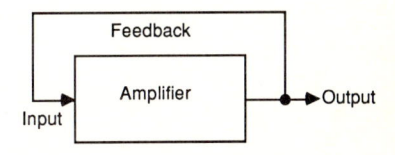

Fig. 12-1 Block Diagram-Basic Oscillator

feedback aids transistor operation and serves to replace the losses present in the circuit maintaining oscillation.

For an oscillator to serve practical purposes, it must operate at a specific frequency. To do this, there must be some type of frequency-determining network or device designed into its circuitry. Actually, the *frequency-determining device* is a form of filter which allows the desired frequency to pass and blocks all others. Without a frequency-determining network or device, the circuit will oscillate at random making it impossible to predict the output frequency.

Again, let's check the three requirements for oscillation:

1. an *amplifier* capable of providing the gain necessary for oscillator operation (amplification [in itself] implies a power source is present.)
2. *feedback* (regenerative) that is of the correct phase and amplitude.
3. a *frequency-determining network* or device that is designed to control the operating frequency

With an understanding of these basic requirements, we can proceed to circuit types and applications. Any circuit that has these three characteristics will oscillate at some frequency. To obtain a signal that has the desired frequency and amplitude, we must design the circuit very carefully. In order to understand these design requirements, we must discuss some of the characteristics that are unique to oscillators.

A good percentage of all operational systems require one or more oscillators in their performance. It is safe to say that, in each case, two things are very important: frequency stability and amplitude stability. *Amplitude stability* refers to the oscillator's ability to supply an output signal that has a constant amplitude. The less the output amplitude varies, the better the circuit's amplitude stability. *Frequency stability* refers to the ability of the oscillator to operate at the same frequency over long periods of time. The less the frequency drift, the better the frequency stability.

To achieve amplitude and frequency stability, care must be taken to prevent variations in gain, load, bias, and component variations. Variations in the load connected to the circuit will alter the oscillator's operation. It is, therefore, necessary to minimize the variations that occur in the load.

As you have already learned, variations in bias affect the transistor's conduction. At best, bias variations will affect the circuit's Q point. At worst, large changes in amplifier gain can result. For this reason, it is necessary to use a well regulated (V_{CC}) power supply and a low-tolerance bias stabilization network. Without these considerations, an output signal that has constant amplitude and frequency is practically impossible.

As components age, it is possible for their values to change. When operating temperature and other environmental conditions change, it is possible that the way components operate can be affected. For these reasons, we must consider component age and tolerances. Solid-state devices are designed to operate at approximately 25° Celsius (77°F) and within certain humidity ranges. For this reason, it is quite often necessary to control both room temperature and humidity in areas where sophisticated electronic equipment is operated.

The power contained in the output signal is also important. In general terms, we can state that high power is obtained at the expense of poorer stability. To maintain oscillator stability and still have the power needed to do a job, we apply the oscillator's output to a power amplifier. This amplifier acts as a *buffer* between the oscillator and load, isolating it from variations in load that would otherwise occur.

Use of a:

Buffer amplifier isolates an oscillator from the variations in load that could occur and which might affect the oscillator's operation.

The buffer amplifier also amplifies the signal, providing sufficient power for the job to be done.

When it is necessary to have an oscillator that not only develops a stable signal but also must develop high power, efficiency becomes a problem. In many cases, the amplifier used in an oscillator operates Class C in order to get higher efficiency. There are many types of oscillators that cannot be operated Class C; therefore, not all oscillators are suited for applications that require high power.

The frequency of a desired signal also imposes certain requirements. For an oscillator to operate at high frequency, its transistor must have *low* interelement capacitances. In fact, some special circuits are used that include the interelement capacitance as part of the circuit design. In this way, the interelement capacitance can be put to good use instead of being viewed as a problem.

Self-Check

Answer each item by inserting the word or words required to correctly complete each statement.

1. The three requirements for oscillation are:

    a. _____
    b. _____
    c. _____

2. The requirements for a good oscillator are _____ stability and _____ stability.
3. Regenerative feedback is _____ phase with the signal that is already present.
4. For an oscillator to operate with a stable Q point, it is necessary to have a _____.
5. Oscillators that must also develop high power are normally operated Class _____.

Sine-Wave Generators

Earlier in this chapter, you were told that RC networks, LC resonant tank circuits, and crystals could be used as frequency determining devices for sine-wave oscillators. Now, we will explore the different types of frequency-determining devices and networks that are used in oscillator design.

RC Networks as Frequency-Determining Networks

To convert an ordinary common-emitter amplifier to a sine-wave oscillator, it is necessary to provide regenerative feedback through an RC network. Figure 12-2(a) contains the block diagram of this type of circuit. Note that the RC network consists of R_1, C_1, R_2, C_2, and R_3, C_3. In this circuit, the RC network provides two things: a path for regenerative feedback and the frequency-determining device. Figure 12-2(b) contains a vector diagram that illustrates the RC network's action.

Assume that the amplifier used in this oscillator is a common-emitter configuration. From this, we know that its output signal is 180° out-of-phase with the base signal. Point A on the block diagram and the vector diagram represents the CE amplifier output signal. Note that in the output signal path, a conductor is connected back to the input through the RC network. Remember—for feedback to be regenerative, it must be in phase with the signal at the point where it is fed back. The common-emitter amplifier shifts the signal's phase 180°. To complete the 360° needed to have the feedback signal and the input signal in phase, the RC network must shift phase by 180°.

When power is first connected to the circuit, the transistor begins to conduct. Because this conduction is not controlled by an input signal, it is classified as *noise*. Noise consists of a large number of random frequencies that carry no intelligence. The variations of each of these frequencies appear at the collector of Q_1 (the output). This point is represented by point A in the two drawings. In order to couple the feedback signal, R_1 and C_1 are connected directly to point A. As the

Wave Generation

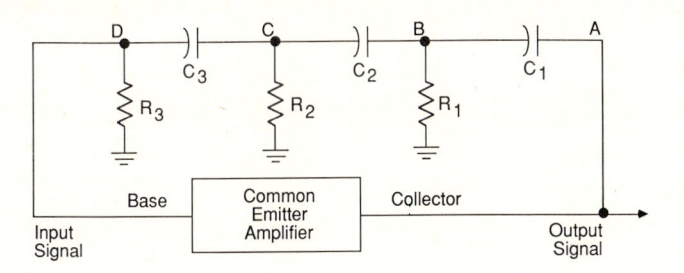

(a) Amplifier with RC Network

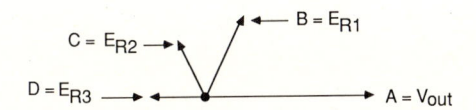

(b) Vector Analysis

Fig. 12-2 RC Oscillator

signal passes (is coupled) through C_1, it causes the capacitor to charge and discharge through R_1. In the process, the signal is shifted in phase (the phase shift that occurs across V_{C1} and V_{R1} in a series RC circuit). The vector located at point B represents this phase shift of about 60°. Note also that vector B is shorter than vector A. This represents the loss in amplitude that occurred across C_1 and R_1. As the signal passes through C_2 and R_2, it is again shifted approximately 60°, and has its amplitude reduced. This is represented by vector C. The signal present at point C has been shifted approximately 120° in relation to point A. To assure that the phase shift and decrease in amplitude are as nearly equal as possible to those of the first RC pair, all resistors and capacitors have equal values. The signal present at point C is passed through C_3 and R_3, where it is again shifted approximately 60° and has its amplitude decreased. This action is represented by vector D. Note that point D is 180° out-of-phase with point A. The signal arriving back at point D has been shifted 360°, 180° across the amplifier and 180° across the RC network.

Notice that point D identifies the base (input) of the common-emitter amplifier. Because of the 360° phase shift, any signal arriving at the input (feedback) is regenerative. Remember, a regenerative signal aids the signal already present. If you look back at the RC network, you should be able to understand that for all the frequencies (noise) present at the collector of the amplifier, this RC network will supply exactly 180° of phase shift for *only* one frequency. It is not required that the phase shift of each capacitor/resistor pair be exactly 60°, as long as the *total* shift is 180°.

For any RC network, only one frequency will receive a phase shift of exactly 180°. In other words, an RC network is frequency selective and the RC time constant determines oscillator frequency of operation. In some texts, vectors are called "phasors." If you consider the fact that vectors represent AC components, you can understand the phase shifts that occur. The length and phase relationship of the vectors shown depend on frequency.

Therefore, the RC network acts as a frequency-determining network because it provides a 180° phase shift for one frequency only. The frequency of oscillation for this circuit is determined by the values of resistance and capacitance included in the RC network. In some applications, these resistors and/or capacitors may be variable, allowing the operator to tune the desired frequency. For the oscillator to provide a sine-wave output, the amplifier must be biased for Class A operation.

LC Tank Circuits as Frequency-Determining Networks

Some sine-wave generators use resonant (parallel LC) networks as their frequency-determination network. In your earlier studies, you learned how resonant circuits store energy during conduction and then transfer energy back and forth during periods when conduction is cut off. It is possible for us to use the sine wave that is produced by a resonant circuit of this type as an output sine wave. The action that occurs during the resonating action of the LC tank circuit is quite often called the **flywheel effect**.

This effect is illustrated in Figure 12-3(a-i). Observe this figure and follow along as we review the action. Note that each section of Figure 12-3 consists of a circuit and the waveform that results during the flywheel action.

When the switch connects the battery (in position A) to the capacitor as shown in Figure 12-3(a), the capacitor charges almost instantly. Notice that there is little resistance in the capacitor's charge path, which makes five time constants (TC = RC) very short. With the switch set to position A, the capacitor is charged to battery potential.

Changing the switch to position B provides the capacitor with a path through which it can discharge. At that time, the capacitor acts like a power source whose negative pole is at the top of the drawing, (see Figure 12-3(b)). As the capacitor discharges, current flows through the coil from the top to the bottom, causing a voltage drop with the polarity

Wave Generation

(a) Capacitor charged

(b) Capacitor discharging, field expanding

(c) Capacitor discharged, field around coil fully expanded and stationary

(d) Field collapsing, capacitor charging

(e) Field collapsed, capacitor charged

(f) Capacitor discharging, field expanding

(g) Capacitor discharged, field maximum and stationary

(h) Field collapsing, capacitor charging

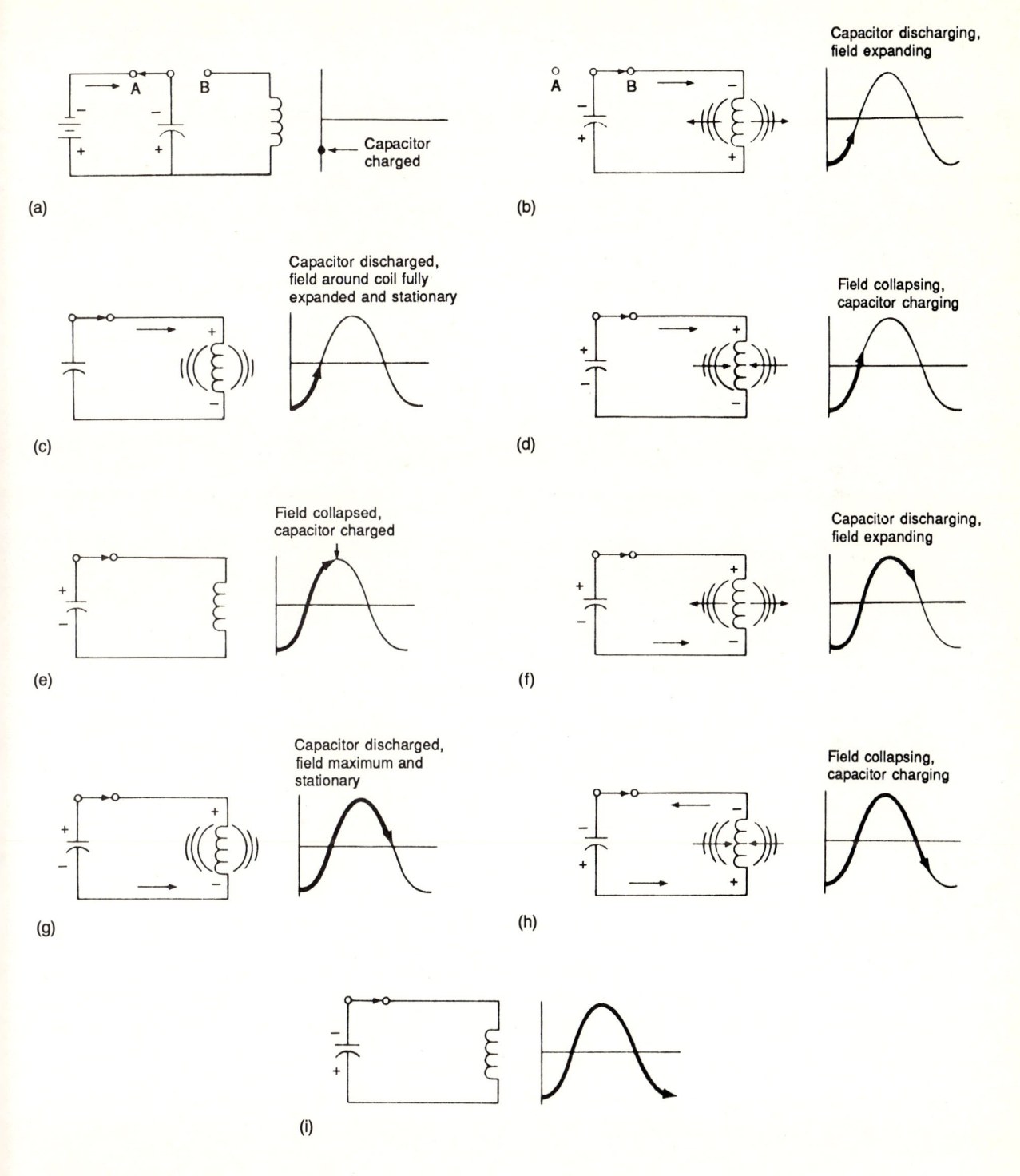

(i)

Fig. 12-3 Flywheel Effect

shown. During this time, an electromagnetic field is expanding around the coil that stores energy. After a period of time, the capacitor has fully discharged, the coil is fully charged, and current tries to stop. The coil senses the change in current and acts to oppose the change.

Refer to Figure 12-3(c). The coil acts to oppose the change in current flow by supplying energy from its electrostatic field. This causes current flow to continue in the same direction. The coil now acts as the power source and charges C_1. Notice that when the coil acts as the power source, its polarity is opposite that present when the coil was charging.

During the next period of time, the coil acts as a power source (see Figure 12-3(d)), and current flows through the capacitor. The capacitor now stores energy in its electrostatic field but note that the polarity is opposite that shown in Figure 12-3(b). Figure 12-3(e) shows the condition that exists when the coil is completely discharged and the capacitor is charged.

See Figure 12-3(f). With the polarity shown (positive at top and negative at bottom), the capacitor again becomes the power source. Current flows through the coil from bottom to top, causing its electromagnetic fields to again expand as is shown in Figure 12-3(f).

Figure 12-3(g) represents the condition that exists when the coil is fully charged and the capacitor is fully discharged. As soon as current tries to stop, the coil senses the change and acts to oppose this change.

At this point, current begins to flow through the capacitor from the top to bottom. This action is represented by Figure 12-3(h). Notice that the coil's polarity has reversed, and it is acting as the power source to recharge the capacitor. The capacitor is now charging with a negative charge at the top and positive at the bottom.

Once the capacitor is fully charged (Figure 12-3(i)), the circuit has returned to the condition that was present when the switch was first moved to position B. The capacitor is fully charged, with the negative potential at its top. During this sequence of operations, the signal that is developed across the coil is a sine wave. If the coil had been a transformer primary, the secondary could have been used to couple this sine wave to the output.

If there were no resistances in the tank circuit, oscillations would continue indefinitely. There is, however, resistance present and this causes power (energy) to be lost through dissipation. This power loss causes each peak of the sine wave to be slightly smaller than the last peak. This loss of amplitude is called **damping**. Damping is illustrated in Figure 12-4.

In Figure 12-4(a), the ideal sine wave is shown. Figure 12-4(b) represents the wave that can be expected from a tank circuit that contains a small internal resistance. Tank circuits that contain high internal resistance produce damped waves more like the one shown in Figure 12-4(c).

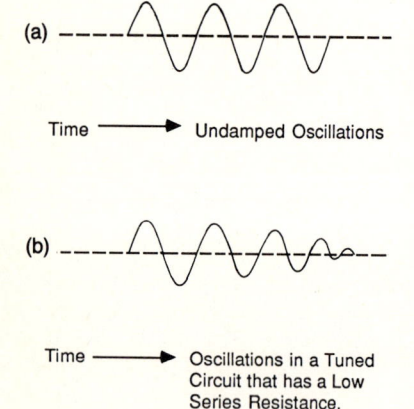

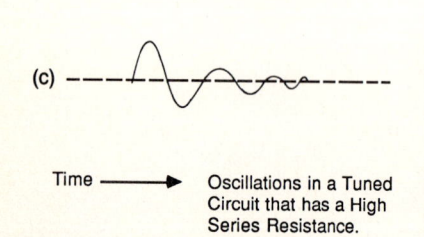

Fig. 12-4 Effects of Damping

(a) Time → Undamped Oscillations

(b) Time → Oscillations in a Tuned Circuit that has a Low Series Resistance.

(c) Time → Oscillations in a Tuned Circuit that has a High Series Resistance.

Before going on, we will discuss one other thing. (Refer back to Figure 12-3(i)). If each time the tank returns to this state it is possible to switch momentarily back to position A, we could recharge the capacitor. If the capacitor could be recharged then, each oscillation would begin with the capacitor charged to the battery potential. The resultant waveshape would be very close to that shown in Figure 12-4(a), with little amplitude being lost during each oscillation.

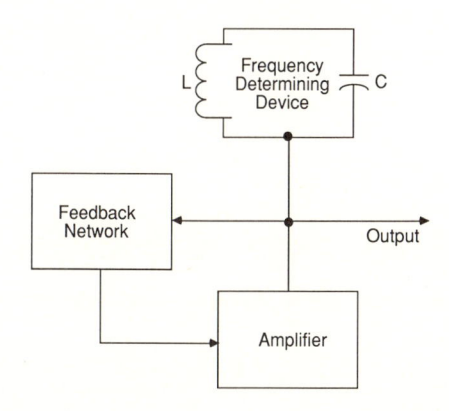

Fig. 12-5 Block Diagram - LC Oscillator

Observe Figure 12-5 which contains the block diagram of a typical **LC oscillator**. An **LC tank** provides the initial signal by generating a sine wave that is fed back to the amplifier input as regenerative feedback. This feedback is sufficient to start and maintain oscillations. The tank LC components generate a sine wave output signal. The LC tank has a high Q and the amplifier operates Class C. When operating Class C, the amplifier is replacing the switch discussed in Figure 12-3 above. For a short period during each oscillation, the amplifier is biased on to provide a quick charge path for the capacitor in the tank circuit.

When an LC tank circuit is used to develop oscillations, the output frequency of the oscillator is determined by the resonant frequency of the tank circuit. The operating frequency of this circuit can be calculated using this formula:

$$f_o = \frac{1}{2\pi\sqrt{LC}}$$

You should realize that this formula is valid for LC tank circuits that have a Q of 10 or greater.

Crystals as Frequency-Determining Devices

Another device that is used to determine oscillator frequency is the crystal. This device may be used with an LC tank circuit or it may be used alone.

Crystals exhibit a characteristic called the **piezoelectric effect**. The piezoelectric effect is:

> **The property of a crystal by which mechanical forces produce electrical voltage and, conversely, electrical voltage produces mechanical forces.**

This creates an oscillation within the crystal that is much like that created within an LC tank circuit.

The piezoelectric effect is present in several crystal substances when they are cut to correct specifications. The most important type of crystal is *quartz*. Quartz has a stronger mechanical strength than other crystals and is, therefore, used almost exclusively. Roselle salt crystals oscillate better but have less mechanical strength. Tourmaline is physically strong like quartz, but it is very expensive to develop into usable crystals. For these reasons, our discussion centers around quartz crystals.

For use in oscillators, the crystals must be a wafer thin sliver of precise thickness, width, and length. The materials from which these wafers are cut may be natural or man made. Once cut, the crystals are mounted in holders which provide physical support and contain the electrodes that can be used to provide excitation. This holder is designed to provide ample room for the mechanical oscillation of the crystal. There are many types of crystal holders. Figure 12-6 contains a drawing of one type of holder.

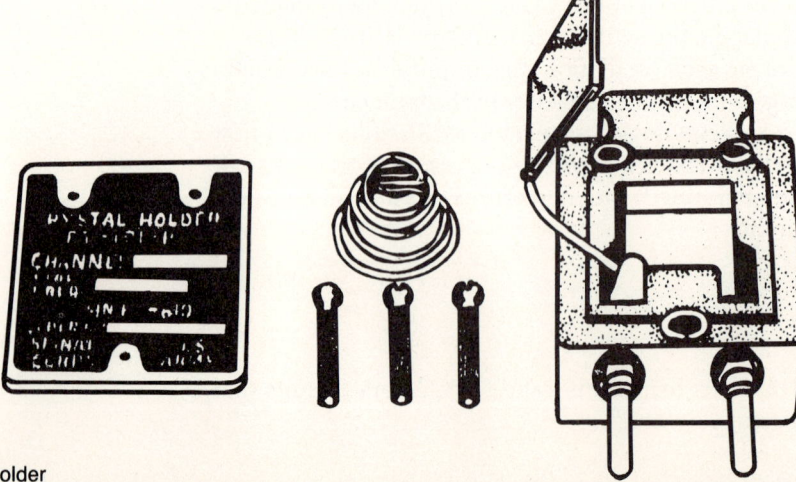

Fig. 12-6 Crystal Holder

Crystals are "cut" to the specific size that will generate a specific frequency. The frequency at which a crystal oscillates is called the **natural resonant frequency** of that device. When small pulses of voltage are applied to the crystal wafer, mechanical vibrations result. The mechanical vibrations produce a sine wave voltage at the crystal's terminals that is at the crystal's resonant frequency. This sine wave is the output signal.

A vibrating crystal can be represented by an equivalent circuit composed of capacitance, inductance, and resistance. Figure 12-7 contains three diagrams that are important to understanding this device. Figure 12-7(a) shows the schematic symbol for a crystal. The crystal's equivalent circuit is shown in Figure 12-7(b).

Placing the crystal in a holder has the same effect as connecting capacitance in parallel with the crystal. This added capacitance is included in the drawing shown in Figure 12-7(c). C_2 represents the capacitance added to the circuit by the holder's metal plates.

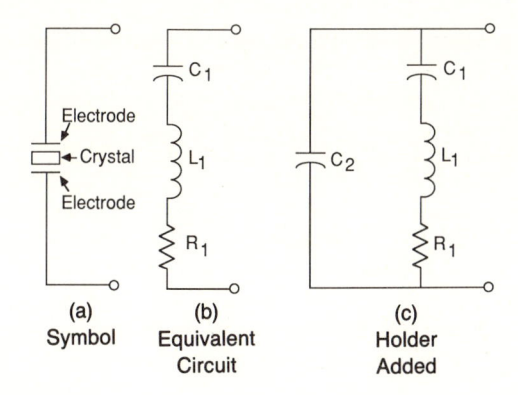

Fig. 12-7 Crystals

Electrical circuits that include crystals can be analyzed by substituting the equivalent circuit for the crystal. Using the equivalent circuit, we can determine the circuit's operational characteristics. This does *not* mean, however, that it is possible to substitute a capacitor and coil designed to oscillate at that frequency and get the same electrical operation as with the crystal.

The Q of a crystal is much higher than that of an LC tank circuit. This high Q results from the fact that the crystal contains very little resistance. Crystals that are available today range in Q from 5,000 to 30,000. This Q indicates that the circuit's reactance is much larger than its resistance. With this high Q, we can expect a crystal controlled oscillator to have a much more stable operating frequency. For this reason crystals are used in oscillators where frequency stability is very important.

Voltage-Controlled Oscillators

Modern circuitry incorporates new tuning techniques for establishing and changing an oscillator's resonant frequency. This is the **voltage-controlled oscillator** (VCO). The VCO contains a **varactor**. Remember this device? It was discussed in Chapter 6, Special Solid-State Devices. A varactor is a special diode. It is designed such that its junction capacitance is proportional to the reverse bias applied to its junction. Changing its reverse bias causes its depletion region to widen or narrow. The two ends of the diode form capacitor plates and the size of the depletion area determines how far the plates are separated and its capacitance. Increasing the reverse bias increases the distance between the plates and reduces capacitance.

In modern circuit design, the varactor's small capacitance is used as part of the oscillator's tank capacitance. This allows the resonant frequency to be controlled by the amount of reverse bias applied to the varactor diode.

Most of the circuits that are in use are of the integrated circuit type. For this reason, specific circuitry will be left for books that address digital circuitry and integrated circuit chips.

Self-Check

Answer each item by inserting the word or words required to correctly complete each statement.

6. For a phase shift oscillator that has three capacitor-resistor pairs, the average phase shift per pair is _____.
7. Three common frequency-determining networks (devices) are:

    a. _____
    b. _____
    c. _____

8. The internal operation of an LC tank circuit is compared to a _____ effect.
9. The most common oscillator crystal is made from _____.
10. Increasing the size of the resistors in an RC feedback network will cause operating frequency to _____.
11. For an amplifier to operate as an oscillator, _____ feedback is required.
12. An oscillator is a device that changes _____ voltage into _____ voltage.
13. When a sine wave's amplitude steadily decreases to zero, the wave is said to be _____.

Wave Generation 385

14. Decreasing the amount of _____ present in an LC tank causes the tank's Q to increase.
15. The Q of crystals varies from _____ to _____.

Nonsinusoidal Wave Generators

Nonsinusoidal Oscillators are normally used to generate square, rectangular, sawtooth and trapezoidal waves. These waveshapes are generated using the same principle that applies when a switch is turned on and off.

Square and Rectangular Wave Generators

Examine Figure 12-8. By manually closing and opening the switch (Figure 12-8(a)) it is possible to create a varying voltage drop across R_1 that has the same characteristics as the square wave shown in Figure 12-8(b).

When Sw_1 is open, the voltage drop on R_1 is zero; but when Sw_1 is closed, V_{R1} quickly rises to the value of the battery voltage and stabilizes. The voltage drop on R_1 remains high until Sw_1 is again opened. At that time, V_{R1} drops back to zero and remains there until the switch is again closed. If the switch is opened and closed at a constant frequency, the output (V_{R1}) has constant amplitude and constant frequency. If the time that Sw_1 is on equals the time it is off, the output is a square wave.

Using the same circuit (Figure 12-8), it is possible to illustrate the generation of a rectangular wave. A **rectangular wave** is a:

> Periodic wave which alternately assumes one of two fixed values, the time of transition being negligible in comparison with the duration of each fixed value.

Two rectangular waves are shown in Figure 12-9(a) and (b). Note that in Figure 12-9(a), Sw_1 is on for a very short period of time as compared to

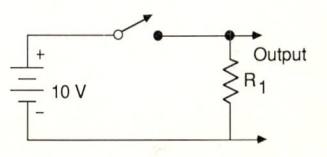

(a) Circuit

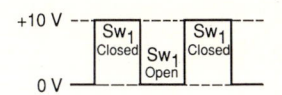

(b) Output Waveshape

Fig. 12-8 Square Waves

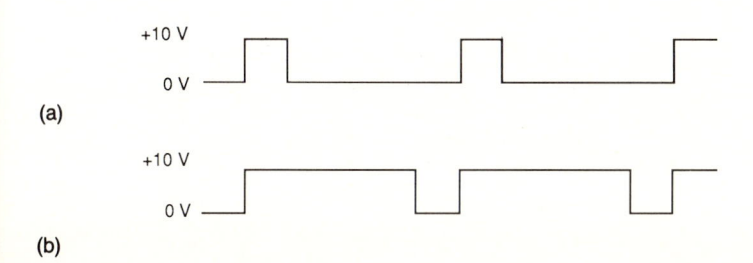

(a)

(b)

Fig. 12-9 Rectangular Waves

the time it is off. In Figure 12-9(b), the opposite is true. Here the switch is in the on position much longer than the period it is off. Note, though, that once the cycle has been established, it is repeated over and over again.

Earlier you learned that the square wave contains:

A fundamental frequency and an infinite number of odd harmonics which have specific amplitude and phase relationships.

A rectangular wave contains:

A fundamental and specific harmonically related frequencies, determined by the time of each alternation.

Because nonsinusoidal waves contain many different frequencies, transistor amplifiers that have *wideband* amplifier characteristics must be used. In other words, the frequency response of the wave generator and its associated circuitry must be large enough to pass each and every frequency contained in the waveform without distortion. Both fundamental and harmonic frequencies must be generated; therefore, the circuit used to generate these waves must be nonlinear. To achieve nonlinearity, the transistor must be operated in the nonlinear portion of its characteristic curve. These are the saturation and cutoff regions of transistor operation. Like sine-wave oscillators, these circuits require regenerative feedback in proper phase and amplitude to maintain circuit operation.

The frequency-determining devices or networks used with these circuits are of two types. If the circuit is *free running*, its frequency is determined internally. If the circuit is *not* a free-running type, the frequency is controlled (triggered) by an external source. In free-running types, the frequency is commonly controlled by an RC network.

Figure 12-10 contains a block diagram representation of a rectangular wave generator. Notice that the block diagram is like that of

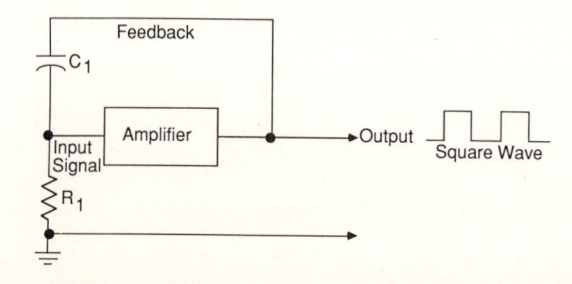

Fig. 12-10 Block Diagram - Square or Rectangular Wave Generator

the sine-wave oscillators. The operation, however, is different because of the difference in amplitude of the feedback. If the PNP common-emitter amplifier is used, the transistor is being reverse biased when its input goes positive. The output signal is supplied as regenerative feedback to the base of the transistor. This drives the transistor into cutoff. The time constant of the RC network determines the amount of time that the transistor remains cut off. When the voltage drop across R_1 decreases to the point where the circuit's bias takes over, the transistor will again conduct (at saturation) and start another cycle of the output. By using different values of resistance (R_1) and capacitance (C_1), the duration of the cut off portion of the waveshape can be varied. Variation of the cutoff time can be used to alter the output frequency. The start and end of each pulse results from either driving the transistor into saturation or cutoff. To obtain a square wave, the transistor's cutoff time is equal to its conduction time. Any other division of time results in a rectangular wave. If the cutoff time is longer than conduction time, the output wave is like the wave shown in Figure 12-9(a). If the cutoff time is shorter than conduction time, the output value is like the wave shown in Figure 12-9(b).

In designing square and rectangular wave generators, you must consider the transistor's switching action. Figure 12-11 contains the load line for a transistor that represents its two extremes (cutoff and saturation). For the output waveshapes to have vertical edges, it is necessary for the transistor to switch from saturation to cutoff and from cutoff to saturation in very small periods of time. The ideal is for switching to take zero time, resulting in edges that are exactly perpendicular. Since this is not possible, we must design circuits that come as close to zero switching time as possible.

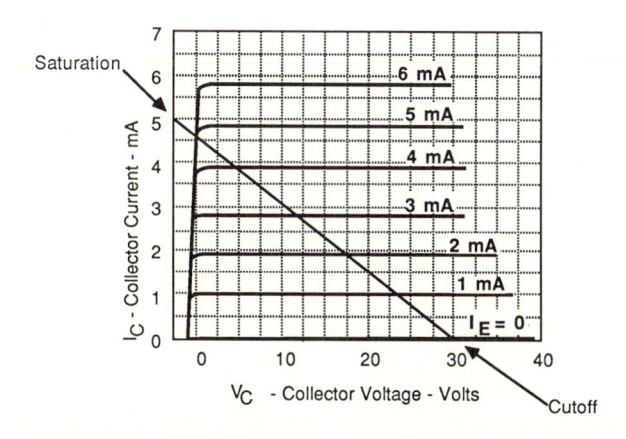

Fig. 12-11 Characteristic Curve Chart

Sawtooth-Wave Generators

Another waveform that proves useful in electronics is the **sawtooth wave** shown in Figure 12-12.

A sawtooth wave can be described as:

> **A periodic wave, the amplitude of which varies linearly between two values. A longer interval is required for one direction of progress than is required for the other direction.**

Observe Figure 12-12 while we examine this waveform. Notice that the amplitude begins at zero and very linearly increases to a maximum during the period t_0 to t_1. Between t_1 and t_2, the amplitude quickly falls back to zero. During the time period covered by t_2 to t_4, the same sequence of events reoccurs. A waveshape of this type is very important to the operation of an oscilloscope. The application of a linearly changing deflection voltage to the deflection plates of the cathode ray tube results in a linear movement of the electron beam across the face of the scope.

To generate the sawtooth wave, we again use circuits that operate on the on/off principle plus the charge discharge characteristics of a capacitor. Figure 12-13 will be used to describe the capacitor action that occurs in sawtooth generators. Review the Universal Exponent Curves chart included in this figure. Remember, this chart is used to illustrate the charge and discharge rate of capacitors and coils by both percentage of charge and time constants. Notice that as the capacitor begins to charge (curve A), the line that represents the first 10% of charge (lower left corner) is almost perfectly linear. This first 10% of charge occurs in

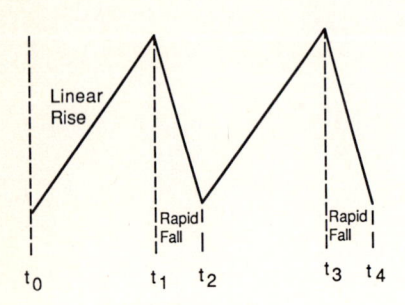

Fig. 12-12 Sawtooth Waveshapes

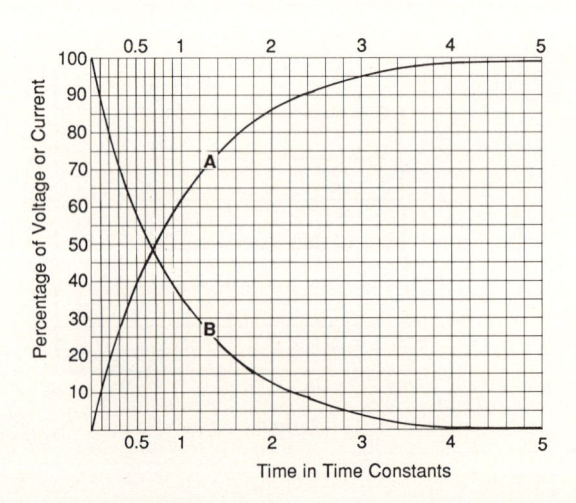

Fig. 12-13 Universal Exponent Curves

one-tenth of one time constant and produces the largest amplitude change of any 10% of charge. Each successive one-tenth of charge becomes less linear and has less amplitude.

The sawtooth generator normally uses the first 10% or less of the capacitor's charge. As soon as the charge voltage reaches the amplitude desired (not to exceed the 10% of charge level), the capacitor quickly discharges back to zero. To illustrate this action, we will discuss the circuit shown in Figure 12-14. Notice that the output is taken across C_1 to ground. When Sw_1 is closed, the capacitor is shunted by a short and can quickly discharge through zero resistance. During this period, the output voltage is held at zero (ground) potential. Opening Sw_1 allows C_1 to begin charging through R_1 to V_t. Because the output is taken across C_1, the output voltage starts at zero and increases at the linear rate of C_1 charge. When the capacitor has charged to the desired voltage level (first 10% of charge or less), Sw_1 is closed. The output voltage drops to zero as the capacitor immediately discharges through the switch. At that point, Sw_1 opens again and the next cycle begins. Each successive cycle repeats this action and provides a sawtooth output wave as shown in Figure 12-15(b).

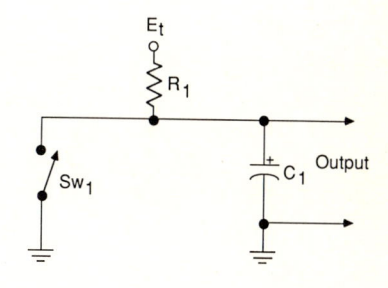

Fig. 12-14 Equivalent Circuit - Sawtooth Generator

(a) Square/Rectangular Waves

(b) Sawtooth Wave

Jump Voltage — Fall or Decay — (c) Trapezoidal Wave

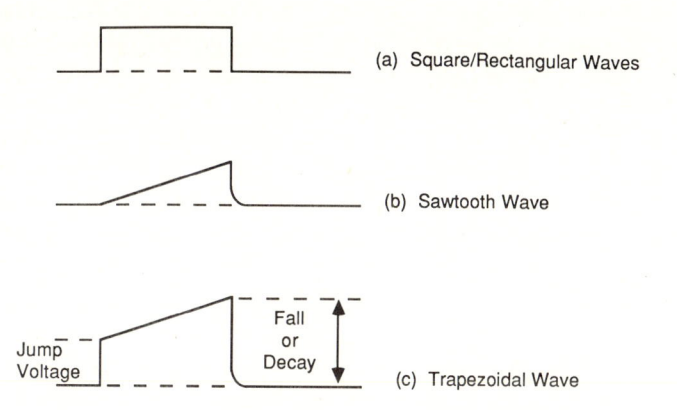

Fig. 12-15

The actual circuit will use a transistor in place of Sw_1. By biasing the transistor at saturation, the voltage drop on the parallel C_1 will be very small. Application of a square wave is used to immediately drive the transistor into cutoff. With the transistor cutoff, C_1 begins to charge. At the end of the desired time, the gate signal is removed from the transistor. This causes output voltage to return to zero and provides a low resistance discharge path (for C_1) through the transistor. The voltage on the capacitor falls to zero and is ready to begin the next cycle.

Trapezoidal-Wave Generators

A **trapezoidal wave** is shown in Figure 12-15(c). Note that it resembles a square wave (Figure 12-15(a)) that has a sawtooth wave (Figure 12-15(b)) sitting on its top.

The leading edge (vertical) of the trapezoidal wave is called **jump voltage**. The linear rise portion of the wave is called the **slope** and the trailing edge is called the **fall** or **decay**. A trapezoidal wave is used to supply the deflection current required to overcome the initial opposition of a deflection coil used in radar or television receivers. The electromagnetic *cathode ray tube* (CRT) uses the electromagnetic fields that are created by charging coils to provide the deflection of the video (picture) on the screen. The jump voltage applies a DC potential that overcomes the coil's initial opposition to a change in current flow. The slope can then cause the deflection current to increase at a linear rate.

Figure 12-16(a) contains a simplified schematic for a trapezoidal wave generator. R_1, R_2, and C_1 are connected across the power source. Notice that when Sw_1 is closed, it places a short in parallel with R_2 and C_1. To illustrate how this circuit operates, assume the following: $E_t = 100$ V, $R_2 = 1$ kΩ, and $R_1 = 99$ kΩ. As shown in Figure 12-16(b), voltage at the output is zero volts when Sw_1 is closed. These conditions are represented by the period between points t_0 and t_1. If Sw_1 is opened at t_1, the following things happen. At the first instant, the capacitor acts like a short. This causes the voltage drop on R_2 to appear as the output voltage. With $R_2 = 1$ kΩ, the voltage drop equals 1 V. This voltage drop ($V_{R2} = 1$ V) is referred to as the jump voltage and is immediately applied to the output.

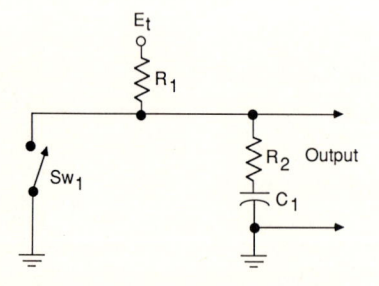

(a) Equivalent Circuit - Trapezoidal Wave Generator

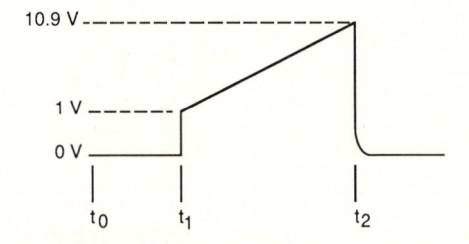

(b) Trapezoidal Waveshape

Fig. 12-16

As the capacitor begins to charge, current flow in R_1 begins to decrease. The voltage on C_1 begins a linear increase. To maintain maximum linearity, C_1 charge is kept within the first 10% of the possible charge. Between times t_1 and t_2, the capacitor is allowed to charge to 10 V. At this time the voltage dropped on the two resistors equals 90 V ($V_{R1} + V_{R2}$). Of this voltage $V_{R2} = 0.9$ V. The output voltage consists of $V_{R2} + V_{C1}$ (0.9 V + 10 V) which provides a peak output of 10.9 V. At time t_2, Sw_1 is closed placing zero volts at the output and providing a low resistance discharge path for C_1. When the voltage on C_1 has fallen to zero, the circuit is ready to begin the next cycle. Each succeeding waveshape will be a trapezoidal wave identical to the one just described.

Circuits that generate sawtooth and trapezoidal waves are often called **time base generators**. They get this name because their outputs are used to generate sweeps that are linear with respect to time.

Self-Check

Answer each item by inserting the word or words required to correctly complete each statement.

16. The four nonsinusoidal waveshapes studied here are:

    a. _____
    b. _____
    c. _____
    d. _____

17. In a square wave, the positive and negative alternations are _____ in duration.
18. To create the slope of a sawtooth wave, the first _____ of the charge of a _____ is used to produce the waveshape.
19. Jump voltage is added to a sawtooth wave in order to overcome the opposition of the _____ used with an electromagnetic cathode ray tube.
20. A rectangular wave has alternations that have _____ durations.

    a. equal   b. unequal

Summary

In this chapter you were introduced to the principles of wave generation. Wave generators can be divided into two broad classifications—sinusoidal and nonsinusoidal.

Three requirements are necessary for oscillations to occur. These are: an amplifier with power source, a frequency-determining network or device, and feedback in the proper phase and amplitude. It is also necessary for the feedback to be regenerative in nature.

Sinusoidal wave generators were discussed from the general standpoint. Three types of frequency determination are used: RC networks, LC resonant tank circuits, and quartz crystals. Each of these can be used in the design of sine-wave generators.

Nonsinusoidal wave generators were discussed. Four output waveshapes and the methods used to generate each wave were covered. The four output waveshapes, a square wave, rectangular wave, sawtooth wave, and trapezoidal wave. In each case, an RC network is used in the generation of the wave. Square and rectangular waves are generated simply by cutting off and saturating a transistor whose collector voltage supplies the output. The generation of sawtooth and trapezoidal waves includes the use of transistors for control of switching, but the output signal is developed across a charging capacitor. To obtain a linear slope for the output signal, the first 10% of charge of an appropriate capacitor is used.

Review Questions and/or Problems

1. It is possible to design wave generating circuits that create signals from as low as _____ to a high of _____ gigahertz.

2. An oscillator is simply an amplifier that _____.

3. An oscillator that produces a sine wave output is classified as a/an _____ oscillator.

4. Three circuits and/or devices are used to control the frequency of sine wave oscillators; these are:

    a. _____
    b. _____
    c. _____

5. Of the three types of frequency determination used with sine wave oscillators the _____ type is the most stable with respect to frequency.

6. The RC phase shift network is designed and used in order to provide _____ to the feedback signal.

7. In oscillators the feedback signal must be of the _____ type.

8. The LC tank oscillator operates on the principle of _____ frequency.

9. In a square or rectangular wave generator, the output frequency is determined by a/an _____.

10. An RC network has four RC pairs and is to be used with an RC oscillator. Each leg of the network is expected to shift the feedback signal's phase by _____.

11. To obtain a higher output power from a sine-wave oscillator, you must design the amplifier to operate Class _____.

12. For high frequency operation, the transistor used in an oscillator circuit must have _____ interelement capacitance.

13. The tank circuit's transfer of energy back and forth between the capacitor and the coil is called the _____ effect.

14. The loss of amplitude on successive alternations within the tank circuit results from _____ and the waveshape that results is called a _____ wave.

15. The piezoelectric effect is the property of a crystal by which the application of _____ vibrations will result in a/an _____ vibration that can be used as a sine wave.

16. A rectangular wave is classified as a _____ waveshape.

17. The square and rectangular waves are generated by switching a transistor between _____ and _____.

18. In a rectangular wave, the positive and negative alternations have _____ time durations.

19. Jump voltage, in a trapezoidal wave, is inserted to overcome the opposition of a _____ to the initial flow of current.

20. The slope included in the sawtooth and trapezoidal waves is used to produce _____.

21. The LC tank oscillator operates on the principle of _____ frequency.

22. A free-running nonsinusoidal wave generator has its frequency determined by _____.

23. Because fundamental and harmonic frequencies must all be amplified, the nonsinusoidal wave generator must be a _____ amplifier.

24. The slope of a trapezoidal wave is generated by the first _____% of charge of a _____.

25. To establish the operating frequency of a non-free running nonsinusoidal wave generator, we use a/an _____.

13
Sine Wave Oscillators

Objectives

1. Identify various types of oscillators by observing their schematic diagrams.
2. State the three requirements for oscillation.
3. Use frequency of operation formulas to calculate the operating frequency of various oscillator circuits.
4. Explain the part that the Q of a resonant circuit plays in oscillator operation.
5. Explain how each of the following circuits operates:

   a. LC tank circuits
   b. Armstrong oscillators
   c. Series-fed Hartley oscillators
   d. Shunt-fed Hartley oscillators
   e. Colpitt's oscillators
   f. Clapp oscillators
   g. Butler oscillators
   h. Phase-shift oscillators
   i. Wein-bridge oscillators
   j. Buffer amplifiers
   k. Frequency multipliers
   l. RC networks

Introduction

In Chapter 12, you were introduced to sine-wave oscillators. In this chapter we will examine different circuits which can produce sine wave signals. Each circuit will be examined for component content and operational characteristics.

A technician spends many hours troubleshooting malfunctioning circuitry. Oscillator circuits are one source of these problems. Because of their importance in communications, computer, and other systems, the oscillator must be thoroughly understood. It is important that you understand how to identify each type, what part each component plays in the circuit's operation, the symptoms that are present when a component malfunctions, and how to locate and replace the defective component.

Remember, an oscillator must generate a signal that has constant amplitude and constant frequency. If either of these is not constant, the circuit is not operating properly.

Basic Oscillator Operations

At some time or another, you have probably attended a concert where the sound system developed a shrill whistle. The whistle that you heard was the result of the amplifier becoming an oscillator whose operating frequency is the same as the shrill audible frequency that you heard. To understand how this can happen, we will discuss the audio system. An understanding of this system will aid you in understanding the explanation of oscillators that follows.

An audio (public address) system consists of three components. These are:

1. an audio amplifier
2. an input device (a microphone)
3. an output device (a loud speaker)

Figure 13-1 contains a block diagram for the public address system discussed above. Notice the positioning of the microphone with respect to the speaker. Mechanical vibrations from the speaker are allowed to

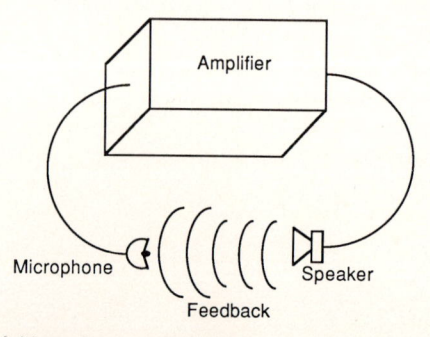

Fig. 13-1 Public Address System Oscillation

strike the diaphragm in the microphone. At first, these bits of noise are small but the amplifier increases their size. The amplified noise arrives at the speaker where it is converted into mechanical vibrations. The vibrations move through the air as waves and strike the microphone. As each cycle of noise is amplified, the speaker and microphone serve to *feedback* a portion of the amplifier's output. This output signal is of such phase and amplitude that it is *regenerative*.

As the amplifier's output signal continues to gain in amplitude, you begin to hear low volume noise on the speaker. Continued amplification of the noise soon overcomes all other operation of the amplifier and you hear a loud, shrill whistle. At this point, the intelligence that you were originally listening to has disappeared and you only hear the frequencies that represent the loud whistle.

When either the speaker or the microphone is removed from the feedpack path, the oscillations cease. Then the amplifier can resume amplification of the intelligence signals that were its original purpose.

Another way to stop the oscillations is to reduce the gain of the amplifier stage. This can be done by turning down the volume. At low gain, the small feedback signal lacks enough amplitude to drive the amplifier into oscillation. In this condition, the amplifier continues to amplify the intelligence that we wish to hear without interruption.

The three basic requirements that are needed for oscillation are:

1. amplification
2. regenerative feedback
3. frequency-determining device or network

A block diagram for an oscillator is shown in Figure 13-2. You must realize that this circuit requires a power source that is capable of meeting its operational needs. In this block diagram, the amplifier provides enough gain so that when the feedback network provides regenerative feedback, oscillations will begin. The frequency-determining network will take over and limit the circuit's operation to a single frequency.

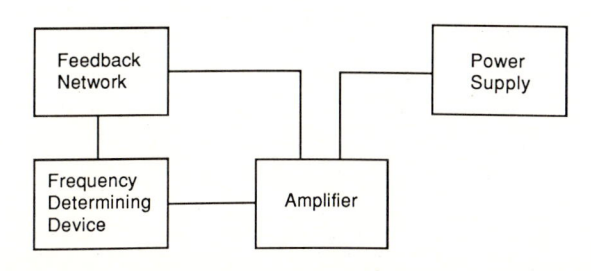

Fig. 13-2 Oscillator Block Diagram

The Amplifier

In an oscillator, the amplifier must develop enough gain to provide for the needs of the output load and regeneration. Regenerative feedback has the same frequency and phase as the signal already present at the amplifier's input. This, then, is the only frequency that is regenerative. The regenerative frequency causes the amplifier to be driven into oscillation at its frequency. Common-emitter, common-base, or common-collector amplifiers can be used in oscillator circuits. Most often, though, a common-emitter amplifier is used. This is because the CE amplifier can develop high power and has input and output impedances that are easy to match.

The Frequency-Determining Device

As the name implies, this circuit, or device, determines the operational frequency for the oscillator. Frequency may be determined by *RC networks*, *LC tank circuits*, or *crystals*.

Probably the most common frequency-determining network is the resonant circuit. In some cases, the resonant circuit may be a series circuit, but our discussion will center around parallel tank circuits. The resonant frequency of tank circuits can be calculated using the formula:

$$f_r = \frac{1}{2\pi\sqrt{LC}}$$

The frequency of this type circuit can be changed by varying the value of either the capacitor or coil. By decreasing either capacitance or inductance, we can cause f_r to increase. To decrease f_r, we must increase the value of either L or C. By adjusting the value of these two components, we can tune the circuit to resonate at a specific frequency.

Figure 13-3 contains the schematic diagram for two parallel LC tank circuits. Figure 13-3(a) contains only series tank resistance, but Figure 13-3(b) has an external load resistor added. In Figure 13-3(a), R_{Int} is one of the factors used to determine the *quality* Q of the tank circuit. In this case, the formula for Q is:

$$Q = \frac{X_L}{R_{Int}}$$

or:

$$Q = \frac{X_C}{R_{Int}}$$

When:

$$X_L = X_C$$

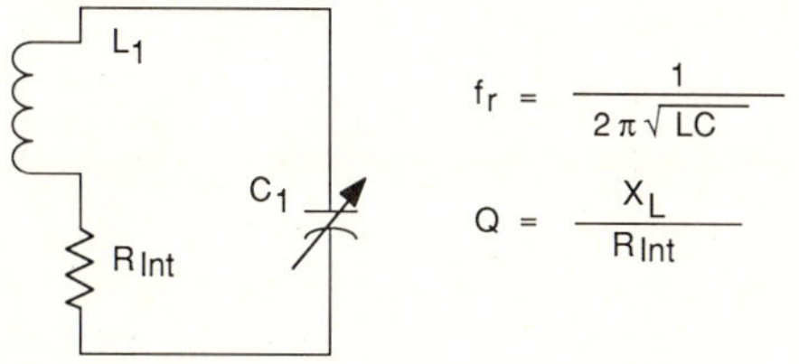

Sine Wave Oscillators

(a) Resonant Circuit

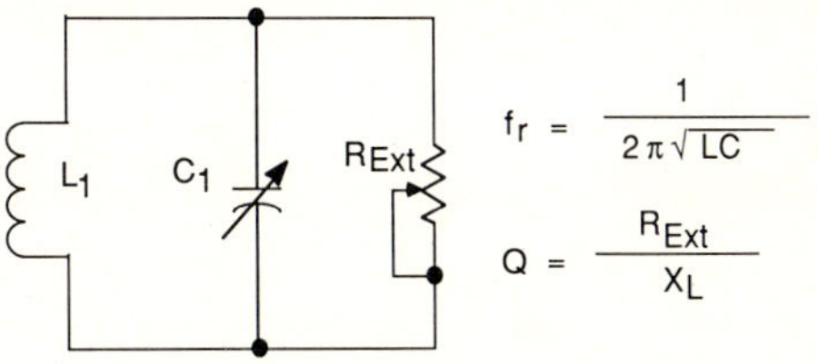

$$f_r = \frac{1}{2\pi\sqrt{LC}}$$

$$Q = \frac{X_L}{R_{Int}}$$

(b) Resonant Circuit With External Loading

$$f_r = \frac{1}{2\pi\sqrt{LC}}$$

$$Q = \frac{R_{Ext}}{X_L}$$

Fig. 13-3 LC Tank Circuit Loading

By examining this formula and the circuit, you can see that Q of the tank and the size of R_{Int} are inversely related. Assume that the circuit is operating at resonance. If the ratio of *reactance* is 20 times larger than *resistance*, then the Q of the circuit is 20. If the ratio is 100:1, the Q is 100.

In Figure 13-3(b), you will notice that an adjustable resistor (R_{Ext}) has been connected in parallel with the tank circuit. In this circuit, R_{Ext} can be used to vary the Q of the tank. Circuit Q for this arrangement is calculated using the formula:

$$Q = \frac{R_{Ext}}{X_L}$$

When R_{Ext} has its resistance decreased, the tank circuit is shunted by a smaller resistance. A smaller shunt resistance causes more losses within the circuit, which reduces the circuit's Q. This is a common method that is used to vary the Q of an LC tank circuit. The limiting factor is that the Q can never be higher than it would have been without R_{Ext} added. For all settings of R_{Ext}, the circuit Q *must* be smaller than if R_{Ext} was not in the circuit.

Using a resonant circuit, the values for C and L will determine the frequency of operation. In most cases, frequency of operation and the tank's resonant frequency are the same. The Q of the tank circuit is important because it determines the feedback requirement of the circuit. If the Q is low, more feedback is required than if the Q is high.

Regenerative Feedback

Feedback is the process of transferring a signal from the high level point in a system to a lower level within the same system. This usually refers to the transfer of part of the output of an amplifier back to the same amplifier's input. When feedback arrives at the input in phase with (aiding) the input signal, it is said to be *regenerative* feedback.

For an oscillator to operate, regenerative feedback is a necessity. Regenerative feedback provides the input. Picture the audio public address system discussed earlier. Amplification of the feedback signal offsets the damping that would otherwise be present in the output sine wave. Remember, all circuits have some losses; therefore, the feedback amplitude must be large enough to overcome the losses of the LC network.

Figure 13-4 contains the block diagram for a PNP common-emitter amplifier that has a feedback network connected between its collector and base. Since a common-emitter amplifier has a 180° phase shift across its base to collector, the feedback network must provide another 180° of phase shift if the feedback is to be regenerative. Various circuits and components are used to provide this 180° of phase shift. If a common-base or common-collector amplifier were used, there would be no need for the phase shift. Remember, these amplifiers have input and output signals that are in phase.

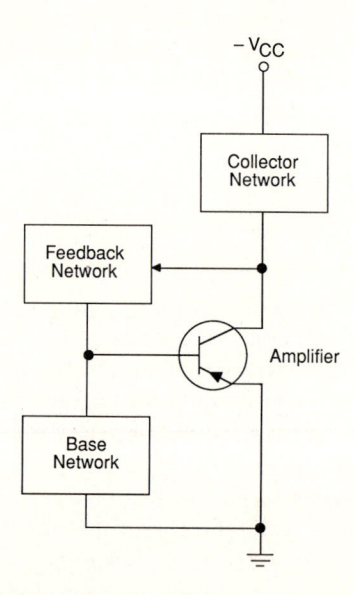

Fig. 13-4 Oscillator Block Diagram

Self-Check

Answer each item by inserting the word or words required to correctly complete each statement.

1. Using an external resistance, we can increase the Q of the tank by _____ the size of R_{Ext}.

Sine Wave Oscillators

2. Three requirements for oscillation are:

   a. _____
   b. _____
   c. _____

3. For sine-wave oscillators, three types of networks or devices are used to establish the operating frequency. These are:

   a. _____
   b. _____
   c. _____

4. The formula for calculating the operating frequency of an LC tank circuit is _____.

5. Figure 13-2 indicates that four things are required for an oscillator circuit's operation. These are:

   a. _____
   b. _____
   c. _____
   d. _____

Armstrong Oscillators

In Chapter 12 we discussed, in general terms, the basic oscillator circuit. To this point, in this chapter, we have reviewed those requirements and have looked at the parts of an oscillator's circuit. Now we will place them all together.

Figure 13-5(a) contains the schematic diagram for a conventional common-emitter amplifier. Forward bias for the circuit is provided by R_B. Capacitor C_C is inserted as the coupling capacitor, and the collector load consists of R_L and L_1. In this configuration, the input signal is shifted 180° between the base (input) and the collector (output) by the transistor. A second 180° occurs because of the **inductive feedback** across the transformer. Inductive feedback is the identifying feature for an Armstrong oscillator.

In Figure 13-5(b), we see a frequency-determining network that contains L_2 and C_1. Notice that C_1 is variable and can, therefore, be used to tune the tank circuit's resonant frequency.

Figure 13-5(c) contains the feedback network. Note that in this drawing, the frequency-determining network and the collector-load network have been joined to form a transformer. The collector-load network acts as the transformer primary and the frequency-determining device as the secondary. This transformer is designed to provide the

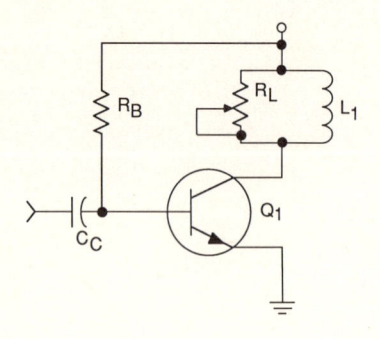

(a) Amplifier

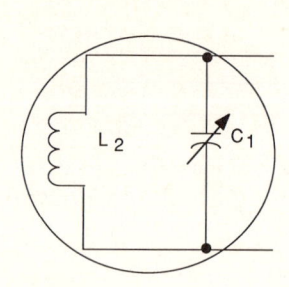

(b) Frequency Determining Device

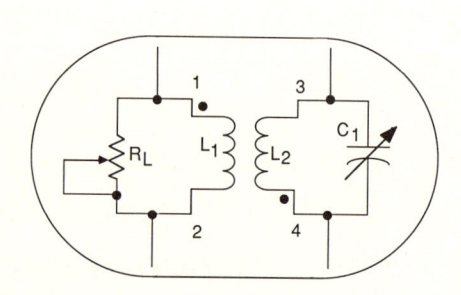

(c) Feedback Network

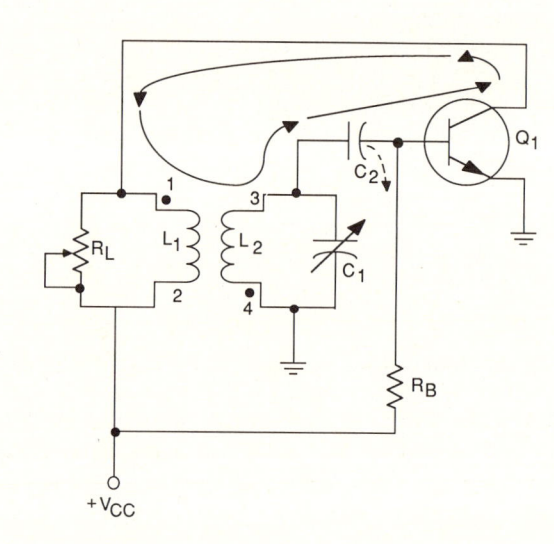

(d) Complete Circuit

Fig. 13-5

180° phase shift needed to assure that the feedback is regenerative. By adjusting R_L, we can control the amount of current that flows through L_1 to V_{CC}. When R_L is set to its maximum resistance, maximum current is flowing through L_1. At this time, the transformer is coupling maximum signal to its secondary (tank circuit). This large signal represents a large regenerative feedback signal. As R_L is adjusted for a smaller resistance, the amount of current flowing in L_1 decreases. This means that less signal is coupled to the tank circuit and that the regenerative feedback has less amplitude. For correct operation, R_L is adjusted so the current flow in L_1 is large enough to couple sufficient feedback to the tank circuit to maintain tank oscillation.

In Figure 13-5(d), we have all sections connected into a complete oscillator circuit. By connecting the feedback network through the coupling capacitor (C_2) to the base of Q_1, we form a **closed loop** for feedback (shown by the solid arrows). We know that this feedback must be regenerative so let's check to see if it is. Assume that the circuit is operating with a positive signal at the base of Q_1. After amplification, the signal arrives at the collector 180° out-of-phase with the base. As collector current flows, a part of the signal is developed across the primary (L_1) of the transformer. Another 180° of phase shift occurs as the signal is coupled across the transformer. (Note the dots on leads 1 and 4 that are used to indicate a 180° phase shift across the transformer.) The negative signal at the collector appears at lead 3 of the transformer as a positive signal. The positive signal is now coupled through C_2 and is developed across R_{eb} of the transistor. The positive signal developed by the feedback is in phase with the positive signal that we assumed when we started this discussion. If the two signals are in phase, the feedback is regenerative. When the input signal at the base goes negative, the feedback signal is also negative. In either case, the feedback signal is in the correct phase to overcome the losses that occur in the frequency-determining device and provides a signal gain of 1. R_1 can be adjusted to assure that the feedback has the correct amplitude.

Figure 13-5(d) can be used to identify all parts of an operational oscillator. Amplification is provided by a common-emitter amplifier. The LC tank circuit acts as the frequency-determining device. Inductive feedback is accomplished by use of the transformer and C_2 forming a closed loop. Regenerative feedback is assured by the 360° phase shift that occurs—180° across the common-emitter amplifier plus 180° across the transformer. The type of oscillator shown in Figure 13-5(d) is called a **tuned base** oscillator because the frequency-determining device is in the base circuit. Should the frequency-determining device (L_2 in parallel with C_1) be in the collector circuit, the oscillator is called a **tuned collector** oscillator. In either case, this type of oscillator is called an **Armstrong** oscillator.

Observe Figure 13-6 as we discuss the *tuned-base Armstrong* oscillator's operation. When V_{CC} is applied to the circuit, a small amount of base current begins to flow through R_B and establishes the Q point for Q_1. The forward bias that is developed in this manner causes current to flow in Q_1 from ground, through the transistor, through L_1, and to V_{CC}. As current flows through L_1, a magnetic field is created which induces a current into the secondary (L_2) of the transformer. L_2 and C_1 form an LC tank circuit that now has a positive voltage applied to the top of L_2 and C_1. At this point, two things happen: first, C_1 charges and stores the energy needed to oscillate the tank circuit; second, capacitor C_2 couples positive going feedback to the base of Q_1. With regenerative feedback at its base, Q_1 conducts harder. This heavier current flows through Q_1 and L_1 causing a larger voltage to be induced into L_2. This causes a larger positive voltage to be coupled through C_2 to the base of Q_1. All of this time, C_1 is storing energy that will later be used to supply energy for oscillation within the tank. Because it is in parallel with L_2, C_1 will charge to the voltage that is induced into L_2.

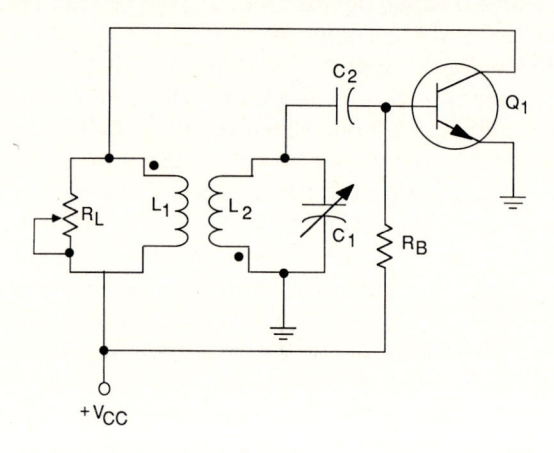

Fig. 13-6 Oscillator Circuit

The conduction of Q_1 increases until the transistor is driven into saturation. At saturation, collector current of Q_1 is maximum and cannot increase further. When current in L_1 is no longer changing, the magnetic field is no longer expanding, and no voltage is induced into L_2.

At this time, C_1 does not have an external voltage applied and it begins to act like a power source. The discharge path of C_1 is through L_2. In the process of C_1 discharging, L_2 stores energy in its electromagnetic field. At this same time, the voltage on C_1 is decreasing. The result is that the *flywheel effect* described in Figure 13-3 begins within the parallel LC tank circuit as soon as Q_1 enters cutoff.

During earlier operation, C_2 has charged to approximately the same voltage as C_1. When C_1 begins to discharge, C_2 also begins to discharge. C_2 discharges through R_B. As soon as C_2 begins its discharge, the forward bias voltage at the base of Q_1 decreases. This causes the collector current of Q_1 to decrease. The decrease in I_C causes the current in L_1 to decrease and the magnetic field around L_1 to begin collapsing. The collapsing field (lines moving in the opposite direction to before) induces a negative voltage into L_2. The negative voltage is coupled through C_2 to the base of Q_1. This reverse bias is applied to the base of Q_1 driving it into cutoff.

With Q_1 cut off, the tank circuit continues to flywheel (oscillate). Oscillation within the LC tank circuit is at the tank's resonant frequency. These oscillations not only form the oscillator's output signal but also aid in keeping Q_1 cut off. Without feedback to overcome the losses in the tank, the amplitudes of the oscillations decrease and, if allowed to continue, soon dampen out completely. To assure that this does not happen, regenerative feedback is applied to the tank circuit for a short period during each cycle of the oscillator.

Regenerative feedback is accomplished as follows: when the voltage on C_1 reaches maximum negative, C_1 begins to discharge toward 0 V with Q_1 still biased below cutoff. Capacitor (C_1) continues its discharge until it reaches 0 V. At that time, the coil begins to oppose the tendency of current to decrease which results from the full discharge of the capacitor. The discharge of L_2 begins to charge C_1 in the opposite polarity—positive at its top plate and negative at its bottom plate. The positive voltage that appears at the top of C_1 is coupled to the base of Q_1 through C_2. This positive voltage is forward bias to Q_1 which causes I_C to start flowing. The current that flows in L_1 creates a magnetic field that expands outward and cuts the windings of L_2, inducing a voltage into the tank circuit. The conduction of Q_1 allows C_1 to charge, quickly replacing any energy lost due to damping. The positive voltage that is fed back immediately drives Q_1 into saturation, which allows flywheel action to start. At this point, Q_1 is again driven into cutoff and remains there until the next positive swing on C_1, which again recharges the tank capacitor.

The basic operation of the Armstrong oscillator is:

1. Power that is applied to Q_1 allows energy to be fed back to the LC tank circuit.
2. The regenerative inductive feedback signal drives Q_1 into cutoff and flywheel action begins.
3. The tank continues to oscillate during the time that Q_1 is cut off.
4. When C_1 is completely discharged and begins to charge in the opposite polarity, Q_1 is biased *on* and recharges the tank capacitor.
5. The regenerative feedback causes Q_1 to conduct for a short period of time during each cycle of the oscillator's output waveshape.
6. Because Q_1 is conducting for only a short portion of the time of one cycle of the output wave, Q_1 is operating Class C. During the conduction time, C_1 quickly recharges to V_{max}.

Class C operation is very efficient in that it uses power for only a small percentage of the operating time. The longer Q_1 is cut off, the more efficient the operation of Q_1 and the less loading that is placed on the frequency-determining device.

Figure 13-7 contains the schematic diagram for an Armstrong oscillator of the type that you are likely to see. Note that R_3 has been added to improve temperature stability. Remember, an unbypassed emitter resistor causes degeneration and results in a loss of circuit gain; therefore, C_3 is placed in the circuit to maintain gain and prevent

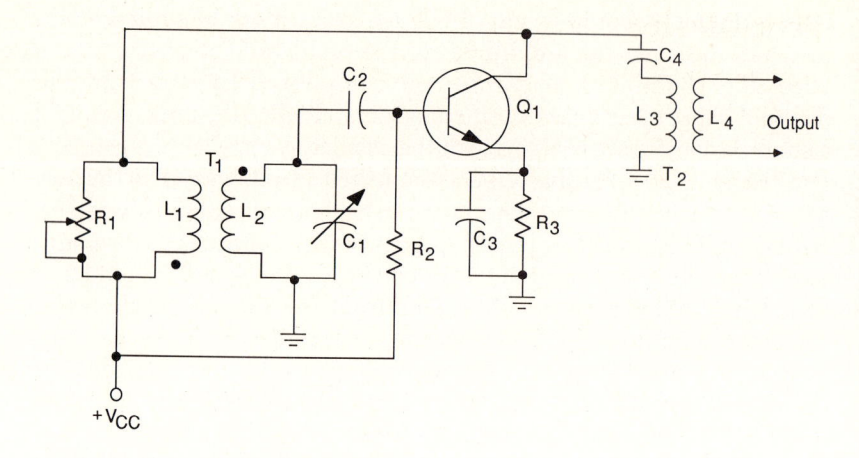

Fig. 13-7 Armstrong Oscillator

degeneration. Capacitor C_4 is placed in the circuit to serve as a coupling capacitor that assures that only AC signals are conducted through the primary of T_2 (L_3). T_2 is inserted into the circuit to provide the output signal across L_4. Use of T_2 assures that the impedance reflected back to Q_1 by the load device is kept to a minimum. R_2 is the base resistor that establishes the Q point and provides a path through which DC base current can flow.

Check Figure 13-8. In this figure we illustrate the relationship that exists between the output signal and the conduction of Q_1. The sine wave shown in Figure 13-8(a) is a true representation of the voltage (V_C) present at the collector of Q_1. Collector current (I_C) of Q_1 is represented by the pulses shown in Figure 13-8(b). You can see that I_C flows for only a very small portion of each cycle of the output signal.

During the time that the LC tank (Figure 13-7) is oscillating, L_1 (primary) and L_2 (secondary) act as a transformer. The signal developed within the tank is applied to the top of L_1. This AC voltage causes current to flow in the primary (L_3) of T_2. Current flow in the primary results in a magnetically coupled signal being present at the secondary (L_4) of T_2. C_4 is placed in the conduction path of L_3 to assure that only AC variations are present in T_2.

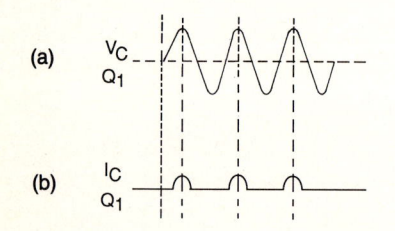

Fig. 13-8 IC versus VC in a Class C Oscillator

How would we check an operational Armstrong oscillator to make sure it is working properly? Some of the ways include:

1. Use an oscilloscope to check for a signal at the secondary of T_2. Normal operation results in a sine wave being observed.
2. Use an oscilloscope to check for the feedback signal.
3. Use a high impedance voltmeter to check the collector voltage (V_C).

Troubleshooting the Armstrong Oscillator

Troubleshooting any oscillator is relatively easy as long as you remember the requirements for oscillation and the characteristics of an amplifier. Oscillators must have a power source, an amplifier, regenerative feedback and a frequency-determining device if they are to operate. Two of these (power source and transistor) are the requirements for an amplifier.

The first question that we must ask is—what will prevent the circuit from oscillating? (Refer to Figure 13-7). If L_1 opens or shorts the regenerative feedback is missing. If either Q_1 or R_3 opens, the DC path for Q_1 is broken. This means that Q_1 cannot amplify and, therefore, the circuit cannot operate. If Q_1 should short, the DC path is present but no amplification can occur, so there is no output.

What can happen within the LC circuit? If either L_2 or C_1 shorts, the tank is shorted and no oscillations can occur. If either L_2 or C_1 should open, the circuit might oscillate at a very high frequency but it is not usable at the design frequency. In this case, we can say that with either C_1 or L_2 open, the circuit does not operate.

Third, what can cause problems with the regenerative feedback path? If C_1, C_2, or L_1 opens all regenerative feedback is lost. Without feedback, the oscillator cannot operate.

Now that we have looked at the three most readily recognizable parts of the oscillator, we will examine the effect of other component malfunctions. If R_1 opens, the output amplitude increases. Remember, the impedance of R_1 and L_1 in parallel is lower than the impedance of the smallest branch. Opening one branch increases the load impedance of the collector which, in turn, increases circuit gain. Shorting either L_1 or R_1 causes the load impedance to decrease to zero and places V_{CC} at the collector of Q_1, causing the gain to decrease to zero. Opening L_1 stops the feedback as explained above.

If R_2 opens, the fixed bias that is applied to Q_1 is removed and the circuit will probably stop oscillating. With some transistors, it is possible that R_2 could open and the circuit still oscillate because the "leakage" current in Q_1 is large enough to start oscillation. If R_2 shorts, the excessive current that flows as I_B will destroy the transistor.

If either R_3 or C_3 should short, the thermal stability of the circuit is lost. However, the circuit may continue to operate if Q_1 does not become too hot. If C_3 opens, the degeneration occurs. The loss of gain that results means that the output signal will have a smaller amplitude and could possibly be zero.

The shorting of C_2 may cause the forward bias applied to Q_1 to decrease to the point that the circuit cannot start oscillating when initially turned on. This is because of the very low DC resistance that is present in L_2.

Should C_4 open, the circuit will continue to operate but there will not be any output signal. This is because C_4 is the coupling capacitor for the output. Opening L_3 or L_4 has the same effect. An output signal cannot be coupled if any one of the three opens.

If either L_3 or L_4 were to short; it is possible that the loss of impedance matching would cause the oscillations to be damped out. The shorting of C_4 places an excessive DC current demand on $+V_{CC}$ (notice L_1 and L_3 would be in series) that would cause the fuse to blow. If, by some chance, the fuse didn't blow, the circuit could not oscillate with the high current flowing in L_1.

There are many different oscillator circuits that can be used to produce sine wave output signals so the troubles discussed here are general in nature. It is possible for the symptoms to vary according to each circuit's design and operation. If, however, you understand the general principles of amplifier operation and the requirements for oscillation (power source, amplifier, feedback, and frequency-determining device), you should find the troubleshooting and oscillators quite simple.

Self-Check

Answer each item by inserting the word or words required to correctly complete each statement.

6. The Armstrong oscillator is a/an _____ oscillator.

   a. LC b. RC

7. Feedback is maintained in the Armstrong oscillator by _____ coupling.

   a. transformer c. capacitive
   b. resistive

8. Refer to Figure 13-6. In this circuit, R_B provides _____ to the transistor.

9. In the Armstrong oscillator, when a positive going signal is present at the base of Q_1, the feedback must have a _____ polarity.

10. Which of the following would be most representative of the time that the transistor conducts in an Armstrong oscillator?

    a. 10%   c. 50%
    b. 25%   d. 100%

Probably the most common oscillator circuits are the Hartley oscillators. Two types (series-fed and shunt-fed) are available. The first discussion will be of the **series-fed** type. A schematic diagram for this type is shown in Figure 13-9.

The Series-Fed Hartley Oscillator

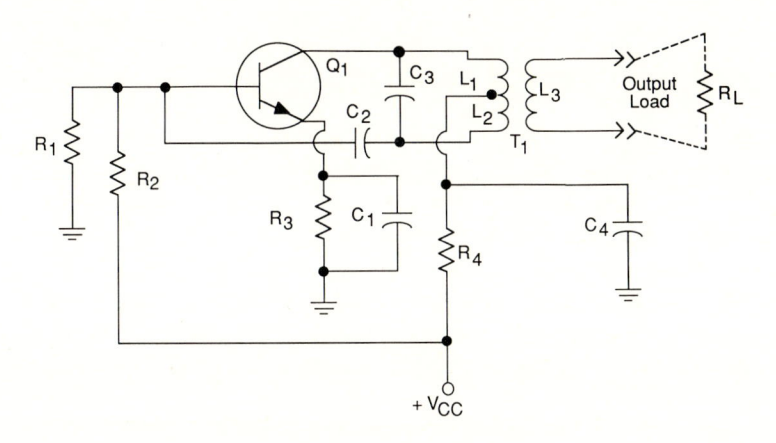

Fig. 13-9 Series-Fed Hartley Oscillator

To understand the purpose of each component, check the following list.

- R_1 and R_2 form the biasing voltage divider.
- R_3 is the temperature stabilization resistor.
- C_1 bypasses R_3 and prevents degeneration.
- C_2 is the feedback coupling capacitor.
- C_3, L_1, and L_2 form the LC tank circuit that acts as the frequency-determining device.
- L_3 provides output coupling and impedance matching for the load circuit.
- R_4 and C_4 form a low pass filter. All frequencies above a specific point are shorted to ground by the low X_C of C_4. As operating frequency increases, the X_C of C_4 decreases to the point that the capacitor acts as a short.
- R_4 also acts as the collector load resistor.

Both Hartley oscillators can be identified by the fact that the primary of T_1 is center-tapped. This oscillator is said to be *series-fed* because DC current flows through a portion of the frequency-determining device. Collector current (I_C) flows from ground, through R_3, through Q_1, through half of the transformer primary (L_1), through R_4, and to $+V_{CC}$. With a part of the tank circuit included in the DC current path, the circuit is said to be *series-fed*. The regenerative feedback from the collector of Q_1 passes through an autotransformer made of L_1 and L_2 and then through C_2 to the base of Q_1. The turns ratio that exists between L_1 and L_2 determines the percentage of the output signal that is fed back to the base of Q_1.

Now to explain how the series-fed Hartley operates, assume that the voltage at the base of Q_1 is going positive and that the circuit is oscillating. A positive at the base of Q_1 causes I_C to increase and V_C to decrease (go negative). This means that the top of L_1 is negative with respect to the center tap. The voltage that is induced into L_2 will be positive at the bottom of L_2 with respect to the center tap. This positive voltage (V_{L2}) is fed back to the base of Q_1 where it arrives as regenerative feedback.

With a negative at the base of Q_1, I_C decreases and V_C increases (goes positive). This places a positive at the top of L_1 with respect to the center tap. L_1 induces a voltage into L_2 that is negative at the bottom of L_2, with respect to the center tap. This negative voltage is coupled through C_2 and back to the base of Q_1. Since both voltages are negative, the feedback is regenerative.

The regenerative feedback path is through L_1, through L_2, through C_2, and to the base of Q_1. The amount (percent) of the output that is fed back can be controlled by selecting the point at which the transformer's primary is center-tapped. If the tap is moved down, L_2 is made smaller while L_1 increases. This causes the feedback amplitude to decrease. Moving the tap upward decreases the size of L_1 and increases L_2. This causes the percentage of feedback to increase. Movement of the center tap does not affect the inductance of the primary coil and will, therefore, not have any affect on the LC tank circuit's resonant frequency.

R_4 and C_4 are placed in the circuit to provide two things. First, the resistor drops V_{CC} to the desired voltage for the transistor. Second, the oscillator signal is isolated from V_{CC} by C_4. The low X_C of C_4 at high frequencies provides a path for shunting RF to ground while DC flows through R_4 without any problem.

The Buffer Amplifier

Observe Figure 13-9. Note that the load resistor (R_L) is "dashed in" across the output terminals. This resistor represents the load that is placed on the frequency-determining device by the next circuit's input impedance. The loading that this circuit provides can affect the oscillator's frequency and amplitude. To prevent this from happening, it is possible to place a **buffer amplifier** between the oscillator output and the circuit that is to receive the oscillator's signal. A buffer amplifier is a circuit that:

> **Isolates the oscillator from the effects of the following circuit.**

The buffer amplifier does this by decreasing the loading effect that would otherwise be placed on the oscillator by reducing the interaction between the load and the oscillator. A schematic diagram for a buffer amplifier is shown in Figure 13-10.

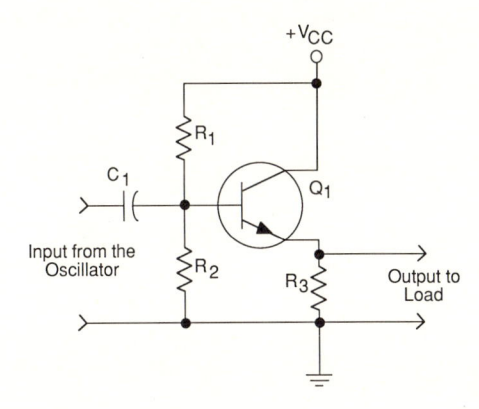

Fig. 13-10 Buffer Amplifier

Notice that the circuit is a common-collector amplifier. You should remember that the common-collector amplifier has a high input impedance and a low output impedance. With the output of the oscillator connected to the buffer's high-impedance input, the buffer has little effect on the operation of the oscillator. The output circuit of the common-collector amplifier is designed to match the impedance of the following circuit. This assures that higher power can be supplied to the buffer output while maintaining efficient operation of the oscillator. Use of a buffer amplifier allows the signal to be passed to the next stage while assuring that output variations in the load are not coupled back to the oscillator.

The Shunt-Fed Hartley Oscillator

Figure 13-11 contains the schematic diagram for a **shunt-fed Hartley** oscillator. Notice the center-tapped winding which identifies a Hartley oscillator. The frequency-determining device for this circuit consists of L_1, L_2, and C_4. In this circuit, the DC path is through R_2, Q_1, R_4, and back to $+V_{CC}$. Notice that DC does not flow through the tank circuit but through a branch that is in parallel with the tank circuit. Because separate parallel paths are used, we say that this circuit is **shunt-fed**.

Each component serves the following purpose(s):

- R_1 and R_3 form the biasing voltage divider for Q_1.
- R_2 provides temperature stabilization.

- C_1 reduces degeneration.
- C_2 is the RF bypass capacitor.
- R_4 is the collector load resistor.
- C_3 couples the output to the tank circuit.
- C_5 provides a feedback path for oscillation.
- L_1, L_2, and C_4 make up the frequency-determining device.
- L_3 provides coupling and impedance matching for the stage.

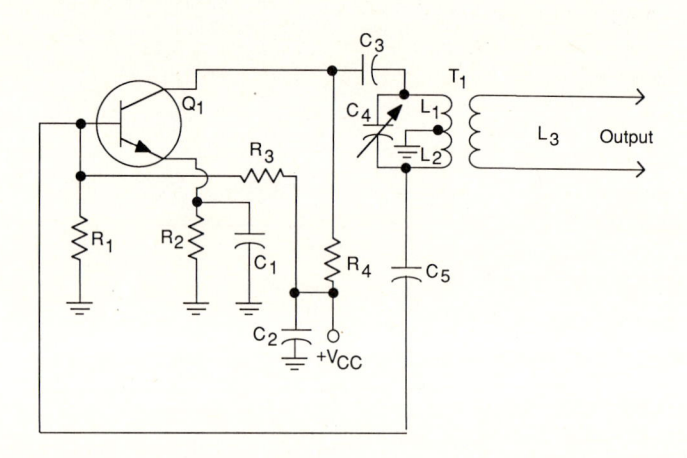

Fig. 13-11 Shunt-Fed Hartley Oscillator

Coupling capacitors C_3 and C_5 serve to "block" DC current from the tank circuit. Regenerative feedback is coupled through C_3, L_1, L_2, and C_5 to the base of Q_1.

Biasing provided by the voltage divider assures that the circuit begins conducting when turned on. Assume that the circuit has been operating for some time and that the signal applied to the base is going positive.

When the positive going signal is applied to the base of Q_1, it causes I_C to increase and V_C to decrease (go in a negative direction). This negative going voltage is coupled through C_3, causing the voltage at the top of L_1 to go negative with respect to the voltage at the primary's center tap. Remember, L_1 and L_2 form an autotransformer. The voltage induced into L_2 is positive at the bottom of L_2 and is applied to the base of Q_1 through C_5 as feedback. With the original input and the feedback being in phase, the feedback is regenerative.

During the negative alternation of the input, I_C decreases causing V_C to increase (go positive). A positive V_C is coupled through C_3, placing a positive at the top of L_1. This means that the voltage at the bottom of L_2 is negative with respect to the center tap. The coupling of C_5 allows this negative voltage to be felt at the base of Q_1. Again we have two signals at the base that are in phase (both negative). This, too, is regenerative feedback.

You can see that the feedback path is through C_3, through L_1, through L_2, through C_5, and back to the base of Q_1. Once again, the position of the center tap on the primary coil determines the percentage of the output signal that is fed back to Q_1.

Use Figure 13-11 as a reference for the following list of troubles that can occur and possible causes for those troubles.

Symptom:
No output signal and $V_C = 0$ V.

Possible Causes:
R_4 open
C_2 shorted
C_3 shorted

Symptom:
No output and V_C normal.

Possible Causes:
C_5 open
C_3 open
Center tap not grounded.

Symptom:
No output and $V_C \cong V_{CC}$.

Possible Causes:
R_2 open
R_3 open
R_4 shorted
C_5 shorted

Self-Check

Answer each item by inserting the word or words required to correctly complete each statement.

11. A series-fed Hartley oscillator _____ have DC current flowing in its tank circuit.

    a. does
    b. does not

12. Check Figure 13-9. R_3 is placed in this circuit to provide _____.
13. A Hartley oscillator can be identified by _____.
14. Explain the purpose of a buffer amplifier.
15. See Figure 13-11. Capacitors _____ and _____ assure that the frequency-determining device is isolated from the DC current.

The Colpitt's Oscillator

Another sine-wave oscillator that is used quite often is the **Colpitt's** oscillator. A schematic diagram for this oscillator is contained in Figure 13-12. A Colpitt's oscillator can be identified by the *split capacitor* (C_3 and C_4) located in the tank circuit. Notice that the feedback loop is tapped into the point that separates the two parts of the tank capacitance. The ratio of the size of C_3 to C_4 determines the percentage of feedback. Components contained in this circuit and their purpose are:

- R_1 – emitter resistor that developes the feedback signal.
- R_2 and R_3 – bias voltage divider
- C_1 – bypass capacitor that places AC ground at the base of Q_1
- C_2 – tank-to-collector coupling capacitor
- C_3, C_4 and L_1 – frequency-determining device
- L_2 – output coupling device

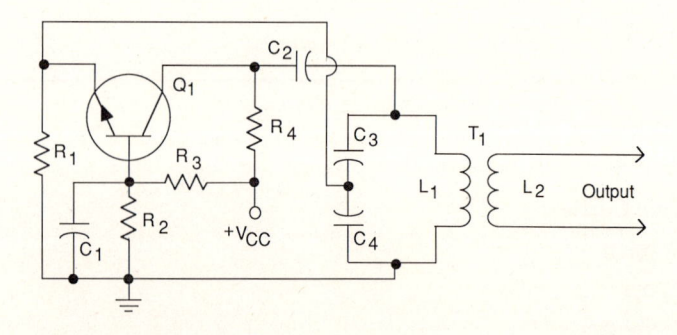

Fig. 13-12 Colpitt's Oscillator

Use of the two capacitors (C_3 and C_4) provide this circuit with capacitive feedback. These two capacitors are said to be *split capacitors* because the capacitance used to determine resonant frequency is "split" into two parts. Before discussing these capacitors, we will establish the regenerative feedback path.

Beginning at the collector of Q_1, feedback passes through C_2 through C_3 of the tank circuit, and back to the emitter of Q_1. Notice that Q_1 is connected as a common-base amplifier—which does not create a phase shift. The tank is constructed such that the feedback does not undergo a phase shift. This means that the signal is in phase at all locations and must, therefore, be regenerative.

Refer to Figure 13-13 while we discuss the purpose of the two capacitors (C_3 and C_4). In this drawing, the emitter circuit of the oscillator has been rearranged for discussion purposes. This allows us to look at the emitter-base circuit with respect to the tank circuit. The distributed (stray) capacitance of the emitter-to-base circuit has been included as C_{eb} in the drawing.

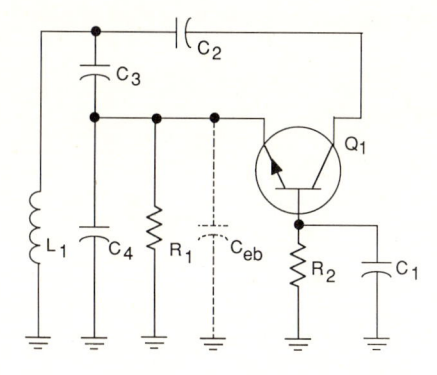

Fig. 13-13 Colpitt's Oscillator with the Emitter Circuit Redrawn

When a tank circuit is connected across a junction of the transistor, the transistor's interelement capacitance (C_{eb}) becomes part of the tank capacitance. A change in C_{eb} causes the tank's resonant frequency to change. Junction capacitance is subject to change in bias or operating temperature, among other things. In the Colpitt's oscillator, C_4 is connected across the emitter-base junction. Placing C_4 in parallel with C_{eb} minimizes the effect of changes in C_{eb}. To accomplish this, the value of C_4 must be quite large. For example, if C_{eb} is 10 pf, C_4 might be 1000 pf. For this example, total parallel capacitance (C_4 and C_{eb}) equals 1010 pf and is part of the tank capacitance. The value of capacitor (C_3) is then selected to assure that the circuit will resonate at the desired frequency. If C_3 equals 100 pf, total capacitance will equal 90.9 pf.

Remember:

Parallel capacitors are treated like resistors in series. Series capacitors are treated like resistors in parallel.

To illustrate how this controls the effect of changes in C_{eb}, consider the following:

Assume that C_{eb} doubles to 20 pf. Now the equivalent value of C_4 and C_{eb} equals:

$C_{eq} = C_4 + C_{eb} = 1000 \text{ pf} + 20 \text{ pf} = 1020 \text{ pf}$

When this is combined with $C_3 = 100$ pf we have:

$C_t = \dfrac{C_3 \times C_{eq}}{C_3 + C_{eq}}$

$C_t = \dfrac{100 \times 1020}{100 + 1020}$

$C_t = \dfrac{102000}{1120}$

$C_t = 91.07 \text{ pf}$

To find the percent of change we do the following:

Amount of C_t change = new C_t − old C_t

Amount of C_t change = 91.07 pf − 90.9 pf

Amount of C_t change = 0.17 pf

% of Change = $\dfrac{\text{Amount of } C_t \text{ change}}{\text{New } C_t} \times 100$

% of Change = $\dfrac{0.17}{91.07} \times 100$

% of Change = .18667 %

This means that if C_{eb} doubles, the actual change in resonant frequency will be less than 0.002 (0.2%). By careful selection of capacitor values, the effect of C_{eb} changes can be kept even smaller than in this example. Using two capacitors reduces the undesirable effect of interelement capacitance that is present in transistor operations. By reducing these changes, we assure that the Colpitt's oscillator operates at a very stable frequency. The two capacitors act as a series capacitive voltage divider whose ratio determines the amplitude of tank voltage that is fed back.

Colpitt's oscillators can be tuned by variations of either tank capacitance or inductance. When capacitance tuning is used, however, both capacitors must be tuned at the same time. This is necessary because C_3 and C_4 form a voltage divider that determines the amplitude of the feedback. Tuning one capacitor and not the other will result in a change in the amplitude of the feedback signal.

The Clapp Oscillator

Figure 13-14 contains the schematic diagram for a **Clapp** oscillator. The split capacitors that are present in the tank circuit identify this circuit as a modified Colpitt's oscillator. Notice that C_5, an adjustable capacitor, has been added. The addition of this capacitor allows the tank frequency to be varied without affecting the feedback ratio that is established by the sizes of C_3 and C_4. The tank circuit now contains C_3, C_4, C_5, and L_1. By placing C_5 in the branch with L_1 provisions are made for tuning while still maintaining control of C_{eb}. Control of the effect of C_{eb} is an advantage of the Colpitt's oscillator. Using a single tuning capacitor (C_5) makes it much easier than if we had to use ganged tuning for C_3 and C_4. Another advantage of this design is the reduction of the **hand capacitance** effect. Hand capacitance refers to the effect that the operator's hand has upon operating frequency. It is defined as:

> **The capacitance introduced into a circuit when one's hand is brought near a tuning capacitor or other insufficiently shielded part of the circuit.**

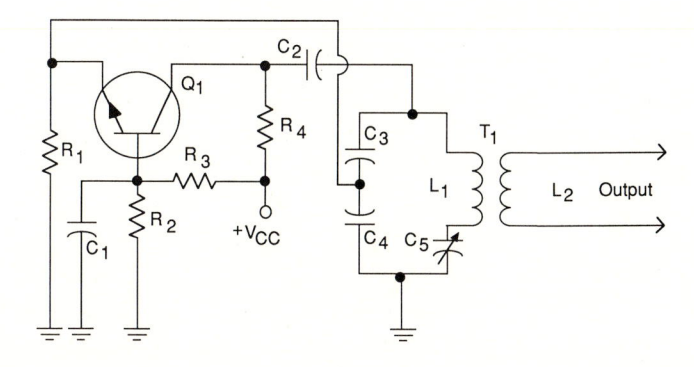

Fig. 13-14 Clapp Oscillator

When you place your hand near the tuning capacitor the capacitance of your hand and/or body, which is grounded, may cause the operating frequency of the oscillator to change without your intending for it to happen. The fact that C_5 has one plate grounded greatly reduces the effect of hand capacitance.

Troubleshooting the Colpitt's and Clapp oscillators is very similar. For that reason, we will discuss only the Clapp oscillator troubleshooting procedures.

Symptom:

No output and V_C is lower than normal.

Possible causes:

C_4 shorted

R_1 shorted

R_2 open

Symptom:

Output frequency lower than normal.

Possible causes:

C_5 shorted

Symptom:

No output and V_C is normal.

Possible causes:

C_2 open

C_3 open

C_4 open

C_5 open

L_1 open

L_2 open

L_1 shorted

L_2 shorted

Sine Wave Oscillators

The Butler Oscillator

The schematic diagram for a **Butler** oscillator is contained in Figure 13-15. This oscillator has two primary identification features. First, two transistors are used; and second, a crystal (CR_1) is connected between the emitters of the two transistors. The purpose for each component is as follows:

- R_1 – Q_1 emitter resistor that develops Q_1 output
- R_2 – Q_2 emitter resistor
- CR_1 – frequency-determining device (a crystal)
- R_4 and R_5 – biasing voltage divider for Q_2
- C_3 – base bypass capacitor for Q_2
- L_1 and C_2 – resonant LC tank circuit that provides the collector load impedance for Q_2
- C_1 – coupling capacitor
- R_3 – biasing resistor for Q_1

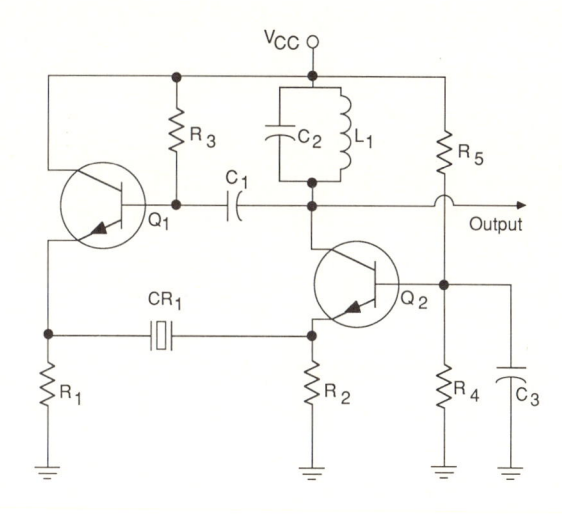

Fig. 13-15 Butler Oscillator

Note that Q_1 is a common-collector amplifier and Q_2 is a common-base amplifier. Regenerative feedback is coupled from the collector of Q_2, through C_1, to the base of Q_1, to the emitter of Q_1, through CR_1, to the emitter of Q_2. For this circuit to operate correctly, the regenerative feedback must pass through CR_1. At its resonant frequency, CR_1 has a very low impedance and passes regenerative feedback easily. At all other frequencies, the impedance of CR_1 is high and effectively blocks the regenerative feedback.

An output signal can be taken directly from the collector of Q_2. Both transistors operate Class C. This allows the tank circuit (L_1 and C_2) time to *flywheel* during the period that Q_2 is cutoff. The flywheel action of the tank circuit produces the output that is present at the collector of Q_2. At the same time, crystal CR_1 is vibrating at its resonant frequency. Because Q_2 is cut off for most of the time, it provides buffer action that isolates the emitter of Q_2 from the collector. With this buffer action, an output can be taken from the collector of Q_2 without affecting the resonant frequency of the oscillator any appreciable amount. This eliminates the need for another buffer amplifier that would isolate the output from the oscillator. Remember:

> **The frequency-determining device for this circuit is the crystal—not the LC tank circuit. For this reason, the tank is usually designed such that its resonant frequency is close to the crystal's resonant frequency.**

Self-Check

Answer each item by inserting the word or words required to correctly complete each statement.

16. A Clapp oscillator can be identified by the fact that an additional _____ has been added to the tank circuit of a Colpitt's oscillator.
17. A Colpitt's oscillator can be identified by its split _____.
18. Refer to Figure 13-14. List two advantages that the insertion of C_5 provides over the Colpitt's oscillator.

   a. _____
   b. _____

19. What is the advantage of having C_4 (Figures 13-13 and 13-14) connected across the emitter-base junction of Q_1?
20. What governs the frequency to which the LC tank circuit of Figure 13-15 is tuned?

The RC Oscillators

Earlier you were introduced to the fact that resistor-capacitor networks could be used as regenerative feedback paths and as frequency-determination devices for oscillators. In this section, we will discuss two different RC types.

The RC Phase Shift Oscillator

An **RC phase shift** oscillator's schematic diagram appears in Figure 13-16. The components that are contained in this circuit serve the following purposes:

- R_1, R_2, R_3, C_1, C_2, and C_3 – these components form the RC frequency-determining network for the circuit and provide the path for regenerative feedback.
- R_3 and R_5 – biasing voltage divider for Q_1.
- R_4 – emitter resistor for Q_1 provides temperature stability.
- C_4 – emitter-resistor bypass capacitor that reduces degeneration.
- R_6 – collector-load resistor for Q_1 that provides control of the feedback amplitude and the gain of Q_1.

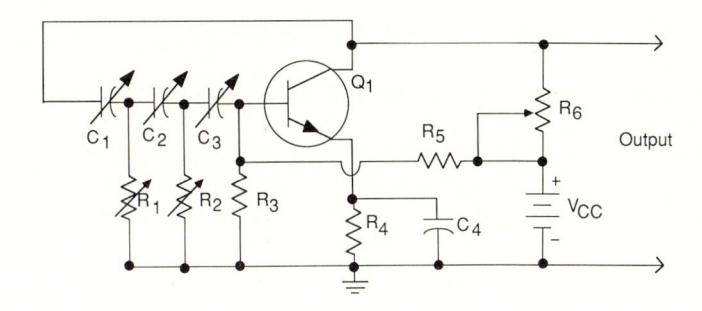

Fig. 13-16 RC Phase-Shift Oscillator

Figure 13-16 has three RC combinations: R_1 and C_1 form combination 1; R_2 and C_2 form combination 2; R_3 and C_3 form combination 3. Resistor R_5 provides forward bias for Q_1 and has a much larger ohmic value than R_3.

The input signal for the RC network comes from Q_1 collector, and the output of the RC network is applied to the base of Q_1. The RC network forms a loop that is used for regenerative feedback. Q_1 is a common-emitter amplifier that provides a 180° phase shift between its base and collector. For the feedback arriving at the base to be regenerative feedback, the RC network must provide an additional 180° of phase shift. This is accomplished by the combined action of the three RC combinations. Each network must provide approximately 60° of feedback for the total to equal 180°. It is possible to use adjustable capacitors and resistors in the RC network, which allows the frequency of operation to be changed. Adjustment of these components changes the time constant and the phase shift of the affected combination. This causes the time required for feedback to change. If the time changes, the frequency of operation must also change. The result is that the collector signal is shifted exactly 180° prior to being applied to the base of Q_1. This assures that the feedback is regenerative.

What about all of the frequencies that are above and below the circuit's frequency of operation (f_o)? You should remember that each resistor-capacitor combination has a charge rate that is constant. Only one frequency is timed such that it receives exactly 180° of phase shift. All other frequencies are shifted more or less than 180°. Frequencies above f_o are shifted less than 180° while all frequencies below f_o are shifted more than 180°. In either of these cases, the feedback signal is not regenerative. Only the one frequency that receives exactly 180° of phase shift is regenerative and can maintain circuit oscillations. In RC oscillators, all amplifiers must operate Class A. This is necessary because there is no tank circuit that can maintain output by its flywheel action. To obtain a sine wave output from an RC oscillator, the transistor must conduct 100% of the time. This assures that 360° of the sine wave is present in the output. If regenerative feedback is too small, the RC oscillator will not oscillate. If regenerative feedback is too large, the transistor will be driven into saturation. In either case, the output signal is distorted and is not a true sine wave. To assure that the regenerative feedback has the correct amplitude, R_6 is placed in the circuit. Adjusting R_6 affects the amplitude of the feedback and the output signal. If the value of R_6 is decreased, the voltage gain of the circuit is decreased and the feedback amplitude is reduced. Increasing the value of R_6 increases both voltage gain and feedback amplitude.

To calculate the frequency of operation for an RC oscillator, we use the formula:

$$f_o = \frac{1}{2\pi RC\sqrt{2N}}$$

where:

N = number of networks

RC oscillators can have three or more phase shift networks. For the circuit with three networks that we have discussed here, the formula is:

$$f_o = \frac{1}{2\pi RC\sqrt{2(3)}}$$

Now that we have considered the circuit's operation, we can take a look at the troubleshooting procedures that apply to its malfunctions.

Symptom:

No output and V_C is normal.

Possible causes:

C_1 open
C_2 open
C_3 open
R_1 open
R_2 open
C_2 shorted
R_1 shorted
R_2 shorted

Symptom:

No output and V_C is higher than normal.

Possible causes:

C_3 shorted.

The Wien Bridge Oscillator

Another oscillator that uses RC networks as its frequency-determining network is the **Wien bridge** oscillator. A schematic diagram for this circuit appears in Figure 13-17. The following list provides an explanation for the part each component plays in circuit operation.

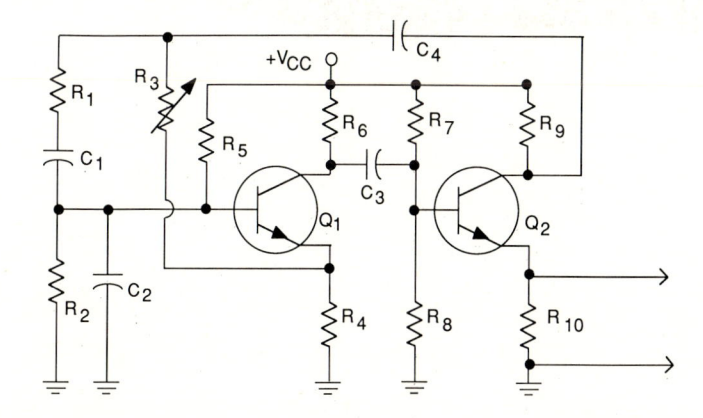

Fig. 13-17 Wien Bridge Oscillator

- R_1, C_1, R_2, and C_2 – these components form the frequency-determining device.
- R_3 and R_4 – degenerative feedback network
- R_2, R_5 and R_7, R_8 – biasing voltage dividers
- R_6 and R_9 – collector load resistors
- C_3 and C_4 – coupling capacitors
- R_{10} – emitter resistor which develops the output signal

The frequency determining device (R_1, R_2, C_1, and C_2) is a series-parallel RC network. Figure 13-18(a) shows this network. Note that R_2, C_2, and the emitter-base circuit of Q_1 (Figure 13-17) are in parallel; therefore, any voltage that is developed across R_2 and C_2 will be applied as the input signal for Q_1. Also notice that R_3 and R_4 form a degenerative feedback network. Any voltage that is developed across R_4 is also felt at the emitter of Q_1. During normal operation of this circuit, it is possible for two different signals to be applied to Q_1. Voltage divider R_3 and R_4 place a degenerative signal across R_4 because R_4 is not capacitor bypassed. Also, the voltage divider (R_3 and R_4) is not frequency selective, meaning that it is not affected by AC signals. Any signal that is coupled from the collector of Q_2 through C_4 will appear at the emitter of Q_1 with an amplitude that is less than that present at the collector of Q_2. The frequency-determining device, however, is frequency selective. The amplitude of the signal at the base of Q_1 depends on the frequency of the signal. Because of the arrangement of these components, one frequency will cause the signal applied to the base of Q_1 to be larger than any other frequency. The frequency at which the largest amplitude occurs becomes the operating frequency for the oscillator circuit.

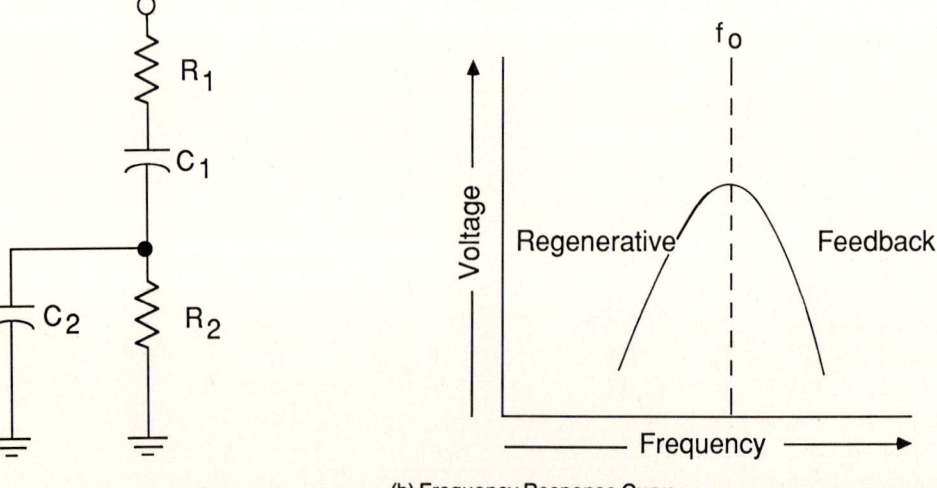

(a) RC Network

(b) Frequency Response Curve

Fig. 13-18 Frequency Determining Device for the Wien Bridge Oscillator

To illustrate the operation of a Wien bridge oscillator frequency-determining device, we will use Figure 13-18(a). First we make the assumptions that R_1 and R_2 have the same ohmic value and that C_1 and C_2 have the same capacitance. Maximum amplitude is developed across parallel branch R_2 and C_2 at the center (resonant) frequency.

Observe Figure 13-18(b) and you see that feedback is maximum at the center frequency of the response curve. When feedback is maximum, the voltage drop across the R_2 and C_2 network must be maximum. As frequency varies either side of center frequency, the amount of voltage developed across R_2 and C_2 decreases. This decrease in voltage results in a decrease in feedback. These decreased values are represented by the response curve of Figure 13-18(b).

The signal across R_2 and C_2 not only varies in amplitude, it also varies in phase. At f_o, current in the series-parallel frequency-determining device leads the applied voltage by $+45°$. The parallel section (R_2 and C_2), however, has an impedance that lags by $-45°$. The result is that after current encounters both of these phase shifts ($+45°$ and then $-45°$), it arrives at the base of Q_1 in phase (regenerative feedback).

At frequencies above and below f_o, the feedback signal will not be of the proper phase to sustain oscillations. So, for given component values within the frequency-determining device, only one frequency (f_o) will have a feedback signal that is the correct phase. Both Q_1 and Q_2 are operating as common-emitter amplifiers. Each amplifier shifts the phase by 180° for a total across the two amplifiers of 360°. This means that Q_2 collector voltage and the voltage applied to the base of Q_1 are in phase.

Maximum regenerative feedback occurs for *only* one frequency. The small amount of degenerative feedback supplied by the voltage divider (R_3 and R_4) increases the frequency stability of the circuit.

Observe Figure 13-19 while we discuss the degenerative feedback. Because R_3 and R_4 are not frequency selective, all frequencies can pass equally well. Note the straight line that represents the degenerative feedback that is superimposed on the regenerative feedback curve. Only at its highest peak does the regenerative feedback become larger than the degenerative feedback. Because regenerative feedback is required for the circuit to operate, operation can occur only at the point or points where regenerative feedback is larger than degenerative feedback.

To tune the circuit, some component within the frequency-determining device must be varied. The method that is most common is to *gang tune* either capacitors C_1 and C_2 or resistors R_1 and R_2.

In Figure 13-17, R_3 is a variable resistor. Adjustment of this resistor varies the amount of degeneration that occurs across R_4. Increasing the size of R_3 reduces degenerative feedback which allows the output signal amplitude to increase and possibly be distorted. Decreasing R_3 causes degenerative feedback to increase which could cause the

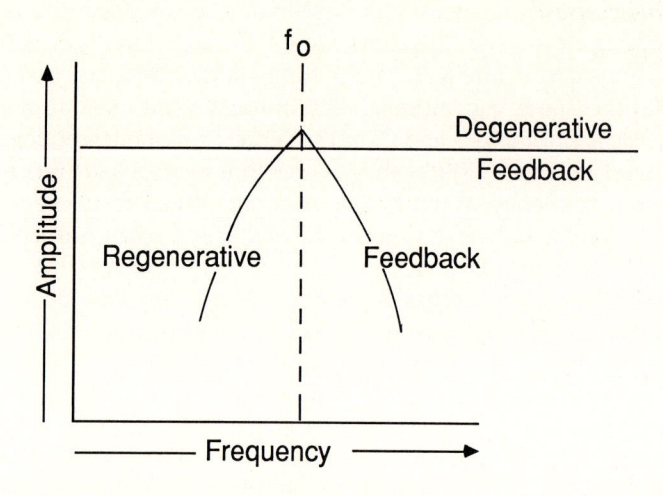

Fig. 13-19 Degenerative Feedback Diagram

oscillator to stop operating or, at best, reduces the output signal amplitude. R_3 is normally adjusted for maximum output amplitude with minimum distortion.

To calculate the frequency of operation for a Wien bridge oscillator, use the formula:

$$f_o = \frac{1}{2\pi\sqrt{(R_1)(R_2)(C_2)(C_3)}}$$

When R_1 equals R_2 and C_1 equals C_2, the following formula can be used. Notice that the square root sign has been removed.

$$f_o = \frac{1}{2\pi RC}$$

Where: R can be either resistor.
C can be either capacitor.

Troubleshooting this oscillator involves recognizing the following symptoms and their possible causes.

Symptom:
No output and V_C of Q_1 is high.

Possible causes:
   R_4 open
   R_5 open
   R_2 shorted
   R_6 shorted
   C_2 shorted

Symptom:
   Q_1 has low V_C.

Possible causes:
   R_2 open
   R_6 open
   R_4 shorted
   R_5 shorted

Symptom:
   Q_2 has low V_C.

Possible causes:
   R_7 shorted
   R_8 open
   R_9 open
   R_{10} shorted
   C_3 shorted

Symptom:
   No output but both collector voltages are normal.

Possible causes:
   R_1 open
   C_1 open
   C_2 open
   C_3 open
   C_4 open
   C_1 shorted
   R_1 shorted
   R_3 shorted

Symptom:
Output frequency and amplitude are unstable.

Possible causes:
R$_3$ open

Self-Check

Answer each item by inserting the word or words required to correctly complete each statement.

21. An RC phase shift oscillator has four RC phase shift networks. This means that each resistor-capacitor pair must shift the phase by _____ degrees.

    a. 45
    b. 90
    c. 180
    d. 360

22. For the circuit in Figure 13-16 to operate correctly total phase shift across the RC networks must equal _____ degrees.

    a. 45
    b. 90
    c. 180
    d. 360

23. The amplifiers for all RC oscillators must operate Class _____.

    a. A
    b. AB
    c. B
    d. C

24. In the Wien bridge oscillator, the operating frequency is established by _____ networks.
25. Refer to Figure 13-17. Regenerative feedback is applied to the _____ of Q$_1$.

Frequency Multipliers

Quite often, we need frequencies higher than these oscillators are capable of producing. In these cases we use circuits called *frequency multipliers*. A frequency multiplier schematic diagram is contained in Figure 13-20. Notice that the output of an oscillator that is operating at 1 MHz is applied to the input of this circuit. As its name implies, the output frequency of the circuit will be a multiple of the input frequency. The circuit shown here is a frequency doubler. It will supply an output signal that is 2 MHz, or double the input frequency. A frequency tripler would supply 3 MHz (triple the input frequency), a quadrupler multiplies the input frequency by four, and so on.

For a circuit to act as a frequency multiplier, the circuit must generate harmonics. Harmonics are generated by amplifiers that operate either Class B or Class C. The frequency doubler shown in Figure 13-20 is operating Class C. With no input signal present, Q_1 is cut off.

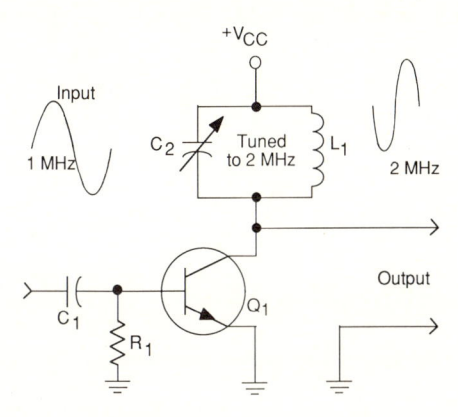

Fig. 13-20 Frequency Multiplier

This is because Q_1 has no forward bias applied. When the positive alternation of the input signal arrives at the base of Q_1 the transistor is forward biased. This causes coupling capacitor C_1 to charge rapidly and Q_1 to conduct. During the conduction of Q_1, capacitor C_2 is charged to the voltage drop of L_1. When the positive alternation (of the input signal) reaches its peak and begins to decrease, the charge on C_1 is larger than the input voltage. At this time, C_1 begins to discharge through R_1 which is a large resistor. This causes the voltage at the top of R_1 to be negative which causes Q_1 to enter cutoff. Because of the large size of R_1, C_1 takes a long time to discharge. The duration of C_1 discharge is much longer than its charge time, meaning that Q_1 is forced to operate Class C. In order to overcome the reverse bias applied to the base by C_1 discharge, the input signal amplitude must be quite large.

With Class C operation, the transistor generates many harmonics of the input signal. By tuning the output tank (L_1 and C_2) to a desired harmonic, that harmonic's frequency can be selected as the output signal. For the frequency doubler, the tank is tuned to the second harmonic; for a tripler, the tank is tuned to the third harmonic, and so on. The amplitude of each higher harmonic decreases rapidly. Because of this lower amplitude, the fourth harmonic (quadrupler) is the highest harmonic for which a frequency multiplier is normally used. If additional multiplication is needed, another stage of multiplication must be added. In fact, two cascade-connected doublers are usually preferred over one quadrupler. The two doublers will provide much greater output power than will a single quadrupler.

If the input frequency and the component values for the tank are known, you can calculate whether the stage is a doubler, tripler, or quadrupler by use of the formula shown below. Assume that we have a multiplier whose input frequency is 10 MHz, tank capacitance is 4 pf and tank inductance is 15.9 µH. We solve for the tank's resonant frequency as follows:

$$f_r = \frac{1}{2\pi\sqrt{LC}}$$

$$f_r = \frac{1}{2(3.1415927)\sqrt{(15.9 \times 10^{-6})(4 \times 10^{-12})}}$$

$$f_r = \frac{1}{2(3.1415927)\sqrt{63.6 \times 10^{-18}}}$$

$$f_r = \frac{1}{2(3.1415927)(8 \times 10^{-9})}$$

$$f_r = \frac{1}{5.0265482 \times 10^{-8}}$$

$f_r = 1.989 \times 10^7$ or approximately 20 MHz

With the 10 MHz input stated above, we can see that the output tank frequency is double the input frequency, or 20 MHz. This means that the circuit is designed to be a frequency doubler for 10 MHz.

Another feature of the frequency multiplier is that it can provide the same characteristics as a *buffer* amplifier. Because the transistor operates Class C, the load of the amplifier does not reflect impedance back to the base circuit. With Class C operation the input impedance at the base is quite high. Remember, the emitter-base junction is reverse biased the majority of the time and acts like an open.

Figure 13-21 contains the schematic diagram for a Butler oscillator that is driving a frequency multiplier. Q_3 and its associated circuitry make up the frequency multiplier.

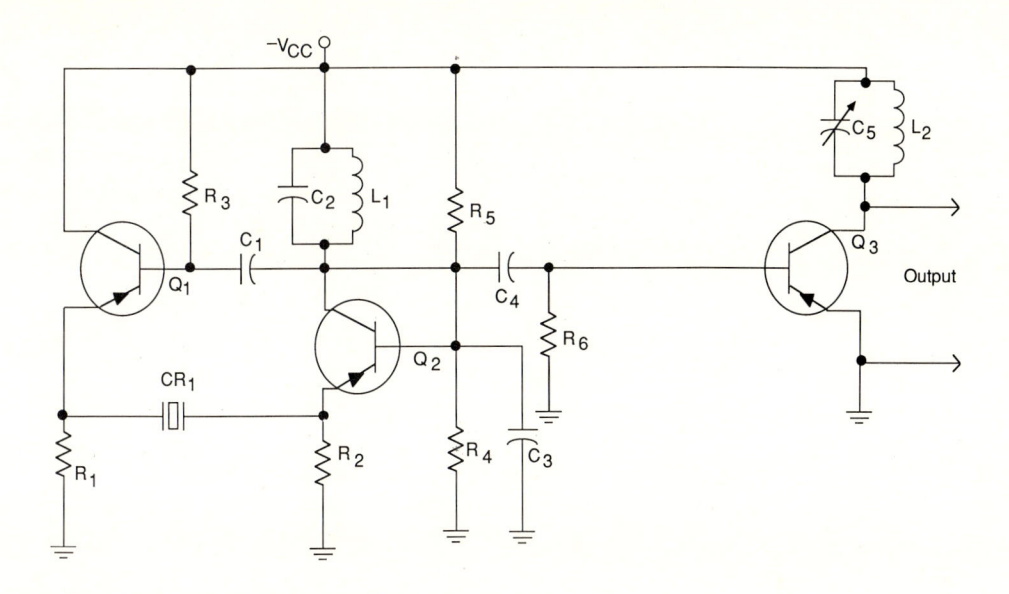

Fig. 13-21 Butler Oscillator Driving a Frequency Multiplier

Self-Check

Answer each item by inserting the word or words required to correctly complete each statement.

26. For a frequency multiplier to act as a multiplier, the _____ circuit must be tuned to a harmonic of the input.
27. The amplifiers used in frequency multipliers must operate Class _____.
28. An added advantage of the frequency multiplier is that it provides the same isolation as a _____ amplifier.
29. The highest harmonic that is used for frequency multipliers is the _____ harmonic.
30. Two frequency _____ will provide much more output power than one frequency _____.

Summary

The three requirements for oscillation are:

1. An amplifier with a suitable power source
2. A frequency-determining device
3. Regenerative feedback.

One type of oscillator is the *sine-wave* oscillator. These are grouped into three categories; LC types, RC, and Crystal types. LC oscillators include the Armstrong, Hartley, Colpitt's and Clapp oscillators. Hartley oscillators can be identified by their tapped transformer winding while Colpitt's and Clapp oscillators can be identified by split capacitors. Further identification between the Colpitt's and Clapp oscillators is that the Clapp has a third capacitor that serves as a tuning device for the tank's resonant frequency. Other than this one feature, the Colpitt's and Clapp oscillators are identical.

Hartley oscillators come in two types; series-fed and shunt-fed. A series-fed Hartley has a portion of the DC current flowing in its tank circuit. In the shunt-fed Hartley, direct current is shunted through a separate path and is not allowed to flow in the tank circuit. All four of these oscillators have a frequency of operation that is determined by the resonant frequency of the tank circuit. In the Hartley oscillators, the regenerative feedback is coupled using *inductive* feedback. In the Colpitt's and Clapp oscillators, regenerative feedback occurs because of *capacitive* feedback.

The Butler oscillator depends upon a quartz crystal for the determination of its frequency of operation. The natural resonant frequency of the crystal establishes the frequency at which regenerative feedback can exist. The Butler oscillator operates Class A while the four LC oscillators operate Class C.

RC oscillators are of the RC Phase Shift and Wien Bridge types. Both depend on RC networks for their frequency of operation. The RC networks provide a 180° phase shift and the amplifier provides another 180° phase shift. The combined 360° phase shift can occur for only one frequency. This frequency becomes the operating frequency.

Buffer amplifiers and frequency multipliers are used with oscillators to provide specific advantages. A buffer amplifier is placed in the output of an oscillator to isolate the oscillator from the effect that might result from connection of a load. Frequency multipliers accept one frequency at their input. A tank circuit that is located in the amplifier's output circuit can be tuned to either the second, third, or fourth harmonic of the input frequency. Operation of the circuit is such that the output tank oscillates at a harmonic of the input signal. Possible output frequencies are second, third, and fourth harmonics. These are called frequency doublers, frequency triplers, and frequency quadruplers. Actually the output of a quadrupler is very low. For that reason it is very seldom used. Frequency doublers and triplers are more likely to be encountered by the technician.

Review Questions and/or Problems

1. Which of the following correctly states the requirements for sustaining oscillations?

    a. power supply, regenerative feedback, and a load
    b. load, frequency-determining device, and an amplifier
    c. frequency-determining device, regenerative feedback and a power supply
    d. frequency-determining device, regenerative feedback and an amplifier

2. Which of the following replaces the energy that is lost as a result of damping?

    a. frequency-determining device
    b. amplifier
    c. load resistor
    d. emitter resistor

3. What is the purpose of an oscillator circuit?

    a. to develop a pulsating DC at its output
    b. to provide an AC signal at the input to its amplifier
    c. to generate a waveform of constant amplitude and frequency
    d. to select one frequency from many frequencies

4. Which of the following establishes an oscillator's frequency of operation?

    a. frequency-determining device
    b. the amplifier
    c. load resistor
    d. emitter resistor

5. A sinusoidal oscillator produces a _____.

    a. square wave of constant amplitude and frequency
    b. square wave of varying amplitude and frequency
    c. sine wave of varying amplitude and frequency
    d. sine wave of constant amplitude and frequency

6. Which of the following is/are methods used to determine the frequency of operation for an oscillator?

   a. crystals
   b. RC networks
   c. LC tank circuits
   d. all of these

7. A resonant LC tank circuit's transference of energy between its coil and capacitor is referred to as the _____ effect.

   a. damping
   b. loading
   c. flywheel
   d. stability

8. When in phase energy is returned from the output back to the input, the feedback is said to be _____ in nature.

   a. generative
   b. regenerative
   c. degenerative
   d. any of these

9. Each successive alternation of the output decreases in amplitude; this is a result of the _____ effect.

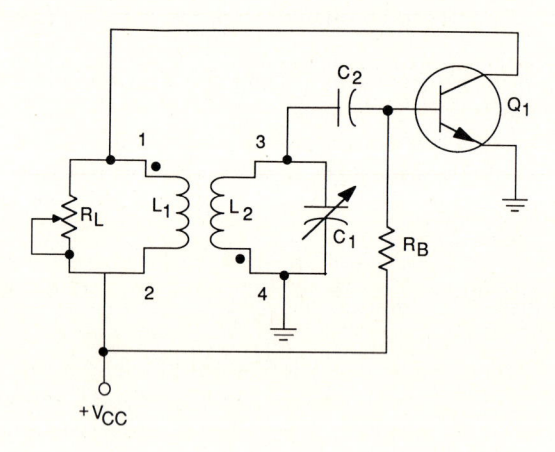

   a. damping
   b. loading
   c. flywheel
   d. stability

Fig. 13-22

10. The oscillator shown in Figure 13-22 is a/an _____ oscillator.

    a. Armstrong
    b. Butler
    c. Colpitt's
    d. Clapp

Sine Wave Oscillators

11. In this circuit (Figure 13-22), frequency of operation is determined by:

    a. L_1, C_1
    b. L_1, C_2
    c. L_2, C_1
    d. L_2, C_2

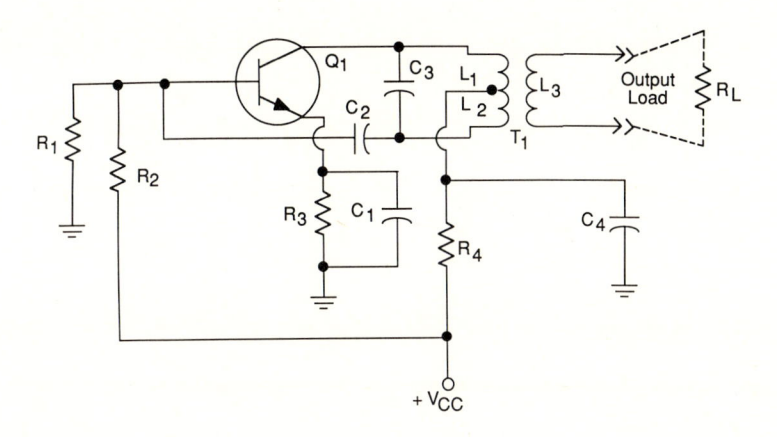

Fig. 13-23

12. The schematic diagram contained in Figure 13-23 represents a/an _____ oscillator.

    a. Armstrong
    b. Colpitt's
    c. series-fed Hartley
    d. shunt-fed Hartley

13. The amount of feedback in this oscillator (Figure 13-23) is determined by the:

    a. size of C_2.
    b. the turns ratio of L_1 to L_2.
    c. the turns ratio of T_1.
    d. ratio of R_1 to R_2.

14. The oscillator (Figure 13-23) does not have an output at the secondary of T_1. Which of the following could cause this?

    a. R_1 open
    b. R_3 shorted
    c. R_4 open
    d. C_4 open

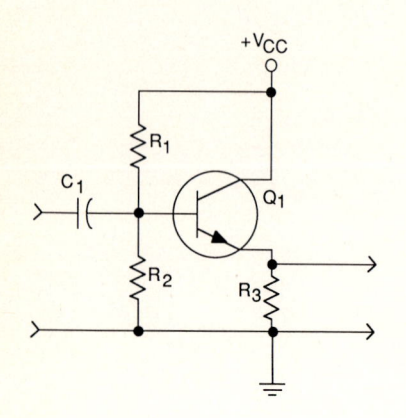

Fig. 13-24

15. Figure 13-24 contains the schematic diagram for a buffer amplifier. This circuit is used to isolate an oscillator from the effects of its load.

   a. True
   b. False

16. In Figure 13-24, the output signal is _____ phase with the input signal.

   a. in
   b. 90° out of
   c. 180° out of
   d. 270° out of

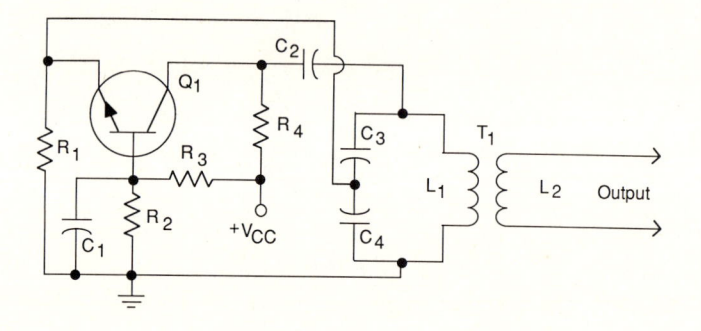

Fig. 13-25

17. The circuit shown in Figure 13-25 is a _____ oscillator and it can be identified by its _____.

   a. Hartley, center-tapped capacitors
   b. Colpitt's, center-tapped capacitors
   c. Hartley, center-tapped inductors
   d. Colpitt's, center-tapped inductors

18. The percent of the output signal (Figure 13-25) that is used as regenerative feedback is determined by:

   a. the turns ratio of L_1 and L_2
   b. the discharge rate of C_4
   c. the ratio of C_3 to C_4
   d. none of the above

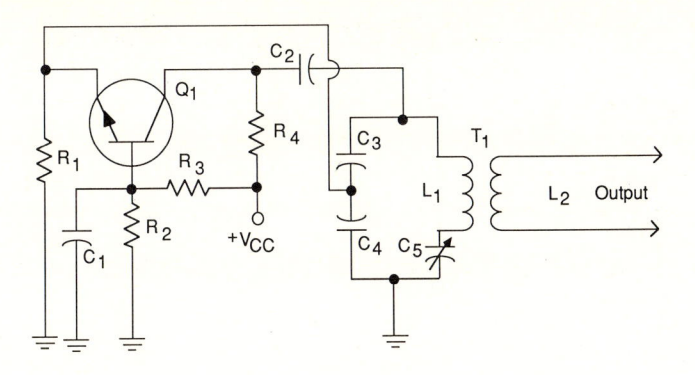

Fig. 13-26

19. The circuit shown in Figure 13-26 has a frequency-determining device that consists of _____.

    a. T_1, C_3 and C_4
    b. L_1 and C_2
    c. L_1, C_3 and C_4
    d. L_1, C_3, C_4 and C_5

20. This circuit (Figure 13-26) is a _____ oscillator.

    a. Hartley
    b. Armstrong
    c. Colpitt's
    d. Clapp

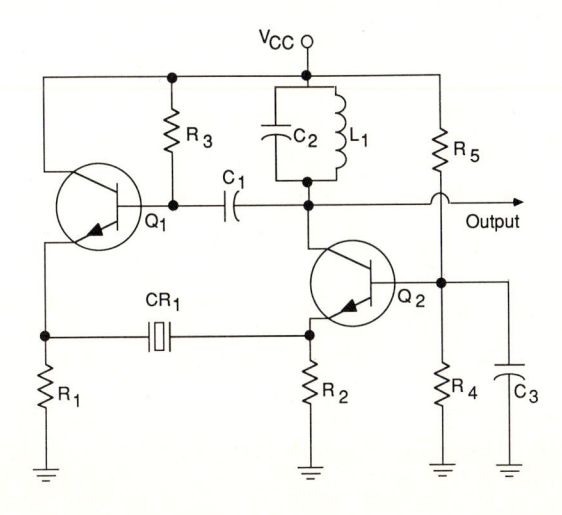

Fig. 13-27

21. In Figure 13-27 the frequency-determining device consists of ____.

   a. L_1 and C_2
   b. C_1 and R_3
   c. CR_1
   d. C_1, C_2, and L_1

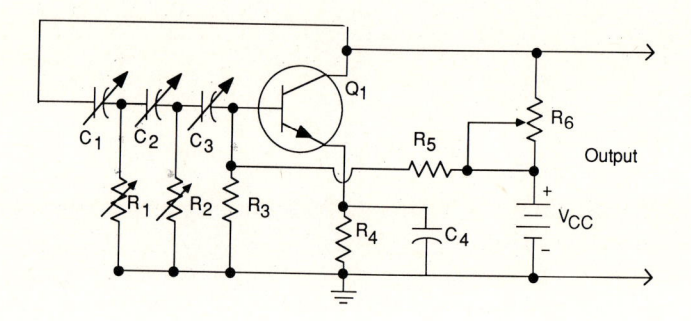

Fig. 13-28

22. The circuit shown in Figure 13-28 has a frequency-determining device that consists of ____.

   a. R_3, R_5 and R_6
   b. R_1, C_1, R_2, C_2, R_3 and C_3
   c. R_4 and C_4
   d. all of these

23. Increasing the value of R_6 (Figure 13-28) will cause gain to ____ and feedback to ____.

   a. increase, increase
   b. increase, decrease
   c. decrease, decrease
   d. decrease, increase

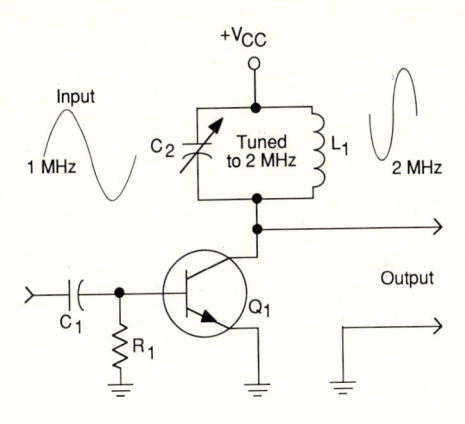

Fig. 13-29

24. The circuit shown in Figure 13-29 is a/an _____.

    a. Hartley oscillator
    b. Armstrong oscillator
    c. buffer amplifier
    d. frequency multiplier

25. Which of the following statements **IS NOT** true regarding a Butler oscillator (Figure 13-27).

    a. Regenerative feedback is developed across C_3.
    b. All frequencies except resonant frequency will be blocked because of the high impedance of CR_1.
    c. CR_1 offers minimum impedance to regenerative feedback at the crystal's natural resonant frequency.
    d. C_1 serves as input coupling capacitor for Q_1.

14
Nonsinusoidal Wave Generators

Objectives

1. Identify the circuit schematics for the following:

    a. Pulsed oscillators
    b. Blocking oscillators
    c. Astable multivibrators
    d. Monostable multivibrators
    e. Bistable multivibrators
    f. Schmitt trigger circuits
    g. Sawtooth wave generators
    h. Trapezoidal wave generators

2. Identify the frequency-determining device for each of the circuits listed in objective 1.
3. Describe the output waveshape for each of the circuits listed in objective 1.
4. Explain the use for each of the circuits listed in objective 1.
5. Define the following terms:

    a. Trigger
    b. Pulse recurrence time (PRT)
    c. Pulse-repetition frequency (PRF)
    d. Rise time
    e. Fall time
    f. Pulse width (PW)
    g. Transit interval

Introduction

The sine-wave (LC) oscillators that were discussed in Chapter 13 are required to provide an output 100% of the time. The sine wave output signal has to be at a predetermined frequency and is generated indefinitely. This is to say that the sine wave oscillator provides a continuous sine-wave output signal.

Many electronic circuits, however, require that an oscillator be turned on for only short periods of time and then remain cut off until a time is reached where the oscillator is again turned on. One type circuit that operates in this fashion is called the **pulsed** oscillator.

A second type, the **blocking** oscillator, is a special type of wave generator. It is used to generate a very narrow pulse that is often called a **trigger**. The circuit may be called a **trigger generator**. Blocking oscillators are often used to provide the timing for complex electronic systems. Other blocking oscillators can be used for many purposes. In this chapter we will discuss the basic blocking oscillator circuit and its waveforms.

A third type of nonsinusoidal oscillator is the **multivibrator**. This circuit operates using an RC network for its timing. It is used to produce square and rectangular waves which can be used to turn other circuits on or off by use of the waveshape's amplitude at either its leading or trailing edges. In this chapter, we will discuss three types of multivibrators and a closely related circuit called a **Schmitt trigger**.

The last pair of circuits discussed are those that are often classified as **time base generators**. *Sawtooth* and *trapezoidal* wave generators fall into this group. They are designed to provide a voltage that has a very linear rise. This rise occurs at a specific rate and in a specific amount. Sawtooth and trapezoidal waves are used to deflect video patterns across a cathode ray tube's screen.

Pulsed Oscillators

There are two primary classifications of pulsed oscillators. These classifications identify the circuit by the type of feedback it uses; *regenerative* or *degenerative*. Both types are discussed in this section.

A pulsed oscillator operates on the same resonating (ringing) principle as did the sine-wave oscillators in Chapter 13. Before going on, we will review the ringing circuit. Simply stated, the **ringing circuit** is:

> An LC tank circuit that is excited
> into oscillation and is then
> allowed to oscillate (ring), without
> feedback, until oscillations cease.

A parallel resonant circuit is shown in Figure 14-1(a) and (b). Note that the circuit is connected to a DC source through Sw_1 and a current limiting resistor (R_1). When Sw_1 is closed, current flows from ground, through R_{Int}, L_1, Sw_1, R_1, to $+V_{CC}$. The DC resistance of L_1 is the internal resistance (R_{Int}) of the tank circuit; therefore, R_{Int} has very small ohmic value. This means that when current flows through L_1, the voltage dropped across L_1 is very small. Because of the current flowing in L_1, however, a magnetic field expands around L_1. As long as current continues to flow in L_1, the magnetic field will exist; therefore, as long as current flows, there will be no flywheel action. At this time C_1 is charged to the parallel voltage of $V_{LI} + V_{Rint}$. This condition is represented by the DC output waveshape illustrated in Figure 14-1(a).

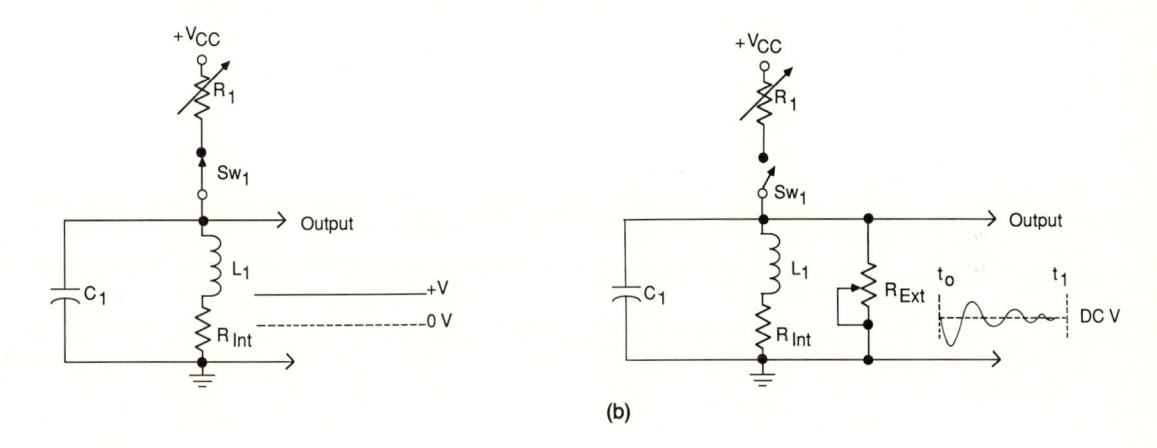

Fig. 14-1 Basic Resonant (Ringing) Circuit

Now, observe Figure 14-1(b). Assume that at t_0, Sw_1 is opened. As soon as current tries to stop, L_1 acts to oppose the change in current. To do this, L_1's magnetic field will collapse returning its energy to the circuit. The action of the magnetic field results in a polarity reversal across L_1. Now current will flow to and charge C_1. When L_1 is completely discharged, the capacitor has maximum charge. At that point, the capacitor begins to supply current for the tank circuit and flywheel action begins. The circuit oscillates at a frequency that matches the resonant frequency of L_1 and C_1. Remember, the formula for calculating resonant frequency is:

$$f_r = \frac{1}{2\pi\sqrt{LC}}$$

Notice that the output waveshape for the circuit shown in Figure 14-1(b) is a damped wave in which the amplitude of each alternation is smaller than the last. The switch is open during the period indicated by t_0 to t_1. During this period, the tank is flywheeling and the tank's resistance (R_{Int}) dissipates a part of the tank's energy during each alternation of oscillation. In this condition, there is no provision for replacing the lost power and after a few cycles of flywheel action, the wave is damped out. The internal (series) resistance (R_{Int}) of the tank has the following relationship to the Q of the tank:

$$Q = \frac{X_L}{R_{Int}}$$

If a variable resistance is connected across the tank as illustrated by R_{Ext}, the rate of damping can be controlled. This is true because now circuit Q and damping have the following relationship:

$$Q = \frac{R_{Ext}}{X_L}$$

From this formula, you can see that the larger R_{Ext} becomes, the greater circuit Q becomes—and damping occurs at a slower rate.

The polarity of the first alternation of the pulsed oscillator output depends on the direction that the initial current flows through L_1. The amplitude of the first alternation depends on the amount of current that flows when Sw_1 is opened and the X_L of L_1 at resonant frequency. Remember, Ohm's law states:

$$E_L = I_t \times X_L$$

If R_1 is small, more current will flow through L_1 causing the output amplitude to be larger. You should be able to see that by adjusting R_1 we can select an amplitude for the output signal that meets our needs.

Figure 14-2 contains the circuit schematic for a pulsed oscillator that contains a ringing circuit in its emitter circuit. The only difference between this circuit and the one shown in Figure 14-1 is that the

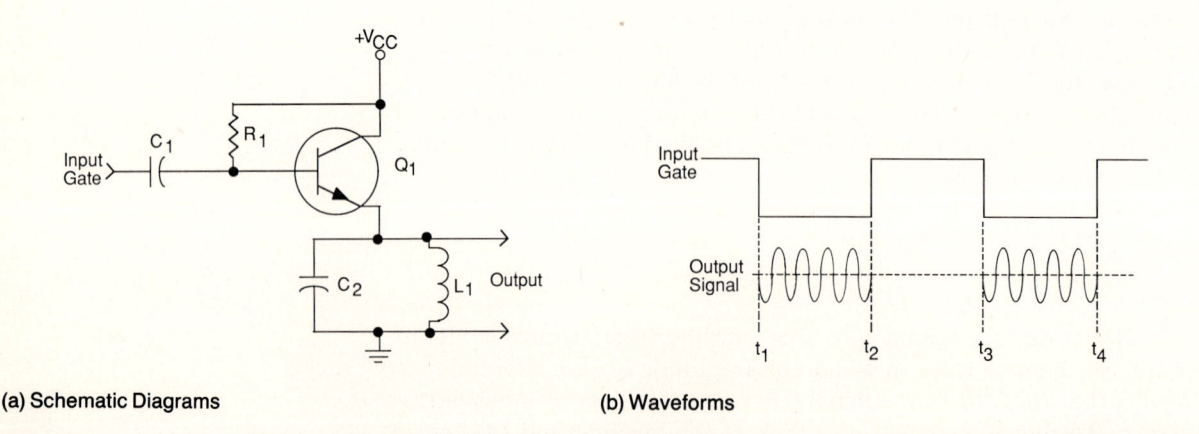

(a) Schematic Diagrams

(b) Waveforms

Fig. 14-2 Emitter - Loaded Pulsed Oscillator

transistor has replaced the switch. By biasing the transistor *on* and *off* at desired intervals, we can control the time that oscillations start and stop and the length of time that the circuit oscillates. Additionally, the amount of forward bias applied to Q_1 determines the amplitude of the first alternation of the output. Remember, increased forward bias on Q_1 results in more I_E flowing through L_1.

Note that the **input gate** is a square wave whose leading edge is negative going. This signal is used to turn Q_1 on and off. In its quiescent state Q_1 is conducting. When the negative (leading) edge of the input signal arrives at the base of Q_1, reverse bias quickly increases to the level required for cutting Q_1 off. At that time, the tank begins to oscillate and will continue to oscillate for the time that Q_1 is cut off. The arrival of the positive (trailing) edge of the input gate allows Q_1 to again become forward biased and to start conducting. Using square or rectangular waves as the input gate allows us to control the operation of the circuit by controlling the time that the negative edge of the input gate is applied. Duration of the oscillations is controlled by the time allowed for the negative going pulse.

The waveforms shown in Figure 14-2(b) show the relationship that exists between the *input gate*, which serves as the input signal, and the *output*. If we assume that the Q of the tank is large enough to prevent damping, we can discuss the circuit. When the input gate (t_1) goes negative, Q_1 enters cutoff. With Q_1 cut off, the tank is placed in a condition where it begins to flywheel. As the tank begins to flywheel, a sine wave output is present at the emitter of Q_1. Notice that the input gate remains negative from times t_1 to t_2. During this entire period (t_1 to t_2) the tank is oscillating and an output is present at Q_1 emitter. Without damping, all alternations have the same amplitude. At t_2 the signal goes positive, biasing Q_1 on and causing I_E to flow through L_1. This stops the tank's oscillation. As long as the input gate remains positive, the tank will not oscillate and the output at Q_1 emitter is a positive DC. From t_3 to t_4 the input gate is again negative, Q_1 is cut off, the tank oscillates, and there is a sine wave output at the emitter of Q_1. When the input gate goes positive again, the transistor is biased on, the tank will not oscillate, and there is no output sine wave.

If the input gate continues to arrive at the base of Q_1, an output sine wave will be present at all times when the input gate is negative. No output is present during the period when the input gate is positive which causes Q_1 to conduct recharging the tank circuit. By adjusting the two periods that are involved, we can vary the amount of time that the output signal is present. The length of time that Q_1 conducts can be varied by the amount of time that the input signal remains positive. During the time that Q_1 conducts, no output sine wave is present.

If we should replace Q_1, with a PNP transistor and $+V_{CC}$ with $-V_{CC}$, the opposite is true. A sine wave is present when the input gate is positive and the sine wave is not present when the input gate is negative.

It is quite simple to calculate the number of cycles of the tank's oscillation present in the output. To do this we must first determine the time that the tank is allowed to oscillate and the number of cycles that can occur during this time.

We will assume that the tank has a resonant frequency of 1 MHz and that the duration of the negative alternation of the input gate is 500 μs. The number of cycles can be determined as follows:

First calculate the time for one cycle of output.

$$t = \frac{1}{f} = \frac{1}{1\,\text{MHz}} = 1\,\mu s$$

If 1 cycle takes 1 μs, then how many cycles occur in 500 μs?

$$\text{No. Cycles} = \frac{\text{total time}}{\text{time 1 cycle}} = \frac{500\,\mu s}{1\,\mu s} = 500$$

Another version of the pulsed oscillator is shown in Figure 14-3. Compare this circuit with Figure 14-2. Notice that this circuit has its tank in the collector circuit. R_1 provides forward bias for Q_1, while C_2 and L_1 form the frequency-determining device. Capacitor C_1 is a coupling capacitor through which the input gate is applied.

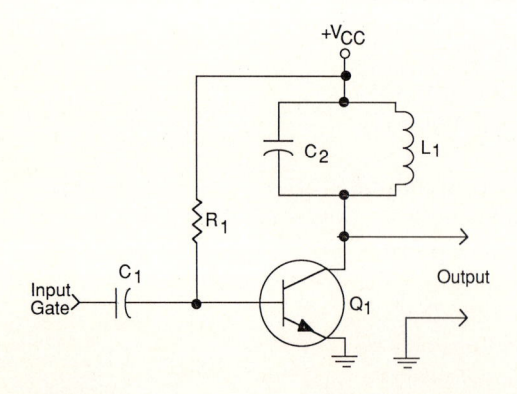

(a) Circuit Diagram

(b) Waveshapes

Fig. 14-3 Collector - Loaded Pulsed Oscillator

Nonsinusoidal Wave Generators

The main difference between this tank circuit and the one in the emitter is the polarity of the first alternation of the oscillation. In this configuration, current flows up through Q_1 and L_1. This causes the bottom of L_1 to be negative with respect to the top of L_1. When Q_1 is cut off by the negative alternation of the input gate, the coil reverses polarity placing the negative at its top, and it begins to charge C_2. While C_2 is charging, L_1 is acting as the power source for the tank. This assures that each time Q_1 (NPN) cuts off, the first alternation of the output will be positive when an NPN transistor is used.

Should the polarity of V_{CC} be reversed and the NPN transistor replaced by a PNP transistor, the first alternation of the output will be negative. Other than this, the PNP circuit operates identically to the NPN operation explained above.

You will observe that both circuits (Figures 14-2 and 14-3) have signals that do not appear to be damped out. Actually, the output of a pulsed oscillator that *does not* have regenerative feedback is a damped wave. The duration of this damped wave is controlled by the duration (time) of the negative alternation of the input gate. Both of the output waveshapes discussed above will have damping as they do not have regenerative feedback.

The second type of pulsed oscillator (Figure 14-4) is the one that has *regenerative* feedback. The output from this pulsed oscillator *will not* have a damped output wave. This circuit diagram is for a **Butler** oscillator that operates Class A. This circuit has been modified so it can be used to produce a pulsed output. Q_2 and Q_3 are the transistors that operate the Butler oscillator. Crystal CR_1 connects the two emitters

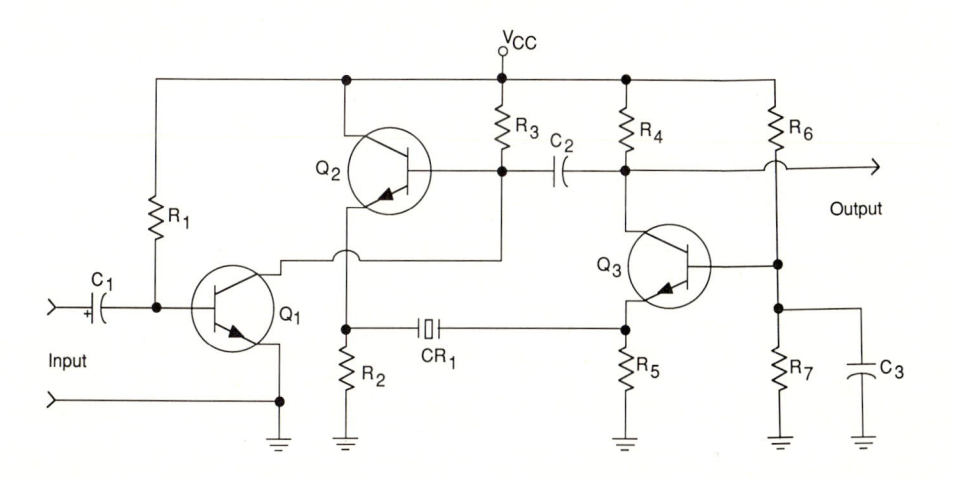

Fig. 14-4 Pulsed Oscillator

together, determines the output frequency, and provides the path for regenerative feedback. Transistor Q_1, R_1, and C_1 have been added to provide a point to which the input gate can be applied. R_1 provides bias for Q_1 and R_3 provides bias for Q_2. R_3 also serves as the collector-load resistor for Q_1. This means that when Q_1 conducts its I_C must flow through R_3. When current flows through R_3, it causes the voltage drop on R_3 to increase, decreasing V_C of Q_1. Because V_C of Q_1 and V_B of Q_2 are the same voltage, we can see that the collector voltage of Q_1 is also bias voltage for Q_2. When current flows in Q_1, the voltage at the base of Q_2 is so small that Q_2 is cut off. When Q_2 is cut off, the regenerative feedback loop for the oscillator is broken and the oscillator will not operate. As long as Q_1 continues to conduct, the oscillator is cut off and there is no AC sine wave output. Reverse biasing of Q_2 places the oscillator in the *off* condition.

When the negative alternation of the input is applied to the base of Q_1, it causes Q_1 to enter cutoff. When Q_1 is cut off, the bias applied to the base of Q_2 is high enough so that Q_2 begins to conduct. This completes the regenerative feedback loop, allowing the oscillator to begin oscillating. A sine wave is present at the output at this time. The oscillator is now operating Class A. Because we are not depending on the flywheel action of a tank circuit, the output amplitude is constant as long as the circuit continues to oscillate.

One advantage of the pulsed oscillator with regenerative feedback is the fact that the output signal is *not damped*. Each alternation of the output has the same peak voltage. When several cycles of the output signal are required, the absence of damping becomes very valuable.

Self-Check

Answer each item by inserting the word or words required to correctly complete each statement.

1. The LC pulsed oscillator uses _____ feedback.

    a. regenerative
    b. degenerative
    c. both

2. In a damped wave, the peak voltage of each alternation _____.

    a. increases
    b. decreases
    c. remains the same

3. A Butler oscillator produces a/an _____ output waveshape.

    a. damped
    b. undamped

Nonsinusoidal Wave Generators

4. In the Butler pulsed oscillator (Figure 14-4), Q_1 is used to _____.
5. Refer to Figure 14-3. In this circuit, Q_1 acts like a _____.

   a. diode
   b. amplifier
   c. switch
   d. all of these

Blocking Oscillators

The **blocking** oscillator can be used to divide frequency, serve as a pulse counter, switch other circuits on or off, and for a number of other purposes. Before we go into the blocking oscillator, however, we must discuss several considerations that enter into their operation.

First, the **timing pulses** of electronic circuits have strict requirements.

A timing pulse is used to trigger a circuit on or off at a specific time.

The time involved may be as small as a few hundredths of a microsecond or as long as several thousand microseconds.

Figure 14-5 contains two timing pulses that are separated by microseconds. The basic requirements for these pulses are:

1. fast rise time
2. a flat top
3. fast fall time
4. specific and controllable frequency

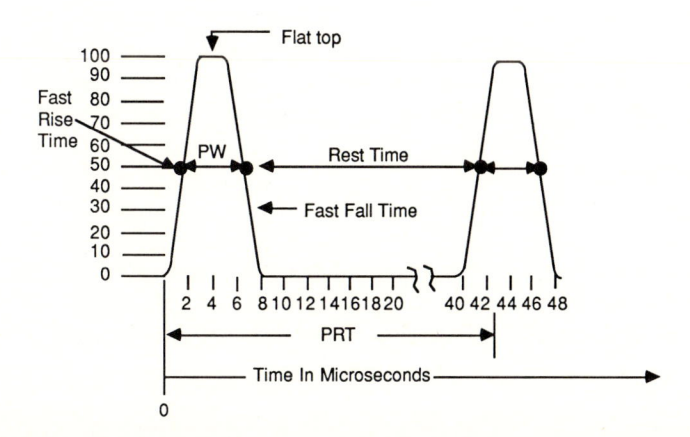

Fig. 14-5 Timing Pulses

Rise-time: The leading edge of each pulse must be as steep as possible. That is, the rise time should be short and the leading edge of the pulse as nearly vertical as possible.

Flat top: The top of each pulse should be as nearly flat as possible. This is especially true of pulses that have long durations.

Fall time: The trailing edge of each pulse should also be as steep as possible. The time required to drop from peak to minimum must be short and the trailing edge as nearly vertical as possible.

Pulse width: (PW) may be thought of as the length, in time, of a pulse that is measured at the point of $0.500 \times E_{peak}$. The time used to measure pulse width is usually microseconds.

Pulse-recurrence time: (PRT) is the time, usually stated in microseconds, between the start of one pulse and the start of the next pulse.

Pulse-repetition frequency: (PRF) states the number of pulses that are present during one second. As with other waves, time and frequency are reciprocals, therefore:

$$PRF = \frac{1}{PRT}$$

$$PRT = \frac{1}{PRF}$$

PW, PRT, and PRF, in a free-running blocking oscillator, are all controlled by the size of the resistors and capacitors used in the oscillator's circuit plus the operational characteristics of the transformer. The primary of the transformer determines the shape and duration of the output pulse. The transformer is very important to blocking oscillator operation. For this reason, we will briefly review transformer action and RL circuit action.

In Figure 14-6(a), you see a transformer that has resistance in both its primary and secondary circuits. If Sw_1 is closed, current flows through R_1 and L_1. As current increases, it causes a magnetic field to expand through L_1, which induces a current and voltage into L_2. Any current that flows in L_2 must pass through R_2. The amount of current and voltage that is induced into L_2 depends upon the turns ratio that exists between L_1 and L_2. Resistor (R_2) acts as the secondary load impedance and, through reflected impedance, will affect the primary impedance. When R_2 is decreased, current flow through L_2 is increased and the current flowing in L_1 also increases. If we decrease R_1, both currents will increase. Thus, any change that affects either side of the transformer is reflected across the transformer where it affects the other side in the same way.

Because of the reflected impedance and the fact that T_1 has inductance, we can simplify this circuit as shown in Figure 14-6(b). Transformer T_1 (L_1 and L_2) is now shown as a single coil and R_1 and R_2 are

(a) Resistance Transformer

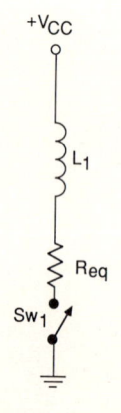

(b) Equivalent Circuit

Fig. 14-6 Series RL Circuits

shown as a single R_{eq} in series with L_1. This circuit is a series RL circuit and can be discussed in those terms.

Observe the series RL circuit of Figure 14-6(b). If Sw_1 is closed, current attempts to flow; however, L_1 opposes the change in current and attempts to keep current at zero. At this point, the coil acts like an open and V_{L1} equals V_{CC}. As current begins to flow, V_{L1} begins to decrease and V_{Req} begins to increase. As I_t continues to increase, V_{L1} continues to decrease while V_{Req} continues to increase. All changes occur at the exponential rate of the Universal Time Constant Chart.

Figure 14-7(a) depicts the curves that govern the charging of L_1 in the circuit. Remember, one time constant of an RL circuit equals:

$$TC = \frac{L}{R}$$

In a period of time that equals five time constants, I_t reaches maximum, V_{L1} decreases to zero volts, and $V_{Req} = V_{CC}$.

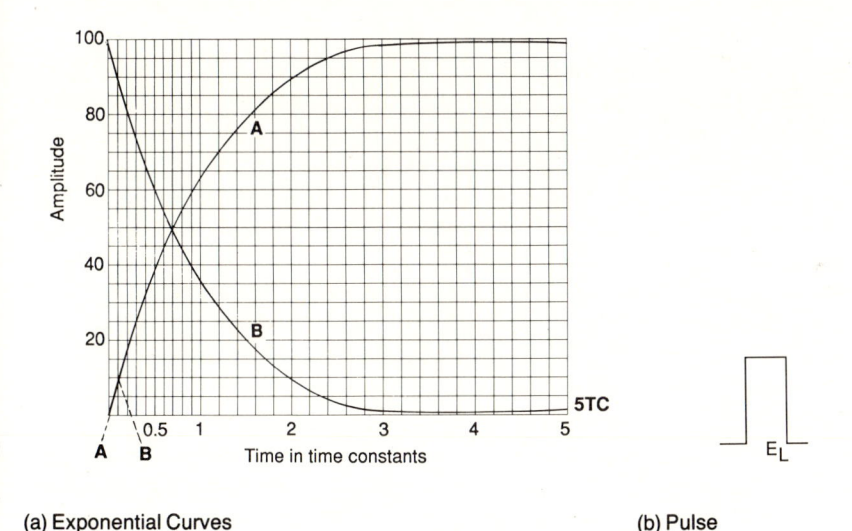

(a) Exponential Curves

(b) Pulse

Fig. 14-7 Voltage Across a Coil

Check Figure 14-6(b) again. If we close Sw_1 in this circuit, the current increases at the rate of curve A of Figure 14-7(a). The time for current to reach maximum depends on the size of R_{eq} and L_1. The first 10% of one time constant (points A to B) of this curve is quite linear. This part of the charge curve (points A to B) provides the largest current change for a given period of time. Also, the smaller the time of charge, the more linear the current change. In the blocking oscillator, a constant current rise in coil L_1 is a key factor.

A basic principle of inductance is that if the rise of current through the coil is linear (that is, the rate of current increase is constant with respect to time), the voltage induced into L_2 of T_1 is constant. This is true of both the primary and secondary of a transformer. Figure 14-7(b) is a waveshape that depicts the voltage across the coil when the current through it increases at a constant linear rate. Notice the similarity between this pulse and the timing pulses shown in Figure 14-5.

Now we can begin our discussion of the blocking oscillator circuit. A **blocking** oscillator is defined as being an:

> **Oscillator which uses inductive regenerative feedback, with output duration and frequency of operation determined by the characteristics of a transformer and the transformer's relationship to the rest of the circuit.**

Figure 14-8 contains the schematic diagram for a simple blocking oscillator. Refer to this circuit diagram for the discussion that follows. R_1 provides forward bias for Q_1, biasing it into conduction. Current flowing through Q_1 and L_1 causes voltage to be induced into L_2. The phasing dots that appear on the transformer symbol tell us that there is a 180° phase shift across T_1. As current flows through L_1, the bottom of the coil is negative and the top is positive. This means that on L_2, the bottom is going positive and the top is going negative. The positive voltage located at the bottom of L_2 is coupled to the base of Q_1 through C_1. This regenerative feedback causes Q_1 to conduct even harder. More voltage is induced into L_2, causing more regenerative feedback to be applied to Q_1. In this way, a very rapid voltage change is applied to the base of Q_1, and Q_1 quickly becomes saturated. Once the base of Q_1 is saturated, it looses control of I_C. At this point, the transistor is conducting so hard that it acts much like a small resistor and has a small voltage (V_C) at its collector. The transistor now acts like a small series resistor connected to a large coil (L_1), forming a series RL circuit.

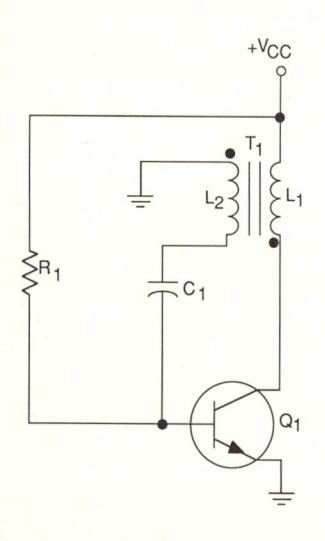

Fig. 14-8 Blocking Oscillator

The operation of the circuit, to this point, has generated the leading edge of the output pulse. Figure 14-9 shows the waveforms that would be viewed at the base and collector of Q_1.

Once the base of Q_1 becomes saturated, the current rise in L_1 is determined by the time constant of L_1 and the total series resistance of the branch. During the time period of t_0 to t_1, current rise in L_1 is linear. This means that the voltage drop on L_2 remains constant as long as the current in L_1 rises at a linear rate. During this period, C_1 has charged to the voltage of L_2.

At time t_1, coil L_1 becomes saturated. With L_1 saturated, the current rise is no longer linear and there is no change in the magnetic field causing it to stop expanding. When the magnetic lines stop expanding, the voltage being induced into L_2 ends. As soon as the voltage across L_2 decreases, C_1 begins to discharge applying a negative voltage to the base of Q_1. This negative voltage immediately drives Q_1 into cutoff. With Q_1 cut off, I_C stops and V_{L1} decreases to 0 V.

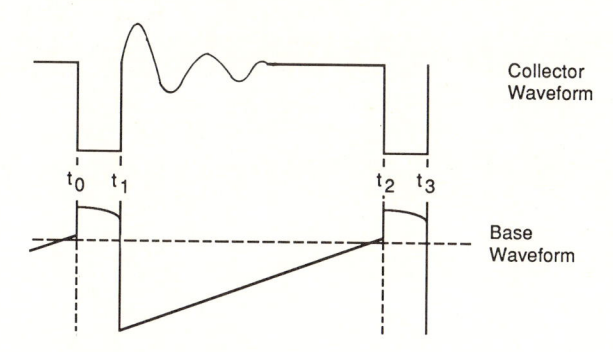

Fig. 14-9 Blocking Oscillator Waveforms

The length of time between t_0 and t_1 determines the pulse width for the output pulse. Pulse width depends, primarily, on the inductance of T_1 and the point at which T_1 saturates. For blocking oscillator purposes, a transformer (T_1) is chosen that will saturate when current reaches approximately 10% of total current. This insures that the current rise in L_1 will be nearly linear. The larger the inductance of L_1, the wider the pulse width. This is because the longer it takes to charge L_1, the longer the linear current rise will exist.

Between times t_1 and t_2 (Figure 14-9), Q_1 of Figure 14-10 is held at cutoff by the discharging of C_1 through R_1. For the time period t_1 to t_2, the oscillator is "blocked" from oscillating. At some point during the discharge of C_1, the voltage drop on R_1 becomes so small that Q_1 can again begin to conduct. At this point (t_2), base current regains control and Q_1 quickly saturates, regenerative feedback is coupled, and the

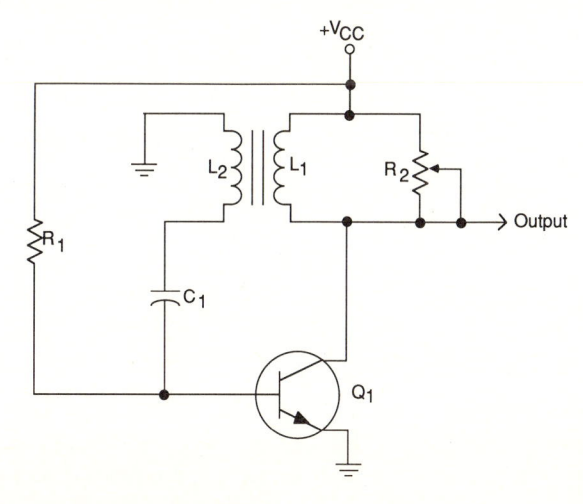

Fig. 14-10 Circuit Damping Compensation

output pulse generation begins all over. From this we can see that the time constant formed by R_1 and C_1 determines the time between pulses (**rest time**). The time for one cycle of output consists of the sum of pulse width plus rest time. In the example shown in Figure 14-5, the PRT equals 40 µs and PRF equals 25000 pulses-per-second (PPS).

It is possible for the collector waveform to have an inductive overshoot at the end of each pulse. This is shown in Figure 14-11.

When Q_1 cuts off (Figure 14-10), I_C in Q_1 stops. This means that I_C is no longer flowing in L_1. L_1, however, has energy stored in its magnetic field which must be returned to the circuit. Because coils oppose a change in current flow, L_1 quickly discharges to V_{CC}. The fast change that occurs creates a positive voltage at the collector of Q_1. These oscillations are not desirable. Quite often, it is necessary to connect a "damping" resistor—like R_2—as shown in Figure 14-10. Variation of R_2 allows us to observe the wave on an oscilloscope and to adjust for critical damping, as shown by the waveforms of Figure 14-11.

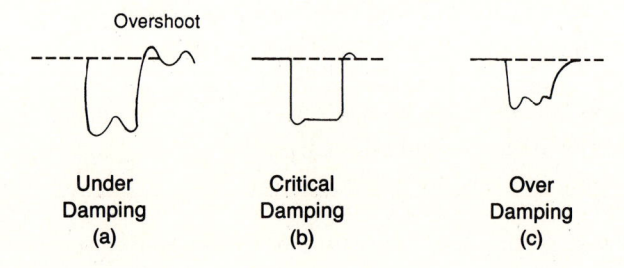

Fig. 14-11 Waveform Damping

Critical damping gives the most rapid transient response with minimum overshoot. This is accomplished by adjusting R_2 for a waveshape like that shown in Figure 14-11(b). The value of R_2 is determined by the Q of the transformer. With critical damping, the oscillations are damped out and overshoot is reduced.

Under damping is shown in Figure 14-11(a). Notice that with under damping, rapid transient response with considerable overshoot is present. This is caused by R_2 being open or too large.

Over damping is caused by the resistance of R_2 being too small. In this case, the transient response is too slow and all overshoot is eliminated. This output wave is not desirable because of the fact that pulse amplitude has probably been reduced as is shown in Figure 14-11(c).

Nonsinusoidal Wave Generators

By over adjustment of R_2, it is possible to reduce damping to the point where the circuit stops oscillating. Adjustment should be set to the point where *critical damping* is observed.

The oscillator that is contained in Figure 14-12(a) is a **free-running synchronized blocking** oscillator circuit. Notice that the oscillator is triggered by a string of input triggers that control its frequency. The control frequency of a blocking oscillator must always be "higher" than the free-run frequency of the oscillator. In the case shown here, the oscillator is triggered by triggers that occur every 200 µs. The discharge rate of C_2 is such that the first trigger that occurs after Q_1 is cut off does not have enough amplitude to trigger Q_1 back on. The arrival of the second input trigger biases Q_1 into conduction at a time that is earlier than would happen if the circuit was free-running.

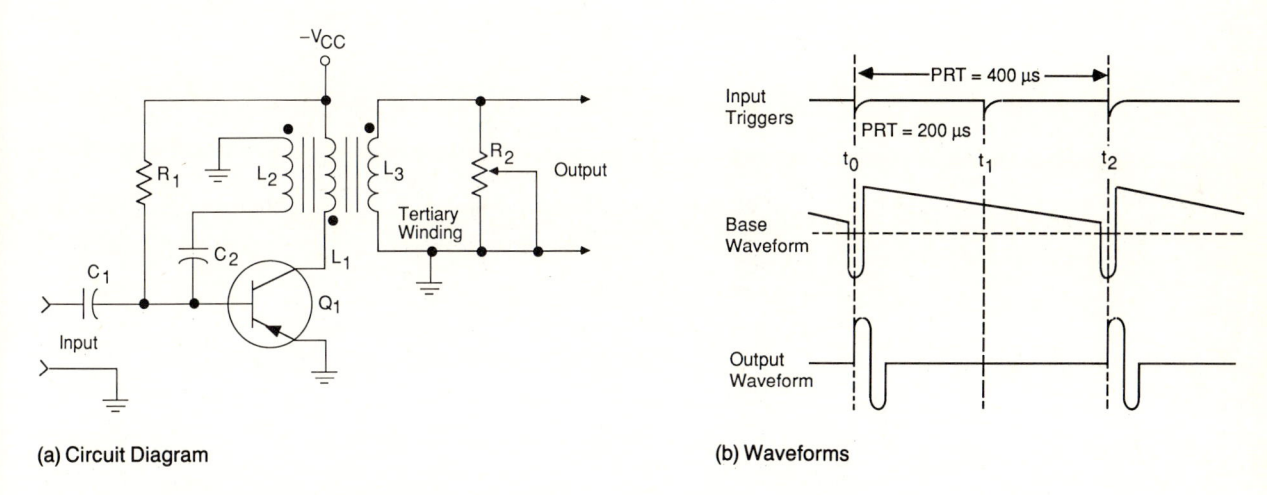

(a) Circuit Diagram

(b) Waveforms

Fig. 14-12 Synchronized Blocking Oscillator

An example of this is an oscillator that has a free-running frequency of 2 kHz, the oscillator produces one output pulse each 500 µs. By applying a string of input triggers that are generated at 2.5 kHz to the base of Q_1, we can force the circuit to supply an output pulse every 400 µs. In this manner, we have increased the operating frequency of the circuit from 2 kHz to 2.5 kHz. When we apply a 5 kHz signal (pulses every 200 µs) to the base of Q_1, every second pulse is too small to trigger Q_1. This provides *frequency division* because the input triggers are applied at the rate of 5 kHz, but the output occurs at the frequency of 2.5 kHz. Should the input frequency be 10 kHz, this circuit could be used as a 4 to 1 divider and continue to produce a 2.5 kHz output.

Refer to Figure 14-12(b). Only those pulses that occur at t_0 and t_2 have enough amplitude to bias Q_1 on. Once the pulse has been generated and Q_1 enters cutoff, Q_1 is maintained in cutoff by C_2's discharge.

Only when the voltage at the bottom of C_1 has decreased low enough can an input pulse trigger the transistor. This *blocking oscillator* is synchronized by a series of input triggers and it is no longer free-running.

Another modification is made to the circuit shown in Figure 14-12(a). This is the addition of a *tertiary* (third) winding to transformer T_1. The third winding is used to couple the output pulse. By use of the tertiary winding, we avoid some of the loading problems that arise from connecting the load directly to the oscillator's collector.

Self-Check

Answer each item by inserting the word or words required to correctly complete each statement.

6. The blocking oscillator uses _____ feedback in the creation of its _____ edge.

   a. degenerative, trailing
   b. degenerative, leading
   c. regenerative, trailing
   d. regenerative, leading

7. Pulse width is determined by _____.

   a. the amplitude of the voltage coupled across T_1.
   b. the time required to saturate Q_1.
   c. the amount of time Q_1 is cut off
   d. all of the above

8. Rest time of the output waveshape coincides with _____.

   a. the discharge time of C_2
   b. the cut off time of Q_1
   c. the reverse bias applied to the base of Q_1
   d. all of the above

9. Explain the purpose of the tertiary output shown in Figure 14-12.
10. In a blocking oscillator, PRF is _____ the free-run frequency of the circuit.

    a. higher than
    b. lower than
    c. the same as

Solid-State Multivibrators

Many electronic circuits are *not* biased to the **on** condition at all times. In computers, for example, circuits must be turned **on** and **off** at specific times and for exact periods of time. The time intervals can vary from thousandths of a microsecond to several thousand microseconds. The best method for providing these exact operating conditions is by application of a square or rectangular wave which performs the switching action.

In this section, we will discuss methods that can be used to generate square and rectangular waves using multivibrators. There are several terms and characteristics that apply to multivibrators. Some of these have already been discussed but we will review them in the discussion that follows. First we will discuss some of the terms that you should already know.

Multivibrator Waveforms

A waveform that begins at a point, undergoes a pattern of changes, and returns to its original value and continually repeats this action, is called a **periodic** waveform. Each completed waveform is called a **cycle** and the time for each cycle is called the **period** of the waveform. The number of cycles of the waveform that occur during one second is called the **frequency**.

Figure 14-13 contains a waveform called a **square wave**. A square wave can be identified by the fact that it is a periodic wave having two voltage levels (negative and positive) that have equal amplitude and durations. Amplitude is the vertical measurement of height and time is the horizontal measurement between vertical lines. To measure the total time (duration or period) of one cycle, we measure the time from the start of one voltage change to the start of the next identical (positive or negative) change. In this figure, period can be measured between either t_0 and t_2, or t_1 and t_3.

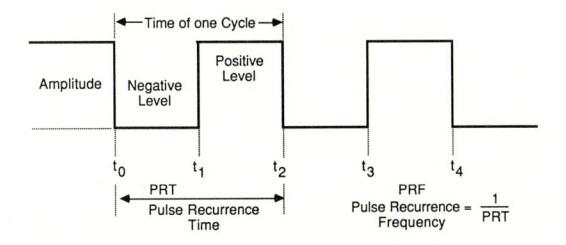

Fig. 14-13 A Typical Square Wave

One period (positive or negative) of the wave is called a **pulse**. The other period of the wave is called the **rest time**. The time for one pulse plus the time for one rest time equals the period and is called the **pulse-recurrence time (PRT). Pulse-repetition frequency (PRF)** is the term used to refer to the number of waveforms that occur during one second. In Figure 14-13, if each *pulse* of the square wave has a duration of 200 μs, the PRT is 400 μs and the PRF is 2.5 kHz. Remember:

$$\text{PRF} = \frac{1}{\text{PRT}}$$

$$\text{PRT} = \frac{1}{\text{PRF}}$$

In Figure 14-14 you see a typical rectangular wave. Notice that this wave also has one positive pulse and one negative pulse. However, you should also note that a rectangular wave has voltage levels (pulses) that have different durations. In this figure the negative voltage level is longest. In many applications, the opposite will be true—the positive pulse will be longer. In the waveform shown here, if the positive pulse has a duration of 100 μs and the negative pulse has a duration of 300 μs, the PRT is 400 μs, and PRF equals 2.5 kHz.

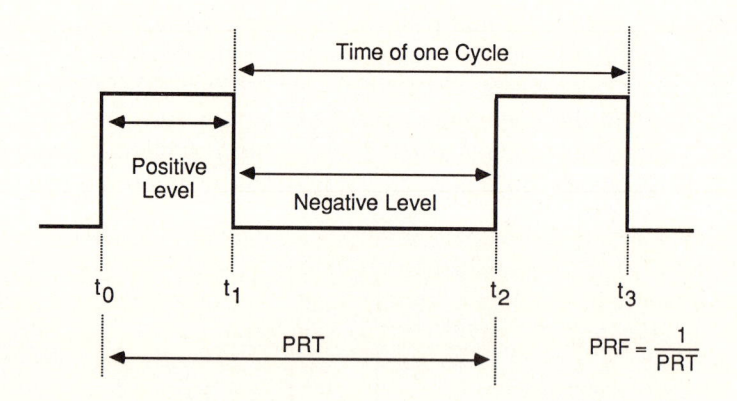

Fig. 14-14 A Typical Rectangular Wave

Another important part of the square and rectangular waves is the **transient interval** shown in Figure 14-15. It is impossible for voltage or current to make instantaneous changes in amplitude; therefore, some time must elapse during the time that the signal is changing amplitude. The transient interval is the time required for voltage or current to change from minimum (10% E_{peak}) to maximum (90% E_{peak}) or maximum (90% E_{peak}) to minimum (10% E_{peak}). Transient intervals occur on the **leading edge** (the change that occurs earliest in time) of the pulse, and on the **trailing edge** (the change that occurs latest in time) of the pulse.

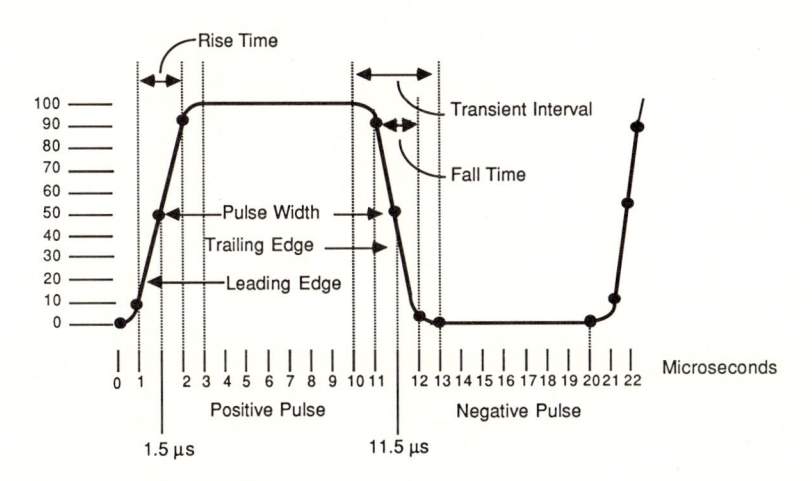

Fig. 14-15 Waveform Characteristics

Other names used here that coincide with transient interval are **rise time** and **fall time**. Rise time is defined as the time required for a waveform to build from the 10% E_{peak} level to the 90% E_{peak} level. Fall time is defined as the amount of time required for the pulse to drop from the 90% E_{peak} level to the 10% E_{peak} level. These two times are not necessarily equal. In fact, there can be drastic variations between rise time and fall time. In Figure 14-15 we have shown rise and fall times that are approximately 1 µs with a transient interval of 3 µs. Total time for generation of the rise time, pulse, and fall time equals 13 µs.

Another term is **pulse width (PW)**. For electronics applications, pulse width indicates the duration of the pulse in microseconds measured at 50% E_{peak} amplitude. In Figure 14-15, pulse width is 10 µs.

Transient interval is a very important consideration in waveform analysis. In Figure 14-16, the first pulse has a PW of approximately 4 µs. Notice that rise and fall time each is 2 µs of the 8 µs total time that elapses between the start of the pulse and the end of the fall time. This is where transient interval must be considered carefully. In some circuitry, pulse width is a fraction of one microsecond. In those cases, transient time must be very short. A pulse of the type shown in Figure 14-16 is called a **trigger**. A trigger is a very narrow pulse that is used to either turn a circuit on or off.

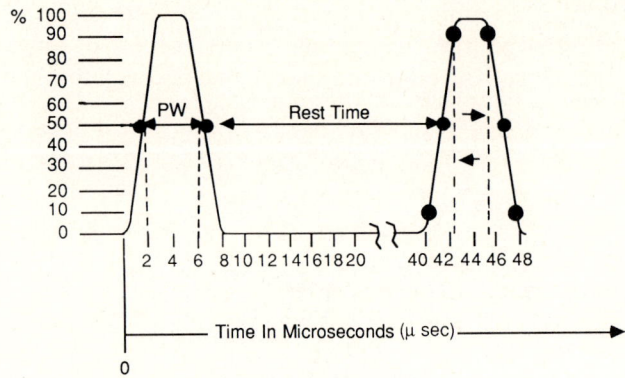

Fig. 14-16 Rectangular Wave with Transient Intervals

Multivibrator Types

Three types of multivibrators will be discussed. These are:

a. Astable
b. Monostable
c. Bistable.

From their names you can probably form some ideas about how each type operates. Each type can be used to generate timing pulses.

The Astable Multivibrator

Figure 14-17 contains the schematic diagram for an **astable multivibrator** and the waveshapes that would be viewed at the base and collector of each transistor. The name *astable* tells us that this circuit is free-running and has *no* stable state. Q_1 and Q_2 alternately switch from saturation to cutoff. When Q_1 is saturated, Q_2 is cut off. When Q_1 cuts

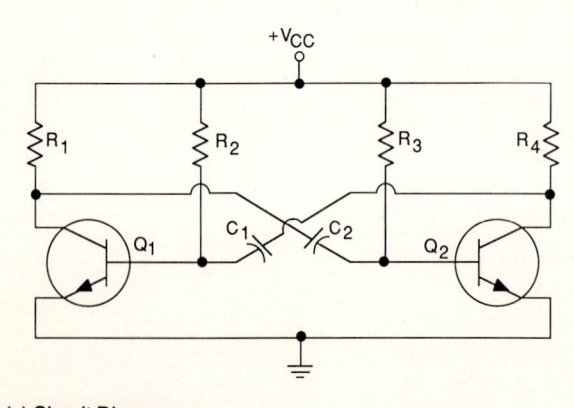

(a) Circuit Diagram

(b) Waveshapes

Fig. 14-17 Astable Multivibrator

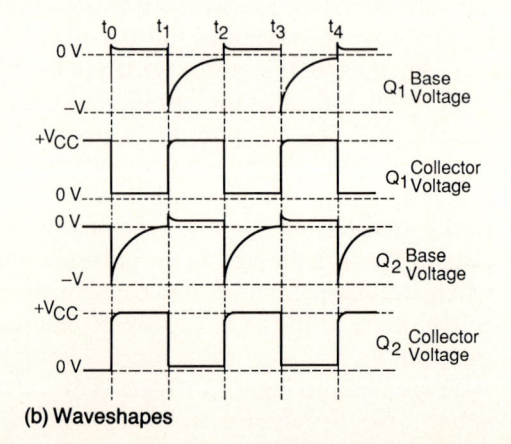

off, Q_2 becomes saturated. Figure 14-17 can be identified as an astable multivibrator by the fact that an RC circuit is connected to the base of each transistor.

For the astable multivibrator of Figure 14-17, notice that a negative signal (Q_1 base voltage—Figure 14-17(b)) is applied to the base of Q_1, keeping it cut off while Q_2 conducts. Q_1 has a low voltage at its collector during the period t_0 to t_1, indicating that it is saturated. This same period provides time for the discharge of C_2 through R_3 which reverse biases Q_2 keeping it cut off. This allows C_1 to charge to a relatively high voltage. At some point, the bias applied to the base of Q_2 reaches the point that Q_2 begins to conduct. As soon as I_C (Q_2) begins to flow, C_1 begins to discharge and reverse biases Q_1. Now Q_2 is saturated ($V_C \cong 0$ V) and Q_1 is cut off ($V_C = V_{CC}$). The base of Q_1 is held below cutoff by the discharge of C_1 through R_2. As the circuit continues to operate, the conduction/cutoff state of the transistors continually alternate. Notice the collector waveshapes; they are identical except that they are 180° out-of-phase.

In this multivibrator, C_2 discharging through R_3 determines how long Q_2 is cut off and Q_1 conducts. C_1 and R_2 determine how long Q_1 is cut off and Q_2 conducts. Cut off time of Q_1 plus the cut off time of Q_2 equals PRT for the circuit. The PRF is controlled by the base RC networks and their time constants. If the values of these capacitors and resistors are not exactly equal, the circuit's output is a rectangular wave. It is possible to take an output from either collector. In the case of a square wave, the only difference is polarity as the two signals are 180° out-of-phase.

Figure 14-18 contains the schematic diagram for a PNP *astable multivibrator*. Except for the PNP transistors, the use of $-V_{CC}$ and the fact that current flows toward ground, this circuit operates exactly like the one shown in Figure 14-17. This PNP circuit is used to discuss troubleshooting of an astable multivibrator. Assume that the circuit is operating with a square wave output being taken from the collector of Q_2.

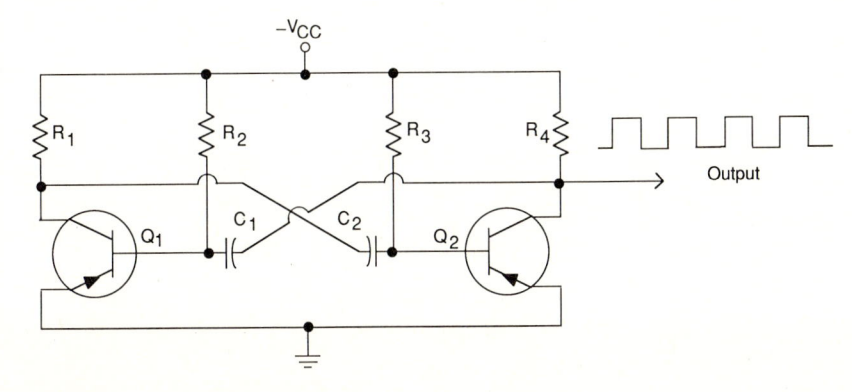

Fig. 14-18 PNP Astable Multivibrator

Symptom:

The negative alternation of the output is longer than the positive alternation.

Possible causes:

The most likely cause is that either R_3 or C_2 has increased in size.

Symptom:

No switching action at the output. V_C of Q_2 is high while V_C on Q_1 is very low. Remember, this is an indication that Q_1 is saturated.

Possible causes:

The most likely causes are R_3 open or Q_2 open.

The Monostable Multivibrator

Figure 14-19 contains the schematic diagram for a **monostable multivibrator**. The monostable multivibrator is often called a **one-shot multivibrator**. Notice that this circuit has an RC network connected to the base of Q_2 only. This assures that the monostable multivibrator always returns to one state and waits for an input trigger to turn it on.

A monostable multivibrator is used when we want a constant frequency (PRF), but desire to have control of pulse width. This circuit is a square or rectangular wave generator that has one stable condition. When no input is applied, Q_2 conducts indefinitely and Q_1 is cut off. When an input trigger is applied to the circuit, Q_2 cuts off and Q_1 conducts. The amount of time that Q_1 conducts is controlled by the RC

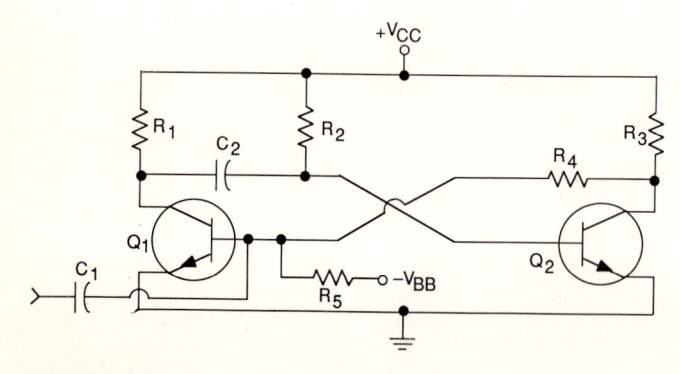

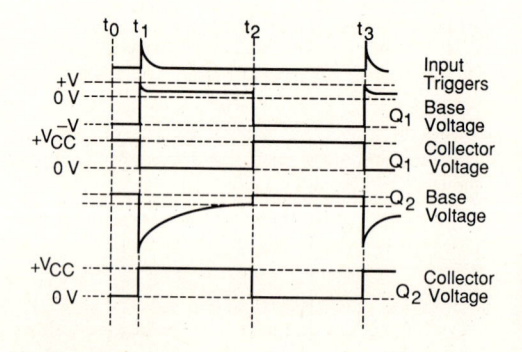

(a) Circuit Diagram (b) Waveshapes

Fig. 14-19 A Monostable Multivibrator

network containing C_2 and R_2. One complete cycle of output is generated for each positive pulse that arrives at the base of Q_1. This means that the circuit's output frequency is controlled by the frequency of the input triggers. Pulse width is determined by the time Q_1 conducts.

To confirm the statements made above, we will discuss the operation of the circuit shown in Figure 14-19. When power is applied to the circuit, Q_2 conducts and Q_1 is cut off. We can prove this by checking the forward bias arrangements for the two transistors. Forward bias for Q_2 is provided by R_2, while forward bias for Q_1 is provided by the voltage divider R_3, R_4, and R_5. Note that the voltage divider is connected between $+V_{CC}$ and $-V_{BB}$. With a $-V_{BB}$, it is possible for the base to have either negative or positive polarity with respect to ground (emitter). When Q_2 conducts it has a V_C that is near zero volts. This allows the $-V_{BB}$ to be applied to the base of Q_1 assuring that it remains cut off as long as Q_2 conducts. This assures that the circuit enters a stable condition of Q_2 conducting and Q_1 cut off as soon as power is applied.

The positive input trigger applied to the base of Q_1 must have large enough amplitude to overcome the $-V_{BB}$ voltage that is present. When this happens, Q_1 begins to conduct and takes control of the circuit. As soon as the voltage at Q_1 collector decreases, C_2 begins to discharge through R_2 cutting Q_2 off. As long as C_2 is discharging, a negative voltage is applied to the base of Q_2 keeping it cut off. However, when C_2 has discharged to a point where the fixed bias can overcome the reverse bias, Q_2 begins to conduct. This returns the $-V_{BB}$ voltage to the base of Q_1 cutting it off. At this point, the circuit has reentered its stable state. All conditions remain stable until the arrival of the next positive input trigger.

A modified version of this circuit results when a variable resistance (rheostat) is placed in series with R_2. This resistor provides control over pulse width. By changing the amount of resistance through which C_2 must discharge, the amount of time Q_2 is cut off will be changed. Since Q_1 conducts and produces the output pulse while Q_2 is cut off, changes of resistance in the RC network affects pulse width. Increasing the size of R_2 increases pulse width. Decreasing the size of R_2 decreases pulse width. Remember, output frequency is determined by the input trigger frequency, and one output pulse is generated for each input trigger. Varying the size of R_2 affects pulse width (PW) only.

Troubleshooting this type of circuit is similar to others we have discussed. Check the following:

Symptoms:

No output waveshape. Q_1 has high collector voltage, and Q_2 has low collector voltage.

Possible causes:

No input triggers
C_1 open
Q_1 open
R_4 open
R_5 open
R_1 shorted

Symptoms:

No output waveshape. Q_1 has low collector voltage, and Q_2 has high collector voltage.

Possible causes:

R_2 open
R_3 open
Q_2 open
R_5 shorted

Symptoms:

No output waveshape. Both collector voltages are low.

Possible causes:

R_1 open
C_1 shorted
Q_1 shorted

The Bistable Multivibrator

As the name implies, the *bistable multivibrator* has two stable states. When a trigger that has the right polarity and amplitude is applied to the input of this circuit, it is steered to the transistor that is cut off and triggers it into conduction. When one transistor begins conducting, the other transistor is cut off. Once cut off, the transistor remains that way until a trigger arrives that can turn it on. It is not necessary for the triggers to have a fixed frequency. In fact, bistable multivibrators often operate where they receive input triggers from several different circuits and are triggered randomly. Versions of this circuit are used extensively in digital circuitry. Now that digital circuitry is used in everything from automobiles to space shuttles, it is necessary for you to understand the operation of this circuit.

A schematic diagram for a **bistable multivibrator** is shown in Figure 14-20. This circuit is often called a **flip-flop**. In this circuit, R_1 and R_7 are collector load resistors. Q_1 has bias provided by R_3, R_6, and R_7. Bias for Q_2 is provided by R_1, R_2, and R_5. In addition, these same voltage dividers form the coupling path between the collectors and bases of the two transistors. With resistors providing the coupling path, we have direct coupling. This is possible because the circuit depends on input triggers for its operation, not RC networks. Notice that both transistors share a common emitter resistor (R_4) which provides emitter coupling. C_1 and C_2 serve as coupling capacitors that couple the input triggers to the bases of the transistors.

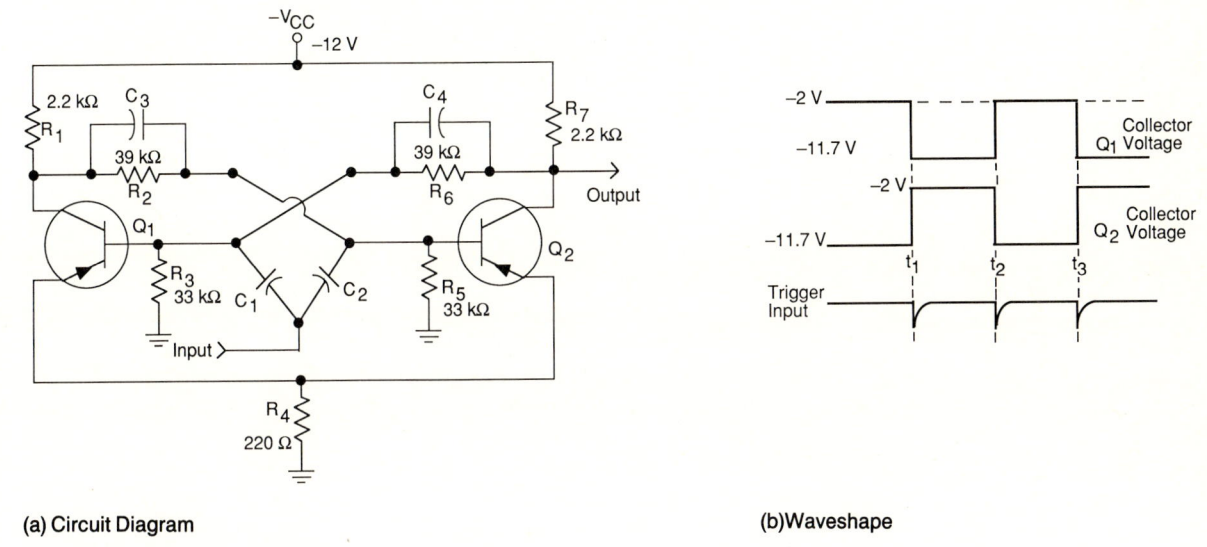

(a) Circuit Diagram   (b) Waveshape

Fig. 14-20 Bistable Multivibrator

Within this circuit, both amplifiers have identical component values. They appear to be balanced. We know, however, that it is not possible to precisely balance a circuit using "off-the-shelf" components. At the first instant after power is applied to the circuit, both transistors have forward bias and begin to conduct.

Because of the slight differences that exist between the two amplifiers, one transistor will conduct harder than the other. We will assume that Q_1 conducts harder than Q_2. Conduction of Q_1 causes Q_1's V_C to decrease (go positive) which decreases the forward bias at the base of Q_2. With Q_2 conducting less, Q_2's V_C goes more negative (nearer cutoff). This negative voltage is applied to the base of Q_1 as forward bias. Q_1 conducts harder, its V_C goes more positive and cuts Q_2 off. At this point, the circuit is in a stable state and will remain in this state until triggered into another state.

At time t_1 of Figure 14-20(b), a negative trigger is applied to the input. This trigger is coupled through C_1 and C_2 to the base of both transistors. Q_1 is saturated and the arrival of a negative pulse has no effect. Q_2, however, is cut off and a negative trigger provides forward bias triggering it into conduction. As Q_2 begins to conduct, its V_C goes positive coupling a positive voltage (reverse bias) to the base of Q_1. As Q_2 begins conduction, Q_1 is cut off. The switching action occurs very quickly and the circuit is now in its other state.

At times t_2 and t_3, negative triggers arrive at the input causing the circuit to change state. In each case, one transistor cuts off and the other is driven into saturation. Notice, though, **the output frequency is one half the input frequency**. In other words, it takes two input pulses to generate one cycle of output.

An output can be taken from the collector of either transistor of this circuit. The waveshapes that can be expected from a circuit of this type are shown in Figure 14-20(b). The voltages noted on the waveshapes are representative of those that can be expected. It is possible to operate bistable multivibrators with voltages much less than 12 V. In fact, the most common voltage is probably 5 V with the circuitry that is used today.

At the high frequencies used with present day circuitry, it is necessary to provide *high frequency compensation* for circuits of this type. The transient interval (switching time) for discrete transistors is controlled by the interelement capacitance of the emitter-base junction (C_{eb}). Switching cannot occur any faster than it takes to charge C_{eb}. Also, this capacitance causes the loss of high frequency components in the output wave. This appears as rounding of the leading and trailing edges.

The charge path for C_{eb} is through the three-resistor voltage divider that supplies bias to the transistor. These three resistors have a total resistance of 74.2 kΩ so charging C_{eb} could take considerable time. However, if we can reduce the amount of resistance in the charge path, we can cause C_{eb} to charge much more quickly and reduce the transient intervals. To compensate for these long transient intervals, we add capacitors C_3 and C_4 as shown in Figure 14-20(a). The addition of these capacitors provide a shunt path for charging C_{eb} that bypasses each of the 39 kΩ resistors. This removes both 39 kΩ resistors from the charge paths. Also, by placing C_3 in series with C_{eb} the total capacitance of the charge path is decreased which shortens the time required to charge C_{eb}.

Refer to Figure 14-20(a) for the troubleshooting of a bistable multivibrator. Assume that all triggers arriving at the input are negative in polarity and have enough amplitude to trigger the circuit.

Symptoms:

Positive trigger waveshape viewed at Q_1 collector. Negative triggers whose frequency coincides with the input trigger frequency are viewed at the collector of Q_2.

Possible causes:

R$_6$ open.

Reason:

Q$_1$ is cut off and Q$_2$ is conducting. This means that the negative triggers are amplified by Q$_1$, their phase is inverted and they are applied to the base of Q$_2$ as forward bias. Q$_2$ cuts off only as long as the trigger is present and then begins conducting again. The circuit again waits for the next input trigger to forward bias Q$_1$—which restarts the process. No output gate is generated because of the missing feedback.

Symptoms:

Voltage between Q$_1$ collector and ground measure 0 V. Positive triggers are viewed at the collector of Q$_2$.

Possible causes:

R$_1$ open.

Reason:

When R$_1$ opens, Q$_1$ has no direct path through which I$_{C1}$ can flow. The base of Q$_2$ is grounded through R$_5$. The negative triggers cause Q$_2$ to conduct but no feedback can be coupled.

The Schmitt Trigger

As square and rectangular waves pass through various points within a system, their edges and tops become rounded as a result of high frequency losses that result from input capacitors and distributed (stray) capacitance. This can affect the accuracy of the time at which circuits that use these waves as triggers are turned on or off. To correct this problem, circuits are used that reshape the wave to its original shape. One such circuit is the **Schmitt trigger.**

Operational characteristics of the Schmitt trigger circuit are very similar to those of multivibrators. For that reason we include them at this point. A schematic diagram for a Schmitt trigger circuit is shown in Figure 14-21.

In this circuit, we see that coupling is provided between the collector of Q$_1$ and the base of Q$_2$. This is similar to the connections used in multivibrators. However, unlike multivibrators, there is no coupling between the collector of Q$_2$ and the base of Q$_1$. A common-emitter resistor (R$_7$) does, however, provide coupling between the two emitters.

First, we will examine the circuit without the input signal being present. Note that the base elements of both transistors are connected

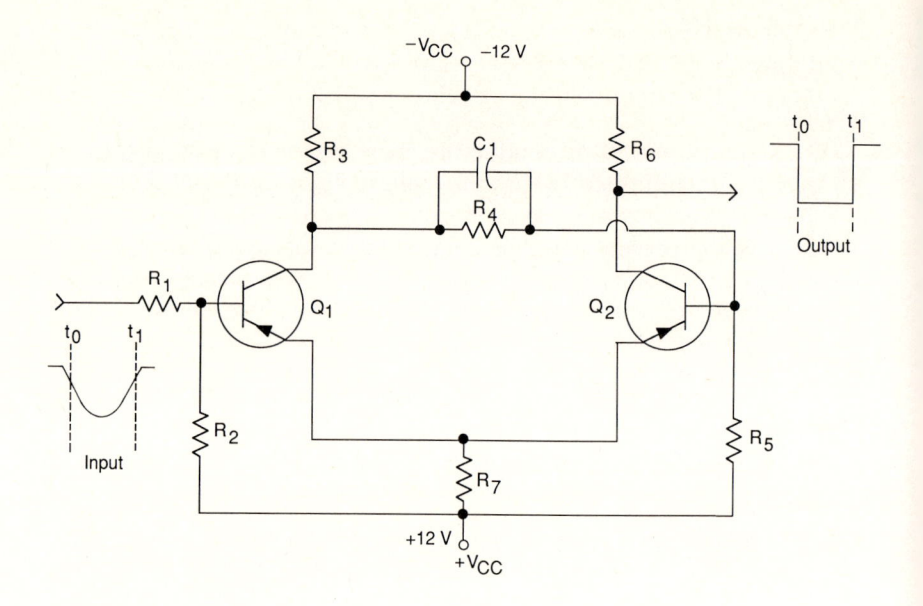

Fig. 14-21 Schmitt Trigger

to +12 V through R_2 and R_5. Both emitters are connected to +12 V through R_7. The collectors are connected to −12 V through resistors R_3 and R_6. With Q_1 base positive and a negative voltage applied to the emitter by conduction of Q_2, Q_1 is cut off. At this point, without an input signal, Q_1 has a V_C that equals −V_{CC}. Q_2, is operating at saturation and has a V_C that is very low.

When a rounded input signal arrives at the base of Q_1, it may be too small at first to trigger Q_1 into conduction. At point t_0, however, input amplitude is large enough to cause Q_1 to begin conduction. Q_1 will quickly saturate. This causes V_C of Q_1 to change from −V_{CC} to approximately zero volts very quickly. This change is coupled to the base of Q_2 where it overcomes the +V_{CC} and cuts Q_2 off. As Q_2 cuts off, its V_C changes from approximately zero volts to −V_{CC}. This is a negative going change which is coupled to the output.

At time t_1, the input signal is small enough that Q_1 will again be cut off. At this time, V_C of Q_1 changes from approximately zero volts to −V_{CC}. This change is coupled to the base of Q_2 allowing it to begin conducting. When Q_2 begins conducting, its V_C changes from −V_{CC} to approximately zero volts. The change in collector voltage of Q_2 is positive going and is coupled to the output. This completes the generation of a negative output pulse like that shown in Figure 14-21. The input and output signals shown on this figure are representative of those present in an actual circuit. The rounded input is reshaped to a square output pulse having approximately the same pulse width as the input signal.

C_1 is added to the circuit to decrease the transient interval. Its operation is identical to that discussed in bistable multivibrators.

Schmitt trigger circuits are also used to sense voltage conditions in circuits that require voltage monitoring. They can be used to signal either an increase or decrease in voltage level. In either case, the signal developed by Q_2 is used to set off an alarm that informs the operators of the voltage change.

Troubleshooting this circuit is much like amplifier troubleshooting. If you have learned those techniques, you will have no problems here.

Symptoms:

No output and V_C of Q_2 remains near zero volts.

Possible causes:

(1) R_6 open. All of V_{CC} is dropped across R_6. This prevents Q_2 from conduction; therefore, V_C of Q_2 cannot change.

(2) R_3 shorted. In this case, V_C of Q_1 would remain at $-V_{CC}$, keeping Q_2 saturated. This would hold V_C of Q_2 near zero volts and no output would be possible.

(3) Either R_4 or C_1 shorted. These two components are in parallel; therefore, a short in either one would cause the same problem. The reduction in resistance that would occur between the collector of Q_1 and the base of Q_2 would increase the forward bias on Q_2 to the point that the input signal could not cut Q_2 off. It is possible that Q_2 could be destroyed by either of these components shorting.

Symptoms:

No output and V_C of Q_2 is at or near $-V_{CC}$.

Possible causes:

(1) When R_6 is shorted, it will drop 0 V. With R_6 dropping 0 V, the voltage at Q_2 collector must equal V_{CC} and V_C cannot change; therefore, no output can be generated.

(2) R_3, R_4, or R_7 open. If one of these resistors opens, there is no current path for Q_2 forward bias current. Q_2 would be cut off and its V_C would equal $-V_{CC}$. This means that no output can be developed.

(3) Q_2 open. All voltage $(-V_{CC})$ is dropped across Q_2. This means that the output would be a constant $-V_{CC}$.

Symptoms:

Collector voltage on Q_1 remains at $-V_{CC}$ and Q_2 collector voltage remains near zero volts.

Possible causes:

(1) R_1 open. Input signal is blocked and will never reach the base of Q_1.

(2) Q_1 open. Q_2 conducts at saturation and the output remains near zero volts.

(3) R_2 shorted. Reverse biasing voltage $(+V_{CC})$ is placed on the base of Q_1. In this case, Q_1 could never be forward biased by the input signal. Q_2 would conduct continuously and the output would be near zero volts.

Symptoms:

High frequency distortion (rounded edges) is present in the output signal.

Possible causes:

C_1 open. C_1 being open increases the time constant for charging C_{eb} and, therefore, increases the transient interval.

Self-Check

Answer each item by inserting the word or words required to correctly complete each statement.

11. Multivibrators are classified into three categories. These are:

    a. _____
    b. _____
    c. _____

12. An astable multivibrator has _____ stable states.

    a. 1   c. 3
    b. 2   d. 0

13. The astable multivibrator can be used to produce a _____ wave output.

    a. square   c. either a or b
    b. rectangular   d. neither a nor b

14. A monostable multivibrator has _____ stable states.

    a. 1
    b. 2
    c. 3
    d. 0

15. The monostable multivibrator produces one output cycle for each _____ input trigger(s).

    a. 1
    b. 2
    c. 3
    d. 4

16. The bistable multivibrator has _____ stable states.

    a. 1
    b. 2
    c. 3
    d. 0

17. A Schmitt trigger circuit produces one output pulse for each _____ input pulse.

    a. 1
    b. 2
    c. 3
    d. 4

18. The bistable multivibrator can be used to produce a _____ wave output.

    a. square
    b. rectangular
    c. either a or b
    d. neither a nor b

19. Refer to Figure 14-20. Capacitors C_3 and C_4 are added to this circuit for what purpose?

20. A Schmitt trigger can be used to _____.

    a. produce a square wave output
    b. produce a rectangular wave output
    c. reshape a rounded pulse
    d. any of the above

Sawtooth and Trapezoidal Wave Generators

Sawtooth and *trapezoidal wave generators* are used to produce very exact waveshapes for use under specific conditions. A sawtooth wave is used to create the sweep on any cathode ray tube that uses *electrostatic* deflection. Oscilloscopes, like those used in your lab, present the signal on the screen through the use of a sawtooth wave. Trapezoidal waves are used to create the sweep on CRTs that use electromagnetic deflection. Most television sets use electromagentic deflection. Trapezoidal waves are used to generate their horizontal sweep.

Sawtooth Wave Generation

As the name implies, a **sawtooth generator** generates an output waveshape that is shaped like the teeth on a saw. This type of signal is used in many applications. The oscilloscope sweep signal is a sawtooth wave. All methods used to develop sawtooth waves involve RC networks.

In Figure 14-22, you will see the schematic diagram for a sawtooth generator and the output waveshape that can be expected. Look at the circuit diagram and disregard the input signal. Transistor Q_1 is biased such that $V_C = -2.5$ V with current flowing from $-V_{CC}$ down through R_1 and Q_1 to ground. As long as there is no input signal, the output will be a constant -2.5 V.

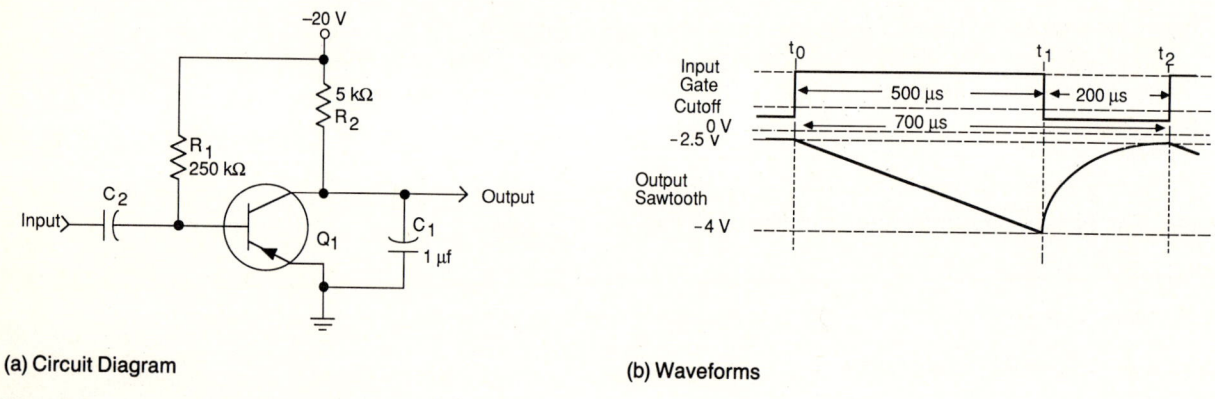

(a) Circuit Diagram

(b) Waveforms

Fig. 14-22 Sawtooth Wave Generator

When a positive input gate arrives at the base of Q_1, it reverse biases Q_1, cutting it off. When Q_1 cuts off, its V_C tries to go from -2.5 V to $-V_{CC}$. This is not possible because C_1 must charge first. The capacitance of C_1 and the resistance of R_2 form an RC time constant that controls the charge of C_1. This means that the voltage at Q_1 collector rises at an exponential rate as shown on the capacitor charge curve of Figure 14-7. At time t_0 of Figure 14-22(b), the output signal starts at -2.5 V and as C_1 begins to charge, the *slope* (charge time) of the output wave begins to form. As time passes, this slope continues until the input gate reverses polarity at time t_1. This causes Q_1 to start conducting. Q_1 provides a low resistance discharge path for C_1 allowing C_1 to quickly revert back to the -2.5 V state from which it started. During the discharge of C_1 (t_1 to t_2), the capacitor discharges, producing the part of the wave shown. This results in a sawtooth wave, like the one shown in Figure 14-22(b), at the output. The PRF of the output is controlled by the PRF of the input.

Observe the output signal. You will see that the slope appears to be linear—which is desirable. You should remember that the first 10% of a capacitor's charge time occurs at a fairly linear rate. For this reason, the time that C_1 is allowed to charge is kept to less than 10% of the applied voltage.

Trapezoidal Wave Generators

Figure 14-23 contains the schematic diagram for a **trapezoidal wave generator** along with the normal input and output waveshapes. This circuit will be discussed briefly. Note the NPN transistor and $+V_{CC}$.

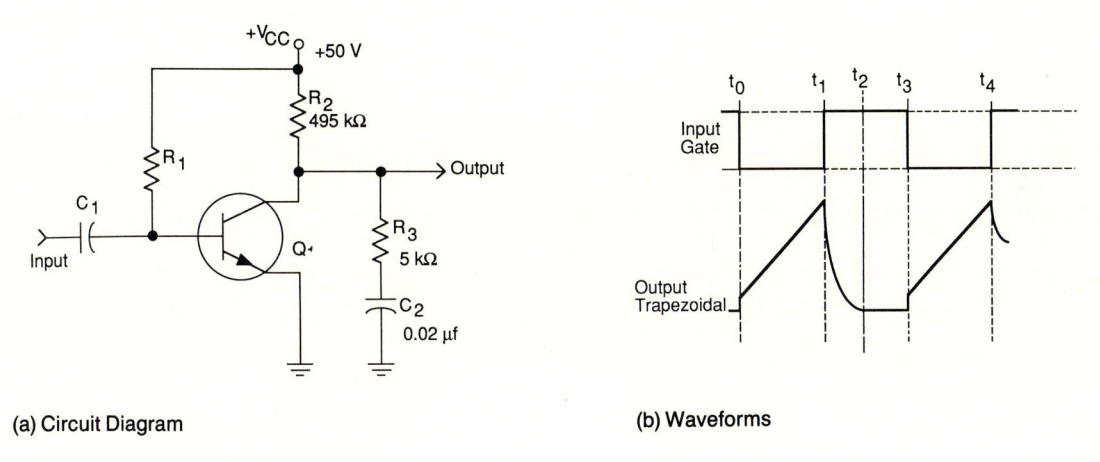

(a) Circuit Diagram

(b) Waveforms

Fig. 14-23 Trapezoidal Wave Generator

The only other difference that exists between this circuit and the sawtooth generator is the addition of R_3 in series with the capacitor (C_2). In this circuit, the positive input biases Q_1 such that the voltage at its collector is low. When the input gate changes to 0 V (t_0), Q_1 is cut off. The charge path for C_2 now contains R_3, C_2, and R_2. At the first instant, C_2 acts like a short and maximum current flows through R_3 and R_2. This means that the entire +50V must be divided by these two resistors. The voltage drop on R_3 appears as part of the output signal and is called **jump voltage**. It gets this name from the fact that when viewing the wave on an oscilloscope, the voltage waveshape appears to "jump" to the level of V_{R3}. As soon as Q_1 cuts off, the jump voltage appears and C_1 begins its charge starting at the voltage level of V_{R3}. From this point on, circuit operation is identical to the sawtooth generator. In the deflection circuit, jump voltage is used to overcome the initial opposition (X_L) of the coil to changes in current.

Self-Check

Answer each item by inserting the word or words required to correctly complete each statement.

21. A sawtooth generator is actually an amplifier that has a/an _____ that controls its collector voltage during cutoff.

22. A sawtooth wave is used with cathode ray tubes that operate using _____ deflection.

    a. electromagnetic   b. electrostatic

23. The output frequency of the trapezoidal wave generator is controlled by the _____.

    a. input gate frequency   c. RC network time constant
    b. free-run frequency   d. none of these
        of the circuit

24. A trapezoidal wave is used with cathode ray tubes that operate using _____ deflection.

    a. electromagnetic   b. electrostatic

25. The output frequency of the sawtooth wave generator is controlled by the _____.

    a. input gate frequency   c. RC network time constant
    b. free-run frequency   d. none of these
        of the circuit

26. In either of these circuits, the slope of the output wave is controlled by the _____.

    a. input gate frequency   c. RC network time constant
    b. free-run frequency   d. none of these
        of the circuit

Summary

In this chapter, we discussed several circuits that generate nonsinusoidal waves. Our intention was to introduce you to these circuits and their uses. We did not try to cover the subject of nonsinusoidal wave generators in its entirety.

Pulsed oscillators were covered early in the chapter. The circuit used for pulsed oscillators is very similar to those used in LC oscillators. With the exception of one crystal-controlled circuit, the main difference is that the transistor is biased off in its quiescent state. An input gate is applied to the base of the transistor which allows the circuit to begin oscillating. Oscillations continue until the input gate is removed. The frequency of the oscillator is controlled by an LC tank circuit.

Blocking oscillators were also discussed. These circuits use LC tank circuits as their frequency-determining devices. Operation of the circuit is controlled by inductive feedback through a transformer located in the collector circuit of the amplifier.

We found that a free-running oscillator can be synchronized by the application of a series of trigger pulses to the base. The frequency of these triggers must be higher than the tank's resonant frequency. In some cases, a tertiary winding is included as part of the transformer and can be used for coupling the output.

Next, you studied multivibrators. The three types discussed were the astable, monostable, and bistable multivibrators. These range from free-running (astable), to one that has one stable state (monostable), to one that is stable in either of two states (bistable). The main difference between the three is the way that biasing and feedback for the transistors is handled. In the monostable and bistable types, input triggers are used to change states. With the monostable, one complete output waveshape is generated for each input trigger. With the bistable circuit, two input pulses are required for each output waveshape.

Even though it is not actually a multivibrator, the Schmitt trigger circuit was also discussed. This circuit is especially valuable in the reshaping of pulses whose leading and trailing edges have become rounded.

Sawtooth and trapezoidal wave generators were discussed briefly. These circuits are used to sweep electron beams across cathode ray tubes. The difference between sawtooth and trapezoidal waves is the addition of jump voltage. The sawtooth wave is used with electrostatic deflection. Trapezoidal waves are used to provide electromagnetic deflection on TV picture tubes and video monitors.

Review Questions and/or Problems

1. In this chapter, we discussed three types of pulse oscillators. The transistor that the input gate is applied to acts as a/an _____ and the transistor must be _____ for a sine wave output to be generated.

    a. switch, conducting
    b. amplifier, conducting
    c. switch, cut off
    d. amplifier, cut off

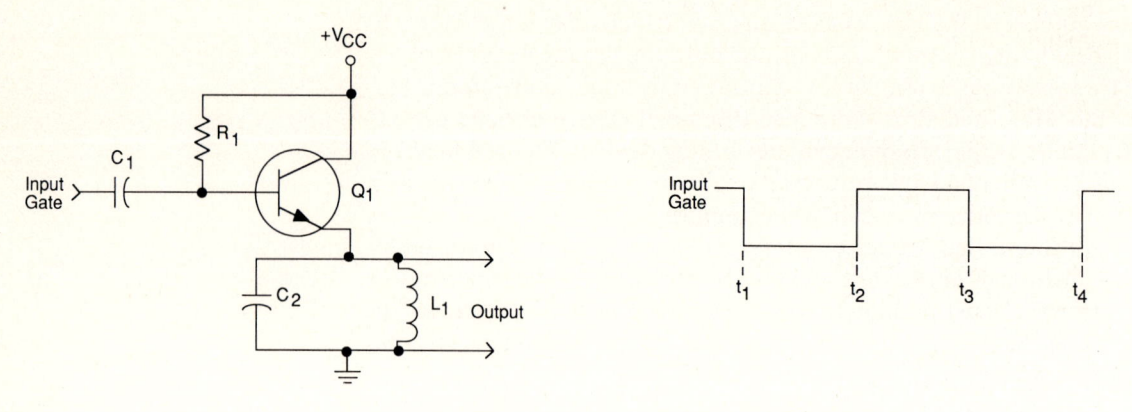

Fig. 14-24

2. Examine Figure 14-24. Sine waves will be present at the output of this circuit during input gate periods:

    a. t_1 to t_2.
    b. t_2 to t_3.
    c. t_1 to t_2 and t_3 to t_4.
    d. t_0 to t_2 and t_3 to t_4.

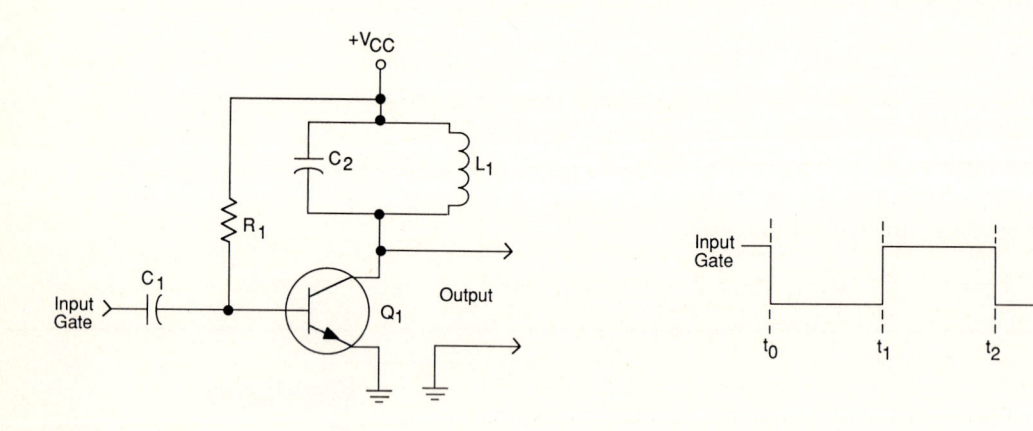

Fig. 14-25

3. Refer to Figure 14-25. The frequency of the output sine wave of this circuit is determined by:

    a. input gate width.
    b. values of L_1 and C_2.

Nonsinusoidal Wave Generators 479

4. The rate at which the output wave (Figure 14-25) is damped out depends on the _____.

   a. amplification factor of the transistor
   b. Q of the tank circuit
   c. amplitude of the feedback
   d. amplitude of the input gate

5. Check Figure 14-25. This is a/an _____ loaded pulse oscillator _____ regeneration.

   a. collector, with
   b. emitter, with
   c. collector, without
   d. emitter, without

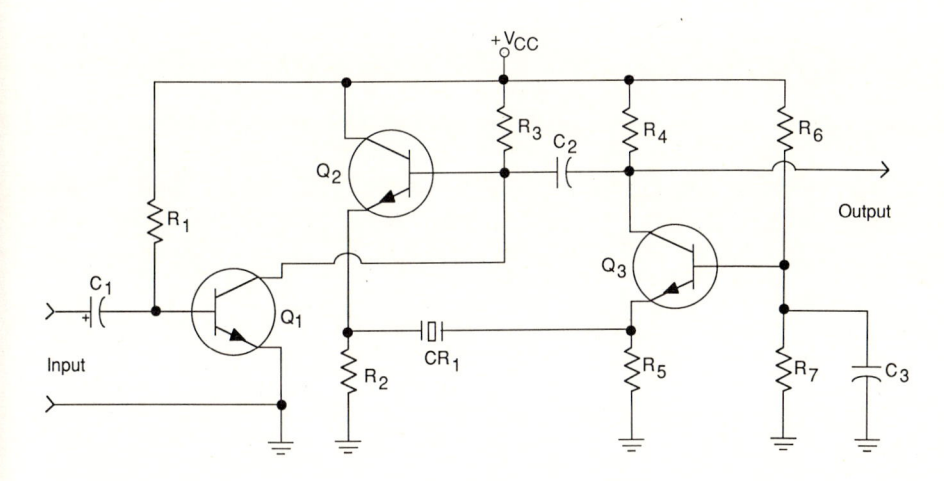

Fig. 14-26

6. Refer to Figure 14-26. The frequency of the sine wave present at the collector of Q_3 is determined by _____.

   a. the width of the negative alternation of the input gate
   b. the width of the positive alternation of the input gate
   c. the RC time constants of the circuit
   d. the resonant frequency of CR_1

7. Check Figure 14-26. Before the input gate is applied to the base of Q_1, transistor Q_2 is being held in the _____ state by the _____ voltage coupled from the collector of Q_1.

   a. conduction, low
   b. cut off, low
   c. conduction, high
   d. cut off, high

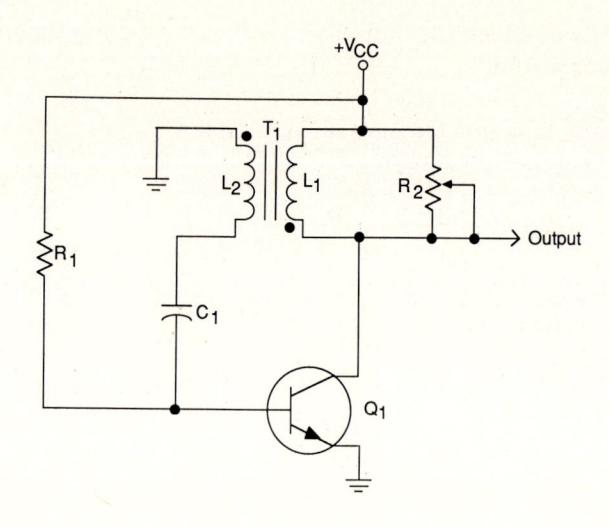

Fig. 14-27

8. Refer to Figure 14-27. Match the statements in the column A to the diagrams in column B.

   a. R_2 resistance too high
   b. R_2 resistance too low
   c. R_2 resistance correct

   (1)
   (2) Overshoot
   (3)

9. Check Figure 14-27. The time that Q_1 is cut off is controlled by _____.

   a. R_1 and C_1 only.
   b. L_1 and L_2 only.
   c. L_1, L_2 and R_2
   d. L_1, L_2, R_1, and C_1

Nonsinusoidal Wave Generators 481

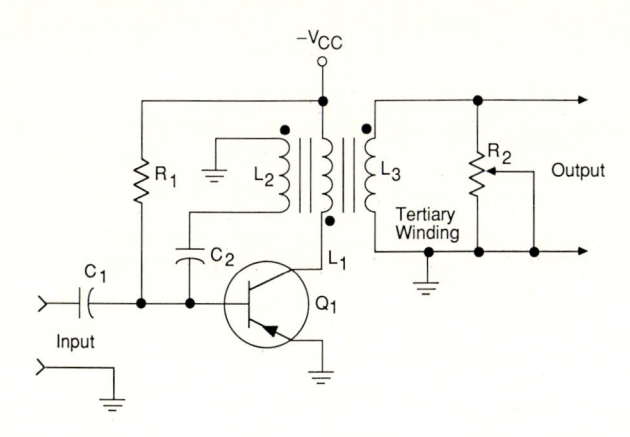

Fig. 14-28

10. Refer to Figure 14-28. This circuit is a _____ blocking oscillator.

    a. free-running   b. synchronized

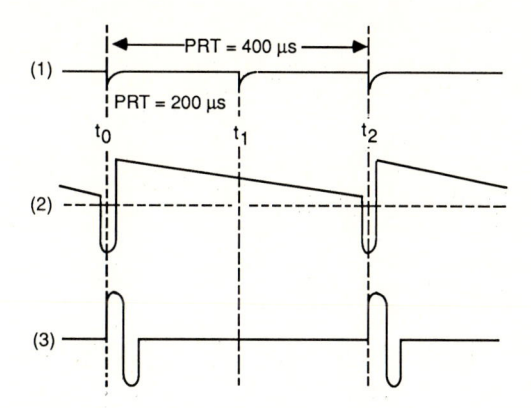

Fig. 14-29

11. Which waveshape in Figure 14-29 depicts the circuit's (Figure 14-28) output waveshape?

    a. 1   c. 3
    b. 2   d. none of these

12. With one input trigger arriving every 200 µs (as shown in Figure 14-29) a complete cycle is developed for each _____ input pulse(s) that has an output frequency of _____.

    a. one, 2 kHz
    b. two, 2 kHz
    c. one, 2.5 kHz
    d. two, 2.5 kHz

13. Refer to Figure 14-28. If C_1 opens, which of the following will be true?

    a. There is no output from the circuit.
    b. The output signal PRF increases.
    c. The output signal PRF decreases.
    d. The output signal PRT decreases.

14. Observe Figure 14-28. A trouble exists in this circuit. Symptoms are no output signal and V_C is approximately equal to V_{CC}. A possible cause is?

    a. R_1 open
    b. C_2 open
    c. Q_1 shorted
    d. L_1 open

15. Check Figure 14-28. Forward bias for Q_1 is provided by the current that flows through _____.

    a. R_1 and the emitter-base junction
    b. L_2 and the emitter-base junction
    c. R_1 and the collector-base junction
    d. L_2 and the collector-base junction

16. Examine Figure 14-29; the period between times t_0 and t_1 is controlled by _____.

    a. R_1 and C_1 only
    b. L_1 and L_2 only
    c. L_1, L_2, and R_2
    d. L_1, L_2, R_1 and C_1

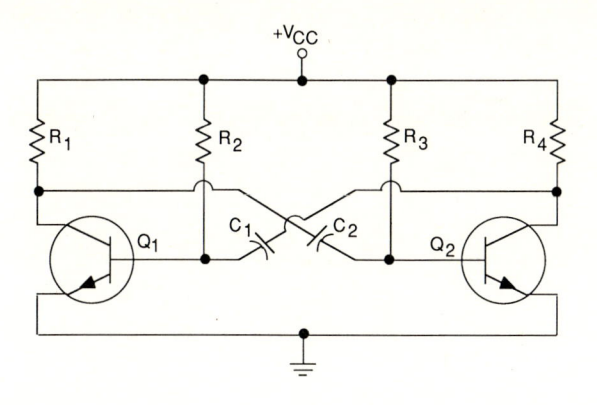

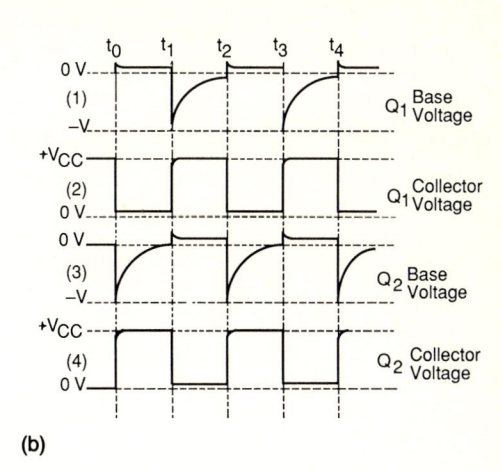

Fig. 14-30

17. Check Figure 14-30. This circuit is a _____ multivibrator of the _____ type.

    a. free-running, monostable
    b. free-running, astable
    c. triggered, monostable
    d. triggered, astable

18. Increasing the value of either R_2 or C_1 (Figure 14-30) causes _____ to remain cutoff longer and _____ to remain saturated longer.

    a. Q_1, Q_1
    b. Q_1, Q_2
    c. Q_2, Q_2
    d. Q_2, Q_1

19. Use Figure 14-30. If Q_1 has zero as its output, which of the following is the probable cause?

    a. R_2 open
    b. R_3 open
    c. C_1 open
    d. Q_2 open

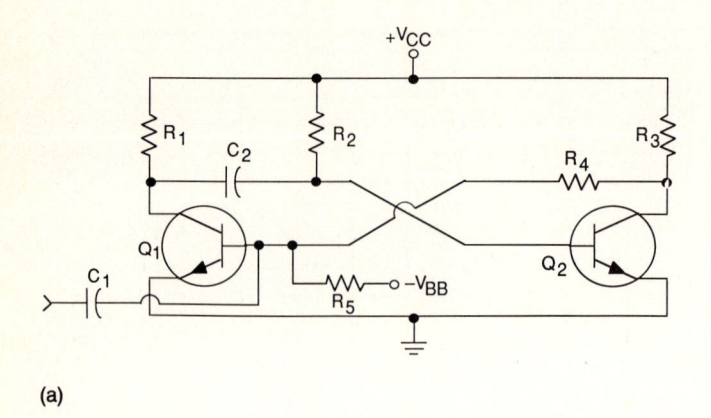

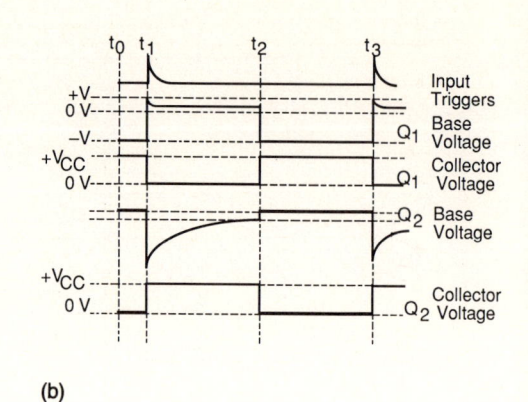

Fig. 14-31

20. Check Figure 14-31. This circuit is a _____ multivibrator of the _____ type.

    a. free-running, monostable
    b. free-running, astable
    c. triggered, monostable
    d. triggered, astable

21. Check Figure 14-31. Triggers are used with this circuit to _____.

    a. increase pulse width
    b. decrease pulse width
    c. establish the output frequency
    d. none of these

22. Observe the waveshapes in Figure 14-31. The period between t_1 and t_3 is called _____.

    a. pulse-repetition frequency
    b. pulse-recurrence time
    c. pulse width

23. Before triggers are applied to this circuit (see Figure 14-31) _____ is cut off and _____ is saturated.

    a. Q_1, Q_1 c. Q_2, Q_2
    b. Q_1, Q_2 d. Q_2, Q_1

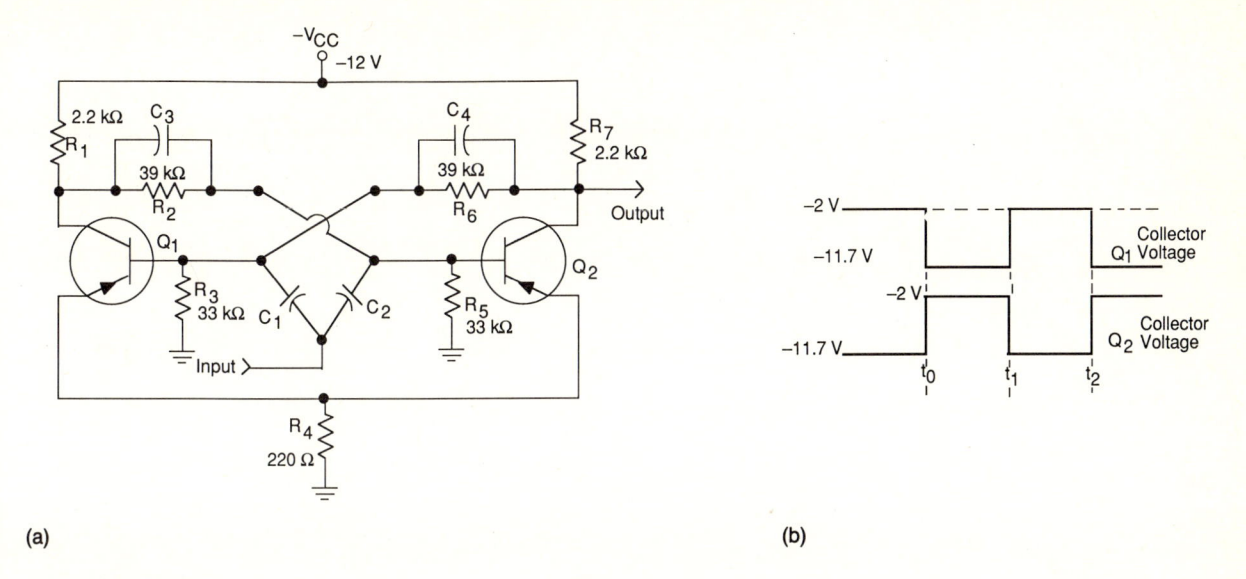

Fig. 14-32

(a)

(b)

24. Check Figure 14-32. This circuit can be identified as a/an _____ multivibrator by the fact that it has a _____ resistor and input triggers are applied to _____ transistor(s).

   a. monostable, emitter, one
   b. bistable, common-emitter, two
   c. monostable, collector, two
   d. bistable, collector, one

25. In Figure 14-32, the input frequency is _____ the output frequency, and the input PRT is _____ the output signal's PRT.

   a. one half, double   c. double, one half
   b. one half, one half   d. double, double

26. There is a problem with the circuit in Figure 14-32. The symptoms are Q_1 collector voltage equals V_{CC}, and Q_2 collector voltage equals V_{CC}. The probable cause is _____.

   a. Q_2 shorted   c. C_2 shorted
   b. Q_1 open   d. R_4 open

27. The two purposes for which Schmitt triggers are used are _____.

   a. voltage shaping and pulse shaping
   b. pulse sensing and voltage shaping
   c. voltage sensing and pulse shaping
   d. none of these

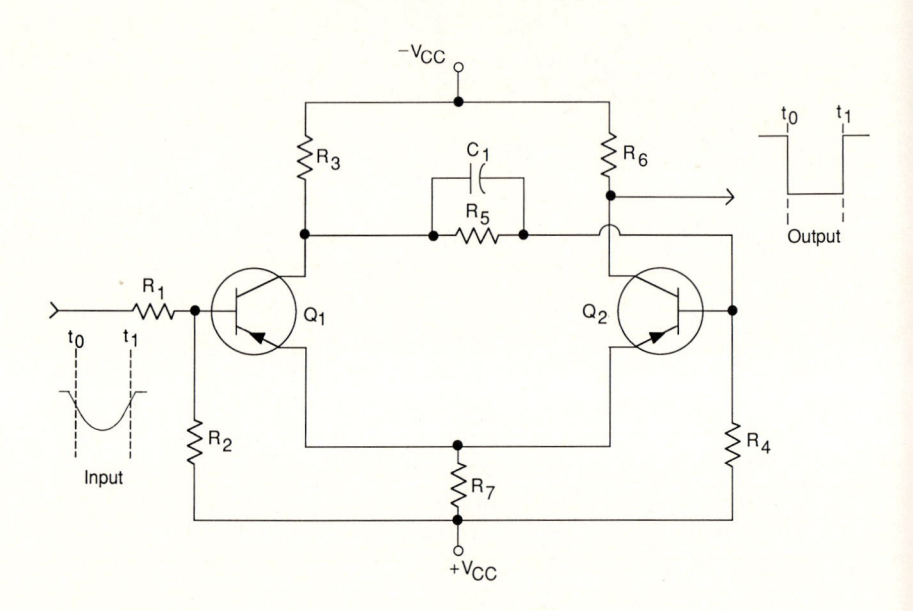

Fig. 14-33

28. Examine Figure 14-33. Before the arrival of an input pulse, _____ is saturated and _____ is cut off.

   a. Q_1, Q_1
   b. Q_1, Q_2
   c. Q_2, Q_2
   d. Q_2, Q_1

29. C_1 is added to the circuit shown in Figure 14-33 in order to prevent _____.

   a. high frequency distortion   c. low frequency distortion
   b. gain   d. amplification

30. Check Figure 14-33. There is a problem in this circuit. The symptom is that the output signal shows high frequency distortion. The probable cause is _____.

   a. C_1 shorted
   b. C_1 is open
   c. Q_2 saturated
   d. input signal is distorted

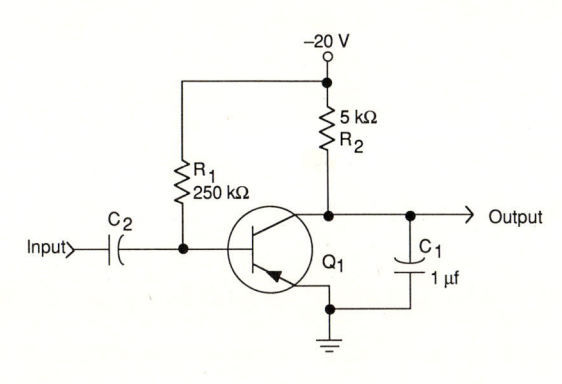

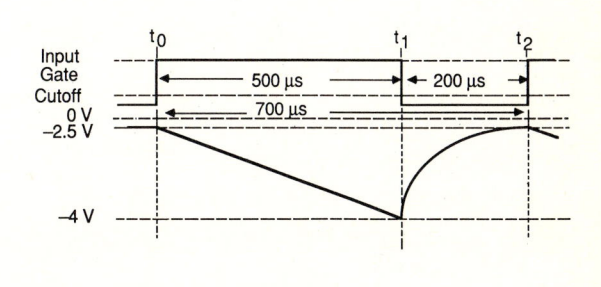

(a)

(b)

Fig. 14-34

31. The circuit shown in Figure 14-34 produces a linear sawtooth because C_1 charges to _____ 10% of total applied voltage.

   a. more than
   b. less than
   c. exactly

32. In the sawtooth generator (Figure 14-34), an output signal is produced when the _____ input gate causes Q_1 to _____.

   a. positive, cut off
   b. positive, conduct
   c. negative, cut off
   d. negative, conduct

33. Decreasing the width of the positive going input gate causes the output wave (Figure 14-34) to _____.

   a. increase in duration
   b. increase in amplitude
   c. decrease in amplitude
   d. none of these

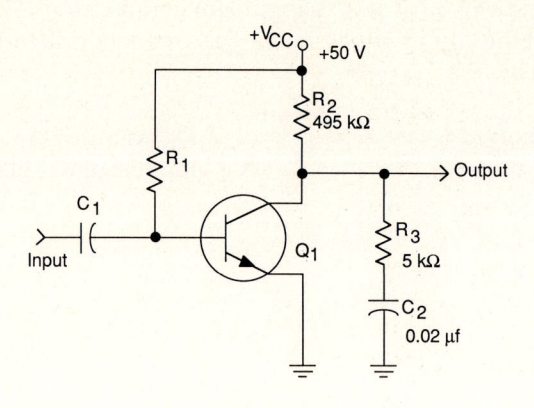

Fig. 14-35

34. In this circuit (Figure 14-35), jump voltage is approximately _____ volts.

    a. 0.005   c. 0.5
    b. 0.05   d. 5

35. In Figure 14-35, the arrival of a negative going input gate causes Q_1 to _____ and the _____ voltage to occur.

    a. saturate, slope   c. cut off, slope
    b. saturate, jump   d. cut off, jump

15
The Junction Field-Effect Transistor (JFET)

Objectives

1. Define the following terms:

    a. P-type junction field-effect transistor
    b. N-type junction field-effect transistor
    c. Drain
    d. Source
    e. Gate
    f. Depletion region
    g. Channel
    h. Channel pinch-off
    i. Transconductance
    j. Pinch-off voltage
    k. Channel saturation current.

2. Explain the effect that gate voltage has on drain current.
3. Identify the schematic diagrams for:

    a. Common-source JFET amplifiers
    b. Common-gate JFET amplifiers
    c. Common-drain JFET amplifiers

4. Describe the circuit operation for the following amplifiers:

    a. Common-source JFET amplifiers
    b. Common-drain JFET amplifiers
    c. Common-gate JFET amplifiers

5. Describe different methods used to provide external bias to JFETs in operational circuits.

Introduction

A type of transistor that operates very differently from any you have studied to this point is the **Field-Effect Transistor (FET)**. These devices provide us with capabilities not available in other components. Field-effect transistors are voltage-controlled devices. Bipolar-junction transistors (BJTs) are current-controlled devices and have current flowing in their input circuits. With this condition (input current), the transistor's input resistance is relatively low. FETs, on the other hand, have very small input currents which means that they have extremely high input resistance. Field-Effect Transistors are grouped into two categories: (1) **Junction Field-Effect Transistors (JFETs)**, (2) **Insulated-Gate Field-Effect Transistors (IGFETs)**. Note that the **Insulated-Gate FET** is also called a **MOSFET (Metal-Oxide-Semiconductor Field-Effect Transistor)** and will be discussed in the next chapter. Further classification of these devices identifies them as either **P-channel** or **N-channel** types.

JFET Construction

Figure 15-1(a) illustrates one method of constructing a JFET. The method shown here is called the **diffusion** method. Starting with a piece of P-type material (substrate), an N-channel is diffused into the P-type substrate. One end of the channel is designated as the *drain* element and the other end as the *source* element. In this device, the source is comparable to the emitter and the drain is comparable to the collector of the bipolar transistor. By precisely injecting N-type semi-

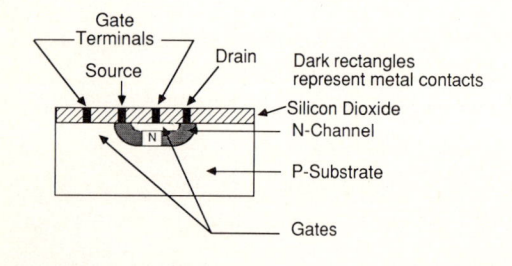

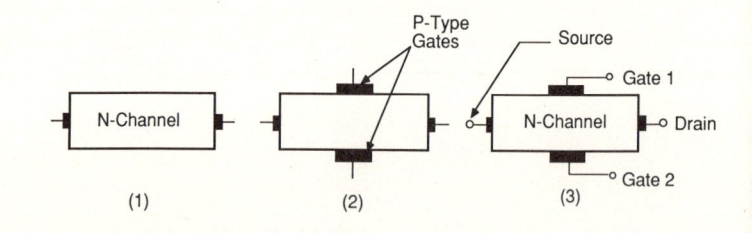

(a) Diffusion Process

(b) Pictorial Diagrams

N-Channel JFET Construction

Fig. 15-1

conductor material onto the P-type substrate, a layer of N-material is constructed—similar to the N-section shown in Figure 15-1(a). Once the N-section is in place, a small layer of P-type material is diffused into the top of the N-section. This P-section forms one part of the *gate* and the substrate serves as the second part of the *gate*. To complete the construction, a layer of silicon dioxide is joined to the top of the semiconductor device to serve as an insulation layer. Metallic contacts are bonded into the silicon dioxide layer and form the electrical connections for connecting the JFET into a circuit. Notice that the gate is made up of two P-sections; one section is the substrate and the other is the P-material that has been diffused into the N-channel. The N-channel is symmetrical and, in this example, the *drain* and *source* can be at either end of the N-channel. Reversing their positions does not change the device's operation. There are, however, many JFETs that *do not* have this symmetrical arrangement. In these, reversing the source and drain will have highly unpredictable results.

In Figure 15-1(b) you will see JFET construction reduced to its simplest form. Figure 15-1(b1) shows a piece of N-type material that has electrodes connected to each of its ends. This is used as the N-channel for the finished JFET. In Figure 15-1(b2) two pieces of P-type material have been attached to the channel. These have leads attached and are used as the gate connections. In Figure 15-1(b3) we see the completed device with the *gate, source,* and *drain* labeled. In most applications the two gates are connected together so that they act as one gate. In some cases, however, there is a need to apply two signals to the JFET. In those cases, one signal is applied to each gate electrode. When used with two input signals, the JFET is said to be **tetrode connected**.

Figure 15-2(a) contains the pictorial diagram for an N-channel JFET and Figure 15-2(b) shows the schematic symbols for both N- and P-type JFETs. Note that the arrow still points to the N-type material.

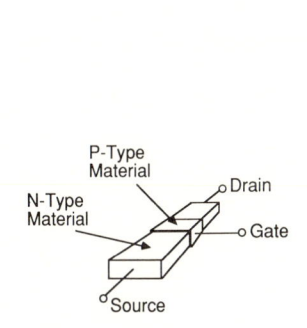

(a) Pictorial Diagram

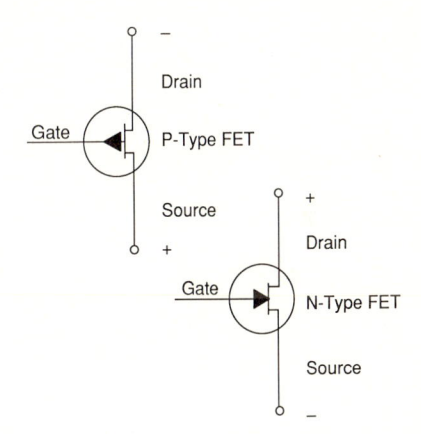

(b) Schematic Symbols

Fig. 15-2 N-Channel JFET

JFET Operational Characteristics

Figure 15-3 contains three pictorial drawings of an N-channel JFET that has two gate electrodes similar to those explained in Figure 15-1(b). Notice that in Figure 15-3(a), the JFET has voltage applied to its *source*

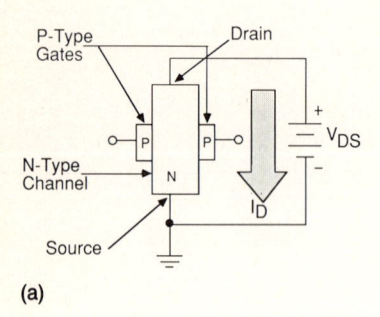

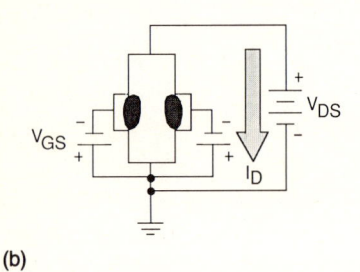

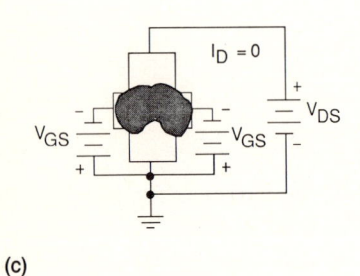

Fig. 15-3 N-Channel JFET Characteristics

and *drain* electrodes only. In Figures 15-3(b) and 15-3(c), the *gate* electrodes also have a voltage (V_{GS}) connected. Figure 15-3(c) has the largest gate-to-source voltage (V_{GS}).

N-channel JFET operation can be explained by examining the operation of these three circuits. The three states we will discuss are zero bias, low bias, and high bias. By examining Figure 15-3(a), you will see that voltage, is applied to the source-drain electrodes of the JFET. Notice that there is no connection to either of the gate electrodes. With voltage connected as shown, *drain current* (I_D) flows through the N-type material. Within the circuit, current leaves the negative side of V_{DS}, enters the JFET at the *source*, passes through the *N-channel*, exits the JFET at the *drain*, and returns to the power source. In this application, the N-type material operates like a resistor. The *drain source voltage* (V_{DS}) is dropped across the N-type material. Assuming that the resistance distribution within the N-type material is linear, the voltage drop is distributed linearly throughout the material.

In Figure 15-3(b), bias has been applied to the gate electrodes in addition to the V_{DS} present in Figure 15-3(a) above. The voltage that is connected between the gate and the source electrodes is called the *gate-source voltage* (V_{GS}). Notice that in the polarity shown, V_{GS} applies a negative voltage to the gate. This means that the gate is negative with respect to the source. Assuming that V_{GS} is a small voltage, we can discuss what occurs within the JFET. A negative voltage applied to the gate (P-type material) reverse biases the PN junction that exists between the gate and the N-channel. The negative voltage acts to repel the electrons from the area adjacent to the gate-channel junction. Note that the gate junction is located at both sides of the pictorial diagram. This means that as negative voltage is applied to the gate, areas are created within the channel that have few electrons. The electrons within these areas have been pushed further into the N-channel, creating a *depletion region*. When an electrical charge is used to create a depletion region, the depleted region is due to the creation of an **electrostatic field**. We can say that application of reverse bias to a junction increases the depletion region that exists near the junction through a process called **electrostatic repulsion**. Remember:

> **To forward bias a PN junction, positive voltage is connected to the P-material and negative voltage is connected to the N-type material.**

and:

> **To reverse biase a PN junction, positive voltage is connected to the N-material and negative voltage is connected to the P-type material.**

When a negative V_{GS} is applied to the gate electrodes, it assures that the PN junction is reverse biased and that an electrostatic field (depletion region) extends into the channel and restricts the area through which current can flow.

Reverse bias applied to the gate causes a depletion region to be formed across the gate-channel junction. Because the bias voltage is low, the depletion region extends only a short distance into the N-channel. The depletion region does, however, form a band completely across the outer edges of the N-channel. Recall that a depletion region is an area where the number of current carriers is greatly reduced. The depletion region acts like an insulator and resists current flow. This means that *drain current* (I_D) must flow through a smaller area (channel) than before. In other words:

The application of reverse voltage to the gate causes channel resistance to increase and drain current (I_D) to decrease.

In Figure 15-3(c) we see the results of applying a high reverse bias (V_{GS}). Notice that as V_{GS} increases the size of the depletion region increases, the channel narrows, and the amount of I_D decreases to zero. At some reverse bias, the depletion regions overlap and the channel is completely blocked. At that point, I_D stops flowing and remains cut off for all reverse biases equal to or larger than this. This condition is illustrated in Figure 15-3(c) where $I_D = 0$.

Notice the similarity that exists between reverse biasing the JFET and the reverse biasing of a bipolar transistor. In both cases, reverse bias can be used to cut off the transistor's conduction. When either transistor is biased at cutoff, the application of an AC signal to the gate causes the transistor to conduct during one alternation of the input and to remain cutoff during the other alternation. If the N-channel JFET input goes positive, it causes the channel to open and allow I_D to flow. When the negative alternation of the input signal is applied, the depletion region keeps the channel blocked.

JFET – Bipolar Transistor Comparisons

Granted, there is a big difference between JFET and bipolar junction transistors. They are similar enough, however, that we can compare their elements. The *drain* is similar to the *collector*. The *source* can be compared to the *emitter* and the *gate* to the *base*. Table 15-1 contains a comparison of the symbols used with JFETs and bipolar transistors.

Table 15-1
JFET and Bipolar Parameter Comparisons

BIPOLAR SYMBOL	JFET SYMBOL	JFET FUNCTION
V_{CE}	V_{DS}	Drain-to-Source Voltage
V_C	V_D	Drain Voltage
V_{BE}	V_{GS}	Gate-to-Source Voltage
V_{EE}	V_{SS}	Source Voltage
I_C	I_D	Drain Current

The JFET operates with *reverse* bias applied to its *gate*, and the bipolar operates with *forward* bias applied to its *base* and current flows across the junction. Because of this, the bipolar transistor can be treated as a *current-controlled* device. Current control in a JFET results from the electrostatic field that is injected into the channel by reverse biasing the junction. Gate-to-source voltage (V_{GS}) is used to do this. The fact that no current flows through the junction and that the electrostatic field is created by voltage (V_{GS}) allows us to make the statement that a JFET is a **voltage-controlled** device.

The P-channel JFET is constructed with a substrate of N-type material diffused with a P-channel. N-type material is then added to complete the gate formation. Both voltages (V_{DS} and V_{GS}) have opposite polarities to those used with the N-channel JFET circuits. Once the P-channel JFET is correctly biased, current (I_D) flows in the opposite direction from the current (I_D) that flows in the N-channel JFET. Assuming that the N-channel JFET and the P-channel JFET have identical construction and voltages, their operation is identical. Both types have their current (I_D) controlled by an electrostatic field whose strength depends on the presence and amplitude of V_{GS}.

The name "field effect" comes from the fact that the depletion regions are controlled by an electrostatic "field". By reverse biasing the gate-channel junctions of a JFET, the electrostatic effect that exists across the junction creates the depletion regions. The result is that the biasing sets up an electrostatic field across the gate-channel junction: thus, the name "field effect". The name "N-channel" or "P-channel" is derived from two things: (1) the name "channel" is given to the narrow opening through which I_D must flow; (2) the "N" or "P" identifies the type of semiconductor material that is diffused into the substrate of the JFET to form the channel.

In a JFET circuit, drain current (I_D) is controlled by V_{GS}. Variations in V_{GS} cause variations in I_D. No usable current flows through the device's junctions. Because I_D is flowing through the channel, only majority current carriers are required to support current flow. In the P-channel JFET, current is supported by holes. In the N-channel JFET, current is supported by electrons.

Operation with $V_{GS} = 0$ Volts

Figure 15-4 is used to show an N-channel JFET that has its source and gate connected to the same potential (ground). Notice the width of the gate sections. Assume that the gate width equals one third of the channel's length. Also assume that 12 V is applied as V_{DS} and that the N-type material used for the channel has a very linear resistance of 3 kΩ. This means that from ground to the bottom of the gate, resistance equals 1 kΩ. The resistance between ground and the top of the gate equals 2 kΩ. With V_{DS} of 12 V applied, the voltage drop to the bottom of the gate equals 4 V (see M_1) and to the top of the gate, it equals 8 V (M_2). Remember, these voltages are developed (dropped) internally within the N-type material that serves as the channel. The potential at the gate electrode is 0 V. The channel-to-gate voltage (indicated by M_1) equals +4 V, and the channel-to-gate voltage (indicated by M_2) equals +8 V. In other words, the reverse bias that exists between the channel and gate varies from +4 V to +8 V over the length of the gate section.

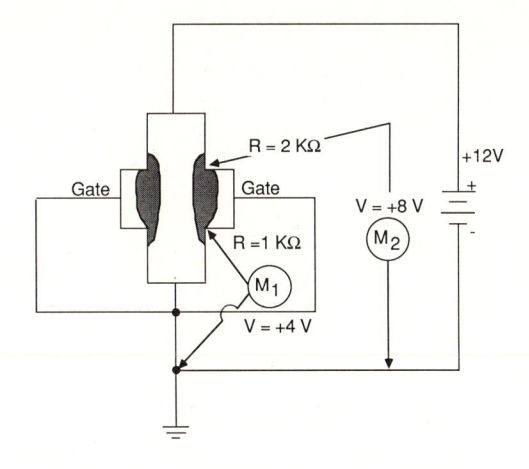

Fig. 15-4 JFET Internal Resistance and Voltage Drop

With each section of the gate having a different reverse bias, the closer that we get to the drain end of the channel, the larger reverse bias becomes. This means that the depletion region becomes larger. In fact, the depletion region forms a cone-shaped funnel within the channel. I_D must be concentrated within the opening at the top of the funnel. The effect is that by use of reverse bias and the resulting depletion regions, current is "channeled" through a small portion of the N-type material.

When both the gate and source electrodes are connected to ground, interesting things happen. Ground places 0 V at the gate terminal. The voltage drop across the junction is reverse bias when V_{DS} is present. As we gradually increase V_{DS} away from zero, I_D gradually increases and the depletion region begins to appear. When V_{DS} is small, the effect of the depletion region is very small. As V_{DS} is increased further, the depletion region continues to expand into the N-type material, making the channel smaller. At the same time, I_D continues to increase at a linear rate. During the early periods when V_{DS} is increasing, the N-type material acts almost like a constant resistance. With each increase in V_{DS}, I_D continues to increase at a linear rate, and the depletion region expands further. This means that the channel is becoming smaller and smaller while I_D continually increases.

Use Figure 15-5 for the discussion that follows. At some V_{DS} level, the channel's resistance begins to be affected more heavily with each increase in V_{DS}. These increases in channel resistance cause the increases in I_D to be smaller than before. The period during which I_D ends its linear increase and levels off is called the **pinch-off region**. Because of the rapid change in channel resistance that occurs in this region, I_D reaches a point where it shows little increase when V_{DS} is increased further. This is the flat portion of the I_D curve on the graph shown in Figure 15-5. In this graph, $I_D \cong 1\,mA$.

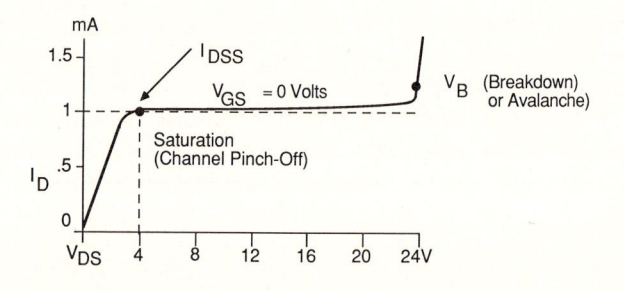

Fig. 15-5 Effect of V_{DS} on I_D

With $V_{GS} = 0$ and further increases in V_{DS} having little effect on I_D, the device is said to have reached *drain-source saturation current* (I_{DSS}) level. This is illustrated in Figure 15-5. At the point where saturation occurs, the depletion region has expanded to the point that the channel has narrowed to a very small opening. The point where the channel becomes saturated is called the **pinch-off point** and the amount of voltage required to reach the pinch-off point is called **pinch-off voltage** (V_P).

If V_{DS} is allowed to continue its increase, the reverse current that flows across the gate-to-channel junctions becomes a problem. The reverse current increases to the point that it beings to *avalanche* as in a Zener diode. The large currents that make up the avalanche current quickly increase I_D and the JFET will probably be destroyed. This is the portion of the chart that is labeled "BREAKDOWN or AVALANCHE."

The following five points should be emphasized.
1. Drain-source-saturation occurs only when V_{GS} equals 0 V and V_{DS} has been increased to the point that I_D has leveled off because of pinch-off.
2. Pinch-off point is normally considered to be the point at which channel current (I_D) stops its linear increase and begins to level off.
3. Pinch-off voltage refers to the amount of voltage required to cause the JFET's drain current to stop increasing and level off.
4. Pinch-off region refers to the region that extends from the point where I_D stops its linear increase to the point where I_D is stabilized.
5. Pinch-off can occur at any point where the sum of gate-source voltage (V_{GS}) and inter-channel voltage drop equals pinch-off voltage (V_P). Assume that $V_{GS} = 0$ V: as V_{DS} is increased, we reach a point where I_D slows its rise and enters the pinch-off region. Continued increases of V_{DS} will cause I_D to increase at a less linear rate. At some higher V_{DS}, drain current is completely pinched-off and I_D stabilizes. The amount of V_{DS} that is present when I_D stabilizes is called *pinch-off voltage* (V_P). This voltage (V_P) also tells us that if we applied the opposite polarity voltage as V_{GS} to a JFET operating with I_{DSS}, the device would enter cutoff. The amount of voltage required to cut the transistor off is called $V_{GS}OFF$. Taking this further, with the value of V_P and $V_{GS}OFF$ known, we can see that any combination of V_{GS} and within-channel voltage drop whose sum equals V_P will cause the drain current to be pinched-off.

The two limits (V_P and $V_{GS}OFF$) represent the saturation and cutoff points of a JFET that compare to the saturation and cutoff points for bipolar transistors. As with bipolar transistors, we can bias the JFET to operate at a point within these limits. One additional consideration enters the picture, though. With JFET operation, we have a single junction. Bias for this junction is comprised of the interchannel voltage drop that results from application of V_{DS} and the voltage applied as V_{GS}. For determination of drain-source saturation current (I_{DSS}), gate-source bias is zero and pinch-off occurs.

It is possible, however, to enter pinch-off with V_{GS} at values other than 0 V. Any bias (V_{GS}) that is applied will have the opposite polarity from the interchannel voltage that caused the drain to saturate. Any combination of V_{GS} other than zero will tend to bias conduction at a point below saturation and as the interchannel voltage drop increases, the sum of the two voltages will equal—and have the same effect as—the pinch-off voltage described above. At the point where V_{GS} and channel voltage drop equal V_P, the device will be pinched-off and current will stabilize at some point below saturation.

Characteristic Curves

A set of characteristic curves for an N-channel JFET is shown in Figure 15-6. Notice that each current curve extends vertically near the right side of the chart. These vertical lines represent the avalanche current and breakdown characteristic of the device.

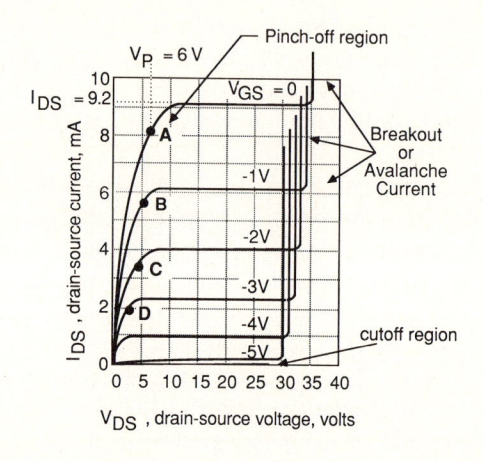

Fig. 15-6 N-Channel JFET Characteristic Curves

First we will discuss the effect of V_{GS} being 0V. For this to happen, the gate and source must be shorted together. You can see, from the chart, that I_D increases at a linear rate until it reaches point A. At point A, I_D begins to level off. Shortly after point A, the current line is practically level. Then comes a region where increases in V_{DS} have little effect on I_D. The point where I_D becomes level is the pinch-off point. As with bipolar transistors, an increase in voltage that does not produce an increase in current indicates that the device is saturated. This current, the maximum possible with the gate and source shorted, is labeled I_{DSS}. At some point, however, increases in V_{DS} cause the JFET to Zener. At that point, the junction will breakdown and *avalanche current* will flow. This is illustrated on the graph (Figure 15-6), by the vertical lines that represent changes in I_D.

> **NOTE:** With $V_{GS} = 0$ V and $V_P = 6$ V, the voltage across the gate-channel junction equals 6 V. This tells us two things:
>
> 1. To cut off the transistor using reverse bias, we must apply -6 V V_{GS}.
> 2. Any combination of V_{GS} ($-$V) and V_{DS} ($+$V) that equals 6 V will cause the channel to be pinched-off. At these points the channel becomes saturated, but I_D is not equal to I_{DSS}. Points B, C, and D (Figure 15-6) are points where this occurs.

Further examination of this chart reveals that as V_{GS} is made more negative, the amount of current (I_D) that flows at pinch-off decreases. This is because negative voltage is reverse bias to the N-channel JFET. Reverse bias causes the depletion region to become larger and, therefore, the channel becomes smaller. You should realize that the curves $V_{GS} = -1$ V to cutoff are supplied by external sources and are in addition to the natural reverse bias that is created by shorting the gates and source. With each of these reverse bias voltages, a point is reached where pinch-off occurs. Once pinch-off occurs, though, it requires approximately the same amount of V_{DS} for avalanche. Remember, avalanche current is reverse current, and in the JFET it flows through the reverse biased gate-source junction. However, avalanche current adds to the drain current causing the fast increase represented by the vertical lines at the right of the chart. When V_{GS} is approximately -6 V, the JFET stops conducting and is cut off.

The amount of V_{DS} that is required to saturate the transistor is called *pinch-off voltage* (V_P). In actuality, this is a difficult point to identify, but we usually consider it to be the point where current ends its linear rise and begins to level off. In Figure 15-6, point A is called the *pinch-off point* and the pinch-off voltage is 6 V. Each JFET has a specific pinch-off voltage just as each bipolar transistor has a specific beta. Remember, any combination of $-V_{GS}$ and internal voltage drop that equals V_P can cause pinch-off.

At the point where V_{GS} equals V_P (pinch-off voltage), we see that the transistor is *cut off*. This point is called the V_{GS}OFF point. The voltage required to reach V_{GS}OFF will always equal V_P. The difference is that they always have opposite polarities; therefore, any combination of V_{GS} and V_{DS} that equals V_P will cause pinch-off to occur.

Characteristic curves and specification sheets can be used for JFETs as they are for bipolar transistors. Once we determine I_{DSS} and either V_P or V_{GS}OFF (remember the two are equal), we can calculate the drain current (I_D) for any value of V_{GS}. To make these calculations, use the following formula:

$$I_D = I_{DSS} \left[1 - \frac{V_{GS}}{V_{GS}OFF}\right]^2$$

To see how this works, we will assume that $I_{DSS} = 5\text{mA}$, $V_{GS} = -4\text{ V}$, and $V_{GS}\text{ OFF} = -8\text{ V}$. With $V_{GS} = -4\text{ V}$, we calculate I_D as follows:

$$I_D = I_{DSS} \left[1 - \frac{V_{GS}}{V_{GS}OFF}\right]^2$$

$$I_D = 5\text{ mA} \left[1 - \frac{-4}{-8}\right]^2$$

$$I_D = 5\text{ mA }[1 - 0.5]^2$$

$$I_D = 5\text{ mA }[0.5]^2$$

$$I_D = 5\text{ mA} \times 0.25$$

$$I_D = 1.25\text{ mA}$$

Sample JFET Specification Sheets can be found in Appendix E. Using one of these sheets, we can find data regarding I_{DSS} and $V_{GS}OFF$ that can be used to develop the transfer characteristics of a specific JFET.

Drain-source saturation current (I_{DSS}) is difficult to predict accurately. This parameter is dependent on the following factors:

1. manufacturing tolerance
2. temperature effects
3. variations in supply voltages

If you examine JFET specification sheets like those contained in Appendix E, you will find that I_{DSS} is stated in different ways. For some JFETs, we find I_{DSS}Min is the only current listed. For others we find that I_{DSS}Min and I_{DSS}Max are listed. A third, and more recent group, have I_{DSS}Min, I_{DSS}Typical, and I_{DSS}Max listed.

This tells us that it is possible for zero voltage-drain current (I_{DSS}) to vary among JFETs that carry the same identification designator. When considering an entire family of JFETs, it is possible that I_{DSS} may vary from I_{DSS}Min to I_{DSS}Max. For a single transistor, though, the typical value will more likely be close to the actual I_{DSS}. When designing circuits for a batch of JFETs, we use minimum and maximum transfer characteristics. The sections that follow illustrate and use these minimum and maximum transfer characteristics.

The formula that was introduced above closely approximates the JFETs transfer characteristics. By using values from the specification sheet that identify I_{DSS}Min and I_{DSS}Max and choosing the appropriate values for V_{GS}OFF voltages, we can build data tables suitable for plotting maximum and minimum transfer characteristics for the JFET. By plotting these data we can actually construct curves on a chart that represent the device's transfer characteristics.

Table 15-2 contains data that were calculated using the values for I_{DSS}Min = 4 mA and various V_{GS} voltages ranging from 0 V to −2.0 V.

> **NOTE:** The JFET used for these calculations was the N-channel, depletion-type JFET, type number 2N5459, designed for general purpose audio and switching purposes. Maximum and minimum saturation current specifications are:
>
> V_{GS} = 0 V, V_PMin = −2 V, I_{DSS}Min = 4 mA.
>
> V_{GS} = 0 V, V_PMax = −8 V, I_{DSS}Max = 16 mA.

Table 15-2
JFET I_{DSS} Minimum Transfer Characteristics

PLOTTING POINT	V_{GS}	I_{DSS}Min	I_D
1	0.0 V	4 mA	4.00 mA
2	−0.5 V	4 mA	2.25 mA
3	−1.0 V	4 mA	1.00 mA
4	−1.5 V	4 mA	0.25 mA
5	−2.0 V	4 mA	0.00 mA

When these points are located on the graph shown in Figure 15-7, they can be connected by a solid line. The line that represents the parameters I_{DSS}Min = 4 mA and V_PMin = −2 V is referred to as the **minimum transfer characteristic curve** for this JFET.

To establish the maximum transfer characteristics, we use the same formula; however, we use the I_{DSS}Max value of 16 mA and V_PMax = −8 V (obtained from the specification sheet). We also use five voltages as values for V_{GS} that range from 0 V to −8 V. The data that result from these calculations are listed in Table 15-3.

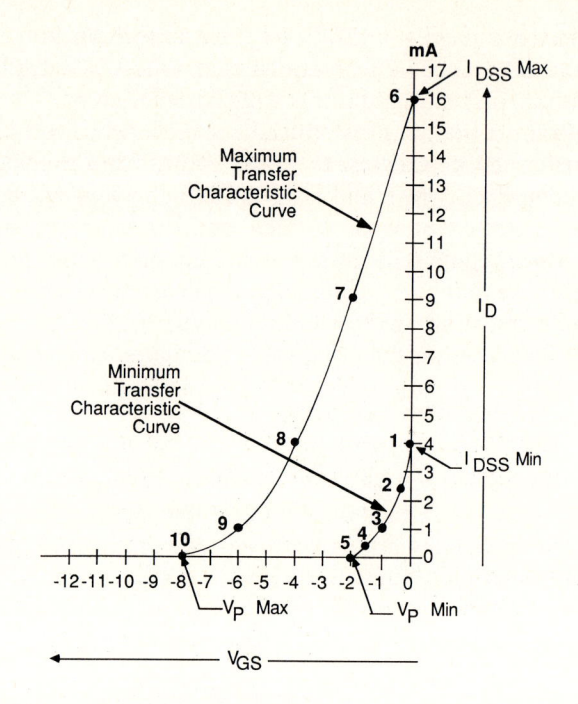

Fig. 15-7 Typical Transfer Characteristics

Table 15-3
JFET I$_{DSS}$ Maximum Transfer Characteristics

PLOTTING POINT	V$_{GS}$	I$_{DSS}$Max	I$_D$
6	0 V	16 mA	16 mA
7	−2 V	16 mA	9 mA
8	−4 V	16 mA	4 mA
9	−6 V	16 mA	1 mA
10	−8 V	16 mA	0 mA

When these points are located on the graph of Figure 15-7 and connected by a solid line, we have the **maximum transfer characteristic curve** for this JFET using parameters I$_{DSS}$Max = 16 mA and V$_P$Max = −8 V.

It is possible for a specific JFET to operate at any combination of V$_{GS}$ and I$_D$ that exists between the minimum and maximum transfer characteristic curves shown on the graph.

Understand that a single JFET will have one transfer characteristic. Another transistor from the same batch may have a completely different transfer characteristic. The maximum and minimum transfer characteristics (Figure 15-7) establish the limits within which all JFETs of this type will fit. This indicates that the selection of a JFET is subject to considerable variations in tolerance.

Transconductance

A parameter that is also valuable in the analysis of JFETs is **transconductance** (g_m). Transconductance is:

> The ratio of the change in drain current (I_D)
> to the change in gate-source voltage (V_{GS})
> for a given level of drain-source voltage (V_{DS}).

Quite often JFET specification charts will show a parameter called *transadmittance* (Y_{fs}) or *transconductance* (g_m). These terms refer to the relative ability of the *gate* to control I_D, and both quantities have a unit of measure, the *Siemens*, symbol **S**. On some specification sheets the unit of measure in micromhos $\mu\mho$.

The term *transconductance* is used because it refers to how a change in V_{GS} is "transferred" to I_D. Transconductance is a very small value, usually stated as micro Siemens (μS). The following formula can be used to calculate transconductance.

$$g_m = \frac{\Delta I_D}{\Delta V_{GS}}$$

We will refer back to transconductance later in this chapter. Values will be substituted for g_m and used in solving problems.

Self-Check

Answer each item by inserting the word or words required to correctly complete each statement.

1. List two types of field-effect transistors.

    a. _____
    b. _____

2. A JFET has a much better _____ than a bipolar transistor.
3. The _____ method is one way in which JFETs are made.
4. A JFET that has a different input connected to each gate is said to be _____ connected.

5. The three elements of a JFET are called:

   a. _____
   b. _____
   c. _____

6. The JFET has _____ junction(s) that exist(s) between the _____ and the _____ that is/are normally _____ biased.
7. Increasing the reverse bias that is applied to the gate-source junction causes the depletion region to _____ and the drain current to _____.
8. The JFET symbol _____ compares to V_{CC} as used in bipolar transistor circuits.
9. The voltage that is used to reverse bias the PN junction in a JFET is referred to by the symbol _____.
10. The voltage point at which I_D stops its linear rise and begins to stabilize is called the _____ voltage.
11. The symbol V_P identifies the _____ voltage.
12. The symbol I_{DSS} identifies the _____ current.
13. The voltage parameter that has the same value as V_P but has opposite polarity is _____.
14. Explain what a chart that contains maximum and minimum transfer characteristic curves represents and how it can be used.
15. Transconductance is often listed on the specification instead of _____ but both are measured in _____.

Biasing JFET Transistors

Before we begin our discussion of JFET biasing, it is important for you to remember that:

> All methods that are used to bias N-channel JFETs work equally well for biasing P-channel JFETs.

Fixed Bias JFET Circuits

Fixed bias is illustrated in the schematic diagram contained in Figure 15-8. Notice how similar this circuit is to the one used with bipolar transistors.

In this circuit, current enters Q_1 at the source, leaves Q_1 at the drain, and flows through R_L to $+V_{DD}$. The negative voltage that is applied at V_G reverse biases the gate-source junction of Q_1. This means that any current that flows across the gate-source junction is reverse current. The amount of reverse current that flows is so small that under normal circumstances it can be ignored. Note that R_G is a large resistor which further limits the amount of I_G that can flow.

The Junction Field-Effect Transistor (JFET)

To better understand the operation of this amplifier, we will examine its DC load line. Consider the fact that when we discuss this device, we are referring to a specific JFET that has a V_{DD} of 24 V. Remember, a load line is specific to a device and a particular applied voltage. Changing the transistor or the applied voltage requires that a new load line be plotted.

Construction of a load line for a JFET is very similar to that of a bipolar transistor. To begin the load line construction, we must first determine the maximum current and voltage limits that exist if Q_1 should open or short. These conditions set the outer limits (I_DMax and V_{DS}Max) of this transistor's operation when +24 V V_{DD} is applied. If the transistor opens, I_D will equal 0 mA, meaning that the entire +24 V (V_{DD}) is dropped by Q_1. Therefore, one limit is:

$$I_D = 0 \text{ mA}$$

and:

$$V_{DS} = V_{DD} = +24 \text{ V}$$

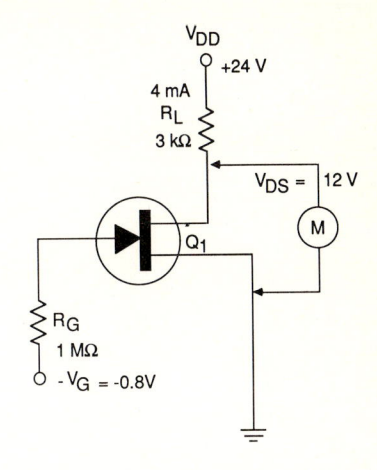

Fig. 15-8 Fixed-Biased JFET Circuit

The other limit is established by replacing Q_1 with a short and then calculating the amount of current that flows through R_L. If Q_1 is shorted, zero volts is present at the transistor's drain; therefore, the other limit is set by the conditions where:

$$V_{DS} = 0 \text{ V}$$

and:

$$I_D = \frac{V_{DD}}{R_L} = \frac{24 \text{ V}}{3 \text{ k}\Omega} = 8 \text{ mA}$$

To locate the limits of the DC load line we locate these two points on the chart shown in Figure 15-9. Point A is located by finding 0 mA (the bottom horizontal line) at the left of the chart and moving right until the 24 V point is located. This spot is marked with a dot that locates Point A.

Now to locate the other limit (Point B). First we locate the point that identifies the V_{DS} voltage of 0 V (the left-most vertical line) and then move up this line until we locate the 8 mA point. This is Point B and we mark it with another dot. A solid line is drawn to connect Points A and B. This line is the DC load line and represents an infinite number of operating points for the circuit shown in Figure 15-8.

Using the DC load line, we can easily establish the values for different bias levels that allow us to operate at different points along the load line. If we want to operate the device at the mid-point of its load line, we locate the +12 V (0.5 × V_{DS}Max) point along the V_{DS} line,

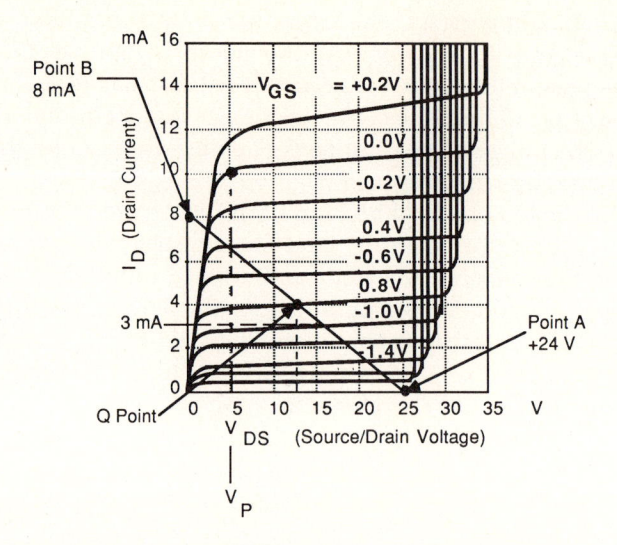

Fig. 15-9 DC Load Line for an N-Channel JFET Common-Source Amplifier

move vertically until we intersect the load line and mark that spot. The V_{GS} bias line that passes through this point is the -0.8 V line as indicated by the Q point. To operate the circuit at this Q point, we would connect -0.8 V bias as $-V_G$ at the end of R_G.

If -1 V is connected as $-V_G$, the transistor would have a Q point that lies where the -1 V V_{GS} crosses the load line. By dropping a vertical line to the V_{DS} line, we find that V_{DS} would be 15 V. If we moved horizontally (from this point of intersection) to the I_D line, we intersect I_D at the 3 mA point. This tells us that with -1 V applied to the $-V_G$ input, the circuit operates at a Q point where $V_{DS} = 15$ V and $I_D = 3$ mA.

When you consult the specification sheet for a particular FET, you will find that *drain-source saturation current* (I_{DSS}) and *pinch-off voltage* are two of the parameters that are listed. By observing the characteristic curve chart shown in Figure 15-9, you can see that several different values of V_{GS} will cause the transistor to enter pinch-off. Each value of V_{GS} has a different pinch-off voltage. Remember, V_P is a voltage value that is the opposite of V_{GS}OFF, and that any combination of V_{GS} and interchannel voltage drop that equals V_P will cause pinch-off. From this, you see that V_P and I_D can vary over considerable ranges, according to the V_{GS} that is applied to the gate. In Figure 15-8, we have used fixed bias in an attempt to have the transistor operate with a fixed V_P and I_D. This is accomplished by connecting a negative voltage $(-V_G)$ to the gate input.

In Figure 15-10, we have included a chart that has the maximum- and minimum-transfer characteristic curves of a specific JFET. Note that I_{DSS}Min = 4 mA, V_{GS}Min = -3 V, I_{DSS}Max = 8 mA, and V_{GS}Max = -6 V. The vertical line represents the -1 V fixed bias applied to the circuit in Figure 15-8. This line is called the **bias line**. The points where the bias line intersects the two transfer curves identify the maximum and minimum values of I_D that are possible with $-1\,V$ fixed bias. Notice that the bias line extends through the $-1\,V$ point of the horizontal (V_{GS}) axis of the chart. By drawing horizontal dashed lines from each of the intersections to the vertical (I_D) axis of the chart, we can determine I_DMax and I_DMin. For the example used here:

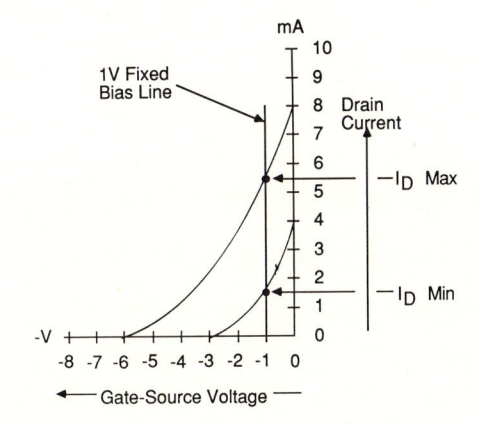

I_DMin $= 1.5\,mA$

I_DMax $= 5.5\,mA$

Fig. 15-10 I_D (Maximum and Minimum) for a Fixed-Bias JFET Circuit

You can see from these values that with fixed bias, the transistor you choose can operate at currents that range from 1.5 mA to 5.5 mA—a 4 mA swing.

These values can be used to make calculations that determine the amount of variation that this swing will cause in V_{DS}. These calculations involve the use of the following formula and procedures:

$V_{RL} = I_D \times R_L$

$V_{DS} = V_{DD} - V_{RL}$

To calculate V_{DS} at I_DMax:

$V_{RL} = 5.5\,mA \times 3\,k\Omega = 16.5\,V$

V_{DS}Min $= 24\,V - 16.5\,V = 7.5\,V$

To calculate V_{DS} at the I_DMin point:

$V_{RL} = 1.5\,mA \times 3\,k\Omega = 4.5\,V$

V_{DS}Max $= 24\,V - 4.5\,V = 19.5\,V$

Remember, V_{DS}Max occurs when I_D is minimum, and V_{DS}Min occurs when I_D is maximum.

From this chart, you should see the use of this type (fixed bias) circuit does not result in a stable circuit. Using -1 V fixed bias, it is possible for the operating point to lie anywhere along the bias line, from the point where $I_D = 1.5$ mA and $V_{DS} = 19.5$ V to a point where $I_D = 5.5$ mA and $V_{DS} = 7.5$ V. It is possible to modify this circuit to get better stability under closely controlled conditions; however, under normal operating conditions the circuit is so unpredictable that it is not suitable for use.

In practical JFET circuits, *self-bias* is used extensively. By placing a resistor (R_S) in series with, and between, the *source* element and ground, it is possible to self-bias the circuit. By adding both self-bias and fixed bias we can have much more accurate control of the circuit's Q point. These biasing methods are discussed in the following section.

Self-Biased JFET Circuits

Three different self-biasing methods are illustrated in Figure 15-11. Notice that each is labeled according to the type of biasing it provides. Each circuit will be discussed in this section.

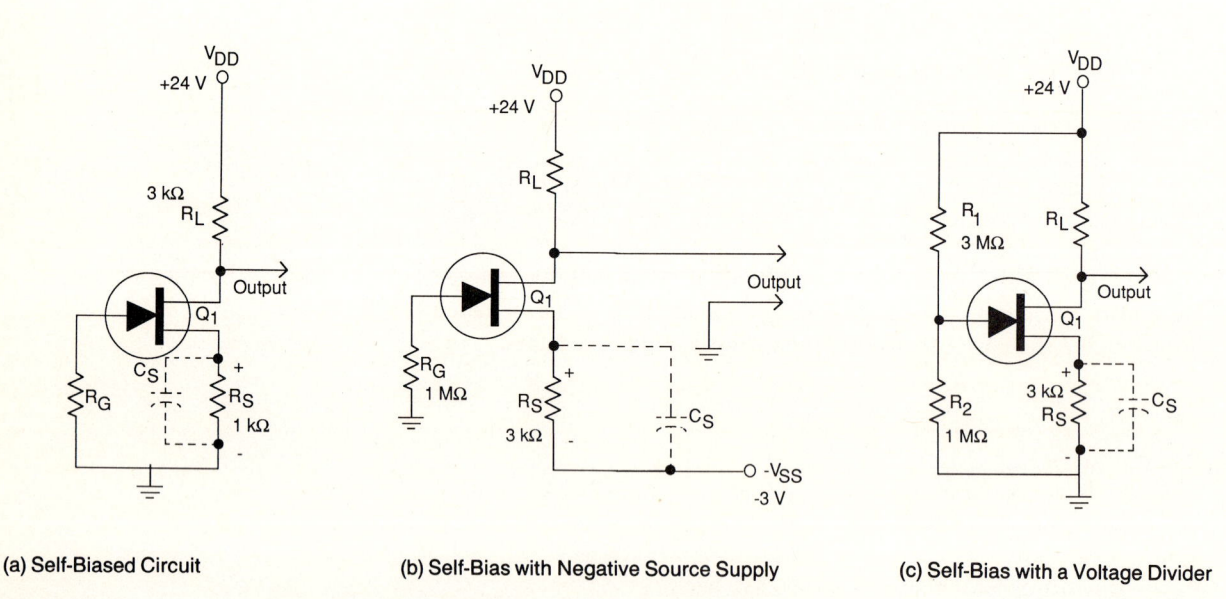

(a) Self-Biased Circuit

(b) Self-Bias with Negative Source Supply

(c) Self-Bias with a Voltage Divider

Fig. 15-11 Three Different Biasing Methods

First, we will consider the circuit that has self-bias only. This circuit is illustrated in Figure 15-11(a). Notice that R_G is connected to ground (0 V) potential. The source is connected to the same (0 V) ground; however, a resistor R_S has been inserted between the source

The Junction Field-Effect Transistor (JFET)

and ground. This means that current flowing away from ground must pass through R_S before arriving at the source of Q_1. Current flowing through R_S causes a voltage (V_{RS}) to be dropped on the resistor. V_{RS} has the polarity shown on the circuit diagram with the positive potential located at the top of R_S.

Remember, the gate-source junction is reverse biased, and any current that flows through it is so small that it can be ignored. With ground (0 V) connected to one end of R_G and no current flowing through R_G, the voltage at the gate is 0 V. The current through R_S drops a voltage, placing a positive voltage at the source. In effect, this combination of voltages (0 V at the gate and +V at the source) reverse biases the gate-source junction.

To graphically analyze this circuit (Figure 15-11(a)), we will use the curves shown in Figure 15-12. Our objective is to determine the variation between I_DMin and I_DMax. The graph shown represents the maximum- and minimum-transfer characteristic curves of a specific JFET. If we select convenient values for I_D, we can construct the bias line for this circuit. To do this, we must use our values for I_D and calculate the values of V_{GS} present for that value of I_D. By using the results of these calculations, we can construct the *bias line* for this circuit.

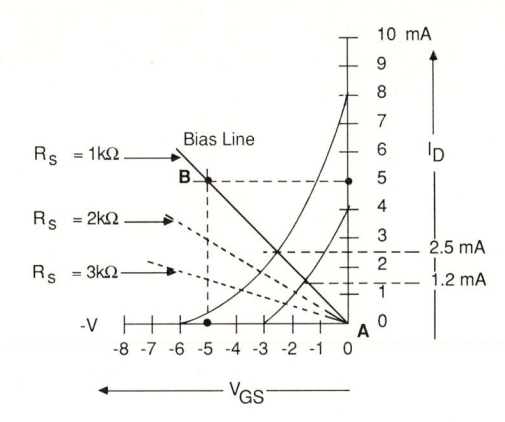

Fig. 15-12 Self-Biased JFET Circuit Determination of I_D Mins and I_D Max

In the analysis that follows, we will use the three formulas shown below:

$$V_{RL} = I_D \times R_L$$

$$V_{RS} = I_D \times R_S$$

$$V_{DS} = V_{DD} - V_{RL} - V_{RS}$$

On the graph, point A represents two things:

1. the point where $I_D = 0$
2. the point where $V_{GS} = 0$.

This point is located at the lower right corner and is labeled 0.0. This point serves as one end of our bias line. Now, to determine the other points that could be used, assume that $I_D = 5$ mA. With this value for I_D and the ohmic value of R_S, we can calculate V_{GS} (difference in potential between gate and source) as follows:

Without external V_G, $V_{GS} = V_{RS}$

therefore:

$$V_{RS} = I_D \times R_S = 5 \text{ mA} \times 1 \text{ k}\Omega = 5 \text{ V}$$

Note that current through R_S is flowing away from ground. This tells us that $V_{RS} = +5$ V. Applying $+5$ V to the source of the JFET has the same effect as applying -5 V to the gate of the same JFET as V_{GS}.

Now we can locate this point (-5 V) on the horizontal (V_{GS}) axis. Mark it with a dot and then construct a vertical line upward from this point. Next we locate the 5 mA point on the vertical (I_D) axis, mark that point with a dot, and construct a horizontal line to the left and delete up to the point where the two dashed lines intersect. We will call the point of intersection point B as indicated on Figure 15-12. Point A (0 V, 0 mA) locates the other end of the *bias line*. To complete the construction of the bias line, points A and B are connected by a solid line as shown in Figure 15-12. To identify the circuit for which the bias line was constructed, we label it $R_S = 1$ kΩ. Bias lines for 2 kΩ and 3 kΩ source resistors are shown using dashed lines.

The points at which the bias line crosses the minimum- and maximum-transfer characteristic curves are the I_DMin and I_DMax points. For $R_S = 1$ kΩ these are: I_DMin $= 1.2$ mA and I_DMax $= 2.5$ mA. By comparing this to the fixed bias method you can see that I_D will have a very small variation compared to fixed bias of -1 V.

Now that we have identified the maximum and minimum values for I_D, we can solve for the values of V_{DS} at each point. Calculations for V_{DS}Max are:

$$V_{RL} = I_D \times R_L = 1.2 \text{ mA} \times 3 \text{ k}\Omega = 3.6 \text{ V}$$

$$V_{RS} = I_D \times R_S = 1.2 \text{ mA} \times 1 \text{ k}\Omega = 1.2 \text{ V}$$

$$V_{DS} = V_{DD} - V_{RL} - V_{RS} = 24 \text{ V} - 3.6 \text{ V} - 1.2 \text{ V} = 19.2 \text{ V}$$

To find V_{DS}Min, we do the following:

$V_{RL} = I_D \times R_L = 2.5 \text{ mA} \times 3 \text{ k}\Omega = 7.5 \text{ V}$

$V_{RS} = I_D \times R_S = 2.5 \text{ mA} \times 1 \text{ k}\Omega = 2.5 \text{ V}$

$V_{DS} = V_{DD} - V_{RL} - V_{RS} = 24 \text{ V} - 7.5 \text{ V} - 2.5 \text{ V} = 14 \text{ V}$

From these calculations we see that the use of self-bias allows much closer control of I_D and V_{DS} than was possible with fixed bias. Examine the $R_S = 2$ kΩ and $R_S = 3$ kΩ lines on Figure 15-12. You can visually determine that the variations in I_D and V_{DS} will be even less for these values of R_S. In fact, the larger the value for R_S, the smaller the variations of I_D and V_{DS}. The use of large values for R_S will, however, reduce the amount of I_D that flows, reduce the gain of the circuit, and cause greater degeneration of the input signal. In many circuits, this is an undesirable result.

As with bipolar transistors, these undesirable effects can be reduced by the insertion of a source capacitor. In each circuit, C_S has been added as a dashed addition. C_S provides a bypass for all AC variations equal to or higher than the circuit's frequency of operation. This places AC ground at the source of Q_1 and returns gain to the level present before R_S was inserted.

An advantage gained from the use of R_S is **temperature stabilization**. JFETs, like other solid state devices, have a negative temperature coefficient. Increases in operating temperature affect transistor operation. By using R_S, we minimize this effect. With any increase in temperature, I_D will increase. When I_D increases, V_{RS} also increases. An increase in V_{RS} increases the junction's reverse bias which, in turn, decreases I_D and V_{RS}, returning operation back to normal.

As with many electronic circuits, obtaining a desired output involves compromise. Biasing of JFET circuits is no different. We have already seen that by using self-bias, we can get better stability. The other circuits we will discuss increase stability even more.

Figure 15-11(b) contains a self-biased JFET circuit. Notice that a negative voltage $(-V_{SS})$ is connected to the bottom of R_S. This circuit is said to have self-bias with external voltage.

A graphical analysis of this circuit is shown in Figure 15-13. In this figure, we see the minimum- and maximum-transfer characteristic curves for a specific JFET. The construction of a bias line for this circuit is similar to that discussed earlier. Use Figure 15-13 as a reference as we analyze this circuit. Formulas that will be used are:

$V_{RS} = I_D \times R_S$

$V_{SS} = V_{RS} + V_{GS}$

$V_{GS} = V_{SS} - V_{RS}$

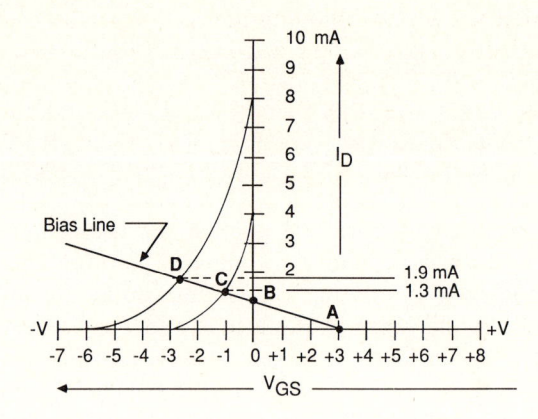

Fig. 15-13 Self-Biased JFET Circuit with V_{SS} Determination of I_D Mins and I_D Max

For this circuit, we see that $I_D = 0$ mA when V_{GS} equals $+3$ V. This is because the gate is at 0 V and $V_{SS} = -3$ V, which means that the gate is $+3$ V with respect to V_{SS}. On the graph, we locate this point at the $+3$ V position on the horizontal (V_{GS}) axis and mark it with a dot. Using these data, we can solve for I_D using the following:

$$I_D = \frac{V_{RS}}{R_S}$$

$$I_D = \frac{3\,V}{3\,k\Omega} = 1\,mA$$

$$V_{GS} = V_{SS} - (I_D \times R_S)$$

$$V_{GS} = 3\,V - (1\,mA \times 3\,k\Omega) = 3\,V - 3\,V = 0V$$

Now we know that $V_{GS} = 0$ V when $I_D = 1$ mA. We indicate this by locating the $I_D = 1$ mA position on the vertical axis and marking this point (B). To construct the bias line, extend a line through point A, point B, and both transfer curves. The points (C and D) where the bias line crosses each transfer curve are the I_DMin and I_DMax points. In this circuit:

I_DMin $= 1.3$ mA

I_DMax $= 1.9$ mA

Note that when using self-bias plus a -3 V V_{SS}, we have narrowed the variation to 0.6 mA (1.9 mA $-$ 1.3 mA). You can see that this combination provides better stability of Q point than either of those discussed earlier.

The Junction Field-Effect Transistor (JFET)

The circuit shown in Figure 15-11(c) is a self-biased circuit with external bias that is supplied by a voltage divider. This type circuit is used quite often in JFET circuits. Figure 15-14 is used for the graphical analysis of this circuit.

This analysis is very similar to the one used for Figure 15-11(b). To solve for V_{GS} when $I_D = 0$, we must use the following formula:

$$V_{GS} = V_{DD} \times \frac{R_2}{R_1 + R_2}$$

$$V_{GS} = 24\,V \times \frac{1\,M\Omega}{3\,M\Omega + 1\,M\Omega} = +6\,V$$

This point is located and marked along the horizontal axis. It has been marked as point A on Figure 15-14.

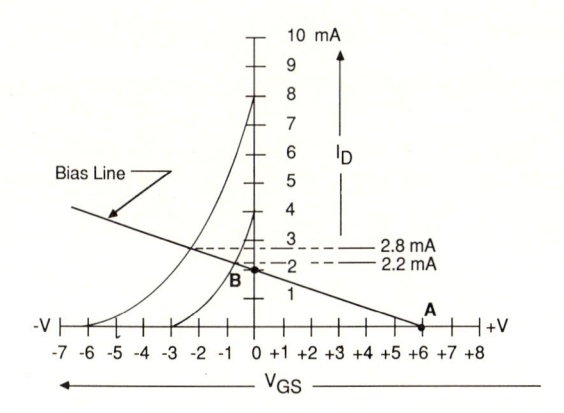

Fig. 15-14 Self-Biased JFET with External Voltage Divider Determination of I_D Min and I_D Max

Next we must calculate the value for I_D when V_{GS} equals 0 V. Remember, for V_{GS} to equal 0 V, V_{RS} must equal V_{GS}; therefore, $V_{RS} = 6\,V$. We calculate I_D as follows:

$$I_D = \frac{6\,V}{R_S} = \frac{6\,V}{3\,k\Omega} = 2\,mA$$

Now that we have the value for I_D we can plot this point by locating the $V_{GS} = 0\,V$ point (at the bottom of the vertical axis) and moving up the vertical (I_D) axis until we arrive at the $I_D = 2\,mA$ point. Mark this spot (point B) with a dot. Now connect points A and B with a solid line that is extended out through both transfer curves. This solid line is the *bias line*.

The points at which the bias line intersects the transfer curves locate I_DMin and I_DMax points for the circuit. In this case these are:

I_DMin = 2.2 mA

I_DMax = 2.8 mA

Notice that in this circuit, the variations of I_D have been reduced to 0.6 mA. It was explained with Figure 15-12 that increasing the size of R_S could increase stability. In that case, though, the increase in R_S limited I_D to values that were not practical. When self-bias plus external voltage (V_{SS} or voltage divider) is used it is possible to use the larger values for R_S without the loss of I_D that occurs with self-bias only.

For the circuits contained in Figure 15-11(b) and (c) we found that the variations in I_D had been reduced to 0.6 mA. This is a great improvement over the fixed and self-biased only circuits.

Design of a JFET Circuit Using Graphic Analysis

To design a JFET circuit that has the desired limits between I_DMin and I_DMax, we must reverse the procedures used for analysis. To do this, we first select convenient values for I_D minimum and maximum. We will use the graph shown in Figure 15-15 in this discussion.

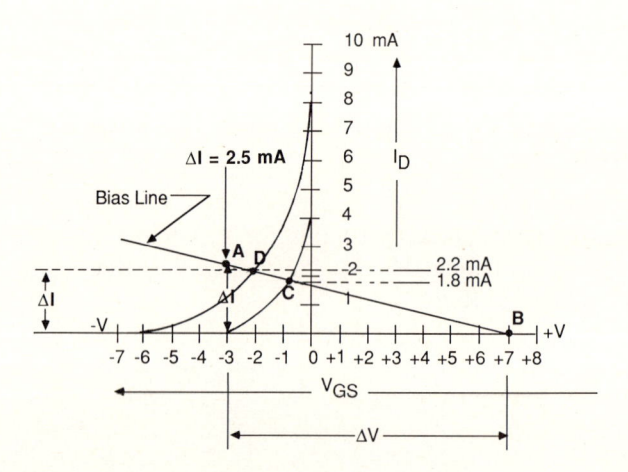

Fig. 15-15 Determining Bias Requirements and Resistor Sizes

The Junction Field-Effect Transistor (JFET)

For this analysis, we will assume the following values for use in the design of the circuit shown in Figure 15-16.

$R_L = 5\ k\Omega$

$I_D Min = 1\ mA$

$I_D Max = 2\ mA$

$V_{DD} = 15\ V$

$V_D =$ will be held to 10 V ± 1 V.

Our purpose is to determine the values for each resistor used in Figure 15-16. For the analysis we will use the transfer characteristic curves shown in Figure 15-15. To make the necessary calculations, we must use the following formulas and calculations:

$V_{RL} = I_D \times R_L = 2\ mA \times 5\ k\Omega = 10\ V$

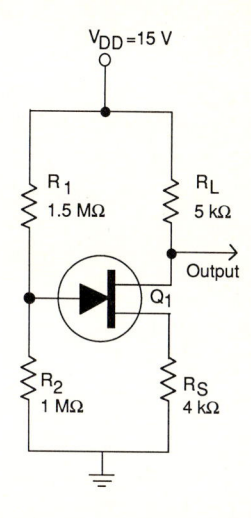

$V_D = V_{DD} - V_{RL} = 15\ V - 10\ V = 5\ V$

$I_D = \dfrac{V_{DD} - V_D}{R_L}$

$I_D = \dfrac{15\ V - 5\ V}{5\ k\Omega}$

$I_D = \dfrac{10\ V}{5\ k\Omega} = 2\ mA$

Fig. 15-16 Self-Bias with External Voltage Divider

Remember, a V_D of 10 V ± 1 V is desired. To calculate the change in I_D that is permissible to maintain this, we perform the following calculation:

$\Delta I_D = \dfrac{\pm 1\ V}{R_L}$

$\Delta I_D = \dfrac{\pm 1\ V}{5\ k\Omega} = 0.2\ mA$

This means that $I_D Min$ and $I_D Max$ can be found as follows:

$I_D Max = I_D + 0.2\ mA = 2\ mA + 0.2\ mA = 2.2\ mA$

$I_D Min = I_D - 0.2\ mA = 2\ mA - 0.2\ mA = 1.8\ mA$

Figure 15-15 can be used to plot the two values ($I_D Min$ and $I_D Max$) on the transfer characteristic curves. These points are identified as points C and D on the graph. Each is marked with a dot. These dots represent two points through which the bias line must extend. When a bias line is drawn that passes through points A and B and the V_{GS} axis, it appears as shown. The point where the bias line intersects the horizontal (V_{GS}) axis represents the externally supplied bias voltage. This voltage is +7V.

Notice that the ends of the bias line are at −3 V and +7 V on the V_{GS} axis. This is a change of 10 V (Δ V). Measured at the −3 V point, the change in current extends from 0 to 2.5 mA. We use this value as our current change (Δ I). The value of R_S that is needed to supply +7 V is the reciprocal of the *slope* of our bias line. The value of R_S can be calculated as follows:

$$R_S = \frac{\Delta V}{\Delta I}$$

$$R_S = \frac{10 \text{ V}}{2.5 \text{ mA}} = 4 \text{ k}\Omega$$

Using these data we can arrive at values for the resistors contained in the voltage divider. Remember, these are large value resistors that limit the gate current. The bias line intersects the horizontal (V_G) axis at +7 V. This means that the voltage divider must provide +7 V at the gate terminal. The ratio can be determined with the following steps:

1. $V_{R2} = 7$ V
2. $V_{DD} = 15$ V is applied to the divider.
3. $V_{R1} = V_{DD} - V_{R2} = 15$ V $- 7$ V $= 8$ V
4. This means that V_{R1} must drop 8 V.
5. The voltage divider ratio is 8 to 7.
6. The ratio for voltage division and resistance division are identical. Therefore, the resistance ratio (R_1 to R_2) is 8 to 7.
7. Ratio $= \dfrac{R_1}{R_2}$

   Ratio $= \dfrac{8}{7} = 1.428571$ to 1

8. Assume that we want to use a 1 MΩ resistor for R_2.
9. To find the value for R_1, do the following:
   $R_1 = R_2 \times 1.428571 = 1$ M$\Omega \times 1.428571 = 1,428,571 \Omega$
10. Therefore, R_1 can be 1.5 MΩ and R_2 can be 1 MΩ

The values that result from these calculations are shown on Figure 15-16. These values take into consideration the fact that resistor tolerances allow the selection of "off-the-shelf" components. A 5% tolerance would be sufficient to allow R_1 to have 1.5 MΩ of resistance. As a result of these solutions we have determined the following resistor values.

$R_S = 4\ k\Omega$

$R_1 = 1\ M\Omega$

$R_2 = 1.5\ M\Omega$

Using this analysis method and the transistor's specification sheet, the biasing arrangement for any JFET can be determined.

Self-Check

Answer each item by inserting the word or words required to correctly complete each statement.

16. Explain any differences between biasing N-channel and P-channel JFETS.
17. The two points that identify the limits of a JFET DC load line are points where the transistor is at _____ and _____.
18. When a JFET is cut off, its V_{DS} = _____ and its I_D = _____.

    a. maximum, maximum   c. minimum, maximum
    b. maximum, minimum   d. minimum, minimum

19. To provide fixed bias for a JFET, we must connect a voltage to the _____.

    a. source   c. gate
    b. drain

20. A DC load line represents _____ Q points for the circuit it represents.

    a. one   c. three
    b. two   d. infinite

21. Each JFET has _____ point(s) where pinch-off can begin.

    a. one   c. three
    b. two   d. several

22. Three things contribute to the fact that within a batch of JFET's I_{DSS} can vary from a minimum to maximum. These are:

    a. _____
    b. _____
    c. _____

23. A bias line is a line that connects the _____ point with _____ point, and it can be extended through the minimum and maximum transfer characteristic curves to locate the points where I_D is minimum or maximum.
24. Best stability of the Q point is obtained when _____ bias is used.

   a. self
   b. fixed
   c. fixed with external voltage.

25. Increasing the size of an unbypassed R_S (Figure 15-11(b)) will cause _____ to decrease, _____ to decrease, and _____ to increase.

JFET Amplifiers

Amplifiers are available that use JFETs as the amplifying device. These amplifiers perform the same functions discussed during your studies of bipolar transistors. As you might expect, the names and even the operation may differ, but the end result is the amplification of an input signal similar to that performed by bipolar transistor amplifiers.

The Common-Source Amplifier

The **common-source amplifier** can be compared to the bipolar common-emitter amplifier. The input to the JFET common-source amplifier is at the *gate*, which compares to the base of a bipolar transistor. The *drain* (collector) is connected to V_{DD} (V_{CC}) and the *source* (emitter) is connected to ground. The output is taken between the *drain* and ground, while the signal is applied between the *gate* and ground. A schematic diagram for a common-source amplifier is contained in Figure 15-17.

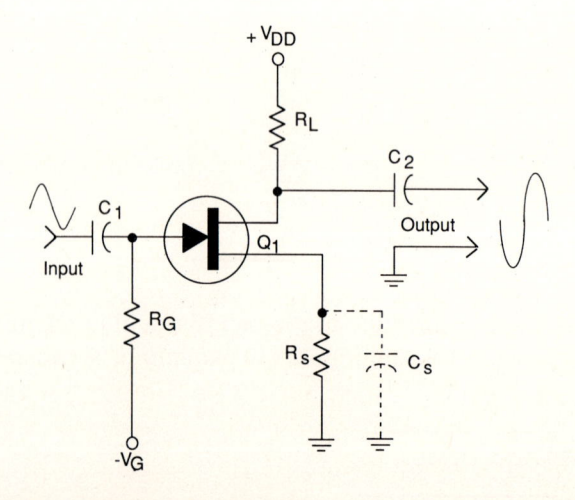

Fig. 15-17 A Common-Source JFET Amplifier

If the input signal goes positive, it reduces the reverse bias present across the gate-channel junction. A reduction in reverse bias reduces the size of the depletion region and enlarges the channel through which I_D flows. With a larger channel, channel resistance is less and I_D will increase. The increase in I_D causes V_{RL} to increase. An increase in V_{RL} causes V_D to decrease (go negative), which creates a 180° phase shift between the input signal and the output signal. Notice that this, too, is similar to the common-emitter amplifier in that a positive input produces a negative going output.

When the input goes negative, reverse bias increases, the depletion region increases, channel resistance increases, I_D decreases, V_{RL} decreases, and V_D increases (goes positive). This provides a 180° phase shift.

Voltage Gain

The voltage gain of any circuit can be calculated by dividing the change (AC pk-pk, or Δ) of the input signal voltage into the change (AC pk-pk, or Δ) of the output voltage. This is expressed by the formula:

$$A_V = \frac{\Delta V_{out}}{\Delta V_{in}}$$

Remember, the amount of drain current (I_D) determines the voltage drop on R_L (V_{RL}). The amount of V_{RL} determines the voltage present at the drain (V_D). Also, we must remember that I_D is controlled by the transconductance (g_m) of the transistor. Voltage gain can also be closely approximated through use of the transistor's specification sheet and the formula:

$$A_V \cong g_m \times R_L$$

We discussed, earlier in this chapter, the effect that R_S has on circuit gain and degeneration. Since this type bias is used widely, we must take R_S into consideration when calculating voltage gain (A_V). To eliminate the effects of R_S, we can install a bypass capacitor across R_S (illustrated by the dashed lines in Figure 15-17). C_S then acts to bypass AC signals around R_S. If R_S is not bypassed, the formula for approximating voltage gain is:

$$A_V \cong \frac{g_m \times R_L}{(1 + g_m) \times R_S}$$

Impedance Values

Closely examine the circuit diagram contained in Figure 15-17. You can see that the AC signal arriving at the gate of Q_1 has two AC paths: (1) down through R_G to ground; (2) into the gate of Q_1 and to ground through R_S. This means that *input impedance* is determined by R_G being in parallel with the gate-source junction and R_S in series. The gate-source junction is reverse biased, which means it will have very high resistance. Because R_G is normally less than 10 MΩ the ratio of the two resistances is so large that we can say that input resistance is approximately equal to R_G. At the very least, we can say that the input impedance is very high. The conclusion is that:

The input impedance of a common-source amplifier is approximately equal to the ohmic value of R_G.

The output signal has three paths: (1) up through R_L as I_D; (2) From the drain to the input of the next stage; (3) through the JFET. This means that the AC signal sees three parallel paths and that the *output impedance* will equal the ohmic value of R_L. Neither of these parameters can be found on the transistor's spec sheet. We will, however, find *output conductance* or *output admittance* listed. You learned earlier that resistance is the reciprocal of conductance or admittance; therefore, to calculate the R_{DS} we would divide either conductance or admittance into one:

$$R_{DS} = \frac{1}{\text{output conductance}}$$

When we perform this calculation, we find that output impedance of the device is very high, probably 600 kΩ or higher. R_L of a common-source circuit will usually be 5 kΩ or smaller. This means that the equivalent of R_{DS} and R_L will be approximately equal to R_L. The conclusion is that:

The output impedance of a common-source amplifier is approximately equal to the ohmic value of R_L.

The fact that a JFET has *high* input impedance and *low* output impedance means that it can be used to *match a high impedance* at its input and to *match a low impedance* in its output circuit. In other words, it can be used to:

Match a high impedance to a low impedance.

Load Line Analysis of a Common-Source Amplifier

To use a DC load line for analysis purposes we must follow the procedures stated earlier. To review the plotting of the load line and how it can be used in circuit design analysis, consider the circuit shown in Figure 15-18(a) and the set of characteristic curves that are included as Figure 15-18(b). To plot the DC load line we must first establish the saturation and cutoff points for the device. At cutoff, conditions in the circuit are $V_{DS} = 25$ V and $I_{DS} = 0$ mA. With Q_1 saturated, the conditions are such that $V_D = 0$ V and I_D is maximum. To determine the actual value for I_D, we treat the circuit as if Q_1 was saturated and R_L the only resistance through which I_D flows. We can calculate I_DMax as follows:

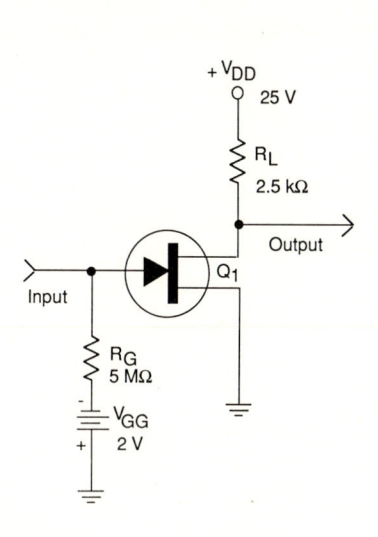

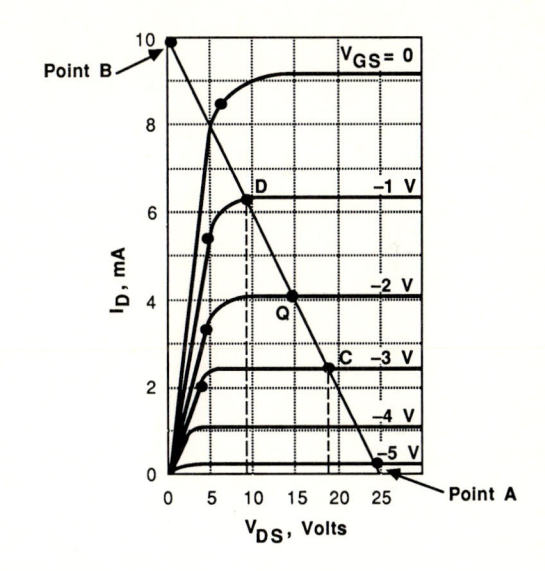

$$I_D\text{Max} = \frac{V_{DD}}{R_L}$$

$$I_D\text{Max} = \frac{25 \text{ V}}{2.5 \text{ k}\Omega} = 10 \text{ mA}$$

(a) Circuit Diagram

(b) Characteristic Curves

Fig. 15-18 Load Line for a Common-Source Amplifier

The conditions to be plotted are:

With Q_1 cut off:

$I_D \text{Min} = 0 \text{ mA}$

$V_{DS} = V_{DD} = 25 \text{ V}$

with Q_1 saturated:

$I_D \text{Max} = 10 \text{ mA}$

$V_{DS} \text{Min} = 0 \text{ V}$

We plot these points by locating their position on the chart and marking them with a dot. For cutoff, we locate the axis (horizontal) that represents V_{DS} and move along it until we locate the 25 V point. This point is marked with a dot as shown at Point A on the chart. Next we locate the point where V_{DS} is 0 V. This is at the bottom of the vertical (I_D) axis. We move up the vertical axis until we find the 10 mA point. This point is marked with a dot and is identified as Point B on the chart. To complete the load line, we connect Points A and B with a solid line as shown on Figure 15-18(b).

With -2 V applied to the gate, we must locate the Q point at the point where our load line and the $V_{GS} = -2$ V line intersect. This spot is labeled "Q" on the drawing.

To determine the amount of swing that occurs at the drain (output) for a given input, we can proceed as follows:

1. Starting at the Q point, move down to the point where you intersect the horizontal axis.
2. Record the value for V_{DS} at this point:
   $V_{DS} = 14.5 \text{ V}$
3. Assume that a 2 V peak-to-peak signal is applied. This means that V_{GS} will vary from -1 V to -3 V.
4. Locate and mark these two points. They are labeled "C" and "D" on the chart.
5. Drop vertical lines to the horizontal axis from these two points and determine the amount of V_{DS} at each point.
   At Point C, $V_{DS} = 19 \text{ V}$
   At Point D, $V_{DS} = 9 \text{ V}$

This tells us that when a 2 V peak-to-peak input signal is applied, the output (V_{DS}) varies from 9 V to 19 V (10 V peak-to-peak). For this application the voltage gain will be:

$$A_V = \frac{\Delta V_{out}}{\Delta V_{in}}$$

$$A_V = \frac{10 \text{ V}}{2 \text{ V}} = 5$$

This completes our explanation of the common-source JFET amplifier. You should be able to see the similarity that exists between this circuit and the bipolar transistor common-emitter amplifier.

The Common-Drain Amplifier

The **common-drain amplifier** is similar to the bipolar transistor common-collector configuration. An input is applied to the gate and the output is taken from the source. Figure 15-19 contains the schematic diagram for a circuit of this type.

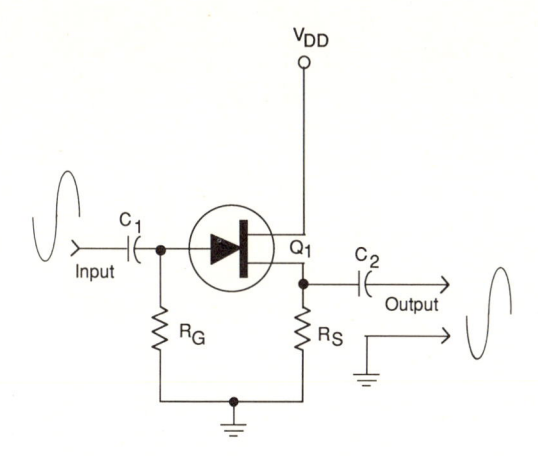

Fig. 15-19 A Common-Drain JFET Amplifier

Notice that the *drain* is connected directly to V_{DD}. The *source* has a resistor (R_S) that acts as the load resistor for the circuit. The output is taken across R_S. As with the common-collector amplifier, the voltage gain (A_V) of this circuit is less than one. The source resistor (R_S) develops self-bias and establishes the Q point. The degeneration present in the source circuit assures minimum distortion of the output signal.

The common-drain circuit has an input impedance that is approximately equal to the value of R_G. Its output impedance can be calculated using the formula:

$$Z_{out} = \frac{R_S}{1 + (g_m \times R_S)}$$

Assuming that g_m = 5000 μS and R_S = 2 kΩ, Z_{out} will equal approximately 200 Ω. This means that it has a high input impedance and a low output impedance. We can use these characteristics to provide impedance matching between a high impedance circuit or device to a low impedance circuit or device.

Again, you should be able to recognize the strong similarity that exists between this circuit and the bipolar transistor common-collector amplifier. It operates in a similar way and is used for purposes similar to the voltage follower, and it is often called a **source follower**.

The Common-Gate Amplifier

Figure 15-20 contains the schematic diagram for a **common-gate amplifier**. Note that the input signal is applied to the *source*, the output is taken from the *drain*, and the *gate* is grounded. In this circuit, R_L is placed in the drain circuit to act as a load resistor. When an input signal is applied to the circuit, it is developed by R_S. With the gate and source both grounded, R_S and the gate-channel junction are placed in parallel. This means that as the input signal is developed across R_S, it is also developed across the parallel branch (the junction).

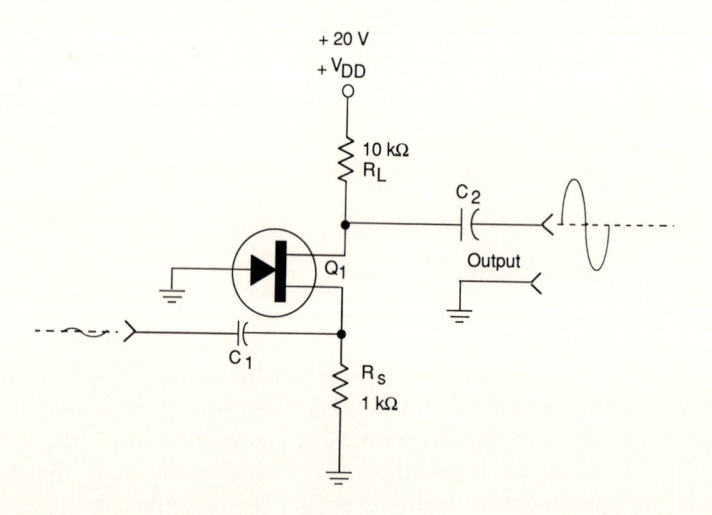

Fig. 15-20 A Common-Gate JFET Amplifier

To analyze this circuit, we will assume the *transconductance* (g_m) = 5000 μS and I_D = 1 mA. With 1 mA flowing V_{RS}, V_{RL} and V_D can be calculated as follows:

$V_{RS} = I_D \times R_S = 1\,mA \times 1\,k\Omega = 1\,V$

$V_{RL} = I_D \times R_L = 1\,mA \times 10\,k\Omega = 10\,V$

$V_D = V_{DD} - V_{RL} = 20 - 10\,V = 10\,V$

We can see that at the top of R_S, the voltage is +1 V; at the drain, V_D = 10 V; and the gate is at 0 V when operating at the Q point. This tells us that the gate is −1 V with respect to the source terminal and is, therefore, reverse biased with $V_{GS} = -1\,V$.

Assume that an input signal (0.2 V peak-to-peak) is applied to the input. This will cause V_{GS} to vary between −0.9 V and −1.1 V. As a result of these changes in V_{GS}, I_D will also be changing above and below 1 mA. For example:

$\Delta I_D = 1\,mA - \left(g_m \times \dfrac{\Delta V_{GS}}{2} \right)$

$\Delta I_D = 1\,mA - \left(5000 \times 10^{-6} \times \dfrac{0.2}{2} \right)$

$\Delta I_D = 1\,mA - (5000 \times 10^{-6} \times 0.1\,V)$

$\Delta I_D = 1\,mA - 0.5\,mA = 0.5\,mA$

This means that I_D is now varying 0.5 mA above and below 1 mA – or a change from 0.5 mA to 1.5 mA. For each alternation of the output, V_{RL} and V_D will change. The change in V_{RL} and V_D can be calculated as follows:

$V_{RL} = I_D \times R_L = 0.5\,mA \times 10\,k\Omega = 5\,V$

$V_{RL}Max = 10\,V + 5\,V = 15\,V$

$V_{RL}Min = 10\,V - 5\,V = 5\,V$

then:

$V_D Max = V_{DD} - V_{RL}Min = 20\,V - 5\,V = 15\,V$

$V_D Min = V_{DD} - V_{RL}Max = 20\,V - 15\,V = 5\,V$

This tells us that the output voltage is swinging between +5 V and +15 V, producing a 10 V peak-to-peak swing in output voltage.

For this circuit, the voltage gain will be:

$$A_V = \frac{\Delta V_{out}}{\Delta V_{in}}$$

$$A_V = \frac{10\text{ V}}{0.2\text{ V}} = 50$$

This completes the discussion of the three configurations used for the design of JFET amplifiers. Using what you learned regarding bipolar amplifiers and what has been presented here, you should have a good basic understanding of how these circuits work.

Advantages and Disadvantages of JFET Circuits

Some of the advantages gained by use of JFET amplifiers are:

1. The JFET's high input impedance and very low input current mean that it is power efficient.
2. JFETs are less noisy than bipolar transistors. Noise is generated in a solid state device by current passing through a junction. In the JFET, very little current flows through the junction. This reduces the noise generated to a very low level.

The major disadvantage encountered in the use of JFETs is the low gain they provide. The bipolar common-emitter amplifier can provide high gain. A common-source amplifier, using a bypassed source resistor, will have an A_V that is considerably lower.

The P-Channel JFET

To this point, our discussion has centered around the N-channel JFET and its applications. Now we will take time to briefly examine the **P-channel JFET**. A P-channel JFET circuit is shown in Figure 15-21. Notice that the channel is made of P-type material and the gates are made of N-type material. In this JFET, current flows from the *drain* to the *source*. Specification sheets for these devices are contained in Appendix E.

For the gate-source junction to be reverse biased, we must either connect a positive voltage to the gate or have a negative voltage present at the *source* terminal of the JFET. The schematic symbol for this circuit is also shown on Figure 15-21. Notice that the arrow points outward, toward the N-type gate.

The minimum- and maximum-transfer characteristics for the P-channel JFET are similar to those discussed for the N-channel JFET.

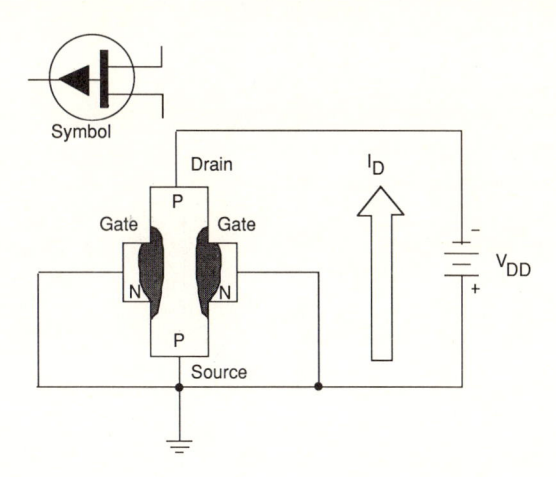

Fig. 15-21 P-Channel JFET

The only difference is that all voltages are negative with respect to ground, and I_D flows from $-V_{DD}$ into the drain, out the source, and to ground.

Saturation current (I_{DSS}) and pinch-off voltage (V_P) were discussed for the N-channel JFET. These parameters are the same for an identical P-channel JFET. Voltage gain (A_V), input impedance (Z_{in}), and output impedance (Z_{out}) also are identical for the two JFETs. The analysis of each configuration is identical except for the different voltage polarities and the direction of current flow.

Because the two have similar circuits, analyses and operations, we will not go into a detailed discussion of P-channel circuitry.

The Use of BI-FET Circuitry

JFET and the bipolar transistors have many similar characteristics. In fact, it is possible to intermix them within the same operational device. An example is shown in Figure 15-22. This circuit is called a **BI-FET amplifier**.

An advantage of this circuit is that it has a very high input impedance and a medium output impedance. This makes it ideal for matching a high-gain common-emitter amplifier to a circuit having a medium input impedance. The overall gain of the two amplifiers is less than if two common-emitter amplifiers were used. The gain is sacrificed to obtain better impedance matching and power efficiency that accompanies the use of a BI-FET circuit.

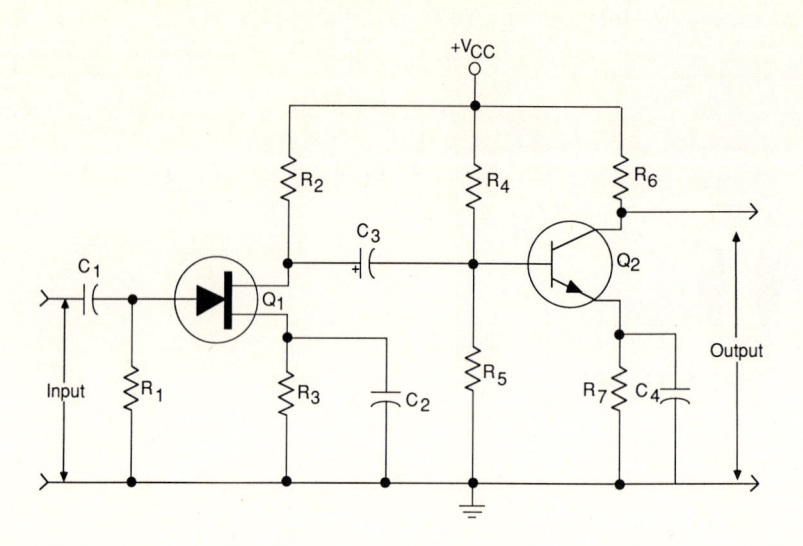

Fig. 15-22 A Two-Stage BI-FET Amplifier

Self-Check

Answer each item by inserting the word or words required to correctly complete each statement.

26. Name the three types of JFET amplifiers and state the comparable bipolar configuration.

    a. _____, _____
    b. _____, _____
    c. _____, _____

27. The main advantage that JFET amplifiers have over the bipolar transistor amplifiers is _____ input impedance and _____ output impedance.

    a. low, high   c. high, high
    b. low, low   d. high, low

28. When compared to bipolar transistor amplifiers, the JFET amplifiers have _____ voltage gain.

    a. high   c. low
    b. medium

The Junction Field-Effect Transistor (JFET)

29. Z_{in} for a common-source amplifier is approximately equal to the ohmic value of _____.

    a. R_S
    b. R_L
    c. R_G
    d. none of these

30. The two formulas for calculating voltage gain are:

    a. _____
    b. _____

31. Z_{out} for a common-source amplifier is approximately equal to the ohmic value of _____.

    a. R_S
    b. R_L
    c. R_G
    d. none of these

32. The input of a JFET amplifier can be used to match the impedance of a circuit or device that has _____ output impedance.

    a. high
    b. low
    c. medium

33. A common-drain circuit has a voltage gain that is _____.

    a. high
    b. low
    c. medium
    d. less than one

34. In the common-gate amplifier, the input is applied to the _____.

    a. gate
    b. source
    c. drain

35. A common-gate circuit has a voltage gain that is _____.

    a. high
    b. low
    c. medium
    d. less than one

Summary

A new device was introduced in this chapter, the *junction field-effect transistor*. This device differs from the bipolar transistor in that the "work" current does not pass through a junction. In the JFET, current is controlled by the electrostatic effect of reverse bias. A channel is provided for current flow. The amount of current is controlled by enlarging or decreasing the depletion region that is induced into the channel. A JFET has a much lower noise figure because "work" current does not pass through a junction, which would inject noise.

JFETs have three elements: *gate, drain*, and *source*. They are manufactured in two types: the P-channel and the N-channel types. They are constructed using a method called *diffusion*. In this process, a piece of semiconductor called a *substrate* has a piece of semiconductor joined to its surface. If the substrate is P-type, N-type material is used for the diffusion. The N-type material is deposited in a narrow strip called a *channel*. A second layer of P-type material is joined to the top of the N-channel, and the two pieces of P-type material serve as the gates.

By biasing the gate-source junction and connecting the JFET into operational circuits, most of the bipolar transistor's capabilities are available using the JFET. Three configurations are available: common-source, common-drain, and common-gate configurations. Each of these parallels a bipolar configuration and can be used for similar purposes.

The JFET has two advantages over the bipolar junction transistor: (1) a much lower noise figure; (2) a very high input impedance which provides good power efficiency. The main disadvantage of the JFET is its low voltage gain.

It is possible to mix JFET and bipolar transistors into the same circuitry. This is an advantage because of the different impedance characteristics of the two. Using combinations of bipolar and JFET transistors, impedance matching is made simpler.

Review Questions and/or Problems

1. In a P-channel JFET the channel is formed of _____-type material and the gates are _____-type material.

    a. P, N
    b. P, P
    c. N, P
    d. N, N

2. The GATE element of an N-channel JFET consists of _____ section(s) of _____-type material.

    a. one, N
    b. two, N
    c. one, P
    d. two, P

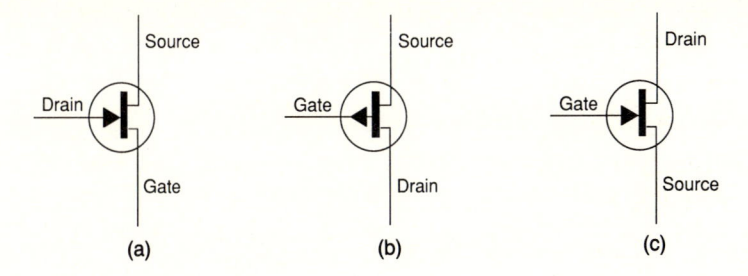

Fig. 15-23

3. Examine Figure 15-23. Which symbol is used to represent a P-channel JFET?

4. It _____ possible to cut off a JFET using self-bias only.

   a. is   b. is not

5. It _____ possible to cut off a JFET using the same source and gate potential.

   a. is   b. is not

6. During the period between V_P and breakdown, I_D remains relatively stable.

   a. True   b. False

7. Each value of V_{GS} has the same PINCH-OFF voltage point.

   a. True   b. False

8. The maximum- and minimum-transfer characteristic curves for a specific JFET have _____ and _____ as their limits.

   a. I_D, V_{DS}
   b. I_G, V_{GS}
   c. I_{DSS}, V_{GS}
   d. I_{DSS}, V_P

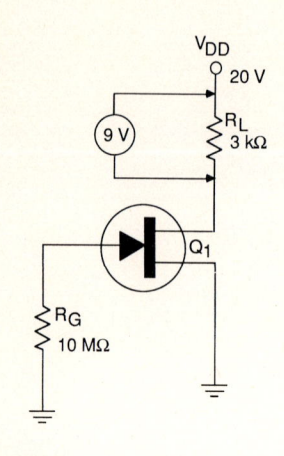

Fig. 15-24

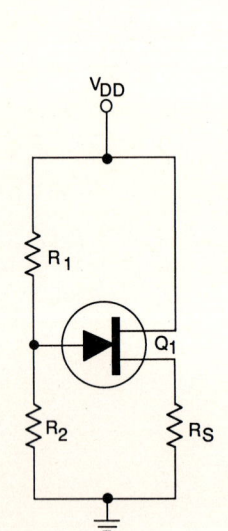

Fig. 15-25

9. Refer to Figure 15-24. What is the drain voltage for this circuit?

   a. 9 V
   b. 12 V
   c. 11 V
   d. 0 V

10. How much drain current (I_D) is flowing in the circuit contained in Figure 15-24?

    a. 3 V
    b. 9 V
    c. 9 mA
    d. 3 mA

11. This circuit (Figure 15-24) is a _____ amplifier.

    a. common-source
    b. common-collector
    c. common-gate
    d. common-drain

12. Adding a resistor between the source (Figure 15-24) and ground will cause A_V to _____.

    a. increase
    b. decrease
    c. remain the same

13. Refer to Figure 15-24. To provide self-bias for this circuit, we would add a _____.

    a. battery between R_G and ground.
    b. resistor between the source terminal and ground.

14. When this circuit (Figure 15-24) is cut off, V_{DS} equals _____.

    a. 20 V
    b. 0 V
    c. 9 V
    d. 11 V

15. The output from this circuit (Figure 15-24) is taken from the _____ element.

    a. source
    b. gate
    c. drain

16. Refer to Figure 15-25. Resistor _____ serves as the load resistor for this circuit.

    a. R_1
    b. R_2
    c. R_S

The Junction Field-Effect Transistor (JFET)

17. Voltage gain for the circuit in Figure 15-25 is _____.

    a. high
    b. low
    c. less than unity

18. Refer to Figure 15-25. Use of R_s causes _____.

    a. regeneration
    b. degeneration

19. The input to this circuit (Figure 15-25) is applied to the _____ and the output is taken from the _____.

    a. gate, drain
    b. gate, source
    c. source, drain
    d. source, gate

20. Check Figure 15-25. Compared to R_S, R_2 will be _____.

    a. very large
    b. large
    c. small
    d. very small

21. Z_{in} for this circuit (Figure 15-25) is approximately equal to the ohmic value of _____.

    a. R_1
    b. R_2
    c. R_S
    d. R_1 and R_2 in parallel

22. The amplifier shown here (Figure 15-25) is a _____ type.

    a. common-source
    b. common-collector
    c. common-gate
    d. common-drain

23. Check Figure 15-26. When Q_1 is saturated, I_D equals _____.

    a. 5 mA
    b. 0 mA
    c. 2.5 mA

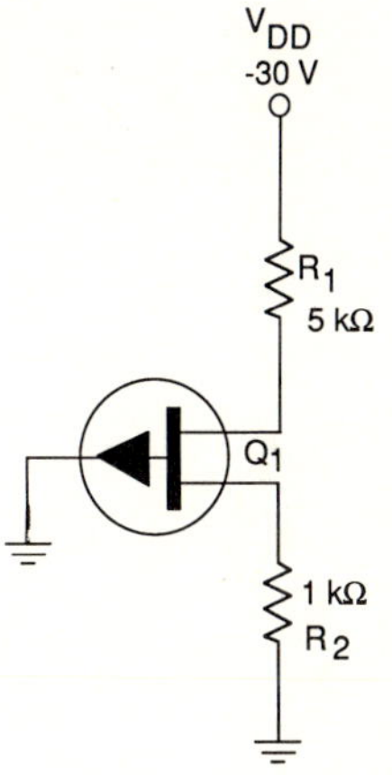

24. In Figure 15-26, current flows from _____ to _____.

    a. ground, V_{DD}
    b. V_{DD}, ground

25. For the circuit in Figure 15-26, the input is applied to the _____ and the output is taken from the _____.

    a. gate, source
    b. drain, source
    c. gate, drain
    d. source, drain

Fig. 15-26

16

MOSFETs — Metal Oxide Semiconductor Field Effect Transistors

Objectives

1. Define the following terms:

    a. MOSFET
    b. Insulated-gate field-effect transistor
    c. Metal-oxide-semiconductor field-effect transistor
    d. Enhancement mode
    e. Depletion-enhancement mode
    f. V-MOSFETs
    g. CMOS
    h. PMOS
    i. NMOS

2. Explain how MOSFETs are biased using:

    a. Fixed bias
    b. Self-bias

3. List the common types of MOSFET amplifier circuits that are used and describe the operation of each type.
4. Describe the construction and uses for V-MOS and CMOS devices.

Introduction

Field-effect transistors were introduced in Chapter 15. FETs are divided into two classes: junction field effect transistors (JFETs) and metal oxide semiconductor field effect transistors (MOSFETs). The difference between the two is that in the **MOSFET** the *gate* is "insulated" from the substrate by the use of a layer of silicon dioxide—the reason these transistors are also called *insulated gate field effect transistors*. Either of the names, *insulated gate FETs* or *MOSFETs*, refers to the devices that you will study in this chapter.

Enhancement- and Depletion-Mode MOSFETs

The physical construction of the MOSFETs is very similar to that used with JFETs in the last chapter. Diffusion processes are used in the manufacture of these devices. A pictorial diagram of a cutaway view of an *enhancement-mode MOSFET* is shown in Figure 16-1. This figure will be used to review the construction methods employed in the fabrication of these devices.

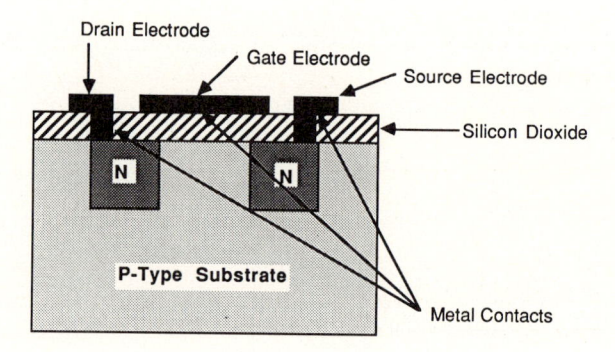

Fig. 16-1 Enhancement Mode MOSFET

     The black areas that appear on the diagram represent metal electrodes. Notice that there are three electrodes and that they are labeled "DRAIN", "GATE", and "SOURCE." Each of these metal contacts is connected to a different region of the semiconductor. This semiconductor device begins as a piece of P-type semiconductor called a **substrate**. After preparation of the substrate, two N-type sections are diffused into the substrate. These will serve as the *drain* and *source*. After these sections are in place, the entire substrate, except for the points where the source and drain electrodes will be mounted, is coated with a layer of *silicon dioxide* that serves two purposes: (1) it provides electrical insulation; (2) it protects the device against damage during further fabrication. After the dioxide is in place, the metal contacts that form the three electrodes are attached to the device. Notice that the source and drain electrodes make physical contact with the two sections of N-type material. The *gate*, however, is insulated from the P-type material by the coat of silicon dioxide. The use of a *metallic* gate and an *insulating* layer of silicon *dioxide* gives rise to the two names (metal-oxide-semiconductor and insulated-gate) that are used to refer to devices of this type.

MOSFETs are available in two classifications: (1) **enhancement mode**; (2) **depletion-enhancement mode**. Both of these are available in types that use either N-channel or P-channel characteristics. In most operational situations, the source and the substrate are connected. To make reference to the two types, it is common to call them **E-MOSFET** or **DE-MOSFET**. The abbreviations should be apparent. Actually, the name DE-MOSFET is usually shortened to **D-MOSFET**.

Construction of the DE-MOSFET is very similar to the E-MOSFET illustrated above. A cutaway pictorial view is shown in Figure 16-2. You will notice that the only difference between the two is the existence of an N-channel that connects the source and drain. This channel is diffused within the substrate prior to application of the silicon dioxide insulation. The N-channel is much more lightly doped than either the source or the drain.

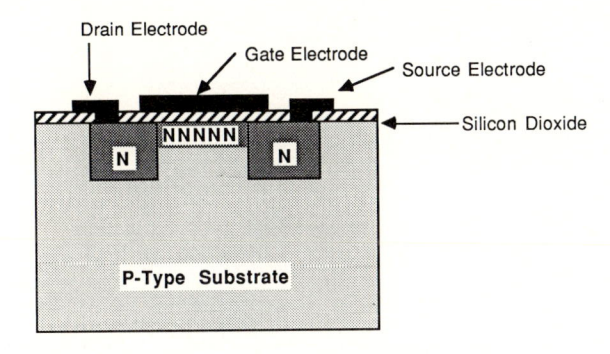

Fig. 16-2 Depletion-Enhancement Mode MOSFET

With this construction the transistor will have current flow at any time when the drain is positive with respect to the source, or vice versa. I_D will flow even when V_{GS} is 0 V. When V_{GS} has a small negative voltage applied, the transistor operates in the *depletion mode*, and when V_{GS} is positive the transistor operates in the *enhancement mode*.

Operation of an E-MOSFET

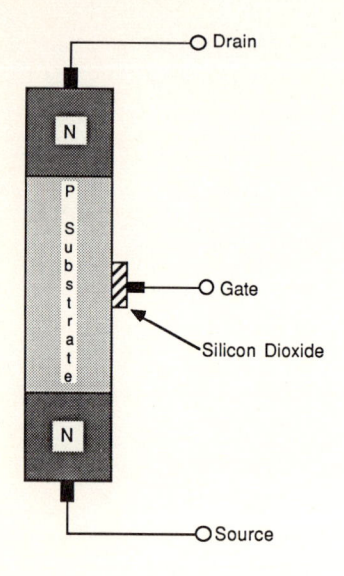

Fig. 16-3 E-MOSFET — Pictorial Diagram

When teaching this subject, many instructors prefer to use the pictorial diagram shown in Figure 16-3 with their explanations. It is common practice to refer to this device as an E-MOSFET. A careful examination of this diagram reveals that it has the same parts as those shown in Figure 16-1. The diagonally shaded area between the P-substrate and the gate electrode is the silicon dioxide insulator. Remember, this is what makes this device different from the JFET. It also gives rise to the names "metal oxide semiconductor" and "insulated gate." The gate is insulated from the semiconductor material, and the gate is metal that is bonded to an oxide insulator that lies between it and the semiconductor.

The device is, however, still a *field effect* device because electron flow between the source and drain is controlled by the effect of an electrostatic field. Certain gate voltages (V_{GS}) can be used to start, increase, and decrease drain current (I_D).

If you check Figure 16-3, you will see that no channel is shown connecting the source and drain. The substrate contains a section of P-type material that separates the two. All current control occurs because of the application of a positive voltage ($+V_{GS}$) to the gate. Observe Figure 16-4(a) where $V_{GS} = 0$ V, and V_{DS} is connected negative to the source and positive to the drain. With this connection the source junction is forward biased and the drain junction is reverse biased. Because of the small current that can flow through a reverse biased junction, I_D will be very small.

(a) $V_{GS} = 0$ V

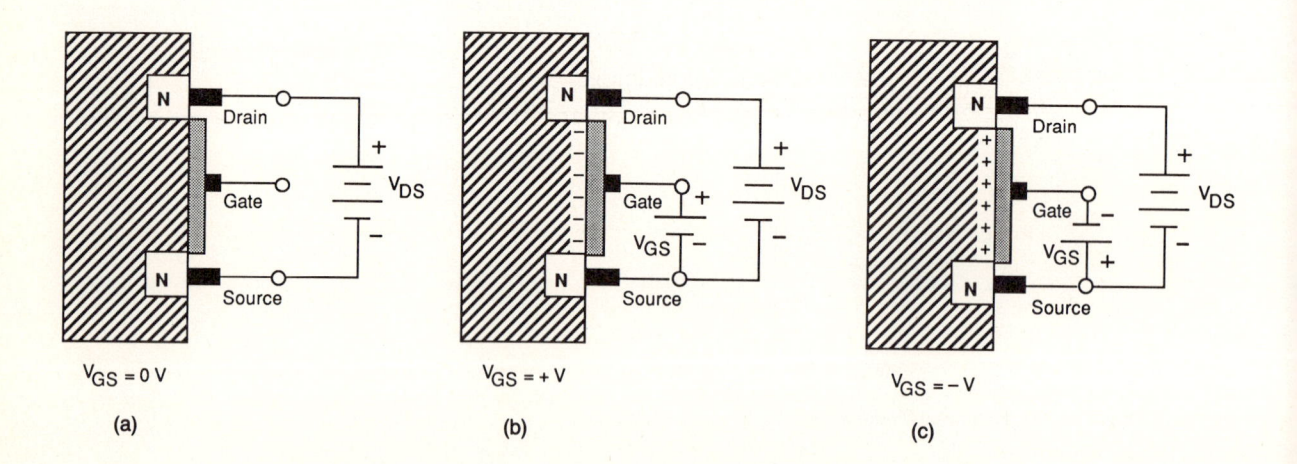

(b) $V_{GS} = +V$

(c) $V_{GS} = -V$

Fig. 16-4 E-MOSFET — Operational Conditions

Now consider Figure 16-4(b), where a $+V_{GS}$ has been connected between the source and gate. The electrostatic action of the $+V_{GS}$ attracts extra electrons into the area opposite the gate electrode. These

MOSFETs—Metal Oxide Semiconductor Field Effect Transistors

electrons form a channel of extra electrons that connect the source and drain. This allows I_D to increase. A further increase in the $+V_{GS}$ attracts even more electrons into the channel region which allows I_D to increase. In effect, both junctions are forward biased when a $+V_{GS}$ is connected. In this mode the number of electrons present in the area between the source and drain is increased *(enhanced)*. For this reason this mode is called the *enhancement mode*. The E-MOSFET will operate well in the enhancement mode.

In Figure 16-4(c), we see that a $-V_{GS}$ has been applied. In this case, the number of electrons that would normally be located near the gate is decreased by the repelling electrostatic field. This causes the small current that was present in Figure 16-4(a) to decrease. In effect, the transistor is cut off and both junctions act as if they were reverse biased. The number of electrons that would normally be available between the source and drain has been *depleted*. Anytime that the electrostatic field reduces the number of electrons present in the current carrying portion of the MOSFET, it is said to be operating in the *depletion mode*. From this you can see that the E-MOSFET cannot operate in the *depletion mode*.

E-MOSFETs are manufactured that have either N-channel (NMOS) or P-channel (PMOS) characteristics. N-channel MOSFETs are referred to as **NMOS** devices, and P-channel MOSFETs are referred to as **PMOS** devices, where NMOS and PMOS refer to the type of construction. Except for voltages that are applied and the direction of current flow, their (PMOS and NMOS) operations are identical. Figure 16-5 contains the schematic symbols for both types. Either type can operate with the source and substrate connected as described here.

(a) Source-Substrate Shorted — (1) N-Type (2) P-Type

(b) Source-Substrate Open — (1) N-Type (2) P-Type

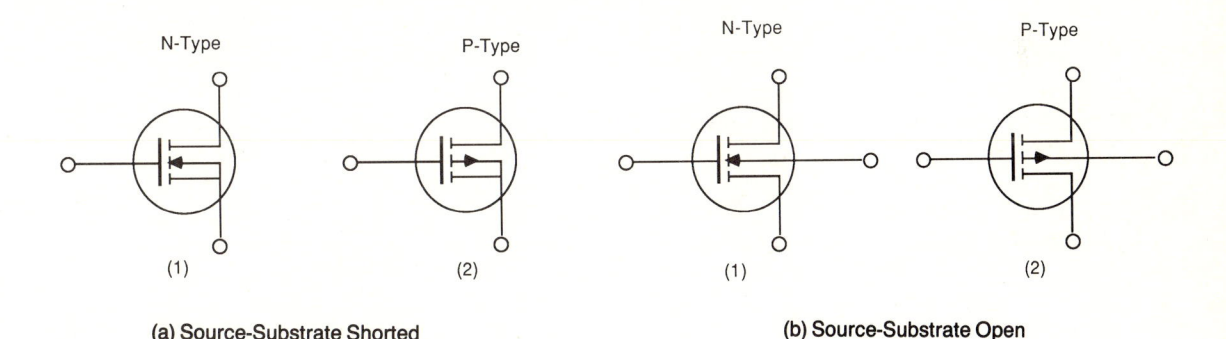

Fig. 16-5 E-MOSFET — Schematic Symbols

Observe that the channel is broken in both schematic symbols. This symbolizes the fact that the channel, within the transistor, is non-existent until enhancement occurs. The space that exists between the gate and channel illustrates the insulation that is provided by the silicon dioxide.

There are, however, other applications where the substrate has its own connection. The symbols for *source-substrate open MOSFETs* are also included.

Operation of a DE-MOSFET

Figure 16-6 contains a pictorial diagram of the DE-MOSFET. If you compare this diagram to Figure 16-2, you will see that both contain the same parts. As stated earlier, the only difference that exists between the DE-MOSFET and the E-MOSFET is that a channel has been diffused into the substrate of the DE-MOSFET. This channel is a lightly doped, thin strip of N-type material while the drain and source are made of heavily doped N-type material. The channel provides a path through which I_D can move more easily once voltage is applied.

Observe Figure 16-7 while we discuss the operation of a DE-MOSFET transistor. In Figure 16-7(a) you will see a diagram that has a battery connected between the source (negative) and drain (positive). Notice that the gate is not connected. This means that $V_{GS} = 0$ V.

When the DE-MOSFET is connected in this way, current flows from the source to drain. This is different from the E-MOSFET where current did not flow when $V_{GS} = 0$ V. The DE-MOSFET will conduct at any time the drain is more positive than the source, even when $V_{GS} = 0$ V.

Figure 16-7(b) contains a diagram that has $+V_{GS}$ connected. Source and drain connections are the same as above. The $+V_{GS}$ attracts electrons from the substrate into the N-channel. This increases the number of electrons available to support I_D. By applying $+V_{GS}$ to the gate, the conduction of I_D is "enhanced" and I_D will increase. The DE-MOSFET is operating in the *enhancement mode*.

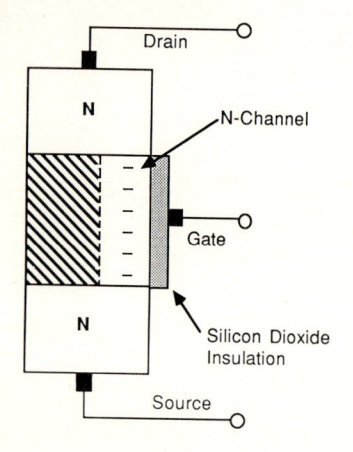

Fig. 16-6 DE-MOSFET — Pictorial Diagram

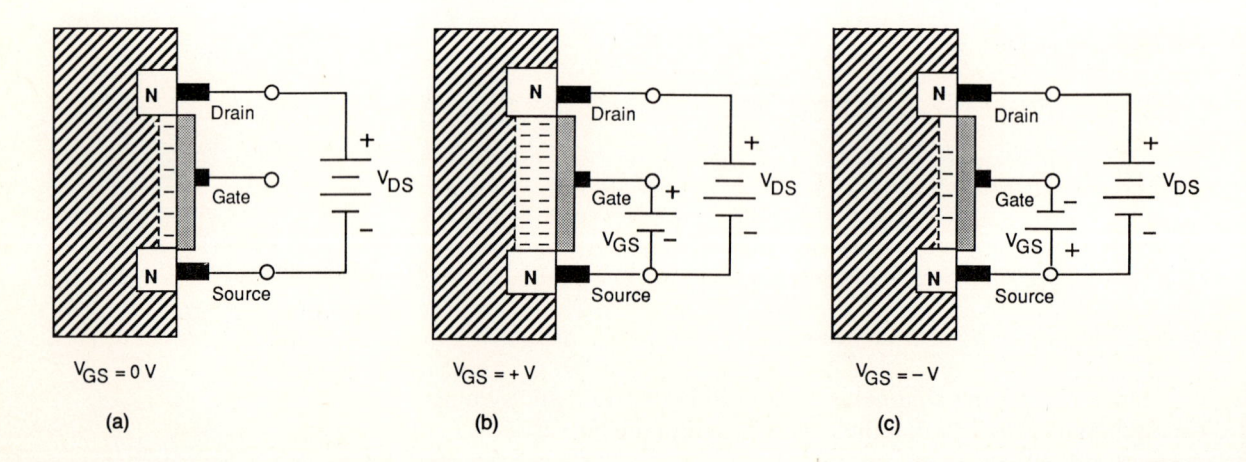

Fig. 16-7 Effect of V_{GS} on DE-MOSFET Operation

In Figure 16-7(c) the connection of V_{GS} has been reversed. In this situation, a $-V_{GS}$ is present at the gate. This causes electrons to be repelled from the N-channel leaving fewer electrons available to support I_D. With these connections, the number of electrons available in the N-channel is "depleted" and I_D decreases. Because of the depletion of available carriers the DE-MOSFET is said to be operating in the *depletion mode*.

When a MOSFET has current flow with $V_{GS} = 0$ V, it has the ability to conduct with either $+V_{GS}$ or $-V_{GS}$ is applied. In one case, the transistor operates in the depletion mode, and in the other, it operates in the enhancement mode. For this reason, we call it a *depletion-enhancement mode MOSFET*. This is shortened to DE-MOSFET for ease of use.

DE-MOSFETs are manufactured with either NMOS or PMOS characteristics. Except for voltages that are applied and the direction of current flow, their operations are identical. Figure 16-8 contains the schematic symbols for both types. Each type has the capability of operating with the source and substrate connected as described here.

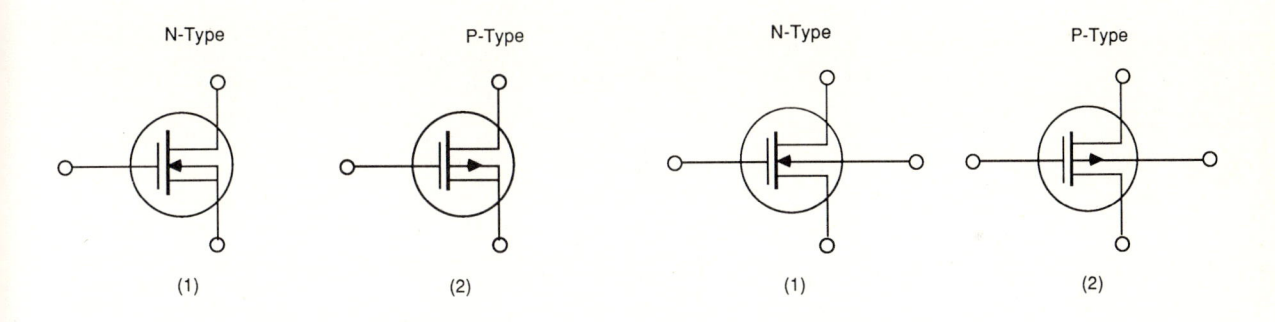

N-Type   P-Type   N-Type   P-Type

(1)   (2)   (1)   (2)

(a) Source-Substrate Shorted   (b) Source-Substrate Open

Fig. 16-8 DE-MOSFET — Schematic Symbols

Notice that the channel is a *solid line* in all of the symbols. This symbolizes the fact that the channel within the transistor is diffused into the substrate and that it connects the source and drain. The space that exists between the gate and channel illustrates the insulation that is provided by the silicon dioxide.

There are, however, other applications where the substrate has its own connection. The symbols for separate source-substrate connections are also included.

Both MOSFETs have many characteristics that are similar. Except for the inclusion of a channel (N or P), they are constructed the same. The addition of this channel causes a drastic change in operation. In the

E-MOSFET, current only flows when V_{GS} is positive. When $V_{GS} = 0\,V$, the transistor will not conduct regardless of the connections made to source and drain terminals. This is called *enhancement mode* operation. The DE-MOSFET will have current flow when $V_{GS} = 0\,V$. When $+V_{GS}$ is applied, it operates in the *enhancement mode*, and when V_{GS} is negative it operates in the *depletion mode*.

MOSFET Characteristic Curves

The characteristic curves for MOSFET devices are very similar to those for the JFETs. There are some differences in transfer characteristics however. Bias line construction and determination of the minimum and maximum transfer characteristics is the same.

E-MOSFET Characteristic Curves

Figure 16-9 contains a typical set of E-MOSFET **drain curves**. The first thing that you should note about this figure is that both V_{GS} and V_{DS} are the same polarity. For the NMOS type MOSFET, both voltages (V_{GS} and V_{DS}) are positive. With PMOS devices, both voltages should be negative. Because V_{DS} and V_{GS} are the same polarity, it is possible to operate this circuit using one power supply and voltage dividers.

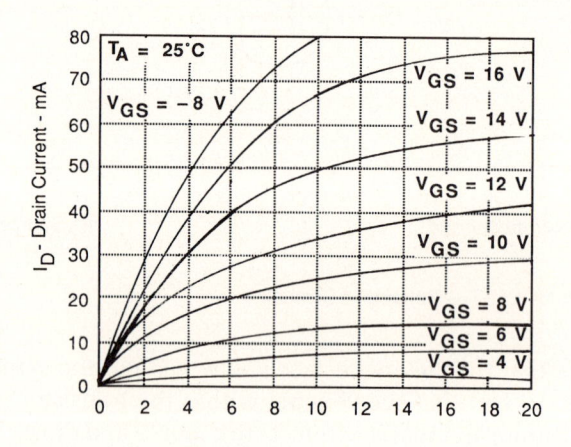

Fig. 16-9 P-Channel E-MOSFET Drain Curves

You should also notice that current does not begin to flow until V_{GS} reaches some positive voltage. In some cases, V_{GS} must be relatively high ($+2$ to $+4\,V$) before the transistor begins to conduct. In the

MOSFETs—Metal Oxide Semiconductor Field Effect Transistors

figure shown, the transistor begins to conduct when V_{GS} is approximately +4 V. The voltage (V_{GS}) required to cause conduction is called the **threshold voltage** ($V_{GS(th)}$). Threshold voltage is defined as:

> **The amount of gate to source voltage
> required to cause current to flow
> between the source and drain.**

Threshold voltage will vary from one E-MOSFET to another. This is a parameter that is usually listed on the manufacturer's specification sheet.

In Figure 16-10 you will see a typical E-MOSFET transistor's transfer characteristic curve. Notice that the transfer characteristic curve has a parabolic shape. This is the same shape as that of JFET transfer characteristic curve. Operating temperature will affect transfer characteristics. The chart shown in Figure 6-10 shows the transfer characteristic for an N-channel (NMOS) E-MOSFET type M-92 at operating temperatures of 25° C, 55° C, and 125° C. Notice that as temperature increases, the transfer characteristic curve more closely approaches a vertical line.

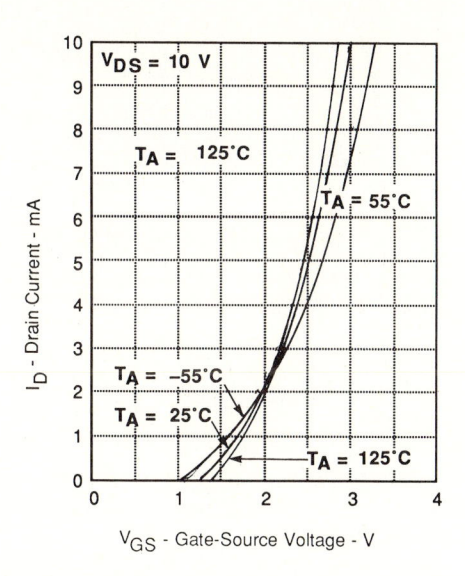

Fig. 16-10 N-Channel E-MOSFET Transfer Characteristics

These data ($V_{GS(th)}$, transfer curves, characteristic curves, and others) are provided in the manufacturer's specification sheet for the device. You should learn how to read these and to apply the data that they include. Appendix D contains a sample of these specification sheets.

DE-MOSFET Characteristic Curves

In Figure 16-11 you will see a typical set of drain curves for a P-channel (PMOS) DE-MOSFET. The curves shown here are for an M-82 as included in Appendix F. The difference between these and NMOS types are reversed voltage polarities and direction of current flow. Looking at this figure, you can see that I_D is flowing when $V_{GS} = 0$ V. In fact, $V_{GS(th)}$ occurs at $V_{GS} = -2$ V. When V_{GS} goes positive I_D increases and when V_{GS} goes negative I_D decreases.

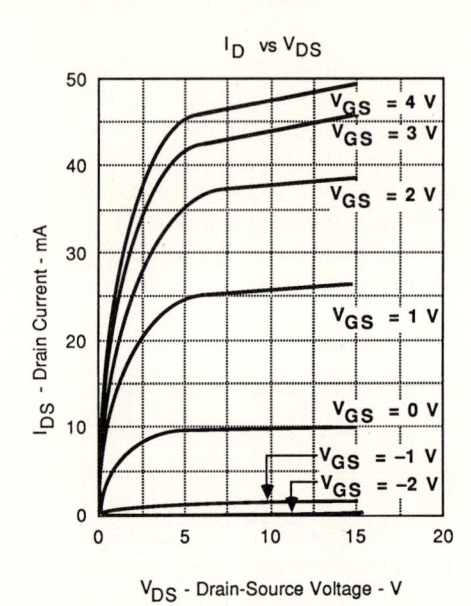

Fig. 16-11 N-Channel DE-MOSFET Drain Curves

Notice that this MOSFET will operate with $V_{GS} = 0$ V, $+V_{GS}$, and $-V_{GS}$. This tells us that it is a *depletion-enhancement* type.

Terms and definitions are the same for MOSFETs as they are for JFETs. I_{DSS} is drain to source saturation. V_{GS}OFF is the amount of bias required to cut off the drain current.

Common values for TRANSCONDUCTANCE of MOSFETs range from 100 μS to 2000 μS. It is not uncommon for R_{in} to be in the range of several millions of ohms.

MOSFETs—Metal Oxide Semiconductor Field Effect Transistors

Answer each item by inserting the word or words required to correctly complete each statement.

Self-Check

1. When the gate's electrostatic field is used to decrease drain current (I_D), the transistor is operating in the _____ mode.

   a. enhancement
   b. depletion-enhancement
   c. depletion
   d. all of these

2. In normal MOSFET operation the _____ and the _____ are shorted together.

   a. source, drain
   b. source, gate
   c. drain, substrate
   d. source, substrate

3. When working with the N-channel (NMOS) MOSFET, V_{GS} must have _____ polarity, and V_{DS} must have _____ polarity.

   a. positive, positive
   b. positive, negative
   c. negative, negative
   d. negative, positive

4. As the operating temperature for a MOSFET increases, the transfer characteristic curve becomes more _____.

   a. rounded
   b. horizontal
   c. vertical

5. The _____ has a diffused channel and the _____ does not have a channel.

   a. E-MOSFET, DE-MOSFET b. DE-MOSFET, E-MOSFET

6. In the enhancement mode, MOSFET current will be _____ when $V_{GS} = 0\,V$.

   a. high
   b. medium
   c. low
   d. zero

7. _____ voltage is the amount of voltage required to start drain current flowing in a MOSFET.

   a. Gate-source
   b. Threshold
   c. Gate-drain
   d. Source-drain

547

8. In MOSFET circuitry analysis, the amount of voltage required to stop the flow of I_D is referred to as _____.

   a. $V_{GS} = 0\,V$
   b. $V_{DS} = 0\,V$
   c. $V_{GS}OFF$
   d. $V_{DS}OFF$

9. The voltage at which drain current begins to flow is called _____ voltage.

   a. V_{GS}
   b. sensing
   c. threshold
   d. emitter

10. When the gate's electrostatic field is used to increase I_D, operation is classified as being in the _____ mode.

    a. enhancement
    b. depletion-enhancement
    c. depletion
    d. all of these

MOSFET Amplifiers

For this discussion of amplifiers, NMOS devices are used. These circuits are used because current flows from the source to drain, with the drain being positive with respect to the source. The circuits used could just as easily have been of the PMOS type. In that case, the drain would be negative with respect to the source, and current would flow from the drain to the source. In all other ways, the explanations that follow can be used for PMOS amplifiers as easily as they are for NMOS amplifiers.

Both E- and DE-MOSFETs can be used in amplifier circuits. One caution: when using the E-MOSFET, you must realize that the threshold voltage is some voltage larger than 0 V; therefore, your designs must take $V_{GS(th)}$ into consideration. If you want to amplify both positive and negative changes, you must bias the transistor high enough to allow for negative swings.

Common-Source Amplifiers

An E-MOSFET that has been connected as a *common-source* amplifier is shown in Figure 16-12. Notice that bias is provided by a voltage divider that contains R_S, R_{G1}, and R_{G2}. The voltage drop on R_{G2} will develop the actual bias voltage. Therefore, the resistance ratio that exists between R_{G1} and R_{G2} must be correct for development of the necessary bias voltage. If Class A operation is desired, the bias developed by R_{G2} must be high enough that negative alternations of the input can be amplified without driving the transistor into cutoff.

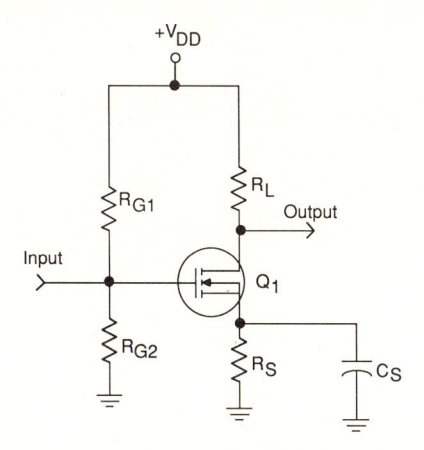

Fig. 16-12 Common-Source N-Channel E-MOSFET Amplifier

Figure 16-13 contains the schematic diagram for a DE-MOSFET that is acting as a *common-source* amplifier. Remember, a DE-MOSFET will conduct when $V_{GS} = 0$ V. Using this biasing arrangement will place 0 V at the gate of Q_1. This circuit can be used as a Class A amplifier.

Input and Output Impedance (Resistance)

Output impedance (Z_{out}) of a MOSFET is approximately the same as that of a JFET. The input impedance (R_{in}) is considerably larger, however. It is not uncommon for R_{in} to be several megohms. In some cases, R_{in} can be as high as 1000 MΩ. The high value of R_{in} results from the use of the insulated gate construction.

A quick way of approximating a MOSFET circuit's R_{in} is to consider the ohmic value of R_G. The transistor's input (gate-source junction) and R_G are in parallel to an incoming signal. We know that the equivalent resistance of parallel branches must be smaller than the smallest resistance of the branches. With the very large ohmic value of the transistor junction, we can say that:

The circuit's R_{in} will be approximately equal to the ohmic value of R_G.

This allows us to make the statement:

All MOSFETs have high input resistance.

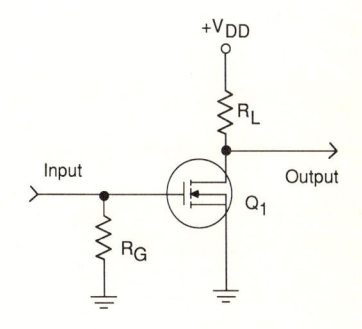

Fig. 16-13 Common-Source N-Channel DE-MOSFET Amplifier

Bipolar–JFET–MOSFET Comparisons

Both JFETs and MOSFETs are voltage controlled while the bipolar transistor is current controlled. The major difference between the JFET and MOSFET is channel area. JFETs and MOSFETs can be used as signal amplifiers and their analyses are quite similar. The extremely low input current (I_{in}) of the MOSFET assures us that it will be affected very little by changes in operating temperature. Both bipolar and JFET types are more likely to be affected by temperature changes. The MOSFET also has a much higher frequency capability than does either the bipolar or JFET devices.

Heat has little effect on MOSFET operation while it can have drastic effect on bipolar and JFET operations. This fact means that the MOSFET can have much smaller size and can be used in circuitry where spacing between elements is much closer than in bipolar or JFET circuitry. This is especially advantageous in *integrated* and *digital* circuitry.

The bipolar transistor can switch (change states) faster than the MOSFET; however, MOSFETs have more applications because they are so small and are less affected by operating temperature. The DE-MOSFET and the JFETs are biased exactly alike. E-MOSFETs differ in their biasing because of their *threshold voltage* requirements. Bipolar transistor biasing is very similar to that used with JFETs and DE-MOSFETs.

V-MOS Devices

A newer member of the MOSFET family is a device called V-MOS. A cutaway pictorial diagram of this device is shown in Figure 16-14. As you can see from the device's construction, it gets its name from the fact that the gate and channel are formed in a "V" shape. This shape provides some improvements in the operation of E-MOSFET devices.

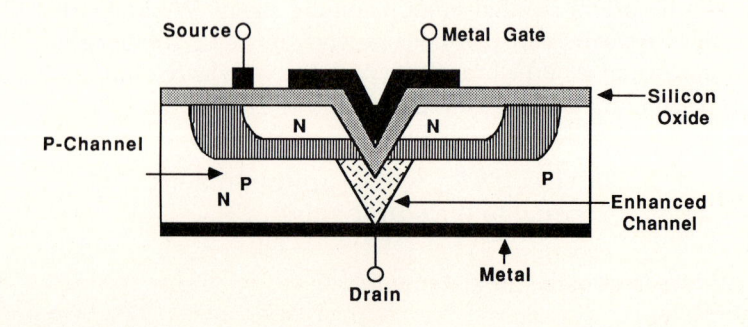

Fig. 16-14 VMOSFET — Pictorial Diagram

Using V-MOSFET devices, higher voltage gains are possible than with the E-MOSFET. Another advantage is that the V-MOSFET has a higher frequency response. It also is capable of conducting higher drain current (I_D) and dissipating the additional heat that results. The higher current- and heat-handling capabilities mean that for the same job, a V-MOSFET can be smaller than the E-MOSFET. Another advantage is that the V-MOSFET can switch states faster than is possible using a conventional E-MOSFET.

CMOS Construction

The term "CMOS" is used to describe a circuit design technique. It identifies a technique where two MOSFETs are constructed on a single substrate. The MOSFETs are interconnected (on the substrate) to form a **complimentary digital circuit**. This is where the CMOS name is derived. CMOS devices provide high-speed switching and have low power dissipation factors. Figure 16-15 is an illustration of a CMOS device. CMOS circuitry can operate at temperatures ranging from −40°C to +85°C. When compared to the operating temperature range (0° to 70°C) for bipolar, PMOS, and NMOS devices, the improvement in temperature range is apparent. An even more dramatic improvement in operating temperature range is available when ceramic-packaged CMOS devices are used as they have a range of −50° to +125°C. Any added temperature capabilities allow the device to be operated in more locations and without external temperature control.

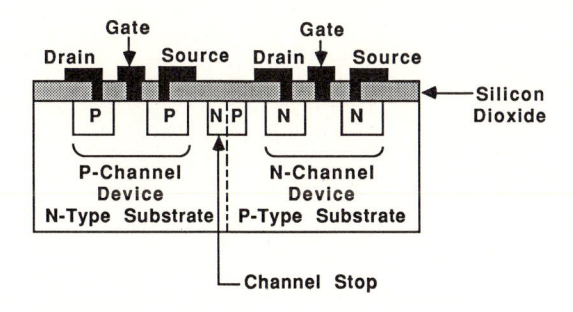

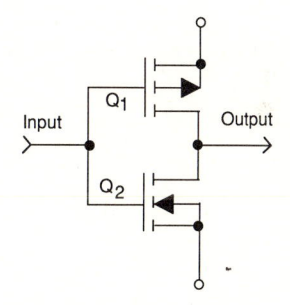

(a) Pictorial Diagram

(b) Schematic Diagram

Fig. 16-15 P- and N-Channel CMOS Integrated Circuit

A second advantage of CMOS devices is their ability to operate under varying conditions. For comparison, consider bipolar devices which require 5 ± 0.25 VDC for their operation. The commonly used 4000 family of CMOS devices can operate with voltages that range from

3 to 18 VDC. Variation anywhere within this range does not appreciably affect the CMOS device's operation. This practically eliminates the need for voltage regulation in systems designed using CMOS devices.

A third advantage is low power consumption. Bipolar devices require much larger currents for their operation which leads to high power dissipation. The increased heat dissipation means that, in many situations, external cooling (fans, blowers, air conditioners, etc.) must be provided. Providing external cooling is expensive as it adds weight and requires added space. CMOS devices, however, operate with currents of only a few microamperes. This means that the power dissipated is much less. Because of this, CMOS devices can operate when mounted more closely together and under more adverse conditions than bipolar devices.

At present, CMOS devices are available that serve the same purposes as most bipolar, PMOS, and NMOS devices. Current trends indicate that CMOS devices will be used in most, if not all, electronic applications of the future.

One disadvantage that is common to PMOS, NMOS, CMOS, VMOS, and other metal oxide devices is that of susceptability to **electrostatic discharge (ESD)**. Possible damage is so critical that the static electricity charge on your body can destroy the device. Special precautions are necessary in all steps from manufacturing to installation and repair. Each person that handles MOS devices **must know and use** the correct procedures for handling these devices. A more complete explanation of this problem is included as **Appendix G**.

Self-Check

Answer each item by inserting the word or words required to correctly complete each statement.

11. V-MOSFETs have an advantage over other devices in that they _____ much faster.

    a. amplify
    b. switch states
    c. conduct
    d. all of these

12. The _____ has a channel that is V-shaped while the _____ has a flat channel.

    a. E-MOSFET, V-MOSFET
    b. DE-MOSFET, V-MOSFET
    c. E-MOSFET, DE-MOSFET
    d. V-MOSFET, E-MOSFET

13. All MOSFETs have the capability of switching faster than bipolar transistors.

    a.   True   b.   False

14. One advantage that the MOSFETs have over bipolar transistors is their _____ power efficiency.

    a.   high   c.   low
    b.   medium

15. The main construction difference between E-MOSFETs and DE-MOSFETs is the existence of a _____ that is diffused into the substrate.

    a.   gate   c.   drain
    b.   source   d.   channel

Summary

In this chapter we discussed the *metal-oxide-semiconductor field effect transistor* (MOSFET). We found that these devices, like all other semiconductor devices, can be designed in either the P or N configuration.

Two types of *enhancement mode MOSFETs* (E-MOSFETs) were discussed. We found that the construction of an E-MOSFET involves the diffusion of source and drain sections into a substrate. These two sections are separated by a large section of the opposite type semiconductor material. The entire device is then insulated using a layer of silicon dioxide. Once the dioxide is in place, a gate terminal is bonded to the dioxide insulator. This effectively insulates the gate from the semiconductor material. When the source and drain electrodes are connected by an external voltage source, I_D will equal 0 mA when $V_{GS} = 0$ V. Increasing V_{GS} to some voltage called the *threshold voltage* will cause the E-MOSFET to begin conduction. Once conducting, an increase in V_{GS} causes I_D to increase.

The DE-MOSFET was also discussed. It differs in construction, from the E-MOSFET by the fact that it has a channel diffused into the substrate that forms a conductor between the source and drain. This channel allows the DE-MOSFET to begin conduction as soon as an external voltage is applied to the source and drain, even with $V_{GS} = 0$ V. Increasing the value of V_{GS} increases I_D, and decreasing V_{GS} causes I_D to decrease. When V_{GS} is decreased, the DE-MOSFET enters a state in which the number of free electrons in the channel has been depleted, causing I_D to decrease. This state is called the *depletion mode*. The DE-MOSFET is able to operate in either mode (*depletion* or *enhancement*).

MOSFETs have several advantages over bipolar transistors and JFETs. The main advantage being higher input resistance, which means low input current and very low power dissipation. The bipolar transistor provides a faster switching time than the MOSFETs. This is offset, to some extent, by the development of the V-MOSFETs. MOS type devices are becoming ever more popular and are being used in a large share of the new circuits that are being designed.

CMOS devices have low power dissipation, tolerate voltage variations, low current requirements, and minimum cooling requirements that make them an improvement over earlier MOS devices. All, however, have a common disadvantage—the delicate handling required to avoid electrostatic discharge.

Review Questions and/or Problems

1. Bipolar transistors are capable of switching at a faster rate than MOSFET transistors.

    a. True b. False

2. One advantage that the MOSFETs have over bipolar transistors is their _____ power efficiency.

    a. good
    b. medium
    c. poor

3. The main construction difference between JFETs and MOSFETs is the existence of a layer of _____ that serves as a/an _____.

    a. silicon dioxide, gate
    b. P-type material, gate
    c. silicon dioxide, insulator
    d. P-type material, insulator

4. When working with the P-channel enhancement-type MOSFET, V_{GS} must have _____ polarity and V_{DS} must have _____ polarity.

    a. positive, positive
    b. positive, negative
    c. negative, negative
    d. negative, positive

5. As the operating temperature for a MOSFET increases, the transfer characteristic curve becomes more _____.

    a. rounded
    b. horizontal
    c. vertical

6. The _____ has a diffused channel and the _____ does not have a channel.

   a. E-MOSFET, DE-MOSFET   b. DE-MOSFET, E-MOSFET

7. In MOSFET circuit analysis the amount of voltage required to stop the flow of I_D is referred to as _____.

   a. $V_{GS} = 0$ V
   b. $V_{DS} = 0$ V
   c. $V_{GS}OFF$
   d. $V_{DS}OFF$

8. Examine Figure 16-16. Which symbol is used to represent a P-channel depletion-enhancement mode MOSFET?

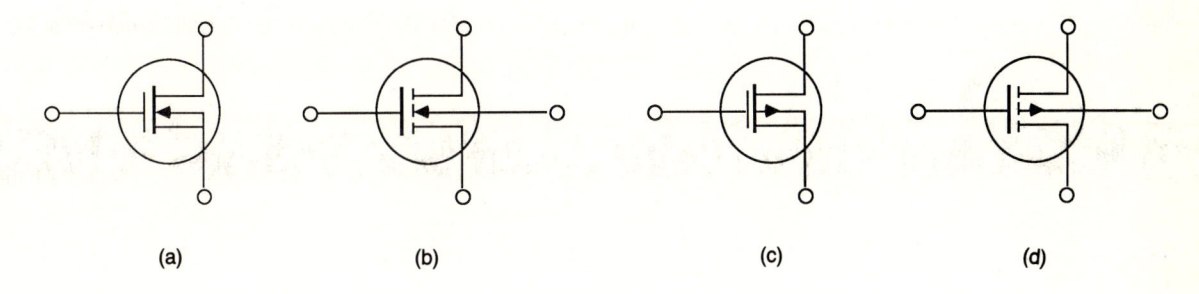

(a)   (b)   (c)   (d)

Fig. 16-16

9. The voltage at which drain current begins to flow is called _____ voltage.

   a. gate-source
   b. sensing
   c. threshold
   d. emitter

10. When the gate's electrostatic field is used to increase drain current (I_D), operation is classified as being the _____ mode.

    a. enhancement
    b. depletion-enhancement
    c. depletion
    d. all of these

11. V-MOSFETs have an advantage over other MOSFETs in that they _____ (change state) much faster.

    a. amplify
    b. switch
    c. conduct
    d. all of these

12. The _____ has a channel that is flat-shaped while the _____ has a channel that is V-shaped.

    a. E-MOSFET, V-MOSFET   c. E-MOSFET, DE-MOSFET
    b. DE-MOSFET, V-MOSFET d. V-MOSFET, E-MOSFET

13. A DE-MOSFET is operating with $V_{GS} = 0V$. Application of a bias causes I_D to decrease. The transistor is operating in the _____ region.
    a. enhancement   c. depletion
    b. depletion-enhancement d. all of these

14. For normal E-MOSFET operation, the _____ and the _____ are shorted together.

    a. source, drain   c. drain, substrate
    b. source, gate   d. source, substrate

15. In the depletion mode, MOSFET current will be _____ when $V_{GS} = 0 V$.

    a. high   c. low
    b. medium d. zero

16. _____ voltage is the amount of voltage required to stop drain current flowing in a MOSFET.

    a. Gate-source off c. Source-drain off
    b. Threshold   d. Source, drain

17
Selected Solid State Devices Operational Characteristics

Objectives

1. Describe the construction and operation of an SCR.
2. List ways in which SCRs can be used.
3. Describe how an SCR can be used as a rectifier.
4. List other types of thyristors and name some of their uses.
5. Explain what occurs when a device operates in its negative resistance region.
6. Describe the construction and operational characteristics of a unijunction transistor and a programmable unijunction transistor.
7. Explain "tunneling". What causes tunneling and how does it effect tunnel diode operation?
8. Describe the construction of:

    a. Photodiode
    b. Phototransistors
    c. Opto couplers
    d. Laser diodes

9. Explain how each of the devices listed in objective 8 operates and each is used.

Introduction

In Chapter 6 we discussed several solid state devices that are designed for special applications. Of those, we have further discussed the *Zener diode* as it is used for *voltage regulation* and the *Varactor diode's* use in a *voltage-controlled oscillator*. In this chapter we will discuss other devices from Chapter 6 in greater depth plus two devices not discussed earlier; the **unijunction transistor (UJT)** and **optoelectronic devices**. Earlier we were concerned with each device's DC characteristics. In this chapter we will discuss both DC and AC applications.

One family of devices that we will discuss is those that are classified as *optoelectronic devices*. Within this family we have both photovoltaic and photoconductive devices. Photovoltaic devices convert light into usable voltage while photoconductive devices undergo predictable resistance changes when exposed to light.

The Silicon Controlled Rectifier

The *silicon controlled rectifier* (SCR) was first introduced in Chapter 6. At that point we discussed DC characteristics. In this chapter we will expand that to cover AC characteristics. Before starting that study, though, we will review the DC characteristics and construction characteristics of the SCR.

The SCR is a member of the *thyristor* family of semiconductor devices. A **thyristor** is a:

**Bistable device that consists of three
or more semiconductor junctions.**

A thyristor is commonly used as a switch. The fact that it is a bistable device makes it ideal for use as a switch. By operating it in one state we can activate another circuit, or by changing the thyristor's state we can turn off the second circuit or device.

Of all thyristors, the SCR is probably the most used. The SCR has three terminals: gate, cathode, and anode. Even though it can be used as a bistable device, it is used as a *unidirectional device* that allows current to flow in one direction only. For this reason it is often referred to as a diode. An SCR can serve the same rectification purpose as a diode, but its construction is much different. Figure 17-1 contains the SCR schematic symbol, a pictorial diagram of SCR construction, and the equivalent circuit for an SCR. The device has four layers of P and N materials that form three junctions. With low voltage applied across the anode to cathode of an SCR, the *anode current* (I_A) that flows in the SCR is zero mA. Application of a small voltage to the gate terminal, however, places all junctions within the SCR in a forward biased condition, allowing heavy current to flow from the cathode to anode. Notice that the *gate* is connected to the *base* of an NPN transistor.

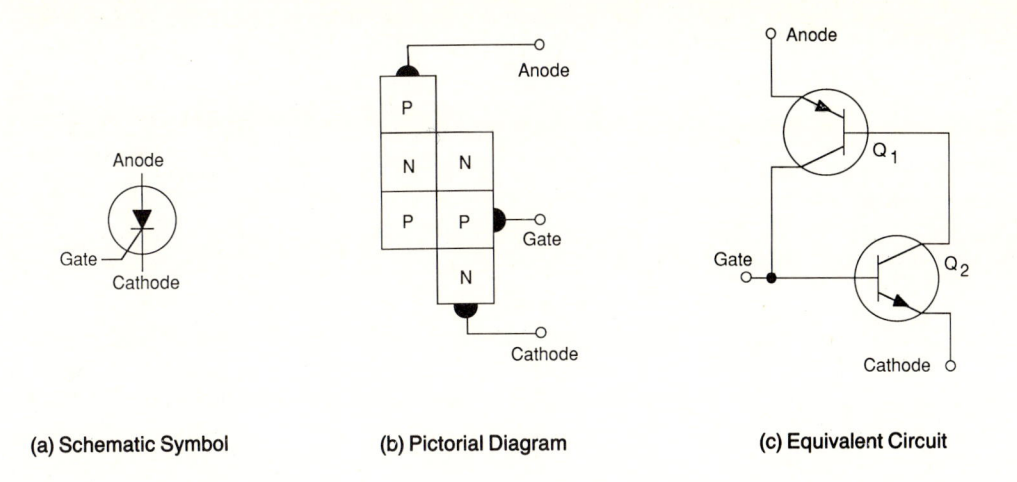

Fig. 17-1 The Silicon Controlled Rectifier

(a) Schematic Symbol (b) Pictorial Diagram (c) Equivalent Circuit

The SCR as a Switch

Consider Figure 17-2 as an example of the SCR being used as a switch. This circuit is typical of the ones that involve the use of an SCR and DC voltage. For reference purposes, we will refer to *anode voltage* as V_A and *gate voltage* as V_G. With Sw_1 open, the SCR is cut off and only reverse current flows. If, however, V_A should be increased to some high value, the junctions will become saturated and the SCR will enter an operational condition called **breakover**. The voltage value at which this happens is called the **breakover potential**. Once the SCR begins to conduct, it has *regenerative* action that keeps the junctions forward biased and anode current (I_A) flowing. To stop I_A, we must either remove the voltage applied to the anode or reverse its polarity.

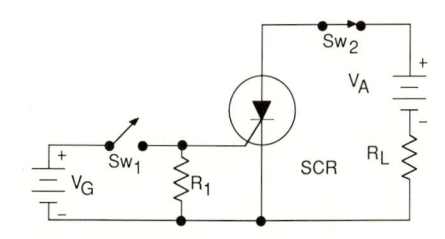

Fig. 17-2 DC Control Using an SCR

For the remainder of this discussion, we will assume that V_A is less than breakover potential and that zero current (I_A) flows when Sw_2 is closed. At this point, the cathode junction is reverse biased, keeping I_A cut off. When Sw_1 is closed, the positive voltage forward biases this junction, allowing I_A to flow. The amount of I_A that flows will be determined by the resistance of the SCR, its load, and the applied voltage. Once I_A begins to flow, Sw_1 can be opened and I_A will continue to flow. In other words, the gate is used to turn the device on and then it loses all control. To cut off I_A we must open Sw_2. Once I_A stops, Sw_2 can be closed and the SCR is back in the cutoff state.

The gate voltage (V_G) performs the same function as closing a switch, and removal of V_A opens the switch. Once V_A has been removed long enough for I_A to stop, the SCR is cut off and awaiting the start of another cycle of operation.

It is also possible for us to use the positive alternation of a sine wave, or other transient wave, as V_G for turning the SCR on. The circuit shown in Figure 17-3 uses the positive alternation of an AC sine wave as a trigger (V_G). Notice that the solid line represents the trigger level. Any positive pulse with an amplitude exceeding this value triggers the SCR into conduction. Remember, as soon as I_A begins to flow, V_G loses control. Therefore, the pulse of V_G can be quite short in duration. As before, the only way to stop I_A is to remove V_A.

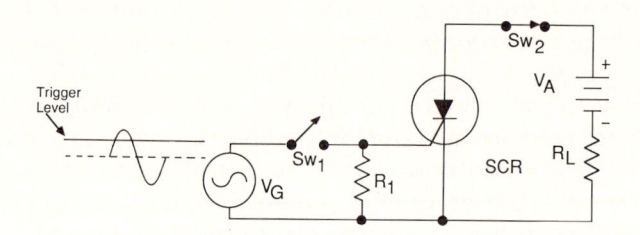

Fig. 17-3 SCR Control Circuit

In Figure 17-4 you see an SCR circuit that has pulsating direct current (PDC) applied as V_G and V_A. Adjustment of the variable resistor (R_2) allows us to have more precise control over the exact point where I_A begins to flow and the duration of its flow. Once V_G has turned the SCR on, I_A is turned off by V_A dropping to zero between pulses. Using this circuit, it is possible to adjust R_2 so that the trigger-on voltage occurs anywhere along the positive slope of the input pulse (V_G). In the output example shown, you can see that by triggering the SCR with the peak of the V_G pulse, the SCR conducts for only half of the positive pulse applied as V_A. By adjusting the amplitude of V_G to other points, the SCR can be turned on earlier during the pulse applied to V_A. This circuit can be used to control operation during a period of 0° to 180° of the positive pulse that is used as V_A.

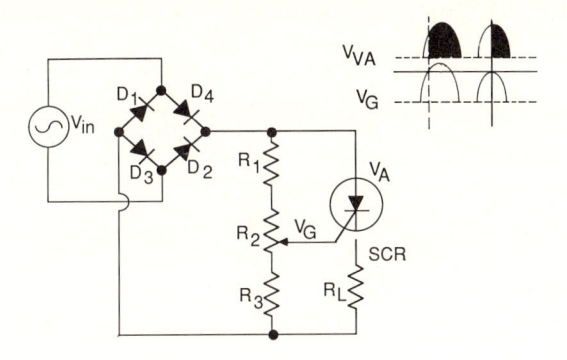

Fig. 17-4 PDC to PDC Switching

DC Power Control Using an SCR

SCRs are used extensively as power control devices. In fact, this is their primary use. An SCR provides the advantage of using a small gate current (I_G) to control a much larger, and more dangerous, cathode-to-anode current (I_A). The gate voltage (V_G) can be any of several combinations of voltages that range from DC to high frequency AC. Triggering can be provided by DC, AC, pulse, and nonsinusoidal waves. Remember, the only purpose served by V_G is to turn the SCR on.

Figure 17-5 contains the schematic diagram for an SCR circuit that uses DC voltage for both V_G and V_A. R_L can represent any device or circuit from a light to a highly complex circuit. In this application the SCR is in series with the load. This means that the entire load current must flow through the SCR. R_1 and R_2 form a voltage divider that determines the amplitude of V_G. V_G will have enough amplitude to turn the SCR on when Sw_1 is closed momentarily, assuming that Sw_2 is closed. Once I_A begins to flow, the SCR's operation can be stopped by opening Sw_2. As soon as I_A has stopped, Sw_2 can be closed, preparing the circuit for another cycle of operation. To begin the next cycle it is only necessary to momentarily close Sw_1.

The circuit shown in Figure 17-6 has an SCR whose V_G is DC voltage and V_A is AC voltage. For this circuit to be placed into operation the positive alternation of the sine wave must be present at the SCR's anode. During that period the application of V_G causes conduction to occur. The SCR continues to conduct until the negative alternation arrives. The application of negative voltage to the anode causes the SCR to cut off.

It is possible to damage the SCR if $+V_G$ is applied when the negative alternation is present at the anode. This could cause reverse current to flow in the SCR and result in structure breakdown. To assure that damage does not occur because of reverse current, we connect a conventional diode (D_1) in series with R_L and the SCR as shown in this

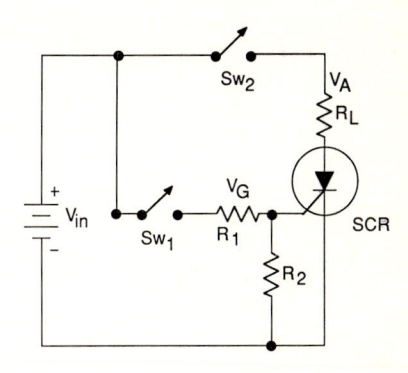

Fig. 17-5 DC to DC Switching

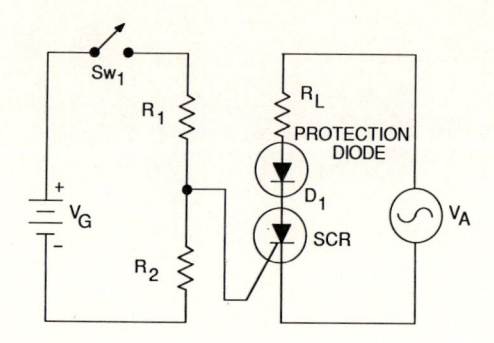

Fig. 17-6 DC to AC Switching

circuit. During the positive alternation, D_1 is forward biased and the circuit operates as if D_1 is not present. During the negative alternation, reverse bias is applied to D_1 which prevents the flow of I_A. Using D_1 allows V_G to remain on and assures that the SCR conducts at any time the positive alternation of V_A is present.

AC Power Control Using an SCR

The circuit contained in Figure 17-7 shows an SCR that has AC applied to both the gate and anode. Notice that the transformer (T_1) provides isolation between the gate and anode. Assume that T_1 does not provide a phase shift. The RC network (R_1 and C_1) does, however provide a phase shift. This provides greater control over the conduction of the SCR than that provided by the circuit in Figure 17-4. Using R_1, it is possible to control the SCR's conduction very closely. In one instance, the SCR can be allowed to conduct for the full 180° of the positive alternation's presence as V_A. The other extreme is to adjust R_1 for zero conduction time with the anode positive. Using this circuit, we can control the flow of I_A to only a few degrees or to much longer periods —up to 180°.

Actually, we have a rectifier circuit whose rectification is controlled to some point less than half-wave. The maximum output would be identical to that of a half-wave rectifier. The minimum output would be only a small portion of a half-wave rectifier. When the anode waveform is viewed on an oscilloscope, you can actually observe the change in conduction that occurs as a result of adjusting R_1.

We have, again, installed a protective diode (D_1) to assure that reverse current cannot flow during the negative alternation and possibly destroy the device.

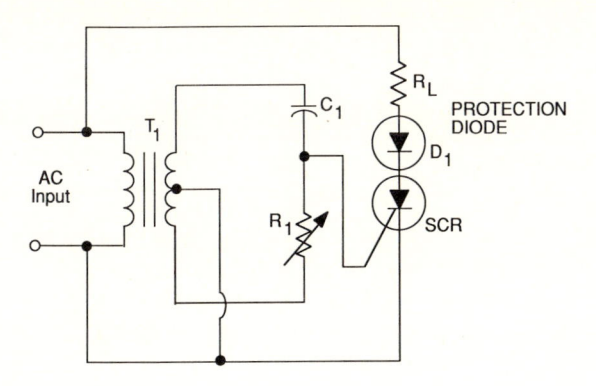

Fig. 17-7 DC to AC Switching

SCR Summary

From this discussion you should be able to see that by use of the correct circuit, we can precisely control the conduction time of a circuit. The current provided to the load can be either DC or pulsating DC. The main advantage gained from the use of SCRs is improved transient time. These devices have very fast turn-on times with switching occurring in as short a period as a few nanoseconds (1×10^{-9} seconds).

Some definitions that you should remember are:

1. **Breakover voltage** – the value of positive anode voltage at which an SCR, with the gate open, will switch into the conductive state.
2. **Transient peak-inverse voltage** – under specified conditions, the maximum allowable instantaneous, short duration, non-repeating, reverse voltage that can be applied to the anode of an SCR with the gate open.
3. **Repetitive peak-reverse voltage** – the maximum allowable instantaneous value of reverse (negative) voltage that can be repeatedly applied to the anode of an SCR with the gate open. This rating should not be exceeded except if the device has a transient rating that allows higher voltages. (See 2 above).
4. **Peak-reverse voltage** – the peak reverse voltage of a non-repeating nature that the anode of an SCR can withstand with the gate open.
5. **Gate trigger voltage/current** – the amount of voltage/current required to trigger the SCR on when positive voltage is present at the anode.
6. **Turn-on time** – the time that elapses between the arrival of a trigger at the gate of an SCR and the time that anode current reaches maximum.

7. **Turn-off time** – the time that elapses between the removal of an SCR's anode voltage and the time that anode current stops flowing.
8. **Average forward current** – the maximum continuous current that can be allowed to flow through an SCR during the conduction state and under stated conditions.

Other thyristor devices that are similar to the SCR are used to perform several functions within electronic circuits. These all have similar construction and operation. Circuits and applications can be found in most books that cover the use of thyristor devices. The number and diversity of operations are too large to cover in this text but the following is a brief summary of these devices.

1. **Four-layer diode (FLD)** – a semiconductor device that has two P-type and two N-type sections arranged such that three junctions are present. This is a reverse-blocking switching thyristor that has low ON-state voltage characteristics. Its uses include triggering, voltage limiting, pulse generation, and timing circuits.
2. **DIAC (DAC)** – a two-terminal, three-layered, bidirectional switching thyristor with a negative resistance characteristic extending over most of its full operating range of currents above its *switching current* (I_S). It is used to provide triggering, phase control, and as a voltage limiter.
3. **Silicon asymmetrical switch (SAS)** – a three-terminal, bidirectional, integrated circuit thyristor with asymmetrical switching characteristics that are determined by the gate bias level. It is used to provide triggering, overvoltage protection, pulse discrimination, and as a lamp driver.
4. **Silicon asymmetrical trigger (SAT)** – a two-terminal, bidirectional, integrated circuit thyristor with asymmetrical switching characteristics. Its uses include application as a light dimmer.
5. **Silicon bilateral switch (SBS)** – a three-terminal, bidirectional, integrated circuit thyristor with symmetrical switching characteristics that are determined by the gate bias level. It is used as a threshold detector, as a trigger generator, and for voltage limiting in AC circuits.
6. **Silicon unilateral switch (SUS)** – a three-terminal multi-layer, reverse-blocking trigger thyristor. The anode-gate terminal bias level determines the forward switching voltage. It is used for timing, triggering and threshold detection circuits.
7. **Gate turnoff thyristor (GTO)** – a three-terminal four-layer, reverse-blocking thyristor which can be turned on with a positive gate voltage and can be turned off by a negative gate voltage. It is used in inverter, pulse generator, chopper and DC switching circuits.

Selected Solid State Devices Operational Characteristics

8. **Light activated SCR (LASCR)** – a three-terminal, four-layer, reverse-blocking thyristor which can be turned on by exceeding its light threshold level which is determined by the gate-bias current. It is used to provide photoelectric control, position monitoring, light coupling of signals, and circuit triggering.
9. **N-gate thyristor (NGT)** – a three-terminal, four-layer, reverse blocking thyristor that is gated by a negative voltage. It is used as a high frequency inverter, chopper or power control device.
10. **Reverse conducting thyristor (RCT)** – a three-terminal, multi-layer, reverse conducting thyristor gated with a positive voltage and capable of conducting large currents in the reverse direction. Its uses include applications as an electroluminescent driver and as an AC bidirectional switching device.
11. **Silicon-controlled assembly (SCA)** – a multi-terminal class of units which include the complete package (wiring and cooling) for systems that require high power operation.
12. **Silicon-controlled bridge (SCB)** – a multi-terminal class of bridge devices which include SCRs as the main control unit. The basic bridge arrangements include single and three-phase devices with several control arrangements. Its uses include single and multi-phase rectification and control for motors, battery chargers, and power inverters.
13. **Silicon-controlled switch (SCS)** – a four-terminal, four-layered, reverse-blocking thyristor which can be gated using a positive voltage applied to the cathode (P-gate), or a negative voltage applied to the anode gate (N-gate). Its applications include use as a lamp driver, counter, alarm circuit, and control circuits.
14. **TRIACS (TAC)** – a three-terminal, multi-layer, bidirectional thyristor with four-quadrant gate turn-on capability (gate voltage positive or negative with respect to anode #1). It is used as a switching device and for phase control of AC power circuits.

Self-Check

Answer each item by inserting the word or words required to correctly complete each statement.

1. The SCR is a _____-layered device that has _____ junctions.

    a. four, three
    b. three, four
    c. three, three
    d. four, four

2. An SCR is operated as a/an _____ device.

    a. unilateral
    b. bilateral

3. In an SCR, the gate is connected to the _____ of a/an _____ transistor.

   a. base, PNP
   b. gate, NPN
   c. base, NPN
   d. gate, PNP

4. Zener diodes are devices that operate with negative resistance regions.

   a. True
   b. False

5. A TRIAC has _____ terminals.

   a. 1
   b. 2
   c. 3
   d. 4

The Unijunction Transistor

In this section we will discuss the Unijunction (UJT). In actuality this device operates more like a switch than a transistor.

Basic construction of a UJT is shown in Figure 17-8(a). Notice that this device has a continuous piece of N-type material that extends from **base 1** to **base 2**. This is similar to the channel of the JFET. The difference is that instead of having two gate sections, the UJT has one small P-type section called the *emitter*. As stated above, the two connections that are made to the N-type material are called "base 1" and "base 2."

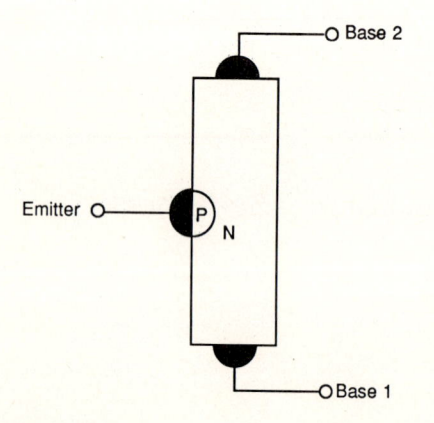

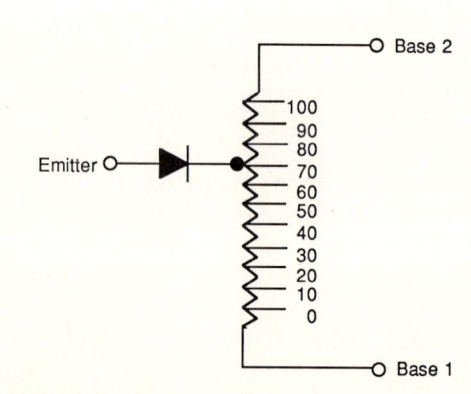

(a) Pictorial Diagram

(b) Resistance Distribution

Fig. 17-8 Unijunction Transistor

Note that the emitter (P-type material) is located closer to base 2 than to base 1. The position of the emitter can range from 50% to 80% of the distance between base 1 and base 2. In this example, the emitter is located approximately seventy percent (70%) along the distance from base 1 to base 2. The UJT shown here is an N-type. If the construction were reversed (base 1 and base 2 connected to P-type material and the emitter made of N-type material) it would be a P-type UJT and would work just as well as the N-type represented here. At the point where the emitter (P-type material) joins the base (N-type material), a PN junction is formed. This junction has the same characteristics and will operate in the same way as all other junctions that we have studied.

Figure 17-8(b) is used to discuss the characteristics of the N-type material that forms the base. As with all semiconductors, this N-type section has a resistive component. This resistance ranges from zero at base 1 to maximum at base 2.

Figure 17-9 represents the equivalent circuit of the interbase resistance and the emitter. Note that the interbase resistance is made of N-type resistive material. The PN junction that exists where the P- and N-type materials join acts much like a PN junction diode. This provides diode protection between the N-type resistor and the emitter. Notice that the diode is pointing toward the N-type material (resistor). The resistance that exists within the N-type material is called the **interbase resistance**. Interbase resistance will normally be within the 5 to 10 kΩ range. Interbase resistance is divided into two parts. We will refer to these as R_{B2} (from the emitter tap to the bottom—base 1) and R_{B1} (from base 2 down to the emitter tap).

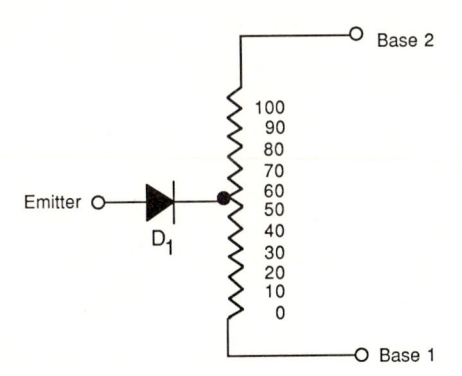

Fig. 17-9 UJT Voltage Distribution

Now we will analyze the operation of an N-type UJT. Notice that Figure 17-10 contains the equivalent circuit for a UJT that has +24 V (V_{BB}) applied to base 2 and base 1 is grounded. With these connections current will flow from ground, up through the resistance, and to the

+24 V. The interbase resistance is manufactured to have a very linear distribution of resistance throughout its length. For example, if the interbase resistance equals 10 kΩ and the resistor is 10 inches long, each one inch of length contains 1 kΩ and each 1/10 inch contains 100 Ω, and so on. This means that the 24 V would be dropped at a very linear rate. If the emitter is located at the 70% distance, then the voltage opposite the emitter is:

$$24 \text{ V} \times 70\% = 24 \text{ V} \times 0.7 = 16.8 \text{ V}$$

You can see that the voltage divider (Figure 17-11) contains resistors R_1 and R_2 each having an ohmic value of 10 kΩ. Notice that the interbase resistance contains 9 sections and is tapped at the sixth (66.67%) section. This means that two-thirds (16 V) of applied voltage is dropped at the tap. With +24 V applied to the divider, each resistor must drop 12 V. The 24 V must divide equally across equal sized resistors. The junction diode (D_1) represents the emitter junction of the UJT. The diode (D_1) has +12 V applied to its anode and +16 V applied to its cathode. This means that D_1 has −4 V reverse bias applied and cannot conduct. The voltage applied to the input diode is often referred to as the **sensing voltage** and may be in the form of a trigger or a gate.

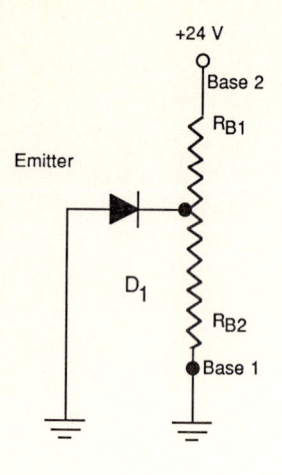

Fig. 17-10 UJT Equivalent Circuit

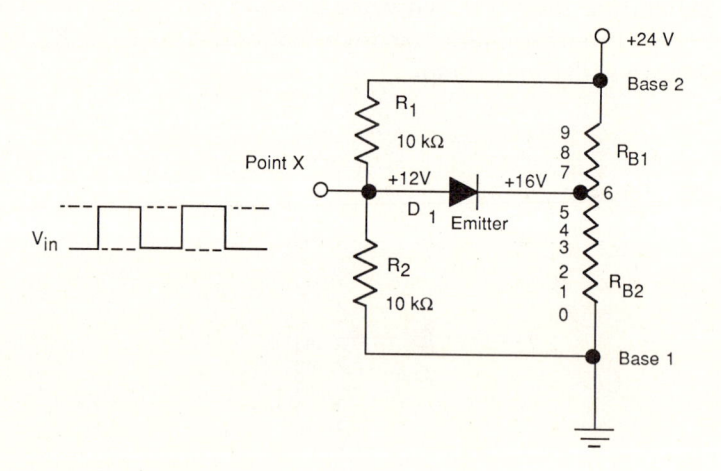

Fig. 17-11 UJT Biasing Circuit

Using these conditions we will describe the circuit's operation. If a square wave (or any alternating signal) is applied to the emitter (point X), the following sequence takes place. When the input signal goes positive, the voltage at point X increases. With a 5 V peak on the square wave, the voltage at point X increases to +17 V. As soon as the voltage

Selected Solid State Devices Operational Characteristics

at the emitter exceeds +16 V, the PN junction is forward biased and its resistance drops to approximately 0 Ω. At this time the resistance that exists between the emitter and base 1 has changed from infinite (reverse biased condition) to zero (shorted condition). With a heavy current flowing through base 1, the emitter, and R_1; the interbase voltage at the emitter drops sharply. The added current through R_1 could cause V_{R1} to be as much as +20 V. This would mean that the voltage from the emitter to ground has dropped to +3 V. The time for the UJT to switch between these conditions is on the order of 1 nanosecond (1×10^{-9} seconds). This makes the device especially useful where fast switching is a necessity.

To explain how the UJT operates in the *on-state*, we will use Figure 17-12. When the input voltage applied to the left side of D_1 exceeded the *interbase voltage* of +16 V, the PN junction was forward biased and emitter current (I_E) began to flow. Because of the emitter junction's low resistance, a high current began to flow from base 1 to the emitter. In the process, the voltage between the emitter and base 1 decreases. Remember, increased current means decreased resistance which means an increase in operating temperature and a decrease in resistance. In the circuit used here, the emitter-to-base 1 voltage drops to approximately 3 V. As long as the input voltage remains positive, the junction is forward biased, V_{R1} is increased, and V_E is approximately 3 V.

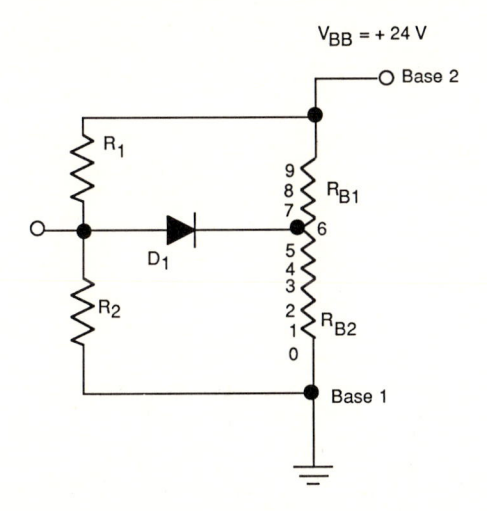

Fig. 17-12 UJT Intrinsic Standoff Ratio

When the input voltage returns to +4 V, the junction is again reverse biased. This causes I_E to cut off and allows the UJT to return to its quiescent conditions with interbase voltage at +16 V and V_E at +12 V. The amount of time that is required for the UJT to switch conditions is called **switching time**. For the UJT the switching time is in the range of 1 nanosecond.

The ratio of V_{RB2} to the voltage (V_{BB}) applied to the UJT is called the **intrinsic stand-off ratio**. The symbol for intrinsic standoff voltage ratio is the greek letter eta (η). This voltage ratio can be expressed by any of the formulas shown below:

$$\eta = \frac{V_{B2}}{V_{BB}}$$

$$\eta = \frac{R_{B2}}{R_{B1} + R_{B2}}$$

$$\eta = \frac{V_{RB2}}{V_{RB1} + V_{RB2}}$$

A check of the manufacturer's specifications will provide a decimal listing of the standoff ratio. It is normal for the ratio to range from 0.5 to 0.8. When we have this specification, it is easy to calculate the value of V_{RB2} using the formula:

$$V_{RB2} = \eta \times V_{BB}$$

When the emitter voltage (V_E) increases to the point that the junction is forward biased, the emitter voltage must equal:

$$V_E = V_{RB2} + V_{in}$$

We will normally call this value V_E peak, symbol (V_P). When V_P is applied, the UJT begins to conduct. The conductivity of the UJT increases rapidly which causes V_E to decrease rapidly to a point that we call *valley voltage*. During the transition from V_P to V_V, the UJT operates in its *negative resistance region*. This region is between points A and B on Figure 17-14(b).

With current flowing away from ground V_{RB2} equals +16 V. If a positive voltage is applied to the emitter—for example +12 V—the junction is reverse biased. Remember, when compared to a +16 V reference, +12 V is −4 V, meaning that the emitter is −4 V with respect to the N-material that is located at the same point. In this case, the junction is normally cut off.

Figure 17-13 contains the schematic symbols for the P-type and N-type UJTs. Notice that the UJT has an emitter lead that is at a 45° angle while the JFET's emitter lead has a 90° angle. This angle is the only difference between the two sets of symbols.

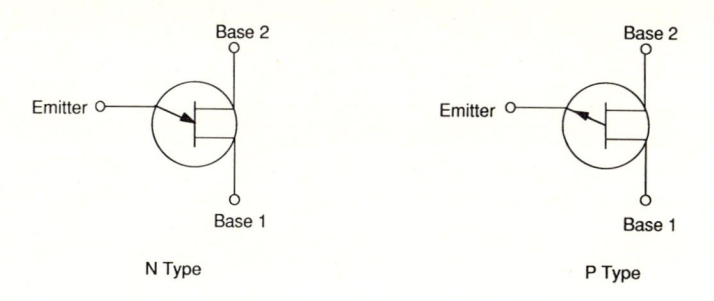

Selected Solid State Devices Operational Characteristics

Fig. 17-13 UJT Schematic Symbols

The UJT Sawtooth Generator

Many circuits can operate more efficiently when a UJT is used to turn the circuit on or off. One application is in the area of waveform generation. Figure 17-14 shows a UJT that is operating as a *sawtooth* generator.

In Chapter 14, you studied nonsinusoidal wave generators. One of the circuits was a bipolar transistor sawtooth wave generator. Remember, one of the things discussed at that time was *transient interval*. You should be able to see the advantages of using the fast

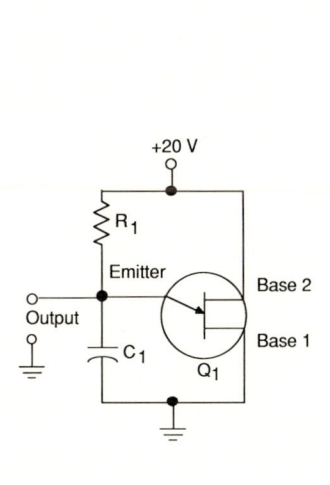

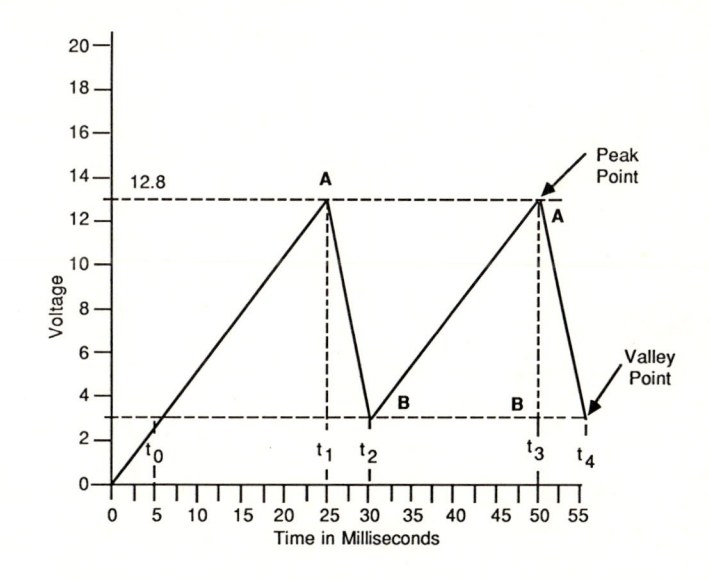

(a) Schematic Diagram   (b) Emitter Waveform

Fig. 17-14 UJT Sawtooth Wave Generator

switching UJT to reduce transient interval where possible. Using the UJT, it is possible to use the circuit shown in Figure 17-14(a) to sweep the trace on an oscilloscope at very high frequencies. The fast switching that UJTs provide allows the sweep to "flyback" in a very short period of time so that another waveshape can be applied.

The circuit operates as follows. At the first instant of operation the capacitor has 0 V charge. When power is applied, the capacitor begins to charge. This part of the output is represented by the period between t_0 and t_1 of Figure 17-14(b). As soon as the capacitor charges to a voltage greater than 12.8 V, the junction becomes forward biased. This stops the charging of the capacitor and provides a low resistance discharge path for it (the capacitor) to discharge. The low-resistance discharge path allows the capacitor's charge to drop quickly to the *valley point*. The capacitor's discharge is represented by the period between t_1 and t_2. This completes the first cycle of operation, but the capacitor will continue to charge and discharge indefinitely. The second cycle of operation is shown in Figure 17-14(b) during the period t_2 to t_4.

It is possible to control the charge time for the capacitor by changing the value of R_1. The charge time can be adjusted such that its duration equals the time required to sweep an electron beam across an oscilloscope or television screen. The discharge time could represent the amount of time for the beam to return to the left side of the screen where it is ready to start the next sweep. The time required for the beam to return is referred to as **flyback time**.

The UJT sawtooth generator operates very similarly to the bipolar sawtooth generator studied earlier. Figure 17-14(a) shows the schematic diagram and Figure 17-14(b) shows the emitter output waveform that is suitable for use as a sawtooth wave. With +20 V applied between base 1 and base 2, the interbase voltage opposite the emitter will be 12.8 V. With some modifications, this circuit can be used to generate a sawtooth or trapezoidal wave suitable for sweeping the CRT in an oscilloscope.

The UJT can be used for several other purposes. These include frequency dividers, pulse generators, phase detectors, and multivibrator circuits (Figure 17-15) that are designed around the use of a UJT.

Note that C_1 and R_2 form an RC network within the circuit. When power is first applied, C_1 begins to charge through R_1. For C_1 to charge, D_1 must be forward biased by the input voltage. During this same period, Q_1 is cut off. The voltage charged on C_1 is also V_E and is used to trigger the UJT. At the point when V_E reaches sufficient amplitude, Q_1 will be triggered on and will begin to conduct. With the low resistance and high conductivity of the interbase resistance (emitter to base 1), V_E drops to the low voltage present at the valley point.

Fig. 17-15 UJT Multivibrator

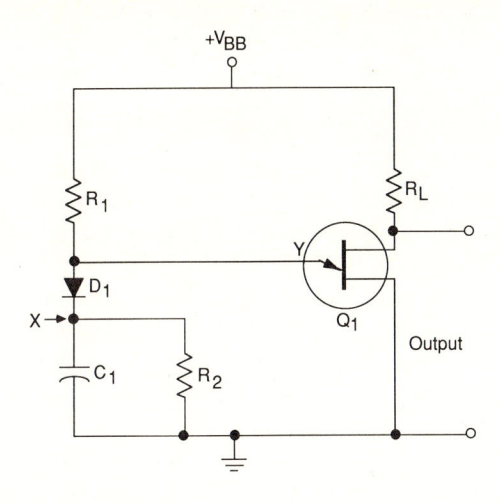

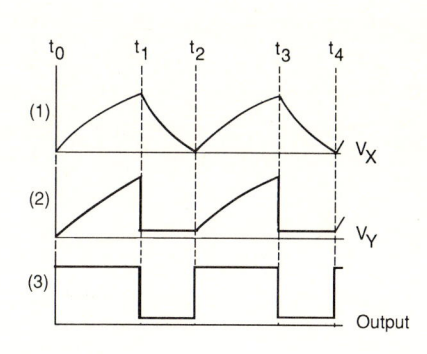

(a) Schematic Diagram
(b) Waveforms

This low voltage is applied to the top of D_1 and causes it to enter cutoff because it is reverse biased. This prevents C_1 from continuing to charge and causes it to discharge through R_2. Q_1 is continuing to conduct at the low (valley) current level with current passing through R_2. This prevents the UJT from being completely cut off. When C_1 has discharged to a voltage near the valley point voltage, Q_1 cuts off. This allows D_1 to be forward biased and C_1 can again charge. At this point another cycle of output begins. The three waveshapes that are present in the circuit are shown in Figure 17-15(b). Notice that waveshape (3) has very good multivibrator output characteristics.

The Programmable UJT (PUT)

It is possible to manufacture the UJT in a form that gives it *programmable* characteristics. When designed and manufactured using these techniques, the device is called a **programmable unijunction transistor (PUT)**. The schematic symbol for a PUT is shown in Figure 17-16. Notice the similarity that exists between this symbol and that of an SCR. The PUT is much like an SCR in both internal construction and operation. Figure 17-17 contains a pictorial diagram of the PUT and its bipolar transistor equivalent circuit. If you compare this figure with Figure 17-1, you will find that the only difference is that this device has the gate connected to the PNP transistor's base, and the SCR has its gate connected to the NPN transistor's base. In fact, the PUT is often called a **complementary SCR**.

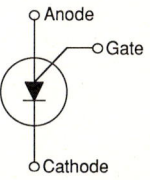

Fig. 17-16 Programmable UJT Schematic Symbol

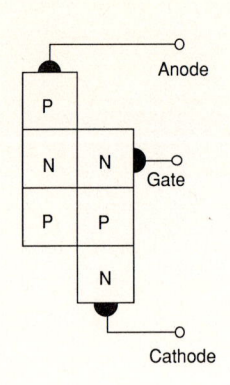

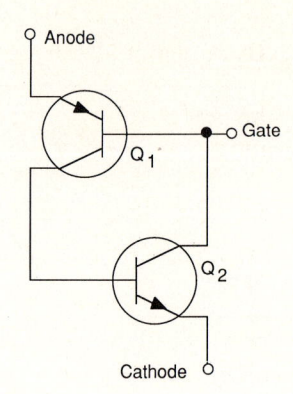

(a) Pictorial Diagram   (b) Equivalent Circuit

Fig. 17-17 Programmable UJT

The PUT can be used to perform the same functions that the UJT performs. The main difference between the PUT and UJT is the fact that the PUT can be programmed for a desired *peak output voltage*. Remember, the UJT has a fixed *intrinsic stand-off voltage ratio* that can be calculated mathematically but is also provided as one of its specifications. To achieve programmability, the PUT is designed and manufactured as a four-layer device. Observe Figure 17-17 for this arrangement.

Notice that the gate of the PUT is actually the base of a PNP transistor. This base is direct coupled to the collector of an NPN transistor. Base voltage for the NPN transistor is also collector voltage for the PNP transistor. This means that:

Collector current of Q_1 depends upon the amount of base current that flows in Q_2. When triggered, the regenerative nature of the gate and Q_1 base junctions will quickly drive the PUT into its negative resistance region.

Switching time for a PUT is approximately 10 times as fast as the UJT. With the PUT, gate voltage is much more stable than with the UJT. Changes in temperature have less effect on the PUT which, in turn, provides a more stable trigger voltage. When the PUT is reverse biased, it has a very small leakage (reverse) current. During conduction it has a very low resistance. All of these characteristics make it more desireable than the UJT.

Figure 17-18 contains a circuit diagram that illustrates a typical biasing arrangement used for PUTs. Notice that the input is applied to the PUT and the Gate is connected to a fixed voltage. This emitter voltage is established by the voltage divider that consists of R_1 and R_2.

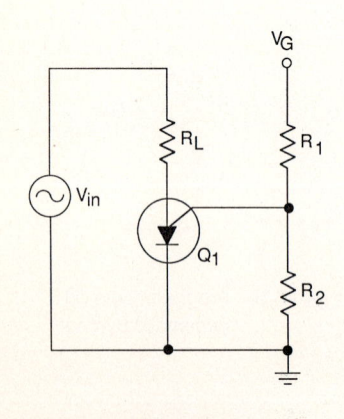

Fig. 17-18 Typical PUT Biasing Network

Selected Solid State Devices Operational Characteristics

Self-Check

Answer each item by inserting the word or words required to correctly complete each statement.

6. The emitter of a UJT can be as little as _____ base length, or as much as _____ of base length away from base 1.

    a. 50%, 100%   c. 80%, 50%
    b. 100%, 50%   d. 50%, 80%

7. For a UJT to begin conduction _____ voltage must exceed _____ voltage.

    a. gate, V_{RB1} c. emitter, V_{RB1}
    b. gate, V_{RB2} d. emitter, V_{RB2}

8. The interbase resistance of a UJT is very linear.

    a. True   b. False

9. When the UJT is conducting, the resistance between the emitter and base 2 is _____.

    a. high   c. low
    b. medium

10. Another name for trigger voltage is _____ voltage.

    a. gate   c. sensing
    b. input   d. all of these

11. A PUT has a switching time that is approximately _____ times _____ than that of the UJT.

    a. 10, slower   c. 5, slower
    b. 10, faster   d. 5, faster

The Tunnel Diode

In Chapter 6 you were introduced to the *tunnel diode*. It was the first device studied that operates with a *negative resistance region* when biased correctly. The tunnel diode's schematic symbol is shown in

Figure 17-19(a), and Figure 17-19(b) contains a voltage current plot for a typical tunnel diode operating with forward bias. Before we go further, we must understand that Ohm's law states:

Current and voltage are directly proportional but current and resistance are inversely proportional.

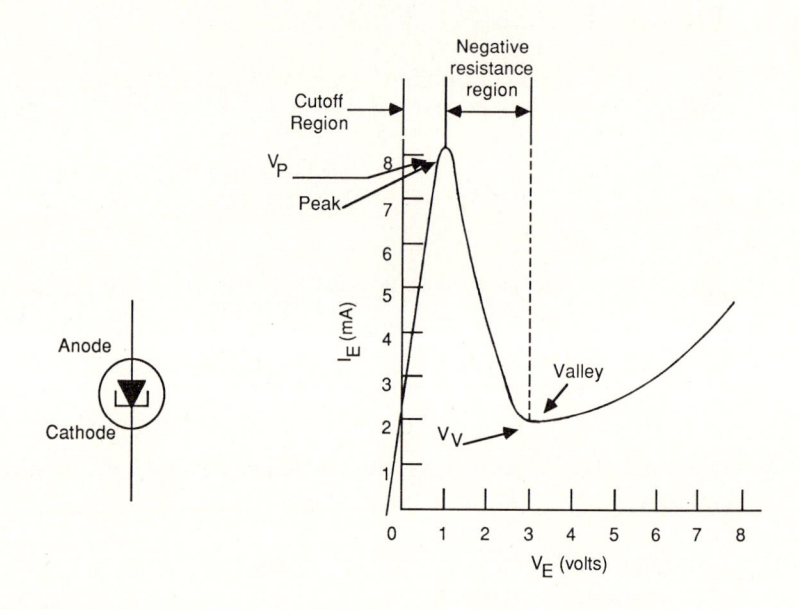

(a) Schematic Symbol   (b) Voltage/Current Plot

Fig. 17-19 Tunnel Diode

In the tunnel diode, both sections (P and N) receive much heavier doping than a conventional diode. This decreases the width of the depletion region to a very narrow barrier. Even at room temperature, electrons are so active that they "tunnel" through areas of low resistance in the junction. The application of forward bias causes the "tunneling" to increase rapidly. The large numbers of electrons that cross the junction cause I_D to increase rapidly. This is represented by the almost vertical line (Figure 17-19(b)) that represents I_D immediately after D_1 begins conducting.

It only takes a relatively small voltage to drive I_D to its *peak* value. Once it reaches the *peak point*, I_D begins to decrease even though V_D continues to increase. This operation is opposite the conditions stated by Ohm's law. Therefore, we say that the tunnel diode is operating within its *negative resistance region*. As forward bias continues to increase, I_D will continue to decrease until the *valley point* is reached. Notice that at the valley, I_D has decreased to a very low amount. This is very close to the current that would flow for a conventional diode operating with the same parameters. Past this (the valley) point the diode operates exactly like a conventional diode. The value of the tunnel diode lies in the opportunities provided by its operation in the negative resistance region.

The Tunnel Diode Oscillator

The circuit diagram for a *tunnel diode oscillator* is contained in Figure 17-20. The frequency-determining network for this circuit consists of the inductance contained in the primary coil of T_1 and C. For sustained oscillations to occur within the tank, the energy lost due to damping must be replaced by D_1 operating in its negative resistance region.

When the tunnel diode is correctly biased, this circuit will oscillate indefinitely. The output frequency will be the resonant frequency of the tank circuit. Tunnel diode oscillators can be designed that operate at ultrahigh frequencies. Frequencies of operation within the 4-5 GHz range are not uncommon.

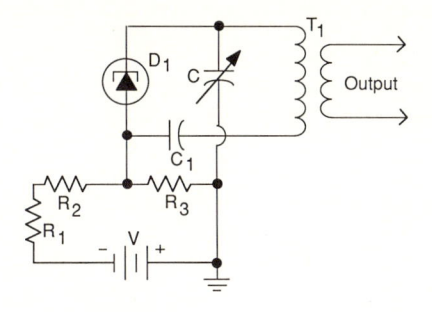

Fig. 17-20 Tunnel Diode Oscillator

The Tunnel Diode Amplifier

It is also possible to use the tunnel diode as a high frequency amplifier. The schematic diagram for a *tunnel diode amplifier* is shown in Figure 17-21. To allow for maximum swing, the diode must be biased so the circuit's Q point is at the center point of its negative resistance region. Notice that R_L is connected in series with D_1.

When an input signal is applied, this circuit operates much like a common-collector or common-drain amplifier. Voltage gain for the circuit is zero while both current and power gain are present. The input signal causes the diode conduction to swing back and forth within the negative resistance region. Power gain from this circuit is in the range

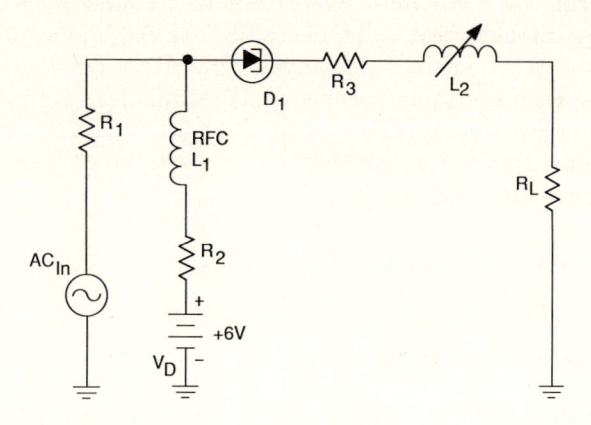

Fig. 17-21 Tunnel Diode Amplifier

of 5 to 1. The main advantage gained from using tunnel diode amplifiers is their ability to operate at ultrahigh frequencies. Frequency of operation can—and will—often exceed 10 GHz.

Self-Check

Answer each item by inserting the word or words required to correctly complete each statement.

12. Examine Figure 17-22. Which of the symbols is used to identify a tunnel diode?

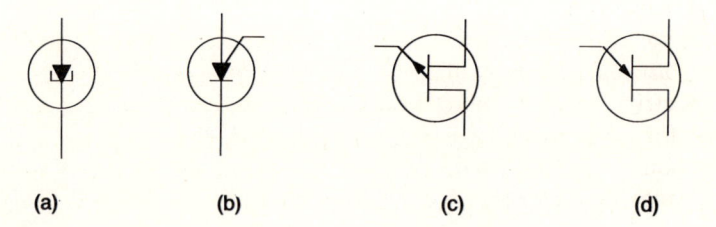

    (a)   (b)   (c)   (d)

Fig. 17-22

13. Negative resistance occurs when a/an _____ in voltage causes a/an _____ in current.

    a. increase, increase   c. decrease, decrease
    b. increase, decrease   d. decrease, increase

14. During negative resistance operation, voltage and current are ____ proportional.

   a. directly
   b. inversely

15. A tunnel diode oscillator uses a/an ____ network as its frequency determining device.

   a. RC
   b. LC

Optoelectronic Devices

Earlier we discussed one optoelectronic device—the *light emitting diode (LED)*. The LED is classified as a photoluminescent device. In the LED, current flow through its junction causes light to be emitted. An advantage of this device is that it provides light without the heat build-up associated with incandescent light sources.

An **optoelectronic** device is:

> Any device that combines a photosensitive material and a controlled light source in the performance of a desired function.

When photosensitive material is exposed to light, it undergoes an internal resistance change. The intensity of the light striking the photosensitive material is inversely proportional to the amount of resistance change that occurs within the material. As light increases, the material becomes more conductive; as light decreases, the material becomes less conductive.

Two types of optoelectronic devices are available: photovoltaic and photoconductive. Photovoltaic devices are used to convert light into electrical energy. Solar cells, like those used to provide electrical power for some satellites, are photovoltaic devices. They act as the power source for the circuits to which they are connected.

Photoconductive devices, on the other hand, operate in conjunction with a power source. When light strikes the photosensitive material of the photoconductive device, the device's internal resistance changes. This causes a change in circuit current flow. One application for this type device that we are all familiar with is the sound track on a movie. Variations in light are used to transmit the sound from the film to the audio amplifiers contained in the sound system.

It is possible for photoconductive devices to respond to frequencies that range from 0 Hz to several hundred kHz. Photoconductive devices can be designed that are sensitive to very small variations in light. Many devices are presently available that operate using optoelectronic principles. Four of the most common are the photodiode, phototransistor, opto coupler, and the LASER diode.

Photosensitive diodes are classified into two categories: photoconductive and photovoltaic. These are defined as follows:

Photoconductive diode - a photoelectric cell, the electrical resistance of which varies inversely with the light that strikes its active material.

Photovoltaic cells - a self-generating semiconductor which converts light into electrical energy.

Photovoltaic devices are currently receiving attention as an alternate energy source that does not drain our natural resources. Imagine a house that uses daylight to store enough electricity to operate during the day and also through the night. Research in the solar cell area appears to be lowering the cost of photovoltaic cells to the point where this may become economically possible in the future.

The photovoltaic device will normally have a large PN junction that is made from silicon. When this junction is exposed to light, a voltage is produced. The voltage produced can be used to provide current for a closed loop as with other power sources.

Photoconductive devices, however, currently find more uses in day-to-day electronics than do photovoltaic cells. They are valuable because of their ability to control current flow. The control that they provide results from the fact that when exposed to light, their internal resistance will vary with the light's intensity.

All semiconductor materials have this sensitivity to light. This explains why conventional diodes, transistors, and other packaged circuits are sealed in light-proof cases. To do otherwise would expose them to light which would alter their operational characteristics as the light changed. The sensitivity of a material depends on its crystalline structure.

The Photodiode

The **photodiode** is a:

Solid-state device similar to an ordinary diode with the exception that light striking its PN junction will cause it to conduct.

A photodiode is normally operated with reverse bias. Then, when light variations occur at its junction, corresponding changes in reverse current will occur. A photodiode that is connected in series with a resistor

Selected Solid State Devices Operational Characteristics

that has a value of 800Ω or less is said to be operating as a short circuit. With a series resistance of 10 kΩ or larger, the circuit acts much like an open.

Photodiodes are also called photoresistive devices. Phototransistors fall within the same classification as photodiodes. Both are, for all practical purposes, semiconductor resistors whose bulk absorbs light energy. This causes the semiconductor's internal resistance to vary with variations in light intensity. It is important to note that they do not depend upon a junction for their operation. It is possible to manufacture some devices that respond to visible light and others that respond to infrared frequencies.

In fabricating the photoconductive device, the packaging must allow maximum light to strike the semiconductor material. In some cases, a lens will be included that maximizes the use of available light. In other cases, a lens is not needed and is not included.

The schematic symbol for a photodiode is shown in Figure 17-23. Notice that the arrows point inward to indicate that the device is light sensitive. Also, note that this device has a PN junction. In many applications this device is used to provide an input to an amplifier. Changes in the photodiode's current can be amplified to a level sufficient to sound an alarm or perform another function.

Photoconductive devices are made from specially doped germanium or silicon. The amount and type of doping allows the device to be designed for operation at specific frequencies. Light striking the reverse biased PN junction causes electron-hole pairs to be generated, which will support reverse current. These variations will be quite small but are large enough to be sensed and amplified by other circuitry. Figure 17-24 contains the block diagram for a circuit using a photodiode as the sensor of an alarm circuit. Many smoke detectors used in homes and offices use a photosensitive device as their smoke sensors.

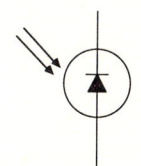

Fig. 17-23 Photodiode Schematic Symbol

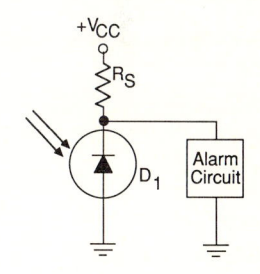

Fig. 17-24 Photodiode Alarm Circuit

The Phototransistor

The phototransistor is similar to a bipolar transistor. It may or may not have a base lead. In many cases, the base lead is eliminated and light variations are used to vary base current which controls both emitter and collector currents. Figure 17-25 contains two schematic diagrams that are used to denote phototransistors. Notice that the only input to Figure 17-25(a) is supplied by light.

In Figure 17-26 you will see a typical phototransistor circuit. Observe that the transistor is reverse biased. If no light strikes the base, the transistor will enter cutoff and collector voltage (V_C) will equal $-V_{CC}$. When light strikes the base, the transistor conducts and the collector voltage drops to less than $-V_{CC}$. In effect, light striking the base causes the collector voltage to go positive. If sufficient light strikes the base, the transistor will saturate causing collector voltage to be a small negative voltage.

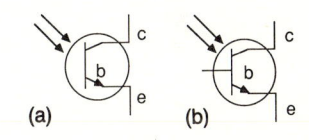

Fig. 17-25 Phototransistor Schematic Symbol

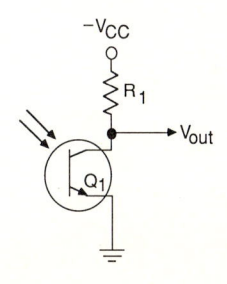

Fig. 17-26 Phototransistor Circuit Diagram

Optoelectronic Couplers

The **optoelectronic coupler** (also called *optically coupled isolator* and *opto coupler*) consists of an LED and a phototransistor (or other light sensitive device). Figure 17-27 contains a schematic diagram of a simple opto coupler. When current flows in the LED, its junction emits light which strikes the base of the phototransistor. With the arrangement shown here, the circuit can be used as a switch. A pulse of current applied to the LED will cause the transistor to switch on. Stopping LED current will cause transistor current to stop. The advantage of this type coupling is that:

> **Optical coupling provides a high degree of isolation between the input circuit and the output circuit.**

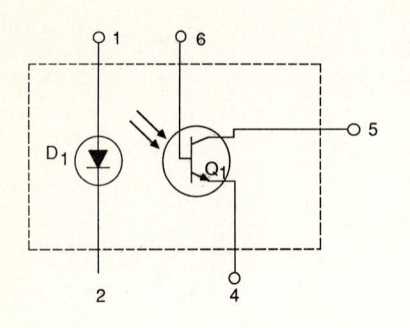

Fig. 17-27 Optical Coupler Diagram

In addition to phototransistors, other optical couplers are available that use SCRs, TRIACs, and diodes as their output circuit.

The LASER Diode

The term **LASER** is defined as follows:

> **Laser is a shortened statement that means: Light amplification by simulated emission of radiation.**

LASERs are used for many purposes. These range from precise distance measurement, to eye, ear, and brain surgery. Surgery is possible because of the concentrated, very narrow, and high powered beam that can be generated by a LASER. This beam can be tuned to a wavelength so that internal surgery can be performed without incision or damage to the external part of the body. Only the area to be repaired is affected.

A LASER emits a highly concentrated, very narrow beam of light that does not spread to a great degree. This allows it to be concentrated in a very small area. With the small beam, small incisions can be made and distances can be measured very accurately.

The LASER diode is made from a sandwich of gallium-arsenide (GaAs). The GaAs may be used by itself or combined with other materials according to the desired use. By forming a sandwich which has its ends covered by light-reflective mirrors, except for a small slot near the center, large amounts of light energy can be accumulated for emission as a LASER beam.

LASER diodes must operate with relatively high current, often several amperes. At lower currents, the diodes will emit light much like that emitted by an LED.

LASER diodes are of two types which operate in either of two modes. These are:

1. Injection LASER diode – a LASER diode that operates in a pulsed pattern.
2. Continuous wave (CW) LASER diode – a LASER diode that operates 100% of the time.

In conclusion, let us emphasize that there are several other optoelectronic devices. Each plays an important part in some circuitry. All, however, operate similarly to the ones discussed here.

Self-Check

Answer each item by inserting the word or words required to correctly complete each statement.

16. Any device that combines a photosensitive device and a controlled light source is called a/an _____ device.
17. Name two types of photosensitive devices.

    a. _____
    b. _____

18. One type optoelectronic device is the LED. It falls within the _____ category.
19. As the light applied to a photosensitive device increases, the material's resistance will _____.
20. An optical coupler can be used to _____ one electrical circuit from another.
21. A LASER diode is used to emit a/an _____.
22. The input to a phototransistor is provided by _____.
23. A solar cell is classified as a _____ device.
24. A photodiode is classified as a _____ device.
25. Operation of a photodiode or phototransistor depends upon the fact that _____ changes as light changes.

Summary

This chapter concentrates on special solid-state devices. Though not used as often as other devices, these devices can provide operational capabilities not otherwise available.

Devices include SCRs, UJTs, PUTs, tunnel diodes, and optoelectronic devices. Except for the tunnel diode, all of these are used as electronic switches. Use of the UJT and PUT provides switching at extremely high frequencies. SCRs are used, primarily, as power control devices. The SCR can also be used as a variable rectifier. Using phase shift techniques, the SCR can be adjusted to rectify any amount from half-wave to a few degrees of the positive alternation. Selection of the correct gate and anode voltages allow the SCR to be used to deliver continuous operation or to supply short, very precise pulses of DC to a load device.

The UJT and the PUT device allows us to design oscillators, multivibrators, and other circuits that operate at frequencies well up into the gigahertz range. This makes circuitry available for operation at frequencies that exceed the capabilities of bipolar and JFET devices.

Tunnel diodes are precisely doped devices that have electron movement through their junctions, without bias being applied. Application of a small forward bias causes the tunnel diode's current to increase quickly until it reaches some peak value (peak point). Increasing bias past this point causes diode current to decrease. When an increase in forward bias results in a decrease in device current, the device is said to be operating in its negative resistance region. At some point, diode current decreases to a low value, called the valley point. Past this point the tunnel diode operates as a conventional diode.

The operation of optoelectronic devices is effected by light striking their surfaces. Two types (photovoltaic and photosensitive) are available. Solar cells are a type of photovoltaic device as they convert light into electricity. LEDs are photoconductive devices in that when light strikes their surface their resistance decreases. This allows current to flow through the device which results in the LED's junction emitting visible light. Photoconductive devices are sensitive to light in that their internal resistance will vary with the amount of light allowed to strike their surface.

Opto couplers can have both photovoltaic and photoconductive devices included in their design. An LED can be used to emit a varying light intensity as results from variations in the LED's current. These light fluctuations are allowed to strike the surface of a photoconductive device that serves as the input to a subsequent circuit. An opto coupler can be used to isolate one electrical circuit from the loading effects of a second electrical circuit and still serve to couple the signal being processed.

Review Questions and/or Problems

1. Zener diodes are used as _____.

    a. current limiters
    b. voltage amplifiers
    c. voltage regulators
    d. all of these

Selected Solid State Devices Operational Characteristics

2. An SCR is triggered ON by a _____ voltage that is applied to the _____ electrode.

   a. positive, anode
   b. negative, gate
   c. positive, gate
   d. negative, anode

3. An SCR is triggered ON by the application of anode voltage. This is called _____.

   a. trigger voltage
   b. trigger current
   c. breakover current
   d. breakover potential

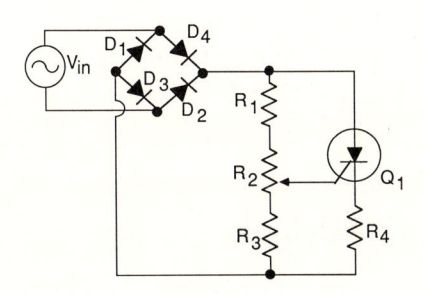

Fig. 17-28

4. Using the circuit shown in Figure 17-28, current can be made to flow not more than _____ degrees or less than _____ degrees.

   a. 180, 0
   b. 360, 180
   c. 180, 90
   d. 180, 360

5. Current flow in an SCR is started by _____ voltage and stopped by removal of _____ voltage.

   a. gate, gate
   b. anode, anode
   c. gate, anode
   d. anode, gate

6. A UJT operates more like a _____ than it does like a _____.

   a. transistor, switch
   b. switch, transistor

7. A UJT is being used as a sawtooth generator. The slope is generated during the charge of C_1.

   a. True
   b. False

8. The output of a UJT sine wave generator is maximum at _____ and minimum at _____.

    a. valley voltage, peak voltage
    b. peak voltage, valley voltage

9. A programmable UJT operates much like an SCR.

    a. True
    b. False

10. UJTs have several advantages over other devices. Which of the following is/are advantages?

    a. fast switching time
    b. ability to operate at high frequencies
    c. operates in the negative resistance region
    d. all of the above

11. Use of a programmable UJT (PUT) allows us to select peak voltage over a small range.

    a. True
    b. False

12. A PUT has a switching time that is approximately _____ times _____ than that of a UJT.

    a. 10, slower
    b. 5, slower
    c. 10, faster
    d. 5, faster

13. Tunneling occurs in the _____ of a tunnel diode because of the _____ of the P and N sections.

    a. junction, reverse biased
    b. junction, forward biased
    c. junction, heavy doping
    d. all of these

14. To get maximum swing from a tunnel diode amplifier, the diode must be _____.

    a. forward biased
    b. reverse biased
    c. biased to the middle of its negative resistance region
    d. none of these

15. Which of the devices discussed in this chapter operate with a negative resistance region?

16. Any device that combines a photosensitive device and a controlled light source is called a/an _____ device.

17. Name two types of photoconductive devices.

    a. _____
    b. _____

18. One type of optoelectronic device is the phototransistor. It falls within the _____ category.

19. As the light applied to a photosensitive device decreases, the material's resistance will _____.

20. An optical coupler can be used to _____ one electrical circuit from another.

21. A LASER diode is used to emit a/an _____.

22. The input to a photovoltaic device is provided by _____.

23. An LED is classified as a/an _____ device.

24. A phototransistor is classified as a _____ device.

25. Operation of a photodiode or phototransistor depends upon the fact that as light changes, _____ will change.

18

Integrated Circuits Fabrication and Applications

Objectives

1. Define each of the following terms:

    a. Discrete circuitry
    b. Microelectronic devices
    c. Thick-film fabrication
    d. Thin-film fabrication
    e. Ceramic printed circuits
    f. Hybrid circuitry
    g. Epitaxial growth
    h. Monolithic circuitry

2. Describe the processes used to design and manufacture each of the following:

    a. Discrete circuitry
    b. Microelectronic devices
    c. Thick-film fabrication
    d. Thin-film fabrication
    e. Ceramic printed circuits
    f. Hybrid circuitry
    g. Epitaxial growth
    h. Monolithic circuitry

3. Identify, when shown examples, the following:

    a. Headers
    b. Flat packs
    c. Dual-In-Line Packages (DIPs)

Introduction

The latest stage in the evolution of miniaturized circuitry is the development and improvement of **integrated circuitry (ICs)**. This field of circuit design has become so highly specialized that it must be performed under specific environmental conditions. Practically all circuits can now be manufactured using ICs due to advances in IC technology.

When compared to other miniaturized circuits ICs are extremely small. Compared to discrete circuitry the small size of these circuits is almost impossible to believe. Steps that show the progress from discrete circuitry to monolithic IC development are traced in this chapter.

Emphasis is placed on the procedures followed during development and manufacture of each type of miniaturized circuitry. Specific types are discussed with attention paid to the applications, advantages, and disadvantages of each type.

Summation of Events and Definitions

For those of us who have been in the electronics field for some time, the changes we have witnessed are almost unbelievable. For many years, our only interest was in vacuum tubes. Today, many textbooks don't even mention these devices. After a very difficult change in thinking, we learned to work with transistors and grew to appreciate their advantages. We had hardly begun to get familiar with transistors when a new technology was upon us. We were told that it was necessary to advance to solid-state *integrated circuit (IC) technology*. In the meantime, we have seen the advancements in IC technology proceed from quite simple to very complex. Operational units, like radios and televisions, are now designed into one small wafer called an integrated circuit. This IC not only contains all of the transistors, it also contains all of the resistors and capacitors. One digital watch contains the equivalent of approximately 20,000 transistor amplifiers. It is no wonder that we continue to be amazed at how far we have come and how fast we are moving ahead.

The challenge of space research, satellite communications, complex commercial electronic systems, and advanced military requirements have required reduction in system size and weight. These needs have placed heavy emphasis on making things smaller and lighter. Along with the move to miniaturization, we have been asked to design systems that are more durable, operate under more varied conditions, and have a longer service life.

The year 1959 saw the formation of the National Aeronautics and Space Administration. The drive to orbit manned space vehicles and to land a man on the moon prior to 1970 gave birth to new research programs. The emphasis placed on this research was the miniaturization of systems that could be used for space exploration. Space and military requirements were the focal point for early attempts at miniaturization, but as it usually does, advanced technology soon spilled over into the civilian community. In 1970, schools that were fortunate enough to have computer systems had to provide large spaces for machines that were not much more capable than today's microcomputers. Computers of this type cost hundreds of thousands of dollars. Today, comparable computer capability can be bought for a few thousand dollars. Advantages that have come about include increased reliability, lower operating cost, easier maintenance, and the design and manufacture of more complex systems that have less weight and provide more capability.

In previous chapters, all circuits have been discussed from the *discrete circuitry* approach. Discrete circuits are defined as:

> **Circuits built from separate components that are individually manufactured, tested, and assembled.**

In this chapter, you will be introduced to electronic circuits that are constructed from extremely small electronic parts that are called **microelectronic** devices. These classifications include:

1. Discrete microcomponents
2. Thick-film or ceramic printed circuits
3. Thin-film fabrication
4. Silicon-integrated circuits
5. Hybrid microcircuitry

Our intent is to provide you with an insight of the entire process—beginning with the design engineer and ending with the operational circuit. It is not our intention to cover the operational capability of many circuits—only to introduce you to these processes. We will begin by spending a short time discussing each of the five classes mentioned above and then investigating these classes in more depth.

Discrete Microcomponents

In this classification we are concerned with the techniques involved in packaging small components into throw-away packages having the same (uniform) shape. When we speak of microcomponents, we are talking about miniature diodes, transistors, resistors, and capacitors. Using these components and a quality manufacturing process we can duplicate the circuit many times. If each package has exactly the same size and operating capability, each can be substituted for the other. By replacing the entire package we can often save money over the cost of isolating and replacing a single component.

Thick-Film or Ceramic Printed Circuits

In this class we are concerned with the techniques used to print circuit patterns on ceramic, or other, substances. These circuits are created through use of paste, or slurry, which is injected in specific patterns through fine mesh wire screens, onto ceramic substrates or other materials. The paste (slurry) used is made by mixing water and an insoluble material such as clay, cement, or soil. Once the paste is deposited on the base substance, the entire package is fired in a kiln, much like the process used in making ceramics. During this firing, the materials undergo a physical change that makes them valuable for use in electronic design.

Thin-Film Fabrication

Thin-film circuits are deposited in, or on thin glass or ceramic substrates. Integrated circuits (ICs) are created using this process. The desired circuit is formed on a single-crystal silicon substrate. The actual circuit is created by use of the *diffusion* process. We referred to diffusion back in Chapter 15 when we discussed JFETs, but we will discuss it in more depth in this chapter. You should realize that of the five microelectronic device classifications listed above, thin-film devices and monolithic silicon integrated circuits (ICs) are fabricated using this method.

Silicon-Integrated Circuits

This technology is a subdivision of thin-film fabrication. Thin-film techniques are used to fabricate circuits on a silicon substrate that have all components (active and passive) that are required for the desired operation. Transistors, MOSFETs, and JFETs are diffused into the silicon substrate along with the required resistors and capacitors.

Hybrid Microcircuitry

This is not a separate technology but a wedding of the others. Hybrid microcircuitry could apply to any combination of discrete and/or thick- or thin-film fabrication. The term *hybrid* indicates that two or more processes are used in designing the circuitry.

Fabrication of circuits using each of these processes requires very specific procedures. Each circuit, however, must be designed to meet specific operational objectives. These objectives are the basis used for selection of the actual process that will be used in fabricating the circuit. In order for us to be able to choose between processes, we must consider the following objectives:

1. Increased functional capabilities
2. Increased (higher) reliability
3. Ease of maintenance
4. Reduced manufacturing cost

Ceramic Printed Circuitry

An early development used to miniaturize electronic circuitry was the use of printed circuitry. In this process a design that includes all conductors needed to connect the discrete components of a circuit is inked onto a copper clad board. The entire board is then treated with an etchant that removes all of the excess copper. The resulting pattern provides all of the conductive paths needed for completion of the circuit. Next, holes are drilled through the board at each location where a component's leads will be installed. Then, each component's leads are inserted through the holes and metal-to-metal connections are made using soldering techniques. This provided a large reduction in circuit size and weight over previous discrete component circuitry.

A more recent approach to the miniaturization of electronic circuitry was the use of **thick-film** fabrication to create **ceramic printed circuits (CPC)**. In this process, all conductive and resistive circuit elements are screened onto a ceramic wafer. All other components that are used in this process are conventional microcomponents that must be soldered in place. CPC can be used with all types of circuitry ranging from the use of conventional discrete components to ICs. In each of these cases, the result is a hybrid circuit.

In the manufacture of CPC circuitry, the first step is to prepare the substrate. A **substrate** is defined as:

> The supporting material on or in which the parts of a circuit are attached or made.

The substrate has an added advantage in that it serves as a heat sink which aids in the dissipation of the heat developed within the operational electronic devices.

Substrates are made from a variety of materials. The most common of these is a high-grade ceramic material that has been mixed with aluminum oxide. If the substrate must have a smooth surface, the ceramic substrate is coated with glass. When glass is applied, care must be taken to assure that the smooth surface has an expansion coefficient that is suitable for the heat expansion that the circuit will require. The ceramic substrate is then cut to the size and shape desired. This is followed by a thorough cleaning using chemical solutions. From this point on, the substrate is handled as little as possible. The intent is to maintain a dry and clean substrate. All of this is done within a temperature-controlled area. A sample substrate is shown in Figure 18-1(a).

Once the substrate is prepared, the next step is the development of circuit *masks* that can be used to transfer the circuit layout to the substrate. A **mask** is defined as:

> A thin sheet of material (usually a thin sheet of metal which contains an open pattern) that is used to shield selected portions of the substrate (base) during a deposition process.

596 Solid State Electronics

(a) Substrate

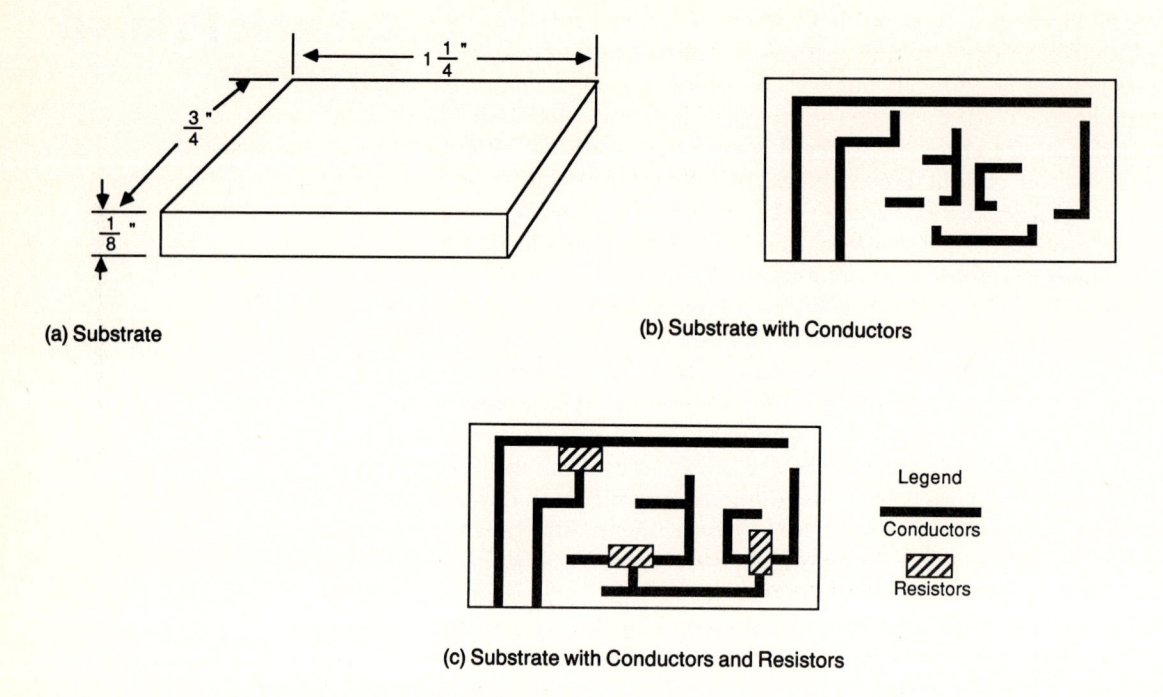

(b) Substrate with Conductors

(c) Substrate with Conductors and Resistors

Legend
Conductors
Resistors

Fig. 18-1 CPC Thick-Film Circuit Fabrication

In the development of these masks, it is common to make a large-scale pictorial diagram of the desired circuit. Then, using photographic methods, the size of the diagram is reduced to the mask size desired. Masks are usually made from stainless steel or other metals. Mesh (openings in the mask) size varies with the type circuit and method that will be used to apply the circuit to the substrate. Once the mask has been prepared, it is duplicated many times so that when it is used, many circuits can be transferred at one time. Because of the way it is made, the mask is a positive image of the circuit that is being fabricated. This means that the paths that will be used as conductors are the areas that are solid, and the nonconductive areas are open.

The next step in CPC fabrication is the coating of the substrate with a photosensitive material called *resist*. Once coated, the substrate is laid aside to dry. During the drying process, the resist hardens into a coating that is sensitive to light. Because of the light sensitivity, much of CPC fabrication must be done under darkroom conditions.

With the hardened resist in place, the mask is placed over the substrate. Once the mask is aligned, the entire package (mask and substrate) is exposed to ultraviolet light. The exposure of the photosensitive areas causes them to harden, but the unexposed (hidden) areas do not harden. Remember, the exposed (hardened) areas *do not* represent conductive areas. When the exposed substrate is rinsed with water, the unexposed photoresist is washed away. The result is that the unhardened areas are an exact reproduction of the desired circuit pattern.

The application of the slurries (pastes) that fabricate circuit components is very similar to the actions of an artist in silk-screening a picture. During this stage of fabrication, the application of components is controlled under very precise conditions and to exact standards.

Usually, the conductors are placed onto the substrate first. Gold, silver, platinum, and palladium (or combinations of these) are the metals used in laying out the conductive paths on the substrate.

The conductive metal is combined with a paste that serves as a binder. This binder assures that the materials being applied will adhere to the substrate and will not be easily broken. Conductors are deposited on the CPC substrate by placing another mask over the substrate that has all conductors traced as open sections within the mask. The conductive paste is applied to the mask with gentle pressure that forces it to pass through the open areas of the mask. Once all conductive patterns have been deposited on the substrate, the mask is removed.

At this point, the conductive pastes are allowed to dry. Afterwards, the entire sandwich is placed in an oven where it is fired at temperatures that range from 600° C to 1100° C. Actual firing temperatures depend on the type of materials being used. During firing, the conductive pastes harden and form a very secure bond to the substrate. A substrate with its conductive pattern is shown in Figure 18-1(b).

After the conductive pattern has been completed, a different mask is used to fabricate the resistive elements. Resistors are formed by the application of a paste designed for its resistive characteristics. Materials that are typically used for resistance applications are silver, palladium, and copper which have been mixed with glass, grit, or other additives. Again, a mask is placed over the device and the resistive substance is applied through holes within the mask. The method used for application is identical to that used with application of the conductive paste. The paste is allowed time for drying and is then fired at temperatures ranging from 600°C to 800° C. A ceramic substrate that has resistors in place is illustrated in Figure 18-1(c).

Electronic circuits require resistors of differing values. In the manufacture of CPC devices, pastes that have varying amounts of resistance are used. The type of paste used determines the amount of resistance deposited. The amount of resistance contained by each paste is determined by the amount of resistance that is exhibited by a square pattern that is 1 mil (0.001 inch) thick. Resistive pastes are divided into four categories; 500 Ω, 3 kΩ, 8 kΩ, and 20 kΩ. If all resistors of one category have the same thickness and size, their resistances will all be equal. This results in resistive inks (pastes) being listed as 500 Ω/square, 3 kΩ/square, and so on. Using the same ink but varying the size of the area covered results in resistors that have different values.

Resistors have a 2:1 Length-to-Width Ratio

(a) Horizontal Resistor

(b) Vertical Resistor

Fig. 18-2 Resistor Size Ratios

Figure 18-2(a) and (b) contain drawings that illustrate changes in length-to-width ratio of resistors attached to a CPC device. Notice that these resistive deposits are rectangular in shape and that their length-to-width ratio is 2:1. When the leads have been attached, as shown (horizontally) in Figure 18-2(a), the value of the resistor will be double the standard used for that type ink. For example, if the ink has a 3 kΩ/square rating, the resistor shown in Figure 18-2(a) will have 6 kΩ value. A resistor like the one (vertical) shown in Figure 18-2(b) will have half the rated value, or 1.5 kΩ.

Resistors that are formed on the CPC will, initially, have a resistive value less than that called for by the circuit design. The resistance of the individual resistor is adjusted to the exact value by shaving off very thin layers of the resistive coating. Remember, smaller conductors have higher resistance than larger conductors made from the same material. Therefore, decreasing the resistance material's thickness increases its ohmic value.

A rule-of-thumb is that the larger the area covered, the higher the resistor's wattage rating. The length-to-width ratio of resistance deposits is 2:1. However, covering the same area with thicker deposits produces different resistance values. Higher power ratings result from spreading equal amounts of the paste over larger areas. Careful control of the amount of paste and the space covered allows both resistance and power rating to be controlled.

The last masking (screening) needed in the fabrication of a ceramic printed circuit (CPC) is the one used to apply an insulating material (ink). Again, a mask is used to assure proper application of the ink. The inked CPC device is again heated to assure that the insulating material bonds to the conductors and resistors. The ink then serves as the insulator that prevents the shorting of board components and discrete components that will be added later. During this period, each resistor receives a final adjustment to bring its ohmic value to within ±1% of the design value. The end result is a ceramic printed circuit that has its conductors and resistors baked into the ceramic material. These components have been adjusted to tolerance, and the final assembly has been cleaned. As a final step, the board has all of its conductors tinned in preparation for the connection of discrete components.

Discrete components, such as diodes, transistors, capacitors, terminal leads, and integrated circuits (ICs) are soldered to the CPC's tinned conductors to complete the circuit. Leads are then attached which allow the CPC to be connected into other circuitry. As a finishing touch, a clear plastic coating is sprayed onto the CPC surface. This coating is then heated enough to assure that it bonds to the entire surface. The coating serves as a barrier that protects the finished circuit from environmental changes. A final check of the circuit includes testing it for operational suitability and ability to meet manufacturing specifications. A finished CPC is shown in Figure 18-3.

Integrated Circuits Fabrication and Applications

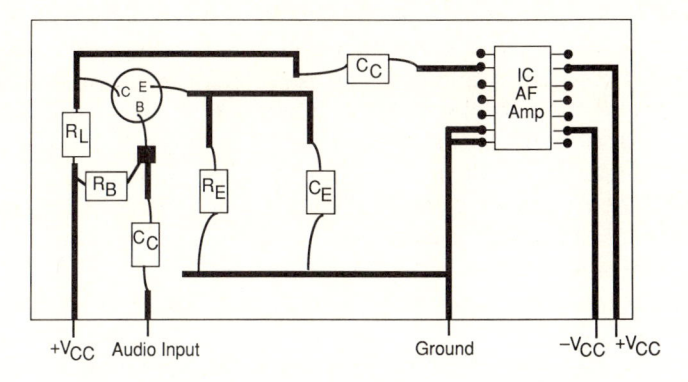

Common-Emitter Amplifier Coupled to an IC Audio Amplifier

Fig. 18-3 A Completed CPC Circuit

Self-Check

Answer each item by inserting the word or words required to correctly complete each statement.

1. Microelectronic devices are classified into five categories. These are:

    a. _____
    b. _____
    c. _____
    d. _____
    e. _____

2. During the manufacture of thick-film devices, _____ and _____ are attached prior to the connection of discrete components.
3. The substrate serves two purposes, these are:

    a. _____
    b. _____

4. Circuits that are placed in a single-crystal silicon substrate are inserted using the _____ process.
5. Hybrid microcircuitry is a combination of _____ or _____ and _____.
6. Shaving a ceramic printed circuit resistor will cause its _____ to decrease and its _____ to increase.
7. A completed ceramic printed circuit could contain the following: _____, _____, _____, _____, and _____.

Thin-Film Circuits

Thin-film fabrication differs considerably from that of thick-film (CPC) fabrication. With thin-film microcircuit fabrication, a substrate is coated with an almost inert substance such as glass or ceramic. The resistors, capacitors, and conductors are deposited on the inert layer using techniques similar to those discussed earlier. Note that in this type of fabrication, capacitors are deposited components—not discrete components. Metal electrodes (called *pads*) are bonded to the circuit board. These pads are usually located near an edge of the board. Active devices (transistors, integrated circuits, hermetically sealed devices, etc.) are connected to these pads by soldering.

The steps involved in the development of thin-film microcircuitry are:

1. Engineers prepare rough, often hand-drawn, layouts of the desired circuit pattern.
2. Drafting personnel produce precision (master) drawings from the engineer's layout that are exact duplicates of the engineer's microcircuit. The finished master drawing is approximately 30 times larger than the size of the final product.
3. The master drawing is then used to make additional drawings from which the masks used in depositing thin-film inks and pastes will be made. The placement of conductors, resistors, and capacitors requires the production of separate, precise drawings called *tracings*.
4. Photographic masters are made for each of the tracings. These photographs must be perfect reproductions of the circuit that show the location of all conductors, or all resistors, or all capacitors.
5. Masks that will be used in the actual depositing of conductors, resistors, and capacitors are made. Using photographic methods, the masters are reduced to the actual size needed for fabrication.
6. Once the masks are available, each mask is used to deposit its conductors or components onto the substrate. The process used for making the deposits is called *vacuum deposition*. This process is defined as follows:

   Vacuum Deposition – A process in which a substance is heated in a vacuum enclosure until the substrate vaporizes and then condenses (deposits) on the surface of another material in the enclosure.

7. Once all deposits are placed on the substrate, the conductive pads are bonded along the edges of the substrate. Small strands of gold or other conductive material are used to attach the pads to the necessary points around the substrate.

8. The final step of assembly requires the mounting and soldering of the discrete components. Metal leads or tapes that are connected to the discrete components are soldered to the pads on the substrate.

A typical thin-film substrate is shown in Figure 18-4(a). Proper preparation of this substrate is very important to the quality of the finished microcircuit. The nature of the substrate and its finish are

(a) Blank Substrate

(b) Conductors Partially Completed

(c) Bottom Plate of C_1 added

(d) Diffused Dielectric for C_1

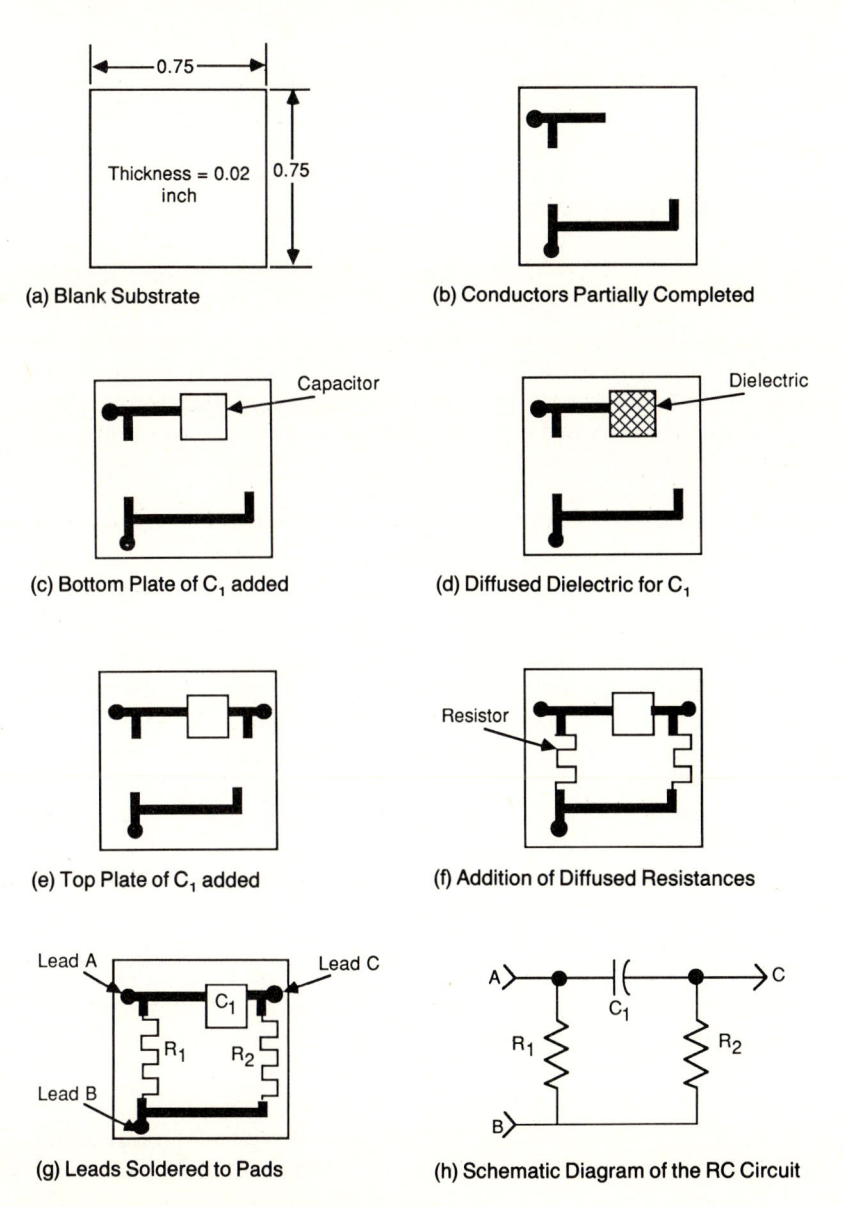

(e) Top Plate of C_1 added

(f) Addition of Diffused Resistances

(g) Leads Soldered to Pads

(h) Schematic Diagram of the RC Circuit

Fig. 18-4 Thin-Film Fabrication Procedure

especially important due to the dependence of the other components on these quantities. (Even the smallest variation in surface smoothness will cause additional material to be deposited which changes component value.) Under ideal conditions, the substrate must have the following properties:

1. A perfectly smooth surface
2. High dielectric (insulating) strength—the ability to insulate one component from another when components are located near to each other
3. High thermal conduction—the ability to remove heat
4. Compatibility with vacuum processing—the substrate must be of a material that will not change properties at the temperatures where the resistive, capacitive, and conductive pastes vaporize
5. Relatively low cost

Substrates can vary in size, but the most common sizes are 0.75 inches by 1.5 inches by 0.02 inches and 0.625 inches by 0.75 inches by 0.02 inches. In Figure 18-4(b), conductive patterns have been added to the substrate shown in Figure 18-4(a). The conductors are formed from several metals that are suitable for use in vacuum deposition processing. Some of these metals are gold, silver, copper, and chromium, or a mixture of these. The type of metal used depends on its electrical properties, its thermal (heat) effect, and how well it bonds to the substrate.

Remember, a capacitor is made from two conductive plates that are separated by a dielectric. To form capacitors, the material used is usually the same as that used for the conductors. If the conductors are aluminum, the capacitor's plates will be aluminum. The dielectric used for capacitors is a type of chemical oxide. The most common of these is silicon monoxide (S_iO). Using this oxide, capacitors can be formed that range from 0.005 μf to 0.02 μf per square centimeter. Figures 18-4(c), (d), and (e) illustrate the steps required to fabricate thin-film capacitors.

Resistors are formed from many different materials. The two most common materials are *nichrome* (nickel-chromium alloy) and *tantalum*. Resistor patterns are shown in Figure 18-4(f). Notice that the resistive deposit arrangement is similar to the resistor symbol used in electronics. Actually, this small square wave is the industrial electronics symbol for resistance. Once the type of resistive material that will be used is determined, resistor values are formed by making the deposits longer, wider, and/or deeper. It is possible to obtain many different values of resistance.

A representation of the completed circuit with pads and conductors attached is shown in Figure 18-4(g). The schematic diagram for the finished product is shown in Figure 18-4(h).

Figure 18-5(a) contains the schematic diagram for a common-emitter NPN amplifier. Figure 18-5(b) shows the same thin-film circuit with the transistor connected. The connection of the transistor and other discrete components requires very delicate soldering techniques. The ideal thin-film circuit is one that contains all of the circuit components as fabricated devices. This requires that active devices other than bipolar transistors be used.

(a) Circuit

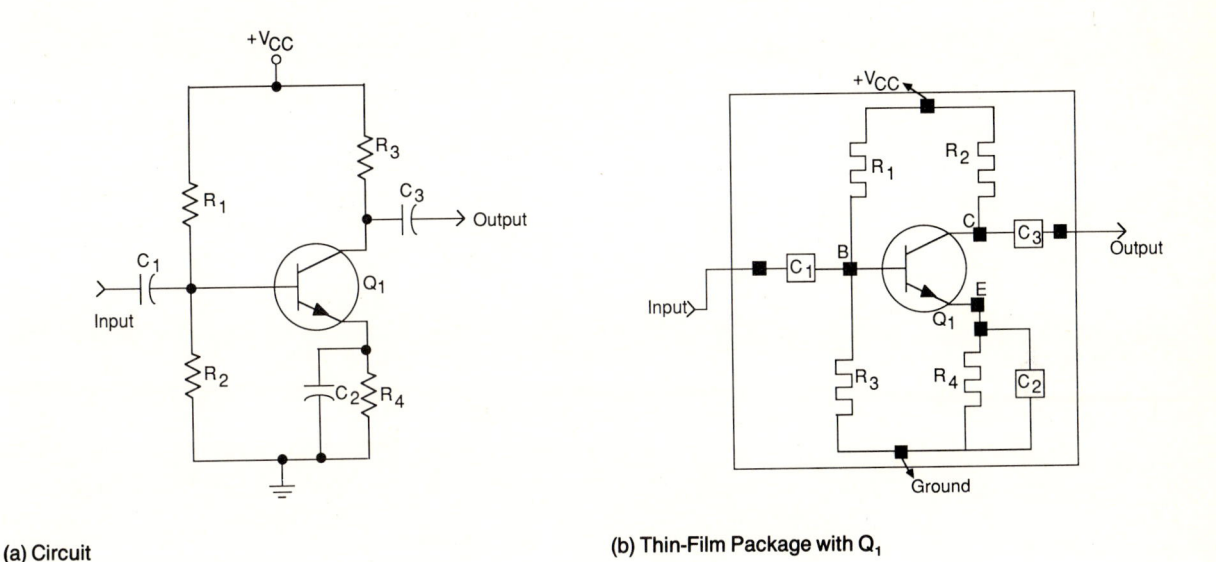

(b) Thin-Film Package with Q_1

Fig. 18-5 Hybrid Thin-Film and Transistor Circuit

One such device is the *field-effect transistor (FET)*, discussed in Chapters 15 and 16. A thin-film version of a FET consists of a *channel* (film of cadmium sulfide) that is deposited on an insulating substrate. Source and drain electrodes that are made of gold and then connected to the cadmium sulfide channel so that they make conductive contact. Next, the gate is deposited on the cadmium sulfide channel. Following this, a layer of silicon monoxide is used to coat the surface of the device with the exception of the source and drain electrodes. The silicon monoxide serves as an effective insulator and as protection against wear. Gate electrodes are inserted through the silicon monoxide. In this way, the monoxide serves as an insulator that isolates the gate and channel electrodes.

Using these techniques FETs can be fabricated that operate in both the *depletion* and *enhancement* modes. Circuit operation is identical to the silicon FETs studied in the earlier lessons.

Tantalum thin-film technology has been the subject of increased interest for the last several years. Advantages gained from tantalum thin-film fabrication are:

1. high annealing (heating and cooling) temperatures

2. good stability of films used
3. high dielectric strength
4. high dielectric constant of tantalum oxide makes it useful for capacitor fabrication.
5. both resistors and capacitors can be fabricated using the same paste

It is possible, using tantalum thin-film fabrication techniques, to fabricate resistors that have such high power dissipation capabilities that heat will shatter the glass substrate before the resistor will fail. Tantalum thin-film circuits have the distinct advantage that both resistors and capacitors can be fabricated using the same materials. Also, much of the fabrication can be done using the photo-etching technique. This allows the circuit to have some advantages over other thin-film fabrication techniques. However, the disadvantage of not being able to fabricate PN junctions (active devices) limits extensive use of these circuits. Diodes and bipolar junction transistors must be discrete components which are soldered to the finished thin-film circuit.

The use of thin-film circuitry has both economical and technical advantages over the use of conventional (discrete) circuitry. When thin-film circuits can be used as passive networks, or as a passive part of a hybrid circuit, they can be produced both economically and in relatively short periods of time. Development time, from schematic drawing to finished circuit, takes approximately 30 days. This fact, alone, makes them ideally suited to some production runs.

The use of thin-film fabrication allows greater flexibility of design than other types of microelectronic circuitry. The number of different materials that are available for use in fabricating thin-film circuitry is larger than those available for CPC fabrication. Also, the best discrete active components can be added to the circuit to complete a highly reliable, thin-film hybrid circuit. It is possible to fabricate a thin-film circuit that is an exact duplicate of a conventional circuit. This is illustrated in Figure 18-5. Since conventional circuits are already available, the design of thin-film versions is greatly simplified. Because of the processes used, it is possible to develop new masks, or to develop new deposit materials, and to change production at any point in the production run.

Circuitry that is fabricated using thin-film techniques is very reliable due to the fact that:

1. Minimum circuit contacts are required.
2. Only a few processes are required for manufacture.
3. The elements used are very stable.

Element stability is such that the estimated failure rate for resistors is 0.0035% per 1000 hours of operating time. For capacitors, the failure rate is 0.0009% per 1000 hours of operating time.

Thin-film and thick-film hybrid circuits can be used in a wide range of electronic equipment. Their adaptation to computer circuitry (analog and digital) has been highly successful. Their high stability and reliability makes them ideally suitable for circuits that operate in remote locations or in space vehicles. They can operate at frequencies ranging from 0 Hz (DC) to approximately 1 GHz.

Circuits fabricated using thin-film techniques are suitable for use where small, compact, and lightweight circuitry is required. This fabrication process also allows these circuits to have some degree of flexibility, if needed. Practically all existing conventional discrete circuits can be duplicated using thin-film fabrication techniques and discrete active components. Figure 18-6(a) contains the drawing for a conventional circuit. Figure 18-6(b) contains the same circuit after thin-film fabrication with discrete components attached.

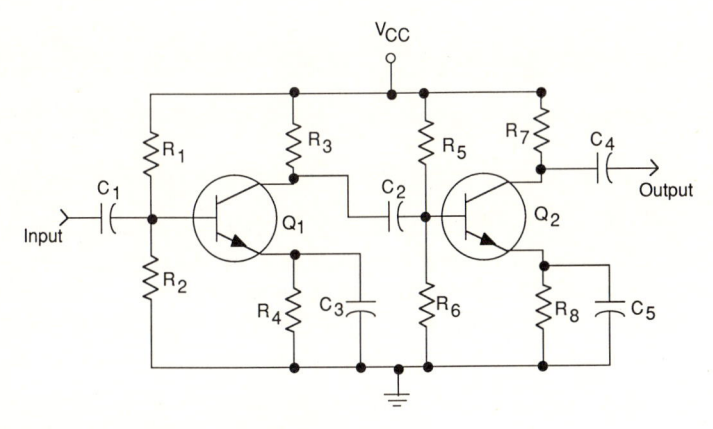

(a) Conventional Circuit Schematic Diagram

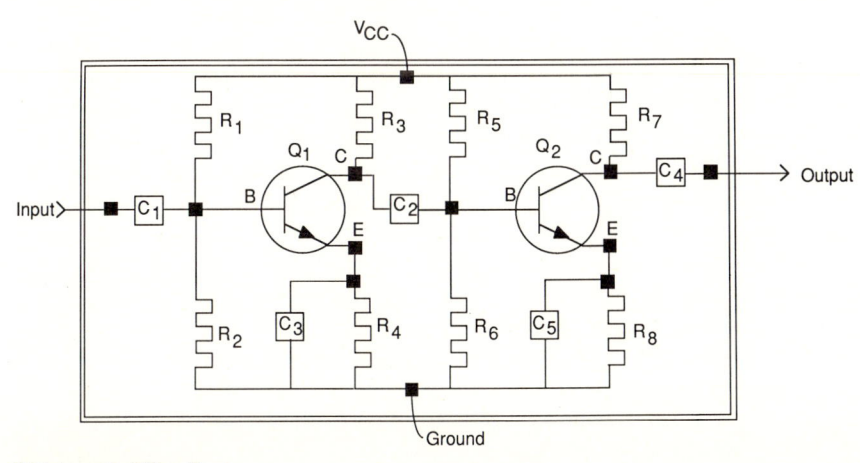

(b) Integrated Circuit

Fig. 18-6 Comparison of Conventional and Integrated Circuits

Self-Check

Answer each item by inserting the word or words required to correctly complete each statement.

8. Thin-film circuits have ____, ____, and ____ deposited on a glass or ceramic substrate.
9. If the conductors of a thin-film circuit are made of aluminum, the capacitor plates will be made of ____.
10. ____ is the dielectric that is most commonly used as a dielectric for thin-film capacitors.
11. In fabricating thin-film resistors, ____ and ____ are the most commonly used materials.
12. It takes an average of ____ days to produce a thin-film circuit from a schematic diagram.
13. The four advantages gained from thin-film fabrication are:

   a. _____
   b. _____
   c. _____
   d. _____

Integrated Circuits

In importance, the development of the *silicon integrated circuits* (ICs) ranks right along with development of the vacuum tube and the transistor as a great advance in the evolution of electronic circuits. Integrated circuits are usually called *ICs* or *chips*. Their use has played a major role in the development of microelectronic circuitry. The definition of **integration** is:

> To form into a whole, or to incorporate into a larger unit.

ICs are designed to contain complete circuits. In fact, many ICs contain hundreds or thousands of different circuits. In the technology of today, ICs contain all elements, active and passive, required to perform all operations of a specific circuit. All components, along with their interconnections, are formed into a single chip of silicon. It is possible for the circuitry that is housed on this chip to perform complex electronic functions. Usually, the chip performs these functions without any outside assistance other than power, input, and output connections. All circuit components, active and passive, are formed from the same semiconductor substrate. For this reason, the completed device is called a **monolithic integrated circuit**.

To better understand IC's, we will begin this discussion with an explanation of their reliability, cost, size, and power requirements. From this discussion, the advantages and disadvantages of ICs will become apparent.

Reliability

To date, no limitation to the application of ICs has been found. On the contrary, we hear almost daily of a new way that ICs are being used or a way that their use has been advanced. The ICs that are available today have been perfected to the point that they can be expected to provide dependable service over long periods of time. For example, a specific IC used in the NASA program has demonstrated a failure rate of 5 failures for every 1 billion hours of operation.

Cost

The cost of ICs varies over a very wide range. Price is determined by the specifications established for the final product. For a device that has high capability, or one that is produced in small quantities, the cost can be hundreds of dollars per unit. As a general rule, the cost of high demand, large run ICs ranges from less than one dollar to about five dollars. The cost of complex computer circuits was drastically reduced with the development of the IC. With conventional circuitry, the manufacturing and operating costs of complex circuitry is high when compared to the same IC circuitry.

Size and Weight

When compared to conventional circuitry, both size and weight of integrated circuitry are much less. In some applications, size and weight may not be that important. It is possible that even though the circuitry is smaller, the overall size of the finished unit may not be smaller. Among the factors that control the size and weight of the finished unit are the power requirements and the required interconnections. To reduce the overall size of operational equipment, new interconnecting methods were designed for use with ICs. Printed circuit boards have been developed that allow the connection of large numbers of ICs in very small spaces. The size of operational units has also been reduced by the use of ICs because of their lower power requirements. In earlier conventional circuits, much power and space had to be devoted to maintaining the temperature of operating equipment within specific ranges. With the IC, we can maintain the operating environment with much less expense.

Power

The factor that influences the use of ICs more than any other is power dissipation requirements. Usefulness and function of any electrical device depends upon its ability to dissipate the heat generated while the circuit is operating. In most cases, power dissipation depends on the physical size of the device. We have, however, devised heat sinks, cooling fins, and ventilation systems that assist in maintaining satisfactory operating temperatures. In ICs the controlling factors are:

1. Physical size of the substrate
2. Material from which the substrate is made
3. The mounting used to house the substrate

Fabrication of the Integrated Circuit (IC)

IC fabrication can be divided into the following major steps:

1. Preparation of the substrate
2. Photoengraving of the substrate
3. Diffusion of conductors and components
4. Oxidation
5. Epitaxy
6. Interconnection, attachment of external leads, and placement of the finished device into a suitable capsule

We will cover each of these steps in detail.

Substrate Preparation

The IC begins as a shaft of silicon material several inches in length. From this shaft, several hundred wafers will be sliced. These wafers are cut from a single crystal of silicon material. A picture of the wafer is included as Figure 18-7.

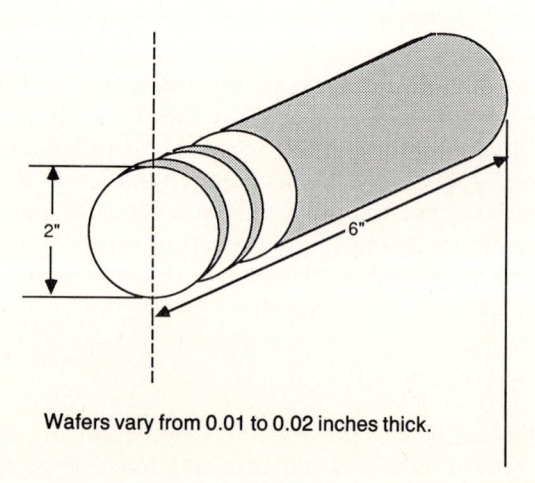

Wafers vary from 0.01 to 0.02 inches thick.

Fig. 18-7 Slicing Wafers from a Silicon Crystal

A wafer is sliced from a long piece of silicon crystal that is sometimes called a *boule*. This wafer must meet certain standards if it is to be suitable for use as the substrate for a group of ICs. Some of these are:

1. Type of base material
2. Type of impurity used for doping the base material
3. Orientation of the crystalline structure
4. Resistivity of the material
5. Etch-pit count
6. Smoothness of surface finish

The substrate can be either P-type or N-type, however, many ICs are fabricated using P-type substrates only. Usually, one type of material is used for all diffusion. This allows a reduction in the number of steps required for fabrication. This also reduces the cost of production. Transistors (both NPN and PNP) are fabricated at the same time using the same substrate. This makes the type of dopant impurity a critical consideration. The base of a PNP transistor is diffused at the same time as the emitter and collector of an NPN transistor is being diffused (see Figure 18-8).

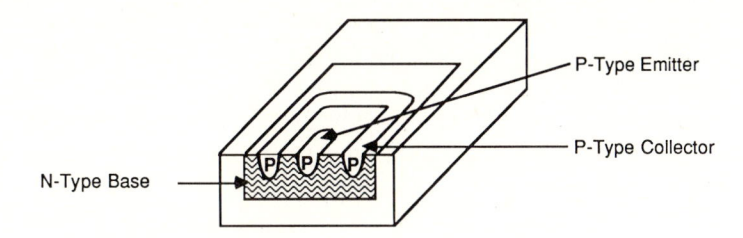

Fig. 18-8 A PNP Lateral Transistor

When the resulting transistor is correctly biased, current flows in the emitter, base, and collector as in discrete components; but in the IC, current flows in parallel with the chip's surface. This type of IC is called the **lateral PNP transistor**.

Crystalline orientation must be considered before the silicon rod is "diced" (cut into wafers). Proper orientation is necessary to assure that everything is correct for the growing of additional crystal wafers. Wafer resistivity is important when parts of the wafer are to be used as passive components (resistors and capacitors) within the circuit.

When a silicon wafer is etched, pits are created in its surface. The number of pits that are formed is a measure of the crystal's purity. The fewer pits that are present, the more perfect the crystal. Remember, variations in thickness can affect component value; therefore, the desired surface is as perfect as possible. A diamond saw is used as one way of slicing a silicon rod into wafers. The action of the saw leaves the two surfaces with a rough finish. The surfaces are then polished using mechanical means or chemical etchants. Another method of slicing wafers is to use a LASER beam. When the LASER is used, both surfaces are very smooth. Once the slicing and polishing phase is completed, the wafers are ready for further processing.

Photoengraving

Masks that are used for photoengraving IC substrates are made using the same processes that were discussed for thick-film and thin-film fabrication. Large drawings are made, photographs are taken, and reduction is performed until a photographic negative suitable for circuit production is available. The original artwork master can be as large as four feet square. After reduction, the mask size will be accurate to approximately 125×10^{-8} inches.

During the photoengraving process, wafer cleanliness is maintained to such a high level that it is almost unbelievable. A microscopic speck of foreign material can result in the destruction of the entire wafer.

Preparation of the wafer begins with the spreading of an even layer of photoresist over the entire surface of the wafer. After allowing the resist to air dry, the wafer is baked at a low temperature prior to printing. To assure accuracy, the patterns to be etched are aligned under a microscope. As several different masks will be used to produce all ICs from a single wafer, each mask must be perfectly aligned before the etching can begin. Once aligned, the mask and wafer are pressed together, and the surface is exposed to ultraviolet light. The exposure to light causes those areas that were not covered by the mask to be sensitized. Once sensitized, the wafer is passed through a developing solution which prepares the exposed areas for etching. After being dried, the wafer is placed in an acid bath where the sensitized areas are etched away, leaving the wafer exposed. Now, other materials can be diffused into the exposed areas to form conductors, resistors, and capacitors. The masking, exposure, developing, drying, and etching steps are repeated over and over until all parts of the circuit have been added through the use of masks and the depositing of correct materials. As a finishing step, the entire surface of the etched side of the wafer is coated with a metallic substance. Now the process is reversed and a mask is applied that identifies the metal electrodes that are to be left. After exposure, all excess metal is removed and the result is a completed circuit that has metal electrodes to which external or internal connections can be made.

Diffusion

The next step in fabrication is called **diffusion**. Diffusion is defined as:

> **A thermally induced process in which one material is absorbed into another.**

In the development of integrated circuits, the silicon dioxide coating on the wafer absorbs the atoms much slower than the etched areas which act as a mask. This assures that diffusion occurs only in the desired areas.

The basic diffusion process depends on the availability of a suitable impurity source. The impurity that is to be diffused will probably be either a phosphorus or boron compound. When exposed to heat these compounds vaporize easily. If the vaporized substance is allowed to settle on the silicon wafer, those exposed areas will allow diffusion to take place. The diffusion of the impurity atoms into the wafer create current-carrying paths. These paths may be either conductors or circuit elements such as resistors and capacitors. To assure that all components are placed at the correct position, multiple etching and diffusion cycles will be repeated.

During the diffusion process, the silicon wafer rests on a thermally flat region within the oven (furnace). Figure 18-9 illustrates this. The entire diffusion process takes place within a hermetically sealed quartz tube. During diffusion, temperatures ranging between 900°C and 1200°C are used.

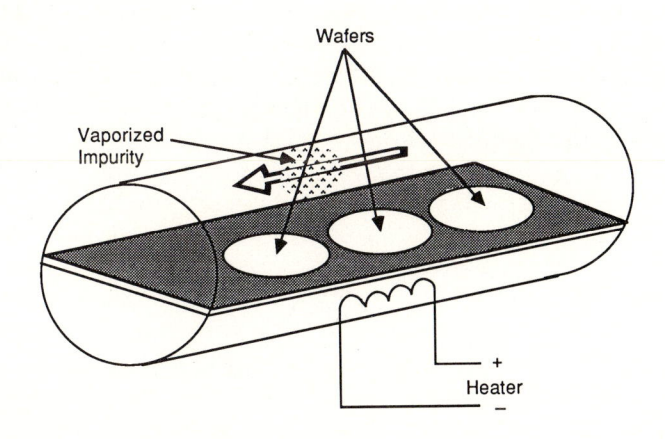

Fig. 18-9 Wafers in Diffusion Oven

A cross-sectional view of a diffused NPN transistor is shown in Figure 18-10. This transistor was fabricated using the same steps that were listed above. It is also possible to diffuse other components into substrates. In Chapter 15, we discussed the way that FETS were diffused. Resistors, capacitors, diodes, and transistors can all be diffused into the same wafer. This results in a complete and operational electronic circuit that is contained in a silicon wafer barely visible to the naked eye.

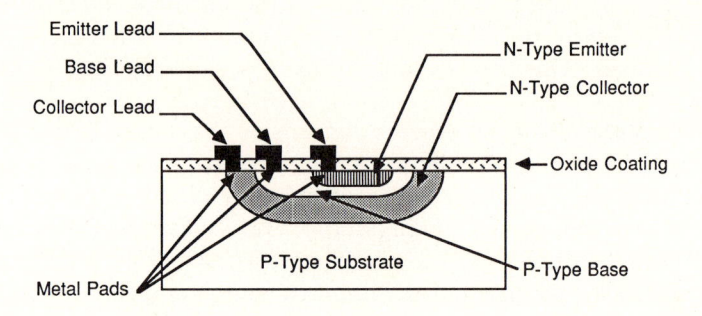

Fig. 18-10 Integrated Circuit - NPN Transistor

A diffused resistor is shown in Figure 18-11. Because all resistors on an IC are fabricated at the same time, the depth of diffusion is identical. The amount of resistance contained in each resistor is controlled by the length and width of the area that is diffused. Earlier, we mentioned that the substrate's resistivity was critical. The wafer is selected because it has the same resistivity throughout its entire structure; therefore, the ratio of depth to resistance (sheet resistance) will be the same for all resistors placed in the IC. It is difficult to precisely control the sheet resistance of a wafer during fabrication. Assume that we wanted four 5 kΩ resistors. Their values might not be exactly 5 kΩ. Assume that instead of 5 kΩ they are actually 4950 Ω. Because of the uniform sheet resistance of the wafer, all of them will be 4950 Ω. This assures us that matched resistors will have exactly the same resistance.

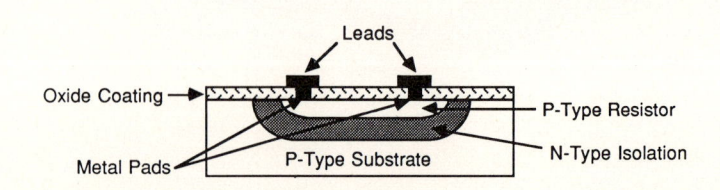

Fig. 18-11 P-Type Resistor Diffused into P-Substrate with N-Type Isolation

Figure 18-12 contains a drawing that represents an integrated capacitor. When diffused into a substrate, these capacitors are of the MOS (metallic-oxide-semiconductor) type. They are called MOS capacitors because of the way they are fabricated. The N-type material is diffused into a P-type substrate to form one plate of the capacitor. An oxide layer is placed over the diffused area and acts as the dielectric. A metal layer is then bonded to the top of the dielectric which acts as the second plate of the capacitor.

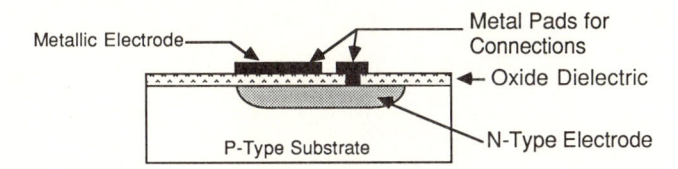

Fig. 18-12 A Diffused (MOS) Capacitor

The value of diffused resistors and capacitors are restricted by the area that is available on the wafer. If values other than those available through diffusion are needed, they must be provided by external connection of discrete components. In many cases, to avoid the use of external components, the engineer will change the circuit design to avoid the use of a capacitor. The result is a circuit that can be fabricated by diffusion that does not require external components.

Oxidation

If you observe Figure 18-10, 11, and 12, you will see that each has an *oxide* layer. The oxide layer is grown on the surface of the wafer and serves five purposes:

1. It prevents diffusion in the areas that it covers
2. It protects the junction from environmental variations—which is very important
3. It serves as insulation between metal conductors
4. It protects the wafer from mechanical abuse during the fabrication process
5. It can serve as the dielectric for capacitors

The oxide layer must have very low conductivity in order for it to be used as a dielectric for the capacitors. Having a high conductivity would allow dielectric losses to occur in the capacitors. Additionally, its temperature and metallurgical properties must be such that it is compatible with the other materials with which it comes in contact.

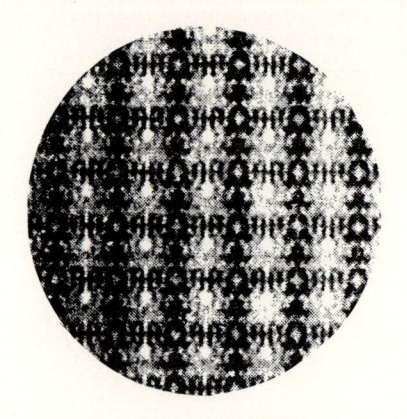

Fig. 18-13

The techniques that are used to place the oxide layer on the wafer may vary. The most common methods are the use of steam, wet oxygen, and dry oxygen. In all cases, the process is to place the wafer into the oven again and to force an oxygen carrying agent (steam or oxygen gas) across the chip's surface. As the carrying agent passes over the surface, some of its oxygen combines (chemically) with the silicon forming the silicon oxide. By careful control of the process, the exact oxide coating desired can be deposited.

Epitaxy

Epitaxy is defined as follows:

> **The controlled growth of a layer of semiconductor material in a suitable substrate.**

In this discussion we will consider epitaxy as it applies to the growth of epitaxial layers on silicon substrates. It is possible for epitaxial layers to be grown that have opposite conductivity and a completely different restivity than the substrate on which they grow. The distribution of impurities is determined solely by diffusion. Impurity distribution, of course, is the determining factor of the device's behavior. As the surface of the silicon substrate is penetrated further, the density of impurities decreases. This condition exists when the types of impurities change. Proper epitaxial techniques allow the reversal of this—placing denser impurity levels below the surface while less density is present at the surface. The most important use of epitaxial layers is to provide a very thin active region that can be used for fabrication of silicon transistors. When epitaxial growth is used to create bipolar transistors, the underlying substrate provides mechanical support only.

A typical system that is used for epitaxial crystal growth consists of an induction header, a quartz tube, a wafer holder, and equipment for handling gasses. The silicon wafer is placed in the quartz tube which is located within the induction heater. The wafer is then heated to approximately 800° to 1200°C. Once the wafer has been heated, it is exposed to hydrogen gas which thoroughly cleans the wafer's surface. Once the wafer has been cleaned, silicon tetrachloride is added to the hydrogen gas which acts as a carrier. When the gas and silicon tetrachloride strike the wafer's surface, they undergo a chemical reaction which forms hydrogen chloride and silicon. The silicon immediately settles on the wafer in the lowest possible energy configuration. In this way, the silicon forms a single crystal layer that has the same orientation as the substrate. It is possible to deposit either P- or N-type silicon, depending on the type impurity that is added to the hydrogen carrier gas. The temperature that exists plus the time allowed for exposure are the determining factors that control the thickness and density of epitaxial layers that are grown in this way.

The major advantage of this process is the ability to precisely control the boundaries of epitaxial layers. These techniques are especially valuable in the fabrication of bipolar junction transistors and diodes on silicon substrates.

Self-Check

Answer each item by inserting the word or words required to correctly complete each statement.

14. Integrated circuits contain both _____ and _____ components.
15. List the four advantages that ICs have over conventional circuits.

   a. _____
   b. _____
   c. _____
   d. _____

16. A wafer of _____ is used as the substrate for IC fabrication.
17. Masks are used during the _____ step of the fabrication process.
18. _____ is used to deposit the different circuit elements onto the substrate during IC fabrication.
19. A layer of _____ is grown on a silicon substrate that serves to protect the wafer from mechanical wear.
20. When a thin layer of silicon material is grown on a silicon substrate, we call this an _____ growth.

Interconnection, Lead Attachment, and Encapsulation

Once the structure on and within the silicon substrate has been completed, it is necessary to make any required internal connections. To make the interconnections, the following procedure is used. Openings are etched into the oxide coating at the locations where metallic contacts are required. The entire device is then placed within a vacuum chamber where aluminum is vaporized and allowed to coat the device's entire surface. In the etched areas, the aluminum easily diffuses into the silicon. In the other areas, the silicon oxide prevents diffusion.

Photoengraving is then used to etch off all metal coating except that needed for internal connections. The silicon wafer is then heated to a temperature at which the aluminum will alloy with the silicon. Once these pads are in place, a second application of aluminum and photoengraving provides aluminum connectors suitable for the internal connections. During this process, quite large pads are placed along the edges of the device which serve as connecting points for the external leads that will be installed later. These leads are formed and attached so that they extend through the final package and provide metallic contact for plugging the device into a socket or soldering it into another circuit.

Once all internal connections have been made and the lead pads are in place, the device is prepared for packaging. A picture of a silicon wafer that has hundreds of identical ICs is shown in Figure 18-13. It is necessary to separate each chip (*die*) from the others before packaging. See Figure 18-14. This is done by scribing and cutting the wafer at very precise points. A laser is usually used to perform this job. Once separated, each die will be mounted in a suitable package. *Note:* a completed chip that contains all active and passive elements is also called a *die*.

Many ICs are mounted in TO-5 cases like the one shown in Figure 18-15(d). Other die are mounted in dual-in-line packages like the ones shown in Figure 18-15(c). We often refer to ICs that are packaged this way as DIP ICs. The packages that are used are referred to as **headers**. Notice the DIP header with the die in place that is shown in Figure 18-15(a). Once the die is mounted to the header, leads that will be used to connect the device into an external circuit are inserted through the header and soldered to the pads mounted on the die. As a final step, a cover is placed over the top of the die and header. Under exact conditions, this cover is then attached to the header forming an environmentally safe package.

Many variations of lead attachment and lead type are used. The most reliable types, however, are those ICs that use gold for all conductive metals. A case in point is the microcomputer manufacturer that used aluminum connectors in an attempt to reduce costs and experienced a drastic increase in lead and connector failures. That company

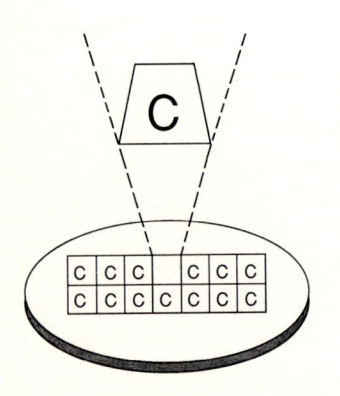

Fig. 18-14 Chip Being Separated from a Wafer

has now returned to all gold or gold-plated conductors. Fabrication techniques that use gold-plated chromium provide almost the same reliability as pure gold. For economical reasons, this type of lead and pad fabrication is being widely adopted.

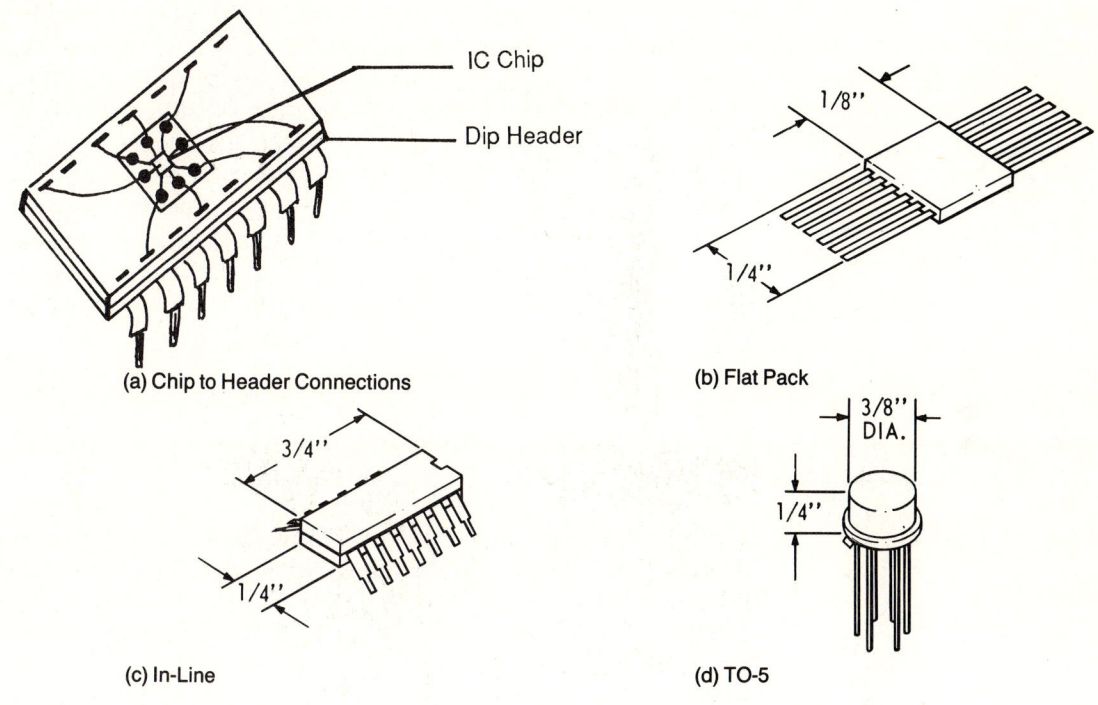

(a) Chip to Header Connections

(b) Flat Pack

(c) In-Line

(d) TO-5

Fig. 18-15

Flat packs are made of **kovar-glass** or glass-ceramic materials. Either of these can be hermetically sealed, and both are inexpensive to manufacture. All package sealing processes take place in a clean, dry, and carefully controlled environment. After being sealed, each package is tested for leaks using helium detector techniques. Each IC is placed in a helium atmosphere that is cooled to $-65°C$. If any leaks are present, the heating causes helium gas to be drawn into the defective packages. They are then immediately removed and placed in heated glycerine. When the packages are placed in glycerine, the gas expands and escapes, causing bubbles to be formed in the glycerine. Another test places the flat packs into a chamber where they are pressurized with nitrogen. Nitrogen will be forced into the chambers of those packages that are leaky. When the leaky package is returned to normal pressures, the nitrogen expands quickly. At best, the nitrogen being expelled from the package can be detected. At worst, the pressure inside the package will cause it to "flip its lid." Those packages that leak cause the IC to be rejected as it would not be suitable for use.

Quality Control

We have intentionally omitted from this discussion the fact that countless tests are being performed at all stages of IC fabrication. Complete electrical testing of each IC is performed at the earliest possible time and continues until the IC is placed in a circuit. In this way, defective devices are identified at the earliest possible moment. From this point on, each defective device is omitted from the fabrication process. The elimination of defective devices can save considerable money and time. This allows the manufacturer to keep the cost per unit as low as possible. Attaching the leads and packaging of the device is a large part of the developmental costs. Eliminating these last two steps for the defective devices can save considerable money. As a last series of tests, each chip is tested in many different environments (hot, cold, humid, dry, pressurized, vibrating, etc.) to establish its operating limits. Only after it passes all these tests is the device ready for use.

Multichip Circuitry

An important variation of the monolithic silicon IC is the **multichip** device. In this device, several chips, each containing different structures, are bonded together so that they share the same substrates and headers. During the packaging process, jumper wires are attached between the individual chips. See Figure 18-16 for an example of a header with a multichip circuit connected. Note that this header contains three chips and the interconnections necessary to complete the electrical connections. External leads are attached at the points needed for connection of inputs, outputs, power, and ground.

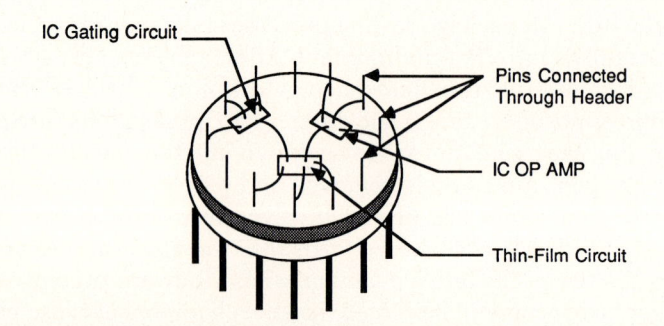

Fig. 18-16 Microchips Connected to TO-5 Heater Microcircuitry

The major advantage gained from this technique is that greater versatility is available. For example, transistors may be on one chip, diodes on another, and resistors on a third. Each chip can be fabricated using the processes that will yield the best results for that particular type chip. Then, the individual chips are mounted and connected as needed. This minimizes the interaction that occurs in monolithic devices because of the isolation of different types of components. Characteristics of multichip circuits more nearly parallel the characteristics of conventional discrete circuits.

With each chip having characteristics that are determined at the point of manufacture, multichip technology is very flexible. The product is improved by pretesting the individual chips, allowing those that are defective to be rejected prior to final assembly.

Multichip silicon integrated circuits have many of the same disadvantages as conventional circuits such as:

1. The necessity to provide conductors that connect the different chips together
2. The increased number of steps required for the fabrication of the different chips. Remember, the monolithic chip combines steps to reduce fabrication time
3. Higher costs for large production-type devices

In summary, the multichip silicon IC is, at present, an interim packaging technique that provides some of the advantages of monolithic silicon ICs. It *does not*, however, provide nearly as many advantages as the monolithic IC.

Self-Check

Answer each item by inserting the word or words required to correctly complete each statement.

21. Individual IC circuits that are cut from a large wafer are called a/an _____.

22. List three packages that are used to package ICs.

   a. _____
   b. _____
   c. _____

23. What is meant by the term *quality control* as it applies to the fabrication of ICs?

24. State the main advantages gained from multichip technology.

25. The part of an IC package on which the die is mounted and through which external leads project is called the _____.

Summary

In this chapter we discussed thick-film, thin-film, and monolithic circuit fabrication techniques. Each of these is a different technique for preparing a durable and reliable microcircuit.

In ceramic printed circuit (thick-film) fabrication, large drawings are made of the circuit, and then through photographic processes, these are reduced to actual circuit sizes. These photopositives are then used to etch circuits into a CPC material. The etched areas become the part where resistors and conductors are deposited using pastes designed to meet the needs of each component. Once the CPC is completed, it is necessary to attach capacitors and active devices to the circuit, creating a *hybrid* circuit.

With thin-film fabrication, the processes are very similar. In this type of fabrication, it is also possible to deposit capacitors as part of the thin-film circuit. Once the thin-film process is completed, it is necessary to mount the active devices (transistors and diodes) to complete the hybrid circuit.

With the *monolithic silicon integrated circuit*, all components—active and passive—can be deposited onto the substrate. This allows entire circuits to be fabricated on a single small silicon substrate. IC circuits range from simple to highly complex.

The last topic discussed was *multichip* technology which involves the use of different chips (diffused transistors, diodes, resistors, etc.) which are wired together much like conventional discrete circuits.

Review Questions and/or Problems

1. An advantage that is gained from the use of integrated circuit is ____.

   a. ease of design
   b. no maintenance
   c. reduced size and weight
   d. all of these

2. All integrated circuits are forms of hybrid circuits.

   a. True
   b. False

3. The use of miniature diodes, transistors, resistors, and capacitors provides an advantage in that identical circuits can be small and completely ____.

4. The process that uses heat to deposit impurities of one material on another material is called ____.

   a. diffusion
   b. photoengraving
   c. etching
   d. masking

5. The term *hybrid microcircuitry* tells us that the final circuit will include _____.

   a. thin-film circuitry
   b. separately attached semiconductor chips
   c. both a and b
   d. neither a nor b

6. All integrated circuit chips are mounted on _____.

   a. TO-5 packages
   b. dual-in-line packages
   c. headers
   d. flat packs

7. Any material that is used to support the conductors and components of an IC is called the _____.

   a. substrate
   b. base material
   c. circuit
   d. any of these

8. In the design of thick-film circuitry, the substrate is most likely a piece of _____.

   a. P-type silicon
   b. N-type silicon
   c. wire mesh
   d. ceramic

9. A _____ allows paste to be applied to desired points of a circuit while protecting all other points.

   a. substrate
   b. mask
   c. screen
   d. both b and c

10. Exposure of a piece of photosensitive substrate to ultraviolet light allows the exposed area to be _____ during etching.

    a. removed
    b. retained

11. During the fabrication of CPC circuitry, the resistors receive final adjustment of their values by _____.

    a. adjusting a shaft
    b. use of a screwdriver
    c. shaving off small slices of material
    d. any of these methods

12. In thin-film fabrication, *pit-count* is important because _____.

13. Which of the following fabrication techniques cannot be used if we want to form transistors and diodes on the substrate?

   a. thin-film fabrication
   b. monolithic IC fabrication

14. Which of the following fabrication processes will provide the largest miniaturization of the finished circuitry?

   a. discrete circuitry
   b. thick-film circuitry
   c. thin-film circuitry
   d. integrated circuitry

15. All circuit components, both active and passive, are deposited on the substrate during _____ fabrication.

   a. CPC
   b. thick-film
   c. monolithic IC
   d. thin-film

16. Which of the following circuit types is most power efficient?

   a. CPC
   b. thick-film
   c. monolithic IC
   d. thin-film

17. The oxide layer that is applied to the surface of the substrate serves as a/an _____.

   a. insulator
   b. wear protector
   c. capacitor's dielectric
   d. all of these

18. In the fabrication of integrated circuits, _____ alignment becomes very critical.

   a. substrate
   b. mask
   c. header
   d. none of these

19. *Sheet resistance* of a substrate refers to the ratio that exists between _____ and _____.

20. Multichip circuitry requires the mounting of _____ chips on a single header.

   a. one
   b. two
   c. three
   d. two or more

19
Operational Amplifiers and Timers

Objectives

1. Define each of the following terms:

    a. Operational amplifier
    b. Inverting input
    c. Non-inverting input
    d. Slew rate
    e. Dual-In-Line package
    f. TO-5 package
    g. Open loop
    h. Closed loop
    i. Offset voltage
    j. Offset voltage drift

2. Recognize various operational amplifier circuits by observing their schematic diagrams.
3. Describe how various operational amplifier circuits operate and the purpose for which they are used.
4. Describe the timing operations provided by circuits built using the 555 timer.

Introduction

Those of us that have been in the electronics field for many years may have been introduced to *operational amplifiers* (**op amps**) through our work with *analog computers*. In those computer systems, it was often necessary to amplify DC voltages. To do this we used DC amplifiers called operational amplifiers. At that time *op amps* were made using vacuum tube circuitry; however, as circuitry became more advanced it became possible to do the same job with semiconductor circuitry. Now the advances have led to a family of integrated circuits called operational amplifiers that can be used for all low-frequency amplification purposes.

The IC op amp is a silicon chip that has all of the transistors, resistors, and capacitors needed to form an amplifier diffused into a monolithic substrate. Op amp circuits are designed to provide large voltage gain along with high input impedance and low output impedance. There are several books on the market that are devoted entirely to the construction and applications of op amps. For the purposes of this book, we will limit our discussion to an introduction to the principles of selection, use, and operation of specific op amps. We will concentrate on applications from the standpoint of inputs and output.

Quite often we must provide accurate timing in the form of triggers or pulses to a circuit or system. One IC that is designed for this purpose is the 555 timer. Using the 555 timer, we can design very precise circuits suitable for controlling the operation of the most complex systems. We will take a brief look at this device and some of its applications.

Operational Amplifier Characteristics

The ideal op amp still has not been developed. Even with the great advances made since the vacuum tube amplifier, some shortcomings still exist. We will explore ways that will allow us to offset some of these deficiencies. Our goal is to operate the op amps as nearly to the "ideal circuit" as possible.

Operational amplifiers are designed and packaged for use. The most common packages are shown in Figure 19-1. Both packages receive wide use in all IC designs. For the operational amplifier, it is common for the package to have either 8, 12, 16, (etc.) pins. The *dual-in-line package* (DIP) is probably the package most used today. It is possible to buy a DIP package that contains a single op amp. It is, however, also possible to purchase a single package that contains several op amps. The cost of op amps is quite low and with just a little study on your part, you can use them to do many jobs where amplification is required.

The typical op amp is constructed using *differential amplifier* stages that are connected in cascade. The first of these differential amplifiers (DIFF AMPs) is considered the *input stage*. With differential amplifier operation, two separate inputs are required. The availability of two inputs provides at least two advantages:

1. Ability to provide either in-phase or 180° out-of-phase output signals
2. "Common-mode" rejection

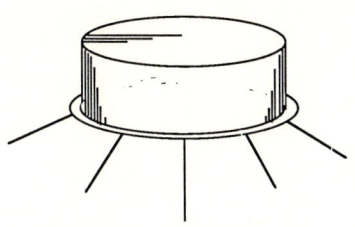

(a) TO-5 Packages are Used for 8, 10 and 12-Pin Integrated Circuits

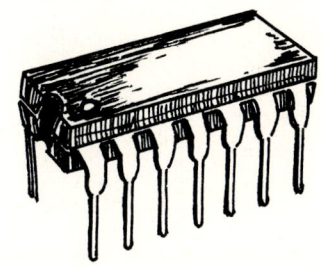

(b) Dual-in-Line Package — Plastic or Ceramic. DIP Packaging is Used for ICs That Have 8, 14, 16, 20, 22, 24, 36 (etc.) Connections

Fig. 19-1 Most Common IC Packaging

Common-mode rejection will be discussed later in this chapter. The availability of two inputs allows the circuit to be designed such that by choosing the correct input terminal, the output signal can be inverted (180° phase shift) or non-inverted (0° phase shift) when compared to the input signal.

The schematic diagram and symbol for a typical op amp (741) is shown in Figure 19-2. Notice that diodes are placed in the input circuit. These diodes protect the two inputs against damage that might occur if the power supply leads were connected to the input terminals. If power is connected to the inputs, one of the diodes will act as a short, causing the power supply circuit breaker to trip. The shorted diode protects the op amp from damage.

Transistors Q_1 and Q_2 form the input differential amplifier. For this operational amplifier, pins 2 and 3 form the inputs for the circuit, pin 6 is the output, pins 4 and 7 are the connections for the two power supply inputs, pins 1 and 5 are the *null offset* connections.

This circuit comes very near to being an ideal op-amp circuit: it's gain is high, its input impedance is high, and its output impedance is low.

Input Circuits

The input circuit for most operational amplifiers is a differential amplifier. This is a part of the op amp and is located within the chip. Observe

Figure 19-2, you will see that the two inputs are connected to the bases of two transistors that operate as a differential amplifier. This type of circuit was discussed in an earlier chapter, but we will review it briefly.

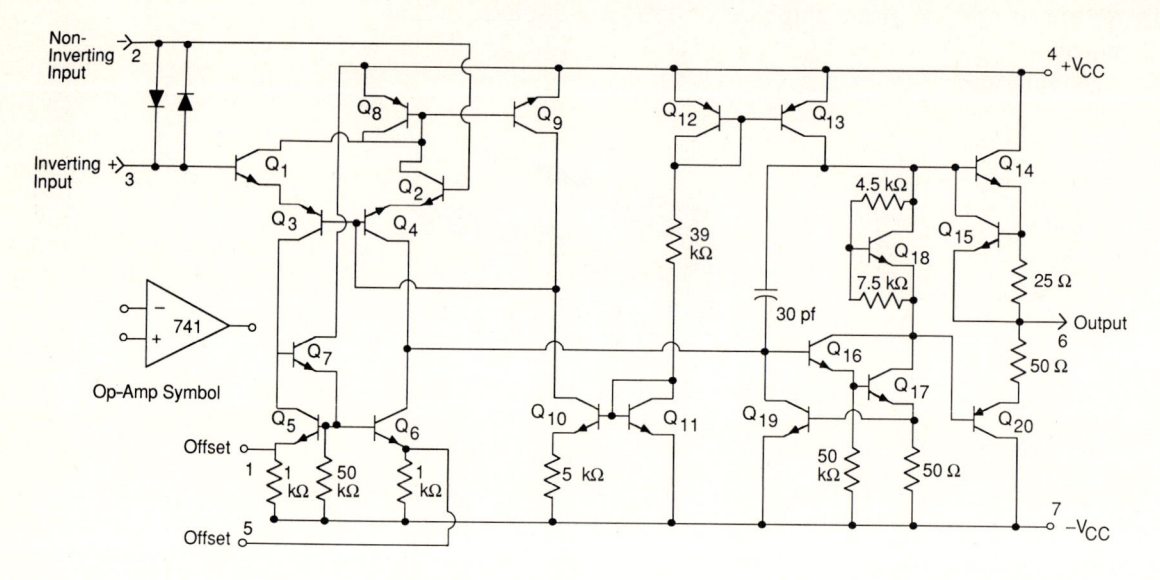

Fig. 19-2 Typical (type 741) Operational Amplifier Schematic Diagram

The circuit diagram that appears in Figure 19-3 is a differential amplifier. You should remember that the differential amplifier's output is taken between V_{C1} and V_{C2}. The output is the difference in amplitude between the two outputs. With a balanced circuit, the conduction of both transistors is such that the collector voltages are equal. With the two voltages being equal, there can be no difference in potential. In operating circuitry, we will design in a variable resistor that can be used to assure that the two stages are balanced and that the output is 0 V. If, however, there is no balance resistor present, the variations in tolerance between the stages will cause one to conduct harder than the other. This results in a small voltage being present in the output. This voltage is called the *offset voltage*, symbol V_{OS}. We can reduce the effect of this voltage by applying a small DC voltage to one of the transistor's emitters. By selecting the correct emitter and the right amount of voltage, we can *offset* the imbalance present in the circuit.

It is very important that V_{OS} be kept to the minimum amount possible. The fact that any offset voltage is present results in a loss of efficiency due to the heat generated. You already know that the operating temperature will affect the operation of solid-state devices. Op amps are no different: temperature effect must be controlled as much as possible. If temperature changes, the offset voltage must change. This voltage change is called the *offset voltage drift*. As a general rule, the drift will be approximately 3 µV per degree of Celsius temperature change for each millivolt of original *offset voltage*. You can see that the more offset voltage present, the greater the amount offset-drift voltage required.

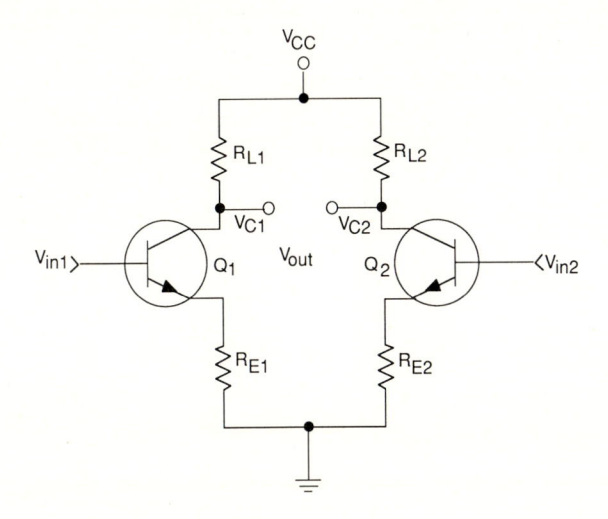

Fig. 19-3 Transistorized Differential Amplifier

In Figure 19-4, you will see the same circuit except that input terminals have been identified as A and B. Assuming that we have a balanced circuit and no need for an offset voltage, we will analyze the circuit. At this point (no input signal), V_{C1} equals V_{C2} and the output is 0 V. With the arrival of a positive input voltage at point A, the voltage at point B will be going negative. This will cause I_C in Q_1 to increase and I_C in Q_2 to decrease. Increased I_C in Q_1 causes V_{RL1} to increase and V_{C1} to decrease. The decreased I_C in Q_2 causes V_{RL2} to decrease, allowing V_{C2} to increase. Notice that V_{C1} decreases while V_{C2} increases. The change in V_{C1} and V_{C2} should be equal. Since they are changing in opposite directions, the change felt in the output is double what would be present from one amplifier only.

An advantage of this type circuit is its *common-mode rejection*. Observe Figure 19-4, assume that a positive spike appears at the input. This positive spike is *common* to both inputs, meaning that it appears at the base of both transistors at the same time as a positive spike. This is characteristic of noise that is generated in a circuit. Because the spike arrives at the input of both transistors at the same time, it causes both transistors to increase conduction. If the spike is negative, both transistors decrease conduction. Because both transistors have the same effect, the outputs (V_{C1} and V_{C2}) both increase or decrease the same amount. This, in effect, cancels (rejects) the effect of the spike. This

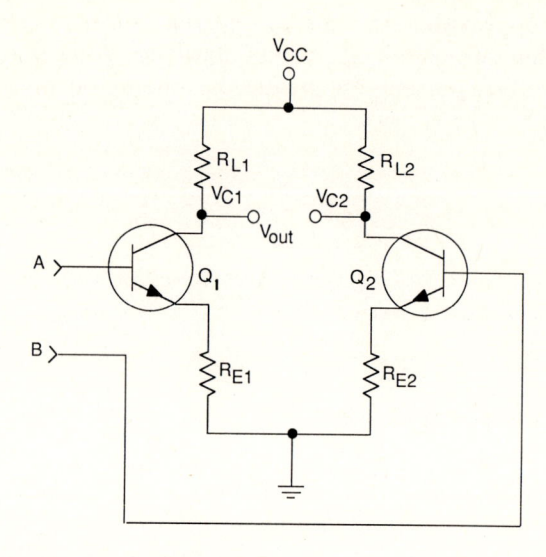

Fig. 19-4 Transistorized Differential Amplifier

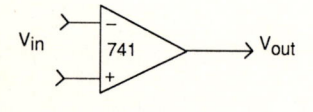

Fig. 19-5 Operational Amplifier Schematic Symbol

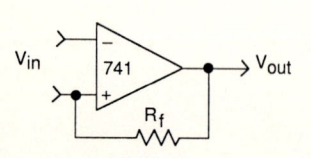

Fig. 19-6 Schematic Symbol with Feedback

cancellation effect is called *common-mode rejection*. As with this differential amplifier, the operational amplifier which contains differential amplifiers will have excellent common-mode rejection.

Figure 19-5 contains the schematic symbol for an operational amplifier. Observe that it has *two* inputs. These are labeled plus (+) and minus (−). You must realize that these are two separate inputs just like inputs A and B of the differential amplifier above. The plus (+) sign indicates that any signal applied to this input **will not** be inverted in the output. In other words, the plus (+) input and output signal will be **in phase**. An input applied to the minus (−) input will be inverted in the output. This tells us that any input applied to the minus (−) input will undergo a 180° phase shift between the input and output.

An "ideal" voltage controlled amplifier would have infinite input impedance. With an infinite input impedance, the amplifier would not drain any current from the input source. This is impossible to accomplish, as we already know. We can, however, provide an even higher input impedance by use of feedback. Observe Figure 19-6 and you will see that a feedback resistor (R_f) has been added to the op amp symbol. The feedback that is supplied by this resistor is degenerative (negative) and opposes the current entering the input. The effect of negative feedback is to reduce the current drain placed on the circuit providing the input signal. In effect, this increases the input impedance of the op amp. Another consideration is *input bias current*. Input bias current is the average of the two currents (I_1 and I_2) that flow in the inputs of an op amp.

Figure 19-7 contains an op amp symbol with I_1 and I_2 inputs shown. Average bias current can be calculated using the formula:

$$I_B = \frac{I_1 + I_2}{2}$$

Input bias current is defined as:

> **The DC current required at the inputs of the amplifier to properly drive the first stage.**

Fig. 19-7 Inverting and Non-Inverting Inputs

Frequency Response

As we start our discussion of frequency response, we must take time to define two terms:

> **Open Loop**—an op amp circuit that does not have feedback of any type.

> **Closed Loop**—an op amp circuit in which part of the output signal is fed back to and provides some control of the input.

Figure 19-8 contains op amp circuit symbols that illustrate both open loop and closed loop conditions.

In their *open loop* configurations, the average op amp has a frequency response that decreases to *unity gain* at approximately 10 MHz. Op amps are seldom operated in open loop configurations.

Even when used in *closed loop* circuits the op amp is highly unstable with an upper frequency response of approximately 1 MHz without compensation. As compensation is added to the circuit the circuit's upper frequency response decreases. The op amp is, however, suitable for use in most of our low frequency applications.

Figure 19-9 is used to show the relationship that exists between voltage *gain* and *frequency response* in op amp circuits.

If you examine Figure 19-9 you will see that as frequency increases, voltage gain decreases. The parameter that compares gain to bandwidth is referred to as the **gain-bandwidth product**. In early lessons you learned that as amplifiers were connected in cascade, the bandwidth decreased. Since the op amp circuit consists of a number of cascaded amplifiers, its gain decreases as frequency increases. The gain-bandwidth product (**GBP**) is numerically equal to the frequency at which amplifier gain equals 1 (unity). The GBP can be used for several purposes. If we know the gain of the amplifier (100) and the bandwidth

(a) Open Loop

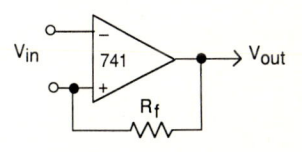

(b) Closed Loop

Fig. 19-8 Open-Closed Loop Symbols

(15 kHz), we can calculate GBP or, as it is frequently called, *unity-gain frequency* (**UGF**) as follows:

$$GBP = \text{Gain} \times BW = UGF$$

$$GBP = 100 \times 15 \text{ kHz} = 1.5 \text{ MHz}$$

Using a transposed form of this equation, we can calculate bandwidth (BW) or gain (A_V) for the amplifier if UGF and either bandwidth (BW) or voltage gain (A_V) are known.

$$BW = \frac{UGF}{A_V}$$

$$BW = \frac{1.5 \text{ MHz}}{100} = 15 \text{ kHz}$$

or:

$$A_V = \frac{UGF}{BW}$$

$$A_V = \frac{1.5 \text{ MHz}}{15 \text{ kHz}} = 100$$

You should remember:

If bandwidth is increased, gain must be sacrificed.

Slew Rate

Another parameter of interest is **slew rate**. Slew rate is the term that is used to describe an operational amplifier's ability to track, or follow, a rapidly changing input signal. This is especially important in circuits where fast changing signals—such as square waves, rectangular waves, or pulses—are present. Terms used to identify op amps that have this capability are *fast* or *high slew rate*. An operational amplifier's slew rate compares to the transient interval of the circuits discussed earlier. Slew rate is defined as:

The maximum rate of change of the output voltage of an amplifier operated within its linear region.

Figure 19-10 contains a chart that illustrates slew rate where slew rate equals:

$$\text{Slew Rate} = \frac{\Delta V_{out}(\text{Max})}{\Delta \text{Time}}$$

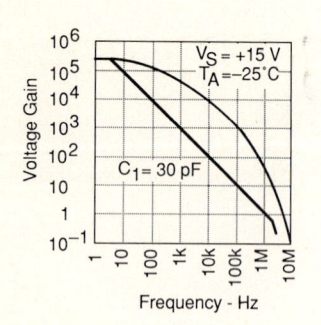

Fig. 19-9 Open-Loop Voltage Gain as a Function of Frequency

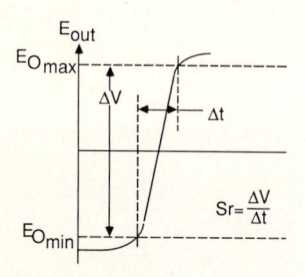

Fig 19-10 Slew Rate

Most circuits require that slew rate be as fast as possible. To obtain a *high slew rate*, there are several things that we can do.

1. All circuit boards used with the op amp must be of the highest quality materials.
2. The pins where the external power supplies connect must have adequate protection. This involves placing external bypass capacitors from the voltage input pins to ground.
3. Distributed capacitance of the output circuit must be kept to a minimum.
4. Resistors are added to the emitters of the input stage. The added resistance reduces first stage current but also improves slew rate. One drawback of this type of compensation is that the added resistance makes it difficult to balance the two stages of the input differential amplifier. To provide for their balance it may be necessary to change the *offset voltage*.
5. Use JFETs or MOSFETs in the op amp design instead of bipolar transistors. FETs have much faster reaction times than do bipolar transistors. Further, FETs have a much better bandwidth capability than does the bipolar transistor.

Large Signal Bandwidth Limitations

When we discuss **large signal behavior**, we are referring to the amplifier's ability to handle signals that produce large swings in output signals. In an op amp, as frequency of operation increases, we reach a point where the slew rate will have an effect on bandwidth. This point can be determined by the formula:

$$f_{UHPP} = \frac{\text{Slew Rate}}{(2\pi)(\Delta V_{out})}$$

where:

f_{UHPP} = frequency at upper halfpower point

V_{out} = peak output voltage

Self-Check

Answer each item by inserting the word or words required to correctly complete each statement.

1. _____ is the most common method used to package ICs.

2. A single package can contain no more than one op amp.

   a. True
   b. False

3. A differential amplifier must have _____ inputs.

   a. one
   b. two
   c. three
   d. four

4. Most op amps have a _____ as their input circuit.

   a. Class A amplifier
   b. Class AB amplifier
   c. Class C amplifier
   d. differential amplifier

5. Any signal that is common to both inputs of the operational amplifier is called a _____ signal.

   a. common-source
   b. common-input
   c. common-mode
   d. common-base

6. If an input signal is applied to the minus (−) input of an operational amplifier, the phase relationship between the input and the output equals _____ degrees.

   a. 0
   b. 60
   c. 180
   d. 360

7. When we apply a small DC voltage to one emitter of a differential amplifier, we reduce the effect of _____.

8. In op amp operations, the feedback signal causes a higher _____ at the input circuit.

   a. gain
   b. frequency
   c. impedance

9. When feedback is provided as part of an op amp circuit, the circuit is said to be a/an _____ loop circuit.

   a. complete
   b. open
   c. closed

10. To calculate slew rate, you must divide the change in _____ by the change in _____.

    a. time, time
    b. time, voltage
    c. voltage, voltage
    d. voltage, time

Sample Operational Amplifier Applications

It is possible to use operational amplifiers for all of the circuits studied in this book with the exception of some wideband amplifiers. Book after book could be written about all of the possible circuits. For that reason, we will introduce you to only a few applications and leave further applications for your future study.

The Inverting Amplifier

An **inverting amplifier** provides the same function as the common-emitter and common-source amplifiers that you studied earlier. The schematic diagram for an inverting amplifier is shown in Figure 19-11. Observe that the offset and DC voltages have been left off of these circuits for simplicity. These connections are generally the same for all circuits using the same type op amp. For that reason they can be omitted from the schematic as long as we realize that they must be part of the finished circuit.

In Figure 19-11, the input signal is applied to the minus **inverting** input. The minus (−) input produces a 180° phase shift between input and output signal. The plus (**non-inverting**) input is grounded and is common to both the input and the output. Negative (degenerative) feedback is coupled from the output back to the input through the feedback resistor (R_f). The ratio of R_1 to R_f will determine, to a large part, the circuits voltage gain. Voltage gain for this circuit can be calculated using the formula:

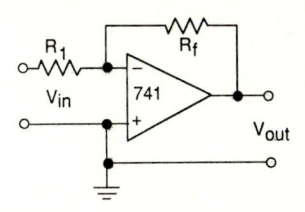

Fig. 19-11 Inverting Amplifier

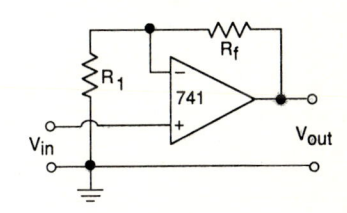

$$A_V = \frac{R_1 + R_f}{R_1}$$

The Non-Inverting Amplifier

Figure 19-12 shows the schematic diagram for a *non-inverting amplifier*. The output is *in phase* with the input. Notice that the input is applied to the *non-inverting* (+) input while the feedback is applied to the *inverting* (−) input. R_1 is connected from the inverting input to the common circuit between input and output. The non-inverting input is always used when we do *not* want the signal to be inverted.

Feedback is applied to the inverting input through resistor R_f which is connected to R_1 and the op amp's inverting input. The ratio of these resistors (R_1 and R_f) has an effect on the circuit's gain. Voltage gain can be determined using the formula:

Fig. 19-12 Non-Inverting Amplifier

$$A_V = \frac{R_1 + R_f}{R_1}$$

The Voltage Follower

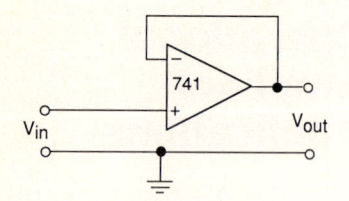

A **voltage follower** circuit operates such that its output is in-phase with the input; there is *no* phase shift between input and output. Figure 19-13 contains the schematic diagram for a voltage follower. The input for this circuit is applied to the non-inverting input. Notice that the entire output signal is fed back to the inverting input. This produces severe degeneration at the input of the input-differential amplifier. This degeneration assures that the output signal will be an exact reproduction of the input—including equal amplitude. The gain of a voltage follower is:

$$A_V = \text{Unity (1)}$$

therefore:

$$V_{out} = V_{in}$$

Fig. 19-13 Voltage Follower

The Summing Amplifier

As stated earlier in this chapter, the operational amplifier can be used to perform many mathematical functions. We will briefly discuss a few of these here. The schematic diagrams for two **summing amplifiers** are shown in Figure 19-14. These circuits are slight variations of the inverting amplifier.

The output of the circuit shown in Figure 19-14(a) is the algebraic sum of all the inputs applied to the inverting input. The output signal will be inverted 180°. Feedback is through R_t which— in conjunction with R_1, R_2, and R_3—controls the amount of output voltage present. Output voltage (V_{out}) for this circuit can be calculated using the formula:

$$V_{out} = \frac{R_f}{R_1}(V_{in1}) + \frac{R_f}{R_2}(V_{in2}) + \frac{R_f}{R_3}(V_{in3})$$

The summing amplifier is especially valuable in the handling of DC voltages. Two or more voltages can be summed using this circuit. An additional input must be provided for each voltage that is to be summed.

The circuit shown in Figure 19-14(a) can be used to provide an inverted output. To provide a non-inverted output, the circuit shown in Figure 19-14(b) would be used.

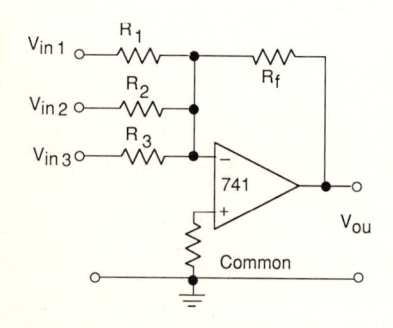

(a) Inverting Type

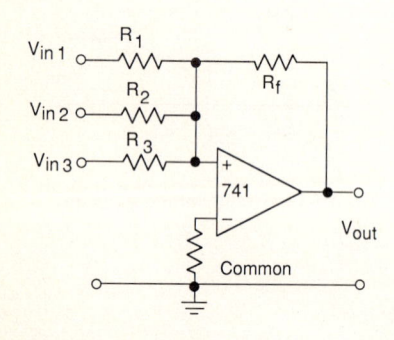

(b) Non-Inverting Type

Fig. 19-14 Summing Amplifiers

The Subtracter

Figure 19-15 shows a circuit that can be used to **subtract** one voltage from another. Degenerative feedback is supplied through R_f. Notice that two inputs are applied, one to the inverting and one to the non-inverting inputs. When connected in this manner, the two voltages cancel and the difference appears in the output. Either AC or DC voltages can be applied to this circuit. Both voltages, however, must be of the same type, that is, both DC or both AC—not intermixed. In this circuit, the mathematical relationship is:

$$V_{out} = V_{in1} - V_{in2}$$

Fig. 19-15 Operational Amplifier — Subtractor

The Adder-Subtracter

It is possible to combine the **summing and subtracting** functions into a single op amp circuit. Figure 19-16 contains the schematic diagram for a circuit of this type. This circuit is called an **adder-subtracter**.

The mathematical expression for V_{out} of this circuit is:

$$V_{out} = (V_{in1} + V_{in2}) - (V_{in3} + V_{in4})$$

In all of these circuits, it is important to note that frequency of operation must be considered when using AC voltages.

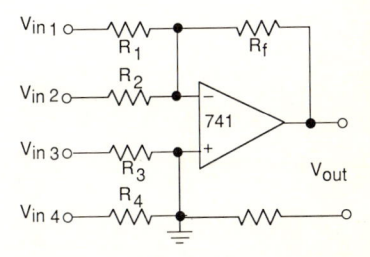

Fig. 19-16 Op-Amp Adder Subtractor

The Differentiator

Before we discuss this circuit, we should emphasize that this circuit has *no* relation to the differential amplifier discussed earlier. You should not confuse the two circuits.

An op amp **differentiator** is shown in Figure 19-17. This circuit produces an output that approaches the pure derivative of the input waveform. The mathematical expression for V_{out} of this circuit is:

$$V_{out} = R_f \times C_1 \times \frac{d}{dt} \times E_{in}$$

> *Note to Students:* The fraction ($\frac{d}{dt}$) shown in this equation is a symbol that is used to indicate the derivative.

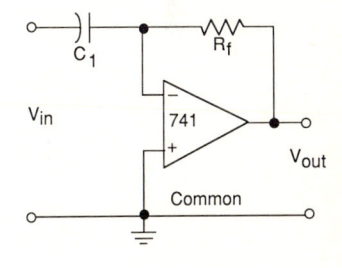

Fig. 19-17 Op-Amp Differentiator

The Integrator

The circuit shown in Figure 19-18 is called an operational amplifier **integrator**. The output of this circuit is the exact integral of the input voltage. Output voltage for this circuit can be represented by the mathematical expression:

$$V_{out} = \frac{1}{R_1 \times C_f} \times \int V_{in} \times dt$$

Note to Students: The $\int$ shown in this equation is a symbol that is used to identify the integration process.

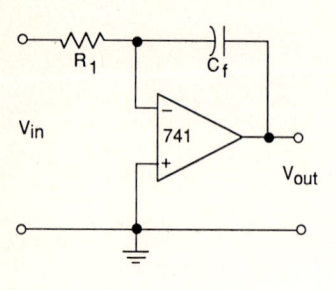

Fig. 19-18 Op-Amp Integrator

An Audio Amplifier

The circuit shown in Figure 19-19 is an *audio amplifier*. This circuit is capable of amplifying all audible frequencies.

If several op amps are connected in cascade, it is possible for the cumulative phase shift to reach 180°. If that condition is allowed to exist, the entire circuit will become an oscillator. To prevent this from happening, compensation in the form of R_3 and C_1 is provided. This feedback results in a decrease in gain at higher operating frequencies. In some op amps, this compensation is provided internally, and external compensation is not required. C_3 and C_4 are added to the circuit to provide a higher slew rate which is useful at higher frequencies (transient waves).

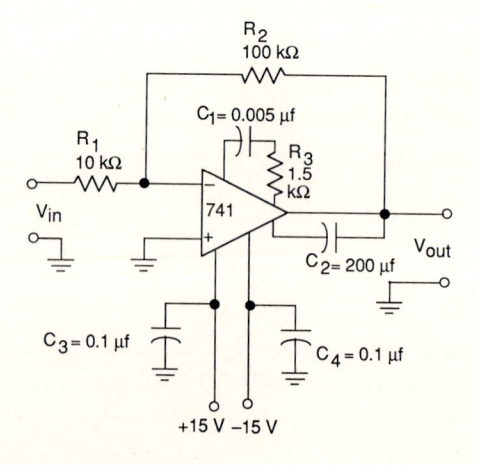

Fig. 19-19 Op-Amp Audio Amplifier

The Differential Amplifier

The operational amplifier can be used as a **differential amplifier**. A schematic diagram is contained in Figure 19-20. Note that inputs are applied to both the inverting and non-inverting inputs of the op amp. Negative feedback is provided by R_2 which reduces gain somewhat. The common-mode rejection provided by this circuit is especially valuable. It is possible to use a circuit of this type to remove the "hum" from an audio signal.

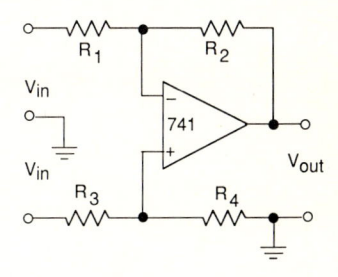

Fig. 19-20 Operational Amplifier Differential Amplifier

The Sine Wave Oscillator

There are many circuits that can be used to generate a sine wave output signal. Figure 19-21 contains the schematic diagram for a simple **RC oscillator**. The frequency-determining device is a Twin-T filter that has been slightly detuned. A perfectly tuned Twin-T filter would act as a *notch filter* at its design frequency and would prevent an output. If we slightly detune the filter from its natural frequency, feedback will be present. At some point on either side of the filter's resonant frequency, we will obtain a free-running sine wave oscillation. When detuning the circuit, the variable resistor should be adjusted using slight changes. As soon as a sine wave is present in the output—stop tuning the resistor. Over tuning will cause the output signal to become distorted. The goal is to obtain a free-running sine wave without distortion.

When the feedback through the filter shifts the output signal 180°, oscillations will begin. This is said to be the *fixed* frequency of the oscillator. Using these procedures, we can design a circuit that has a *single frequency* of operation and that cannot be further tuned.

The frequency of operation at that point will equal:

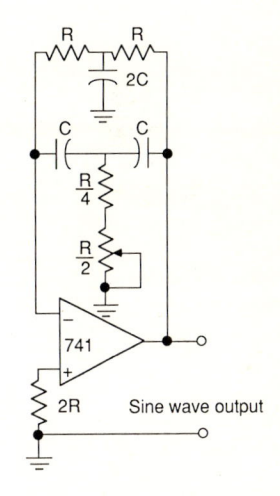

$$f_r = \frac{1}{2\pi RC}$$

Fig. 19-21 Sine Wave RC Oscillator

Another commonly used operational amplifier sine wave oscillator is the *wien bridge oscillator*. A circuit diagram for this type of oscillator is shown in Figure 19-22. It too is an RC oscillator that operates at a "fixed" frequency. To operate at more than one frequency, these oscillators require a switching network that allows the RC frequency-determining device to be changed for each "fixed" frequency.

The Wien bridge oscillator is capable of operating at frequencies ranging from approximately 100 Hz to approximately 6 kHz. Frequency of operation can be determined using the same formula shown above.

Notice that this circuit has a bulb (B_1) inserted in the non-inverting input. This bulb acts as a current ballast. As current increases, the bulbs resistance changes, maintaining output stability.

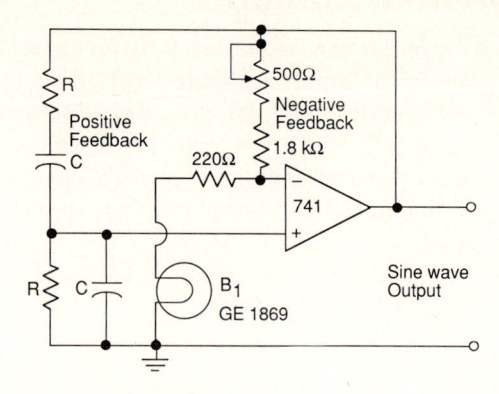

Fig. 19-22 Wein-Bridge Sine Wave Oscillator

Self-Check

Answer each item by inserting the word or words required to correctly complete each statement.

11. If the input and output of an op amp are 180° out-of-phase, the circuit is called a/an _____ amplifier.

    a. inverting
    b. non-inverting

12. When the op amp is used as a voltage follower, we can expect the circuit to have _____ voltage gain.

    a. high
    b. medium
    c. low

13. The op amp circuit that is used to add two or more voltages is called a _____ amplifier.

    a. summing
    b. non-inverting
    c. subtracting
    d. none of these

14. It is impossible for an operational amplifier to be used to add and subtract within the same circuit.

    a. True
    b. False

15. An op amp that is being used as a differential amplifier can be used to remove _____ from an audio signal.

    a. gain
    b. noise
    c. frequency
    d. all of these

16. It is possible to use an op amp as an oscillator.

    a. True
    b. False

The 555 Timer

In advanced electronic and digital circuitry, it is often necessary to provide a precise timing signal for control of a circuit's operation. One integrated circuit that is used extensively for this purpose is the 555 timer. The pin layout, DIP package, and block diagram for this monolithic IC are shown in Figure 19-23.

(a) package

(b)

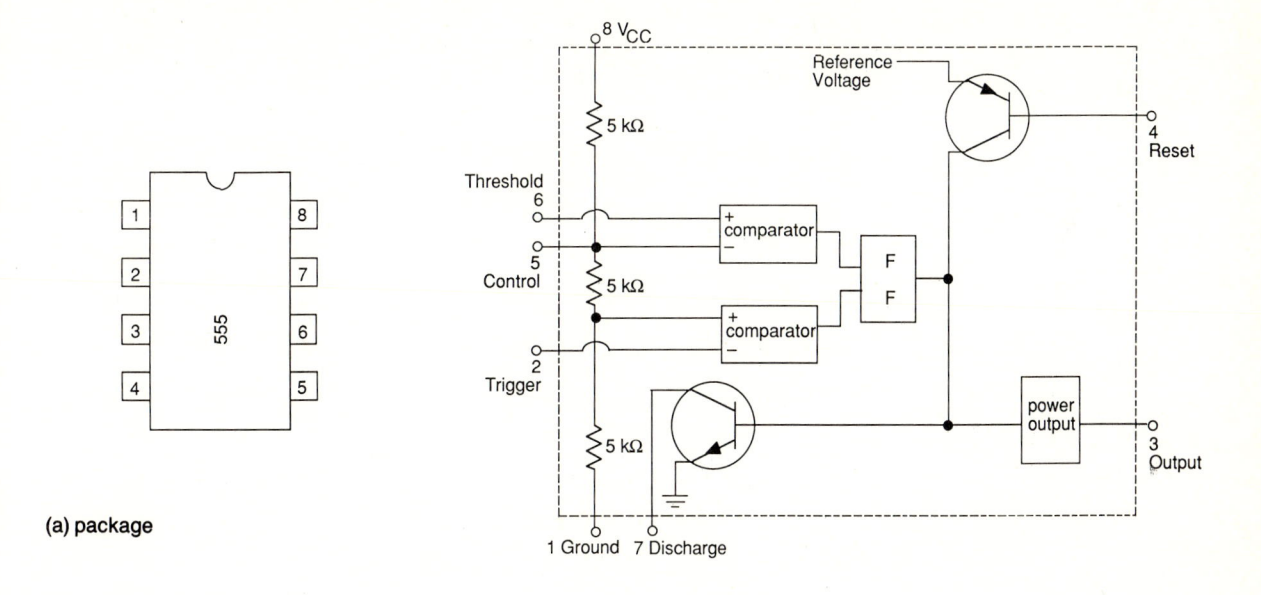

Fig. 19-23 555 Timer-Major Operations Block Diagram

Using this integrated circuit and a few external components, we can design many circuits that can provide timing triggers and pulses, such as:

1. Astable (free running) multivibrators
2. Monostable multivibrators
3. Trigger generators
4. Frequency dividers

The 555 Timer is produced by several manufacturers. It has received such wide usage that several books have been devoted entirely to its uses. For that reason we will limit our discussion to a few popular applications.

Astable Multivibrator Operation

The **astable (free-running) multivibrator** is used in many applications. Using the 555 timer, a power source, two resistors and two capacitors, we can construct an excellent astable multivibrator. A diagram of this circuit is contained in Figure 19-24.

This is a simple version of the astable circuit. Note that pins 4 and 5 are not used but are shorted to V_{CC} and ground. In this circuit R_1, R_2, and C_1 form an RC network that determines the circuit's frequency of operation and output pulse width.

When operating, this circuit provides a rectangular wave output. By modifying the circuit, we can approach a perfect square wave output signal. One way to approach square wave operation is to bypass R_2 with a diode as is shown in dotted lines on the schematic diagram contained in Figure 19-24. With D_1 inserted, the output signal will approximate a square wave. More elaborate modifications using the control and discharge functions of the IC can produce a perfect square wave output signal.

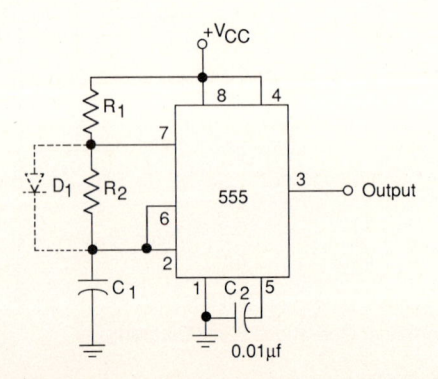

Fig. 19-24 Astable Multivibrator

Monostable Multivibrator Operation

Figure 19-25 contains the block diagram for a 555 timer that is connected to operate as a **monostable multivibrator**. Note that as with other monostable circuits, an input trigger is required to control circuit operation. When a trigger appears at the input, the circuit switches ON and begins to produce an output pulse. The pulse width is controlled by the RC time constant of the RC network made up of R_2 and C_2. When the trigger arrives, C_2 begins to charge. Once C_2 charges to a sufficient voltage, it biases the input circuit OFF. Once the circuit returns to its original OFF state, C_2 can quickly discharge and it then awaits the arrival of the next pulse.

Variation of the sizes of R_2 and C_2 can vary the pulse width of the output signal. Remember, though, frequency of operation is determined by the frequency of the input triggers. This means that *pulse-recurrence time* (PRT) is established by the frequency of the input triggers.

The 555 timer is a valuable IC. It can be used to fabricate countless useful circuits quickly and easily. It would be to your advantage to secure manuals devoted to this chip for use in your career as a technician.

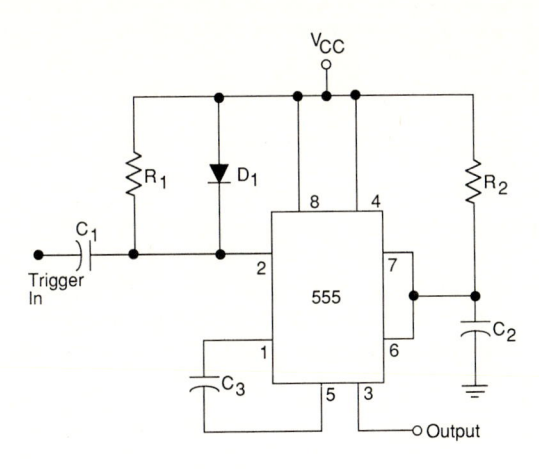

Fig. 19-25 555 Monostable Multivibrator

Self-Check

Indicate whether each statement is True or False.

17. The 555 Timer is an 8-pin monolithic integrated circuit.

    a. True   b. False

18. Timing circuits are easily fabricated using the 555 Timer.

    a. True   b. False

19. Addition of a diode in parallel with R_1 of the astable (555) multivibrator will cause its output to be a more perfect square wave.

    a. True   b. False

20. In the monostable (555) multivibrator, increasing the size of C_2 will cause PRF to increase.

    a. True   b. False

Summary

In this chapter you have been introduced to *operational amplifiers* (OP AMPs). The approach was to introduce you to the way these devices are made, packaged, and used.

Op amps make up one family of integrated circuits. These devices are available in different packages and in single or multiple op amps per package. The most common package is probably the Dual-In-Line Package (DIP).

Compensation methods used to stabilize frequency, increase slew rate, and to balance the input differential amplifier were discussed. You were introduced to the terms INVERTING INPUT, NON-INVERTING INPUT, OPEN LOOP, CLOSED LOOP, SLEW RATE, OFFSET VOLTAGE, and OFFSET VOLTAGE DRIFT.

Op amps can be used to perform the same types of amplification as bipolar transistors and JFETs. Also, they can be used to perform mathematical functions, as differential amplifiers, and as sine-wave oscillators.

The 555 timer monolithic IC can be used to easily design and construct circuits stable enough to provide timing signals for complex systems.

Review Questions and/or Problems

1. Examine Figure 19-26. In this symbol, the plus (+) input is used to denote the _____ input and the minus (−) input is the _____ input.

    a. inverting, non-inverting
    b. non-inverting, inverting

2. R_f (Figure 19-26) is used to provide _____ feedback to the input.

   a. degenerative   b. regenerative

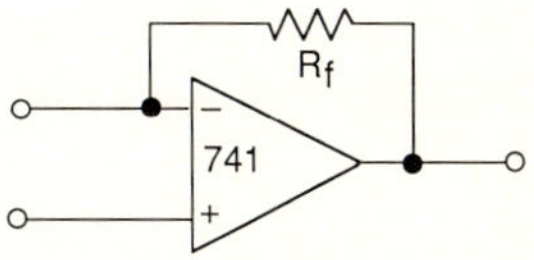

   Fig. 19-26

3. TO-5 packaging is the most common method used to package ICs.

   a. True   b. False

4. A single package can contain two or more op amps.

   a. True   b. False

5. A differentiator amplifier must have _____ inputs.

   a. one   c. three
   b. two   d. four

6. Audio frequency op amps operate as a _____ amplifier.

   a. Class A   c. Class C
   b. Class AB   d. differential

7. Any signal that is common to both inputs of the operational amplifier is called a _____ signal.

   a. common-source   c. common-mode
   b. common-input   d. common-base

8. If an input signal is applied to the plus (+) input of an operational amplifier, the phase relation between the input and the output will equal _____ degrees.

   a. 0   c. 180
   b. 60   d. 360

9. When we apply a small DC voltage to one base of a differential amplifier, the output signal will equal _____.

10. In op amp operations, a larger feedback resistor causes a lower _____.

    a. gain   c. impedance
    b. frequency

11. When feedback is *not* part of an op amp circuit, the circuit is said to be a/an _____ loop circuit.

    a. complete
    b. open
    c. closed

12. To calculate voltage gain, you must divide the change in _____ by the change in _____.

    a. output voltage, input voltage
    b. input voltage, output voltage

13. If the input and output of an op amp are in phase, the circuit is called a/ an _____ amplifier.

    a. inverting
    b. non-inverting

14. When the op amp is used as a voltage amplifier, we can expect the circuit to have _____ voltage gain.

    a. high
    b. medium
    c. low

15. One of the op amp circuits that was discussed is used to subtract two or more voltages. This circuit is called a _____ amplifier.

    a. summing
    b. non-inverting
    c. subtracter
    d. none of these

16. It is possible for an operational amplifier to be used to add and subtract within the same circuit.

    a. True
    b. False

17. An op amp that is being used as a notch filter can be used to remove _____ from the output.

    a. gain
    b. noise
    c. frequencies
    d. both b and c

18. It is impossible to use an op amp as an oscillator.

    a. True
    b. False

19. Diodes are connected between the two input circuits to prevent damage caused by _____.

    a. improper connection of power supplies
    b. overly large input signals
    c. reversing the connections of the two input signals
    d. all of the above

20. A typical operational amplifier has _____.

    a. high output impedance, high gain, and high input impedance
    b. low input impedance, high gain, and high output impedance
    c. high output impedance, high gain, and low input impedance
    d. low output impedance, high gain, and high input impedance

21. The 555 timer is a 14-pin monolithic integrated circuit.

    a. True   b. False

22. Timing circuits are easily fabricated using the 555 timer.

    a. True   b. False

23. Addition of a diode in parallel with R_2 of the astable (555) multivibrator will cause its output to be a more perfect square wave.

    a. True   b. False

24. In the Monostable (555) Multivibrator, increasing the size of C_2 will cause PRF to decrease.

    a. True   b. False

25. Which of the following *is not* a circuit that can be made using a 555 timer?

    a. Monostable multivibrator
    b. Astable multivibrator
    c. Trigger generator
    d. Wideband amplifier

Glossary

ACCEPTOR IMPURITY - Also called *acceptor*, an impurity that has too few valence electrons to complete the crystalline structure when covalent bonded with silicon or germanium. P-type material results.

AC LOAD LINE - A line plotted on a set of characteristic curves that takes into consideration the effect of AC on emitter-base junction resistance.

ACTIVE - 1. The conducting state of a control device. 2. Controlling power from a separate supply. 3. Requiring a power supply separate from the controls.

ADDER-SUBTRACTER - A circuit that performs two mathematical functions (addition and subtraction) simultaneously.

A_I - Symbol used to denote power gain.

ALPHA (α) - In a transistor, the ratio that exists between emitter and collector gain. Symbol for current gain in a common-base amplifier. In a junction transistor, alpha is always less than unity (1).

ANALOG - Angular movement.

ANALOG COMPUTER - A computer that operates with both input and output circuits that are continuously varying in amplitude.

ANALOG VOLTMETER - A voltmeter whose indications are presented by the angular movement of a pointer.

ANALOG MULTIMETER - An analog measuring device that can be used to measure two or more parameters.

ANODE - The positive point of an active device to which electrons are attracted.

ANODE GATE - One of the gate electrodes of a silicon bilateral switch.

APPROXIMATION METHOD - A method used to analyze transistor circuits where small amounts of current and/or resistance are ignored in order to simplify the analyses.

A_P - Symbol used to denote power gain.

ARMSTRONG OSCILLATOR - An inductive feedback oscillator that consists of a tuned-base circuit and an untuned output coil either in the collector or base circuit.

ASTABLE MULTIVIBRATOR - A circuit that has two momentarily stable states, between which it continually alternates. Time spent in each fixed state and switching time are controlled by circuit parameters.

A_V - Symbol used to denote power gain.

AVALANCHE CURRENT - A rapid increase in current that occurs in a reverse biased junction that can cause structural breakdown.

BARRIER REGION - See DEPLETION REGION.

BASE - 1. The region between the emitter and collector of a transistor which receives minority carriers injected from the emitter. 2. On a printed circuit board, the base is the material that supports the printed pattern.

BASE CURRENT - The current that flows in the base of a transistor. Base current will normally range between 2% and 8% of emitter current.

BETA (β) - Also called *current transfer ratio*. Current gain within a common-emitter amplifier. Current gain is calculated by comparison of the change in collector current to the change in emitter current. The formula for beta is:

$$\text{Beta } (\beta) = \frac{I_C}{I_E}$$

BIAS - DC voltage or current applied to a solid-state device that establishes the operating (Q) point.

BIAS VOLTAGE - Voltage applied to a diode or transistor that establishes the quiescent conditions.

BIDIRECTIONAL - 1. Responsive in opposite directions. 2. Capable of supporting current in two directions.

BI-FET CIRCUITRY - Circuitry that contains both bipolar and field effect transistor devices.

BILATERAL - See **BIDIRECTIONAL**.

BIPOLAR TRANSISTOR - A transistor in which both positive and negative current carriers are used to support current flow. All NPN and PNP transistors are this type.

BIPOLAR JUNCTION TRANSISTOR - See **BIPOLAR TRANSISTOR**.

BISTABLE MULTIVIBRATOR - A circuit having two stable states; it will remain in either state indefinitely unless triggered by a voltage of correct polarity and amplitude, at which time it will immediately switch states.

BLOCKING OSCILLATOR - A transistor oscillator that operates intermittently as its base bias increases during oscillation to a point where oscillations stop, and then decreases until oscillation resumes.

BREAKDOWN - The phenomena that occurs in a reverse biased semiconductor diode where current behaves as if the circuit contained negative resistance.

BREAKDOWN POINT - The point at which conduction of a semiconductor device enters the negative resistance region.

BREAKDOWN POTENTIAL - See **BREAKDOWN VOLTAGE**.

BREAKDOWN VOLTAGE - The voltage that causes conduction within a semiconductor device when it enters the negative resistance region.

BREAKOVER - The start of current flow in a silicon controlled rectifier.

BREAKOVER POINT - The point at which current begins to flow in a silicon controlled rectifier.

BREAKOVER VOLTAGE - The voltage required to cause breakover to occur in a silicon controlled rectifier.

BUFFER AMPLIFIER - An amplifier designed to isolate a preceeding circuit from the effects of a following circuit.

CATHODE - General name used to denote the negative electrode of an active device.

CATHODE GATE - One of the gate electrodes of a silicon bilateral switch.

CB AMPLIFIER - See **COMMON-BASE AMPLIFIER**.

cb JUNCTION - The collector-base junction of a bipolar transistor.

CC AMPLIFIER - See **COMMON-COLLECTOR AMPLIFIER**.

CE AMPLIFIER - See **COMMON-EMITTER AMPLIFIER**.

CELSIUS TEMPERATURE - Also called *centigrade temperature*. A temperature scale based on freezing = 0° and boiling = 100° when both exist under normal atmospheric conditions.

CENTRIFUGAL FORCE - The force which acts on a rotating body and which tends to throw the body farther from its axis of rotation.

CENTRIPETAL FORCE - The force which compels a rotating body to move inward toward the center of rotation.

CERAMIC PRINTED CIRCUITS - A printed circuit that is fabricated on a ceramic substrate.

CHANNEL PINCH-OFF - The reduction of channel current in a field-effect transistor to its smallest amount possible.

CHARACTERISTIC CURVE - A graph plotted to show the relationship that exists between changing values. An example is: collector current change as compared to collector voltage change.

CHEMICAL ACTIVITY - The tendency of atoms to combine with other atoms and form molecules.

CLAPP OSCILLATOR - A modified Colpitt's oscillator that has an added capacitor that provides ease of tuning.

CLOSED LOOP CIRCUIT - A circuit in which the output is continuously fed back to the input for constant comparison.

CMOS - See **COMPLEMENTARY MOS**.

COLLECTOR - The element of a transistor in which the majority of emitter current flows.

COLLECTOR CURRENT - The amount of total transistor current that flows in the collector element. Collector current ranges between 92% and 98% of total (emitter) current.

COLLECTOR FEEDBACK BIAS - See **SELF BIAS**.

COLPITT'S OSCILLATOR - An oscillator that has a tuned tank circuit connecting the collector and base. The tank capacitance consists of "split" capacitors which determine the operating frequency and the ratio of output to feedback coupled to the base.

COMMON-BASE AMPLIFIER - Also called a *grounded-base amplifier*. A transistor amplifier in which the base element is common to both the input and output circuit.

COMMON-BASE CONFIGURATION - See **COMMON-BASE AMPLIFIER**.

COMMON-COLLECTOR AMPLIFIER - Also called an *emitter follower*. A transistor amplifier in which the collector element is common to both the input and output circuit.

COMMON-COLLECTOR CONFIGURATION - See **COMMON-COLLECTOR AMPLIFIER**.

COMMON-DRAIN AMPLIFIER - An amplifier configuration used with field-effect transistors whose operation is similar to that of the bipolar transistor common-emitter amplifier.

COMMON-DRAIN CONFIGURATION - See **COMMON-DRAIN AMPLIFIER**.

COMMON-EMITTER AMPLIFIER - Also called a *grounded-emitter amplifier*. A transistor amplifier in which the emitter element is common to both the input and output circuit.

COMMON-EMITTER CONFIGURATION - See **COMMON-EMITTER AMPLIFIER**.

COMMON-GATE AMPLIFIER - An amplifier configuration used with field-effect transistors whose operation is similar to that of the bipolar transistor common-base amplifier.

COMMON-GATE CONFIGURATION - See **COMMON-GATE AMPLIFIER**.

COMMON-MODE REJECTION - Also called *in-phase rejection*. A measure of how well a differential amplifier ignores a signal which appears simultaneously and in phase at both input terminals.

COMMON-SOURCE AMPLIFIER - An amplifier configuration used with field-effect transistors whose operation is similar to that of the bipolar transistor common-base amplifier.

COMMON-SOURCE CONFIGURATION - See **COMMON-SOURCE AMPLIFIER**.

COMPLEMENTARY MOS - Pertaining to N- and P-channel enhancement-mode devices fabricated compatibly on a silicon chip and connected into push-pull complementary circuits.

COMPLEMENTARY SCR - See **PROGRAMMABLE UJT**.

COVALENT BONDING - A type of linkage between atoms. Each atom provides one electron to a shared pair that constitutes an ordinary chemical bond.

CONDUCTION BAND - A partially filled energy band in which electrons can move freely, allowing the material to act as a conductor of electron current.

CONDUCTOR - A material that has been designed to act as a carrier for electron flow.

CONVENTIONAL FLOW - An explanation of current based on the belief that current in an external circuit will flow from the positive pole of a battery and return to the negative pole of the battery.

CRITICAL DAMPING - The value of damping that provides the most rapid transient response with minimum overshoot.

CRYSTAL - A solid in which the available atoms are arranged with some degree of geometric regularity.

CURRENT GAIN - The ratio of change in output current to the change in input current.

CUTOFF - Minimum value of bias which stops all current flow in an active device. An operating condition that is called cutoff and during which zero current flows.

DAMPING - The reduction of an oscillator's output power that results from the dissipation that occurs because of internal resistance.

DC LOAD LINE - A line that is plotted on a device's characteristic curves that represents an infinite number of Q points ranging from cutoff to saturation.

DEGENERATION - Also called *negative feedback*. A process by which a portion of the output signal of an amplifier is fed back to the input circuit 180° out-of-phase, thereby decreasing the amplification of the circuit.

DE-MOSFET - A MOSFET device that can operate in both the depletion and enhancement modes depending on the polarity of the bias applied to the device.

DEPLETION LAYER - Also called *barrier layer*. In a semiconductor, the region in which the mobile-carrier charge density is insufficient to neutralize the net fixed charge density of donors or acceptors.

DEPLETION MODE - A MOSFET device that operates in the mode where bias is used to "deplete" the number of carriers present in the channel.

DEPLETION REGION - The region, extending on both sides of a reverse-biased semiconductor junction, in which all carriers are swept away from the vicinity of the junction: that is, the area is depleted of carriers.

DEPOSITION PROCESS - The application of a material to a substrate through the use of chemical, vapor, electrical, vacuum, or other processes.

DIAC - A four-layered, two-lead device that is used as a switch for control of electrical circuitry.

DIFF AMP - See **DIFFERENTIAL AMPLIFIER**.

DIFFERENTIAL AMPLIFIER - An amplifier having two similar input circuits so connected that they respond to the difference between two voltages or currents, but effectively reject common mode voltages or currents.

DIFFERENTIATOR - Also called *differentiating circuit*. A circuit in which the output voltage is substantially in proportion to the rate of change of the input voltage or current.

DIFFUSION - A thermally induced process in which one type of material permeates another.

DIGITAL - Using numbers expressed in digits and in certain scale of notation to represent all the variables that appear in a problem.

DIGITAL CIRCUITRY - Circuitry which operates like a switch: i.e., it is either "on" or "off."

DIGITAL MULTIMETER - A measuring instrument that is capable of checking two or more parameters and that provides its indication in digital form.

DIGITAL VOLTMETER - A voltmeter whose indications are presented in digital form.

DIODE - A two-element semiconductor device that makes use of the rectifying characteristics of a PN junction (junction diode).

DIP - See **DUAL-IN-LINE**.

DISCRETE CIRCUITRY - Circuits built from separate components that are individually manufactured, tested, and assembled.

DONOR IMPURITY - Also called *donor*. An impurity which tends to release a free electron and thereby affect the conductivity of the crystal.

DOPING - The addition of impurities to a semiconductor crystal in order to control its resistivity and to determine whether a P- or N-type material results.

DRAIN - The element of field-effect transistors that performs the same function as the collector in a bipolar transistor.

DRAIN CURRENT - Current that flows in the drain circuit of a field-effect transistor that compares to collector current in a bipolar transistor.

DRAIN-SOURCE SATURATION CURRENT - Amount of current that is required to saturate the drain element of a field-effect transistor, when gate-source voltage (V_{GS}) equals zero volts.

DUAL BATTERY BIAS - See **DUAL SOURCE BIAS**.

DUAL-IN-LINE (DIP) - A type of package used to contain integrated circuits.

DUAL SOURCE BIAS - Transistor biasing that involves the use of separate batteries for the biasing of each junction.

DUAL SUPPLY BIAS - See **DUAL SOURCE BIAS**.

eb JUNCTION - Emitter-base junction within a bipolar transistor.

ELECTRON - The negative particle contained in all atoms that orbits the nucleus in specific energy levels.

ELECTRON FLOW - The movement of electrons from the negative pole of a power source to the positive pole of the source.

ELECTRON-HOLE PAIR - A positive current carrier (hole) and a negative carrier (electron), considered together as one entity.

ELECTROSTATICS - The branch of physics concerned with electricity at rest.
EMITTER - The transistor element through which 100% of the current flows.
EMITTER CURRENT - The direct current flowing in the emitter of a bipolar transistor. Emitter current = 100% of transistor current.
EMITTER FOLLOWER - See **COMMON-COLLECTOR AMPLIFIER**.
E-MOSFET - A MOSFET that operates because of the increase in (enhancement of) the number of carriers available in the channel resulting from correct biasing.
ENERGY LEVEL - The electrical power contained in an orbiting electron.
ENHANCEMENT MODE - See **E-MOSFET**.
EPITAXY - The controlled growth of a layer of semiconductor material in a suitable substrate.
EPITAXIAL GROWTH - A semiconductor fabrication process in which single-crystal N or P material is deposited and grows on the surface of the substrate.
ESAKI DIODE - See **TUNNEL DIODE**.
ETCH - The selective removal of a solid material, usually metal, as in the preparation of printed circuit boards.
ETCHANT - A chemical substance that is used to etch.
EXTRINSIC - Not perfect, contains impurities.
FALL TIME - The amount of time required for a pulse to drop from 90% of peak to 10% of its maximum amplitude.
FEEDBACK - The return of a portion of an output signal to the input of the same stage.
FET - See **FIELD-EFFECT TRANSISTOR**.
FIELD-EFFECT TRANSISTOR - A semiconductor device that operates using the electrostatic effect caused by bias changes to control the amount of current that flows.
FIXED BIAS - A constant value of DC bias voltage.
FLAT PACKS - A flat, rectangular integrated circuit or hybrid-circuit package with coplanar leads.
FLAT TOP - The horizontal portion of a square or rectangular wave that results from the low frequency components that make up the wave.
FLIP FLOP - See **BISTABLE MULTIVIBRATOR**.
FLYWHEEL EFFECT - The maintaining of oscillations in an LC resonant tank circuit during intervals between pulses of excitation energy.
FORBIDDEN BAND - The region that exists between the valence band and the conduction band of an element.
FORBIDDEN REGION - See **FORBIDDEN BAND**.

FORWARD BIAS - An external voltage applied in the conducting direction of a PN junction.
FORWARD BREAKOVER POTENTIAL - Voltage required to cause breakover in an SCR.
FORWARD CURRENT - Current that flows across a PN junction that has forward bias applied.
FOUR-LAYER DIODE - A semiconductor device that has three junctions and four layers of semiconductor material. Connections are made to the anode and cathode with current flow in one direction only.
FREQUENCY-DETERMINING DEVICE (NETWORK) - The device or network that determines the frequency of operation for an oscillator.
FREQUENCY DIVIDER - A circuit or device whose output is at a lower frequency than the frequency of the input.
FREQUENCY MULTIPLIER - A device or circuit that delivers an output signal whose frequency is a multiple of the input frequency.
GAIN - Any increase in signal power that results from being processed by an amplifier.
GAIN-BANDWIDTH PRODUCT - The product of the closed-loop gain of an operational amplifier and its corresponding closed-loop bandwidth.
GAMMA (γ) - Ratio of collector voltage to collector supply voltage.
GATE - One of the electrodes of several semiconductor devices.
GATE TRIGGER CURRENT - Amount of current required to start gate current flow in a four-layer device such as an SCR.
GATE TRIGGER VOLTAGE - Amount of voltage required to start gate current flow in a four-layer device such as an SCR.
GBP - See **GAIN-BANDWIDTH PRODUCT**.
GERMANIUM - A brittle, grayish-white metallic element having semiconductor properties that is used in the fabrication of transistors and PN junction diodes. Atomic number – 32, valence – 4, symbol – Ge.
HAND CAPACITANCE EFFECT - The capacitance introduced when one's hand is brought near a tuning capacitor or other insufficiently shielded part of a circuit or receiver.
HEADERS - The part of a sealed component or assembly that provides support and insulation for the leads passing through the walls of a package.
HOLDING CURRENT - The value of average forward current (with the gate open) below which a silicon controlled rectifier will cut off after having been conducting in the forward direction.
HOLE FLOW - Conduction in a semiconductor where electrons move into holes and thereby create new holes. The holes appear to move toward the negative terminal of the battery, giving the equivalent of current movement from positive to negative.

HOLES - In the electronic valence structure of a semiconductor, a mobile vacancy which acts like a positive electronic charge with a positive mass.

HOT CARRIER DIODE - A diode in which a closely controlled, metal-semiconductor junction provides virtual elimination of charge storage. This device has excellent diode forward and reverse characteristics, extremely fast turn-on and turn-off times, low noise characteristics and a wide dynamic range. Abbreviation HCD.

HYBRID CIRCUITRY - 1. A circuit which combines thin-film and semiconductor technologies. 2. An integrated circuit combining parts made by a number of techniques, such as diffused monolithic portions, thin-film elements, and discrete devices.

I_B - See **BASE CURRENT**.

IC - See **INTEGRATED CIRCUIT**.

I_C - See **COLLECTOR CURRENT**.

I_E - See **EMITTER CURRENT**.

IMPURITY - A material such as boron, phosphorus, or arsenic that is added to a semiconductor such as germanium or silicon to produce either P-type or N-type material.

INPUT SIGNAL - The current, voltage, power, or other driving force that is applied to a circuit or device.

INSULATED-GATE FIELD-EFFECT TRANSISTOR - In general, any field-effect transistor that has an insulated gate regardless of the fabrication process. Abbreviated IGFET.

INSULATOR - A material which has its valence electrons tightly bound to the atom and they are not free to move.

INTEGRATED CIRCUIT - A combination of interconnected circuit elements inseparably associated on or within a continuous substrate.

INTEGRATOR - A device or circuit whose output is proportionate to the integral of the input signal.

INTERBASE RESISTANCE - Resistance that is arranged between base 1 and base 2 of a unijunction transistor.

INTRINSIC MATERIAL - A semiconductor material in which there are equal numbers of holes and electrons (no impurities).

INTRINSIC STANDOFF VOLTAGE - The voltage drop that exists across the bottom portion (tap to ground) of the interbase resistance of a UJT.

INVERTING AMPLIFIER - An op amp whose output signal is 180° out-of-phase with the input signal.

JFET - See **JUNCTION FIELD-EFFECT TRANSISTOR**.

JUMP VOLTAGE - Voltage within a trapezoidal wave whose function is to overcome the X_L of the CRT's deflection coils.

JUNCTION - 1. A contact between two sections of semiconductor material as in a diode. 2. A region of transistor P- and N-type semiconductor material.

JUNCTION BARRIER - The opposition to the diffusion of majority carriers across a PN junction due to the charge of the fixed donor and acceptor ions.

JUNCTION CAPACITANCE - Capacitance that exists within the junction region of a semiconductor device.

JUNCTION FIELD-EFFECT TRANSISTOR - A transistor made up of a gate region that is diffused into a channel region. When a control voltage is applied to the gate, the channel is depleted or enhanced, and the current between the source and the drain is thereby controlled. No current flows when the channel is "pinched off."

JUNCTION RESISTANCE - Resistance that exists within the junction region of a semiconductor device.

JUNCTION TRANSISTOR - A transistor having three alternate sections of P-type and N-type semiconductor material.

KINETIC ENERGY - Energy which a system possesses by virtue of its motion.

LARGE SIGNAL BEHAVIOR - Reaction of a circuit to a signal having high amplitude.

LATCHED - Locked into a condition awaiting an opposite signal to turn it off.

LCD - See **LIQUID CRYSTAL DISPLAY**.

LC OSCILLATOR - An oscillator whose operating frequency is controlled by parallel-connected inductance and capacitance that is operating at the resonant frequency.

LC TANK - A parallel-connected inductance and capacitance that, through its resonance, controls the frequency of operation of an LC oscillator.

LEADING EDGE - The transition of a pulse which occurs first.

LEAKAGE CURRENT - A current that flows in a reverse biased semiconductor junction that is supported by minority carriers.

LED - See **LIGHT-EMITTING DIODE**.

LIGHT-EMITTING DIODE - A PN junction that emits light when biased in the forward direction.

LIQUID CRYSTAL DISPLAY - A crystalline substance whose molecules are normally translucent but when an electronic current is introduced, molecules become agitated forming a surface that will reflect light.

LOAD LINE - A line that is drawn on the collector characteristic curves of a transistor which represents an infinite number of operating points for that transistor with specific collector resistance and collector voltage.

MAJORITY CARRIER - The predominant carrier in a semiconductor; electrons in N-type and holes in P-type materials.

MAJORITY CURRENT - Current in a semiconductor that is supported by the predominant carriers.

MASK - A device (usually a thin sheet of metal which contains an open pattern) used to shield selected portions of a base during a deposition process.

MAXIMUM AVERAGE FORWARD CURRENT - Maximum average forward current that a diode can conduct without damage at an operating temperature of 25°C or 77° F.

MAXIMUM AVERAGE REVERSE CURRENT - Maximum average reverse current that a diode can conduct without damage at an operating temperature of 25°C or 77°F.

MAXIMUM POWER DISSIPATION - The maximum power that a diode is capable of dissipating when operating at a temperature of 25°C or 77°F.

MAXIMUM SURGE CURRENT - Maximum current that can be allowed to flow in a diode without damage at an operating temperature of 25°C or 77°F.

METAL-OXIDE-SEMICONDUCTOR FIELD-EFFECT TRANSISTOR - Abbreviated MOSFET - A device consisting of diffused source and drain regions on either side of a P and N channel region, and a gate electrode insulated from the channel by silicon oxide.

MICROCOMPONENTS - Those components that are smaller than existing discrete components by several orders of magnitude.

MICROELECTRONICS - Also called *microsystems electronics*. The entire body of electronic art which is connected with, or applied to, the realization of electronic systems from extremely small electronic parts.

MINORITY CARRIER - The less predominant carrier in a semiconductor; electrons in P-type and holes in N-type materials.

MINORITY CURRENT - Current in a semiconductor that is supported by the less predominant carriers.

MONOLITHIC IC - An integrated circuit that is fabricated on a substrate made from a single semiconductor material.

MONOSTABLE MULTIVIBRATOR - Also called a *one-shot multivibrator*. A circuit having only one stable state, from which it can be triggered to change state, but only for a predetermined interval, after which it returns to the original state.

MOSFET - Abbreviation for *metal-oxide-semiconductor field-effect transistor*.

MULTICHIP CIRCUITRY - A microcircuit in which discrete, miniature, active electronic devices (diodes and transistors) and thin-film or diffused passive components or component clusters are interconnected by thermocompression bonds, alloying, soldering, welding, chemical deposition, or metalization.

MULTIVIBRATOR - A relaxation oscillator in which the in-phase feedback voltage is obtained from two transistors and whose operating frequency and pulse durations are controlled by RC networks.

N-CHANNEL - The resistive bar (channel) in a FET that is constructed from N-type material.

NEGATIVE CARRIER - Electrons.

NEGATIVE IONS - Ions that have an excess of electrons.

NEGATIVE-RESISTANCE REGION - The operating region within which an increase in applied voltage results in a decrease in current.

NEGATIVE TEMPERATURE COEFFICIENT - The amount of reduction in the value of a quantity, such as capacitance or resistance, for each degree of increase in operating temperature.

NEUTRON - One of the three major particles that exist within an atom. The neutron possesses a neutral electrical charge and is found in the nucleus of the atom.

NPN TRANSISTOR - A junction transistor formed using two sections of N-type material and one section of P-type material. The emitter and collector are N-type and the base is P-type.

NON-INVERTING AMPLIFIER - An operational amplifier whose output is in-phase with the input.

NMOS - MOSFET devices that have an N-channel.

NONSINUSOIDAL - Any waveform that is not a sine wave. It, therefore, contains a sine wave plus harmonics.

N-TYPE MATERIAL - Semiconductor material in which electrons are the majority carriers and holes are the minority carriers.

OFFSET - The measure of imbalance between halves of a symmetrical circuit.

OFFSET VOLTAGE DRIFT - The amount of voltage drift that occurs between the balanced inputs of an operational amplifier.

ONE-SHOT MULTIVIBRATOR - See **MONOSTABLE MULTIVIBRATOR**.

OP AMP - See **OPERATIONAL AMPLIFIER**.

OPEN LOOP GAIN - The gain present in an operational amplifier that does not have feedback.

OPERATIONAL AMPLIFIER - An amplifier that performs various mathematical operations. Also useable as a DC amplifier.

OPERATING POINT - Also called *quiescent point* and *Q point*. The point along a load line where an active device operates with only DC voltages applied.

ORBIT - The path that an electron takes around the nucleus of an atom.

OSCILLATOR - An electronic circuit which generates AC power at a frequency determined by the values of certain constants in its circuit. An oscillator may be considered an amplifier that has positive (regenerative) feedback.

OUTPUT SIGNAL - The current, voltage, power, or other driving force that is present at the output terminals of a circuit or device.

OVER DAMPING - Any periodic damping greater than the amount required for critical damping.

PACKAGE - The case that contains and protects a transistor or other device.
PACKAGING - The physical process of locating, connecting, and protecting devices, components, etc.
P-CHANNEL - The resistive bar (channel) in a FET that is constructed from P-type material.
PEAK INVERSE VOLTAGE - The peak AC voltage which a rectifying cell or PN junction will withstand in the reverse direction.
PEAK POINT - In a Zener diode or other thermistor, the maximum forward current that exists prior to entering the negative resistance region.
PEAK RECURRENT CURRENT - Maximum peak current that a diode can conduct when a varying current is present as in AC operations.
PEAK REVERSE VOLTAGE - See **PEAK INVERSE VOLTAGE**.
PENTAVALENT - Having five valence electrons.
PERMISSIBLE ENERGY LEVEL An energy level at which an electron can exist as it orbits an atom.
PIV - See **PEAK INVERSE VOLTAGE**.
PHASE - The angular relationship between current and voltage in AC circuits.
PHOTODIODE - A solid-state device similar to an ordinary diode with the exception that when light strikes the junction, the diode conducts.
PINCH-OFF REGION - Region at which channel current in an FET is minimum.
PINCH-OFF VOLTAGE - Amount of bias voltage present when an FET enters the pinch-off region of its operation.
PIN DIODE - A diode whose junction acts like a variable resistor and whose resistance can be controlled by junction bias.
PMOS - MOSFET devices that have a P-channel.
PN JUNCTION - The region of a transistor between P-type and N-type material in a single semiconductor crystal.
PN JUNCTION DIODE - See **DIODE**.
PNP TRANSISTOR - A junction transistor formed using two sections of P-type material and one section of N-type material. Emitter and collector are P-type and the base is N-type.
POSITIVE CARRIER - Holes.
POSITIVE IONS - Ions that have a deficiency of electrons.
POTENTIAL ENERGY - Energy due to the position of one body with respect to another or to the relative parts of the same body.
POTENTIAL HILL - Forward bias required to start current in a PN junction.
POWER GAIN - The ratio of output power to input power.
PPS - Pulses-per-second.

PRF - See **PULSE REPETITION FREQUENCY**.

PRINTED CIRCUITS - Circuits in which the interconnecting wires have been replaced by conductive strips that are printed, etched, etc., onto an insulating board or substrate.

PROGRAMMABLE UJT - A four-layer switch that operates at very low current levels that has only one anode gate for control of triggering.

PROTON - The positively charged particle that exists in the nucleus of all atoms.

PRT - See **PULSE RECURRENCE TIME**.

PRV - See **PEAK INVERSE VOLTAGE**.

P-TYPE MATERIAL - Semiconductor material in which holes are the majority carriers and electrons are the minority carriers.

PULSE - 1. The variation of a quantity having a normally constant value. This variation is characterized by a rise and decay of a finite value. 2. An abrupt change in voltage, either positive or negative, which conveys information to a circuit.

PULSED OSCILLATOR - An oscillator in which oscillations are sustained by either self-generated or external pulses.

PULSE-REPETITION FREQUENCY (PRF) - The rate (usually given in frequency or pulses-per-second) at which a series of pulses occur.

PULSE-RECURRENCE TIME (PRT) - The time that elapses between the start of one pulse to the start of the next pulse.

PULSES PER SECOND (PPS) - Number of pulses that occur in one second.

PULSE WIDTH (PW) - The time interval between the points at which the instantaneous value on the leading and trailing edges bear a specified relationship to the peak pulse amplitude.

Q POINT - See **OPERATING POINT**.

QUIESCENT POINT - See **OPERATING POINT**.

RC OSCILLATOR - An oscillator in which the operating frequency is determined by the resistance-capacitance elements.

READOUT - The manner in which information is presented.

RECOMBINATION - The simultaneous elimination of both an electron and a hole in a semiconductor.

RECTANGULAR WAVE - A periodic wave which alternately assumes one of two fixed values, the time of transition being negligible in comparison with the duration of each fixed value and each fixed value having a different duration.

REGENERATION - The gain in power achieved by coupling part of the output signal back to the input in phase with the incoming signal.

REPETITIVE PEAK-INVERSE VOLTAGE - The maximum allowable instantaneous value of reverse voltage that may be repeatedly applied to the anode of an SCR with the gate open.

RESIST - The material placed on the surface of a copper-clad board to prevent the removal by etching of the conductive layer beneath the area covered.

RESONANCE - The circuit condition that exists when X_C equals X_L. This is usually represented by a single frequency or very narrow band of frequencies.

RESONANT FREQUENCY - The center frequency at which a circuit or device resonates.

REST TIME - The time between the end of one pulse and the beginning of the next pulse.

REVERSE BIAS - An external voltage applied to a PN junction that reduces the flow of current through the junction.

REVERSE CURRENT - See **MINORITY CURRENT**.

RISE TIME - The amount of time required for a pulse to rise from 10% of peak to 90% of its maximum amplitude.

SAWTOOTH WAVE - A periodic wave, the amplitude of which varies linearly between two values.

SBS - See **SILICON BILATERAL SWITCH**.

SCHMITT TRIGGER - A bistable pulse generator in which an output pulse of a constant amplitude exists only as long as the input voltage exceeds a certain DC value.

SCR - See **SILICON CONTROLLED RECTIFIER**.

SCS - See **SILICON CONTROLLED SWITCH**.

SELF BIAS - The bias developed by current flow through a resistor connected in series with the emitter of a transistor.

SEMICONDUCTOR - An element that is neither a good conductor nor a good insulator.

SERIES-FED HARTLEY OSCILLATOR - A Hartley oscillator that has a portion of its tank circuit in series with the DC path.

SHEET RESISTANCE - The resistance per square centimeter that exists within a substrate or other piece of conductive material.

SHELL - See **ENERGY LEVEL**.

SHUNT-FED HARTLEY OSCILLATOR - A Hartley oscillator which has a parallel branch for DC current flow, and thereby prevents the flow of DC current in its tank circuit.

SIGNAL - Intelligence placed into and/or extracted from an active device such as an amplifier.

SILICON - A metallic element that in its pure state is used as a semiconductor.

SILICON CONTROLLED RECTIFIER (SCR) - A four-layer PNPN device that, when in the normal state, blocks applied voltage in either direction. Application of the correct voltage to a gate terminal enables the device to conduct in the forward direction.

SILICON CONTROLLED SWITCH (SCS) - A four-terminal PNPN semiconductor switching device that can be triggered into conduction by the application of either positive or negative pulses.

SILICON BILATERAL SWITCH (SBS) - A device that has characteristics similar of the SCS but that is capable of conduction in either direction.

SINUSOIDAL - Varying in proportion to the sine of an angle or time function. Normally a sine wave.

SLEW RATE - The maximum rate of change of the output voltage of an amplifier operated within its linear region.

SOLID-STATE - Pertaining to circuits and components made of or using semiconductors.

SOURCE - The element of a field-effect transistor that compares to the collector of a bipolar transistor.

SPACE CHARGE REGION - The region where the net charge density of a semiconductor differs significantly from zero. (Also see **DEPLETION LAYER**.)

SQUARE WAVE - A periodic wave which alternately assumes two fixed values for equal lengths of time, the transition being negligible compared to the duration of each fixed value.

SUBSTRATE - The supporting material on or in which the parts of a semiconductor device or integrated circuit are made or attached.

SUBTRACTER - An operational amplifier that performs the mathematical function of subtraction.

SUMMING AMPLIFIER - An operational amplifier that performs the mathematical operation of addition.

SURGE CURRENT - Sudden current change in a circuit.

TERTIARY WINDING - The third winding of a transformer, normally an output winding.

THERMAL RUNAWAY - A condition in which the dissipation in a transistor or other device increases so rapidly with higher temperature that the temperature continues to rise.

THICK-FILM CIRCUITRY - A microcircuit in which passive components of a ceramic-metal composition are formed on a suitable substrate by screening or firing.

THIN-FILM CIRCUITRY - A circuit consisting of a passive substrate on which various passive elements (resistors and capacitors) are deposited in the form of thin-patterened films of conductive or nonconductive materials and to which active components are attached separately.

THYRISTOR - A bistable device comprising three or more junctions.

THRESHOLD VOLTAGE - The level of voltage at which a PN junction begins to conduct.

TIMER - An assembly of electric circuits and associated equipment which provides the following: trigger pulses, sweep voltages, intensifier pulses, gate voltages and blanking voltages.

TIMING PULSE - A pulse used to synchronize a circuit.

TRAILING EDGE - The transition of a pulse that occurs last.

TRANSISTOR - An active semiconductor device, usually made of silicon or germanium, having three or more electrodes.

TRANSIT INTERVAL - The amount of time required for a pulse to rise from zero to maximum value or to fall from maximum value to zero.

TRAPEZOIDAL WAVE - A trapezoidal-shaped waveform.

TRIAC - A five-layer NPNPN device that is equivalent to two SCRs that can be used to switch either polarity of applied voltage. Operation can be controlled in either polarity from the single gate electrode.

TRIGGER - A pulse that starts an action.

TRIGGER GENERATOR - A pulsed oscillator designed to generate single pulses at a predetermined frequency.

TRIVALENT - Having three valence electrons.

TUNED-BASE OSCILLATOR - An oscillator whose tuned circuit is connected in the base circuit.

TUNED-COLLECTOR OSCILLATOR - An oscillator whose tuned circuit is connected in the collector circuit.

TUNNEL DIODE - A highly doped PN diode which, when forward biased, exhibits a region of negative restance as part of its operational characteristics.

TUNNEL EFFECT - Piercing of a rectangular potential barrier in a semiconductor by a particle that does not have enough energy to go over the barrier.

TUNNELING - See **TUNNEL DIODE**.

TURN-OFF TIME - The time that a switching circuit takes to stop the flow of current in the circuit it is controlling.

TURN-ON TIME - The time that a switching circuit takes to start the flow of current in the circuit it is controlling.

UJT - See **UNIJUNCTION TRANSISTOR**.

UNDER DAMPING - The condition where damping is so small that excessive overshoot is present.

UNIJUNCTION TRANSISTOR (UJT) - A three-terminal semiconductor device exhibiting stable, open-circuit, negative-resistance characteristics.

UNILATERAL - Conductivity in one direction only.

UNITY GAIN - A gain (amplification factor) of one (1).

UNITY-GAIN FREQUENCY - The frequency at which gain equals unity (1).

VACUUM DEPOSITION - A process in which a substance is heated in vacuum enclosure until the substance vaporizes and condenses (deposits) on the surface of another material in the enclosure.

VALENCE - A number that represents the chemical activity of an element.

VALENCE BAND - The area, within an element, just below the conduction band.

VALENCE ORBIT - The outermost orbit in an atom.
VALENCE SHELL - The electrons which form the outermost shell of an atom.
VALLEY POINT - In a tunnel diode, the point on the characteristic curve at which junction current is minimum prior to beginning its rise and after having passed through the negative resistance region.
VALLEY VOLTAGE - In a tunnel diode, the voltage at which junction current is minimum prior to beginning its rise and after having passed through the negative resistance region.
VARACTOR - A diode whose junction capacitance can be controlled by a reverse bias voltage.
V_{CB} - Voltage drop – collector to base.
V_{CE} - Voltage drop – collector to emitter.
VCO - See **VOLTAGE CONTROLLED OSCILLATOR**.
V_{EB} - Voltage drop emitter to base.
VMOS - Any MOS device whose gate is V-shaped.
VOLTAGE-CONTROLLED OSCILLATOR - An oscillator whose frequency can be controlled by a changing voltage.
VOLTAGE DIVIDER BIAS - Fixed bias that is established by use of a voltage divider.
VOLTAGE FOLLOWER - An operational amplifier circuit that operates much like a common-collector amplifier.
VOLTAGE GAIN - Ratio that exists between the change in output voltage and the change in input voltage of an amplifier.
ZENER CURRENT - See **AVALANCHE CURRENT**.
ZENER DIODE - Also called a *voltage regulator diode*. A two-layer device that, when reverse biased, reaches a point where current increases rapidly with little change in voltage. Used as a voltage regulator.

Bibliography

Badrhhan, K. S. and Larky, N. D. 1984, *Electronics principles and applications*. Cincinnati: South-Western

Bannon, E. 1975, *Operational amplifiers:* Theory and Servicing. Reston, VA: Reston.

Boylestad, R. and Nashelsky, L. 1977, *Electricity, electronics, and magnetism*. Engelwood Cliffs, NJ: Prentice-Hall

Brazee, J. G. 1968, *Semiconductor and tube electronics*: An Introduction. 1968. Holt, Rinehart, and Winston

Buban, P. and Schmitt, M. L. 1972, *Technical electricity and electronics*. New York: McGraw-Hill

Churchman, L. W. 1971, *Survey of electronics*. San Francisco: Rinehart

Cirovic, M. M. 1977, *Integrated circuits: a user's handbook*. Reston, VA: Reston

Cooper, W. D. and Weisbecker, H. B. 1982, *Solid-state devices and integrated circuits*. Reston, VA: Reston

Crozier, P. 1983, *Introduction to electronics*. North Scituate, Mass.: Breton

Davis, C. A. 1973, *Industrial electronics: design and application*. Columbus, OH: Merrill Publishing

Department of Defense, *Miscellaneous electronics training materials*. Washington, DC: U. S Government Printing Office

Gerrish, H. H. and Dugger Jr., W. E. 1979, *Transistor electronics*. South Holland, Illinois: Goodheart-Willcox

Graf, R.F., ed. *Dictionary of electronic terms*, 1972. 4th ed, Fort Worth, TX: Tandy Corporation

Grob, B. 1977, *Basic electronics*. New York: McGraw-Hill

Kiver, M.S. 1972, *Transistor and integrated electronics*. New York: McGraw-Hill

Lenk, J.D. 1973, *Manual for integrated circuits*. Reston, VA: Reston

Malvino, A. P. 1973, *Electronic principles*. New York: McGraw-Hill

Malvino, A. P. 1973, *Transistor circuit approximations*. New York: McGraw-Hill

Manera, A. S. 1973, *Solid-state electronic circuits: for engineering technology*. New York: McGraw-Hill

Matthews, J. I. 1972, *Solid-state electronics concepts*. New York: McGraw-Hill

Prensky, S. D. and Seidman, A. H. 1981, *Linear integrated circuits: practice and applications*. Reston, VA: Reston

RCA Corporation, 1984, *RCA SK Replacement guide*. Deptford, NJ: RCA Corporation

Robinson, V. 1975, *Solid-state circuit analysis*. Reston, VA: Reston

Tepper, I. 1972, *Solid-state devices: Volume 1/theory*. Reading, MA: Addison-Wesley

Tepper, I. 1974, *Solid-state devices: volume 2/applications*. Reading, MA: Addison-Wesley

Texas Instruments, Inc. 1973, *The transistor and diode data book*. Dallas: Texas Instruments, Inc.

Traister, R. J. 1984, *The experimenter's guide to integrated circuits*. Englewood Cliffs, NJ: Prentice-Hall

Veatch, H. C. 1978, *Electrical circuit action*. Chicago: Science Research Associates

Zeines, B. 1976, *Transistor circuit analysis and application*. Reston, VA: Reston

A
Appendix
General Purpose, Rectifier, and Varactor Diodes

Appendix A

TYPES 1N645 THRU 1N649, 1N645A
SILICON GENERAL PURPOSE DIODES
BULLETIN NO. DL-S 739125, OCTOBER 1966—REVISED MARCH 1973

225 V to 600 V • 400 mA AVERAGE
- Rugged Double-Plug Construction

mechanical data

Double-plug construction affords integral positive contact by means of a thermal compression bond. Moisture-free stability is ensured through hermetic sealing. The coefficients of thermal expansion of the glass case and the dumet plugs are closely matched to allow extreme temperature excursions. Hot-solder-dipped leads are standard.

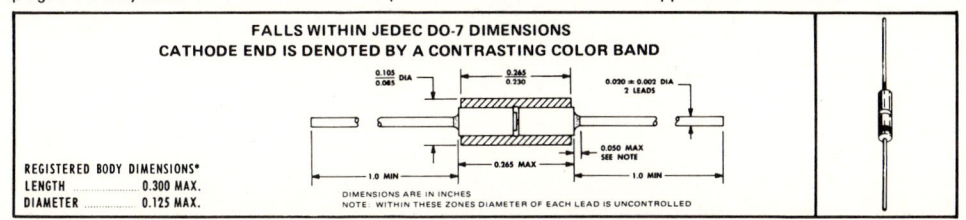

FALLS WITHIN JEDEC DO-7 DIMENSIONS
CATHODE END IS DENOTED BY A CONTRASTING COLOR BAND

REGISTERED BODY DIMENSIONS*
LENGTH .... 0.300 MAX.
DIAMETER .... 0.125 MAX.

DIMENSIONS ARE IN INCHES
NOTE: WITHIN THESE ZONES DIAMETER OF EACH LEAD IS UNCONTROLLED

*absolute maximum ratings at 25°C free-air temperature (unless otherwise noted)

		1N645	1N645A	1N646	1N647	1N648	1N649	UNIT
$V_{RM(wkg)}$	Working Peak Reverse Voltage over Operating Free-Air Temperature Range	225	225	300	400	500	600	V
I_O	Average Rectified Forward Current at (or below) 25°C Free-Air Temperature (See Note 1)	400						mA
I_O	Average Rectified Forward Current at 150°C Free-Air Temperature	150						mA
$I_{FM(surge)}$	Peak Surge Current, One Second, at 25°C to 150°C Free-Air Temperature (See Note 2)	3						A
P	Continuous Power Dissipation at (or below) 25°C Free-Air Temperature (See Note 3)	600						mW
$T_{A(opr)}$	Operating Free-Air Temperature Range	−65 to 150						°C
	Altitude at Rated Working Peak Reverse Voltage	100 000						ft

*electrical characteristics at 25°C free-air temperature (unless otherwise noted)

PARAMETER		TEST CONDITIONS	1N645 MIN MAX	1N645A MIN MAX	1N646 MIN MAX	1N647 MIN MAX	1N648 MIN MAX	1N649 MIN MAX	UNIT
$V_{(BR)}$	Reverse Breakdown Voltage	$I_R = 100\ \mu A$, $T_A = 100°C$	275	275	360	480	600	720	V
I_R	Static Reverse Current	$V_R =$ Rated $V_{RM(wkg)}$	0.2	0.2	0.2	0.2	0.2	0.2	μA
		$V_R =$ Rated $V_{RM(wkg)}$, $T_A = 100°C$	15	15	15	20	20	25	μA
		$V_R = 60\ V$		0.05					μA
		$V_R = 60\ V$, $T_A = 125°C$		10					μA
V_F	Static Forward Voltage	$I_F = 400\ mA$	1	1	1	1	1	1	V
C_T	Total Capacitance	$V_R = 12\ V$, $f = 1\ MHz$	6 typ	6 typ	6 typ	6 typ	6 typ	6 typ	pF

NOTES:
1. These values may be applied continuously under single-phase 60-Hz half-sine-wave operation with resistive load. Derate linearly to 150 mA at 150°C free-air temperature at the rate of 2 mA/°C.
2. These values apply for a one-second square-wave pulse with the device at nonoperating thermal equilibrium immediately prior to the surge.
3. Derate linearly to 200 mW at 150°C free-air temperature at the rate of 3.2 mW/°C.

*JEDEC registered data.

This information is out of date and parts could possibly be absolete, reprint for educational purposes only. Courtesy of Texas Instruments, Inc.

TYPES 1N4305, 1N4444, 1N4454
SILICON SWITCHING DIODES

BULLETIN NO. DL-S 699266, OCTOBER 1966—REVISED AUGUST 1969

FAST SWITCHING DIODES

- **Rugged Double-Plug Construction**
  Electrical Equivalents
  1N4305 . . . 1N3063 . . . 1N4532
  1N4454 . . . 1N3064

mechanical data

Double-plug construction affords integral positive contact by means of a thermal compression bond. Moisture-free stability is ensured through hermetic sealing. The coefficients of thermal expansion of the glass case and the dumet plugs are closely matched to allow extreme temperature excursions. Hot-solder-dipped leads are standard.

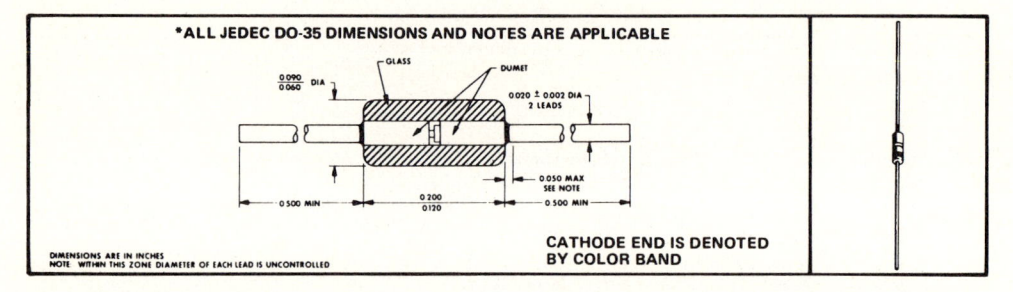

*ALL JEDEC DO-35 DIMENSIONS AND NOTES ARE APPLICABLE

CATHODE END IS DENOTED BY COLOR BAND

*absolute maximum ratings at 25°C free-air temperature (unless otherwise noted)

		1N4305	1N4444	1N4454	UNIT
V_{RM}	Peak Reverse Voltage	75		75	V
$V_{RM(wkg)}$	Working Peak Reverse Voltage		50		V
P	Continuous Power Dissipation at (or below) 25°C Free-Air Temperature (See Note 1)	500			mW
T_{stg}	Storage Temperature Range	−65 to 200			°C
T_L	Lead Temperature 1/16 Inch from Case for 10 Seconds	300			°C

*electrical characteristics at 25°C free-air temperature (unless otherwise noted)

	PARAMETER	TEST CONDITIONS	1N4305 MIN	1N4305 MAX	1N4444 MIN	1N4444 MAX	1N4454 MIN	1N4454 MAX	UNIT
$V_{(BR)}$	Reverse Breakdown Voltage	$I_R = 5\ \mu A$	75		70		75		V
I_R	Static Reverse Current	$V_R = 50\ V$		0.1		0.05		0.1	μA
		$V_R = 50\ V,\ T_A = 150°C$		100		50		100	μA
V_F	Static Forward Voltage	$I_F = 0.1\ mA$			0.44	0.55			V
		$I_F = 0.25\ mA$	0.505	0.575					V
		$I_F = 1\ mA$	0.55	0.65	0.56	0.68			V
		$I_F = 2\ mA$	0.61	0.71					V
		$I_F = 10\ mA$	0.70	0.85	0.69	0.82		1	V
		$I_F = 100\ mA$			0.85	1			V
α_{VF}	Forward Voltage Temperature Coefficient	$I_F = 10\ \mu A$ to 10 mA, See Note 2				3			mV/°C
C_T	Total Capacitance	$V_R = 0,\ f = 1\ MHz$		2		2		2	pF

NOTES: 1. Derate linearly to 200°C at the rate of 2.85 mW/°C.
2. Temperature coefficient, α_{VF}, is determined by the following formula:

$$\alpha_{VF} = \frac{V_F\ @\ 150°C - V_F\ @\ -55°C}{150°C - (-55°C)}$$

* JEDEC registered data

TYPES 1N4305, 1N4444, 1N4454
SILICON SWITCHING DIODES

*operating characteristics at 25°C free-air temperature

PARAMETER	TEST CONDITIONS	1N4305 MIN	1N4305 MAX	1N4444 MIN	1N4444 MAX	1N4454 MIN	1N4454 MAX	UNIT
t_{rr} Reverse Recovery Time	$I_F = 10$ mA, $I_{RM} = 10$ mA, $i_{rr} = 1$ mA, $R_L = 100\ \Omega$, See Figure 1, Condition 1		4		7		4	ns
	$I_F = 10$ mA, $V_R = 6$ V, $i_{rr} = 1$ mA, $R_L = 100\ \Omega$, See Figure 1, Condition 2		2				2	ns
$V_{FM(rec)}$ Forward Recovery Voltage	$I_F = 100$ mA, $R_L = 50\ \Omega$, See Figure 2						3	V
η_r Rectification Efficiency	$V_r = 2$ V, $R_L = 5$ kΩ, $C_L = 20$ pF, $Z_{source} = 50\ \Omega$, $f = 100$ MHz	45 %						

*PARAMETER MEASUREMENT INFORMATION

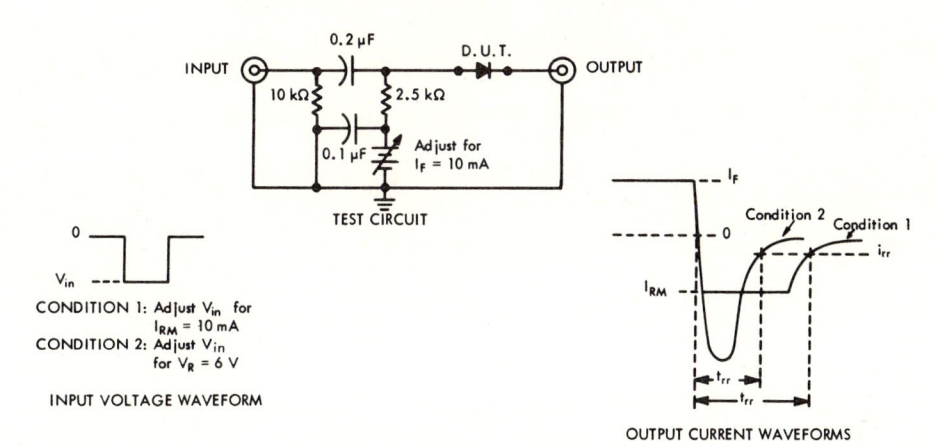

FIGURE 1 – REVERSE RECOVERY TIME

NOTES: a. The input pulse is supplied by a generator with the following characteristics: $Z_{out} = 50\ \Omega$, $t_r \leq 0.5$ ns, $t_p = 100$ ns.
b. Output waveforms are monitored on an oscilloscope with the following characteristics: $t_r \leq 0.6$ ns, $Z_{in} = 50\ \Omega$.

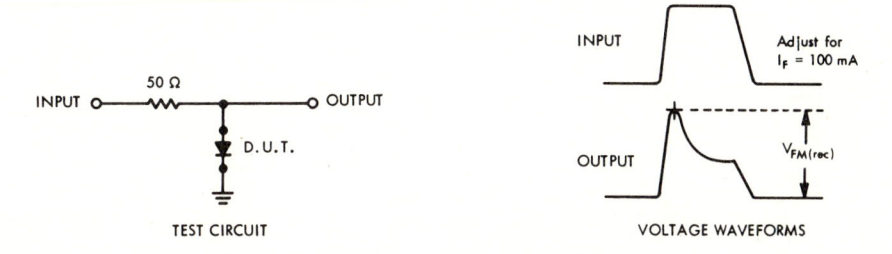

FIGURE 2 – FORWARD RECOVERY VOLTAGE

NOTES: c. The input pulse is supplied by a generator with the following characteristics: $Z_{out} = 50\ \Omega$, $t_r \leq 30$ ns, $t_p = 100$ ns, PRR = 5 to 100 kHz.
d. The output waveform is monitored on an oscilloscope with the following characteristics: $t_r \leq 15$ ns, $R_{in} \geq 1$ MΩ, $C_{in} \leq 5$ pF.

* JEDEC registered data

This information is out of date and parts could possibly be obsolete, reprint for educational purposes only. Courtesy of Texas Instruments, Inc.

TYPES TIV24, TIV25
SILICON VOLTAGE-VARIABLE-CAPACITANCE DIODES

BULLETIN NO. DL-S 7211743, JUNE 1972

VHF TUNING DIODES

- Small Size, Double-Plug Construction
- Extremely Stable and Reliable
- Available in Matched Sets†

mechanical data

Double-plug construction affords integral positive contact by means of a thermal compression bond. Moisture-free stability is ensured through hermetic sealing. The coefficients of thermal expansion of the glass case and the dumet plugs are closely matched to allow extreme temperature excursions. Hot-solder-dipped leads are standard.

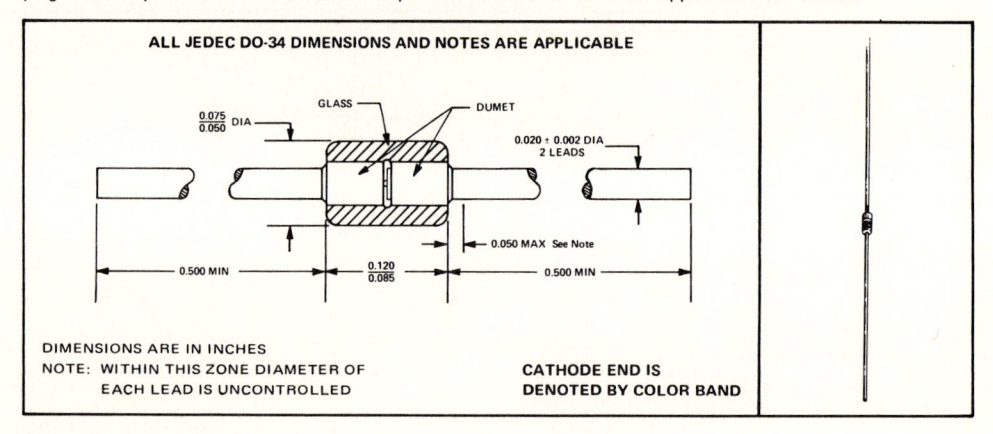

ALL JEDEC DO-34 DIMENSIONS AND NOTES ARE APPLICABLE

DIMENSIONS ARE IN INCHES
NOTE: WITHIN THIS ZONE DIAMETER OF EACH LEAD IS UNCONTROLLED
CATHODE END IS DENOTED BY COLOR BAND

absolute maximum ratings at 25°C free-air temperature (unless otherwise noted)

Peak Reverse Voltage	30 V
Continuous Power Dissipation at (or below) 25°C Free-Air Temperature (see Note 1)	250 mW
Storage Temperature Range	−65°C to 150°C
Lead Temperature 1/16 Inch from Case for 10 Seconds	260°C

electrical characteristics at 25°C free-air temperature

PARAMETER		TEST CONDITIONS	TIV24 MIN	TIV24 MAX	TIV25 MIN	TIV25 MAX	UNIT
$V_{(BR)}$	Breakdown Voltage	$I_R = 10\,\mu A$	30		30		V
I_R	Reverse Current	$V_R = 25\,V$		100		100	nA
C_t	Total Capacitance	$V_R = 3\,V,\ f = 1\,MHz$	22	34	23	34	pF
		$V_R = 25\,V,\ f = 1\,MHz$	5.2	7.5	4.2	6.5	
Q	Figure of Merit (See Note 2)	$V_R = 3\,V,\ f = 100\,MHz$	80		80		
$\dfrac{C_{t1}}{C_{t2}}$	Capacitance Ratio	$V_1 = 3\,V,\ V_2 = 25\,V,\ f = 1\,MHz$	3.5	6	4.5	6	

†The capacitance of diodes in matched sets is matched at all voltages between 3 and 25 volts to within 1.5 % or 0.1 pF, whichever is greater. For ordering matched sets, add dash number to basic part number to indicate the quantity of diodes in the set. For example, TIV24-4 indicates a matched set of 4 diodes.

NOTES: 1. Derate linearly to 150°C at the rate of 2 mW/°C.
2. Figure of Merit, Q, is defined by the equation $Q = \dfrac{1}{2\pi f\, C_t\, r_s}$ where r_s is the equivalent series resistance.

This information is out of date and parts could possibly be absolete, reprint for educational purposes only. Courtesy of Texas Instruments, Inc.

TYPES TIV24, TIV25
SILICON VOLTAGE-VARIABLE-CAPACITANCE DIODES

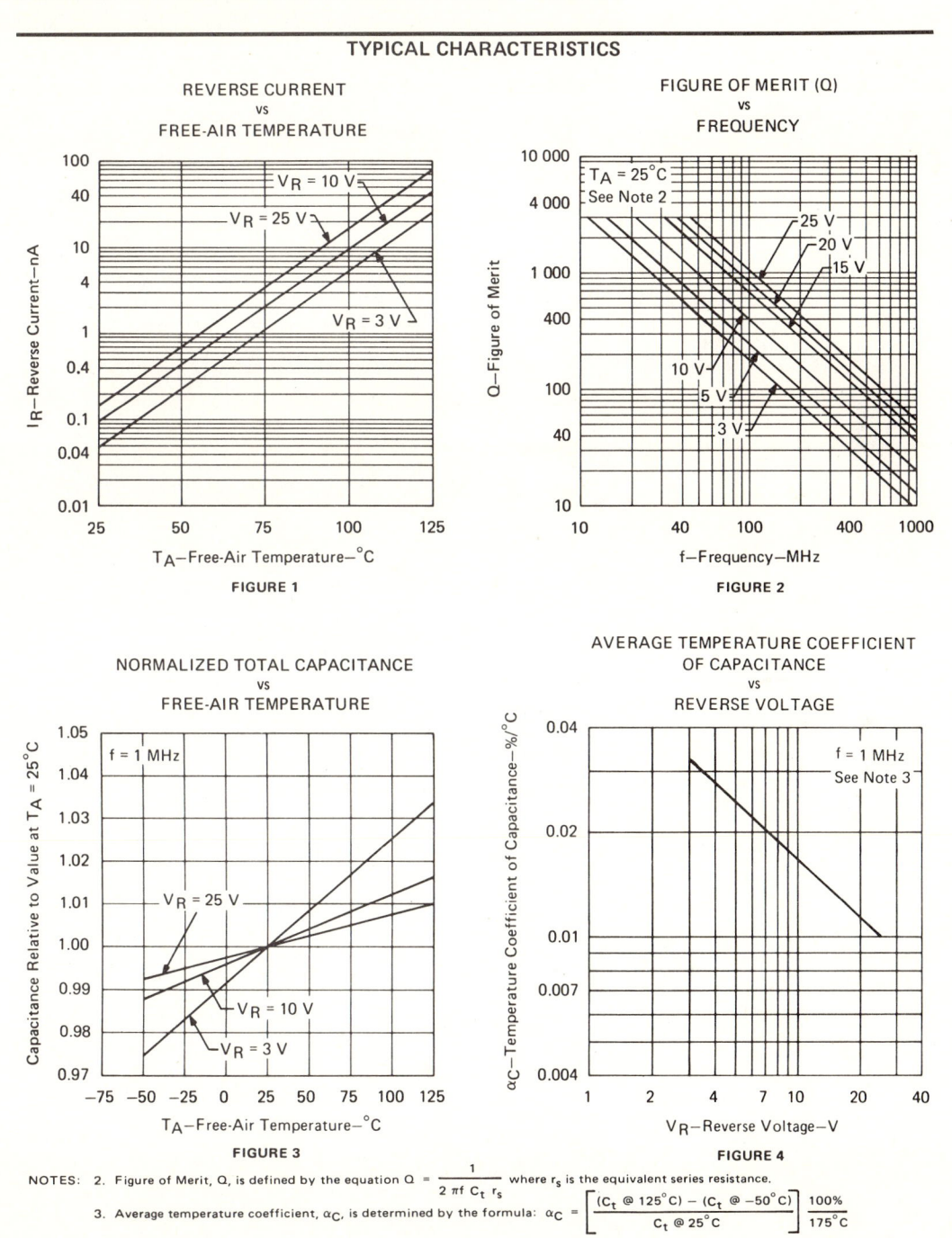

TYPICAL CHARACTERISTICS

REVERSE CURRENT vs FREE-AIR TEMPERATURE

FIGURE 1

FIGURE OF MERIT (Q) vs FREQUENCY

FIGURE 2

NORMALIZED TOTAL CAPACITANCE vs FREE-AIR TEMPERATURE

FIGURE 3

AVERAGE TEMPERATURE COEFFICIENT OF CAPACITANCE vs REVERSE VOLTAGE

FIGURE 4

NOTES: 2. Figure of Merit, Q, is defined by the equation $Q = \dfrac{1}{2\pi f\, C_t\, r_s}$ where r_s is the equivalent series resistance.

3. Average temperature coefficient, α_C, is determined by the formula: $\alpha_C = \left[\dfrac{(C_t\ @\ 125°C) - (C_t\ @\ -50°C)}{C_t\ @\ 25°C}\right] \dfrac{100\%}{175°C}$

This information is out of date and parts could possibly be absolete, reprint for educational purposes only. Courtesy of Texas Instruments, Inc.

Appendix B
Zener Diodes (Voltage Regulator Diodes)

TYPES 1N5226 THRU 1N5257, 1N5226A THRU 1N5257A, 1N5226B THRU 1N5257B
SILICON VOLTAGE-REGULATOR DIODES

BULLETIN NO. DL-S 7311944, MARCH 1973

V_Z ... 3.3 V to 33 V
P_D ... 500 mW

- Available with 5%, 10% and 20% Tolerances
- Rugged Double-Plug Construction

description and mechanical data

These voltage regulator diodes have been designed using the best of both silicon material processing and packaging technologies. The silicon die is a planar oxide-passivated structure which has additional true-glass passivation over the junction. The double-plug package, proven by years of volume production, ensures the best in mechanical integrity and the lowest possible junction temperature when compared to the thermal characteristics of whisker packages. Because of this rugged double-plug (heat-sink) package, these devices offer very conservatively rated power dissipation capabilities.

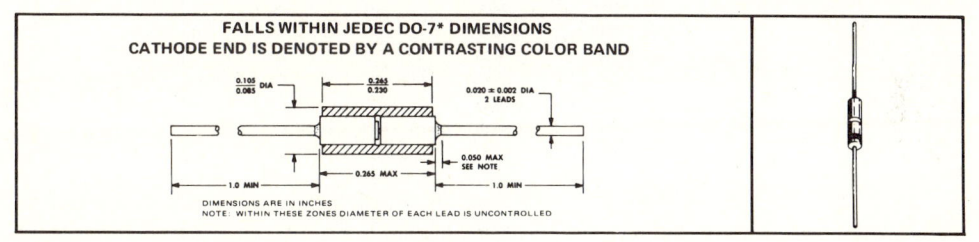

FALLS WITHIN JEDEC DO-7* DIMENSIONS
CATHODE END IS DENOTED BY A CONTRASTING COLOR BAND

*absolute maximum ratings at specified lead temperature

Steady-State Regulator Current, I_{ZM}, at (or below) 75°C	See Table 2
Continuous Power Dissipation at (or below) 75°C (See Note 1)	500 mW
Peak Nonrepetitive Reverse Surge Power at 55°C (See Note 2)	10 W
Operating Lead Temperature Range	−65°C to 200°C
Lead Temperature 1/16 Inch from Case for 10 Seconds	230°C

TABLE 1—STEADY-STATE REGULATOR CURRENT

TYPE	I_{ZM}† (mA)	TYPE	I_{ZM}† (mA)	TYPE	I_{ZM}† (mA)	TYPE	I_{ZM}† (mA)
1N5226, A, B	138	1N5234, A, B	73	1N5242, A, B	38	1N5250, A, B	23
1N5227, A, B	126	1N5235, A, B	67	1N5243, A, B	35	1N5251, A, B	21
1N5228, A, B	115	1N5236, A, B	61	1N5244, A, B	32	1N5252, A, B	19.1
1N5229, A, B	106	1N5237, A, B	55	1N5245, A, B	30	1N5253, A, B	18.2
1N5230, A, B	97	1N5238, A, B	52	1N5246, A, B	28	1N5254, A, B,	16.8
1N5231, A, B	89	1N5239, A, B	50	1N5247, A, B	27	1N5255, A, B	16.2
1N5232, A, B	81	1N5240, A, B	45	1N5248, A, B	25	1N5256, A, B	15.1
1N5233, A, B	76	1N5241, A, B	41	1N5249, A, B	24	1N5257, A, B	13.8

†The nominal I_{ZM} currents shown are applicable for devices having regulator voltages approximately 10% above the nominal V_Z values shown under electrical characteristics. These values do not represent absolute limits. The actual steady-state current-voltage product must not exceed the power rating.

NOTES: 1. Derate linearly to 200°C lead temperature at the rate of 4 mW/°C.
2. This value applies for an 8.3-ms square-wave pulse with the device at nonoperating thermal equilibrium immediately prior to the surge.

This information is out of date and parts could possibly be absolete, reprint for educational purposes only. Courtesy of Texas Instruments, Inc.

TYPES 1N5226 THRU 1N5257, 1N5226A THRU 1N5257A, 1N5226B THRU 1N5257B
SILICON VOLTAGE-REGULATOR DIODES

1N5226A THRU 1N5257A AND 1N5226B THRU 1N5257B

*electrical characteristics at 25°C lead temperature (unless otherwise noted)

PARAMETER	V_Z Regulator Voltage	αV_Z Temperature Coefficient of Regulator Voltage	z_z Small-Signal Regulator Impedance	z_{zk} Small-Signal Regulator Knee Impedance	I_R Static Reverse Current	V_F Static Forward Voltage	$I_{Z(T)}$	$V_{R(T)}$ 1N5226A thru 1N5257A	$V_{R(T)}$ 1N5226B thru 1N5257B
TEST CONDITIONS	$I_R = I_{Z(T)}$, See Note 3	See Note 4	$I_R = I_{Z(T)}$, $I_r = 10\% I_{Z(T)}$, f = 60 Hz	$I_{ZK} = 250\,\mu A$, $I_{zk} = 25\,\mu A$, f = 60 Hz	$V_R = V_{R(T)}$	$I_F = 200$ mA			
LIMIT	NOM§	MAX	MAX	MAX	MAX	MAX			
UNIT	V	%/°C	Ω	Ω	µA	V	mA	V	V
1N5226A, B	3.3	−0.070	28	1600	25	1.1	20	0.95	1.0
1N5227A, B	3.6	−0.065	24	1700	15	1.1	20	0.95	1.0
1N5228A, B	3.9	−0.060	23	1900	10	1.1	20	0.95	1.0
1N5229A, B	4.3	±0.055	22	2000	5	1.1	20	0.95	1.0
1N5230A, B	4.7	±0.030	19	1900	5	1.1	20	1.9	2.0
1N5231A, B	5.1	±0.030	17	1600	5	1.1	20	1.9	2.0
1N5232A, B	5.6	+0.038	11	1600	5	1.1	20	2.9	3.0
1N5233A, B	6.0	+0.038	7	1600	5	1.1	20	3.3	3.5
1N5234A, B	6.2	+0.045	7	1000	5	1.1	20	3.8	4.0
1N5235A, B	6.8	+0.050	5	750	3	1.1	20	4.8	5.0
1N5236A, B	7.5	+0.058	6	500	3	1.1	20	5.7	6.0
1N5237A, B	8.2	+0.062	8	500	3	1.1	20	6.2	6.5
1N5238A, B	8.7	+0.065	8	600	3	1.1	20	6.2	6.5
1N5239A, B	9.1	+0.068	10	600	3	1.1	20	6.7	7.0
1N5240A, B	10	+0.075	17	600	3	1.1	20	7.6	8.0
1N5241A, B	11	+0.076	22	600	2	1.1	20	8.0	8.4
1N5242A, B	12	+0.077	30	600	1	1.1	20	8.7	9.1
1N5243A, B	13	+0.079	13	600	0.5	1.1	9.5	9.4	9.9
1N5244A, B	14	+0.082	15	600	0.1	1.1	9.0	9.5	10
1N5245A, B	15	+0.082	16	600	0.1	1.1	8.5	10.5	11
1N5246A, B	16	+0.083	17	600	0.1	1.1	7.8	11.4	12
1N5247A, B	17	+0.084	19	600	0.1	1.1	7.4	12.4	13
1N5248A, B	18	+0.085	21	600	0.1	1.1	7.0	13.3	14
1N5249A, B	19	+0.086	23	600	0.1	1.1	6.6	13.3	14
1N5250A, B	20	+0.086	25	600	0.1	1.1	6.2	14.3	15
1N5251A, B	22	+0.087	29	600	0.1	1.1	5.6	16.2	17
1N5252A, B	24	+0.088	33	600	0.1	1.1	5.2	17.1	18
1N5253A, B	25	+0.089	35	600	0.1	1.1	5.0	18.1	19
1N5254A, B	27	+0.090	41	600	0.1	1.1	4.6	20	21
1N5225A, B	28	+0.091	44	600	0.1	1.1	4.5	20	21
1N5226A, B	30	+0.091	49	600	0.1	1.1	4.2	22	23
1N5257A, B	33	+0.092	58	700	0.1	1.1	3.8	24	25

§V_Z tolerance is ±10% for 1N5226A thru 1N5257A series; ±5% for 1N5226B thru 1N5257B series. See preceding page for 20%-tolerance devices.

NOTES: 3. V_Z is measured with the device at thermal equilbrium while held in clips at least 3/8 inch from the case in still air at 25°C.

4. Temperature Coefficient $\alpha_{VZ} = \left[\dfrac{(V_Z @ 125°C) - (V_Z @ 25°C)}{V_Z @ 25°C} \right] \times \dfrac{100\%}{125°C - 25°C}$

For determining α_{VZ}, V_Z is measured at 7.5 mA for 1N5226A/1N5226B thru 1N5242A/1N5242B and at I_{ZT} for 1N5243A/1N5243B thru 1N5257A/1N5257B.

This information is out of date and parts could possibly be absolute, reprint for educational purposes only. Courtesy of Texas Instruments, Inc.

TYPES 1N5226 THRU 1N5257, 1N5226A THRU 1N5257A, 1N5226B THRU 1N5257B SILICON VOLTAGE-REGULATOR DIODES

1N5226 THRU 1N5257

*electrical characteristics at 25°C lead temperature

PARAMETER	V_Z Regulator Voltage	I_R Static Reverse Current	V_F Static Forward Voltage	TEST CURRENT AND VOLTAGE	
TEST CONDITIONS	$I_R = I_{Z(T)}$, See Note 3	$V_R = V_{R(T)}$	$I_F = 200$ mA	$I_{Z(T)}$	$V_{R(T)}$
LIMIT	NOM‡	MAX	MAX		
UNIT	V	µA	V	mA	V
1N5226	3.3	100	1.1	20	0.95
1N5227	3.6	100	1.1	20	0.95
1N5228	3.9	75	1.1	20	0.95
1N5229	4.3	50	1.1	20	0.95
1N5230	4.7	50	1.1	20	1.9
1N5231	5.1	50	1.1	20	1.9
1N5232	5.6	50	1.1	20	2.9
1N5233	6.0	50	1.1	20	3.3
1N5234	6.2	50	1.1	20	3.8
1N5235	6.8	30	1.1	20	4.8
1N5236	7.5	30	1.1	20	5.7
1N5237	8.2	30	1.1	20	6.2
1N5238	8.7	30	1.1	20	6.2
1N5239	9.1	30	1.1	20	6.7
1N5240	10	30	1.1	20	7.6
1N5241	11	30	1.1	20	8.0
1N5242	12	10	1.1	20	8.7
1N5243	13	10	1.1	9.5	9.4
1N5244	14	10	1.1	9.0	9.5
1N5245	15	10	1.1	8.5	10.5
1N5246	16	10	1.1	7.8	11.4
1N5247	17	10	1.1	7.4	12.4
1N5248	18	10	1.1	7.0	13.3
1N5249	19	10	1.1	6.6	13.3
1N5250	20	10	1.1	6.2	14.3
1N5251	22	10	1.1	5.6	16.2
1N5252	24	10	1.1	5.2	17.1
1N5253	25	10	1.1	5.0	18.1
1N5254	27	10	1.1	4.6	20
1N5255	28	10	1.1	4.5	20
1N5256	30	10	1.1	4.2	22
1N5257	33	10	1.1	3.8	24

‡V_Z tolerance is ±20% for 1N5226 thru 1N5257. See next page for 5%-tolerance and 10%-tolerance devices.
NOTE 3: V_Z is measured with the device at thermal equilibrium while held in clips at least 3/8 inch from the case in still air at 25°C.
*JEDEC registered data

This information is out of date and parts could possibly be absolete, reprint for educational purposes only. Courtesy of Texas Instruments, Inc.

C
Appendix
Bipolar Transistors

Appendix C 681

TYPES 2N2604, 2N2605
P-N-P SILICON TRANSISTORS
BULLETIN NO. DL-S 7311966, MARCH 1973

FOR LOW-LEVEL, LOW-NOISE, HIGH-GAIN AMPLIFIER APPLICATIONS

- For Complementary Use with 2N929, 2N930, 2N2483, 2N2484, and 2N2586
- Guaranteed h_{FE} at 10 μA, −55°C and 25°C
- Low Noise Characteristics
- Usable at Collector Currents as Low as 1 μA

*mechanical data

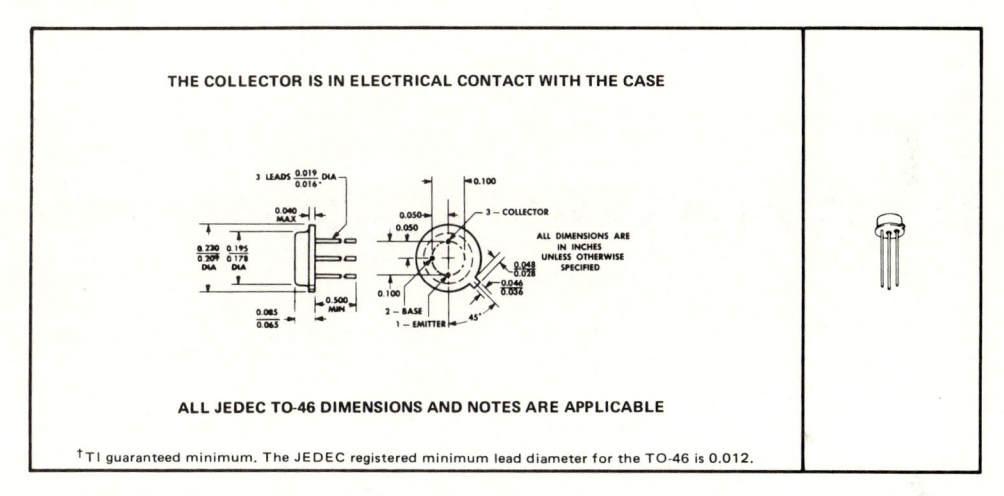

THE COLLECTOR IS IN ELECTRICAL CONTACT WITH THE CASE

ALL DIMENSIONS ARE IN INCHES UNLESS OTHERWISE SPECIFIED

3 – COLLECTOR
2 – BASE
1 – EMITTER

ALL JEDEC TO-46 DIMENSIONS AND NOTES ARE APPLICABLE

†TI guaranteed minimum. The JEDEC registered minimum lead diameter for the TO-46 is 0.012.

*absolute maximum ratings at 25°C free-air temperature (unless otherwise noted)

Collector-Base Voltage	−60 V
Collector-Emitter Voltage (See Note 1)	−45 V
Emitter-Base Voltage	−6 V
Continuous Collector Current	−30 mA
Continuous Device Dissipation at (or below) 25°C Free-Air Temperature (See Note 2)	400 mW
Storage Temperature Range	−65°C to 200°C
Lead Temperature 1/16 Inch from Case for 10 Seconds	230°C

NOTES: 1. This value applies between 0 and 10 mA collector current when the base-emitter diode is open-circuited.
2. Derate linearly to 200°C free-air temperature at the rate of 2.28 mW/°C.

This information is out of date and parts could possibly be absolete, reprint for educational purposes only. Courtesy of Texas Instruments, Inc.

TYPES 2N2604, 2N2605
P-N-P SILICON TRANSISTORS

*electrical characteristics at 25°C free-air temperature (unless otherwise noted)

PARAMETER		TEST CONDITIONS	2N2604 MIN	2N2604 MAX	2N2605 MIN	2N2605 MAX	UNIT
$V_{(BR)CBO}$	Collector-Base Breakdown Voltage	$I_C = -10\,\mu A$, $I_E = 0$	−60		−60		V
$V_{(BR)CEO}$	Collector-Emitter Breakdown Voltage	$I_C = -10\,mA$, $I_B = 0$, See Note 3	−45		−45		V
$V_{(BR)EBO}$	Emitter-Base Breakdown Voltage	$I_E = -10\,\mu A$, $I_C = 0$	−6		−6		V
I_{CBO}	Collector Cutoff Current	$V_{CB} = -45\,V$, $I_E = 0$		−10		−10	nA
I_{CES}	Collector Cutoff Current	$V_{CE} = -45\,V$, $V_{BE} = 0$		−10		−10	nA
		$V_{CE} = -45\,V$, $V_{BE} = 0$, $T_A = 170°C$		−10		−10	μA
I_{EBO}	Emitter Cutoff Current	$V_{EB} = -5\,V$, $I_C = 0$		−2		−2	nA
h_{FE}	Static Forward Current Transfer Ratio	$V_{CE} = -5\,V$, $I_C = -10\,\mu A$	40	120	100	300	
		$V_{CE} = -5\,V$, $I_C = -10\,\mu A$, $T_A = -55°C$	10		20		
		$V_{CE} = -5\,V$, $I_C = -500\,\mu A$	60		150		
		$V_{CE} = -5\,V$, $I_C = -10\,mA$, See Note 3		350		600	
V_{BE}	Base-Emitter Voltage	$I_B = -0.5\,mA$, $I_C = -10\,mA$, See Note 3	−0.7	−0.9	−0.7	−0.9	V
$V_{CE(sat)}$	Collector-Emitter Saturation Voltage	$I_B = -0.5\,mA$, $I_C = -10\,mA$, See Note 3		−0.5		−0.5	V
h_{ib}	Small-Signal Common-Base Input Impedance	$V_{CB} = -5\,V$, $I_E = 1\,mA$, $f = 1\,kHz$	25	35	25	35	Ω
h_{rb}	Small-Signal Common-Base Reverse Voltage Transfer Ratio			10×10^{-4}		10×10^{-4}	
h_{ob}	Small-Signal Common-Base Output Admittance			1		1	μmho
h_{fe}	Small-Signal Common-Emitter Forward Current Transfer Ratio		60	350	150	600	
$|h_{fe}|$	Small-Signal Common-Emitter Forward Current Transfer Ratio	$V_{CE} = -5\,V$, $I_C = -500\,\mu A$, $f = 30\,MHz$	1		1		
C_{obo}	Common-Base Open-Circuit Output Capacitance	$V_{CB} = -5\,V$, $I_E = 0$, $f = 1\,MHz$		6		6	pF
$h_{ie(real)}$	Real Part of Small-Signal Common-Emitter Input Impedance	$V_{CE} = -5\,V$, $I_C = -1\,mA$, $f = 100\,MHz$		200		200	Ω

*operating characteristics at 25°C free-air temperature

PARAMETER		TEST CONDITIONS	2N2604 MIN	2N2604 MAX	2N2605 MIN	2N2605 MAX	UNIT
$\overline{F}$	Average Noise Figure	$V_{CE} = -5\,V$, $I_C = -10\,\mu A$, $R_G = 10\,k\Omega$, Noise Bandwidth = 15.7 kHz, See Note 4		4		3	dB

NOTES: 3. These parameters must be measured using pulse techniques. $t_w = 300\,\mu s$, duty cycle $\leq 2\%$.
4. Average Noise Figure is measured in an amplifier with response down 3 dB at 10 Hz and 10 kHz and a high-frequency roll-off of 6 dB/octave.

This information is out of date and parts could possibly be absolete, reprint for educational purposes only. Courtesy of Texas Instruments, Inc.

Appendix C 683

TYPES 2N3704 THRU 2N3706, A8T3704 THRU A8T3706
N-P-N SILICON TRANSISTORS

BULLETIN NO. DL-S 7311771, JANUARY 1973

SILECT† TRANSISTORS‡

- For Medium-Power Amplifiers, Class B Audio Outputs, Hi-Fi Drivers
- Also Available in Pin-Circle Versions ... 2N5449, 2N5451
- For Complementary Use with 2N3702, 2N3703 or A8T3702, A8T3703

mechanical data

These transistors are encapsulated in a plastic compound specifically designed for this purpose, using a highly mechanized process developed by Texas Instruments. The case will withstand soldering temperatures without deformation. These devices exhibit stable characteristics under high-humidity conditions and are capable of meeting MIL-STD-202C, Method 106B. The transistors are insensitive to light.

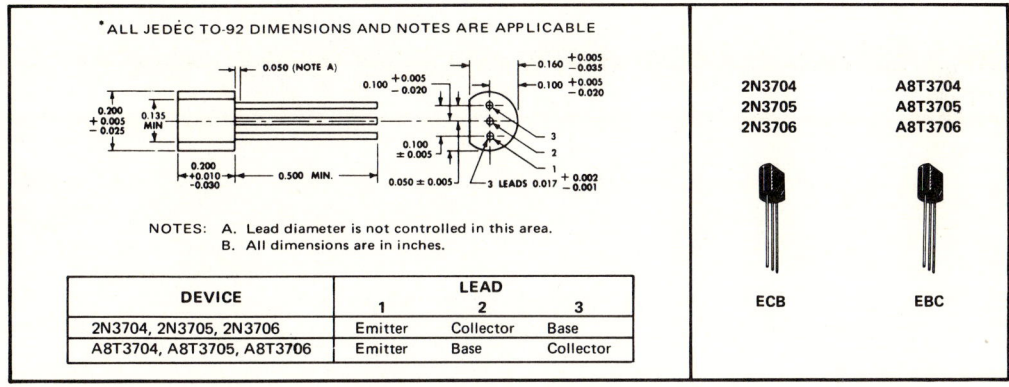

*ALL JEDEC TO-92 DIMENSIONS AND NOTES ARE APPLICABLE

NOTES: A. Lead diameter is not controlled in this area.
B. All dimensions are in inches.

2N3704   A8T3704
2N3705   A8T3705
2N3706   A8T3706

ECB   EBC

DEVICE	LEAD 1	LEAD 2	LEAD 3
2N3704, 2N3705, 2N3706	Emitter	Collector	Base
A8T3704, A8T3705, A8T3706	Emitter	Base	Collector

absolute maximum ratings at 25°C free-air temperature (unless otherwise noted)

	2N3704 / 2N3705 / A8T3704 / A8T3705	2N3706 / A8T3706
Collector-Base Voltage	50 V*	40 V*
Collector-Emitter Voltage (See Note 1)	30 V*	20 V*
Emitter-Base Voltage	5 V*	5 V*
Continuous Collector Current	← 800 mA →	
Continuous Device Dissipation at (or below) 25°C Free-Air Temperature (See Note 2)	← 625 mW§ / 360 mW* →	
Continuous Device Dissipation at (or below) 25°C Lead Temperature (See Note 3)	← 1.25 W§ / 500 mW* →	
Storage Temperature Range	← −65°C to 150°C* →	
Lead Temperature 1/16 Inch from Case for 10 Seconds	← 260°C* →	

NOTES: 1. These values apply when the base-emitter diode is open-circuited.
2. Derate the 625-mW rating linearly to 150°C free-air temperature at the rate of 5 mW/°C. Derate the 360-mW (JEDEC registered) rating linearly to 150°C free-air temperature at the rate of 2.88 mW/°C.
3. Derate the 1.25-W rating linearly to 150°C lead temperature at the rate of 10 mW/°C. Derate the 500-mW (JEDEC registered) rating linearly to 150°C lead temperature at the rate of 4 mW/°C. Lead temperature is measured on the collector lead 1/16 inch from the case.

*The asterisk identifies JEDEC registered data for the 2N3704, 2N3705, and 2N3706 only. This data sheet contains all applicable registered data in effect at the time of publication.
†Trademark of Texas Instruments.
‡U.S. Patent No. 3,439,238.
§Texas Instruments guarantees these values in addition to the JEDEC registered values which are also shown.

This information is out of date and parts could possibly be absolete, reprint for educational purposes only. Courtesy of Texas Instruments, Inc.

TYPES 2N3704 THRU 2N3706, A8T3704 THRU A8T3706
N-P-N SILICON TRANSISTORS

*electrical characteristics at 25°C free-air temperature

PARAMETER		TEST CONDITIONS	2N3704 A8T3704 MIN	MAX	2N3705 A8T3705 MIN	MAX	2N3706 A8T3706 MIN	MAX	UNIT
$V_{(BR)CBO}$	Collector-Base Breakdown Voltage	$I_C = 100\ \mu A$, $I_E = 0$	50		50		40		V
$V_{(BR)CEO}$	Collector-Emitter Breakdown Voltage	$I_C = 10$ mA, $I_B = 0$, See Note 4	30		30		20		V
$V_{(BR)EBO}$	Emitter-Base Breakdown Voltage	$I_E = 100\ \mu A$, $I_C = 0$	5		5		5		V
I_{CBO}	Collector Cutoff Current	$V_{CB} = 20$ V, $I_E = 0$		100		100		100	nA
I_{EBO}	Emitter Cutoff Current	$V_{EB} = 3$ V, $I_C = 0$		100		100		100	nA
h_{FE}	Static Forward Current Transfer Ratio	$V_{CE} = 2$ V, $I_C = 50$ mA, See Note 4	100	300	50	150	30	600	
V_{BE}	Base-Emitter Voltage	$V_{CE} = 2$ V, $I_C = 100$ mA, See Note 4	0.5	1	0.5	1	0.5	1	V
$V_{CE(sat)}$	Collector-Emitter Saturation Voltage	$I_B = 5$ mA, $I_C = 100$ mA, See Note 4		0.6		0.8		1	V
f_T	Transition Frequency	$V_{CE} = 2$ V, $I_C = 50$ mA, See Note 5	100		100		100		MHz
C_{obo}	Common-Base Open-Circuit Output Capacitance	$V_{CB} = 10$ V, $I_E = 0$, $f = 1$ MHz		12		12		12	pF

NOTES: 4. These parameters must be measured using pulse techniques. $t_w = 300\ \mu s$, duty cycle ≤ 2%.
5. To obtain f_T, the $|h_{fe}|$ response with frequency is extrapolated at the rate of −6 dB per octave from $f = 20$ MHz to the frequency at which $|h_{fe}| = 1$.

*The asterisk identifies JEDEC registered data for the 2N3704, 2N3705, and 2N3706 only.

TYPICAL CHARACTERISTICS

2N3705, A8T3705
STATIC FORWARD CURRENT TRANSFER RATIO vs COLLECTOR CURRENT

FIGURE 1

BASE-EMITTER VOLTAGE vs COLLECTOR CURRENT

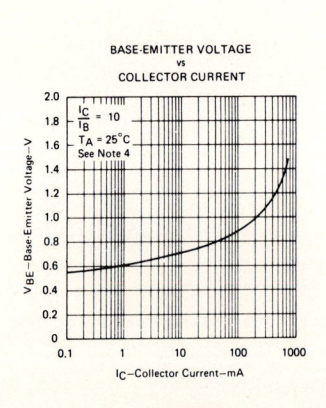

FIGURE 2

COLLECTOR-EMITTER SATURATION VOLTAGE vs COLLECTOR CURRENT

FIGURE 3

This information is out of date and parts could possibly be absolete, reprint for educational purposes only. Courtesy of Texas Instruments, Inc.

D

Appendix

Appendix D

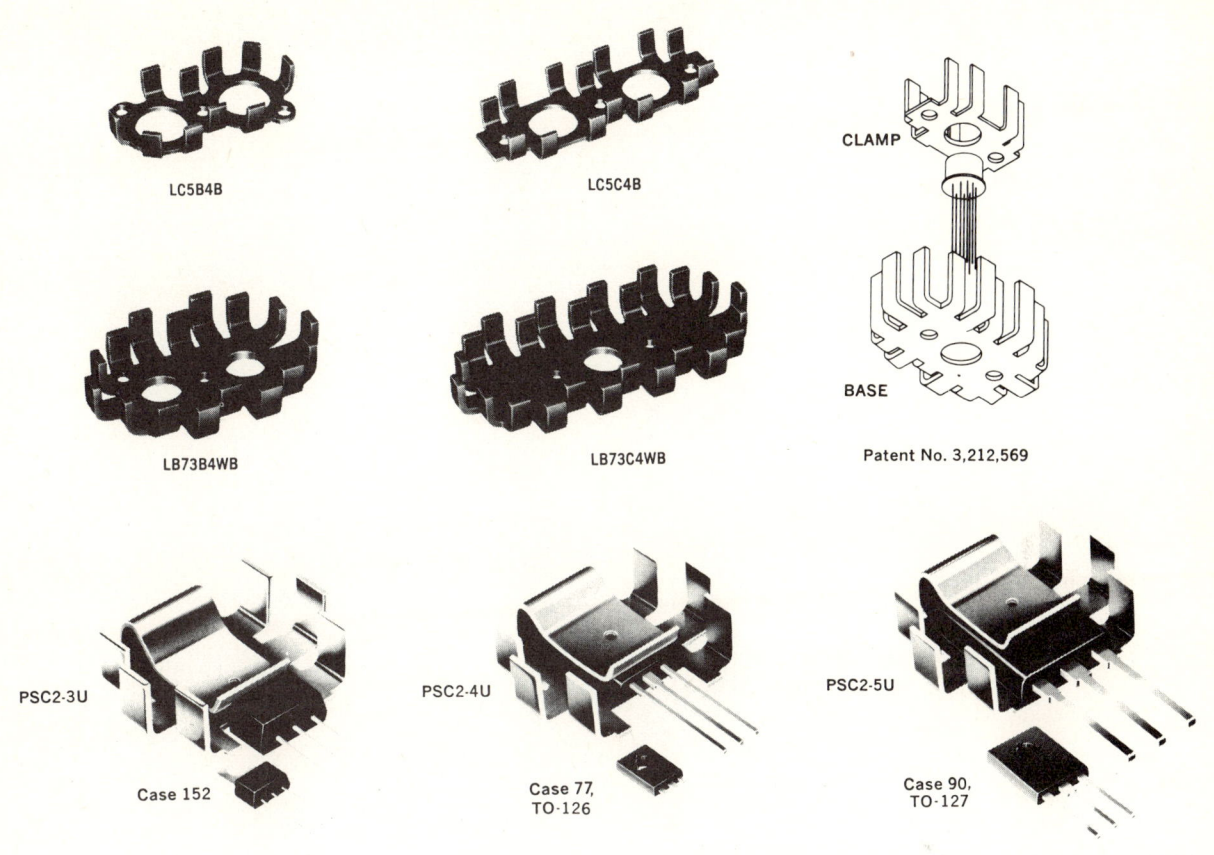

Fig. D-1 Typical Heat Sinks

Heat is one of the biggest problems that must be overcome in the field of solid-state electronics. In many cases, it is necessary to take extreme measures to protect solid-state devices from damage due to overheating. The illustrations shown above are typical of *heat-dissipation devices* used for these purposes. These are only a few of the heat-dissipation devices, commonly called *heat sinks*, that are available.

A heat sink is used to enclose a device so that heat generated in the device flows into the metal heat sink. Once in the metal of the heat sink, the heat is spread over a much greater surface and will not build up to the high temperatures that could exist within the device alone. When heat is a special problem, it may be that two or more heat sinks will be used with a single device as is illustrated by the *clamp* and *base* drawing in the figure above.

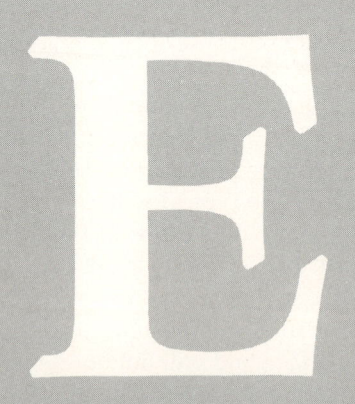

Appendix E
Junction Field-Effect Transistors

TYPE 2N3820
P-CHANNEL SILICON JUNCTION FIELD-EFFECT TRANSISTOR

BULLETIN NO. DL-S 687947, AUGUST 1965–REVISED JULY 1968

SILECT† FIELD-EFFECT TRANSISTOR‡
For Industrial and Consumer Small-Signal Applications

mechanical data

This transistor is encapsulated in a plastic compound specifically designed for this purpose, using a highly mechanized process developed by Texas Instruments. The case will withstand soldering temperatures without deformation. The device exhibits stable characteristics under high-humidity conditions and is capable of meeting MIL-STD-202C, Method 106B. The transistor is insensitive to light.

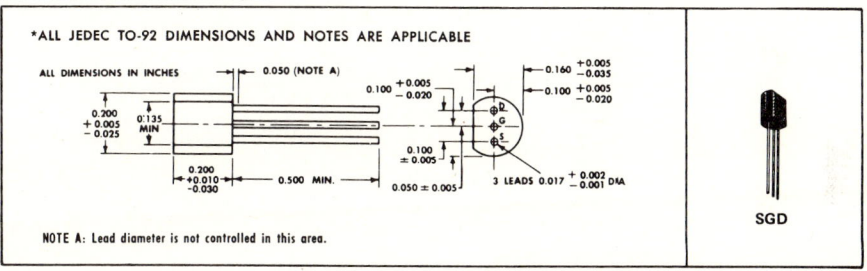

*absolute maximum ratings at 25°C free-air temperature (unless otherwise noted)

Drain-Gate Voltage	−20 v
Drain-Source Voltage	−20 v
Reverse Gate-Source Voltage	20 v
Gate Current	−10 ma
Continuous Device Dissipation at (or below) 25°C Free-Air Temperature (See Note 1) . . .	360 mw
Storage Temperature Range	−65°C to +150°C
Lead Temperature 1/16 Inch from Case for 10 Seconds	260°C

*electrical characteristics at 25°C free-air temperature (unless otherwise noted)

PARAMETER		TEST CONDITIONS	MIN	MAX	UNIT
$V_{(BR)GSS}$	Gate-Source Breakdown Voltage	$I_G = 10\ \mu a$, $V_{DS} = 0$	20		v
I_{GSS}	Gate Cutoff Current	$V_{GS} = 10$ v, $V_{DS} = 0$		20	na
		$V_{GS} = 10$ v, $V_{DS} = 0$, $T_A = 100°C$		2	μa
I_{DSS}	Zero-Gate-Voltage Drain Current	$V_{DS} = -10$ v, $V_{GS} = 0$, See Note 2	−0.3	−15	ma
V_{GS}	Gate-Source Voltage	$V_{DS} = -10$ v, $I_D = -30\ \mu a$	0.3	7.9	v
$V_{GS(off)}$	Gate-Source Cutoff Voltage	$V_{DS} = -10$ v, $I_D = -10\ \mu a$		8	v
$|y_{fs}|$	Small-Signal Common-Source Forward Transfer Admittance	$V_{DS} = -10$ v, $V_{GS} = 0$, $f = 1$ kc, See Note 2	800	5000	μmho
$|y_{os}|$	Small-Signal Common-Source Output Admittance	$V_{DS} = -10$ v, $V_{GS} = 0$, $f = 1$ kc, See Note 2		200	μmho
C_{iss}	Common-Source Short-Circuit Input Capacitance	$V_{DS} = -10$ v, $V_{GS} = 0$, $f = 1$ Mc		32	pf
C_{rss}	Common-Source Short-Circuit Reverse Transfer Capacitance			16	pf
$|y_{fs}|$	Small-Signal Common-Source Forward Transfer Admittance	$V_{DS} = -10$ v, $V_{GS} = 0$, $f = 10$ Mc	700		μmho

NOTES: 1. Derate linearly to 150°C free-air temperature at the rate of 2.88 mw/°C.
        2. These parameters must be measured using pulse techniques. $t_w \approx 100$ ms, duty cycle $\leqslant 10\%$.

*JEDEC registered data
†Trademark of Texas Instruments
‡U.S. Patent No. 3,439,238

This information is out of date and parts could possibly be absolete, reprint for educational purposes only. Courtesy of Texas Instruments, Inc.

TYPES 2N3821 THRU 2N3824
N-CHANNEL SILICON JUNCTION FIELD-EFFECT TRANSISTORS

BULLETIN NO. DL-S 7311919, MARCH 1973

2N3821, 2N3822
FOR SMALL-SIGNAL APPLICATIONS

- Low I_{GSS}: ⩽100 pA
- Low C_{iss}: ⩽6 pF
- High y_{fs}/C_{iss} Ratio (High-Frequency Figure-of-Merit)

2N3823
FOR VHF AMPLIFIER AND MIXER APPLICATIONS

- Low Noise Figure: ⩽2.5 dB at 100 MHz
- Low C_{rss}: ⩽2 pF
- High y_{fs}/C_{iss} Ratio (High-Frequency Figure-of-Merit)

2N3824
FOR HIGH-SPEED COMMUTATOR AND CHOPPER APPLICATIONS

- Low $r_{ds(on)}$: ⩽250 Ω
- Low $I_{D(off)}$: ⩽100 pA
- Low C_{rss}: ⩽3 pF

*mechanical data

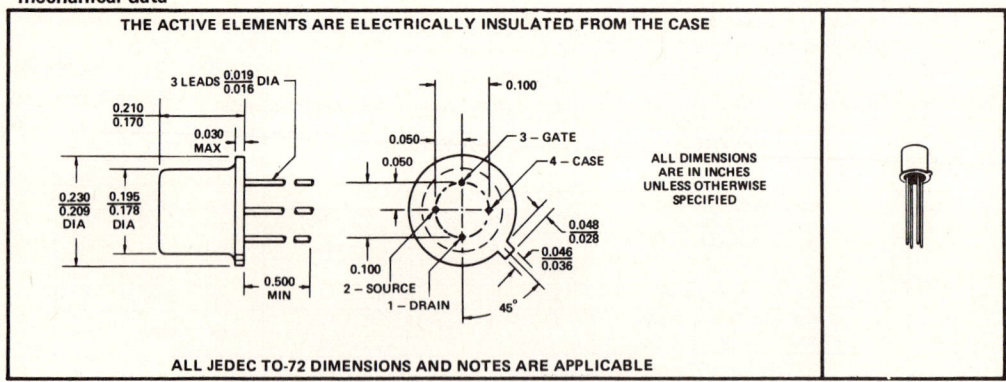

THE ACTIVE ELEMENTS ARE ELECTRICALLY INSULATED FROM THE CASE

ALL DIMENSIONS ARE IN INCHES UNLESS OTHERWISE SPECIFIED

ALL JEDEC TO-72 DIMENSIONS AND NOTES ARE APPLICABLE

*JEDEC registered data. This data sheet contains all applicable registered data in effect at the time of publication.

This information is out of date and parts could possibly be absolete, reprint for educational purposes only. Courtesy of Texas Instruments, Inc.

TYPES 2N3821 THRU 2N3824
N-CHANNEL SILICON JUNCTION FIELD-EFFECT TRANSISTORS

*absolute maximum ratings at 25°C free-air temperature (unless otherwise noted)

	2N3821 2N3822 2N3824	2N3823
Drain-Gate Voltage	50 V	30 V
Drain-Source Voltage	50 V	30 V
Reverse Gate-Source Voltage	−50 V	−30 V
Continuous Forward Gate Current	← 10 mA →	
Continuous Device Dissipation at (or below) 25°C Free-Air Temperature (See Note 1)	← 300 mW →	
Storage Temperature Range	−65°C to 200°C	
Lead Temperature 1/16 Inch from Case for 10 Seconds	← 300°C →	

2N3821, 2N3822

*electrical characteristics at 25°C free-air temperature (unless otherwise noted)

PARAMETER		TEST CONDITIONS†	2N3821 MIN MAX	2N3822 MIN MAX	UNIT
$V_{(BR)GSS}$	Gate-Source Breakdown Voltage	$I_G = -1\ \mu A$, $V_{DS} = 0$	−50	−50	V
I_{GSS}	Gate Cutoff Current	$V_{GS} = -30\ V$, $V_{DS} = 0$	−0.1	−0.1	nA
		$V_{GS} = -30\ V$, $V_{DS} = 0$, $T_A = 150°C$	−0.1	−0.1	μA
$V_{GS(off)}$	Gate-Source Cutoff Voltage	$V_{DS} = 15\ V$, $I_D = 0.5\ nA$	−4	−6	V
V_{GS}	Gate-Source Voltage	$V_{DS} = 15\ V$, $I_D = 50\ \mu A$	−0.5 −2		V
		$V_{DS} = 15\ V$, $I_D = 200\ \mu A$		−1 −4	
I_{DSS}	Zero-Gate-Voltage Drain Current	$V_{DS} = 15\ V$, $V_{GS} = 0$, See Note 2	0.5 2.5	2 10	mA
$|y_{fs}|$	Small-Signal Common-Source Forward Transfer Admittance	$V_{DS} = 15\ V$, $V_{GS} = 0$, $f = 1\ kHz$, See Note 2	1500 4500	3000 6500	μmho
$|y_{os}|$	Small-Signal Common-Source Output Admittance	$V_{DS} = 15\ V$, $V_{GS} = 0$, $f = 1\ kHz$, See Note 2	10	20	μmho
C_{iss}	Common-Source Short-Circuit Input Capacitance	$V_{DS} = 15\ V$, $V_{GS} = 0$, $f = 1\ MHz$	6	6	pF
C_{rss}	Common-Source Short-Circuit Reverse Transfer Capacitance		3	3	pF
$|y_{fs}|$	Small-Signal Common-Source Forward Transfer Admittance	$V_{DS} = 15\ V$, $V_{GS} = 0$, $f = 100\ MHz$	1500	3000	μmho

*operating characteristics at 25°C free-air temperature

PARAMETER		TEST CONDITIONS†	2N3821 2N3822 MAX	UNIT
$\overline{F}$	Average Noise Figure	$V_{DS} = 15\ V$, $V_{GS} = 0$, $R_G = 1\ M\Omega$, $f = 10\ Hz$, Noise Bandwidth = 5 Hz	5	dB
V_n	Equivalent Input Noise Voltage	$V_{DS} = 15\ V$, $V_{GS} = 0$, $f = 10\ Hz$, Noise Bandwidth = 5 Hz	200	$nV/\sqrt{Hz}$

NOTES: 1. Derate linearly to 175°C free-air temperature at the rate of 2 mW/°C.
2. These parameters must be measured using pulse techniques. $t_w = 100\ ms$, duty cycle ≤ 10%.

*JEDEC registered data
†The fourth lead (case) is connected to the source for all measurements.

This information is out of date and parts could possibly be absolute, reprint for educational purposes only. Courtesy of Texas Instruments, Inc.

TYPES 2N3821 THRU 2N3824
N-CHANNEL SILICON JUNCTION FIELD-EFFECT TRANSISTORS

2N3823

*electrical characteristics at 25°C free-air temperature (unless otherwise noted)

PARAMETER		TEST CONDITIONS†	2N3823 MIN	2N3823 MAX	UNIT
$V_{(BR)GSS}$	Gate-Source Breakdown Voltage	$I_G = -1\ \mu A$, $V_{DS} = 0$	−30		V
I_{GSS}	Gate Cutoff Current	$V_{GS} = -20\ V$, $V_{DS} = 0$		−0.5	nA
		$V_{GS} = -20\ V$, $V_{DS} = 0$, $T_A = 150°C$		−0.5	μA
$V_{GS(off)}$	Gate-Source Cutoff Voltage	$V_{DS} = 15\ V$, $I_D = 0.5$ nA		−8	V
V_{GS}	Gate-Source Voltage	$V_{DS} = 15\ V$, $I_D = 400\ \mu A$	−1	−7.5	V
I_{DSS}	Zero-Gate-Voltage Drain Current	$V_{DS} = 15\ V$, $V_{GS} = 0$, See Note 2	4	20	mA
$|y_{fs}|$	Small-Signal Common-Source Forward Transfer Admittance	$V_{DS} = 15\ V$, $V_{GS} = 0$, $f = 1$ kHz, See Note 2	3500	6500	μmho
$|y_{os}|$	Small-Signal Common-Source Output Admittance	$V_{DS} = 15\ V$, $V_{GS} = 0$, $f = 1$ kHz, See Note 2		35	μmho
C_{iss}	Common-Source Short-Circuit Input Capacitance	$V_{DS} = 15\ V$, $V_{GS} = 0$, $f = 1$ MHz		6	pF
C_{rss}	Common-Source Short-Circuit Reverse Transfer Capacitance			2	pF
$|y_{fs}|$	Small-Signal Common-Source Forward Transfer Admittance	$V_{DS} = 15\ V$, $V_{GS} = 0$, $f = 200$ MHz	3200		μmho
g_{is}	Small-Signal Common-Source Input Conductance			800	μmho
g_{os}	Small-Signal Common-Source Output Conductance			200	μmho

*operating characteristics at 25°C free-air temperature

PARAMETER		TEST CONDITIONS†	2N3823 MAX	UNIT
F	Common-Source Spot Noise Figure	$V_{DS} = 15\ V$, $V_{GS} = 0$, $R_G = 1$ kΩ, $f = 100$ MHz	2.5	dB

2N3824
electrical characteristics at 25°C free-air temperature (unless otherwise noted)

PARAMETER		TEST CONDITIONS†	2N3824 MIN	2N3824 MAX	UNIT
*$V_{(BR)GSS}$	Gate-Source Breakdown Voltage	$I_G = -1\ \mu A$, $V_{DS} = 0$	−50		V
*I_{GSS}	Gate Cutoff Current	$V_{GS} = -30\ V$, $V_{DS} = 0$		−0.1	nA
		$V_{GS} = -30\ V$, $V_{DS} = 0$, $T_A = 150°C$		−0.1	μA
*$I_{D(off)}$	Drain Cutoff Current	$V_{DS} = 15\ V$, $V_{GS} = -8\ V$		0.1	nA
		$V_{DS} = 15\ V$, $V_{GS} = -8\ V$, $T_A = 150°C$		0.1	μA
I_{DSS}	Zero-Gate-Voltage Drain Current	$V_{DS} = 15\ V$, $V_{GS} = 0$, See Note 2	12	24	mA
*$r_{ds(on)}$	Small-Signal Drain-Source On-State Resistance	$V_{GS} = 0$, $I_D = 0$, $f = 1$ MHz		250	Ω
*C_{iss}	Common-Source Short-Circuit Input Capacitance	$V_{DS} = 15\ V$, $V_{GS} = 0$, $f = 1$ MHz		6	pF
*C_{rss}	Common-Source Short-Circuit Reverse Transfer Capacitance	$V_{DS} = 0$, $V_{GS} = -8\ V$, $f = 1$ MHz		3	pF

NOTE 2: These parameters must be measured using pulse techniques. $t_w = 100$ ms, duty cycle ≤ 10%.
*JEDEC registered data
†The fourth lead (case) is connected to the source for all measurements.

This information is out of date and parts could possibly be absolete, reprint for educational purposes only. Courtesy of Texas Instruments, Inc.

F

Appendix
Metal-Oxide-Semiconductor Field-Effect Transistors

CHIP TYPE MN82
N-CHANNEL DEPLETION-TYPE
INSULATED-GATE FIELD-EFFECT TRANSISTORS

- MN82 is a 19 X 19-mil, epitaxial, planar, expanded-contact MOS silicon chip
- Available in TO-72 packages
- For use in VHF amplifier circuits

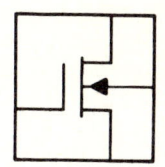

electrical and operating characteristics at 25°C free-air temperature

PARAMETER		CONDITIONS			LOW	TYP	HIGH	UNIT
$V_{(BR)DSV}$	Drain-Source Breakdown Voltage	$I_D = 10\ \mu A,$	$V_{GS} = -8\ V$		20♦	28		V
I_{GSSF}	Forward Gate-Terminal Current	$V_{GS} = 8\ V,$	$V_{DS} = 0$			<1		pA
I_{GSSR}	Reverse Gate-Terminal Current	$V_{GS} = -8\ V,$	$V_{DS} = 0$			−<1	−50	pA
$V_{GS(off)}$	Gate-Source Cutoff Voltage	$V_{DS} = 15\ V,$	$I_D = 50\ \mu A$		−0.8	−1.5	−8	V
I_{DSS}	Zero-Gate-Voltage Drain Current	$V_{DS} = 15\ V,$	$V_{GS} = 0,$	See Note 1	5	10	30	mA
$\|y_{fs}\|$	Small-Signal Common-Source Forward Transfer Admittance	$V_{DS} = 15\ V,$	$I_D = 5\ mA,$	$f = 1\ kHz$	5	10	12	mmho
$\|y_{os}\|$	Small-Signal Common-Source Output Admittance					0.25		mmho
C_{iss}	Common-Source Short-Circuit Input Capacitance	$V_{DS} = 15\ V,$	$I_D = 5\ mA,$	$f = 1\ MHz,$ See Note 2		4		pF
C_{rss}	Common-Source Short-Circuit Reverse Transfer Capacitance					0.3	0.35	pF
C_{oss}	Common-Source Short-Circuit Output Capacitance					1.6		pF
g_{is}	Small-Signal Common-Source Input Conductance					0.2		mmho
b_{is}	Small-Signal Common-Source Input Susceptance					4.5		
g_{fs}	Small-Signal Common-Source Forward Transfer Conductance	$V_{DS} = 15\ V,$	$I_D = 5\ mA,$	$f = 200\ MHz$		10		mmho
b_{fs}	Small-Signal Common-Source Forward Transfer Susceptance					−2		
g_{rs}	Small-Signal Common-Source Reverse Transfer Conductance					0.05		mmho
b_{rs}	Small-Signal Common-Source Reverse Transfer Susceptance					−0.4		
g_{os}	Small-Signal Common-Source Output Conductance					0.25		mmho
b_{os}	Small-Signal Common-Source Output Susceptance					2		
F	Spot Noise Figure	$V_{DS} = 15\ V,$	$I_D = 5\ mA,$	$f = 200\ MHz$			5	dB

♦This value does not modify guaranteed limits for specific devices and does not justify operation in excess of absolute maximum ratings.
CAUTION: The measurement of $V_{(BR)DSV}$ may be destructive.
NOTES: 1. This parameter was measured using pulse techniques. $t_w = 300\ \mu s$, duty cycle ≤ 2%.
2. Capacitance measurements were made using chips mounted in TO-72 packages.

This information is out of date and parts could possibly be absolete, reprint for educational purposes only. Courtesy of Texas Instruments, Inc.

CHIP TYPE MN82
N-CHANNEL DEPLETION-TYPE
INSULATED-GATE FIELD-EFFECT TRANSISTORS

TYPICAL CHARACTERISTICS

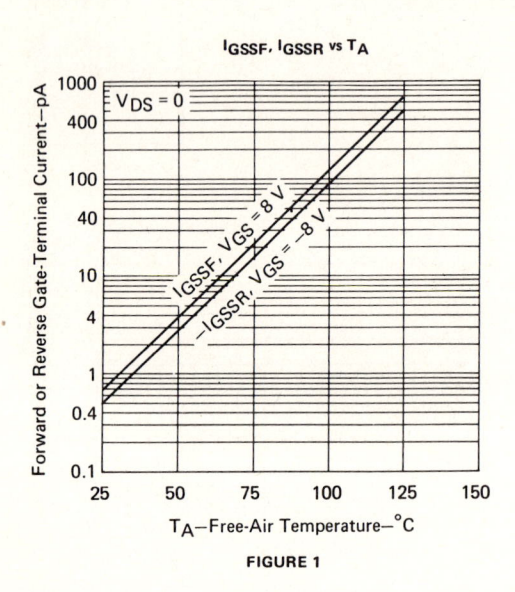

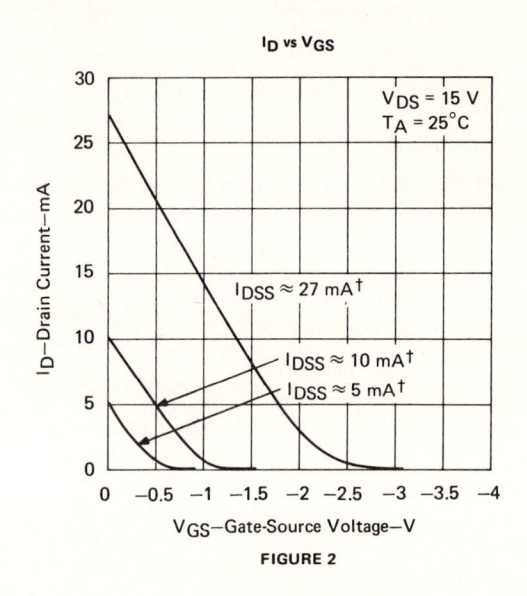

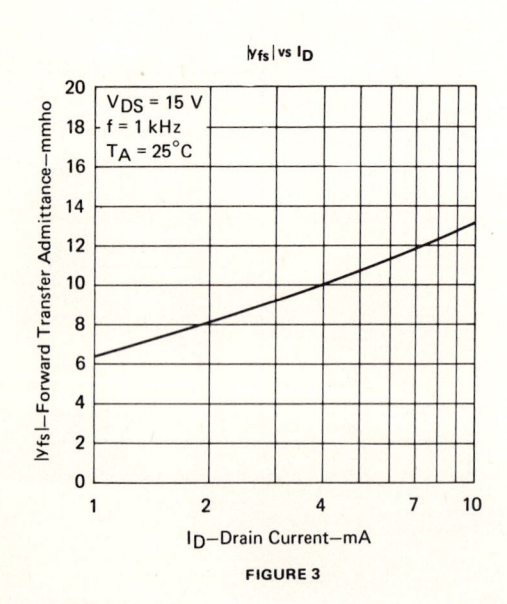

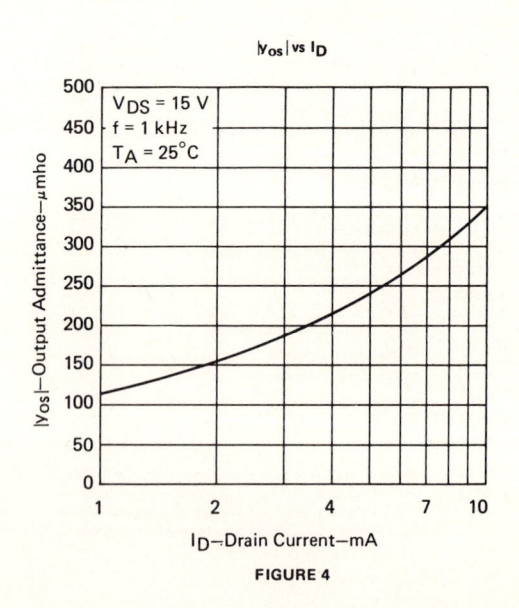

FIGURE 1 — I_{GSSF}, I_{GSSR} vs T_A ($V_{DS} = 0$); curves for I_{GSSF}, $V_{GS} = 8$ V and $-I_{GSSR}$, $V_{GS} = -8$ V.

FIGURE 2 — I_D vs V_{GS} ($V_{DS} = 15$ V, $T_A = 25°C$); curves for $I_{DSS} \approx 27$ mA†, $I_{DSS} \approx 10$ mA†, $I_{DSS} \approx 5$ mA†.

FIGURE 3 — $|y_{fs}|$ vs I_D ($V_{DS} = 15$ V, $f = 1$ kHz, $T_A = 25°C$).

FIGURE 4 — $|y_{os}|$ vs I_D ($V_{DS} = 15$ V, $f = 1$ kHz, $T_A = 25°C$).

†Data is for devices having the indicated value of I_{DSS} at $V_{DS} = 15$ V, $V_{GS} = 0$, and $T_A = 25°C$.

This information is out of date and parts could possibly be absolete, reprint for educational purposes only. Courtesy of Texas Instruments, Inc.

Appendix F

CHIP TYPE MN82
N-CHANNEL DEPLETION-TYPE
INSULATED-GATE FIELD-EFFECT TRANSISTORS

TYPICAL CHARACTERISTICS

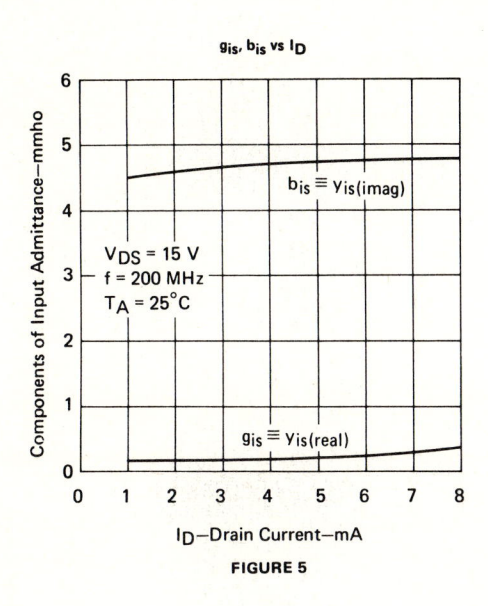

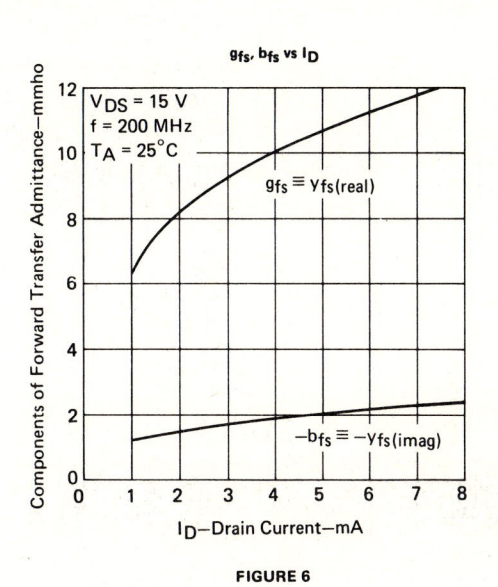

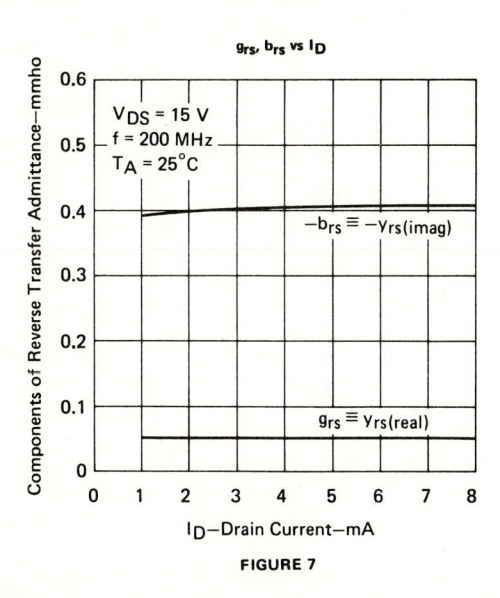

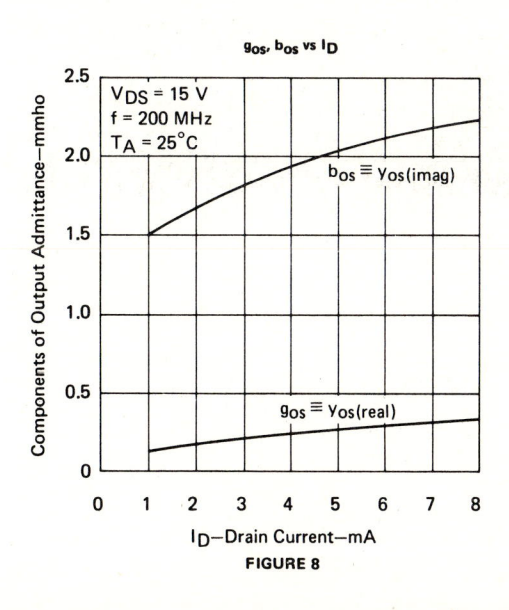

g_{is}, b_{is} vs I_D

$b_{is} \equiv y_{is}(imag)$

V_{DS} = 15 V
f = 200 MHz
T_A = 25°C

$g_{is} \equiv y_{is}(real)$

I_D—Drain Current—mA

FIGURE 5

g_{fs}, b_{fs} vs I_D

V_{DS} = 15 V
f = 200 MHz
T_A = 25°C

$g_{fs} \equiv y_{fs}(real)$

$-b_{fs} \equiv -y_{fs}(imag)$

I_D—Drain Current—mA

FIGURE 6

g_{rs}, b_{rs} vs I_D

V_{DS} = 15 V
f = 200 MHz
T_A = 25°C

$-b_{rs} \equiv -y_{rs}(imag)$

$g_{rs} \equiv y_{rs}(real)$

I_D—Drain Current—mA

FIGURE 7

g_{os}, b_{os} vs I_D

V_{DS} = 15 V
f = 200 MHz
T_A = 25°C

$b_{os} \equiv y_{os}(imag)$

$g_{os} \equiv y_{os}(real)$

I_D—Drain Current—mA

FIGURE 8

This information is out of date and parts could possibly be absolete, reprint for educational purposes only. Courtesy of Texas Instruments, Inc.

CHIP TYPE MN82
N-CHANNEL DEPLETION-TYPE
INSULATED-GATE FIELD-EFFECT TRANSISTORS

TYPICAL CHARACTERISTICS

g_{is}, b_{is} vs f

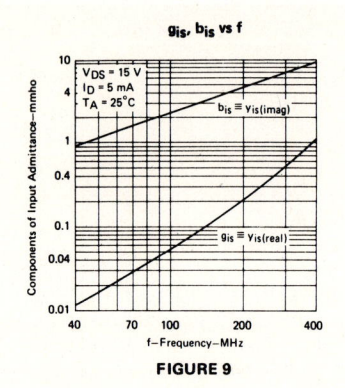

V_{DS} = 15 V
I_D = 5 mA
T_A = 25°C

$b_{is} \equiv y_{is(imag)}$
$g_{is} \equiv y_{is(real)}$

Components of Input Admittance—mmho
f—Frequency—MHz

FIGURE 9

g_{fs}, b_{fs} vs f

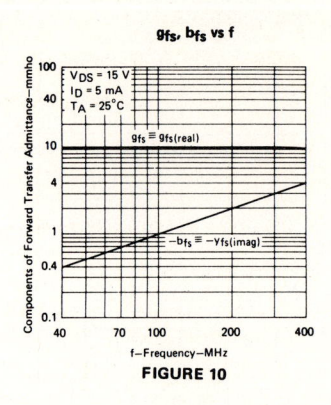

V_{DS} = 15 V
I_D = 5 mA
T_A = 25°C

$g_{fs} \equiv y_{fs(real)}$
$-b_{fs} \equiv -y_{fs(imag)}$

Components of Forward Transfer Admittance—mmho
f—Frequency—MHz

FIGURE 10

g_{rs}, b_{rs} vs f

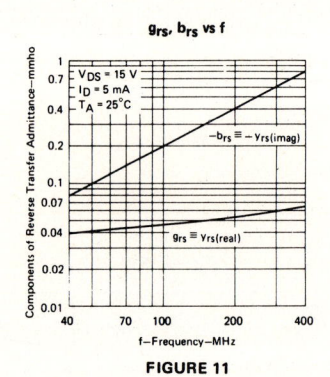

V_{DS} = 15 V
I_D = 5 mA
T_A = 25°C

$-b_{rs} \equiv -y_{rs(imag)}$
$g_{rs} \equiv y_{rs(real)}$

Components of Reverse Transfer Admittance—mmho
f—Frequency—MHz

FIGURE 11

g_{os}, b_{os} vs f

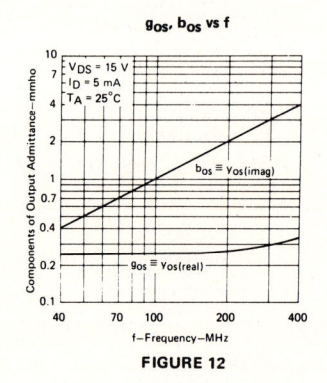

V_{DS} = 15 V
I_D = 5 mA
T_A = 25°C

$b_{os} \equiv y_{os(imag)}$
$g_{os} \equiv y_{os(real)}$

Components of Output Admittance—mmho
f—Frequency—MHz

FIGURE 12

C_{iss} vs I_D

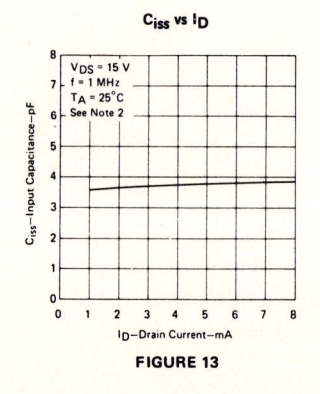

V_{DS} = 15 V
f = 1 MHz
T_A = 25°C
See Note 2

C_{iss}—Input Capacitance—pF
I_D—Drain Current—mA

FIGURE 13

C_{rss} vs I_D

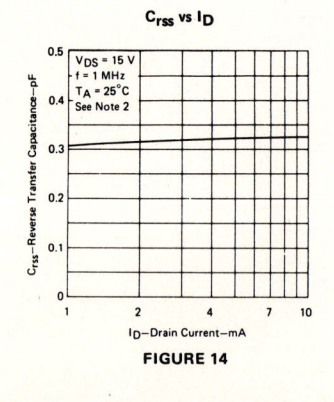

V_{DS} = 15 V
f = 1 MHz
T_A = 25°C
See Note 2

C_{rss}—Reverse Transfer Capacitance—pF
I_D—Drain Current—mA

FIGURE 14

C_{oss} vs I_D

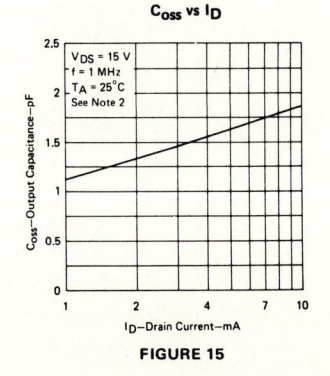

V_{DS} = 15 V
f = 1 MHz
T_A = 25°C
See Note 2

C_{oss}—Output Capacitance—pF
I_D—Drain Current—mA

FIGURE 15

NOTE 2: Capacitance measurements were made using chips mounted in TO-72 packages.

This information is out of date and parts could possibly be obsolete, reprint for educational purposes only. Courtesy of Texas Instruments, Inc.

CHIP TYPE MN83
N-CHANNEL ENHANCEMENT-TYPE
INSULATED-GATE FIELD-EFFECT TRANSISTORS

- MN83 is a 21 X 21-mil, epitaxial, planar, expanded-contact MOS silicon chip
- Available in TO-72 packages
- For use in switching and chopper circuits

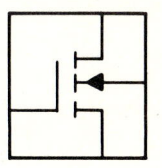

electrical and operating characteristics at 25°C free-air temperature

PARAMETER		CONDITIONS			OBSERVED VALUES LOW	TYP	HIGH	UNIT
$V_{(BR)DSS}$	Drain-Source Breakdown Voltage	$I_D = 10\ \mu A$,	$V_{GS} = 0$		25♦	40		V
I_{GSSF}	Forward Gate-Terminal Current	$V_{GS} = 35\ V$,	$V_{DS} = 0$			<1	10	pA
I_{GSSR}	Reverse Gate-Terminal Current	$V_{GS} = -35\ V$,	$V_{DS} = 0$			−<1	•−10	pA
I_{DSS}	Zero-Gate-Voltage Drain Current	$V_{DS} = 10\ V$,	$V_{GS} = 0$			<1	10	nA
$V_{GS(th)}$	Gate-Source Threshold Voltage	$V_{DS} = 10\ V$,	$I_D = 10\ \mu A$		0.5	1	3	V
$I_{D(on)}$	On-State Drain Current	$V_{DS} = 10\ V$,	$V_{GS} = 10\ V$,	See Note 1	10	150	400	mA
$r_{ds(on)}$	Small-Signal Drain-Source On-State Resistance	$V_{GS} = 10\ V$,	$I_D = 0$,	$f = 1\ kHz$		15	200	Ω
C_{iss}	Common-Source Short-Circuit Input Capacitance	$V_{DS} = 10\ V$, See Note 2	$V_{GS} = 0$,	$f = 1\ MHz$,		4.5	6	pF
C_{rss}	Common-Source Short-Circuit Reverse Transfer Capaciatnce	$V_{DS} = 0$, See Note 2	$V_{GS} = 0$,	$f = 1\ MHz$,		1.1	1.5	pF
$t_{d(on)}$	Turn-On Delay Time	$V_{DD} = 10\ V$,	$I_{D(on)} \approx 10\ mA$,	$R_L = 800\ \Omega$,		1		ns
t_r	Rise Time	$V_{GS(on)} = 10\ V$,	$V_{GS(off)} = 0$,	Figure 1 Circuit		2		
$t_{d(off)}$	Turn-Off Delay Time					3		
t_f	Fall Time					12		

♦This value does not modify guaranteed limits for specific devices and does not justify operation in excess of absolute maximum ratings.
CAUTION: The measurement of $V_{(BR)DSS}$ may be destructive.
NOTES: 1. This parameter was measured using pulse techniques. $t_w = 300\ \mu s$, duty cycle ≤ 2%.
2. Capacitance measurements were made using chips mounted in TO-72 packages.

This information is out of date and parts could possibly be absolete, reprint for educational purposes only. Courtesy of Texas Instruments, Inc.

CHIP TYPE MN83
N-CHANNEL ENHANCEMENT-TYPE
INSULATED-GATE FIELD-EFFECT TRANSISTORS

PARAMETER MEASUREMENT INFORMATION

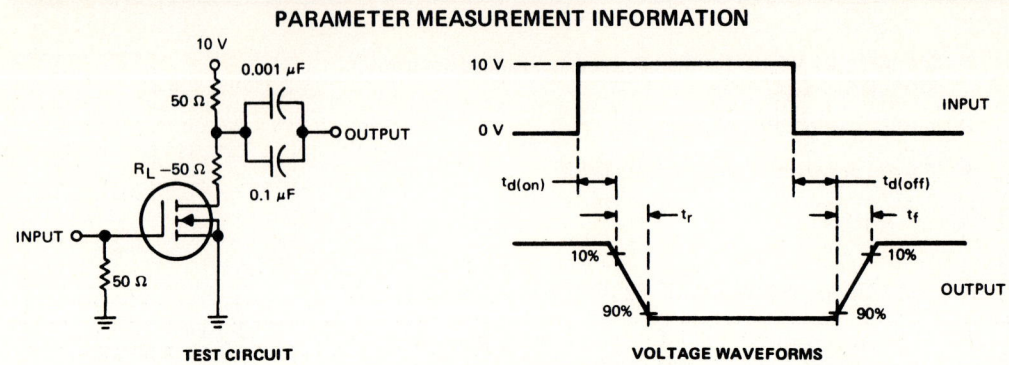

TEST CIRCUIT

VOLTAGE WAVEFORMS

NOTES: a. The input waveform is supplied by a generator with the following characteristics: Z_{out} = 50 Ω, duty cycle ≤ 1%, t_r ≤ 0.33 ns, t_f ≤ 0.33 ns, t_w ≈ 100 ns.
b. Waveforms are monitored on an oscilloscope with the following characteristics: t_r ≤ 0.4 ns, R_{in} = 50 Ω, C_{in} ≤ 2 pF.

FIGURE 1

TYPICAL CHARACTERISTICS

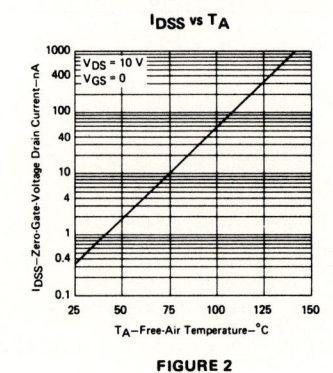

I_{DSS} vs T_A

V_{DS} = 10 V
V_{GS} = 0

I_{DSS} — Zero-Gate-Voltage Drain Current — nA
T_A — Free-Air Temperature — °C

FIGURE 2

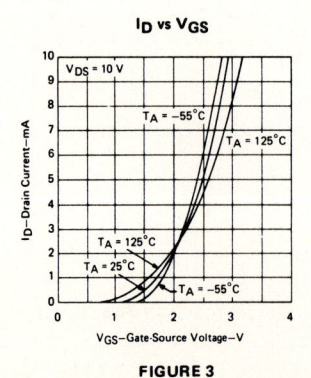

I_D vs V_{GS}

V_{DS} = 10 V

T_A = −55°C
T_A = 125°C
T_A = 125°C
T_A = 25°C
T_A = −55°C

I_D — Drain Current — mA
V_{GS} — Gate-Source Voltage — V

FIGURE 3

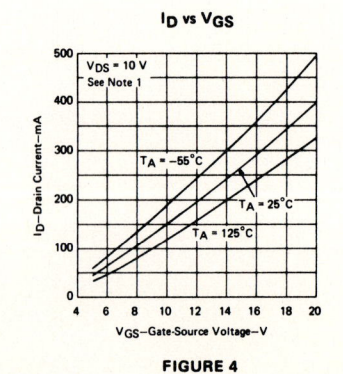

I_D vs V_{GS}

V_{DS} = 10 V
See Note 1

T_A = −55°C
T_A = 25°C
T_A = 125°C

I_D — Drain Current — mA
V_{GS} — Gate-Source Voltage — V

FIGURE 4

NOTE 1: This parameter was measured using pulse techniques. t_w = 300 μs, duty cycle ≤ 2%.

This information is out of date and parts could possibly be absolete, reprint for educational purposes only. Courtesy of Texas Instruments, Inc.

Appendix F 701

**CHIP TYPE MN83
N-CHANNEL ENHANCEMENT-TYPE
INSULATED-GATE FIELD-EFFECT TRANSISTORS**

TYPICAL CHARACTERISTICS

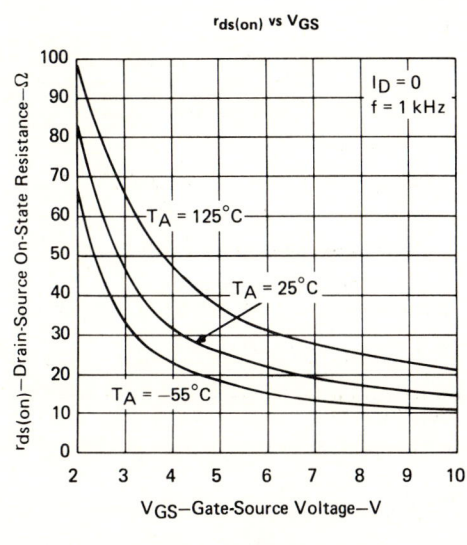

FIGURE 5 — $r_{ds(on)}$ vs V_{GS} ($I_D = 0$, $f = 1$ kHz; curves for $T_A = 125°C$, $T_A = 25°C$, $T_A = -55°C$)

FIGURE 6 — $t_{d(on)}$, t_r, $t_{d(off)}$, t_f vs R_L ($V_{DD} = 10$ V, $V_{GS(on)} = 10$ V, $V_{GS(off)} = 0$ V, $T_A = 25°C$, Figure 1 Circuit)

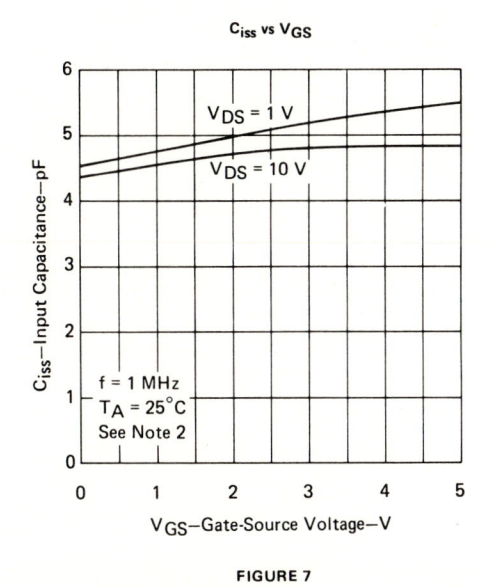

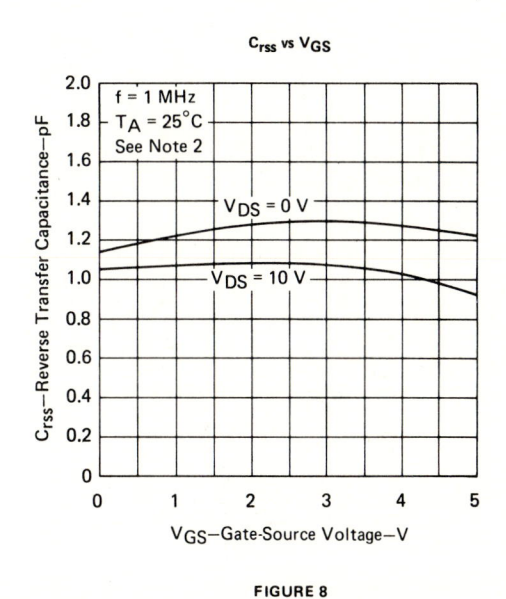

FIGURE 7 — C_{iss} vs V_{GS} ($f = 1$ MHz, $T_A = 25°C$, See Note 2; curves for $V_{DS} = 1$ V and $V_{DS} = 10$ V)

FIGURE 8 — C_{rss} vs V_{GS} ($f = 1$ MHz, $T_A = 25°C$, See Note 2; curves for $V_{DS} = 0$ V and $V_{DS} = 10$ V)

NOTES: 1. This parameter was measured using pulse techniques. $t_w = 300$ μs, duty cycle ⩽ 2%.
2. Capacitance measurements were made using chips mounted in TO-72 packages.

This information is out of date and parts could possibly be absolete, reprint for educational purposes only. Courtesy of Texas Instruments, Inc.

CHIP TYPE MN84
N-CHANNEL DEPLETION-TYPE
INSULATED-GATE FIELD-EFFECT TRANSISTORS

- MN84 is a 21 X 21-mil, epitaxial, planar, expanded-contact MOS silicon chip which has integrated back-to-back diodes between the gate and the substrate
- Available in TO-72 packages
- For low-power switching and chopper circuits

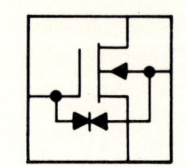

electrical and operating characteristics at 25°C free-air temperature

PARAMETER		CONDITIONS			OBSERVED VALUES LOW	TYP	HIGH	UNIT
$V_{(BR)GSSF}$	Forward Gate-Source Breakdown Voltage	I_G = 1 mA, See Note 1	V_{DS} = 0,	V_{US} = 0,	7♦	10		V
$V_{(BR)GSSR}$	Reverse Gate-Source Breakdown Voltage	I_G = −1 mA, See Note 1	V_{DS} = 0,	V_{US} = 0,	−7♦	−35		V
I_{GSSF}	Forward Gate-Terminal Current	V_{GS} = 7 V,	V_{DS} = 0,	V_{US} = 0		<0.1	10	nA
I_{GSSR}	Reverse Gate-Terminal Current	V_{GS} = −7 V,	V_{DS} = 0,	V_{US} = 0		−<0.1	−10	nA
$I_{S(off)}$	Source Cutoff Current	V_{SD} = 12 V,	V_{GD} = −6 V,	V_{UD} = 0		<0.1	1000	nA
		V_{SD} = 12 V,	V_{GD} = −6 V,	V_{UD} = −6 V		<0.1	1000	
$I_{D(off)}$	Drain Cutoff Current	V_{DS} = 12 V,	V_{GS} = −6 V,	V_{US} = 0		<0.1	100	nA
		V_{DS} = 12 V,	V_{GS} = −6 V,	V_{US} = −6 V		<0.1	100	
I_{USS}	Substrate Reverse Current	V_{US} = −20 V,	V_{DS} = 0,	V_{GS} = 0		−<0.1	−10	nA
$V_{GS(off)}$	Gate-Source Cutoff Voltage	V_{DS} = 12 V,	I_D = 10 μA,	V_{US} = 0	−0.1	−0.75	−1.5	V
I_{DSS}	Zero-Gate-Voltage Drain Current	V_{DS} = 12 V,	V_{GS} = 0,	V_{US} = 0	1	5	12	mA
$I_{D(on)}$	On-State Drain Current	V_{DS} = 3 V, See Note 2	V_{GS} = 6 V,	V_{US} = −6 V	50	100		mA
$r_{ds(on)}$	Small-Signal Drain-Source On-State Resistance	V_{GS} = 6 V, f = 1 kHz	I_D = 0,	V_{US} = 0		18	70	Ω
C_{iss}	Common-Source Short-Circuit Input Capacitance	V_{DS} = 12 V, f = 1 MHz,	V_{GS} = −6 V, See Note 3	V_{US} = 0,		5.5	7	pF
C_{rss}	Common-Source Short-Circuit Reverse Transfer Capacitance	V_{DS} = 0, f = 1 MHz,	V_{GS} = −6 V, See Note 3	V_{US} = 0,		1.4	2	pF
C_{ds}	Drain-Source Capacitance	V_{DS} = 12 V, f = 1 MHz,	V_{GS} = −6 V, See Notes 3 and 4	V_{US} = 0,		3.5	5	pF
$t_{d(on)}$	Turn-On Delay Time	V_{DD} = 12 V, $V_{GS(on)}$ ≈ 6 V, $V_{GS(off)}$ ≈ −2 V, Figure 1 Circuit	$I_{D(on)}$ ≈ 55 mA,	R_L = 200 Ω, V_{US} = −6 V,		1.4		ns
t_r	Rise Time					0.7		
$t_{d(off)}$	Turn-Off Delay Time					2.5		
t_f	Fall Time					4		

♦This value does not modify guaranteed limits for specific devices and does not justify operation in excess of absolute maximum ratings.

NOTES: 1. Both gate breakdown voltages are measured while the device is conducting rated gate current. This ensures that the gate-voltage-limiting network is functioning properly.
2. This parameter was measured using pulse techniques. t_w = 300 μs, duty cycle ≤ 2%.
3. Capacitance measurements were made using chips mounted in TO-72 packages.
4. C_{ds} measurement employs a three-terminal capacitance bridge incorporating a guard circuit. The gate and case are connected to the guard terminal of the bridge.

This information is out of date and parts could possibly be absolete, reprint for educational purposes only. Courtesy of Texas Instruments, Inc.

Appendix G
Electrostatic Discharge Problems

Electrostatic discharge (ESD) is a phenomenon in which an electrostatic charge will accumulate on a person's body or on a particular type of material and discharges to ground. In some cases a person's body can act as a capacitor. In this state it is possible for body capacitance to charge to several thousand volts. It is possible for the current pulse that is present when this charge is discharged to equal several hundred milliamperes and to be present for a few tenths of a microsecond. If this current is allowed to flow through a metallic-oxide semiconductor device, it is highly likely that the device will be destroyed.

Problems that are associated with ESD have long been recognized. In earlier years ESD was not nearly the problem that it is today. With the evolution of MOS devices this problem has grown tremendously. Additionally, the use of plastic cases has added to the problem. It is possible for plastic cases to accumulate an ESD charge of several thousand volts. The cost from ESD damage has become quite large.

The electronics industry has been working very hard to establish guidelines that will assist in minimizing the effect of ESD discharges. It is possible for ESD discharge damage to occur at any time. Many home computer owners have experienced this trouble. A sudden ESD discharge can do extensive damage to equipment of this type. Static control blankets have been developed that allow the operator and computer to be sitting on a static-free carpet. Technicians who work with ground straps attached to their body and their workbenches are common. Many efforts are being made to assure that all components of the electronics manufacturing, and maintenance operations are conscious of ESD and act to minimize its effect.

The approach you take toward ESD must be influenced by three considerations. These are especially important to the technician who repairs MOS-type circuitry.

First, static electricity is always present. The amount of static electricity present can be influenced by our day-to-day actions. Factors that must be considered are shown in Table G-1.

Table G-1

Means of Static Generation	Charges (in volts) Relative Humidity Low 10-20%	High 65-90%
Walking across carpet	35,000	1500
Walking across vinyl floor	12,000	250
Workers at bench	6,000	100
Vinyl envelopes used to hold working instructions	7,000	600
Common poly bag picked up from bench	20,000	1200
Work chair or stool padded with urethane padding	18,000	1500

Most of us have little knowledge regarding the static hazard associated with our actions. Just walking across a room can create a charge large enough to destroy an expensive MOS device. These charges can build up and discharge into an electronic device where it causes damage. This damage may not be immediately apparent but results in a change in operating parameters that cause operations below desired standards and maybe eventual breakdown.

Second, we must consider the work environment. Many of the materials we work with and around every day can influence static electricity conditions. Some of these are shown in Table G-2.

Table G-2

Work Surfaces	Melamine Laminates (waxed or highly resistive) Finished Wood Synthetic Mats
Floors	Waxed Vinyl
Chairs	Fiberglass Finished Wood Vinyl
Clothing	Common clean room smocks Personal Clothing (other than virgin cotton) Non-conductive shoes
Packaging and Handling	Common-plastic - bags, wraps, envelopes Common bubble pack, foam Common plastic trays, plastic tote boxes, vials, parts bins, etc.
Assembly, Cleaning, Test, and Repair Areas	Spray cleaners Common solder suckers Common solder irons Solvent brushing (synthetic brushes) Cleaning, Drying sand blasting Temperature chambers Cryogenic sprays Heat guns and blowers Electrostatic copiers

Third, many of the components we work with can be damaged by levels of static electricity that are below our sensing level. For humans, the sensing level is approximately 3000 volts of static electricity. Table G-3 illustrates the susceptibility of some devices to ESD damage.

Table G-3

MOSFET	100 V – 200 V
JFET	140 V – 10,000 V
CMOS	250 V – 2,000 V
SCHOTTKY Diodes, TTL	300 V – 2,500 V
Bipolar Transistors	380 V – 7,000 V
SCR	680 V – 1,000 V

Remember, even the packing and handling of the devices can cause static build-up. We must take extra precautions to protect sensitive components at all steps in the process. Manufacturers conduct all their operations under conditions specifically designed to minimize ESD problems. We must do the same with our laboratories and repair facilities.

Many of the failures we credit to shelf life or manufacturing defect are quite likely caused by ESD during handling. Many ESD failures are not recognized as such because they may have been masked by some other problem.

Unfortunately, our problems with ESD are not getting fewer, nor are they likely to in the near future. Many facilities incur the high cost of spare parts and repairs which could be greatly reduced with an effective ESD control program. Without careful attention, the ESD problem is expected to grow during the next few years. As devices grow smaller, the ESD problems associated with them will grow. The use of reasonable ESD prevention techniques must become part of the engineer's and technician's work habits.

RCA cites four "cardinal *thou shall* rules" that must be followed to minimize the effect of electrostatic discharge. Thou shall:

1. Assume that *all* components are sensitive to ESD damage.
2. Not touch a sensitive component unless properly grounded.
3. Not transport, store, or handle sensitive components except in a "static free" environment.
4. Test static-sensitive components only in ESD protected areas.

With these four cardinal rules in mind, the following general guidelines should be used to perform service and/or field repairs.

1. Turn off **all** power to the equipment being serviced.
2. Do not use just any canned coolant to isolate troubles. Use antistatic spray coolant **only**.
3. Touch the grounded chassis with your hand prior to removing or inserting any ESD sensitive device.

4. **Do not** probe or test ESD sensitive devices with a voltmeter unless absolutely necessary. When using test probes for in-circuit testing, ground the meter and test leads prior to touching the terminals of the device.
5. Ground the ESD protective package containing replacement devices to the grounded chassis before opening the package. This will dissipate any accumulated charge that exists on the package.
6. Remove the ESD sensitive device from its package and install it in the equipment. **Avoid touching parts, connections, and circuitry**.

Warning: Testing of solid-state devices should be performed by experienced persons using test equipment designed for such purposes. Typical test instruments include ohmmeters, transistor testers and curve tracers. Line voltage should not be applied directly to a device undergoing a test.

Further Operating Considerations

COS/MOS Operations

Operating Voltage
During operation near the maximum supply voltage limit, care should be taken to avoid or suppress power supply turn-on and turn-off transients, power supply ripple, or ground noise.

Input Signals
To prevent damage to the input protection circuit, input signals should never exceed the value for V_{DD} nor less than the value of V_{SS}. Input currents must not exceed 10 mA—even when the power supply is off.

Unused Inputs
A connection must be provided at every input terminal. All unused input terminals must be connected to either V_{DD} or V_{SS}, whichever is appropriate.

Output Short Circuits
Shorting of outputs to V_{DD} or V_{SS} may damage COS/MOS devices by exceeding the maximum device dissipation.

THIS APPENDIX WAS COMPILED FROM INFORMATION PROVIDED BY RCA Corporation, Solid-State Division, Route 202, Sommerville, NJ 08876.

H Appendix
PN Unijunction Transistors and Programmable Unijunction Transistors

Appendix H

TYPES 2N4851, 2N4852, 2N4853
P-N UNIJUNCTION SILICON TRANSISTORS

BULLETIN NO. DL-S 7311967, MARCH 1973

PLANAR UNIJUNCTION TRANSISTORS SPECIFICALLY CHARACTERIZED FOR A WIDE RANGE OF MILITARY, SPACE, AND INDUSTRIAL APPLICATIONS

- Planar Process Ensures Low Leakage, High-Performance With Low Driving Currents, and Greatly Improved Reliability

*mechanical data

Package outline is same as JEDEC TO-18 except for lead position. All TO-18 registration notes also apply to this outline.

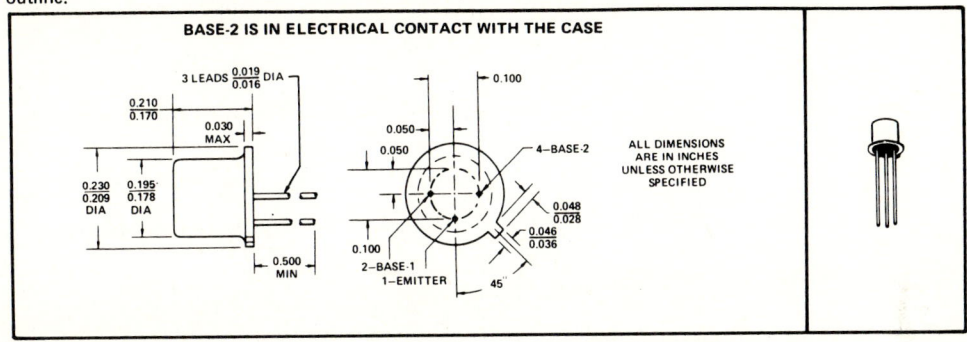

BASE-2 IS IN ELECTRICAL CONTACT WITH THE CASE

*absolute maximum ratings at 25°C free-air temperature (unless otherwise noted)

Emitter–Base-Two Reverse Voltage	–30 V
Interbase Voltage (See Note 1)	35 V
Continuous Emitter Current	50 mA
Peak Emitter Current (See Note 2)	1.5 A
Continuous Device Dissipation at (or below) 25°C Free-Air Temperature (See Note 3)	300 mW
Storage Temperature Range	–65°C to 200°C
Lead Temperature 1/16 Inch from Case for 10 Seconds	260°C

*electrical characteristics at 25°C free-air temperature (unless otherwise noted)

	PARAMETER	TEST CONDITIONS	2N4851 MIN	2N4851 MAX	2N4852 MIN	2N4852 MAX	2N4853 MIN	2N4853 MAX	UNIT
r_{BB}	Static Interbase Resistance	V_{B2B1} = 3 V, I_E 0	4.7	9.1	4.7	9.1	4.7	9.1	kΩ
α_{rBB}	Interbase Resistance Temperature Coefficient	V_{B2B1} = 3 V, I_E = 0, T_A = –65°C to 125°C, See Note 4	0.2	0.8	0.2	0.8	0.2	0.8	%/°C
η	Intrinsic Standoff Ratio	V_{B2B1} = 10 V, See Figure 3	0.56	0.75	0.7	0.85	0.7	0.85	
I_{EB2O}	Emitter Reverse Current	V_{EB2} = 30 V, I_{B1} = 0		100		100		50	nA
I_P	Peak-Point Emitter Current	V_{B2B1} = 25 V		2		2		0.4	µA
I_V	Valley-Point Emitter Current	V_{B2B1} = 25 V	2		4		6		mA
V_{OB1}	Base-One Peak Pulse Voltage	See Figure 4	3		5		6		V
f_{max}	Maximum Frequency of Oscillation	See Figure 5	1		1		1		MHz

NOTES:
1. The interbase voltage rating is based upon allowable power dissipation: $V_{B2B1} = \sqrt{r_{BB} \cdot P_T}$.
2. The peak emitter current rating is based on the capability of the transistor to operate safely in the circuit of Figure 4.
3. Derate linearly to 125°C free-air temperature at the rate of 3 mW/°C.
4. Temperature coefficient α_{rBB} is determined by the following formula:

$$\alpha_{rBB} = \left[\frac{(r_{BB} @ 125°C) - (r_{BB} @ -65°C)}{r_{BB} @ 25°C}\right] \frac{100\%}{190°C}$$

To obtain r_{BB} for a given temperature $T_{A(2)}$, use the following formula:

$$r_{BB(2)} = [r_{BB} @ 25°C] \, [1 + (\alpha_{rBB}/100\%)(T_{A(2)} - 25°C)]$$

*JEDEC registered data. This data sheet contains all applicable data in effect at the time of publication.

This information is out of date and parts could possibly be absolete, reprint for educational purposes only. Courtesy of Texas Instruments, Inc.

TYPES 2N4851, 2N4852, 2N4853
P-N UNIJUNCTION SILICON TRANSISTORS

*PARAMETER MEASUREMENT INFORMATION

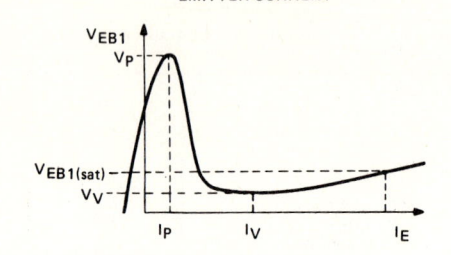

FIGURE 1—GENERAL STATIC EMITTER CHARACTERISTIC CURVE

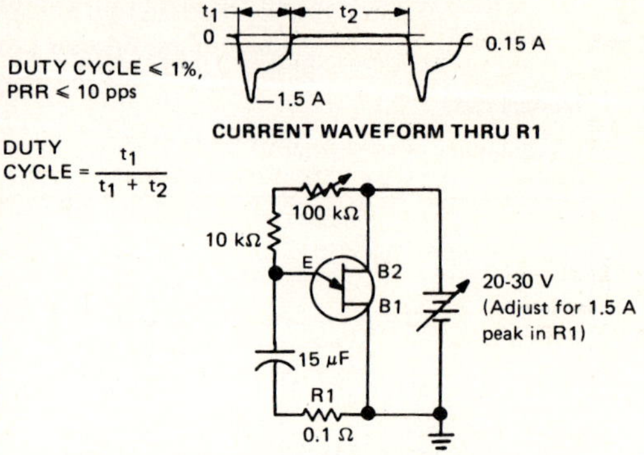

DUTY CYCLE ≤ 1%, PRR ≤ 10 pps

$$\text{DUTY CYCLE} = \frac{t_1}{t_1 + t_2}$$

CURRENT WAVEFORM THRU R1

FIGURE 2—PEAK-EMITTER-CURRENT TEST CIRCUIT AND WAVEFORM

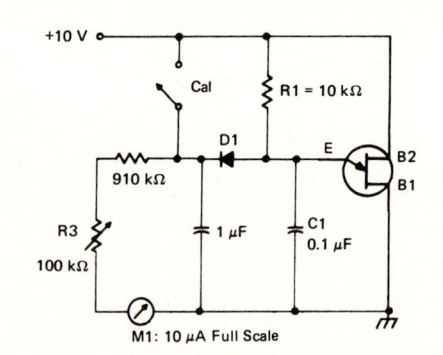

η—Intrinsic Standoff Ratio—This parameter is defined in terms of the peak-point voltage, V_P, by means of the equation: $V_P = \eta \, V_{B2B1} + V_F$, where V_F is about 0.49 volt at 25°C and decreases with temperature at about 2 millivolts/°C.

The circuit used to measure η is shown in the figure. In this circuit, R1, C1 and the unijunction transistor form a relaxation oscillator, and the remainder of the circuit serves as a peak-voltage detector with the diode D1 automatically subtracting the voltage V_P. To use the circuit, the "cal" button is pushed, and R3 is adjusted to make the current meter M1 read full scale. The "cal" button then is released and the value of η is read directly from the meter, with N = 1 corresponding to full-scale deflection of 10 µA.
D1: 1N457, or equivalent, with the following characteristics:
V_F = 0.49 V at I_F = 10 µA
I_R ≤ 2 µA at V_R = 20 V

FIGURE 3—TEST CIRCUIT FOR INTRINSIC STANDOFF RATIO (η)

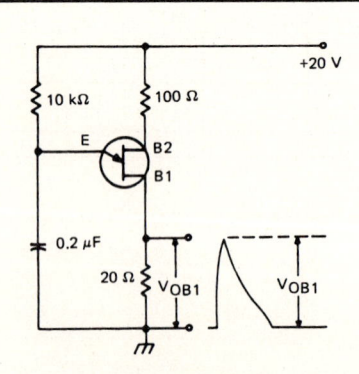

FIGURE 4—V_{OB1} TEST CIRCUIT

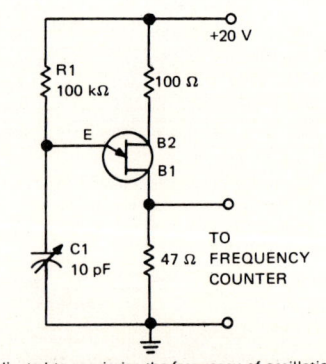

R1 and C1 are adjusted to maximize the frequency of oscillation.
FIGURE 5—f_{max} TEST CIRCUIT

*JEDEC registered data

This information is out of date and parts could possibly be absolete, reprint for educational purposes only. Courtesy of Texas Instruments, Inc.

Appendix H 713

TYPES 2N6116, 2N6117, 2N6118
P-N-P-N SILICON PROGRAMMABLE UNIJUNCTION TRANSISTORS
BULLETIN NO. DL-S 7211776, DECEMBER 1972

- For Use in Pulse, Timing, Sweep, Trigger, and Oscillator Circuits
- Features Low Peak-Point Current and Low Forward Voltage
- Programmable η, r_{BB}, I_P, and I_V

mechanical data

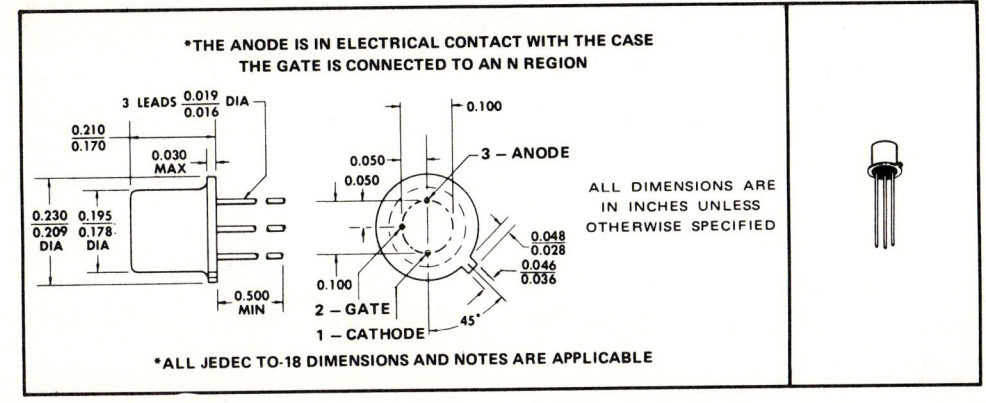

*THE ANODE IS IN ELECTRICAL CONTACT WITH THE CASE
THE GATE IS CONNECTED TO AN N REGION

ALL DIMENSIONS ARE IN INCHES UNLESS OTHERWISE SPECIFIED

3 – ANODE
2 – GATE
1 – CATHODE

*ALL JEDEC TO-18 DIMENSIONS AND NOTES ARE APPLICABLE

*absolute maximum ratings at 25°C free-air temperature (unless otherwise noted)

Anode-Cathode Voltage	±40 V
Gate-Anode Voltage	40 V
Gate-Cathode Voltage: (Positive Limit)	40 V
(Negative Limit)	−5 V
Continuous Anode Current at (or below) 25°C Free-Air Temperature (See Note 1)	200 mA
Repetitive Peak Anode Current: (t_W = 100 μs, Duty Cycle ≤ 1%)	1 A
(t_W = 20 μs, Duty Cycle ≤ 1%)	2 A
Nonrepetitive Peak Anode Current: (t_W = 10 μs, Duty Cycle = 0)	5 A
Continuous Gate Current	±20 mA
Continuous Device Dissipation at (or below) 25°C Free-Air Temperature (See Note 2)	250 mW
Storage Temperature Range	−65°C to 200°C
Lead Temperature 1/16 Inch from Case for 10 Seconds	260°C

*electrical characteristics at 25°C free-air temperature (unless otherwise noted)

PARAMETER		TEST CONDITIONS	2N6116 MIN	2N6116 MAX	2N6117 MIN	2N6117 MAX	2N6118 MIN	2N6118 MAX	UNIT
I_{GAO}	Gate Reverse Current	V_{GA} = 40 V, I_K = 0		5		5		5	nA
		V_{GA} = 40 V, I_K = 0, T_A = 75°C		75		75		75	
I_{GKS}	Gate Reverse Current	V_{GK} = 40 V, V_{AK} = 0		50		50		50	nA
V_P–V_S	Offset Voltage	V_S = 10 V, R_G = 10 kΩ	0.2	0.6	0.2	0.6	0.2	0.6	V
		V_S = 10 V, R_G = 1 MΩ	0.2	1.6	0.2	0.6	0.2	0.6	
I_P	Peak-Point Current	V_S = 10 V, R_G = 10 kΩ		5		2		1	μA
		V_S = 10 V, R_G = 1 MΩ		2		0.3		0.15	
I_V	Valley-Point Current	V_S = 10 V, R_G = 10 kΩ	70		50		50		μA
		V_S = 10 V, R_G = 1 MΩ	50		50		25		
V_F	Anode-Cathode On-State Voltage	V_S = 10 V, R_G = 10 kΩ, I_F = 50 mA		1.5		1.5		1.5	V

NOTES: 1. Derate linearly to 125°C free-air temperature at the rate of 2 mA/°C.
2. Derate linearly to 125°C free-air temperature at the rate of 2.5 mW/°C.

*JEDEC registered data. This data sheet contains all applicable registered data in effect at the time of publication.

USES CHIP U41

This information is out of date and parts could possibly be absolute, reprint for educational purposes only. Courtesy of Texas Instruments, Inc.

TYPES 2N6116, 2N6117, 2N6118
P-N-P-N SILICON PROGRAMMABLE UNIJUNCTION TRANSISTORS

*operating characteristics at 25°C free-air temperature

PARAMETER		TEST CONDITIONS	2N6116 MIN MAX	2N6117 MIN MAX	2N6118 MIN MAX	UNIT
V_{OM}	Peak Output Voltage	V_{AA} = 20 V, C1 = 0.2 µF, See Figure 4	6	6	6	V
t_r	Output Pulse Rise Time		80	80	80	ns

*PARAMETER MEASUREMENT INFORMATION

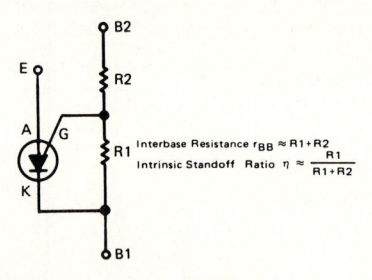

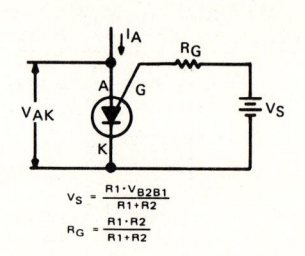

Interbase Resistance $r_{BB} \approx R1 + R2$
Intrinsic Standoff Ratio $\eta \approx \dfrac{R1}{R1+R2}$

FIGURE 1—PROGRAMMABLE UNIJUNCTION CIRCUIT

$V_S = \dfrac{R1 \cdot V_{B2B1}}{R1+R2}$

$R_G = \dfrac{R1 \cdot R2}{R1+R2}$

FIGURE 2—EQUIVALENT CIRCUIT USED FOR TESTING

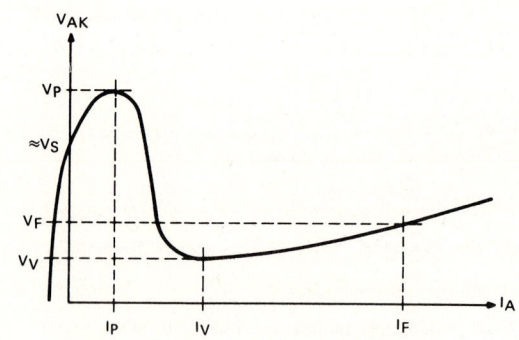

FIGURE 3—GENERAL ANODE CHARACTERISTICS

TEST CIRCUIT

OUTPUT VOLTAGE WAVEFORM

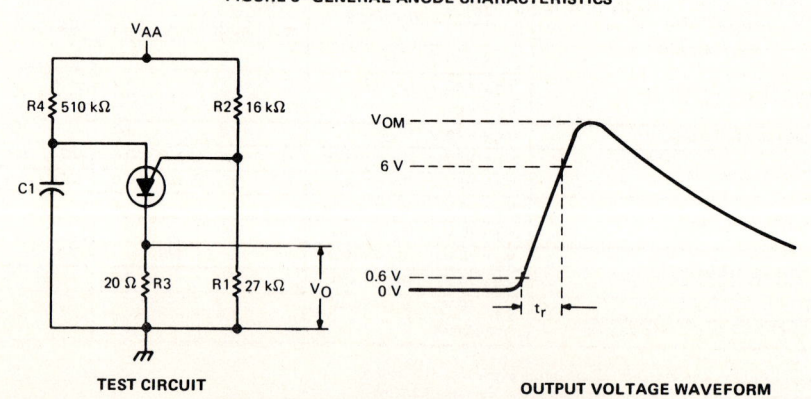

FIGURE 4—TESTING OPERATING CHARACTERISTICS

*JEDEC registered data

This information is out of date and parts could possibly be absolete, reprint for educational purposes only. Courtesy of Texas Instruments, Inc.

Appendix H 715

TYPES A7T6027, A7T6028
P-N-P-N SILICON PROGRAMMABLE UNIJUNCTION TRANSISTORS

BULLETIN NO. DL-S 7311796, JANUARY 1973

SILECT† TRANSISTORS‡ FOR USE IN PULSE, TIMING, SWEEP, TRIGGER, AND OSCILLATOR CIRCUITS

- Plug-in Replacements for 2N6027, 2N6028 (TO-98 Package)
- Low Peak-Point Current and Low Forward Voltage
- Programmable η, r_{BB}, I_P, and I_V

mechanical data

These transistors are encapsulated in a plastic compound specifically designed for this purpose, using a highly mechanized process developed by Texas Instruments. The case will withstand soldering temperatures without deformation. These devices exhibit stable characteristics under high-humidity conditions and are capable of meeting MIL-STD-202C, Method 106B. The transistors are insensitive to light.

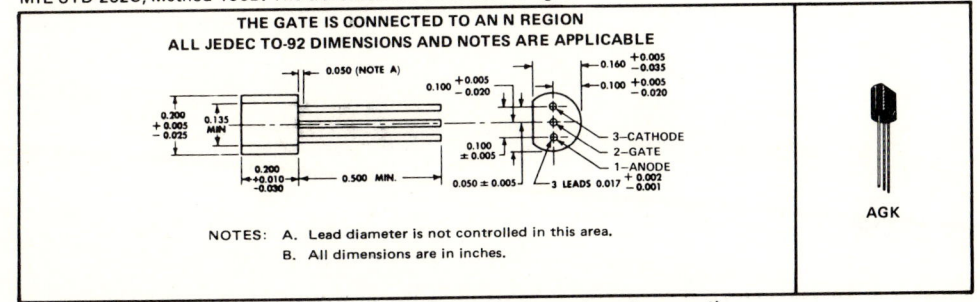

THE GATE IS CONNECTED TO AN N REGION
ALL JEDEC TO-92 DIMENSIONS AND NOTES ARE APPLICABLE

3—CATHODE
2—GATE
1—ANODE

AGK

NOTES: A. Lead diameter is not controlled in this area.
B. All dimensions are in inches.

absolute maximum ratings at 25°C free-air temperature (unless otherwise noted)

Anode-Cathode Voltage	±40 V
Gate-Anode Voltage	40 V
Gate-Cathode Voltage: (Positive Limit)	40 V
(Negative Limit)	−5 V
Continuous Anode Current	150 mA
Repetitive Peak Anode Current: (t_w = 100 μs, Duty Cycle ≤ 1%)	1 A
(t_w = 20 μs, Duty Cycle ≤ 1%)	2 A
Nonrepetitive Peak Anode Current: (t_w = 10 μs, Duty Cycle = 0)	5 A
Continuous Gate Current	±50 mA
Continuous Device Dissipation at (or below) 25°C Free-Air Temperature (See Note 1)	300 mW
Storage Temperature Range	−65°C to 150°C
Lead Temperature 1/16 Inch from Case for 60 Seconds	260°C

electrical characteristics at 25°C free-air temperature

PARAMETER		TEST CONDITIONS	A7T6027 MIN	A7T6027 MAX	A7T6028 MIN	A7T6028 MAX	UNIT
I_{GAO}	Gate Reverse Current	V_{GA} = 40 V, I_K = 0		10		10	nA
		V_{GA} = 40 V, I_K = 0, T_A = 75°C		100		100	
I_{GKS}	Gate Reverse Current	V_{GK} = 40 V, V_{AK} = 0		100		100	nA
$V_P - V_S$	Offset Voltage	V_S = 10 V, R_G = 10 kΩ	0.2	0.6	0.2	0.6	V
		V_S = 10 V, R_G = 1 MΩ	0.2	1.6	0.2	0.6	
I_P	Peak-Point Current	V_S = 10 V, R_G = 10 kΩ		5		1	μA
		V_S = 10 V, R_G = 1 MΩ		2		0.15	
I_V	Valley-Point Current	V_S = 10 V, R_G = 200 Ω	1500		1000		μA
		V_S = 10 V, R_G = 10 kΩ	70		25		
		V_S = 10 V, R_G = 1 MΩ		50		25	
V_F	Anode-Cathode On-State Voltage	V_S = 10 V, R_G = 10 kΩ, I_F = 50 mA		1.5		1.5	V

NOTE 1: Derate linearly to 125°C free-air temperature at the rate of 3 mW/°C.
†Trademark of Texas Instruments
‡U.S. Patent No. 3,439,238

This information is out of date and parts could possibly be obsolete, reprint for educational purposes only. Courtesy of Texas Instruments, Inc.

Answers to Self-Check

Chapter 1

1. True 2. False 3. True 4. True 5. False 6. True 7. False 8. True 9. True 10. True 11. False 12. True 13. True 14. False 15. False

Chapter 2

1. True 2. False 3. False 4. True 5. True 6. True 7. False 8. True 9. True 10. True 11. False 12. False 13. False 14. True 15. True 16. True 17. False 18. True 19. False 20. True 21. True 22. False 23. True 24. True 25. False

Chapter 3

1. two, one 2. a. emitter, b. base, c. collector 3. base 4. emitter, N-type material, direction of current flow 5. emitter 6. recombination 7. holes 8. direction opposite to 9. barrier 10. base 11. collector-base, emitter-base 12. P-type, N-type 13. 0.7 V 14. increase 15. P-type 16. $I_E = 100\%$, $I_C = 92\%$ to 98%, $I_B = 2\%$ to 8% 17. low, high 18. base 19. current 20. directly 21. minority, collector-base, reverse 22. temperature 23. positive 24. voltage divider 25. positive

Chapter 4

1. False 2. False 3. True 4. True 5. False 6. False 7. False 8. True 9. True 10. False 11. Beta (β) 12. negative 13. increase 14. $I_B + I'_{RB}$ 15. Alpha (α) 16. negative 17. I_E and R_{eb} 18. $V'_{RB} - V_{eb}$ 19. collector, ground 20. 180° out of phase 21. points along the load line 22. V_{CE} and I_C (reverse order ok.) 23. V_{CC} or R_L 24. saturation 25. maximum $I_C = 8$ mA

$$I_C = \frac{20 \text{ V}}{2.5 \text{ k}\Omega}$$

26. fixed 27. bridge 28. R_B in parallel with R_{cb} 29. R_B in parallel with $R_L + R_{cb}$ 30. are not

Chapter 5

1. collector, ground 2. emitter, collector, base 3. R_{cb} 4. small, large 5. collector to ground 6. a. alpha 7.

$$\frac{I_C}{I_E}$$

8. emitter, collector 9. V_{CB} 10. I_C 11. in 12. increases 13. increase 14. b. saturation 15. increasing 16. emitter 17. less than 1 18. I_E, I_B 19. less 20. V_{in}, V_{EB} 21. cutoff, saturation. 22. False 23. emitter-base I_C, I_E 24. emitter follower 25. common-emitter

Chapter 6

1. PIN 2. Zener 3. photo 4. not been doped 5. tunnel diode 6. reverse, capacitance 7. battery powered, low current requirements. 8. reverse, Zener (negative resistance) 9. light 10. Hot carrier, metal (gold, silver or aluminum) 11. latched on 12. cathode, anode 13. breakover point 14. source (anode) 15. four-layer 16. anode 17. both 18. anode to cathode 19. SCRs, gate 20. one direction only

Chapter 7

1. diode 2. reverse 3. high 4. either $0.318 \times E_{pk}$ or $0.45 \times E_{eff}$ 5. equal to 6. center-tapped secondary 7. two 8. 800 PPS 9. 95.45 V 10. 200 V 11. four (4) 12. double 13. two times 14. equal to 15. higher voltage at higher current 16. It regulates the output voltage, which reduces ripple voltage. 17. It increases average DC out and reduces ripple voltage. 18. An RC-pi filter has capacitance and resistance; an LC-pi filter has capacitance and inductance. RC filter does not provide current regulation of the LC filter. 19. It maintains a more stable load current. 20. No effect as E_{pk} remains the same. 21. No output 22. Blow fuse, stops all output. 23. Places short to ground on diode output

causing blown fuse or diode. **24.** Filtering decreases, ripple voltage increases, average DC out decreases. **25.** All output stops. **26.** parallel (shunt) **27.** a. reverse **28.** a. reverse **29.** both **30.** d. all of these **31.** reference **32.** zero **33.** +10 V **34.** resistor, diode **35.** long

Chapter 8

1. emitter current, collector current **2.** inversely **3.** resistor **4.** b. decreases **5.** C_1 blocks DC voltages, present at the preceeding stage from biasing the transistor. C_2 acts as a bypass capacitor that places AC ground at the transistor emitter and thereby increases voltage gain. **6.** emitter current (I_E) **7.** beta + 1 times **8.** R_B, R_B', and R_{ib} **9.** short **10.** impedance, X_C **11.** 20 kHz **12.** bypass, resistor **13.** 100 V peak-to-peak **14.** R_L and/or bypassing R_E with a capacitor. **15.** I_CMax, V_{CE}Max **16.** Q point **17.** I_C^*, V_{CE}^* **18.** higher **19.** impedance **20.** Z_{in} of the next stage **21.** V_{CC} **22.** high fidelity **23.** symmetrical **24.** voltage gain (A_V) **25.** change **26.** decrease **27.** current gain (A_I), voltage gain (A_V) **28.** increase **29.** AC ground **30.** larger **31.** high input impedance (Z_{in}) and a low output impedance (Z_{out}) **32.** voltage, current **33.** input, output **34.** increase input impedance (Z_{in}) and to better develop the input signal. **35.** less **36.** positive **37.** larger **38.** positive **39.** emitter **40.** collector **41.** larger

Chapter 9

1. one, 180° **2.** phase splitter **3.** differential **4.** negative **5.** grounded **6.** A **7.** B **8.** overdriven **9.** AB **10.** C **11.** negative **12.** double-diode **13.** capacitor, AC ground **14.** lowest **15.** degenerative **16.** 100 **17.** common-collector **18.** common-emitter **19.** common-collector **20.** common-emitter **21.** common-collector **22.** common-collector **23.** common-emitter **24.** less than one **25.** common-collector **26.** Zener, resistor **27.** voltage, current **28.** series **29.** R_L **30.** increase

Chapter 10

1. direct **2.** output, input **3.** direct **4.** True **5.** impedance, coupling capacitor. **6.** impedance **7.** higher **8.** distributed (stray) capacitance **9.** coils **10.** transformer **11.** output, input **12.** $A_{V_1} \times A_{V_2} \times \ldots A_{V_n}$ **13.** V_{CC} **14.** No input signal applied to Q_2, and no output from Q_2. **15.** zero volts **16.** 20 Hz to 20 kHz **17.** maximum (high) (large) etc. **18.** percentage **19.** True **20.** 20 kHz **21.** left **22.** a heat sink **23.** last **24.** one **25.** smaller **26.** two **27.** (a) transformer (b) phase splitter (Answers may be in reverse order and still be correct.) **28.** 100% **29.** phase shift **30.** double **31.** biasing **32.** efficient **33.** crossover **34.** AB **35.** opposite types (PNP and NPN) of **36.** reference **37.** They provide a discharge path for coupling capacitors. **38.** A **39.** two, same, 180° **40.** Class B

Chapter 11

1. loopstick, ferrite-rod **2.** selectivity **3.** narrower **4.** 10 kHz **5.** tuned stages **6.** decrease **7.** transformer **8.** transformer, tuned **9.** signal-to-noise ratio **10.** gang **11.** There should be zero output. **12.** saturate **13.** There would be zero output. **14.** R_1 shorted, Q_1 shorted, R_{B2} shorted **15.** reduce degeneration **16.** distributed (stray) capacitance **17.** 4 MHz **18.** increase **19.** increase **20.** False **21.** True **22.** remain the same (no effect). **23.** an infinite number of odd harmonics. **24.** high frequency or shunt **25.** remain the same (no effect)

Chapter 12

1. (a) amplifier, (b) regenerative feedback, (c) frequency determining device or network **2.** amplitude, frequency **3.** in **4.** regulated V_{CC} and low tolerance bias networks. **5.** C **6.** 60° **7.** (a) RC networks, (b) LC tank circuits, (c) crystals **8.** flywheel **9.** quartz **10.** decrease **11.** regenerative **12.** DC, AC **13.** damped **14.** resistance **15.** 5000 to 30,000 **16.** (a) square wave, (b) rectangular wave, (c) sawtooth wave, (d) trapezoidal wave **17.** equal **18.** 10%, capacitance **19.** coils **20.** unequal

Chapter 13

1. increasing 2. (a) amplifier, (b) regenerative feedback, (c) frequency-determining device or network 3. (a) RC networks, (b) LC tank circuits, (c) crystals 4.

$$f_O = \frac{1}{2\pi\sqrt{LC}}$$

5. (a) feedback network, (b) frequency-determining device, (c) amplifier, (d) power source 6. LC 7. transformer 8. bias 9. positive 10. 10% 11. does 12. temperature stabilization 13. center-tapped transformer 14. It isolates an oscillator from its load. 15. C_3 and C_5 16. capacitor 17. capacitance 18. (a) reduces hand capacitance effect (b) provides tuning capability 19. It reduces the effect of emitter-base interelement capacitance. 20. the crystal 21. 45 22. 180 23. Class A 24. RC 25. base 26. output 27. either b or c 28. buffer 29. fourth 30. doublers, quadrupler

Chapter 14

1. degenerative 2. decreases 3. undamped 4. provide a point that can accept the input gate. This cuts Q_2 off stopping output oscillations. 5. switch 6. regenerative, leading 7. the time required to saturate Q_1. 8. all of these 9. The tertiary winding couples the output to the next stage. 10. higher than 11. (a) astable, (b) monostable, (c) bistable 12. (d) 0 13. either a or b 14. 1 15. 1 16. 2 17. 2 18. either a or b 19. C_3 and C_4 reduce the time constant of the emitter-base junction capacitances, which will shorten the transient interval. 20. any of the above. 21. capacitor 22. electrostatic 23. input gate frequency. 24. electromagnetic 25. input gate frequency 26. RC network time constant

Chapter 15

1. (a) JFET, (b) MOSFET 2. noise factor 3. diffusion 4. tetrode 5. (a) source, (b) drain, (c) gate 6. one, channel, gate, reverse 7. increase, decrease 8. V_{DD} 9. V_{GS} 10. pinch-off 11. pinch-off 12. drain-source saturation 13. V_{GS}OFF 14. Since JFETs are affected by temperature, manufacturing tolerances, and power source tolerances, it is impossible to state an exact I_{DSS} and V_P for JFET. The parameters (I_{DSS}Min I_{DSS}, V_PMin, and V_PMax) are used to construct characteristic curves that allow us to design a JFET amplifier, with some degree of certainty. 15. transadmittance, Siemens 16. For P-channel JFETs all voltages must have their polarity reversed from those of the N-channel type. 17. I_DMax, V_{DS}Max 18. maximum, minimum 19. gate 20. infinite 21. several 22. (a) manufacturing tolerance, (b) temperature effects, (c) bias supply tolerances 23. I_D, V_{GS} 24. self 25. I_D, gain, degeneration 26. (a) common-source, (b) common-drain, (c) common-gate 27. high, low 28. low 29. R_G 30. (a) $A_V = g_m \times R_L$,

(b) $A_V = \dfrac{\Delta V_{out}}{\Delta V_{in}}$

31. R_L 32. high 33. less than one 34. source 35. medium

Chapter 16

1. depletion 2. source, gate 3. negative, positive 4. vertical 5. DE-MOSFET, E-MOSFET 6. zero 7. threshold 8. V_{GS}OFF 9. threshold 10. enhancement 11. switch states 12. V-MOS, E-MOSFET 13. True 14. high 15. channel

Chapter 17

1. four, three 2. unilateral 3. base, NPN 4. True 5. three 6. 50%, 80% 7. emitter, V_{RB2} 8. True 9. low 10. sensing 11. 10, faster
12.

13. increase, decrease 14. inversely 15. LC 16. optoelectronic 17. (a) photovoltaic, (b) photoconductive 18. luminescent 19. decrease 20. isolate

21. highly concentrated and powerful light beam. 22. light 23. photovoltaic 24. photoconductive 25. resistance

Chapter 18

1. (a) Discrete microcomponents, (b) Thick-film (Ceramic Printed) circuitry, (c) Thin-film circuitry, (d) Hybrid microcircuitry, (e) Silicon integrated circuits 2. resistors, conductors 3. (a) serves as a base for mounting components, (b) acts as a heat dissipater (heat sink) 4. diffusion 5. discrete, thick-film, thin-film 6. size, resistance 7. diodes, transistors, integrated circuits, capacitance, and leads 8. resistance, capacitance, conductors 9. aluminum 10. Silicon monoxide 11. nichrome, tantalum 12. 30 13. (a) circuit flexibility, (b) allows the use of the best discrete components, (c) reliability, (d) It is possible to duplicate all discrete circuits. 14. passive, active 15. (a) reliability, (b) cost, (c) size, (d) power efficiency 16. silicon 17. photoengraving 18. Diffusion 19. oxide 20. epitaxial 21. die 22. (a) Dual-In-Line (DIP), (b) TO-5, (c) Flat Packs 23. continual testing 24. parallel duplication of discrete circuits, flexibility, and quality control. 25. header

Chapter 19

1. Dual-In-Line (DIP) 2. False 3. two 4. differential amplifier 5. common-mode 6. 180 7. offset drift 8. impedance 9. closed 10. voltage, time 11. (a) inverting 12. low 13. summing 14. False 15. noise 16. True 17. True 18. True 19. True 20. False

Answers to Review Questions and/or Problems

Chapter 1
1. b. negative ion **3.** d. all of these **5.** b. less than **7.** c. electrons, holes **9.** b. wide **11.** a. positive **13.** b. insulator **15.** a. positive ion **17.** a. pentavalent **19.** a. True

Chapter 2
1. c. junction **3.** d. either a or b **5.** a. a continuous crystalline structure must exist. **7.** d. all of these are correct **9.** a. barrier **11.** a. reverse biased **13.** a. A, B **15.** c. avalanche **17.** b. 9.3 V **19.** c. active **21.** d. 0.3 V **23.** a. unilateral **25.** b. infinity

Chapter 3
1. a. base b. emitter c. collector **3.** emitter, type **5.** forward, reverse **7.** low, high **9.** 100 **11.** a. base-emitter voltage **13.** a. increase
15.

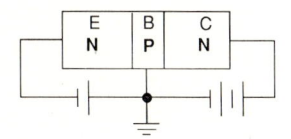

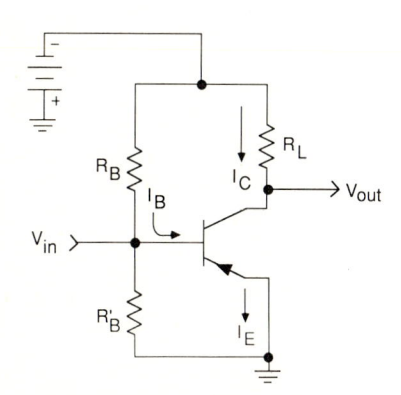

17.

19. a.

Chapter 4
1. negative **3.** NPN **5.** resistance of the emitter-base junction **7.** voltage divider **9.** b. self **11.** a. True **13.** dual-source **15.** maximum power dissipation limits **17.** 6 mA, mark 6 mA point on the vertical axis **19.** $I_C \cong 4.5$ mA and $V_{CE} \cong 6$ V

Chapter 5
1. base **3.** positive going **5.** collector, emitter **7.** I_C, I_E **9.** 0.95 **11.** collector **13.** base, emitter **15.** reverse the polarity of V_{CC} **17.** I_E, I_B **19.** smaller

Chapter 6
1. reverse **3.** variable capacitance, reverse **5.** breakover potential **7.** a. Increase in source voltage causes SCR current to increase. b. Decrease in source voltage causes SCR current to decrease. c. At some point the decreasing voltage cause I_A to drop below the holding current level. d. Removal, or reverse in phase of source voltage causes the SCR to enter cutoff. **9.** a. forward b. reverse c. forward **11.** 2, 3 **13.** Light Emitting Diode (LED) **15.** unilateral **17.** diode junction, light pipes **19.** metal **21.** both (two) **23.** current to flow in one direction only **25.** d, e

Chapter 7
1. AC to Pulsating DC (PDC) **3.** d. Junction resistance decreases and junction current increases **5.** 60 ($f_{out} = f_{in}$) **7.** 143 V (3.18 x $E_{pk\ sec}$) **9.** 450 V ($E_{pk\ out} = E_{pk\ sec}$) **11.** 200 ($f_{ripple} = 2 \times f_{input}$) **13.** 90 V $\left(0.636 \times \dfrac{E_{pk\ sec}}{2}\right)$

15. 282 V ($E_{Ave \cdot DC} = 0.636 \times E_{pk\ out}$) ($E_{AveDC} = 0.636 \times 170$ V) ($E_{AveDC} = 108.12$ V) **17.** a. output is positive **19.** reverse the connections to all four diodes. **21.** 170 V ($E_{pk\ out} = E_{pk\ secondary}$) **23.** output waveshape is a positive Half-wave PDC. **25.** a. T_1 Primary open **27.** voltage **29.** blown fuse **31.** negative, positive **33.** reverse **35.** b. False **37.** negative, positive **39.** positive, negative

Chapter 8
1. d. it blocks DC and couples input signals above a specific frequency. **3.** a. increase **5.** 20 μA, **7.** 20Ω **9.** 21 Hz **11.** 1000 **13.** a. J_1 **15.** 360 **17.** decouple AC (Place AC ground at the collector) **19.** b. J_2

Chapter 9
1. a. I_C flows 100% of the time. **3.** R'_B and R_E **5.** b. decrease **7.** in phase **9.** 180° out **11.** common-emitter **13.** differential amplifier **15.** a comparison of the two output signals. **17.** So that it will act as a common-base amplifier that reacts to the changes felt across R_4. **19.** 180° **21.** double-diode **23.** thermistor **25.** 9-26 **27.** 7.5 V to 17.5 V (10 V peak-to-peak) **29.** maximum (5 mA) **31.** a. Q_1 **33.** d. at the top of Z_1 **35.** b. down

Chapter 10
1. d. any of the above. **3.** X_C, Z_{in} **5.** d. all of these **7.** d. 300:1 **9.** c. impedance **11.** b. low **13.** d. 360° **15.** a. A **17.** c. the amount of bias applied to Q_1 and Q_2 **19.** d. both b and c

Chapter 11
1. 10 **3.** d. choose a frequency or small band of frequencies **5.** magnetic shield **7.** a. selectivity **9.** distributed (stray) **11.** b. decrease **13.** a. increase **15.** low, removed **17.** d. below, above **19.** c. high frequency loss

Chapter 12
1. 10 Hz, several thousand gigahertz **3.** sinusoidal **5.** quartz crystal **7.** regenerative **9.** RC network **11.** C **13.** flywheel **15.** mechanical, electrical **17.** cutoff and saturation **19.** coil **21.** resonant **23.** wideband **25.** external trigger

Chapter 13
1. d. frequency-determining device, regenerative feedback, and an amplifier **3.** c. to generate a waveform of constant amplitude and frequency **5.** d. sine wave of constant amplitude and frequency **7.** c. flywheel **9.** a. damping **11.** c. L_2, C_1 **13.** b. ratio of the inductances L_1 to L_2 **15.** a. True **17.** b. Colpitt's, split capacitance **19.** d. L_1, C_3, C_4, and C_5 **21.** c. CR_1 **23.** a. increase, increase **25.** a. regenerative feedback is developed across C_3

Chapter 14
1. c. switch, cut off **3.** b. L_1 and C_2 **5.** a. collector with **7.** b. cut off, low **9.** a. C_1 and R_1 **11.** c. (3) **13.** c. PRF decreases **15.** a. R_1 and eb junction **17.** b. free-running, astable **19.** a. R_2 open **21.** c. establish the output frequency **23.** b. Q_1, Q_2 **25.** c. double, one half **27.** c. voltage sensing and pulse shaping **29.** a. high frequency distortion **31.** c. exactly **33.** c. decrease in amplitude **35.** c. cut off, slope

Chapter 15
1. a. PN **3.** b. **5.** a. is **7.** b. False **9.** c. 11 V **11.** a. common-source **13.** b. resistor between the source and ground **15.** c. drain **17.** c. less than unity (1) **19.** b. gate, source **21.** d. R_1 and R_2 in parallel **23.** a. 5 mA **25.** d. source, drain

Chapter 16
1. a. True **3.** c. silicon dioxide, insulator **5.** c. vertical **7.** c. $V_{GS}OFF$ **9.** threshold **11.** b. switch **13.** c. depletion **15.** c. low

Chapter 17

1. c. voltage regulators **3.** d. breakover potential **5.** c. gate, anode **7.** a. True **9.** a. True **11.** a. True **13.** c. heavy doping **15.** tunnel diode, UJT **17.** a. photovoltaic b. photoconductive **19.** increase **21.** high powered, narrow beam of light. **23.** photoconductive **25.** internal resistance

Chapter 18

1. d. all of these **3.** interchangeable **5.** c. both a and b **7.** a. substrate **9.** d. both b and c **11.** c. shaving off small slices of material. **13.** a. thin-film **15.** c. monolithic **17.** d. all of these **19.** depth, resistance

Chapter 19

1. b. non-inverting, inverting **3.** b. False **5.** b. two **7.** c. common-mode **9.** the difference between the outputs of the two amplifiers **11.** b. open **13.** b. non-inverting **15.** c. subtractor **17.** d. both b and c **19.** a. improper connection of the power supply. **21.** b. False **23.** a. True **25.** d. wideband amplifier

Index

Index

Acceptor impurity, 9, 20
AC load line, 226-231
Active, 40, 122
AC voltage gain, 223-226
Adder-subtracter, 637
A_I, See Current gain.
Alpha, 81, 82, 121, 122, 124
Alpha cutoff frequency, 301
Ambient temperature, 33
Amplification factor, 273
Amplifier
  buffer, 375, 412-413
  common-base amplifier, 74, 75, 114, 117, 240-244
  common-collector amplifier, 74, 75, 114, 125, 126, 127, 128, 212-240
  common-drain amplifier, 525-526
  common-emitter amplifier, 74, 75, 245-253, 256,
  common-gate amplifier, 526-528
  common-source amplifier, 520-525
  complimentary-symmetry, 325-330
  coupling, 295
  differential, 258-262, 639
  narrow-band, 344-354
  overdriven, 266
  paraphase, 323
  power, 294, 312-331
  single-ended, 257, 314, 315
  stagger tuned, 362-364
  wideband, 354-359
Amplitude distortion, 263
Amplitude stability, 374
Anode, 150
Anode current, 560
Anode gate, 153
Anode voltage, 561
A_P, See Power gain.
Approximation method, 90, 212
Atom, 2
A_V, See Voltage gain
Avalanche current, 35, 137
Avalanche point, 35, 499
Average DC output voltage, 168, 172, 175, 176, 193, 194

Bandwidth, 346, 349
Barrier, 23, 54
Barrier region, 23
Base current, 81, 212
Beta, 81, 82, 115
Beta cutoff frequency, 301
Bias, 20, 26, 27, 29, 56
  dual source, 76, 78
  fixed bias, 76, 506-510
  forward, 26, 27, 32, 37, 43, 56, 66, 74, 163
  reverse, 29, 32, 38, 43, 56, 66, 74, 163
  self bIas, 76, 78, 90, 510-519
  voltage divider, 76, 78
Bidirectional device, 13, 33, 154

BI-FET, 529-530
Bilateral, 13, 154, 155
Bipolar junction transistor, 52, 54
Bipolar transistor, 52, 54, 74
Bistable multivibrator, 466-469, 643
Bleeder resistor, 164, 182, 187, 188
Blocking oscillator, 444
Breakdown,
  point, 35, 136
  potential, 32, 136, 499
  reverse, 39, 277
  voltage, 32, 136, 499
Breakover, 149, 561
Breakover point, 149
Breakover voltage, 149, 561, 565
Buffer amplifier, 375

Capacitive filter, 179
Carriers,
  current, 139
  majority, 25, 30, 77, 139
  minority, 25, 30
  negative, 9
  positive, 9
Cascade connected amplifiers, 307, 308
Cathode, 31, 150
Cathode gate, 153
CB amplifier, See Common-base amplifier.
CC amplifier, See Common-collector amplifier.
CE amplifier, See Common-emitter amplifier.
Ceramic printed circuits, 593-594-599
Characteristic curve, 92, 93, 97, 120, 500-505, 546
Clampers, 200-203
Clapp oscillator, 419-420
Class of amplifiers, 311, 312
  by frequency, 311
  by type of operation, 312
  by use, 312
Class of operation, 263-265, 282
  Class A, 263, 264, 282
  Class AB, 263, 265, 282
  Class B, 263, 265, 282
  Class C, 263, 265, 282, 375, 407
Closed loop circuit, 631
CMOS, See Complementary MOS
Collector, 31, 52
Collector-base junction, 74
Collector current, 77, 81, 212
Collector feedback bias, See Self bias.
Colpitt's oscillator, 416-419
Common-base amplifier, 74, 75, 114, 117, 212, 245-253
Common-base configuration, See Common-base amplifier.
Common-collector amplifier, 74, 75, 114, 125, 212, 240-244
Common-collector configuration, See Common-collector amplifier.

Common-drain amplifier, 525-526
Common-drain configuration, See Common-drain amplifier.
Common-emitter amplifier, 74, 75, 212-240
Common-emitter configuration, See Common-emitter amplifier.
Common-gate amplifier, 526-528
Common-gate configuration, See Common-gate amplifier.
Common ground, 114
Common-mode rejection, 627
Common-source amplifier, 520-525
Common-source configuration, See Common-source amplifier.
Complementary SCR, See Programmable UJT.
Complementary MOS, 551, 552
Complementary-symmetry amplifier, 325-330
Complementary transistor, 235
Conduction band, 5
Conventional flow, 11, 55
Coupling capacitor, 300
Coupling circuits, 301, 362
Conduction, 5
Covalent bonding, 6, 7
Crossover distortion, 319, 327
Crystal, 8, 400, 422
Current,
  base, 60, 61, 81
  Carriers, 11
  collector, 60, 61, 77, 81
  conventional flow, 11, 55
  Electron flow, 55,
  emitter, 60, 61, 121
  majority, 11, 20, 28
  minority, 11, 20, 24, 25, 28
  reverse, 25, 136, 146
Current controlled device, 64, 121
Current gain, 115, 116, 117, 127, 294
Cutoff, 40, 120, 263, 499

Damping, 380, 456
DC load line, 93, 120, 228
Decay, 390
Decoupling capacitor, 345
Degeneration, 269
Degenerative feedback, 373
Delta, 237
DE-MOSFET, See Metal-Oxide-Semiconductor Field-Effect Transistor - depletion type.
Depletion layer, 23
Depletion mode, See DE-MOSFET
Depletion region, 23, 26, 54, 136, 494
DIAC, 155
DIFF AMP, See Differential amplifier.
Differential amplifier, 639
Differentiator, 637
Diffusion, 492, 611-613

Digital multimeter, 39
Diode, 20,
   discharge, 324
   clampers, 162
   LASER, 584-585
   limiters, 162
   load lines, 40
DIP, See Dual-in-line.
Direct coupling, 295-299
Discharge diodes, 324
Discrete circuitry, 593
Distortion, 263
Distributed capacitance,
   299, 301
Donor impurity, 9, 20
Doping, 8, 136, 139
Drain, 492, 494, 495, 538
Drain current, 494
Drain curves, 544
Drain-source saturation current, 498-499
Drain-source voltage, 494, 499
Driver stage, 322
Dual battery bias, See Dual source bias.
Dual-In-Line (DIP), 616-617
Dual source bias, 76, 78
Dual supply bias, See Dual source bias.

EB junction, See Emitter-base junction.
Effective voltage, 166,
Electron, 2, 20
   free, 5, 9, 25
   flow, 13
Electron-hole pair, 9, 25
Electronic voltage regulator, 138, 139,
   276-281
Electronic voltmeter
   (EVOM), 39
Electrostatic, 22
Electrostatic discharge, 550, Appendix G
Element stability, 4
Emitter, 31, 52
Emitter-base junction, 74
Emitter-bypass capacitor, 216
Emitter current, 81
Emitter follower, 114, 125, 127
E-MOSFET, See Metal-Oxide-
   Semiconductor Field-Effect Transistor,
   enhancement type.
Energy
   bands, 5
   diagram, 6, 23
   kinetic, 2
   level, 3, 5
   permissible level, 2, 5
   potential, 2
ENHANCEMENT MODE, See
   E-MOSFET.
Epitaxy, 614-615
Error amplifier, 280
Error voltage, 280
ESAKI DIODE, See TUNNEL DIODE.
EVOM, See Electronic Voltmeter.

Fall time, 390, 451, 452, 461
Feedback,
   degenerative, 373
   inductive, 403, 454
   regenerative, 373, 399-402, 407, 421, 454
Feedback signal, 377
FET, See Field effect transistor.
Fidelity, 263
Field-effect transistor,
   Chapter 15
Filter, 178-190
   capacitive input, 182, 183
   choke, 188, 189
   inductive input, 182
   L-type, 182, 183
   pi-type, 183
   RC-Type, 184
Final amplifier, 294
Fixed bias, 76, 506-510
Flip flop, 466
Flywheel effect, 378-380
Forbidden band, 5, 6
Forbidden region, See Forbidden band.
Force,
   centrifugal, 2
   centripetal, 2
Forward bias, 26, 27, 32, 56, 66, 163
Forward breakover potential, 151
Four-layer device, 148
Free-running, 386, 462
Frequency-determining devices
   (networks), 376, 382, 400-401
Frequency multiplier, 430-433
Frequency response, 220-223
Frequency stability, 374, 383
Full-wave rectifier, 169-177, 175, 192-193
   conventional full-wave rectifier, 169-173
   full-wave bridge rectifier, 173-177, 175,
   194-195
Fundamental wave, 320, 386

Gain,
   current, 115, 116, 117, 127, 273, 294
   power, 273, 294
   voltage, 115, 117, 273, 294, 521
Gain-bandwidth product, 631
Gamma, 97, 98, 118, 119
Ganged tuning, 345
Gate, 150, 152, 492, 494, 538
Gate-source voltage, 494, 497
Gate voltage, 561
GBP, See Gain-bandwidth product.
Germanium, 3

Half-Power points, 347
Half-wave rectifier, 163 - 169, 175, 179, 191,
   192
Harmonics, 320, 321, 322
   even, 320, 321, 322
   odd, 320
Heat sink, 313, Appendix D
High fidelity, 233,
High frequency compensation, 358, 361,
   362, 468

High frequency loss, 365
Holding current, 152, 153, 154
Hole, 9, 11, 20, 25
Hole flow, 10
Hot carrier diode, 147
Hybrid circuitry, 594

I_B, See Base current.
IC, See Integrated circuit.
I_C, See Collector current.
I_E, See Emitter current.
Impedance coupling, 303-305
Impurity, 8
Inductive filter, 180, 181
Input circuit, 627
Input impedance, 214-220, 522, 549
Input signal, 114
Input stage, 626
Input voltage, 128
Insulated-Gate Field-Effect Transistor, See
   Metal-Oxide-Semiconductor Field-
   Effect Transistor.
Integrated circuit, 606-618
Integrator, 637-638
Interbase resistance, 569
Interbase voltage, 569
Interelement capacitance, 299
Inverting amplifier, 635
Ions,
   negative, 8, 11
   positive, 8, 11

JFET, See Junction Field Effect Transistor.
Jump voltage, 389-390, 475
Junction,
   barrier, 23
   field, 23
   photosensitive, 147
   transistor, 54

Large signal amplifier, 233
Latching device, 150
Lateral PNP transistors, 609
LCD, See Liquid Crystal Diode.
LC oscillator, 381
LC tank, 378, 381, 400
Leading edge, 461
Leakage current, 64, 65
LED, See Light Emitting Diode.
Light Emitting Diode, 143
Light pipe, 144
Limiters, 196-200
Liquid Crystal Display, 145
Load, 165, 166, 167
   heavy, 166
   light, 166
Load impedance, 231, 233
Load line,
   AC, 226-231
   DC, 40, 93, 120, 228, 523-525
Load resistor, 76
Low frequency compensation, 360, 361,
   362
Low frequency loss, 365

Index

Low frequency response, 302

Majority carrier, 11, 20, 28, 77
Majority current, 28
Mask, 595
Matched pair transistors, 325
Maximum average forward current, 32
Maximum collector dissipation, 312
Maximum power dissipation, 32, 92, 312
Maximum surge current, 32
Metal-Oxide Semiconductor Field-Effect Transistor, 492, Chapter 16
    CMOS, 551-553
    depletion type, 538-539, 542-544
    enhancement type, 538-539
    N-Channel type, 494
    P-Channel type, 496, 528-529
    V-MOS, 550
Microcomponents, 593
Microelectronics, 593
Minority carrier, 11, 20, 24, 28
Minority current, 25, 146
Monolithic integrated circuit, 606
Monostable multivibrator, 464-466
MOSFET, Abbreviation for Metal-Oxide Semiconductor Field-Effect Transistor.
Multichip circuitry, 618-619
Multivibrator, 444, 459-469
    astable, 462-464, 642
    bistable, 466-469, 643
    monostable, 464-466

Narrow-band amplifier, 344-354
Negative resistance region, 140, 570
Negative temperature coefficient, 12, 269
Negative voltage, 188
Non-inverting amplifier, 635
Nonsinusoidal, 344, 354, 372
Nonsinusoidal wave generators, 365-391
NPN transistor, 54
N-type material, 9, 11, 13, 24

Offset, 628
Offset voltage drift, 626
One-shot multivibrator, 464
OP AMP, See Operational Amplifier.
Open, 190, 192, 195
Open loop gain, 631
Operating point, 40, 83, 95, 96, 98, 234
Operational amplifiers, 626-641
    adder-subtracter, 637
    audio amplifier, 638
    differential amplifier, 639
    differentiator, 637
    integrator, 637
    inverting amplifier, 635
    non-inverting amplifier, 635
    voltage follower, 636
    sine-wave oscillator, 639-640
    summing amplifier, 636
Optoelectronic devices, 581-585
Orbit, 3,

Oscillators, Chapter 12
    Armstrong, 403-410
    blocking, 451-458
    Butler, 421-422, 449
    Clapp, 410-420
    Colpitt's, 416-419
    phase shift (RC), 422-424
    pulsed, 444-450
    series-fed Hartley, 411-413
    shunt-fed Hartley, 413-415
    Wien bridge, 425-430
Output impedance, 215, 232, 241, 522, 549
Output resistance, 274
Output signal, 114, 299
Output stage, 294
Output voltage, 128
Overdriven, 263
Oxidation, 613-615

Package, 52, 53, 146
Packaging, 617
PDC, See Pulsating Direct Current.
Peak Inverse Voltage, 32, 167, 171, 174, 186
Peak Output Voltage, 169, 170
Peak point, 140
Peak Recurrent Current, 32
Peak-to-peak voltage, 166
Peak voltage, 166
Pentavalent elements, 8
Phase relationship,
    common-base amplifier, 115, 124, 128
    common-collector amplifier, 126, 127, 128
    common-drain amplifier, 525-526
    common-emitter amplifier, 84, 128
    common-gate amplifier, 526-528
    common-source amplifier, 520-525
Phase splitter, 257-258, 322, 323, 324
Piezoelectric effect, 382
Photoconductive, 581
Photo diode, 146, 582-583
Photoengraving, 610
Photosensitive, 146, 147
Phototransistor, 583
Photovoltaic, 581
Pinch off, 499
Pinch off point, 498
Pinch off voltage, 499
PIN diode, 147
PIV, See Peak Inverse Voltage.
P_{MAX} curve, 312, 313
PN junction, 20, 136
PN junction diode, 20, 31 - Also see Diode.
PNP transistor, 54
Positive voltage, 188
Potential hill, 24, 54
Power amplifier, 294, 312-331
Power source, 165
Power supply, 160, 181
Power supply filters, 178-190
PRF, See Pulse Repetition Frequency.
Programmable UJT, 575-577
PRT, See Pulse Reccurrence Time.

PRV, See Peak Inverse Voltage.
P-type material, 9, 11, 13, 24
Pulsating direct current, 162, 178,
Pulse, 460
Pulsed oscillator, 444-450
Pulse Reccurrence Time, 452, 460-461
Pulse Repetition Frequency, 452, 460-461
Pulse width, 452, 455, 461
Push-pull amplifiers, 316-324
    Class A, 317, 318, 378
    Class AB, 319, 320
    Class B, 318, 319

Q point, See Operating point.
Quiescent point, See Operating point.

RC network, 376, 377, 400
RC oscillator, 422-424
Recombination, 22, 25, 54
Rectangular wave, 386
Rectifier, 162, 175, 179, 180
    conventional full wave, 180, 186
    half wave, 163-168
    full wave bridge, 180
    troubleshooting, 190-195
Reference voltage, 279, 280
Regenerative feedback, 373, 399-402, 407, 421
Regulators,
    electronic, 276-281
    series, 280
    Zener diode, 277, 278
Relaxation oscillators, 372
Resist, 596
Resistance,
    input, 274
    output, 274
Resistivity, 612
Resonant frequency, 383
Rest time, 456, 460
Reverse bias, 26, 27, 29, 56, 66, 163
Reverse breakdown, 39
Reverse current, 25, 30, 31, 146 - Also see Minority current.
Ringing circuit, 444-446
Ripple frequency, 168, 171, 173, 192, 193, 195
Ripple voltage, 178, 179, 192, 193, 195
Rise time, 451, 452, 460-461

Saturation, 40, 120, 263, 499
Sawtooth wave, 388
Sawtooth wave generator, 388-389, 473-475, 573-575
SBS, See Silicon Bilateral Switch.
Schmitt trigger, 444, 468
SCR, See Silicon controlled rectifier.
SCS, See Silicon Controlled Switch.
Selectivity, 343
Self bias, 76, 78, 90, 510-519
Semiconductor, 6,
    extrinsic, 8,
    intrinsic, 147

N-type, 9,
P-type, 9
Sensing voltage, 570
Sensitivity, 343
Series compensation, 359
Series peaking coil, 359
Seven-segment display, 144
Sheet resistance, 612
Shell, See Orbit
Short, 190, 192, 193, 195
Shunt compensation, 359
Shunt peaking coil, 359
Signal-to-noise ratio, 349, 350
Silicon, 3
Silicon bilateral switch, 154
Silicon-controlled rectifier, 150, 560-567
Silicon-controlled switch, 153
Silicon integrated circuits, 606-618
Single-ended power amplifier, 314, 315
Sinusoidal waves, 372
Sine wave oscillators, 398-430, 639-640
Single source biasing, 66, 67
Slew rate, 632-633
Slope, 390
Small signal amplifier, 233
Source, 494-495, 538
Space charge region, 23
Square wave, 354, 386, 459
Stagger tuned amplifiers, 362-364
Static, 349
Steady DC, 178
Stray capacitance, 299
Substrate, 538, 595, 608-610
Subtractor, 636
Summing amplifier, 636
Surge current, 185
Surge suppressors, 185

Temperature, ambient, 33
Temperature stabilization, 267-273, 511
  emitter resistor, 267-269
  diode, 271-273

thermistor, 269-271, 270
Thermal runaway, 39, 65, 267
Thick-film fabrication, 593-599
Thin-film fabrication, 594, 600-605
Threshold voltage, 545
Thyristor, 560
Time base generator, 390, 444
Timer (555), 641-643
Timing pulse, 451
Trailing edge, 461
Transfer characteristics,
  maximum, 502-505
  minimum, 502-505
Transformer coupling, 305, 306, 362
Transient interval, 460-461, 573
Transient peak-inverse voltage, 565
Transistor,
  NPN, 54
  PNP, 54
  field-effect, Chapter 15
  metallic-oxide-semiconductor field-effect, Chapter 16
  unijunction, 569-576
Trapezoidal wave, 389-390, 473-477
TRIAC, 155
Trigger, 444
Trigger generator, 444
Trivalent, 20
Troubleshooting,
  common-emitter amplifiers, 309, 310
  oscillators, 409-410, 420, 428-430
  multivibrators, 463-464, 465-466, 468-469
  narrow-band amplifier, 350-353
  push-pull amplifiers, 330, 331
  rectifier circuits, 190-195
  Schmitt trigger, 471-472
  wideband amplifier, 364-365
Tuned base oscillator, 405
Tuned collector oscillator, 405
Tunnel diode, 139, 577-580

amplifier, 579-580
oscillator, 579
Tunnel effect, 139
Tunneling, See Tunnel diode.
Turn-off time, 566
Turn-on time, 565

UJT, See Unijunction Transistor.
Unidirectional device, 33, 34, 155, 560
Unijunction transistor, 569-576
Unilateral, Also See Unidirectional.
Unity-gain frequency, 632
Upper frequency cutoff, 222, 223, 301, 303

Vacuum deposition, 600
Valence, 6
  electrons, 3
  shell, 3
Valley point, 140
Valley voltage, 572
Varactor, 141, 142, 384
VCO, See Voltage Controlled Oscillator.
V-MOS, 550
Voltage,
  breakdown, 499
  breakover, 149, 561
  drain-source, 494, 499
  gate-source, 494, 497
  emitter-follower, 62
  interchannel, 499
  negative, 328
  pinch-off, 499
  positive, 328
Voltage controlled oscillator, 384
Voltage divider bias, 76, 78
Voltage follower, 636
Voltage gain, 115, 117, 273, 294

Wideband amplifier, 355-356

Zener current, See Avalanche current.
Zener diode, 136, 138